Motion Control

Edited by

Federico Casolo

Motion Control
Edited by Federico Casolo

Republished by InTech
Janeza Trdine 9, 51000 Rijeka, Croatia

Additional hard copies can be obtained from orders@intechopen.com

Motion Control, Edited by Federico Casolo
p. cm.
ISBN 978-953-7619-55-8

For EU product safety concerns:
IN TECH d.o.o., Prolaz Marije Krucifikse Kozulić 3, 51000 Rijeka, Croatia,
info@intechopen.com or visit our website at intechopen.com.

Meet the editor

Federico Casolo, mechanical engineer (general construction engineer) in charge of the Polytechnic of Milan since 1983, first as a researcher, then as a professor of Applied Mechanical Engineering (ing.ind / 13). He held courses in Applied Mechanics in Machines and Biomechanics at the Polytechnic of Milan and at the University of Parma. His current academic role includes, in addition to courses in applied mechanics, also a course of Mechanics of human-machine systems within the doctoral school of the Polytechnic. The main research sectors he deals with mainly refer to biomechanics and mechatronics: the main applications concern artificial limbs for amputees, systems for the rehabilitation of disabled people and sports equipment. He also worked on mechanical systems for household appliances and for the handling of aggregates. He had the opportunity to deepen his studies in the field of biomechanics as a visiting scientist (following a CNR-NATO competition) at the Biomechanics Laboratory of Pennsylvania State University, which he attended for one year (under the guidance of Prof. P. Cavannagh). From 2000 to 2004 he was president of the Tecnical Group on Computer Simulation, organ of the International Society of Biomechanics. Currently he is a member of the Mechanics Department, a mechatronics department and is part of the Faculty of Systems Engineering. Regarding the other institutional positions covered by his role: for two terms he was elected a member of the Board of the Polytechnic of Milan and a member of the extended Academic Senate. He is a member of the Order of Engineers of the Province of Milan.

Motion Control
Edited by Federico Casolo

Republished by InTech
Janeza Trdine 9, 51000 Rijeka, Croatia

Additional hard copies can be obtained from orders@intechopen.com

Motion Control, Edited by Federico Casolo
p. cm.
ISBN 978-953-7619-55-8

Preface

This book deals with a crucial aspect of mechanical systems design: the synthesis and control of movements of machine components. The scientific research in this area is strongly linked to the evolution of robotics and mechatronics and is focused on various aspects of motion dynamics. Purpose of this volume is to provide researchers with updated information, useful from the point of view of application as well as of theory. For this reason, most chapters describe the research in depth and show all the required details to implement the methods toward other practical applications. The need of motor control studies is still growing in numerous research fields; therefore, our selection shows applications from very diverse areas. They range from terrestrial, aerial, underwater and space vehicles control, to the precision control of industrial components in manufacturing systems and include the control of structures connected to the human body and transmitting to it noise and vibrations.

The book reveals many different aspects of motion control, and a wide multiplicity of approaches to the problem as well. Despite the number of examples, however, this volume is not meant to be exhaustive: it intends to offer some original insights for all researchers that hopefully, in the future, will make their experience available for a forthcoming publication on the subject.

Editor

Federico Casolo

Politecnico di Milano,
Department of Mechanics
Italy

Contents

1

Dynamics and Control of Multibody Systems

Marek Vondrak[1], Leonid Sigal[2] and Odest Chadwicke Jenkins[1]
[1]*Brown University*,
[2]*Univesity of Toronto*
[1]*U.S.A.*,
[2]*Canada*

1. Introduction

Over the past decade, physics-based simulation has become a key enabling technology for variety of applications. It has taken a front seat role in computer games, animation of virtual worlds and robotic simulation. New applications are still emerging and physics is becoming an integral part of many new technologies that might have been thought of not being directly related to physics. For example, physics has been recently used to explain and recover the motion of the subject from video (Vondrak et al., 2008). Unfortunately, despite the availability of various simulation packages, the level of expertise required to use physical simulation correctly is quite high. The goal of this chapter is thus to establish sufficiently strong grounds that would allow the reader to not only understand and use existing simulation packages properly but also to implement their own solutions if necessary. We choose to model world as a set of constrained rigid bodies as this is the most commonly used approximation to real world physics and such a model is able to deliver predictable high quality results in real time. To make sure bodies, affected by various forces, move as desired, a mechanism for controlling motion through the use of constraints is introduced. We then apply the approach to the problem of physics-based animation (control) of humanoid characters.

We start with a review of unconstrained rigid body dynamics and introduce the basic concepts like body mass properties, state parameterization and equations of motion. The derivations will follow (Baraff et al., 1997) and (Erleben, 2002), using notation from (Baraff, 1996). For background information, we recommend reading (Eberly, 2003; Thornton et al., 2003; Bourg, 2002). We then move to Lagrangian constrained rigid body dynamics and show how constraints on body accelerations, velocities or positions can be modeled and incorporated into simpler unconstrained rigid body dynamics. Various kinds of constraints are discussed, including equality constraints (required for the implementation of "joint motors"), inequality constraints (used for the implementation of "joint angle limits") and bounded equality constraints (used for implementation of motors capable of generating limited motor forces). We then reduce the problem of solving for constraint forces to the problem of solving linear complementarity problems. Finally, we show how this method can be used to enforce body non-penetration and implement a contact model, (Trinkle et al., 1997; Kawachi et al., 1997).

Lastly, we illustrate how before mentioned constraints can be used to implement composite articulated bodies and how these bodies can be actuated by generating appropriate motor torques at joints, following (Kokkevis, 2004). Various kinds of convenient joint parameterizations with different degrees of freedom, together with options for their actuation, are discussed.

Fig. 1. Examples of constrained rigid body systems. Constraints glue bodies together at designated points, actuate the structures or enforce non-penetration.

1.1 Related work

While physical simulation is conceptually well understood, control of articulated high degree of freedom bodies (or characters) remains a challenging problem. On the simulation side there currently exist a number of commercial and open source engines that deliver robust and computationaly efficient performance (e.g., Crisis, Havoc, Newton, Open Dynamics Engine (ODE), PhysX). Quantitative analysis of performance among some of these and other popular choices are discussed in (Boeing et al., 2007). However, control over the motion of characters within these simulators is still very limited. Those packages that do provide means for building user defined dynamic controllers (e.g., Euphoria by NaturalMotion and Dynamic Controller Toolbox (Shapiro et al., 2007)) still lack fidelity and ability to model stylistic variations that are important for producing realistic motions.

In this chapter, we describe trajectory-based control (either in terms of joint angles or rigidly attached points) implemented in the form of constraints. This type of the control is simple, general, stable, and is available (or easy to implement) within any simulator environment that supports constraints (e.g., Crisis, ODE, Newton). That said, other control strategies have also been proposed and are applicable for appropriate domains and tasks. For example, where modeling of high fidelity trajectories is hard, one can resort to sparse set of key-poses with proportional derivative (PD) control (Hodgins et al., 1995); such controllers can produce very stable motions (e.g., human gait (Yin et al., 2007)) but often look artificial or robotic. Locomotion controllers with stable limit cycle behavior are popular and appealing choices for various forms of cyclic gates (Laszlo et al, 1996); particularly in the robotics and biomechanics communities (Goswami et al., 1996).

At least in part the challenges in control stem from the high dimensionality of the control space. To that end few approaches have attempted to learn low-dimensional controllers through optimization (Safonova et al., 2004). Other optimization-based techniques are also popular, but often require initial motion (Liu et al., 2005) or existing controller (Yin et al., 2008) for adaptation to new environmental conditions or execution speed (McCann et al., 2006). Furthermore, because it is unlikely that a single controller can produce complex motions of interest, approaches that focus on building composable controllers (Faloutsos et al., 2001) have also been explored. Alternatively, controllers that attempt to control high degree-of-freedom motions using task-based formulations, that allow decoupling and composing of controls required to complete a particular task (e.g., maintain balance) from controls required to actuate redundant degrees of freedom with respect to the task, are also appealing (Abe et al., 2006). In robotics such strategies are known as operational space control (Khatib, 1987; Nakamura et al., 1987).

Here we discuss and describe trajectory-based control that we believe to strike a balance between the complexity and effectiveness in instances where desired motion trajectories are available or easy to obtain. Such control has been illustrated to be effective in the emerging applications, such as tracking of human motion from video (Vondrak et al., 2008).

2. Rigid body dynamics

Rigid bodies are solid structures that move in response to external forces exerted on them. They are characterized by mass density functions describing their volumes ("mass properties"), positions and orientations ("position information") in the world space and their time derivatives ("velocity information").

2.1 Body space, mass properties, position, orientation

Properties of rigid bodies are derived from an assumption that rigid bodies can be modeled as particle systems consisting of a large (infinite) number of particles constrained to remain at the same relative positions in the body spaces. Internal spatial interaction forces prevent bodies from changing their shapes and so as a result, any rigid body can only translate or rotate with respect to a fixed world frame of reference. This allows one to associate local coordinate frames with the bodies and define their shapes/volumes in terms of local *body spaces* that map to the world reference frame using rigid transformations.

We describe a volume of a rigid body by a *mass density function* $\rho: \boldsymbol{R}^3 \mapsto \boldsymbol{R}^+$ that determines the body's mass distribution over points $\vec{r}^b$ in the body space. The density function is non-zero for points forming the body's shape and zero elsewhere and its moments characterize the body's response to the exerted forces. We are namely interested in *total mass* $m = \int \rho(\vec{r}^b)\,\mathrm{d}\vec{r}^b$, *center of mass* $\vec{r}^b_{cm} = \int \frac{\vec{r}^b \rho(\vec{r}^b)}{M}\,\mathrm{d}\vec{r}^b$, *principal moments of inertia* $I_{xx} = \int \left(\left(\vec{r}^b_y\right)^2 + (\vec{r}^b_z)^2\right)\rho(\vec{r}^b)\,\mathrm{d}\vec{r}^b$, $I_{yy} = \int ((\vec{r}^b_x)^2 + (\vec{r}^b_z)^2)\rho(\vec{r}^b)\,\mathrm{d}\vec{r}^b$, $I_{zz} = \int \left((\vec{r}^b_x)^2 + \left(\vec{r}^b_y\right)^2\right)\rho(\vec{r}^b)\,\mathrm{d}\vec{r}^b$ and *products of inertia* $I_{xy} = \int \left(\vec{r}^b_x \vec{r}^b_y\right)\rho(\vec{r}^b)\,\mathrm{d}\vec{r}^b$, $I_{xz} = \int (\vec{r}^b_x \vec{r}^b_z)\rho(\vec{r}^b)\,\mathrm{d}\vec{r}^b$, $I_{yz} = \int \left(\vec{r}^b_y \vec{r}^b_z\right)\rho(\vec{r}^b)\,\mathrm{d}\vec{r}^b$ that we record into *inertia matrix*

$$I_{body} = \begin{pmatrix} I_{xx} & -I_{xy} & -I_{xz} \\ -I_{xy} & I_{yy} & -I_{yz} \\ -I_{xz} & -I_{yz} & I_{zz} \end{pmatrix}.$$

To place a rigid body's volume in the world, we need to know the *mapping from the body space to the world space*. For that, we assume that the body's center of mass lies at the origin of the body space, $\vec{r}^{\,b}_{cm} = \vec{0}$, and construct a mapping $[R, \vec{x}]$ so that a point $\vec{p}^{\,b}$ in the body space will get mapped to the world space point $\vec{p}$ by applying a *rotation* R, represented by a 3×3 rotation matrix mapping body space axes to the world space axes (*orientation of the body in the world space*), followed by applying a *translation* $\vec{x}$ that corresponds to the world space position of the body's center of mass *(position of the body in the world space)*, $\vec{p} = R \cdot \vec{p}^{\,b} + \vec{x}$.

2.2 Velocity

Having placed the body in the world coordinate frame, we would like to characterize the motion of this body over time. To do so we need to compute time derivatives of the position and orientation of the body, i.e. $\frac{\partial}{\partial t}[R, \vec{x}]$. We decompose instantaneous motion over infinitesimally short time periods to the translational (*linear*) motion of the body's center of mass and a rotational (*angular*) motion of the body's volume. We first define *linear velocity* $\vec{v} = \dot{\vec{x}}$ as the time derivative of the rigid body's position $\vec{x}$, characterizing the instantaneous linear motion and describing the direction and speed of the body translation. Next, we describe the rotational motion as a rotation about a time varying axis that passes through the center of mass. We define *angular velocity* $\vec{\omega}$ as a world-space vector whose direction describes the instantaneous rotation axis and whose magnitude $[rad \cdot s^{-1}]$ defines the instantaneous rotation speed. Linear and angular velocities are related such that they can describe velocities of arbitrary points or vectors attached to the body. For example, if $\vec{r} = \vec{p} - \vec{x}$ is a vector between the point on the body, $\vec{p}$, the center of mass of the body, $\vec{x}$, then $\dot{\vec{r}} = \vec{\omega} \times \vec{r}$ and $\dot{\vec{p}} = \vec{v} + \vec{\omega} \times \vec{r}$. This can be used to derive a formula for $\dot{R}$ that says $\dot{R} = \vec{\omega}^* \cdot R$, where $\vec{\omega}^*$ is a "cross-product matrix" such that $\vec{\omega}^* \cdot \vec{r} = \vec{\omega} \times \vec{r}$. It is worth noting that because $\vec{p}$ is fixed in the body centric coordinate frame, so is the vector $\vec{r}$.

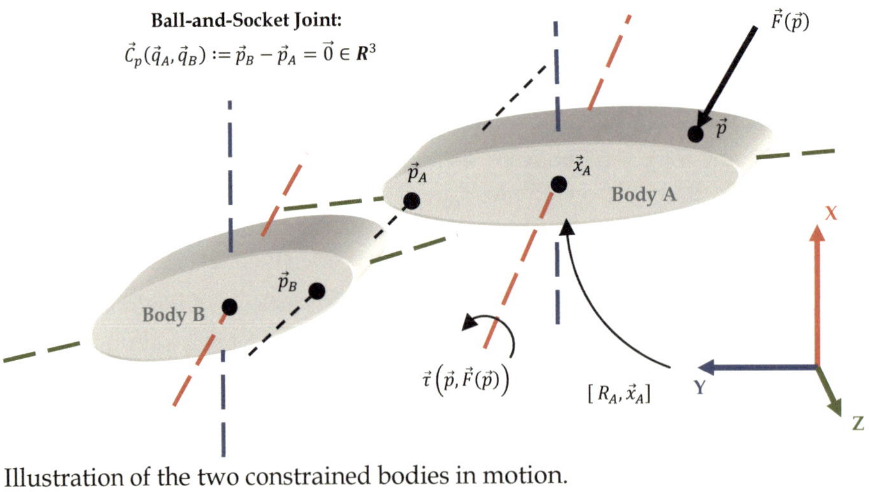

Fig. 2. Illustration of the two constrained bodies in motion.

2.3 Force

From previous section we have $\frac{\partial}{\partial t}[R, \vec{x}] = [\vec{\omega}^* \cdot R, \vec{v}]$ relating changes of the position and orientation to the values of the body's linear and angular velocities. Now, we would like to characterize how the linear and angular velocities of a rigid body change in response to forces exerted on the body. Intuitively, these changes should depend on the location where the force is applied as well as mass distribution over the body volume. So we need to know not only the directions and magnitudes of the exerted forces, but also the points at which these forces are applied.

To capture the effects for a single force $\vec{F}(\vec{p})$ acting at a world space point $\vec{p}$, we define a *force-torque pair* $\left[\vec{F}(\vec{p}),\ \vec{\tau}\left(\vec{p}, \vec{F}(\vec{p})\right)\right]$, where $\vec{\tau}\left(\vec{p}, \vec{F}(\vec{p})\right) = (\vec{p} - \vec{x}) \times \vec{F}(\vec{p})$ is the *torque* due to the force $\vec{F}(\vec{p})$. The torque can be imagined as a scale of the angular velocity $\vec{\omega}$ that the rigid body would gain if $\vec{F}(\vec{p})$ was the only force acting on the body and the force was exerted at $\vec{p}$. To capture the overall effects of all force-torque pairs $[\vec{F_i}, \vec{\tau_i}]$ due to all forces acting on the body, it is sufficient to maintain only the corresponding aggregate statistics: *total force* $\vec{F}_{\text{total}} = \sum_i \vec{F}_i$ and *total torque* $\vec{\tau}_{total} = \sum_i \vec{\tau}_i$ about the center of mass of the body, $\vec{x}$.

Now, we express the body's linear and angular velocities in the form of linear and angular momentums whose instantaneous changes can be directly related to the values of the total forces and torques acting on the body. The reason for doing so is that it is actually the momentums that remain unchanged when no forces act on the body, not the velocities. We define *linear momentum* $\vec{P} = m \cdot \vec{v}$ and *angular momentum* $\vec{L} = I \cdot \vec{\omega}$ where $I = R \cdot I_{body} \cdot R^T$. The relation between the velocity and force information is then given by derivatives of linear and angular momentum with respect to time, $\dot{\vec{P}} = \vec{F}_{total}$ and $\dot{\vec{L}} = \vec{\tau}_{total}$.

2.4 Equations of motion

We are now ready to present complete equations describing motion of a set of rigid bodies in Newtonian dynamics under the effect of forces. The equations are *first order ordinary differential equations* (ODEs). To simulate the system, one has to numerically integrate the equations of motion, which can be done by using standard numerical ODE solvers. We explore several formulations of the equations of motion below.

2.4.1 Momentum form

We start with the momentum form that makes the linear and angular momentum a part of a rigid body's state and builds directly upon the concepts presented in earlier sections. To make the body's state complete, only the position and orientation information has to be added to the state. Therefore, the *state* is described by a vector $\vec{y}$, $\vec{y} = \left(\vec{x}, R, \vec{P}, \vec{L}\right)$, where $\vec{x}$ is the position of the body's center of mass, R is the orientation of the body and $\vec{P}$ and $\vec{L}$ are the body's linear and angular momentums. The *equation of motion for the rigid body in the momentum form* is then given by $\frac{\partial \vec{y}}{\partial t} = (\vec{v}, \vec{\omega}^* \cdot R, \vec{F}_{total}, \vec{\tau}_{total})$, where $\vec{F}_{total}$ and $\vec{\tau}_{total}$ are the total external force and torque exerted on the body and $\vec{v}$ and $\vec{\omega}$ are auxiliary quantities derived from the state vector $\vec{y}, \vec{v} = m^{-1} \cdot \vec{P}, I = R \cdot I_{body} \cdot R^T, I^{-1} = R \cdot I_{body}^{-1} \cdot R^T, \vec{\omega} = I^{-1} \cdot \vec{L}$.

If there are n rigid bodies in the system, the individual ODE equations are combined into a single ODE by concatenating the body states $\vec{y}_1, \cdots, \vec{y}_n$ into a single state vector $\vec{y} = (\vec{y}_1, \cdots, \vec{y}_n)$ and letting $\frac{\partial \vec{y}}{\partial t} = \left(\frac{\partial \vec{y}_1}{\partial t}, \cdots, \frac{\partial \vec{y}_n}{\partial t}\right)$.

2.4.2 Velocity form

As a conceptually more common alternative, the equations of motion can be reformulated so that linear and angular momentums in the state vector are replaced with linear and angular velocities. In this formulation, the state vector $\vec{y}$ is defined as

$$\vec{y} = (\vec{x}, R, \vec{v}, \vec{\omega}) \tag{1}$$

To formulate the right-hand-side vector of the ODE, we need know time derivatives of the linear and angular velocities and relate them to external forces and torques. We define *linear acceleration* $\vec{a}$ of a rigid body as the acceleration of the body's center of mass, that is, $\vec{a} = \dot{\vec{v}} = \ddot{\vec{x}}$, and because $\dot{\vec{P}} = \vec{F}_{total}$ we immediately get $\vec{a} = m^{-1} \cdot \vec{F}_{total}$. For the angular motion, we define *angular acceleration* $\vec{\alpha}$ as the time derivative of the body's angular velocity, $\vec{\alpha} = \dot{\vec{\omega}}$, and it can be shown that $\vec{\alpha} = I^{-1} \cdot (\vec{\tau}_{coriolis} + \vec{\tau}_{total})$, where $\vec{\tau}_{coriolis} = (I \times \vec{\omega}) \times \vec{\omega}$ is an implicit internal inertial (coriolis) torque due to body rotation and $\vec{\tau}_{total}$ is the total external torque applied on the body. This way we get the *equation of motion for a single*[1] *rigid body in the velocity form*

$$\frac{\partial \vec{y}}{\partial t} = \left(\vec{v}, \vec{\omega}^* \cdot R, m^{-1} \cdot \vec{F}_{total}, I^{-1} \cdot \left((I \times \vec{\omega}) \times \vec{\omega} + \vec{\tau}_{total}\right)\right) \tag{2}$$

2.4.3 Generalized form

We now elaborate on the velocity-form of the equation of motion, define the notion of *generalized velocities and forces* and the concept of *mass matrices* for rigid bodies, which will allow us to treat rigid bodies as a kind of particles moving in $\boldsymbol{R}^6$, simplifying many equations. We will call any block vector consisting of a block due to a linear quantity and a block due to the corresponding angular quantity a *generalized quantity*. That way, we obtain *generalized velocity* $\vec{v}_{gen} = (\vec{v}, \vec{\omega})$, *generalized acceleration* $\vec{a}_{gen} = (\vec{a}, \vec{\alpha})$, *generalized total external force* $\vec{F}_{gen}^{total} = (\vec{F}_{total}, \vec{\tau}_{total})$ and *generalized coriolis force* $\vec{F}_{gen}^{coriolis} = (\vec{0}, \vec{\tau}_{coriolis})$. In addition, we define *generalized position* $\vec{q} = (\vec{x}, \vec{R})$ that encodes both position of the body's center of mass and orientation in $3D$ space.

We now define the *mass matrix* M of a rigid body which is a 6×6 time-dependent matrix consisting of four 3×3 blocks encoding the body's mass properties,

$$M = \begin{pmatrix} m \cdot E & 0 \\ 0 & I \end{pmatrix}, \tag{3}$$

and E is a 3×3 identity matrix. From the previous section, we know that $m \cdot \vec{a} = \vec{F}_{total}$ and $I \cdot \vec{\alpha} = \vec{\tau}_{total} + \vec{\tau}_{coriolis}$ which can be rewritten using the mass matrix simply as $M \cdot \vec{a}_{gen} = \vec{F}_{gen}^{total} + \vec{F}_{gen}^{coriolis}$. Let's assume that the generalized coriolis force $\vec{F}_{gen}^{coriolis}$ is implicitly incorporated into to the total generalized external force $\vec{F}_{gen}^{total}$ and, to improve readability, let's remove the $_{gen}$ subscripts and omit the "generalized" adjective whenever it is clear that the generalized notation is used. This lets us write

$$M \cdot \vec{a} = \vec{F}_{total} \tag{4}$$

[1] As for the momentum form, equation of motion for a set of n bodies is obtained by "cloning" the equation for a single body n-times.

which yields a relation between the total force $\vec{F}_{total}$ and the total acceleration $\vec{a}$. Because the relation is linear, this equation also holds for *any force* $\vec{F}$ acting on the body and the corresponding *acceleration* $\vec{a} = M^{-1} \cdot \vec{F}$ the body would gain in *response to the application* of $\vec{F}$[2]. The relation resembles Newton's Second Law for particles and rigid bodies can thus be imagined as special particles with time-varying masses M that move in $\boldsymbol{R}^6$.

3. Constraints

One of the challenges one has to face in physical simulation is how to generate appropriate forces so that rigid bodies would move as desired. Instead of trying to generate these forces directly, we describe desired motion in terms of *motion constraints* on accelerations, velocities or positions of rigid bodies and then use *constraint solver* to solve for the forces. We still use the same equations of motion (and numerical solvers) to drive our bodies like before, but this time, we introduce *constraint forces* that implicitly act on constrained bodies so that given motion constraints are enforced. We study the approach of *Lagrange multiplier method* that handles each constraint in the same uniform way and allows to combine constraints automatically. Examples of constrained rigid bodies are given in Fig. 1.

In general, the motion constraint on the position or orientation of a body will subsequently result in the constraints on its velocity and acceleration (to ensure that there is no velocity or acceleration in the constrained direction, leading to violation of constraint after integration of the equations of motion); similarly a constraint on velocity will impose a constraint on the acceletation. We will discuss these implications in the following section. A first-order rigid body dynamics with impulsive formulation of forces (discussed in Section 3.3.1) allows one to ignore the acceleration constraints in favor of simplicity, but at expense of inability to support higher-order integration schemes.

3.1 Example: point-to-point equality constraint

Let's start with a motivational example. Imagine we are given two bodies and we want to enforce a position constraint that stipulates that point $\vec{p}_1 = \vec{x}_1 + \vec{r}_1$attached to the first body is to coincide with a point $\vec{p}_2 = \vec{x}_2 + \vec{r}_2$ attached to the second body (see Fig. 2 where the two bodies are denoted as A and B), making the two bodies connected at $\vec{p}_1 = \vec{p}_2$ and preventing them from tearing apart. We can express this *position-level constraint* as a vector equation $\vec{C}_p(\vec{q}_1, \vec{q}_2) := \vec{p}_2 - \vec{p}_1 = \vec{0} \in \boldsymbol{R}^3$, defined in terms of generalized positions $\vec{q}_1, \vec{q}_2$ of the two bodies, such that all valid position pairs, for which the constraint is maintained, correspond to a manifold $\vec{C}_p(\vec{q}_1, \vec{q}_2) = \vec{0}$. Granted the constraint is maintained already, the goal is to compute an appropriate constraint force so that $(\vec{q}_1, \vec{q}_2)$ stays on the manifold during the state update. Given the total external forces $\vec{F}_1^{total}$and $\vec{F}_2^{total}$ acting on the two bodies, we will construct a constraint force such that it would cancel exactly those components of the $\vec{F}_1^{total}$ and $\vec{F}_2^{total}$ vectors that would make the bodies accelerate away from the manifold. To do this, we will reformulate our position-level constraint to a constraint on body accelerations and from that derive the constraint force. Our constraint formulation will give us a set of basis vectors that need be combined to get the constraint

---

[2] If $\vec{F}$ refers to the total external force exerted on the body, coriolis force is assumed to be included in $\vec{F}$.

force. Appropriate coefficients of this combination are computed by solving a system of linear equations.

Let's assume that at the current time instant the bodies are positioned so that the constraint is maintained, that is, $\vec{C}_p = \vec{0}$. To make sure the constraint will also be maintained in the future, we have to enforce $\dot{\vec{C}}_p = \vec{0}$. Let's have a look at what $\dot{\vec{C}}_p$ looks like, $\dot{\vec{C}}_p = \frac{\partial}{\partial t}(\vec{p}_2 - \vec{p}_1) = \frac{\partial}{\partial t}(\vec{x}_2 + \vec{r}_2 - \vec{x}_1 - \vec{r}_1) = \dot{\vec{x}}_2 + \vec{\omega}_2 \times \vec{r}_2 - \dot{\vec{x}}_1 - \vec{\omega}_1 \times \vec{r}_1 = \dot{\vec{x}}_2 - \vec{r}_2 \times \vec{\omega}_2 - \dot{\vec{x}}_1 + \vec{r}_1 \times \vec{\omega}_1 = \dot{\vec{x}}_2 - \vec{r}_2^* \cdot \vec{\omega}_2 - \dot{\vec{x}}_1 + \vec{r}_1^* \cdot \vec{\omega}_1 = (-E \quad \vec{r}_1^*) \cdot \vec{v}_1 + (E \quad -\vec{r}_2^*) \cdot \vec{v}_2 = (-E \quad \vec{r}_1^* \quad E \quad -\vec{r}_2^*) \cdot \begin{pmatrix} \vec{v}_1 \\ \vec{v}_2 \end{pmatrix} = (J_1 \quad J_2) \cdot \begin{pmatrix} \vec{v}_1 \\ \vec{v}_2 \end{pmatrix}$, where J_1 and J_2 are 3×6 matrices called the Jacobian matrices due to the position constraint $\vec{C}_p$ and the first and the second body. So we need to enforce another *constraint* $\vec{C}_v(\vec{v}_1, \vec{v}_2) := J_1 \cdot \vec{v}_1 + J_2 \cdot \vec{v}_2 = \vec{0}$, this time formulated in terms of generalized *velocities* $\vec{v}_1, \vec{v}_2$. This is good because we were able to reformulate the original constraint specified in terms of generalized positions to a constraint specified in terms of generalized velocities.

Let's assume that the velocity constraint also holds, that is, $\vec{C}_v = \vec{0}$, and let's guarantee the velocity constraint will be maintained in the future by requesting $\dot{\vec{C}}_v = \vec{0}$ (this will also guarantee that the original position-level constraint will be maintained, because $\vec{C}_p = \vec{0}$ at the current time instant). We have $\dot{\vec{C}}_v = \frac{\partial}{\partial t}(J_1 \cdot \vec{v}_1 + J_2 \cdot \vec{v}_2) = J_1 \cdot \vec{a}_1 + J_2 \cdot \vec{a}_2 + \dot{J}_1 \cdot \vec{v}_1 + \dot{J}_2 \cdot \vec{v}_2$ and so we obtain a constraint $\vec{C}_a(\vec{a}_1, \vec{a}_2) := J_1 \cdot \vec{a}_1 + J_2 \cdot \vec{a}_2 - \vec{c} = \vec{0}$, where J_1 and J_2 are the Jacobian matrices defined above, $\dot{J}_1$and $\dot{J}_2$ are their time derivatives and $\vec{c} = -\dot{J}_1 \cdot \vec{v}_1 - \dot{J}_2 \cdot \vec{v}_2$. This constraint is formulated directly in terms of generalized *accelerations* $\vec{a}_1, \vec{a}_2$ and because we already know the relation between accelerations and forces, this constrains the forces that can act on the two bodies. To complete the formulation of $\vec{C}_a$, we need to get the value of $\vec{c}$. It is usually easier to compute $\vec{c}$ directly from $\dot{\vec{C}}_v$ rather than by computing the time derivatives of the Jacobian matrices. We can for example do, $\ddot{\vec{C}}_p = \dot{\vec{C}}_v = \frac{\partial}{\partial t}(-\dot{\vec{x}}_1 - \vec{\omega}_1 \times \vec{r}_1) + \frac{\partial}{\partial t}(\dot{\vec{x}}_2 - \vec{\omega}_2 \times \vec{r}_2) = \left(-\ddot{\vec{x}}_1 - \dot{\vec{\omega}}_1 \times \vec{r}_1 - \vec{\omega}_1 \times (\vec{\omega}_1 \times \vec{r}_1)\right) + \left(\ddot{\vec{x}}_2 + \dot{\vec{\omega}}_2 \times \vec{r}_2 + \vec{\omega}_2 \times (\vec{\omega}_2 \times \vec{r}_2)\right) = \left(-\ddot{\vec{x}}_1 + \vec{r}_1^* \cdot \dot{\vec{\omega}}_1 - \vec{\omega}_1 \times (\vec{\omega}_1 \times \vec{r}_1)\right) + \left(\ddot{\vec{x}}_2 - \vec{r}_2^* \cdot \dot{\vec{\omega}}_2 + \vec{\omega}_2 \times (\vec{\omega}_2 \times \vec{r}_2)\right) = (-E \quad \vec{r}_1^* \quad E \quad -\vec{r}_2^*) \cdot \begin{pmatrix} \vec{a}_1 \\ \vec{a}_2 \end{pmatrix} - \vec{\omega}_1 \times (\vec{\omega}_1 \times \vec{r}_1) + \vec{\omega}_2 \times (\vec{\omega}_2 \times \vec{r}_2)$ and obtain $\vec{c} = \vec{\omega}_1 \times (\vec{\omega}_1 \times \vec{r}_1) - \vec{\omega}_2 \times (\vec{\omega}_2 \times \vec{r}_2)$.

So given our original constraint $\vec{C}_p(\vec{q}_1, \vec{q}_2) := \vec{p}_2 - \vec{p}_1 = \vec{0}$ and assuming $\vec{C}_p = \vec{0}$ and $\dot{\vec{C}}_p = \vec{0}$ we were able to reduce the problem of maintaining $\vec{C}_p = \vec{0}$ to the problem of enforcing $\ddot{\vec{C}}_p = \vec{0}$ which is an *acceleration-level constraint* with $J_1 = (-E \quad \vec{r}_1^*)$, $J_2 = (E \quad -\vec{r}_2^*)$ and $\vec{c} = \vec{\omega}_1 \times (\vec{\omega}_1 \times \vec{r}_1) - \vec{\omega}_2 \times (\vec{\omega}_2 \times \vec{r}_2)$. We now need to compute the generalized constraint forces $\vec{F}_1^c$ and $\vec{F}_2^c$ to be applied to the first and second body, respectively. *Lagrange multiplier method* computes these forces as a linear combination of the rows of the Jacobian matrices (that are *known apriori*), $\vec{F}_1^c = J_1^T \cdot \vec{\lambda}$, $\vec{F}_2^c = J_2^T \cdot \vec{\lambda}$, and solves for the *unknown* coefficients (multipliers) $\vec{\lambda}$ in the combination so that $J_1 \cdot \vec{a}_1 + J_2 \cdot \vec{a}_2 = \vec{c}$ after the external forces $\vec{F}_1^{total}$ and $\vec{F}_2^{total}$ and constraint forces $\vec{F}_1^c$ and $\vec{F}_2^c$ were applied to the bodies. This can be imagined as follows. Each row of the three rows in $J_1 \cdot \vec{a}_1 + J_2 \cdot \vec{a}_2 = \vec{c} \in \boldsymbol{R}^3$ defines a hypersurface in the space of points $(\vec{a}_1, \vec{a}_2)$ and the $(\vec{a}_1, \vec{a}_2)$ acceleration is valid if $(\vec{a}_1, \vec{a}_2)$ lies on each of these hypersurfaces. Now, the normal of the j-th hypersurface equals the j-th row of $(J_1 \quad J_2)$ and so in order to project $(\vec{a}_1, \vec{a}_2)$ onto the j-th hypersurface, the force $\lambda_j \cdot (J_1)_j$ has to be applied to the first body and $\lambda_j \cdot (J_2)_j$ has to be applied to the second body.

Let's solve for the multipliers $\vec{\lambda}$. For that, let's concatenate individual vectors and matrices into global vectors and matrices characterizing the whole rigid body system, we get $\vec{a} = (\vec{a}_1, \vec{a}_2)$, $J = (J_1 \quad J_2)$, $\vec{F}_{total} = \left(\vec{F}_1^{total}, \vec{F}_2^{total}\right)$, $\vec{F}_c = J^T \cdot \vec{\lambda} = \left(\vec{F}_1^c, \vec{F}_2^c\right)$, $M = \begin{pmatrix} M_1 & 0 \\ 0 & M_2 \end{pmatrix}$ and $J \cdot \vec{a} = \vec{c}$. From the section on equations of motion, we get that the acceleration $\vec{a}$ of the rigid body system after the total external force $\vec{F}_{total}$ and constraint force $\vec{F}_c$ are added to the system equals $\vec{a} = M^{-1} \cdot \left(\vec{F}_{total} + \vec{F}_c\right) = M^{-1} \cdot \left(\vec{F}_{total} + J^T \cdot \vec{\lambda}\right) = M^{-1} \cdot \vec{F}_{total} + M^{-1} \cdot J^T \cdot \vec{\lambda}$. This acceleration has to satisfy the constraint $J \cdot \vec{a} = \vec{c}$ and so $J \cdot M^{-1} \cdot \vec{F}_{total} + J \cdot M^{-1} \cdot J^T \cdot \vec{\lambda} = \vec{c}$, $(J \cdot M^{-1} \cdot J^T) \cdot \vec{\lambda} + \left(J \cdot M^{-1} \cdot \vec{F}_{total} - \vec{c}\right) = \vec{0}$, finally producing a system of linear equations $A \cdot \vec{\lambda} + \vec{b} = \vec{0}$, where $A = J \cdot M^{-1} \cdot J^T$ is a 3×3 matrix, $\vec{b} = J \cdot M^{-1} \cdot \vec{F}_{total} - \vec{c}$ is a 3×1 vector and $\vec{\lambda} \in \boldsymbol{R}^3$ are the multipliers to be solved for. Once $\vec{\lambda}$ are known, constraint force $\vec{F}_c = J^T \cdot \vec{\lambda} = \left(\vec{F}_1^c, \vec{F}_2^c\right)$ is applied to the bodies.

3.2 Acceleration constraints

We will now generalize the approach from the previous section for c constraints and n bodies. The index i will be used to index constraints, $i = 1, \dots, c$, and the index j will be used to index bodies, $j = 1, \dots, n$. Vectors $\vec{q} = (\vec{q}_1, \dots, \vec{q}_n)$, $\vec{v} = (\vec{v}_1, \dots, \vec{v}_n)$ and $\vec{a} = (\vec{a}_1, \dots, \vec{a}_n)$ will refer to the generalized position, velocity and acceleration of the rigid body system, $\vec{F}_{total} = (\vec{F}_1^{total}, \dots, \vec{F}_n^{total})$ will refer to the total external force exerted on the system and $\vec{F}_c = \left(\left(\vec{F}_c\right)_1, \dots, \left(\vec{F}_c\right)_n\right)$ will refer to the total constraint force exerted on the system due to all constraints.

Let M_j be the mass matrices of the individual bodies in the system. We then have $M_j \cdot \vec{a}_j = \vec{F}_j^{total}$ and so if M is a square block diagonal matrix with the individual matrices M_j on the diagonal, which we call the *mass matrix of the rigid body* system, we can relate the system acceleration $\vec{a}$ due to the application of $\vec{F}_{total}$ by $M \cdot \vec{a} = \vec{F}_{total}$, where

$$M = \begin{pmatrix} M_1 & 0 & \dots & 0 \\ 0 & M_2 & \dots & 0 \\ \vdots & \vdots & \ddots & \vdots \\ 0 & 0 & \dots & M_n \end{pmatrix}.$$

Constraint i acts on two bodies A_i and B_i, has a dimensionality m_i and removes m_i degrees of freedom (DOFs) from the system. For example, if the two bodies are connected by a 3D revolute joint - $m_i = 3$, because the joint constrains position of body A_i with respect to B_i such that the two are affixed at the joint location (see Fig. 2). Note that while the constraint removes only 3 degrees of freedom, it affects both linear and angular properties of the system. A hinge joint will remove additional 2 degrees of freedom, allowing only rotational motion about a single axis with respect to the joint, resulting in a constraint of dimension $m_i = 5$, etc.

The constraint is characterized by a $m_i \times 6n$ matrix J_i of rank m_i called the constraint's Jacobian matrix consisting of n $m_i \times 6$ blocks due to individual bodies and a constraint equation right-hand-side vector $\vec{c}_i$ of length m_i. J_i has only two non-zero blocks, one due to the first constrained body A_i and one due to the second constrained body B_i, referred to by J_{i,A_i} and J_{i,B_i}. According to the Lagrange multiplier approach, the constraint is enforced by applying a constraint force $\vec{F}_c^i = J_i^T \cdot \vec{\lambda}_i = \left(\left(\vec{F}_c^i\right)_1, \dots, \left(\vec{F}_c^i\right)_n\right)$ to the rigid body system, determined by the values of m_i multipliers $\vec{\lambda}_i$. Each row $k = 1, \dots, m_i$ of J_i removes one DOF from the system and contributes to the constraint force $\vec{F}_c^i$ by exerting a force $\left(\vec{\lambda}_i\right)_k \cdot (J_i)_k$ on the system. Due to the way J_i is defined, $\left(\vec{F}_c^i\right)_{A_i} = J_{i,A_i}^T \cdot \vec{\lambda}_i$ and $\left(\vec{F}_c^i\right)_{B_i} = J_{i,B_i}^T \cdot \vec{\lambda}_i$ are the only non-zero blocks of $\vec{F}_c^i$ and $\left(\vec{F}_c^i\right)_{A_i}$ is the constraint force applied to the first body and $\left(\vec{F}_c^i\right)_{B_i}$ is the constraint force applied to the second body.

Let's stack the individual $m_i \times 6n$ Jacobian matrices J_i by rows to a single $m \times 6n$ Jacobian matrix J, where $m = \sum_i m_i$ is the total number of DOFs removed from the system. J is then a block matrix with $c \times n$ blocks whose non-zero blocks are given by J_{i,A_i} and J_{i,B_i}. Then the total constraint force $\vec{F}_c$ exerted on the system equals $\vec{F}_c = \sum_i \vec{F}_c^i = J^T \cdot \vec{\lambda}$, where $\vec{\lambda} = (\vec{\lambda}_1, \dots, \vec{\lambda}_c)$ is a $m \times 1$ vector of Lagrange multipliers due to all constraints. Because constraints should not be conflicting, J is assumed to have full rank.

Let $A = J \cdot M^{-1} \cdot J^T$, $\vec{c} = (\vec{c}_1, \dots, \vec{c}_c)$ and $\vec{b} = J \cdot M^{-1} \cdot \vec{F}_{total} - \vec{c}$. Matrix A is a $m \times m$ matrix and can be treated as if it consisted of $c \times c$ blocks due to individual constraint pairs such that the value of the (i_1, i_2)-th block of size $m_{i_1} \times m_{i_2}$ due to the i_1-th constraint and the i_2-th constraint is given by $A_{i_1,i_2} = \sum_j J_{i_1,j} \cdot M_j^{-1} \cdot \left(J_{i_2,j}\right)^T$. Because the individual matrices M_j and M_j^{-1} are positive definite, M and M^{-1} are positive definite and so because J is assumed to have full rank, A is also positive definite. We will use A_i (with slight abuse of notation) to denote the i-th block row of A due to constraint i. Vector $\vec{b}$ is a vector of length m consisting of c blocks due to the individual constraints. We use $\vec{b}_i$ to refer to the i-th block of $\vec{b}$ of length m_i due to constraint i.

We will now discuss specific types of constraints. Each constraint i will generate a constraint force of the same form $\vec{F}_c^i = J_i^T \cdot \vec{\lambda}_i$ but different constraint types will lead to different conditions on the legal values of the multipliers $\vec{\lambda}$, essentially constraining the directions the constraint force can act along (can it push, can it pull or can it do both?).

3.2.1 Equality constraints

We define *acceleration level equality constraint* i as follows. The constraint acts on two bodies A_i and B_i, has a dimensionality m_i and is specified by two $m_i \times 6$ matrices J_{i,A_i} and J_{i,B_i} and a right-hand-side vector $\vec{c}_i$ of length m_i. The constraint requests that $J_{i,A_i} \cdot \vec{a}_{A_i} + J_{i,B_i} \cdot \vec{a}_{B_i} = \vec{c}_i$ for accelerations $\vec{a}_{A_i}$ and $\vec{a}_{B_i}$.

The J_{i,A_i} and J_{i,B_i} matrices are called the Jacobian blocks due to the first and the second body and are supposed to have full rank. This terminology stems from the fact that if the acceleration-level constraint implements a position-level constraint $\vec{C}_p(\vec{q}_{A_i}, \vec{q}_{B_i}) = \vec{0}$ or a velocity-level constraint $\vec{C}_v(\vec{v}_{A_i}, \vec{v}_{B_i}) = \vec{0}$ then $J_{i,A_i} = \frac{\partial \vec{C}_p}{\partial \vec{q}_{A_i}}$ and $J_{i,B_i} = \frac{\partial \vec{C}_p}{\partial \vec{q}_{B_i}}$ or $J_{i,A_i} = \frac{\partial \vec{C}_v}{\partial \vec{v}_{A_i}}$ and $J_{i,B_i} = \frac{\partial \vec{C}_v}{\partial \vec{v}_{B_i}}$. The constraint is an equality constraint because it is described by a linear equality.

Let's derive conditions on $\vec{\lambda}$ due to the acceleration level equality constraint i. Using our rigid body system dynamics equation, we get that the system acceleration $\vec{a}$ after the total external force $\vec{F}_{total}$ and total constraint force $\vec{F}_c = J^T \cdot \vec{\lambda}$ are applied to the system equals $\vec{a} = M^{-1} \cdot (\vec{F}_{total} + J^T \cdot \vec{\lambda})$. The constraint equation requests that $J_i \cdot \vec{a} - \vec{c}_i = \vec{0}$ which means that $J_i \cdot \vec{a} - \vec{c}_i = (J \cdot \vec{a} - \vec{c})_i = \left(J \cdot M^{-1} \cdot \vec{F}_{total} + J \cdot M^{-1} \cdot J^T \cdot \vec{\lambda} - \vec{c}\right)_i = \left((J \cdot M^{-1} \cdot J^T) \cdot \vec{\lambda} + (J \cdot M^{-1} \cdot \vec{F}_{total} - \vec{c})\right)_i = \left(A \cdot \vec{\lambda} + \vec{b}\right)_i = A_i \cdot \vec{\lambda} + \vec{b}_i = \vec{0}$. Hence we get that equality constraint i requires that

$$A_i \cdot \vec{\lambda} + \vec{b}_i = \vec{0} \tag{5}$$

which is an equality constraint on the values of $\vec{\lambda}$.

3.2.2 Inequality constraints

Let's think of enforcing a different kind of constraint such that the equality sign = in the constraint's formulation is replaced with either a greater-than-or-equal sign ≥ or a less-than-or-equal sign ≤. For example, if $C_p(\vec{q}_1, \vec{q}_2)$ measures a distance of a ball from the ground plane, we might want to enforce a one-dimensional position constraint $C_p(\vec{q}_1, \vec{q}_2) \geq 0$ requesting that the ball lies above the ground. Assuming that both $C_p(\vec{q}_1, \vec{q}_2) = 0$ and $\dot{C}_p(\vec{q}_1, \vec{q}_2) = 0$ (the ball rests on the ground), the constraint can be implemented by maintaining $\ddot{C}_p(\vec{q}_1, \vec{q}_2) \geq 0$, which is an acceleration-level greater-or-equal constraint.

3.2.2.1 Greater-or-equal constraints

We define *acceleration level greater-or-equal constraint* i as follows. The constraint acts on two bodies A_i and B_i, has a dimensionality m_i and is specified by two $m_i \times 6$ matrices J_{i,A_i} and J_{i,B_i} and a right-hand-side vector $\vec{c}_i$ of length m_i. The constraint requests that $J_{i,A_i} \cdot \vec{a}_{A_i} + J_{i,B_i} \cdot \vec{a}_{B_i} \geq \vec{c}_i$ for accelerations $\vec{a}_{A_i}$ and $\vec{a}_{B_i}$.

Let's present conditions on $\vec{\lambda}$ due to the acceleration level greater-or-equal constraint i. Similarly to the equality case, $J_{i,A_i} \cdot \vec{a}_{A_i} + J_{i,B_i} \cdot \vec{a}_{B_i} \geq \vec{c}_i$ can be rewritten as (1) $J_{i,A_i} \cdot \vec{a}_{A_i} + J_{i,B_i} \cdot \vec{a}_{B_i} - \vec{c}_i = J_i \cdot \vec{a} - \vec{c}_i = A_i \cdot \vec{\lambda} + \vec{b}_i \geq \vec{0}$, which is an inequality greater-or-equal constraint on the values of $\vec{\lambda}$. Now, let's recall that in *Lagrange multiplier approach*, the goal of $\vec{F}_c^i$ is to cancel those components of $\vec{F}_{total}$ that would make the bodies accelerate towards invalid states. In the case of an equality constraint, the bodies were restricted to remain on the intersections of the hypersurfaces due to the constraint's DOFs and $\vec{F}_c^i$ cancelled accelerations along the directions of the hypersurface normals. In the case of a greater-or-equal constraint, however, the bodies can move away from a hypersurface along the direction of the hypersurface's normal, but not in the opposite direction. In other words, positive accelerations along the positive directions of the normals are unconstrained and therefore (2) $\vec{\lambda}_i \geq \vec{0}$ (the constraint force can not pull the bodies back to the hypersurface). In addition, (3) if the bodies are already accelerating to the front of the hypersurface k, $\left(J_{i,A_i} \cdot \vec{a}_{A_i} + J_{i,B_i} \cdot \vec{a}_{B_i} - \vec{c}_i\right)_k > 0$, then the constraint force due to that hypersurface must vanish, that is $\left(\vec{\lambda}_i\right)_k = 0$, so that no energy would be added to the system (constraint force is as "lazy" as possible). These conditions can be restated in terms of the i-th block row of matrix A and the i-th block of vector $\vec{b}$ as follows,

$$A_i \cdot \vec{\lambda} + \vec{b_i} \geq \vec{0}$$
$$\vec{\lambda_i} \geq \vec{0} \tag{6}$$
$$\left(A_i \cdot \vec{\lambda} + \vec{b_i}\right) \cdot \vec{\lambda_i} = 0,$$

where $\left(A_i \cdot \vec{\lambda} + \vec{b_i}\right) \cdot \vec{\lambda_i} = \sum_{k=1}^{m_i} \left(A_i \cdot \vec{\lambda} + \vec{b_i}\right)_k \cdot \left(\vec{\lambda_i}\right)_k = 0$ in fact means that $\left(A_i \cdot \vec{\lambda} + \vec{b_i}\right)_k \cdot \left(\vec{\lambda_i}\right)_k$ for $1 \leq k \leq m_i$ because both the products have to be positive. It is said that the components of $A_i \cdot \vec{\lambda} + \vec{b_i}$ are *complementary* to the corresponding components of $\vec{\lambda_i}$.

3.2.2.2 Less-or-equal constraints

We define *acceleration level less-or-equal constraint i* as follows. The constraint acts on two bodies A_i and B_i, has a dimensionality m_i and is specified by two $m_i \times 6$ matrices J_{i,A_i} and J_{i,B_i} and a right-hand-side vector $\vec{c_i}$ of length m_i. The constraint requests that $J_{i,A_i} \cdot \vec{a}_{A_i} + J_{i,B_i} \cdot \vec{a}_{B_i} \leq \vec{c_i}$ for accelerations $\vec{a}_{A_i}$ and $\vec{a}_{B_i}$.

Analogously to the previous case, we obtain the following set of conditions on multipliers $\vec{\lambda}$ due to the acceleration level less-or-equal constraint i. In addition to the condition $J_i \cdot \vec{a} - \vec{c_i} = A_i \cdot \vec{\lambda} + \vec{b_i} \leq \vec{0}$, multipliers due to constraint i have to be negative and complementary to $\vec{\lambda_i}$,

$$A_i \cdot \vec{\lambda} + \vec{b_i} \leq \vec{0}$$
$$\vec{\lambda_i} \leq \vec{0} \tag{7}$$
$$\left(A_i \cdot \vec{\lambda} + \vec{b_i}\right) \cdot \vec{\lambda_i} = 0.$$

Less-or-equal constraints i can trivially be converted to greater-or-equal constraints by negating the Jacobian blocks and the right-hand-side vector $\vec{c_i}$ and so they do not have to be handled as a special case.

3.2.3 Bounded equality constraints

Let's suppose we want to implement a one-dimensional constraint that would behave like an equality constraint $J_i \cdot \vec{a} = \vec{c_i}$ such that the constraint would break if the magnitude $\| J_i^T \| \cdot \left| \left(\vec{\lambda_i}\right)_1 \right|$ of the constraint force $\vec{F_c^i} = J_i^T \cdot \vec{\lambda_i}$ required to maintain the constraint exceeds a certain limit. Such a capability could, for example, be used for the implementation of various kinds of motors with limited power. Now, because $\| J_i^T \|$ is known, limiting the force magnitude (in this case) is equivalent to specifying the lower and upper bound on the value of the multiplier $\left(\vec{\lambda_i}\right)_1$. Hence, without loss of generality we can assume the bounds on $\vec{\lambda_i}$ are given instead. In the general case of a multi-dimensional constraint, we assume that each multiplier has its own bounds, independent of the values of other multipliers, so that the problem of solving for $\vec{\lambda}$ remains tractable.

We define *acceleration level bounded equality constraint i* as follows. The constraint acts on two bodies A_i and B_i, has a dimensionality m_i and is specified by two $m_i \times 6$ matrices J_{i,A_i} and J_{i,B_i}, a right-hand-side vector $\vec{c_i}$ of length m_i and $\vec{\lambda_i}$ bounds $\vec{\lambda}_i^{lo} \leq \vec{0}$ and $\vec{\lambda}_i^{hi} \geq \vec{0}$. The constraint requests that $\left(\vec{\lambda}_i^{lo}\right)_k \leq \left(\vec{\lambda_i}\right)_k \leq \left(\vec{\lambda}_i^{hi}\right)_k$ and implements the equality constraint $J_{i,A_i} \cdot \vec{a}_{A_i} + J_{i,B_i} \cdot \vec{a}_{B_i} = \vec{c_i}$ for accelerations $\vec{a}_{A_i}$ and $\vec{a}_{B_i}$ subject to constraint force limits given by $\vec{\lambda}_i^{lo}$ and $\vec{\lambda}_i^{hi}$.

We will now elaborate on what constraint force limits due to the acceleration level bounded equality constraint i really mean and what the corresponding conditions on $\vec{\lambda}$ look like. Following up on the hypersurface interpretation of the equality constraint $J_i \cdot \vec{a} - \vec{c}_i = \vec{0}$, if the bodies are to move off the hypersurface k due to the k-th constraint DOF in the direction of the surface normal, a negative $(\vec{\lambda}_i)_k$ is required to cancel the acceleration. Now, if the value of $(\vec{\lambda}_i)_k$ required to fully cancel the acceleration is less than the allowed lower limit $(\vec{\lambda}_i^{lo})_k$, clamped $(\vec{\lambda}_i)_k \geq (\vec{\lambda}_i^{lo})_k$ would not yield a constraint force strong enough to cancel the prohibited acceleration and in the end $J_i \cdot \vec{a} - \vec{c}_i > \vec{0}$. Similarly, if the bodies are to move off the hypersurface in the opposite direction, a positive $(\vec{\lambda}_i)_k$ is required to cancel the acceleration. If $(\vec{\lambda}_i)_k$ is clamped such that $(\vec{\lambda}_i)_k \leq (\vec{\lambda}_i^{hi})_k$ and the acceleration is not cancelled fully then $J_i \cdot \vec{a} - \vec{c}_i < \vec{0}$. Putting this discussion into equations and assuming $(\vec{\lambda}_i^{lo})_k \leq 0$ and $(\vec{\lambda}_i^{hi})_k \geq 0$, we get

$$
\begin{gathered}
(\vec{\lambda}_i^{lo})_k \leq (\vec{\lambda}_i)_k \leq (\vec{\lambda}_i^{hi})_k \\
(\vec{\lambda}_i)_k = (\vec{\lambda}_i^{lo})_k \Rightarrow (A_i \cdot \vec{\lambda} + \vec{b}_i)_k \geq 0 \\
(\vec{\lambda}_i)_k = (\vec{\lambda}_i^{hi})_k \Rightarrow (A_i \cdot \vec{\lambda} + \vec{b}_i)_k \leq 0 \\
(\vec{\lambda}_i^{lo})_k < (\vec{\lambda}_i)_k < (\vec{\lambda}_i^{lo})_k \Rightarrow (A_i \cdot \vec{\lambda} + \vec{b}_i)_k = 0.
\end{gathered}
\tag{8}
$$

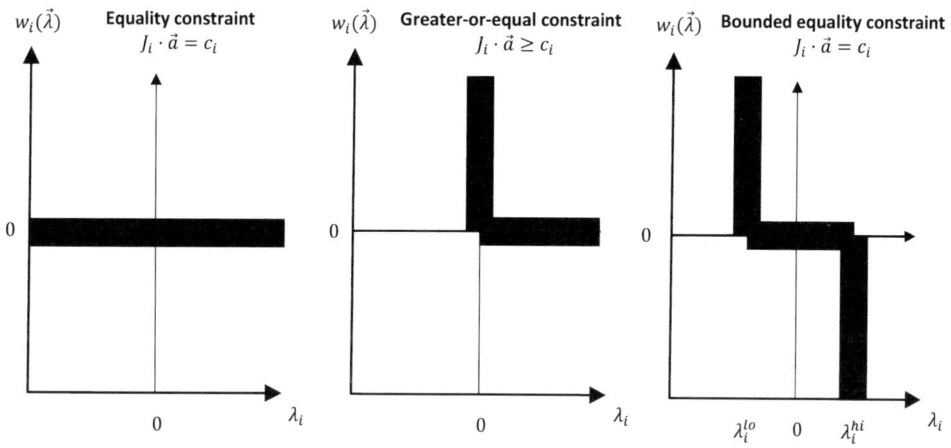

Fig. 3. Visualization of complementarity conditions on the pairs $(\lambda_i, w_i(\vec{\lambda}))$ due to different kinds of one dimensional constraints i, where $w_i(\vec{\lambda}) := A_i \cdot \vec{\lambda} + b_i = J_i \cdot \vec{a} - c_i$. Thick lines indicate permissible values for the $(\lambda_i, w_i(\vec{\lambda}))$ pairs. As can be seen, equality constraint requests $w_i(\vec{\lambda})$ to be zero and lets λ_i take an arbitrary value. Greater-or-equal constraint requests both $w_i(\vec{\lambda})$ and λ_i to be non-negative and complementary to each other. Bounded equality constraint generalizes the two previous cases by introducing explicit limits $\lambda_i^{lo} \leq 0$ and $\lambda_i^{hi} \geq 0$ on the values of λ_i. For improved readability, $\overrightarrow{\ }$ accents have been removed from one-dimensional vectors related to the constraint i.

Bounded equality constraints are generalization of both inequality and equality constraints. For example, if we set $\vec{\lambda}_i^{lo} = \vec{0}$ and $\vec{\lambda}_i^{hi} = \vec{\infty}$ then the bounded equality constraint i turns to a greater-or-equal constraint i with the same Jacobian blocks and right-hand-side vector $\vec{c}_i$. Similarly, by setting $\vec{\lambda}_i^{lo} = -\vec{\infty}$ and $\vec{\lambda}_i^{hi} = \vec{0}$, the constraint turns to a less-or-equal constraint. Finally, by setting $\vec{\lambda}_i^{lo} = -\vec{\infty}$ and $\vec{\lambda}_i^{hi} = \vec{\infty}$, the constraint turns to an unbounded equality constraint.

3.2.4 Reduction to LCP

In the previous section we have discussed several constraint types and showed what conditions on the multipliers $\vec{\lambda}$ they impose. Our goal is now to solve for $\vec{\lambda}$ obeying the presented conditions so that the constraint force $\vec{F}_c = J^T \cdot \vec{\lambda}$ could be exerted on the system. As it turns out, the problem of solving for $\vec{\lambda}$ is equivalent to solving of specific kinds of *linear complementarity problems (LCPs)* for which efficient algorithms exist and so we can compute $\vec{\lambda}$ by using a LCP solver, (Smith, 2004; Vondrak, 2006; Cline, 2002). To simplify the discussion, we assume that every inequality and bounded equality constraint i is one-dimensional, $m_i = 1$. As a result, we can simply write λ_i instead of $\left(\vec{\lambda}_i\right)_1$, etc.

If all the constraints are unbounded equalities, the corresponding conditions on $\vec{\lambda}$ are given by $A \cdot \vec{\lambda} + \vec{b} = \vec{0}$ which is a linear system that can be solved efficiently by standard factorization techniques. If all constraints are greater-or-equal constraints, we get a *pure linear complementarity problem* of the form $A \cdot \vec{\lambda} + \vec{b} \geq \vec{0}$, $\vec{\lambda} \geq \vec{0}$, $\vec{\lambda} \cdot \left(A \cdot \vec{\lambda} + \vec{b}\right) = \vec{0}$, which can be solved by a standard LCP solver. If there are k unbounded equality constraints and $c - k$ greater-or-equal constraints, we get a *mixed linear complementarity problem* $A_{eq} \cdot \vec{\lambda} + \vec{b}_{eq} = \vec{0}, A_{ineq} \cdot \vec{\lambda} + \vec{b}_{ineq} \geq \vec{0}, \vec{\lambda}_{ineq} \geq \vec{0}, \vec{\lambda}_{ineq} \cdot \left(A_{ineq} \cdot \vec{\lambda} + \vec{b}_{ineq}\right) = 0$, where $A_{eq}, \vec{b}_{eq}$ denotes the rows of $A, \vec{b}$ due to equality constraints and $A_{ineq}, \vec{b}_{ineq}$ denotes the rows of $A, \vec{b}$ due to inequality constraints. Mixed LCPs can be solved by mixed LCP solvers. Finally, if there are k unbounded equality constraints and $c - k$ bounded equality-constraints (including inequality constraints i with appropriately set $\vec{\lambda}_i$ limits), we get a *lo-hi linear complementarity problem* $A_{eq} \cdot \vec{\lambda} + \vec{b}_{eq} = \vec{0}, \lambda_i^{lo} \leq \lambda_i \leq \lambda_i^{hi}, \lambda_i = \lambda_i^{lo} \Rightarrow A_i \cdot \vec{\lambda} + b_i \geq 0, \lambda_i = \lambda_i^{hi} \Rightarrow A_i \cdot \vec{\lambda} + b_i \leq 0, \lambda_i^{lo} < \lambda_i < \lambda_i^{hi} \Rightarrow A_i \cdot \vec{\lambda} + b_i = 0$, where i indexes unbounded equality and inequality constraints. This is the most general form that can handle all constraint forms we have discussed and can also be solved efficiently.

3.3 Velocity constraints

So far we have discussed how constraints can be implemented on the accelerations. It is useful, however, to specify constraints on the velocities as well. Let's recall the example with the ball and the ground plane where the goal is to enforce a one-dimensional position-level constraint $C_p(\vec{q}_1, \vec{q}_2) \geq 0$ stipulating that the ball has to stay above the ground. Now, if $C_p(\vec{q}_1(t), \vec{q}_2(t)) = 0$ and $\dot{C}_p(\vec{q}_1(t), \vec{q}_2(t)) < 0$ at the current time t (the ball strikes the ground plane) then $C_p(\vec{q}_1(t + \epsilon), \vec{q}_2(t + \epsilon)) < 0$ at the time instant $t + \epsilon$ regardless of accelerations at time t for a sufficiently small ϵ. In order to ensure that the constraint is maintained at $t + \epsilon$, velocities at time t have to change so that $\dot{C}_p(\vec{q}_1(t), \vec{q}_2(t)) \geq 0$. This, however, is a constraint on the velocity.

3.3.1 Impulsive dynamics

We will now outline the concept of impulsive forces and first-order rigid body dynamics. With regular forces, the effects of forces on positions and orientations of rigid bodies are determined by second-order (Newtonian) dynamics in which velocities change through the integration of forces while positions change through the integration of velocities. With impulsive forces, the effects of forces on positions and orientations are determined by first-order (impulsive) dynamics in which velocities change directly through the application of impulsive forces and positions change through the integration of velocities.

We postulate *impulsive force* $\vec{J}_F$ as a force with "units of momentum". If $\vec{P}$ and $\vec{L}$ are the linear and angular momentums of a rigid body and $\vec{J}_F$ is applied to the body at the world space position $\vec{r}$, then the linear momentum $\vec{P}$ changes by the value $\Delta\vec{P} = \vec{J}_F$ and the angular momentum $\vec{L}$ changes by the value $\Delta\vec{L} = \vec{J}_\tau$, where $\vec{J}_\tau = (\vec{r} - \vec{x}) \times \vec{J}_F$ is *impulsive torque* due to the impulsive force $\vec{J}_F$. Impulsive forces and torques can be seen as "ordinary" forces and torques that directly change the body's linear and angular momentums, instead of affecting their time derivatives.

Similarly to the second-order dynamics, we couple linear and corresponding angular quantities to generalized quantities. That way, we obtain *generalized momentum* $\vec{F}_{imp}^{total} = (\vec{P}, \vec{L})$ and *generalized impulsive force (impulse)* $\vec{F}_{imp} = (\vec{J}_F, \vec{J}_\tau)$. Then if M is the mass matrix of the rigid body and $\vec{v}$ is the body's generalized velocity, we immediately get $M \cdot \vec{v} = \vec{F}_{imp}^{total}$ from the definition of the linear and angular momentum. Moreover, our momentum update rules state that the change $\Delta\vec{v}$ of generalized velocity $\vec{v}$ due to the application of the generalized impulse $\vec{F}_{imp}$ equals $\Delta\vec{v} = M^{-1} \cdot \vec{F}_{imp}$. Therefore the first-order dynamics relating velocities $\vec{v}$ to impulses $\vec{F}_{imp}$ is given by

$$M \cdot \vec{v} = \vec{F}_{imp} \tag{9}$$

and $\vec{F}_{imp}^{total}$ can be seen as a generalized *total external impulse* acting on the body that consists of the only term - the inertial term $(\vec{P}, \vec{L})$. This directly compares to the case of second-order dynamics that relates accelerations $\vec{a}$ to forces $\vec{F}$ by $M \cdot \vec{a} = \vec{F}$.

If we have a set of n rigid bodies with mass matrices $M_1, \dots, M_n$, generalized velocities $\vec{v}_1, \dots, \vec{v}_n$ and total external impulses $\left(\vec{F}_{imp}^{total}\right)_1, \dots, \left(\vec{F}_{imp}^{total}\right)_n$ then the first-order dynamics of the system is given by $M \cdot \vec{v} = \vec{F}_{imp}^{total}$, where M is a mass matrix of the system made of $M_1, \dots, M_n$, $\vec{v} = (\vec{v}_1, \dots, \vec{v}_n)$ and $\vec{F}_{imp}^{total} = \left(\left(\vec{F}_{imp}^{total}\right)_1, \dots, \left(\vec{F}_{imp}^{total}\right)_n\right)$. Analogously to the acceleration case, we call $\vec{v}$ the velocity of the system and $\vec{F}_{imp}^{total}$ the total external impulse exerted on the system (system momentum).

3.3.2 Constraints

We can now transfer everything we know about acceleration-level constraints, defined with respect to accelerations and forces, to the realm of velocity-level constraints, defined with respect to velocities and impulsive forces. There is no need to do any derivations because acceleration-level formulation of rigid body dynamics exactly corresponds to the velocity-level formulation of the impulsive dynamics. The only differences are due to the fact that we will now work with system velocities $\vec{v}$, impulsive constraint forces $\vec{F}_{imp}^{c}$ and

momentums $\vec{F}_{imp}^{total}$ instead of accelerations $\vec{a}$, constraint forces $\vec{F}_c$ and total external forces $\vec{F}_{total}$. In consequence, the same algorithms can be used to implement velocity constraints.
We define *velocity level constraint* i as follows. The constraint acts on two bodies A_i and B_i, has a dimensionality m_i and is specified by two $m_i \times 6$ matrices J_{i,A_i} and $J_{i,B_i}$and a right-hand-side vector $\vec{k}_i$ of length m_i. The constraint requests either $J_{i,A_i} \cdot \vec{v}_{A_i} + J_{i,B_i} \cdot \vec{v}_{B_i} = \vec{k}_i$, $J_{i,A_i} \cdot \vec{v}_{A_i} + J_{i,B_i} \cdot \vec{v}_{B_i} \leq \vec{k}_i$ or $J_{i,A_i} \cdot \vec{v}_{A_i} + J_{i,B_i} \cdot \vec{v}_{B_i} \geq \vec{k}_i$ and is implemented by exerting a constraint impulse $\left(\vec{F}_c^i\right)_{imp} = J_i^T \cdot \vec{\lambda}_i$ determined by the values of multipliers $\vec{\lambda}_i$. In addition, if bounds on the valid multiplier values $\vec{\lambda}_i^{lo} \leq \vec{0}$ and $\vec{\lambda}_i^{hi} \geq \vec{0}$ are provided, then the constraint describes a *bounded equality constraint* i that requests $\left(\vec{\lambda}_i^{lo}\right)_k \leq \left(\vec{\lambda}_i\right)_k \leq \left(\vec{\lambda}_i^{hi}\right)_k$ and implements the equality constraint $J_{i,A_i} \cdot \vec{v}_{A_i} + J_{i,B_i} \cdot \vec{v}_{B_i} = \vec{k}_i$ for velocities $\vec{v}_{A_i}$ and $\vec{v}_{B_i}$ subject to constraint impulse limits given by $\vec{\lambda}_i^{lo}$ and $\vec{\lambda}_i^{hi}$. Multipliers $\vec{\lambda}$ can be computed by solving the same LCP problems like before. If there are c constraints, we will get $A = J \cdot M^{-1} \cdot J^T$ and $\vec{b} = J \cdot M^{-1} \cdot \vec{F}_{imp}^{total} - \vec{k}$, where $\vec{k} = \left(\vec{k}_1, \dots, \vec{k}_c\right)$.

3.4 Position constraints

Motion control constraints are most often specified on the position level because it is the natural way of expressing desired motion. In the earlier section, we have already discussed how position level constraints can be implemented either on the acceleration or velocity level, but this time, we will do it more thoroughly and will also show how prior constraint errors due to numerical inaccuracies could be reduced during simulation.
We never enforce constraints directly on the position level. Position level enforcement would require use of custom equations of motion specific to the set of constraints. As a result equations would have to change each time the constraint set is updated. For the rest of the section, we will assume we have n rigid bodies and c position-level constraints.
We define *position level constraint* i as follows. The constraint acts on two bodies A_i and B_i, has a dimensionality m_i and is specified by a function $\vec{C}_p^i\left(\vec{q}_{A_i}, \vec{q}_{B_i}\right) \in \boldsymbol{R}^{m_i}$ that is differentiable with respect to time so that its velocity level and acceleration level formulations (consistent with our prior definitions) can be obtained by differentiation. *Position level equality constraint* i requests that $\vec{C}_p^i\left(\vec{q}_{A_i}, \vec{q}_{B_i}\right) = \vec{0}$ for generalized positions $\vec{q}_{A_i}$ and $\vec{q}_{B_i}$ and the value of $\vec{C}_p^i\left(\vec{q}_{A_i}, \vec{q}_{B_i}\right)$ can intuitively be thought of as a measurement of the position error for bodies at the position configuration $\left(\vec{q}_{A_i}, \vec{q}_{B_i}\right)$. *Position level greater-or-equal constraint* i requests that $\vec{C}_p^i\left(\vec{q}_{A_i}, \vec{q}_{B_i}\right) \geq \vec{0}$ and *position level less-or-equal constraint* i requests that $\vec{C}_p^i\left(\vec{q}_{A_i}, \vec{q}_{B_i}\right) \leq \vec{0}$.

3.4.1 Acceleration or velocity level

We use constraint forces to implement position level constraints in an incremental way. We start from an initial state that is consistent with the constraint formulation (such that positions and velocities are valid with respect to the position level and velocity level formulations of the constraints) and then apply constraint forces to ensure that the velocity level and position level constraints remain maintained. Alternatively, we start from a state that is consistent with the position level formulations and then apply constraint impulses to ensure that the position level constraints remain maintained.
Please note that whenever an impulse is applied to a body, its velocity changes. In consequence, conditions that have to be met so that a particular constraint could be

implemented on the acceleration level need no longer be valid after the impulse is applied and so it cannot be reliably determined in advance which constraints can be implemented on the acceleration level. To address this issue, we implement all constraints on the velocity level *whenever there is at least one position constraint that has to be implemented on the velocity level.*

3.4.2 Equality constraints with stabilization

Consider the position level equality constraint $\vec{C}_p^i(\vec{q}_{A_i}, \vec{q}_{B_i}) = \vec{0}$. By differentiating $\vec{C}_p^i(\vec{q}_{A_i}, \vec{q}_{B_i}) = \vec{0}$ with respect to time, we get a corresponding velocity level formulation of the position constraint in the form of $\vec{C}_v^i(\vec{v}_{A_i}, \vec{v}_{B_i}) = \vec{0}$, where $\vec{C}_v^i(\vec{v}_{A_i}, \vec{v}_{B_i}) = \frac{\partial}{\partial t}\vec{C}_p^i(\vec{q}_{A_i}, \vec{q}_{B_i}) = J_{i,A_i} \cdot \vec{v}_{A_i} + J_{i,B_i} \cdot \vec{v}_{B_i}$. By differentiating this velocity constraint, we get a corresponding acceleration level formulation $\vec{C}_a^i(\vec{a}_{A_i}, \vec{a}_{B_i}) = \vec{0}$, where $\vec{C}_a^i(\vec{a}_{A_i}, \vec{a}_{B_i}) = \frac{\partial}{\partial t}\vec{C}_v^i(\vec{v}_{A_i}, \vec{v}_{B_i}) = J_{i,A_i} \cdot \vec{a}_{A_i} + J_{i,B_i} \cdot \vec{a}_{B_i} - \vec{c}_i$ and $\vec{c}_i = -\dot{J}_{i,A_i} \cdot \vec{v}_{A_i} - \dot{J}_{i,B_i} \cdot \vec{v}_{B_i}$. The position level constraint i $\vec{C}_p^i = \vec{0}$ can thus be implemented *incrementally* either (1) on the acceleration level, by starting from a state where $\vec{C}_p^i = \dot{\vec{C}}_p^i = \vec{0}$ and applying constraint forces so that $\ddot{\vec{C}}_p^i = \vec{0}$ or (2) on the velocity level, by starting from a state where $\vec{C}_p^i = \vec{0}$ and applying constraint impulses so that $\dot{\vec{C}}_p^i = \vec{0}$. In the first case, constraint forces are applied under the assumption that $\vec{C}_p^i = \dot{\vec{C}}_p^i = \vec{0}$, while in the second case, constraint impulses are applied under the assumption that $\vec{C}_p^i = \vec{0}$. In practice, however, these assumptions often do not hold for various pragmatic reasons. For example, the numerical solver that integrates the equations of motion incurs an integration error or constraint forces are computed with an insufficient precision.

Let's assume we implement the position level constraint i on the velocity level. If the constraint is currently broken, that is $\vec{C}_p^i \neq \vec{0}$, we want to generate a constraint impulse so that the constraint error $\vec{C}_p^i$ will be driven towards a zero vector. This is called *constraint stabilization*. Fortunately, simple stabilization can be implemented by following a procedure suggested in (Cline, 2002). Instead of requiring that $\dot{\vec{C}}_p^i = 0$, we can require that

$$\dot{\vec{C}}_p^i = -\vec{C}_p^i \cdot \alpha \,, \tag{10}$$

where α is a small positive value (dependent on the integration step size) that determines the speed with which the constraint is stabilized. Then, if t is the current time, we have $\vec{C}_p^i(t + \Delta t) \approx \vec{C}_p^i(t) + \Delta t \cdot \dot{\vec{C}}_p^i(t) = \vec{C}_p^i(t) \cdot (1 - \Delta t \cdot \alpha)$ and so we can reduce the position error by simply biasing the request on the desired velocity.

Analogously to the previous case, if we implement the position level constraint i on the acceleration level, we need to reduce both the position error $\vec{C}_p^i$ as well as velocity error $\dot{\vec{C}}_p^i$. That could be done by biasing the request on the desired acceleration $\ddot{\vec{C}}_p^i$. Instead of requiring that $\ddot{\vec{C}}_p^i = \vec{0}$ we can require

$$\ddot{\vec{C}}_p^i = -\vec{C}_p^i \cdot \alpha - \dot{\vec{C}}_p^i \cdot \beta, \tag{11}$$

where α and β are positive constants. Because $\dot{\vec{C}}_p^i = J_i \cdot \vec{v}$ we get $\ddot{\vec{C}}_p^i = -\vec{C}_p^i \cdot \alpha - J_i \cdot \vec{v} \cdot \beta$. Plugging these equations into our constraint definitions, we can therefore implement the position level equality constraint i with stabilization by submitting either the velocity level equality constraint $J_{i,A_i} \cdot \vec{v}_{A_i} + J_{i,B_i} \cdot \vec{v}_{B_i} = -\vec{C}_p^i \cdot \alpha$ or the acceleration level equality constraint $J_{i,A_i} \cdot \vec{a}_{A_i} + J_{i,B_i} \cdot \vec{a}_{B_i} = -\vec{C}_p^i \cdot \alpha - J_i \cdot \vec{v} \cdot \beta - \dot{J}_i \cdot \vec{v}$. Moreover, if we want to implement powering limits, these constraints can be submitted as bounded equality constraints with appropriate force limits.

3.4.3 Inequality constraints with stabilization

Let's assume for simplicity that we work with a one-dimensional position level greater-or-equal constraint $C_p^i \geq 0$, like the one that would stipulate that the ball lies above the ground. This constraint is fundamentally different from a equality constraint because no velocity or acceleration constraints are actually imposed until the constrained bodies reach the boundary $\{(\vec{q}_{A_i}, \vec{q}_{B_i}) \mid C_p^i(\vec{q}_{A_i}, \vec{q}_{B_i}) = 0\}$ of the set of valid positions $\{(\vec{q}_{A_i}, \vec{q}_{B_i}) \mid C_p^i(\vec{q}_{A_i}, \vec{q}_{B_i}) \geq 0\}$ in the position space. We will now discuss three cases, depending on whether $C_p^i > 0$, $C_p^i = 0$ or $C_p^i < 0$ at the current state.

If $C_p^i > 0$ then the bodies did not reach the boundary of the set of valid positions and so no constraints will be imposed. This corresponds to the case when the ball is in the air and does not touch the ground.

If $C_p^i = 0$ then the bodies lie at the boundary of the set of valid positions (the ball touches the ground) and $\dot{C}_p^i \geq 0$ has to be enforced in order to maintain $C_p^i \geq 0$. This $\dot{C}_p^i \geq 0$ constraint can be implemented (1) on the acceleration level by submitting $\ddot{C}_p^i \geq 0$ if $\dot{C}_p^i = 0$, (2) on the acceleration level by ignoring the constraint if $\dot{C}_p^i > 0$ (the velocity-level constraint will hold regardless of the value of $\ddot{C}_p^i$ if a sufficiently small integration step is taken) or (3) on the velocity level by directly requesting $\dot{C}_p^i \geq 0$. The constraint must be implemented on the velocity level if $\dot{C}_p^i < 0$. In practice, we make use of comparisons with tolerances when discriminating between the cases (1) - (3) and add extra terms to the velocity level and acceleration level formulations so that the original position level constraint would be stabilized, i.e. we use $J_{i,A_i} \cdot \vec{a}_{A_i} + J_{i,B_i} \cdot \vec{a}_{B_i} \geq -\vec{C}_p^i \cdot \alpha - J_i \cdot \vec{v} \cdot \beta - \dot{J}_i \cdot \vec{v}$ and $J_{i,A_i} \cdot \vec{v}_{A_i} + J_{i,B_i} \cdot \vec{v}_{B_i} \geq -\vec{C}_p^i \cdot \alpha$.

If $C_p^i < 0$ then the system state is invalid (ball penetrates the ground) and should be rejected. To handle this case, we would like to locate the state when $C_p^i = 0$ so that we could fall-back onto the previous case $C_p^i = 0$. For that, we can either (1) roll the simulation state back to the previous state and then use a bisection-like algorithm to locate the latest valid state $C_p^i = 0$ or (2) we can ignore the position error and act as if $C_p^i = 0$ (this way, we would only rely on the constraint stabilization mechanism to recover $C_p^i \geq 0$).

3.5 Contact

In this section we briefly discuss the problem of enforcing body non-penetration and modeling friction. We will account for these phenomena through constraints that will be associated with contacts reported by a collision detection library. Given the system state at time t, we will use the library to find which shapes are in contact and then formulate non-penetration and friction constraints specific to time t to constrain relative body motion at the contacting points. We informally define *contacts* as relevant coinciding points of contacting pairs of body shapes where contact forces or impulses should act in order to prevent penetration. We assume collision detection library describes contacts by vectors

$(A_i, B_i, \vec{p}_i, \vec{n}_i, \dot{\vec{n}}_i)$ such that contact i involves bodies A_i and B_i contacting at the world space point $\vec{p}_i = \vec{p}_{A_i} = \vec{p}_{B_i}$ ($\vec{p}_{A_i}$ and $\vec{p}_{B_i}$ are the corresponding points attached to A_i and B_i), the contact surface normal at $\vec{p}_i$ is given by a unit vector $\vec{n}_i$ pointing towards B_i and $\dot{\vec{n}}_i$ is the time derivative of $\vec{n}_i$. Let us denote $\vec{r}_{A_i} = \vec{p}_{A_i} - \vec{x}_{A_i}$ and $\vec{r}_{B_i} = \vec{p}_{B_i} - \vec{x}_{B_i}$.

We can then request *non-penetration at contact i* by stipulating that $C_i^n := \vec{n}_i \cdot (\vec{p}_{B_i} - \vec{p}_{A_i}) \geq 0$, which is a one-dimensional position level greater-or-equal constraint. Let's have a look at its time derivatives to retrieve the constraint's Jacobian blocks and the right-hand-side vector. We have $\dot{C}_i^n = \dot{\vec{n}}_i \cdot (\vec{p}_{B_i} - \vec{p}_{A_i}) + \vec{n}_i \cdot (\dot{\vec{p}}_{B_i} - \dot{\vec{p}}_{A_i})$, $\ddot{C}_i^n = \ddot{\vec{n}}_i \cdot (\vec{p}_{B_i} - \vec{p}_{A_i}) + \dot{\vec{n}}_i \cdot (\dot{\vec{p}}_{B_i} - \dot{\vec{p}}_{A_i}) + \dot{\vec{n}}_i \cdot (\dot{\vec{p}}_{B_i} - \dot{\vec{p}}_{A_i}) + \vec{n}_i \cdot (\ddot{\vec{p}}_{B_i} - \ddot{\vec{p}}_{A_i}) = \ddot{\vec{n}}_i \cdot (\vec{p}_{B_i} - \vec{p}_{A_i}) + 2 \cdot \dot{\vec{n}}_i \cdot (\dot{\vec{p}}_{B_i} - \dot{\vec{p}}_{A_i}) + \vec{n}_i \cdot (\ddot{\vec{p}}_{B_i} - \ddot{\vec{p}}_{A_i})$. Because $\vec{p}_{A_i} = \vec{x}_{A_i} + \vec{r}_{A_i}$ and $\vec{p}_{B_i} = \vec{x}_{B_i} + \vec{r}_{B_i}$ we get $\dot{\vec{p}}_{A_i} = \vec{v}_{A_i} + \vec{\omega}_{A_i} \times \vec{r}_{A_i}$, $\dot{\vec{p}}_{B_i} = \vec{v}_{B_i} + \vec{\omega}_{B_i} \times \vec{r}_{B_i}$, $\ddot{\vec{p}}_{A_i} = \vec{a}_{A_i} + \vec{\alpha}_{A_i} \times \vec{r}_{A_i} + \vec{\omega}_{A_i} \times (\vec{\omega}_{A_i} \times \vec{r}_{A_i})$, $\ddot{\vec{p}}_{B_i} = \vec{a}_{B_i} + \vec{\alpha}_{B_i} \times \vec{r}_{B_i} + \vec{\omega}_{B_i} \times (\vec{\omega}_{B_i} \times \vec{r}_{B_i})$. Considering that $\vec{p}_{A_i} = \vec{p}_{B_i}$ at the contact, we then have $\dot{C}_i^n = \vec{n}_i \cdot (\dot{\vec{p}}_{B_i} - \dot{\vec{p}}_{A_i}) = \vec{n}_i \cdot \vec{v}_{B_i} + \vec{n}_i \cdot (\vec{\omega}_{B_i} \times \vec{r}_{B_i}) - \vec{n}_i \cdot \vec{v}_{A_i} - \vec{n}_i \cdot (\vec{\omega}_{A_i} \times \vec{r}_{A_i}) = \vec{n}_i \cdot \vec{v}_{B_i} + (\vec{r}_{B_i} \times \vec{n}_i) \cdot \vec{\omega}_{B_i} - \vec{n}_i \cdot \vec{v}_{A_i} - (\vec{r}_{A_i} \times \vec{n}_i) \cdot \vec{\omega}_{A_i}$ and $\ddot{C}_i^n = 2 \cdot \dot{\vec{n}}_i \cdot (\dot{\vec{p}}_{B_i} - \dot{\vec{p}}_{A_i}) + \left(\vec{n}_i \cdot \vec{a}_{B_i} + (\vec{r}_{B_i} \times \vec{n}_i) \cdot \vec{\alpha}_{B_i} - \vec{n}_i \cdot \vec{a}_{A_i} - (\vec{r}_{A_i} \times \vec{n}_i) \cdot \vec{\alpha}_{A_i} + \vec{n}_i \cdot \left(\vec{\omega}_{B_i} \times (\vec{\omega}_{B_i} \times \vec{r}_{B_i})\right) - \vec{n}_i \cdot \left(\vec{\omega}_{A_i} \times (\vec{\omega}_{A_i} \times \vec{r}_{A_i})\right)\right)$. By comparing these equations against our constraint definitions, we obtain $J_{i,A_i} = \left(-\vec{n}_i^T \quad -(\vec{r}_{A_i} \times \vec{n}_i)^T\right)$, $J_{i,B_i} = \left(\vec{n}_i^T \quad (\vec{r}_{B_i} \times \vec{n}_i)^T\right)$ and $\dot{J}_i \cdot \vec{v}_{gen} = 2 \cdot \dot{\vec{n}}_i \cdot (\vec{v}_{B_i} + \vec{\omega}_{B_i} \times \vec{r}_{B_i} - \vec{v}_{A_i} - \vec{\omega}_{A_i} \times \vec{r}_{A_i}) + \vec{n}_i \cdot \left(\vec{\omega}_{B_i} \times (\vec{\omega}_{B_i} \times \vec{r}_{B_i}) - \vec{\omega}_{A_i} \times (\vec{\omega}_{A_i} \times \vec{r}_{A_i})\right)$.

The non-penetration constraint has the following direct interpretation. The value of C_i^n measures *relative body separation* at contact i, $\dot{C}_i^n$ measures *normal velocity* at the contact and $\ddot{C}_i^n$ measures *normal acceleration*. If λ_i^n is the Lagrange multiplier due to the acceleration level formulation of the constraint $\ddot{C}_i^n := J_i \cdot \vec{a}_{gen} - c_i \geq 0$ then λ_i^n equals the *normal force magnitude* and $-\vec{n}_i \cdot \lambda_i^n$ force is exerted on A_i at $\vec{p}_{A_i}$ and $\vec{n}_i \cdot \lambda_i^n$ is exerted on B_i at $\vec{p}_{B_i}$. Our conditions from equation (6) due to non-penetration constraints thus say that both the normal acceleration and the normal force magnitude have to be non-negative and complementary to each other.

If λ_i^n is the Lagrange multiplier due to the velocity level formulation of the constraint $\dot{C}_i^n := J_i \cdot \vec{v}_{gen} \geq 0$ then $-\dot{C}_i^n$ measures the relative approaching body velocity and λ_i^n equals the normal impulse magnitude required to stop the bodies at $\vec{p}_i$ so that $\dot{C}_i^n \geq 0$. This fact can be utilized to model *impacts*. Instead of stopping the bodies, we would like to take the velocity $-\dot{C}_i^n$ and have it opposed so that the bodies would bounce. This effect can be achieved by a minor modification of the original velocity level non-penetration constraint, (Baraff, 1997). Given a *coefficient of restitution* $\epsilon_i \in [0,1]$ at the contact i determining how bouncy the contacting surface is, we replace the non-penetration velocity level constraint $J_i \cdot \vec{v}_{gen} \geq 0$ with a constraint $J_i \cdot \vec{v}_{gen} \geq -\epsilon_i \cdot \dot{C}_i^n$.

Friction is usually modeled according to Coulomb friction law, (Trinkle et al., 1997). Coulomb law introduces friction forces acting along the contact's tangential plane. It extends the complementarity conditions on the normal acceleration and normal force magnitude by adding conditions on the direction and magnitude of the friction force, by relating the direction and magnitude to the relative tangential velocity at the contact and the corresponding normal force magnitude. The relation is quadratic and is most often linearized by considering two separate friction directions that the friction force can act along, (Trinkle et al., 1997).

In this linearization, friction at contact i is approximated by two additional *friction bounded equality constraints* that constrain relative body motion in two tangential directions perpendicular to each other and the contact normal. That is, if $\vec{n}_i$ is the contact normal at contact i and $\vec{t}_i^x$ and $\vec{t}_i^y$ are two unit vectors such that $\vec{t}_i^x \cdot \vec{t}_i^y = 0$ and $\vec{t}_i^x \times \vec{t}_i^y = \vec{n}_i$ then it is requested that $C_i^x := \vec{t}_i^x \cdot (\vec{p}_{B_i} - \vec{p}_{A_i}) = 0$ and $C_i^y := \vec{t}_i^y \cdot (\vec{p}_{B_i} - \vec{p}_{A_i}) = 0$ subject to force limits. Friction constraints are special because their force limits are functions of the normal forces. If the corresponding non-penetration constraint $C_i^n \geq 0$ is implemented on the acceleration level then the friction constraints are implemented on the acceleration level as well by bounded equality constraints $\ddot{C}_i^x = 0$ and $\ddot{C}_i^y = 0$ with force limits $(\lambda_i^x)^{lo} = (\lambda_i^y)^{lo} = -\mu_i \cdot \lambda_i^n$ and $(\lambda_i^x)^{hi} = (\lambda_i^y)^{hi} = \mu_i \cdot \lambda_i^n$, where $\mu_i \geq 0$ is a static friction coefficient and λ_i^n is the normal force magnitude at contact i. Otherwise, the constraints are implemented on the velocity level by $\dot{C}_i^x = 0$ and $\dot{C}_i^y = 0$ with force limits $(\lambda_i^x)^{lo} = (\lambda_i^y)^{lo} = -\mu_i \cdot \lambda_i^n$ and $(\lambda_i^x)^{hi} = (\lambda_i^y)^{hi} = \mu_i \cdot \lambda_i^n$, where $\mu_i \geq 0$ is an impulsive friction coefficient and λ_i^n is the normal impulse magnitude at contact i, (Kawachi et al., 1997).

Note that for the friction constraints, force limits are not constant and depend on the values of λ_i^n. This implies we have to use a specialized solver to solve for $\vec{\lambda}$ or estimate λ_i^n first (e.g., we might ignore friction constraints and solve for an estimate of $\vec{\lambda}$, then fix the friction force limits using the estimated values of λ_i^n), (Smith, 2004). Friction constraints have analogous Jacobians and equation right-hand-side vectors as non-penetration constraints. It is just that $\vec{n}_i$ vectors in the corresponding formulations are replaced with $\vec{t}_i^x$ and $\vec{t}_i^y$ vectors.

Fig. 4. Contact. Each contact point illustrated by a white spike yields a non-penetration acceleration constraint $\ddot{C}_i^n \geq 0$ and two friction bounded equality constraints $\ddot{C}_i^x = 0$ and $\ddot{C}_i^y = 0$.

4. Articulated bodies and motion control

In this section we will illustrate some examples of useful constraints that can be used to build articulated structures. In particular, we look at construction of joints (implemented with "anchors") that connect individual rigid bodies by virtual hinges and restrict rotations about specific axes. Anchors are formally defined as position level equality constraints (implemented either on the velocity or acceleration level) that restrict relative body positions and/or orientations. They implement various virtual joints (e.g., ball-and-socket or hinges) that connect bodies. This way, complex articulated jointed structures could be implemented.

4.1 Ball-and-socket joint

We start with a ball-and-socket joint that was discussed earlier (in the form of "point-to-point" constraints) in Section 3. The joint is defined in terms of two anchor points fixed in the coordinate frames of the first and the second constrained body. Ball-and-socket joint requires the anchors to occupy the same position in the world coordinate frame. As such, it removes 3 degrees of freedom from the system.

If i is the constraint index, A_i and B_i are the indices of the constrained bodies and $\vec{r}^{\,b}_{A_i}$ and $\vec{r}^{\,b}_{B_i}$ are the positions of the two anchors on the first and second body, expressed in the corresponding body-centric coordinate frames, then $\vec{p}_{A_i} = \vec{x}_{A_i} + R_{A_i} \cdot \vec{r}^{\,b}_{A_i}$ and $\vec{p}_{B_i} = \vec{x}_{B_i} + R_{B_i} \cdot \vec{r}^{\,b}_{B_i}$ are the world space positions of the anchors and the goal is to ensure that $\vec{p}_{A_i} = \vec{p}_{B_i}$. Let us denote $\vec{r}_{A_i} = \vec{p}_{A_i} - \vec{x}_{A_i}$ and $\vec{r}_{B_i} = \vec{p}_{B_i} - \vec{x}_{B_i}$. These vectors are fixed in the coordinate frame of the first and the second body respectively, resulting in $\dot{\vec{r}}_{A_i} = \vec{\omega}_{A_i} \times \vec{r}_{A_i}$ and $\dot{\vec{r}}_{B_i} = \vec{\omega}_{B_i} \times \vec{r}_{B_i}$. Following the derivation from the previous section, our constraint can then be formulated in the form of equations $\vec{C}^i_p := \vec{p}_{B_i} - \vec{p}_{A_i} = \vec{x}_{B_i} + \vec{r}_{B_i} - \vec{x}_{A_i} - \vec{r}_{A_i} = \vec{0}$, $\dot{\vec{C}}^i_p := \vec{v}_{B_i} + \vec{\omega}_{B_i} \times \vec{r}_{B_i} - \vec{v}_{A_i} - \vec{\omega}_{A_i} \times \vec{r}_{A_i} = \vec{0}$ and $\ddot{\vec{C}}^i_p := \vec{a}_{B_i} - \vec{r}^{\,*}_{B_i} \cdot \vec{\alpha}_{B_i} + \vec{\omega}_{B_i} \times (\vec{\omega}_{B_i} \times \vec{r}_{B_i}) - \vec{a}_{A_i} + \vec{r}^{\,*}_{A_i} \cdot \vec{\alpha}_{A_i} - \vec{\omega}_{A_i} \times (\vec{\omega}_{A_i} \times \vec{r}_{A_i}) = \vec{0}$ and so we obtain $J_{i,A_i} = \begin{pmatrix} -E & \vec{r}^{\,*}_{A_i} \end{pmatrix}$, $J_{i,B_i} = \begin{pmatrix} E & -\vec{r}^{\,*}_{B_i} \end{pmatrix}$, $\dot{J}_i \cdot \vec{v}_{gen} = \vec{\omega}_{B_i} \times (\vec{\omega}_{B_i} \times \vec{r}_{B_i}) - \vec{\omega}_{A_i} \times (\vec{\omega}_{A_i} \times \vec{r}_{A_i})$, $\vec{C}^i_p = \vec{p}_{B_i} - \vec{p}_{A_i}$, $m_i = 3$. These terms can be directly substituted into the acceleration level or velocity level formulations of the position level equality constraint.

4.2 Universal joint

Universal joint i attaches two bodies A_i and B_i like a point-to-point joint that additionally removes one rotational degree of freedom so that the constrained bodies can only rotate about two remaining axes. The two axes are defined explicitly and are perpendicular to one another. The first axis is attached to the first body A_i and the second axis is attached to the second body B_i.

The joint is defined by the positions of the two anchors $\vec{r}^{\,b}_{A_i}$ and $\vec{r}^{\,b}_{B_i}$ attached to the first and the second body and the directions of the two joint axes $\vec{u}^{\,b}_{A_i}$ and $\vec{u}^{\,b}_{B_i}$attached to the first and the second body. Given the world space positions $\vec{p}_{A_i} = \vec{x}_{A_i} + R_{A_i} \cdot \vec{r}^{\,b}_{A_i}$ and $\vec{p}_{B_i} = \vec{x}_{B_i} + R_{B_i} \cdot \vec{r}^{\,b}_{B_i}$ of the anchors and the world space directions $\vec{u}_{A_i} = R_{A_i} \cdot \vec{u}^{\,b}_{A_i}$ and $\vec{u}_{B_i} = R_{B_i} \cdot \vec{u}^{\,b}_{B_i}$ of the axes, the constraint requires the anchors to occupy the same world space position $\vec{p}_{A_i} = \vec{p}_{B_i}$ and requests the axes to be perpendicular in world space such that $\vec{u}_{A_i} \cdot \vec{u}_{B_i} = 0$.

The constraint removes four degrees of freedom, $m_i = 4$, and is defined by four constraint rows $(\vec{C}^i_p)_k = 0$, $k = 1, \dots, 4$ of which the first three rows are due to the ball-and-socket joint discussed earlier and the fourth row is given by $(\vec{C}^i_p)_4 := -\vec{u}_{A_i} \cdot \vec{u}_{B_i} = 0$. The unstabilized velocity level formulation of the constraint is obtained by differentiating $\vec{C}^i_p$ with respect to time. We have already seen this formulation for the first three rows and so only need to consider $(\dot{\vec{C}}^i_p)_4$. We have $(\dot{\vec{C}}^i_p)_4 := \frac{\partial}{\partial t}(-\vec{u}_{A_i} \cdot \vec{u}_{B_i}) = -(\vec{\omega}_{A_i} \times \vec{u}_{A_i}) \cdot \vec{u}_{B_i} - \vec{u}_{A_i} \cdot (\vec{\omega}_{B_i} \times \vec{u}_{B_i}) = \vec{\omega}_{B_i} \cdot (\vec{u}_{A_i} \times \vec{u}_{B_i}) - \vec{\omega}_{A_i} \cdot (\vec{u}_{A_i} \times \vec{u}_{B_i}) = 0$ and thus $(J_{i,A_i})_4 = \begin{pmatrix} 0 & -(\vec{u}_{A_i} \times \vec{u}_{B_i})^T \end{pmatrix}$ and $(J_{i,B_i})_4 =$

$\begin{pmatrix}0 & (\vec{u}_{A_i} \times \vec{u}_{B_i})^T\end{pmatrix}$. Note that $\vec{\omega}_{B_i} \cdot (\vec{u}_{A_i} \times \vec{u}_{B_i}) - \vec{\omega}_{A_i} \cdot (\vec{u}_{A_i} \times \vec{u}_{B_i}) = (\vec{u}_{A_i} \times \vec{u}_{B_i}) \cdot (\vec{\omega}_{B_i} - \vec{\omega}_{A_i})$ equals the relative angular velocity (rotation speed) about the axis $\vec{u}_{A_i} \times \vec{u}_{B_i}$ and so the velocity-level constraint $\left(\dot{\vec{C}}_p^i\right)_4 = 0$ explicitly prohibits relative body rotation about $\vec{u}_{A_i} \times \vec{u}_{B_i}$. In turn, the two constrained bodies can only rotate about $\vec{u}_{A_i}$ and $\vec{u}_{B_i}$.

The unstabilized acceleration level formulation of the constraint is obtained by differentiating the velocity level formulation with respect to time. The first three rows of the unstabilized acceleration constraint are again the same as in the ball-and-socket joint case and the fourth row is given by $\left(\ddot{\vec{C}}_p^i\right)_4 := \frac{\partial}{\partial t}\left((\vec{u}_{A_i} \times \vec{u}_{B_i}) \cdot (\vec{\omega}_{B_i} - \vec{\omega}_{A_i})\right) = \left(\frac{\partial \vec{u}_{A_i}}{\partial t} \times \vec{u}_{B_i} + \vec{u}_{A_i} \times \frac{\partial \vec{u}_{B_i}}{\partial t}\right) \cdot (\vec{\omega}_{B_i} - \vec{\omega}_{A_i}) + (\vec{u}_{A_i} \times \vec{u}_{B_i}) \cdot (\vec{\alpha}_{B_i} - \vec{\alpha}_{A_i}) = \left((\vec{\omega}_{A_i} \times \vec{u}_{A_i}) \times \vec{u}_{B_i} + \vec{u}_{A_i} \times (\vec{\omega}_{B_i} \times \vec{u}_{B_i})\right) \cdot (\vec{\omega}_{B_i} - \vec{\omega}_{A_i}) + (\vec{u}_{A_i} \times \vec{u}_{B_i}) \cdot (\vec{\alpha}_{B_i} - \vec{\alpha}_{A_i}) = 0$ because $\vec{u}_{A_i}$ is attached to the first body A_i and $\vec{u}_{B_i}$ is attached to the second body B_i. We thus obtain $(J_i \cdot \vec{v}_{gen})_4 = \left((\vec{\omega}_{A_i} \times \vec{u}_{A_i}) \times \vec{u}_{B_i} + \vec{u}_{A_i} \times (\vec{\omega}_{B_i} \times \vec{u}_{B_i})\right) \cdot (\vec{\omega}_{B_i} - \vec{\omega}_{A_i})$.

4.3 Hinge joint

Hinge joint i attaches two bodies like a ball-and-socket joint but additionally removes two more rotational degrees of freedom so that the constrained bodies A_i and B_i can only rotate about a single common axis (hinge axis). The joint is defined by the positions of the two anchors $\vec{r}_{A_i}^b$ and $\vec{r}_{B_i}^b$ attached to the first and the second body and the directions of the axes $\vec{u}_{A_i}^b$ and $\vec{u}_{B_i}^b$attached to the first and the second body. Given the world space positions $\vec{p}_{A_i} = \vec{x}_{A_i} + R_{A_i} \cdot \vec{r}_{A_i}^b$ and $\vec{p}_{B_i} = \vec{x}_{B_i} + R_{B_i} \cdot \vec{r}_{B_i}^b$ of the anchors and the world space directions $\vec{u}_{A_i} = R_{A_i} \cdot \vec{u}_{A_i}^b$ and $\vec{u}_{B_i} = R_{B_i} \cdot \vec{u}_{B_i}^b$ of the axes, the constraint requires the anchors to occupy the same position in the world coordinate frame $\vec{p}_{A_i} = \vec{p}_{B_i}$ and requests the hinge axes to align in world space such that $\vec{u}_{A_i} = \vec{u}_{B_i}$.

This time, we will show the velocity level formulation of the constraint directly because such a formulation allows to naturally express conditions on what rotation axes the constrained bodies cannot rotate about. The constraint removes 5 degrees of freedom, $m_i = 5$, and is defined by five velocity constraint rows $\left(\dot{\vec{C}}_p^i\right)_k = 0$, $k = 1, \dots, 5$ of which the first three rows are due to the point-to-point joint discussed earlier. We will now define the remaining two constraint rows $\left(\dot{\vec{C}}_p^i\right)_4 = 0$ and $\left(\dot{\vec{C}}_p^i\right)_5 = 0$.

Let $\vec{g}$ and $\vec{h}$ be two world space vectors perpendicular to $\vec{u}_{A_i}$ and assume that $\vec{u}_{A_i} = \vec{u}_{B_i}$. To make sure that the two bodies can rotate only about $\vec{u}_{A_i}$, the relative angular velocity $(\vec{\omega}_{B_i} - \vec{\omega}_{A_i}) \cdot \vec{w}$ about any axis $\vec{w}$ perpendicular to $\vec{u}_{A_i}$ (relative rotation speed about $\vec{w}$) must be zero. This can be enforced by simply requiring that the relative angular velocity about $\vec{g}$ and $\vec{h}$ is zero. We thus get $\left(\dot{\vec{C}}_p^i\right)_4 := \vec{g} \cdot \vec{\omega}_{B_i} - \vec{g} \cdot \vec{\omega}_{A_i} = 0$ and $\left(\dot{\vec{C}}_p^i\right)_5 := \vec{h} \cdot \vec{\omega}_{B_i} - \vec{h} \cdot \vec{\omega}_{A_i} = 0$ and so can write $(J_{i,A_i})_4 = \begin{pmatrix}\vec{0}^T & -\vec{g}^T\end{pmatrix}$, $(J_{i,B_i})_4 = \begin{pmatrix}\vec{0}^T & \vec{g}^T\end{pmatrix}$ and $(J_{i,A_i})_5 = \begin{pmatrix}\vec{0}^T & -\vec{h}^T\end{pmatrix}$, $(J_{i,B_i})_5 = \begin{pmatrix}\vec{0}^T & \vec{h}^T\end{pmatrix}$.

Now, if the constraint is broken and the hinge axes are not aligned, we can express the orientation error by a vector $\vec{u}_{B_i} \times \vec{u}_{A_i}$. To stabilize the constraint and eventually bring the axes back into alignment, we want to rotate the bodies about $\vec{z} = \frac{\vec{u}_{B_i} \times \vec{u}_{A_i}}{\left\|\vec{u}_{B_i} \times \vec{u}_{A_i}\right\|}$ with a speed

proportional to $\|\vec{u}_{B_i} \times \vec{u}_{A_i}\|$. Because $\vec{z}$ lies in the plane perpendicular to $\vec{u}_{A_i}$, the stabilization can be decomposed to requests on rotation speeds about $\vec{g}$ and $\vec{h}$ and thus directly incorporated into the constraint equations as $\left(\dot{\vec{C}}_p^i\right)_4 := \vec{g} \cdot \vec{\omega}_{B_i} - \vec{g} \cdot \vec{\omega}_{A_i} - \vec{g} \cdot (\vec{u}_{B_i} \times \vec{u}_{A_i}) \cdot \alpha = 0$ and $\left(\dot{\vec{C}}_p^i\right)_5 := \vec{h} \cdot \vec{\omega}_{B_i} - \vec{h} \cdot \vec{\omega}_{A_i} - \vec{h} \cdot (\vec{u}_{B_i} \times \vec{u}_{A_i}) \cdot \alpha = 0$, where α is the stabilization constant. We can thus write $\left(\vec{C}_p^i\right)_4 \approx \vec{g} \cdot \left(\vec{u}_{A_i} \times \vec{u}_{B_i}\right)$ and $\left(\vec{C}_p^i\right)_5 \approx \vec{h} \cdot \left(\vec{u}_{A_i} \times \vec{u}_{B_i}\right)$.
The unstabilized acceleration level formulation of the constraint is obtained by differentiating the velocity level formulation with respect to time. The first three rows of the unstabilized acceleration constraint are the same like in the point-to-point case and the remaining two rows are given by $\left(\ddot{\vec{C}}_p^i\right)_4 := \left(\vec{\omega}_{A_i} \times \vec{g}\right) \cdot \left(\vec{\omega}_{B_i} - \vec{\omega}_{A_i}\right) + \vec{g} \cdot \left(\vec{\alpha}_{B_i} - \vec{\alpha}_{A_i}\right) = 0$ and $\left(\ddot{\vec{C}}_p^i\right)_5 := \left(\vec{\omega}_{A_i} \times \vec{h}\right) \cdot \left(\vec{\omega}_{B_i} - \vec{\omega}_{A_i}\right) + \vec{h} \cdot \left(\vec{\alpha}_{B_i} - \vec{\alpha}_{A_i}\right) = 0$ because $\vec{g}$ and $\vec{h}$ can be assumed to be attached to the first body A_i. We thus obtain $\left(\dot{J}_i \cdot \vec{v}_{gen}\right)_4 = \left(\vec{\omega}_{A_i} \times \vec{g}\right) \cdot \left(\vec{\omega}_{B_i} - \vec{\omega}_{A_i}\right)$ and $\left(\dot{J}_i \cdot \vec{v}_{gen}\right)_5 = \left(\vec{\omega}_{A_i} \times \vec{h}\right) \cdot \left(\vec{\omega}_{B_i} - \vec{\omega}_{A_i}\right)$. By now we have defined all the terms J_{i,A_i}, J_{i,B_i}, $\dot{J}_i \cdot \vec{v}_{gen}$, $\vec{C}_p^i$ and m_i required to implement hinge joint either on the velocity or acceleration level with stabilization.

5. Motion control and motors

In this section we show how to actuate the articulated body defined in the previous section. In particular, we show how constraints can be used to define motors that can be applied at the joints of articulated body, similarly to (Kokkevis, 2004). Motors control angles or displacements along certain axes; they also control rotational or translational speed.
Motors provide mechanisms to actuate body pairs by controlling their relative linear and angular properties. Given desired relative positions, orientations or velocities that need be followed by a motor, the motor formulates appropriate position level or velocity level constraints. These constraints are then implemented on the velocity or acceleration level by submitting corresponding (bounded) equality or inequality constraints with stabilization. Most often, motors are combined with anchors so that bodies can be actuated about the valid (originally unconstrained) degrees of freedom at joints, like a hinge axis or universal joint's axes.
Anchors discussed earlier were implemented by equality constraints that were always "active". Once constrained bodies are connected by a joint, the corresponding anchor constraint becomes effective and is incrementally maintained. With motors, the corresponding constraint changes in accordance with desired motion. The motor constraint is not permanent and the system state is often inconsistent with the constraint formulation when the constraint is requested to be followed. As a result, the constraint has to be stabilized and it is the constraint stabilization mechanism and/or impulses due to first-order dynamics that generate the energy needed to actuate the bodies. For example, when a position motor, following a desired position of the second body relative to the position of the first body, is implemented on the acceleration level by submitting an acceleration constraint, both constraint position and velocity errors have to be stabilized and the necessary actuation energy is generated by the constraint's stabilization mechanism. Stabilization parameters determine how long it takes the motor to approach the desired position.

We will now discuss various types of motors. Each particular type will define a motor-specific measure of relative body state relating the state of one actuated body to the state of the other actuated body. Then, given the desired (intended) value and the current value of this measure (e.g., relative angle, relative orientation, relative position in space, relative rotation speed, relative angular velocity, relative linear velocity), the motor formulates motion control constraints that would drive the measure towards its desired value.
We distinguish between linear and angular motors. Pure linear motors affect only linear properties of the bodies while pure angular motors affect only angular properties. As a convention, we assume that the measure used by motor i actuating bodies A_i and B_i always relates the state of the second body B_i to the state of the first body A_i. In addition, for the sake of convenience, we assume that the position level measure is defined with respect to the initial state at time t_0 (in other words the position measure is always equal to a zero vector at t_0).

5.1 3-DOF linear motor

We start with a simple *3-DOF velocity level linear motor* i that controls relative linear velocity of A_i and B_i along world space axes. The motor velocity measure $\dot{\vec{u}}(t) := \vec{v}_{B_i}(t) - \vec{v}_{A_i}(t)$ is equivalent to relative body velocity in the world coordinate frame and is independent of relative angular properties. Then, given a desired value $\dot{\vec{u}}_d(t)$ of this measure, we impose a three dimensional velocity level constraint, $m_i = 3$, $\vec{C}_v^i := \dot{\vec{u}}(t) - \dot{\vec{u}}_d(t) = \vec{0}$. This constraint affects only linear properties of the two bodies and can be implemented either directly on the velocity level or the acceleration level by requesting $\dot{\vec{C}}_v^i := \vec{a}_{B_i}(t) - \vec{a}_{A_i}(t) = -\vec{C}_v^i \cdot \beta$, where β is a velocity stabilization parameter. The formulations produce $J_{i,A_i} = (-E \quad 0)$, $J_{i,B_i} = (E \quad 0)$ and $\dot{J}_i \cdot \vec{v}_{gen} = 0$.
Now, if we are to implement a *3-DOF position level linear motor* i that controls relative position of two bodies A_i and B_i in the world coordinate frame, we can define the motor position measure $\vec{u}(t)$ as a vector of current relative body displacements along the world space axes, that is, $\vec{u}(t) := \left(\vec{x}_{B_i}(t) - \vec{x}_{A_i}(t)\right) - \left(\vec{x}_{B_i}(t_0) - \vec{x}_{A_i}(t_0)\right)$ and then, given a desired value $\vec{u}_d(t)$ of the measure, we can impose a three dimensional position level constraint, $m_i = 3$, $\vec{C}_p^i(t) := \vec{u}(t) - \vec{u}_d(t) = \vec{0}$. We get $\dot{\vec{C}}_p^i = \dot{\vec{u}}(t) - \dot{\vec{u}}_d(t) \approx \dot{\vec{u}}(t) = \vec{v}_{B_i}(t) - \vec{v}_{A_i}(t)$ and can once again implement the position level constraint on the velocity level by requesting $\vec{C}_v^i := \vec{v}_{B_i}(t) - \vec{v}_{A_i}(t) = -\vec{C}_p^i \cdot \alpha$, where α is a position stabilization constant. This lets us implement the position level linear motor by using the corresponding velocity level linear motor.

5.2 3-DOF angular motor

Let's first present a *3-DOF velocity level angular motor* i that directly controls relative angular velocity of A_i and B_i in the world coordinate frame, both in terms of the rotation direction and speed. This motor is an angular analog to the 3-DOF linear velocity motor presented earlier and hence we can simply define the three dimensional motor velocity measure $\dot{\vec{u}}$(t) as $\dot{\vec{u}}(t) := \vec{\omega}_{B_i}(t) - \vec{\omega}_{A_i}(t)$. Then, given a desired value $\dot{\vec{u}}_d(t)$ of the velocity, we impose a three dimensional velocity level constraint, $m_i = 3$, $\vec{C}_v^i := \dot{\vec{u}}(t) - \dot{\vec{u}}_d(t) = \vec{0}$ that can be implemented either directly on the velocity level or the acceleration level by requesting

$\dot{\vec{C}}_v^i = -\vec{C}_v^i \cdot \beta$, where β is a velocity stabilization parameter. We then get $J_{i,A_i} = (0 \quad -E)$, $J_{i,B_i} = (0 \quad E)$ and $\dot{J}_i \cdot \vec{v}_{gen} = 0$.

We now present a *3-DOF position level angular motor* i, implemented according to (Vondrak, 2006), that controls relative orientation of the two bodies. The motor constraint will be formulated directly on the velocity level. This simplifies the formulation but still allows us to directly specify the world space axis that the actuated bodies have to rotate about and how fast in order to reach the desired relative orientation. We define *orientation of B_i relative to the orientation of A_i at time* t as $R_{A_i}^{-1}(t) \cdot R_{B_i}(t)$.

To formulate the constraint, we first need to define the motor measure, relative orientation, $U(t)$. Following our convention, $U(t)$ measures orientation of B_i relative to A_i relative to the initial relative orientation $R_{A_i}^{-1}(t_0) \cdot R_{B_i}(t_0)$ and hence $U(t)$ is defined as $U(t) := R_{A_i}^{-1}(t) \cdot R_{B_i}(t) \cdot R_{B_i}^{-1}(t_0) \cdot R_{A_i}(t_0)$. When we are given a desired value $U_d(t)$ for the relative orientation, we need to find the world space rotation axis $\vec{w}$ and an angle z that the two bodies have to rotate about so that $U = U_d$ after the rotation is applied. For that, we define relative B_i to A_i rotational error $U_e(t) := R_{A_i}^{-1}(t) \cdot R_{B_i}(t) \cdot R_{B_i}^{-1}(t_0) \cdot R_{A_i}(t_0) \cdot U_d^{-1}(t)$ which we then decompose to a rotation axis $\vec{w}_{A_i}^b$ (attached to A_i) and an angle $z \in [0, \pi)$. We then transform $\vec{w}_{A_i}^b$ to world coordinate frame obtaining the axis $\vec{w} = R_{A_i}(t) \cdot \vec{w}_{A_i}^b$. Because the relative rotation about $\vec{w}$ by z is the only valid relative rotation consistent with our orientation request, we can finally constrain relative body angular velocity by stipulating that $\vec{C}_v^i := \vec{\omega}_{B_i} - \vec{\omega}_{A_i} = -\vec{w} \cdot z \cdot \alpha$, where α is a position stabilization constant. This lets us implement the position level angular motor by using the corresponding velocity level angular motor that follows $\dot{\vec{u}}_d(t) = -\vec{w} \cdot z \cdot \alpha$.

5.3 1-DOF motors

We will now present a template for 1-DOF linear and/or angular position and velocity motors that control relative displacement or angle (generally called the offset) and their time derivatives along or about given fixed axes attached to the first body in the actuated body pair. Unlike the previous cases, our 1-DOF motor will not define its position-level motor measure $u(t)$, the offset of B_i from A_i relative to the initial offset at time t_0, explicitly. In contrast, it will be assumed that at any time t the value of $u(t)$ specific to the motor could be determined. That is because 1-DOF motors will be most often combined with anchors and the value of $u(t)$ will be specific to the anchor type. For example, the hinge joint's hinge axis could be controlled by a 1-DOF motor and the value of $u(t)$ in that case would equal the rotation angle of B_i about the hinge axis relative to the initial relative orientation of B_i and A_i. The motors will additionally provide support for enforcing one-dimensional joint displacement and angle limits because the enforcement in the 1-DOF case can be implemented by using simple one-dimensional constraints, given the value of $u(t)$.

5.3.1 1-DOF angular motor

Let's consider the *1-DOF angular position and velocity motor* i that actuates bodies A_i and B_i and is specified by a rotation axis $\vec{w}_{A_i}^b$ attached to the first body A_i, motor angle measure $u(t)$ and motor angle limits z_i^{lo} and z_i^{hi}. Given the world space rotation axis $\vec{w}_i = R_{A_i} \cdot \vec{w}_{A_i}^b$, the value of the angle measure $u(t)$ is supposed to equal the angle the second body B_i is rotated about $\vec{w}_i$ with respect to the first body A_i, relative to the initial relative body orientation. We then define velocity measure $\dot{u}(t)$ as the time derivative of the angle measure $u(t)$. The velocity measure equals the relative rotation speed about $\vec{w}_i$ and so is given by $\dot{u}(t) = \vec{w}_i \cdot (\vec{\omega}_{B_i} - \vec{\omega}_{A_i})$. We will say the motor's axis is limited, if the angle limits are in effect and we will say the axis is powered if the motor is to follow either a desired angle $u_d(t)$ or a desired rotation speed $\dot{u}_d(t)$.

We now show velocity level constraints with stabilization implementing the motor. The constraints ensure that angle limits are obeyed and a desired angle or angular velocity is followed. If the axis is limited and $u(t) \leq z_i^{lo}$ then it is requested that $C_v^i := \vec{w}_i \cdot (\vec{\omega}_{B_i} - \vec{\omega}_{A_i}) - \left(z_i^{lo} - u(t)\right) \cdot \alpha \geq 0$, where α is a position level stabilization constant. If the axis is limited and $u(t) \geq z_i^{hi}$ then $C_v^i := \vec{w}_i \cdot (\vec{\omega}_{B_i} - \vec{\omega}_{A_i}) - \left(z_i^{hi} - u(t)\right) \cdot \alpha \leq 0$ is requested. If the motor is to follow a desired angle $u_d(t)$ then it is requested that $C_v^i := \vec{w}_i \cdot (\vec{\omega}_{B_i} - \vec{\omega}_{A_i}) - (u_d(t) - u(t)) \cdot \alpha = 0$. If the motor is to follow a desired angular velocity $\dot{u}_d(t)$ then we request $C_v^i := \vec{w}_i \cdot (\vec{\omega}_{B_i} - \vec{\omega}_{A_i}) - \dot{u}_d(t) = 0$. Each constraint produces the same Jacobian blocks $J_{i,A_i} = (\vec{0}^T \quad -\vec{w}_i^T)$ and $J_{i,B_i} = (\vec{0}^T \quad \vec{w}_i^T)$. By differentiation, we obtain $\dot{J}_i \cdot \vec{v}_{gen} = (\vec{\omega}_{A_i} \times \vec{w}_i) \cdot (\vec{\omega}_{B_i} - \vec{\omega}_{A_i})$.

5.3.2 1-DOF linear motor

To formulate a *1-DOF linear position and velocity motor i* we proceed similarly to the 1-DOF angular motor case. The motor is specified by a translation axis $\vec{w}_{A_i}^b$ attached to the center of mass of A_i, motor displacement measure $u(t)$ of the second body's center of mass from the first body's center of mass along $\vec{w}_i$ (relative to the initial displacement) and motor displacement limits z_i^{lo} and z_i^{hi}. Because $\vec{w}_{A_i}$ is attached to A_i, the relative displacement $u(t)$ is a function of $R_{A_i}(t)$ and so both linear and angular properties have to be constrained. Here, we constrain the linear properties only and assume the associated anchor will impose additional constraints to maintain proper body orientation, if necessary. This lets us formulate the same set of constraints as in the previous section, but this time, linear velocities are constrained instead of angular velocities. We get $J_{i,A_i} = (-\vec{w}_i^T \quad \vec{0}^T)$, $J_{i,B_i} = (\vec{w}_i^T \quad \vec{0}^T)$ and $\dot{J}_i \cdot \vec{v}_{gen} = (\vec{\omega}_{A_i} \times \vec{w}_i) \cdot (\vec{v}_{B_i} - \vec{v}_{A_i})$.

5.4 Trajectory motor

When controlling motion of loop-free articulated structure it is sometimes convenient to specify desired position of selected points (markers), rigidly attached to the segments, instead of specifying desired angles at joints. Given the desired positions of markers, one could use inverse kinematics to solve for desired angles. These desired angles can then be submitted as constraints for motors, resulting in the motion of the body that would implicitly satisfy desired positions of the markers.

We propose to use a different approach, previously used in (Vondrak et al., 2008), based on first-order inverse dynamics - given the desired positions of markers, we solve for desired angular velocities that should be followed by the structure's angular motors so that markers would reach their desired positions. Using inverse dynamics is simpler in our case, because we already formulated how to solve for velocities that satisfy a given set of constraints (by solving for the impulses).

Let's assume our articulated structure is made up of n rigid body segments connected by $n-1$ actuated joints of various kinds. We assume the root segment r has index 1 and that the segment $i = 2, \ldots, n$ is connected to its parent segment by joint i. In addition, let's connect the root segment $r = 1$ to an external fixed body (i.e., the world, implemented as a rigid body with infinite mass) by an additional 6-DOF motor r that is a combination of the 3-DOF linear and the 3-DOF angular motors presented earlier. The external fixed body establishes a coordinate frame in world space and so the position measure $\vec{q}_r \in \boldsymbol{R}^6$ of the combined 6-DOF motor defines the global position and orientation of the structure's root segment in world space. The other motor measures $\vec{q}_i$, $i = 2, \ldots, n$, recursively parameterize the positions and orientations of the remaining segments in world space and so we can define the kinematic state $\vec{q}$ of the structure as a vector $\vec{q} = (\vec{q}_1, \vec{q}_2, \ldots, \vec{q}_n)$. We also define the structure's velocity $\dot{\vec{q}} = (\dot{\vec{q}}_1, \dot{\vec{q}}_2, \ldots, \dot{\vec{q}}_n)$ which is a concatenation of the structure's motor velocity measures.

Now, if there are m markers $j = 1, \ldots, m$ attached to the kinematic structure such that the world space position of marker j at state $\vec{q}$ is given by $\vec{z}^j(\vec{q})$ and its corresponding desired world space position is given by $\vec{z}_d^j$, the inverse kinematics approach would solve for desired position measures $\vec{q}_d$ so that (1) $\vec{z}^j(\vec{q}_d) = \vec{z}_d^j$ for all $j = 1, \ldots, m$ and (2) $(\vec{q}_d)_1 = \vec{q}_1$. The first constraint requests the markers to reach their desired positions at state at $\vec{q}_d$ while the second constraint fixes the position and orientation of the root segment to ensure that requests (1) cannot be satisfied by simple translation or rotation of the root segment. Once $\vec{q}_d$ values are known, structure's motors would be programmed to follow $\vec{q}_d$.

With the inverse dynamics approach, we want to solve for desired velocity measures $\dot{\vec{q}}_d$ so that (1) $\dot{\vec{z}}^j = -(\vec{z}^j(\vec{q}) - \vec{z}_d^j) \cdot \alpha$, for all $j = 1, \ldots, m$, where α is a position stabilization constant and (2) $(\dot{\vec{q}}_d)_1 = \dot{\vec{q}}_1$. The first constraint requests the markers to be moved towards their desired positions while the second constraint ensures that constraints (1) cannot be solved by forced translation or rotation of the root segment. Because each equation (1) is actually a constraint on relative velocities of two points (anchors), in practice, the value of $\dot{\vec{q}}_d$ can be computed by (1) attaching m ball-and-socket joints between anchors $\vec{z}^j(\vec{q})$ on the structure parts and anchors $\vec{z}_d^j$ on the world body, (2) programming the root segment's 6-DOF motor to follow the current values of the velocity measures $\dot{\vec{q}}_1$ and taking an auxiliary simulation step from the current time t to a time $t + \Delta t$ to retrieve the figure velocities $\dot{\vec{q}}(t + \Delta t)$ (if we need to enforce additional constraints, e.g. due to joint angle limits or angle-based actuation of other joints, we can just let those constraints be active at this time). These velocities can then be used as the desired velocities $\dot{\vec{q}}_d$. We can thus let $\dot{\vec{q}}_d := \dot{\vec{q}}(t + \Delta t)$. Once the $\dot{\vec{q}}_d$ values are known, we roll the simulation state back to t and reprogram the structure's motors to follow $\dot{\vec{q}}_d$.

5.5 Control example

In this section, we illustrate the use of joint and motor constraints to build a simple humanoid character that will be animated using a physical simulation. Motion control constraints will be set up such that desired motion specified by given motion capture data will be followed. We assume that the desired motion is physically valid.

We compose the character out of rigid bodies corresponding to the character's body parts (in lieu of Section 4) and then connect these parts by actuated ball-and-socket joints that have 3-DOF angular motors associated with them so that the character's motion could be controlled. We use 3-DOF linear and 3-DOF angular motors to directly control the position and orientation of the character's root segment in world space. The direct control of the root segment (while not entirely physically plausible) makes control simple, such that we don't have to worry about character's balance and/or unanticipated collisions with the environment. In addition, to allow the motion of the character to adjust to external disturbances, such as dragging of body parts by a mouse, in a plausible way, we set the angular motors actuating the character's upper body, head and arms to generate bounded constraint forces[3]. To produce the final animation, we program the character's motors to follow desired positions and joint orientations given by motion capture and simulate the rigid body system forward in time.

Figure 5 shows animation results generated by (Vondrak, 2006) using the described character model. The top row shows the animation when no external disturbances are present. The middle row illustrates effects of attaching a basket to the character's right hand by a hinge joint. The axis motor is programmed to maintain zero rotation speed subject to force limits so that friction between the hand and the basket handle could be modeled. The bottom row extends the previous case by replacing the basket with an umbrella. The umbrella is pulled to the front of the character's head by a ball-and-socket joint attached to the right wrist and an anchor in front of the head. In addition, two 1-DOF angular motors are attached to the umbrella and the external world body to keep the umbrella's "pitch" and "bank" angles close to zero so that the umbrella will be held upright regardless of the character's pose.

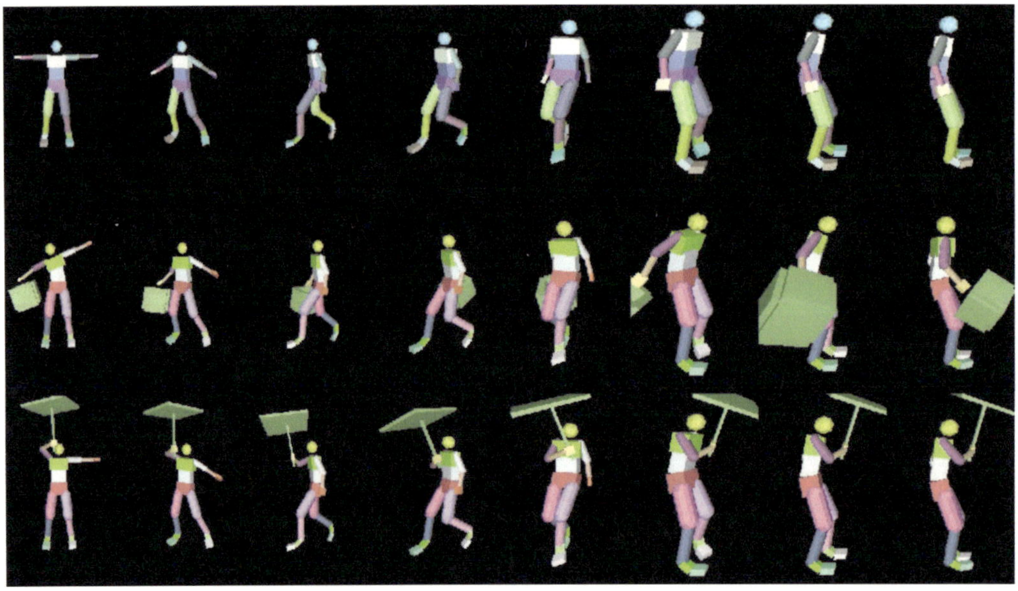

Fig. 5. Physics-based animation of an articulated character under the effects of different external disturbances. Top row shows the original motion of a subject preparing to make a jump when no external disturbances are present, the other rows show adjusted motion when objects are attached to the character's right hand.

[3] Motors actuating lower body and legs are programmed to generate almost unbounded forces so that the direct control of the character's (root) position and orientation is consistent with the actual motion of the lower body.

6. Discussion and conclusions

Physics-based simulation has emerged as a popular approach for realistic animation and analysis of rigid and articulated bodies in motion. This chapter briefly reviewed basic principles of unconstrained rigid body mechanics and then focused on the more challenging constrained rigid body mechanics principles. We have outlined basic concepts of rigid and articulated body modeling and simulation and advocated a constraint-based motion control that is based on motors implemented by constraints imposed on the position, velocity and/or acceleration of joint angles or points rigidly attached to the bodies.

The formulated approach to control is simple and accurate within the context of readily available physics-based engines. That said, general control of complex articulated models, such as humanoids, is very challenging, especially in absence of trajectories that constrain all or most parts of the body over time. In particular, design of controllers that reproduce dynamics and energetics of human motion as well as can model dynamic variations due to the physical morphology or style of the individual remains an open issue. A variety of other approaches to motion control exist (see Section 1.1). For instance, task-based control (where the user specifies the task instead of joint angles or trajectories, e.g., pick up a mug from the table) has been emerging as the new alternative direction in the control and has a number of appealing properties from the point of view of animators and game designers. Discussing these alternative approaches falls outside the scope of this chapter. Lastly, we also do not consider numerical and performance aspects of the constraint-based motion control method and do not discuss various integration methods that clearly affect the quality (and the speed) of resulting simulations (see (Boeing et al., 2007) for discussion).

That said, constraint-based motion control has become the standard approach for animating virtual worlds with stunning realism. This approach is versatile enough to model distinct phenomena like body articulation, joint actuation and contact in a uniform way; it is also capable of producing stable high quality simulations with predictable results in real time. Consequently, constraint-based control has become the default motion control strategy employed by all major commercial and open-source simulation packages.

7. References

Abe, Y.; Popovic, J. (2006). Interactive animation of dynamic manipulation. In: Eurographics/ACM SIGGRAPH Symposium on Computer Animation

Baraff, D. (1996). Linear-Time Dynamics using Lagrange Multipliers, In: SIGGRAPH'96, pages 137 - 146, New Orleans

Baraff, D.; Witkin, A. & Kass M. (1997). An Introduction to Physically-Based Modeling, In: SIGGRAPH'97. Course Notes

Bourg, D. (2002). Physics for Game Developers, O'Reilly & Associates, Inc.

Boeing, A. &Braunl, T (2007). Evaluation of real-time physics simulation systems. In: International Conference on Computer Graphics and Interactive Techniques (Graphite), pages 281 - 288

Cline, M. (2002). Rigid Body Simulation with Contact and Constraints, In: Master's Thesis, The University of British Columbia

Eberly, D. (2003). Game Physics (Interactive 3d technology series), MorganKaufman

Erleben, K. (2002). Rigid Body Simulation, In: Lecture Notes http://www.diku.dk/forskning/image/teaching/Courses/e02/RigidBodySimulation/

Faloutsos, P.; van de Panne, M. & Terzopoulos, D. (2001). Composable controllers for physics-based character animation. In: ACM Transactions on Computer Graphics (SIGGRAPH)

Goswami, A.; Espiau, B. & Keramane, A (1996). Limi cycles and their stability in a passive bipedal gait, In: IEEE Robotics and Automation

Havok Inc. (2007). Havok behavior. http://www.havok.com/

Hodgins, J.; Wooten, W.; Brogan, D. & O'Brien, J. (1995). Animating human athletics. In: ACM Transactions on Computer Graphics (SIGGRAPH)

Kawachi, K.; Suzuki, H. & Kimura, F. (1997). Simulation of Rigid Body Motion with Impulsive Friction Force, In: Proceedings of IEEE International Symposium on Assembly and Task Planning, pages 182 - 187

Khatib, O. (1987). A unified approach to motion and force control of robot manipulators: The operational space formulation, In: IEEE Journal on Robotics and Automation 3, 1 (February), pages 43–55

Kokkevis, E. (2004). Practical Physics for Articulated Characters

Laszlo, J. F.; van de Panne, M. & Fiume, E. (1996). Limit Cycle Control and its Application to the Animation of Balancing and Walking, In: ACM Transactions on Computer Graphics (SIGGRAPH)

Liu, C. K.; Hertzmann, A. & Popovic, Z. (2005). Learning physics-based motion style with nonlinear inverse optimization, In: ACM Transactions on Graphics, 24(3):1071.1081

McCann, J.; Pollard, N. S. & Srinivasa, S. (2006). Physics-Based Motion Retiming, In: ACM Transactions on Computer Graphics (SIGGRAPH)/Eurographics Symposium on Computer Animation

Nakamura, Y.; Hhanafusa, H. & Yoshikawa, T. (1987). Task-priority based redundancy control of robot manipulators, In: International Journal of Robotics Research, 6(2), pages 3-15

PhysX. NVIDIA. http://www.nvidia.com/object/physx_new.html

Safonova A.; Hodgins J. K. & Pollard N. S. (2004). Synthesizing Physically Realistic Human Motion in Low-Dimensional, Behavior-Specific Spaces, In: ACM Transactions on Graphics 23(3), SIGGRAPH

Shapiro, A.; Chu, D.; Allen, B. & Faloutsos, P. (2007). A Dynamic Controller Toolkit, In: ACM Transactions on Graphics Video Game Symposium (Sandbox)

Smith, R. (2004). Open Dynamics Engine, In: http://www.ode.org/

Thornton, S. & Marion, J. (2003). Classical Dynamics of Particles and Systems, Books Cole, 5th Edition

Trinkle, J.; Pang, J.; Sudarsky, S. & Lo, G. (1997). On Dynamic Multi-Rigid-Body Contact Problems with Coulomb Friction, In: Zeitschrift für Angewandte Mathematik, pages 267 - 279

Vondrak, M. (2006). Crisis Physics Engine, In: http://crisis.sourceforge.net/

Vondrak, M.; Sigal, L. & Jenkins, C. (2008). Physical Simulation for Probabilistic Motion Tracking, In: Computer Vision and Pattern Recognition (CVPR 2008), pages 1 - 8

Yin, K.; Coros, S.; Beaudoin, P. & van de Panne, M. (2008). Continuation Methods for Adapting Simulated Skills, In: ACM Transactions on Graphics (SIGGRAPH)

Yin, K.; Loken, K. & van de Panne, M. (2007). SIMBICON: Simple Biped Locomotion Control, In: ACM Transactions on Graphics (SIGGRAPH)

2

Intelligent Control

Maouche Amin Riad
Houari Boumedien University of Algiers
Algeria

1. Introduction

This chapter presents a novel family of intelligent controllers. These controllers are based on semiphysical (or gray-box) modeling. This technique is intended to combine the best of two worlds: knowledge-based modeling and black-box modeling.

A knowledge-based (white-box) model is a mathematical description of the phenomena that occur in a process, based on the equations of physics and chemistry (or biology, sociology, etc.); typically, the equations involved in the model may be transport equations, equations of thermodynamics, mass conservation equations, etc. They contain parameters that have a physical meaning (e.g., activation energies, diffusion coefficients, etc.), and they may also contain a small number of parameters that are determined through regression from measurements.

Conversely, a black-box model is a parameterized description of the process based on statistical learning theory. All parameters of the model are estimated from measurements performed on the process; it does not take into account any prior knowledge on the process (or a very limited one). Very often, the devices and algorithms that can learn from data are characterized as intelligent. The human mental faculties of learning, generalizing, memorizing and predicting should be the foundation of any intelligent artificial device or smart system (Er & Zhou, 2009). Even if we are still far away from achieving anything similar to human intelligence, many products incorporating Neural Networks (NNs), Support Vector Machines (SVMs) and Fuzzy Logic Models (FLMs) already exhibit these properties. Among the smart controller's intelligence is its ability to cope with a large amount of noisy data coming simultaneously from different sensors and its capacity to plan under large uncertainties (Kecman, 2001).

A semiphysical (or gray-box) model may be regarded as a tradeoff between a knowledge-based model and a black-box model. It may embody the entire engineer's knowledge on the process (or a part thereof), and, in addition, it relies on parameterized functions, whose parameters are determined from measurements. This combination makes it possible to take into account all the phenomena that are not modeled with the required accuracy through prior knowledge (Dreyfus, 2005).

A controller based on gray-box modeling technique is very valuable whenever a knowledge-based model exists, but is not fully satisfactory and cannot be improved by further analysis, or can only be improved at a very large computational cost (Maouche & Attari, 2008a; 2008b). Physical systems are inherently nonlinear and are generally governed by complex equations with partial derivatives. A dynamic model of such a system, to be used in control design, is by nature an approximate model. Thus, the modeling error

introduced by this approximation influences the performance of the control. Choosing an adaptive control based on neural network, allows dealing with modeling errors and makes it possible to compensate, until a certain level, physical phenomena such as friction, whose representation is difficult to achieve (Maouche & Attari, 2007).
We will consider as an application to this type of control, a robot manipulator with flexible arms. Flexible manipulators are a good example of complex nonlinear systems difficult to model and to control.
In this Chapter we describe a hybrid approach, based on semiphysical modelling, to the problem of controlling flexible link manipulators for both structured and unstructured uncertainties conditions (Maouche & Attari, 2008a; 2008b). First, a neural network controller based on the robot's dynamic equation of motion is elaborated. It aims to produce a fast and stable control of the joint position and velocity, and to damp the vibration of each arm. Then, an adaptive neural controller is added to compensate the unknown nonlinearities and unmodeled dynamics, thus enhancing the accuracy of the control. The robustness of the adaptive neural controller is tested under disturbances and compared to a classical nonlinear controller. Simulation results show the effectiveness of the proposed control strategy.

2. Lightweight flexible manipulators

The demand for increased productivity in industry has led to the use of lighter robots with faster response and lower energy consumption. Flexible manipulator systems have relatively smaller actuators, higher payload to weight ratio and, generally, less overall cost. The drawbacks are a reduction in the stiffness of the robot structure which results in an increase in robot deflection and poor performance due to the effect of mechanical vibration in the links.
The modeling and control of non-rigid link manipulator motion has attracted researchers attentions for almost three decades. A non-rigid link in a manipulator bears a resemblance to a flexible (cantilever) beam that is often used as a starting point in modeling the dynamics of a non-rigid link (Book, 1990). Well-known approaches such as Euler-Lagrange's equation and Hamilton's principle are commonly used in modeling the motion of rigid-link manipulators and to derive the general equation of motion for flexible link manipulators. The infinite-dimensional manipulator system is commonly approximated by a finite-dimensional model for controller design. The finite element method is used in the derivation of the dynamical model leading to a computationally attractive form for the displacement bending.
The motion control of a flexible manipulator consists of tracking the desired trajectory of the rigid variables which are the angular position and velocity. But due to the elasticity of the arms, it has also to damp the elastic variables which are, in our case, deflection and elastic rotation of section of the tip. The main difficulty in controlling such a system is that unlike a rigid manipulator, a flexible manipulator is a system with more outputs to be controlled (rigid and elastic variables) then inputs (applied torques), that involves the presence of dynamic coupling equations between rigid and elastic variables.
Moreover, the dynamic effect of the payload is much larger in the lightweight flexible manipulator than in the conventional one.
However, most of the control techniques for non-rigid manipulators are inspired by classical controls. A multi-step control strategy is used in (Book et al., 1975; Hillsley & Yurkovitch,

1991; Ushiyama & Konno, 1991; Lin & Lee, 1992; Khorrami et al., 1995; Azad et al., 2003; Mohamed et al., 2005) that consists of superimposing to the control of the rigid body, the techniques of shaping or correction of the elastic effects. Other algorithms use the technique of decoupling (De Shutter et al., 1988; Chedmail & Khalil, 1989), others are based on the method of the singular perturbation approach (Siciliano & Book, 1988; Spong, 1995; Park et al., 2002), use noncollocated feedback (Ryu et al., 2004) or use model-based predictive control for vibration suppression (Hassan et al., 2007).

Neural network-based controllers were also used as they reduce the complexity and allow a faster computation of the command (Kuo & Lee, 2001; Cheng & Patel, 2003; Tian & Collins, 2004; Tang et al., 2006).

With recent developments in sensor/actuator technologies, researchers have concentrated on control methods for suppressing vibration of flexible structures using smart materials such as Shape Memory Alloys (SMA) (Elahinia & Ashrafiuon, 2001), Magnetorheological (MR) materials (Giurgiutiu et al., 2001), Electrorheological (ER) materials (Leng & Asundi, 1999), Piezoelectric transducers (PZT) (Shin & Choi, 2001; Sun et al., 2004; Shan et al., 2005), and others.

The use of knowledge-based modeling, whereby mathematical equations are derived in order to describe a process, based on a physical analysis, is important to elaborate effective controllers. However, this may lead to a complex controller design if the model of the system to be controlled is more complex and time consuming.

Therefore, we propose a controller based on artificial neural networks that approximate the dynamic model of the robot. The use of artificial neural networks, replacing nonlinear modeling, may simplify the structure of the controller and, reduce its computation time and enhance its reactivity without a loss in the accuracy of the tracking control (Maouche & Attari, 2008a; 2008b). This is important when real time control is needed.

The main advantage of neural networks control techniques among others is that they use nonlinear regression algorithms that can model high-dimensional systems with extreme flexibility due to their learning ability.

Using dynamic equations of the system to train the neural network presents many advantages. Data (inputs/outputs set) are easily and rapidly obtained via simulation, as they are not tainted with noise, and they can be generated in sufficient number that gives a good approximation of the model. Moreover, it is possible to generate data that have better representation of the model of the system.

To reduce the modeling error between the actual system and its representation, we propose to add an adaptive neural controller. Here, the neural network is trained online, to compensate for errors due to structured and unstructured uncertainties, increasing the accuracy of the overall control.

The control law presented here has several distinguished advantages. It is easy to compute since it is based on artificial neural network. This robust controller design method maximizes the control performance and assures a good accuracy when regulating the tip position of the flexible manipulator in the presence of a time-varying payload and parameter uncertainties.

3. Dynamic modeling

The system considered here consists of two links connected with a revolute joint moving in a horizontal plane as shown in Figure 1. The first and the second link are composed of a flexible beam cantilevered onto a rigid rotating joint. It is assumed that the links can be bent freely in the horizontal plane but are stiff in the vertical bending and torsion. Thus, the Euler-Bernoulli beam theory is sufficient to describe the flexural motion of the links. Lagrange's equation and model expansion method can be utilized to develop the dynamic modeling of the robot.

As shown in Figure 1, $\{O_0\,\vec{x}_0\,\vec{y}_0\}$ represents the stationary frame, $\{O_1\,\vec{x}_1\,\vec{y}_1\}$ and $\{O_2\,\vec{x}_2\,\vec{y}_2\}$ are the moving coordinate frames with origin at the hubs of links 1 and 2, respectively. $\vec{y}_1$ and $\vec{y}_2$ are omitted to simplify the figure. θ_1 and θ_2 are the revolving angles at the hub of the two links with respect to their frames. f_1, α_1, f_2 and α_2 are the elastic displacements, they describe the deflection and the section rotation of the tip for the first and the second arm, respectively.

The motion of each arm of the manipulator is described by one rigid and two elastic variables:

$$\mathbf{q} = [\mathbf{q_r}\ \mathbf{q_e}]^{\mathrm{T}} \tag{1}$$

where $\mathbf{q_r} = [\theta_1\ \theta_2]^{\mathrm{T}}$ and $\mathbf{q_e} = [f_1\ \alpha_1\ f_2\ \alpha_2]^{\mathrm{T}}$.

The torques applied to the manipulator joints are given by:

$$\mathbf{\Gamma} = [\Gamma_1\ \Gamma_2]^{\mathrm{T}} \tag{2}$$

Let consider an arbitrary point M_i on the link i ($i = 1,2$). The kinetic energy of the link i is given by:

$$T_i = \frac{1}{2}\rho_i \int_0^{L_i}\int_0^{S_i} V(M_i)^2\, ds\, dx \tag{3}$$

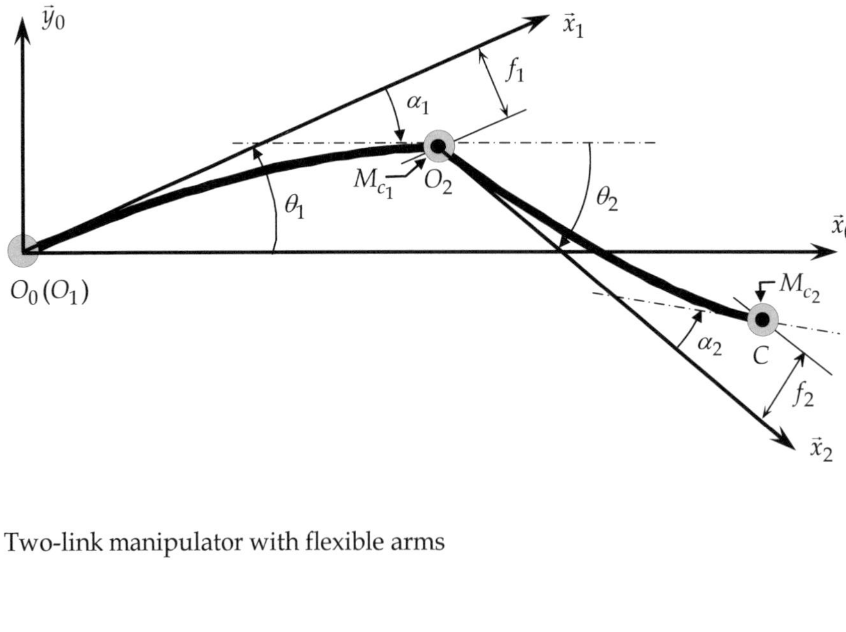

Fig. 1. Two-link manipulator with flexible arms

where $V(M_i)$ is the velocity of M_i on the flexible link i. L_i, S_i and ρ_i are the length, the section and the mass density of link i ($i = 1,2$), respectively.
Now, the total kinetic energy T can be written as (Ower & Van de Vegt, 1987):

$$T = T_1 + T_2 + \frac{1}{2}J_{A_1}\dot{\theta}_1^2 + \frac{1}{2}J_{B_1}(\dot{\theta}_1+\dot{\alpha}_1)^2 + \frac{1}{2}J_{A_2}(\dot{\theta}_1+\dot{\alpha}_1+\dot{\theta}_2)^2$$

$$+ \frac{1}{2}J_{B_2}(\dot{\theta}_1+\dot{\alpha}_1+\dot{\theta}_2+\dot{\alpha}_2)^2 + \frac{1}{2}M_{C_1}V(O_2)^2 + \frac{1}{2}M_{C_2}V(C)^2 \quad (4)$$

where J_{A_i} and J_{B_i} are, respectively, the mass moment of inertia at the origin and at the end of the link i ($i = 1,2$). Note that the first and the second terms on the right-hand side in (4) are kinetic energy of the flexible links 1 and 2, respectively. The third term is due to moment of inertia of the portion of the mass of the first actuator relative to link 1. The fourth and the fifth terms are due to moment of inertia of the portion of the mass of the second actuator in relation to link 1 and portion of the mass of the second actuator in relation to link 2, respectively. The sixth term is due to moment of inertia of mass at C (payload). The seventh and the eighth terms are kinetic energy of mass at O_2 and C respectively.
The potential energy U can be written as:

$$U = \frac{1}{2}\mathbf{q}^{\mathrm{T}}\mathbf{K}\mathbf{q} \quad (5)$$

with $\mathbf{K}=\begin{bmatrix}\mathbf{0} & \mathbf{0}\\ \mathbf{0} & \mathbf{K_e}\end{bmatrix}$, $\mathbf{K_e}=\begin{bmatrix}\mathbf{K_{E_1}} & \mathbf{0}\\ \mathbf{0} & \mathbf{K_{E_2}}\end{bmatrix}$ and $\mathbf{K_{E_i}}=\begin{bmatrix}12E_iI_i/L_i^3 & -6E_iI_i/L_i^2\\ -6E_iI_i/L_i^2 & 4E_iI_i/L_i\end{bmatrix}$ ($i = 1,2$).

The term on the right-hand side in (5) describes the potential energy due to elastic deformation of the links. Note that the term relative to the gravity is not present here as the manipulator moves on a horizontal plane. $\mathbf{K}$ is the stiffness matrix. The first two rows and columns of $\mathbf{K}$ are zeros as U does not depend on $\mathbf{q_r}$. E_i is the Young modulus and I_i the quadratic moment of section of the considered link.
The dynamic motion equation of the flexible manipulator can be derived in terms of Lagrange-Euler formulation:

$$\frac{\mathrm{d}}{\mathrm{d}t}\left[\frac{\partial \ell}{\partial \dot{q}_r(i)}\right] - \left[\frac{\partial \ell}{\partial q_r(i)}\right] = \Gamma_i \quad (i = 1,2) \quad (6a)$$

$$\frac{\mathrm{d}}{\mathrm{d}t}\left[\frac{\partial \ell}{\partial \dot{q}_e(j)}\right] - \left[\frac{\partial \ell}{\partial q_e(j)}\right] = 0 \quad (j = 1,4) \quad (6b)$$

where ℓ is the Lagrangian function and $\ell = T - U$.
Substituting (4) and (5) into (6a) and (6b) yields to:

$$\mathbf{L_r\Gamma} = \mathbf{A(q)\ddot{q}} + \mathbf{B(q)}[\mathbf{\dot{q}\dot{q}}] + \mathbf{C(q)}\left[\mathbf{\dot{q}}^2\right] + \mathbf{Kq} \quad (7)$$

where $\mathbf{A(q)}$ is the $(n \times n)$ inertia matrix, $\mathbf{B(q)}$ is the $(n \times (n^2-n)/2)$ matrix of Coriolis terms and $[\dot{\mathbf{q}}\dot{\mathbf{q}}]$ is an $((n^2-n)/2 \times 1)$ vector of joint velocity products given by: $[\dot{q}_1\dot{q}_2, \dot{q}_1\dot{q}_3, \dot{q}_1\dot{q}_4, ..., \dot{q}_{n-1}\dot{q}_n]^T$, $\mathbf{C(q)}$ is the $(n \times n)$ matrix of centrifugal terms and $\left[\dot{\mathbf{q}}^2\right]$ is an $(n \times 1)$ vector given by: $\left[\dot{q}_1^{\,2}, \dot{q}_2^{\,2}, ..., \dot{q}_n^{\,2}\right]^T$, $\mathbf{K}$ is the $(n \times n)$ stiffness matrix and $\mathbf{L_r\Gamma}$ is the n torque vector $[\Gamma_1 ... \Gamma_r, 0...0]^T$ applied to the joints. n is the total number of variables: $n_r + n_e$ (rigid and elastic, respectively) of the system, in our case, $n = 6, n_r = 2, n_e = 4$.
If we suppose known the length of the two links, we define (Pham et al., 1991):

$$\mathbf{X} = [J_{A_1} + J_{B_1}, J_{B_1}, M_{C_1} + M_{C_2}, \rho_1 l_1, M_1, E_1 I_1, J_{A_2} + J_{B_2}, J_{B_2}, M_{C_2}, \rho_2 l_2, M_2, E_2 I_2]^T \quad (8)$$

where $\mathbf{X}$ is the vector of the robot dynamic parameters.

4. Nonlinear control

This control is a generalization of the classically known 'computed torque' used to control rigid manipulator (Slotine & Li, 1987). It consists of a proportional and derived (PD) part completed by a reduced model which contains only the rigid part of the whole nonlinear dynamic model of the flexible manipulator (Pham, 1992). Let:

$$\mathbf{h}(\mathbf{q},\dot{\mathbf{q}}) = \mathbf{B}(\mathbf{q})[\dot{\mathbf{q}}\dot{\mathbf{q}}] + \mathbf{C}(\mathbf{q})\left[\dot{\mathbf{q}}^2\right] \quad (9)$$

Then, the model can be reduced to:

$$\mathbf{L_r\Gamma} = \mathbf{A}(\mathbf{q})\ddot{\mathbf{q}} + \mathbf{h}(\mathbf{q},\dot{\mathbf{q}})\dot{\mathbf{q}} + \mathbf{K}\mathbf{q} \quad (10)$$

or even:

$$\begin{bmatrix} \mathbf{\Gamma} \\ 0 \end{bmatrix} = \begin{bmatrix} \mathbf{A_r} & \mathbf{A_{re}} \\ \mathbf{A_{er}} & \mathbf{A_e} \end{bmatrix} \begin{bmatrix} \ddot{\mathbf{q}}_\mathbf{r} \\ \ddot{\mathbf{q}}_\mathbf{e} \end{bmatrix} + \begin{bmatrix} \mathbf{h_r} & \mathbf{h_{re}} \\ \mathbf{h_{er}} & \mathbf{h_e} \end{bmatrix} \begin{bmatrix} \dot{\mathbf{q}}_\mathbf{r} \\ \dot{\mathbf{q}}_\mathbf{e} \end{bmatrix} + \begin{bmatrix} 0 & 0 \\ 0 & \mathbf{K_e} \end{bmatrix} \begin{bmatrix} \mathbf{q_r} \\ \mathbf{q_e} \end{bmatrix} \quad (11)$$

we deduce from (11) that:

$$\mathbf{\Gamma} = \mathbf{A_r}\,\ddot{\mathbf{q}}_\mathbf{r} + \mathbf{h_r}\,\dot{\mathbf{q}}_\mathbf{r} + \mathbf{A_{re}}\,\ddot{\mathbf{q}}_\mathbf{e} + \mathbf{h_{re}}\,\dot{\mathbf{q}}_\mathbf{e} \quad (11a)$$

$$0 = \mathbf{A_{er}}\ddot{\mathbf{q}}_\mathbf{r} + \mathbf{h_{er}}\dot{\mathbf{q}}_\mathbf{r} + \mathbf{A_e}\,\ddot{\mathbf{q}}_\mathbf{e} + \mathbf{h_e}\dot{\mathbf{q}}_\mathbf{e} + \mathbf{K_e}\mathbf{q_e} \quad (11b)$$

we can then use the following control law:

$$\mathbf{\Gamma_{NL}} = \mathbf{A_r}(\mathbf{q_r},\mathbf{q_e})\ddot{\mathbf{q}}_\mathbf{r}^\mathbf{d} + \mathbf{h_r}(\mathbf{q_r},\mathbf{q_e},\dot{\mathbf{q}}_\mathbf{r},\dot{\mathbf{q}}_\mathbf{e})\dot{\mathbf{q}}_\mathbf{r}^\mathbf{d} + \mathbf{K_{pr}}\tilde{\mathbf{q}}_\mathbf{r} + \mathbf{K_{vr}}\dot{\tilde{\mathbf{q}}}_\mathbf{r} \quad (12)$$

where, $\mathbf{q_r^d}$, $\dot{\mathbf{q}}_\mathbf{r}^\mathbf{d}$ and $\ddot{\mathbf{q}}_\mathbf{r}^\mathbf{d}$ define the desired angular trajectory. $\tilde{\mathbf{q}}_\mathbf{r} = \mathbf{q_r^d} - \mathbf{q_r}$, $\dot{\tilde{\mathbf{q}}}_\mathbf{r} = \dot{\mathbf{q}}_\mathbf{r}^\mathbf{d} - \dot{\mathbf{q}}_\mathbf{r}$ are angular position and velocity errors. $\mathbf{K_{vr}}$ and $\mathbf{K_{pr}}$ are positive definite matrices of gain. If we consider the ideal case where no errors are made while evaluating the dynamic parameters X, a Lyapunov stability analysis of this control law is presented on Appendix A.

5. Adaptive neural control

The control system structure proposed in this paper is a combination of two controllers. The construction of the first controller is based on the approximation of the nonlinear functions in (12) by neural network to reduce the computation burden. The second controller is based on adaptive neural network. Here the network is trained online, to compensate for errors due to structured and unstructured uncertainty, increasing the precision of the overall controller.

5.1 Reducing the computation burden using Neural Network

The nonlinear law presented in (12) has some major advantages as it uses information extracted from the dynamic motion equation of the system to control the manipulator. Physical characteristics like the passivity of the system can then be used to elaborate a stable controller (Kurfess, 2005).

The drawback is that, using dynamic motion equation of the system in the construction of the controller can lead to a complex controller. Computing such a controller can be time consuming. This is mainly the case with flexible manipulators as they are governed by complex equations which lead generally to a huge model. Using such a model can be incompatible with real time control.

To avoid this problem we propose to approximate the part of the model which is used in the controller with neural networks. The main feature that makes neural network ideal technology for controller systems is that they are nonlinear regression algorithms that can model high-dimensional systems and have the extreme flexibility due to their learning ability. In addition their computation is very fast.

The functions $\mathbf{A_r}(\mathbf{q_r},\mathbf{q_e})$ and $\mathbf{h_r}(\mathbf{q_r},\mathbf{q_e},\dot{\mathbf{q}}_\mathbf{r},\dot{\mathbf{q}}_\mathbf{e})$ are approximated with the artificial neural networks $\mathbf{A_rNN}$ and $\mathbf{h_rNN}$. We will then use their outputs in addition to the PD part of (12) to elaborate the first controller:

$$\mathbf{\Gamma_{NN}} = \mathbf{A_rNN}\,\ddot{\mathbf{q}}_\mathbf{r}^\mathbf{d} + \mathbf{h_rNN}\,\dot{\mathbf{q}}_\mathbf{r}^\mathbf{d} + \mathbf{K_{pr}}\,\tilde{\mathbf{q}}_\mathbf{r} + \mathbf{K_{vr}}\,\dot{\tilde{\mathbf{q}}}_\mathbf{r} \tag{13}$$

In the neural network design scheme of $\mathbf{A_rNN}$ and $\mathbf{h_rNN}$, there are three-layered networks consisting of input, hidden and output layers. We use sigmoid functions in the hidden layer and linear functions in the output layer.

The back-propagation algorithm is adopted to perform supervised learning (Gupta et al., 2003). The two distinct phases to the operation of back-propagation learning include the forward phase and the backward phase.

In the forward phase the input signal propagate through the network layer by layer, producing the response $\mathbf{Y}$ at the output of the network:

$$\mathbf{Y} = f_o(f_h(\mathbf{Xi}\cdot\mathbf{Wij})\cdot\mathbf{Wjk}) \tag{14}$$

where $\mathbf{Xi}$ is the input signal, $\mathbf{Y}$ is the actual output of the considered neural network. In this control scheme, the input signals of the input layer for $\mathbf{A_rNN}$ are the rigid and elastic position of the two links: $[\theta_1, \theta_2, f_1, \alpha_1, f_2, \alpha_2]^\mathrm{T}$. For $\mathbf{h_rNN}$ the inputs are rigid and elastic position and velocity of the two links: $[\theta_1, \theta_2, f_1, \alpha_1, f_2, \alpha_2, \dot{\theta}_1, \dot{\theta}_2, \dot{f}_1, \dot{\alpha}_1, \dot{f}_2, \dot{\alpha}_2]^\mathrm{T}$. $\mathbf{Xi}\cdot\mathbf{Wij}$ is the weighted sum of the outputs of the previous layer, Wij_{ij} and Wjk_{jk} denote the weights between units i and j in the input layer to the hidden layer and between units j and k in the hidden layer to the output layer, respectively.

In this paper, the function f_o is a linear function and f_h is a tangent sigmoid function expressed by:

$$f_h(x) = \frac{2}{1+e^{-2x}} - 1 \tag{15}$$

The actual responses of $\mathbf{A_rNN}$ and $\mathbf{h_rNN}$ so produced are then compared with the desired responses of $\mathbf{A_r}$ and $\mathbf{h_r}$ respectively. Error signals generated are then propagated in a backward direction through the network.

In the backward phase, the delta rule learning makes the output error between the output value and the desired output value change weights and reduce error.

The training is made off line so that it does not disturb the real time control. The free parameters of the network are adjusted so as to minimize the following error function:

$$\mathbf{E_{NN}} = \frac{1}{2}(\mathbf{Y^d} - \mathbf{Y})^2 \tag{16}$$

where $\mathbf{Y^d}$ and $\mathbf{Y}$ are the desired and actual output of the considered neural network ($\mathbf{A_rNN}$ or $\mathbf{h_rNN}$).

The connect weight Wjk_{jk} is changed from the error function by an amount:

$$\Delta Wjk_{jk} = \gamma \cdot \delta k_k \cdot H_j \tag{17}$$

where γ is the learning rate and H_j is the jth hidden node.

The connect weight Wij_{ij} is changed from the error function by an amount:

$$\Delta Wij_{ij} = \gamma \cdot \delta j_j \cdot Xi_i \tag{18}$$

Delta rule learning for the units in the output layer is given by:

$$\delta k_k = Y_k^d - Y_k \tag{19}$$

Delta rule learning for the units in the hidden layer is given by:

$$\delta j_j = (1 - H_j^{\,2}) \cdot \sum_k (\delta k_k \cdot Wjk_{jk}) \tag{20}$$

Neural networks corresponding to $\mathbf{A_rNN}$ and $\mathbf{h_rNN}$ have been trained over different trajectories (training set). The stop criterion is a fundamental aspect of training. We consider, that the simple ideas of capping the number of iterations or of letting the system train until a predetermined error value are not recommended. The reason is that we want the neural network to perform well in the test set data; i.e., we would like the system to perform well in trajectories it never saw before (good generalization) (Bishop, 1995).

The error in the training set tends to decrease with iteration when the neural network has enough degrees of freedom to represent the input/output map. However, the system may

be remembering the training patterns (overfitting or overtraining) instead of finding the underlying mapping rule. To overcome this problem we have used the 'Cross Validation' method.
To avoid overtraining, the performance in a validation set (data set from trajectories that the system never saw before) must be checked regularly during training. Here, we performed once every 50 passes over the training set. The training should be stopped when the performance in the validation set starts to decrease, despite the fact that the performance in the training set continues to increase.

5.2 Construction of the adaptive neural controller

Let us consider now the case where the estimated parameters $\hat{\mathbf{X}}$ used in the dynamic equations to model the system are different from the actual parameters $\mathbf{X}$ of the manipulator. This will introduce an error in the estimation of the torque.
In addition to the structured uncertainties, there are also unstructured uncertainties due to unmodeled phenomena like frictions, perturbations etc. A more general equation of motion of the horizontal plane flexible robot is given by:

$$\mathbf{L_r \Gamma} = \mathbf{A(q)\ddot{q}} + \mathbf{B(q)\dot{q}\dot{q}} + \mathbf{C(q)\dot{q}^2} + \mathbf{Kq} + \mathbf{F(q,\dot{q})} \tag{21}$$

where $\mathbf{F(q,\dot{q})}$ is the unstructured uncertainties of the dynamics, including frictions (viscous friction and dynamic friction) and other disturbances.
We will then add a second controller to the system based on adaptive neural network in order to compensate the errors induced by the structured and unstructured uncertainties.
The basic concept of the adaptive neural network used in the second controller is to produce an output that forms a part of the overall control torque that is used to drive manipulator joints to track the desired trajectory.
The errors between the joint's desired and actual position/velocity values are then used to train online the neural controller.
In the adaptive neural network design scheme there are also three layers. Sigmoid and linear functions are used in the hidden and the output layer respectively.
The input signals of the input layer are angular position and velocity: $[\theta_1, \theta_2, \dot{\theta}_1, \dot{\theta}_2]^T$ at the hub 1 and 2, and the output signals $\mathbf{Y}$ of the output layer is the torque $\mathbf{\Gamma_{AN}}$ = $[\Gamma_{AN_1}, \Gamma_{AN_2}]^T$.
Training is made online and the parameters of the network are adjusted to minimize the following error function:

$$\mathbf{E_{AN}} = \frac{1}{2}(\mathbf{Y^d} - \mathbf{Y})^2 = \frac{1}{2}(\mathbf{K_{pn}}\tilde{\mathbf{q}}_r + \mathbf{K_{vn}}\dot{\tilde{\mathbf{q}}}_r)^2 \tag{22}$$

where $\mathbf{Y^d}$ and $\mathbf{Y}$ are the desired and actual output of the neural network, $\mathbf{K_{pn}}, \mathbf{K_{vn}}$ are positive definite matrices of gain.
As the training of the adaptive neural controller is made online, we must minimize its computational time. The learning rate is designed relating the network learning, local minimum, and weight changes which can be overly large or too small in the neural network learning. A momentum factor is then used to help the network learning (Krose & Smagt, 1996).

The formulation of the weight change is then given by:

$$\Delta \mathbf{W}(t+1) = \gamma \cdot \frac{\partial \mathbf{E}}{\partial \mathbf{W}} + \eta \cdot \Delta \mathbf{W}(t) \tag{23}$$

where $\mathbf{W}$ designates $\mathbf{Wij}$ or $\mathbf{Wjk}$, t indexes the presentation number and η is a constant which determines the effect of the previous weight change.

When no momentum term is used, it takes a long time before the minimum has been reached with a low learning rate, whereas for the high learning rates the minimum is never reached because of the oscillations. When adding the momentum term, the minimum will be reached faster. This will drive the adaptive neural controller to produce a faster response. A better control can then be achieved.

The overall robotic manipulator control system proposed is shown in Figure. 2. It can be written:

$$\mathbf{\Gamma} = \mathbf{\Gamma_{NN}} + \mathbf{\Gamma_{AN}} \tag{24}$$

where $\mathbf{\Gamma}$ is the overall controller output (torque); $\mathbf{\Gamma_{NN}}$ is the first controller output based on the neural model of the robot, as defined in (13); $\mathbf{\Gamma_{AN}}$ is the second controller output based on the adaptive neural network.

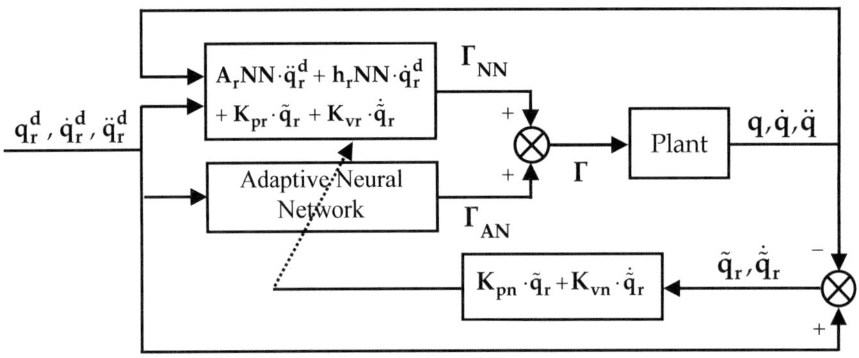

Fig. 2. The overall control system

6. Simulation results

Performance of the control strategy proposed is tested using a dynamic trajectory having 'Bang-Bang' acceleration, with a zero initial and final velocity:

$$\ddot{\theta}_1^d(t) = \ddot{\theta}_2^d(t) = \begin{cases} \dfrac{16\pi}{T^2} & \text{for } t \in \left[0, \dfrac{T}{4}\right] \\ -\dfrac{16\pi}{T^2} & \text{for } t \in \left[\dfrac{T}{4}, \dfrac{3T}{4}\right] \\ \dfrac{16\pi}{T^2} & \text{for } t \in \left[\dfrac{3T}{4}, T\right] \end{cases} \tag{25}$$

with $\theta_1^d(0) = \theta_2^d(0) = 0$ and $\dot{\theta}_1^d(0) = \dot{\theta}_2^d(0) = 0$.

To avoid the destabilization of the control induced by fast dynamics, we choose $T = 30$ *sec*. The maximum angular velocity is reached for $t = T/4$ and for $t = 3T/4$ and its absolute value is $4\pi / T$ *rad/sec* or 24 *deg/sec*.
The gain matrices are adjusted as follows:

- in the nonlinear control law (12), $\mathbf{K_{pr}} = \begin{bmatrix} 1 & 0 \\ 0 & 0.6 \end{bmatrix}$ and $\mathbf{K_{vr}} = \begin{bmatrix} 4 & 0 \\ 0 & 0.8 \end{bmatrix}$
- in the error function (23), $\mathbf{K_{pn}} = \begin{bmatrix} 0.8 & 0 \\ 0 & 0.4 \end{bmatrix}$ and $\mathbf{K_{vn}} = \begin{bmatrix} 4.8 & 0 \\ 0 & 1.5 \end{bmatrix}$.

Let us suppose now that the actual values of the parameters of the robot are such as specified in Table 1. To test the robustness of the proposed control strategy, we consider the extreme case where the estimation error on the dynamic parameters $\mathbf{X}$ is:

$$\hat{\mathbf{X}} = \frac{\mathbf{X}}{100} \tag{26}$$

then, we use these values ($\hat{\mathbf{X}}$) in the training of $\mathbf{A_rNN}$ and $\mathbf{h_rNN}$. This will drive the first controller to produce an incorrect torque. We will see how the second controller deals with this error and how it will correct it.
Our goal here is to simulate an important error due to a bad estimation of the dynamic parameters (or ignorance of some of them). We can suppose that if the hybrid controller can handle this important error, it can a fortiori handle a small one.
For simplicity on the simulation tests, dynamic parameters are equally bad estimated in (26). Even if it is not always the case on practice, this will not affect the adaptive neural controller, which is in charge of compensating these errors, because the adaptive neural controller considers the global error (the resultant of the sum of all errors).
In order to better appreciate the effectiveness of the overall adaptive neural controller we compare its results with the nonlinear controller given in (12).
Figures 3 to 12 illustrate the results obtained with the adaptive neural controller applied to the two-link flexible manipulator. They describe the evolution of: angular position, error on position, deflection, angular velocity and error on the angular velocity, for the joints 1 and 2, respectively.
Results of the nonlinear control are reported in dashed line for comparison. The desired trajectory (target) is reported on Figures 3, 6, 8 and 11 in dotted line.
Table 2 and Table 3 presents the maximum error and the Root Mean Square error (RMS) of the angular position and velocity obtained with the two types of control strategy used.
The desired trajectory imposes a fast change of acceleration on moment $t = T/4 = 7.5$ *sec* and $t = 3T/4 = 22.5$ *sec*. This radical change from a positive to a negative acceleration for the first moment and from a negative to positive acceleration for the second one stresses the control. We can see its impact on the control of the angular velocity in Figure 6 and Figure 11.
However, the trajectory following obtained with the adaptive neural control is good and the error induced is acceptable. Whereas, the nonlinear control strongly deviates from the target.
We can see from Table 2 and Table 3 that the error on velocity, obtained with the conventional nonlinear control, reaches 0.19 *rad/sec* (10.9 *deg/sec*) for the first joint and 0.13 *rad/sec* (7.7 *deg/sec*) for the second one. With the adaptive neural control, results are

significantly better with an velocity error lower than 0.01 *rad/sec* (lower than 0.5 *deg/sec*) for the two joints.
For the position control (see Figures 3 and 8), we notice that the angular trajectory obtained with the adaptive neural controller matches perfectly the target, with an error of no more than 0.003 *rad*, (0.2 *deg*) for the first and the second links, whereas it exceeds 0.34 *rad* (20 *deg*) with the nonlinear controller for the two links (see Table 2 and Table 3).
The hybrid controller proposed deals well with the flexibility of the link as the deflection is lessened (see Figure 5 and Figure 10). However, results obtained with the nonlinear control alone are slightly better.
The deflection of the first flexible link, shown in Figure 5, is within ± 0.055 *m* with the hybrid control where as it is lower than 0.036 *m* with the nonlinear control. For the second flexible link and as shown in Figure 10, the deflection reaches 0.017 *m* with the hybrid control where as it is lower than 0.011 *m* with the nonlinear control
We notice also from Figure 5 and Figure 10, the appearance of vibrations with the hybrid control. However, their amplitude is lessened.
Therefore, we can make the following conclusion. On the one hand, the use of the nonlinear model based controller ($\boldsymbol{\Gamma}_{\mathbf{NN}}$) alone reduces the precision of the control in the presence of structured and unstructured uncertainties. But, on the other hand, the use of the adaptive part of the neural controller ($\boldsymbol{\Gamma}_{\mathbf{AN}}$) alone increases the deflection of the links and no damping of vibrations is achieved which can lead to an unstable system.
Combining these two control technique schemes gives a good compromise between stability and precision. Simulation results show the effectiveness of the control strategy proposed.

Physical parameters	**Link 1**	**Link 2**
Length (m)	$L_1 = 1.00$	$L_2 = 0.50$
Moment of inertia at the Origin of the link (kg m²)	$J_{A_1} = 1.80\ 10^{-3}$	$J_{A_2} = 1.85\ 10^{-4}$
Moment of inertia at the end of the link (kg m²)	$J_{B_1} = 4.70\ 10^{-2}$	$J_{B_2} = 0.62$
Mass of the link (kg)	$M_1 = 1.26$	$M_2 = 0.35$
Mass at the tip (kg)	$M_{C_1} = 4.0$	$M_{C_2} = 1.0$
Mass density (kg/m³)	$\rho_1 = 7860$	$\rho_2 = 7860$
Young's modules	$E_1 = 1.98\ 10^{11}$	$E_2 = 1.98\ 10^{11}$
Quadratic moment of section (m⁴)	$I_1 = 3.41\ 10^{-11}$	$I_2 = 6.07\ 10^{-12}$

Table 1. Manipulator characteristics

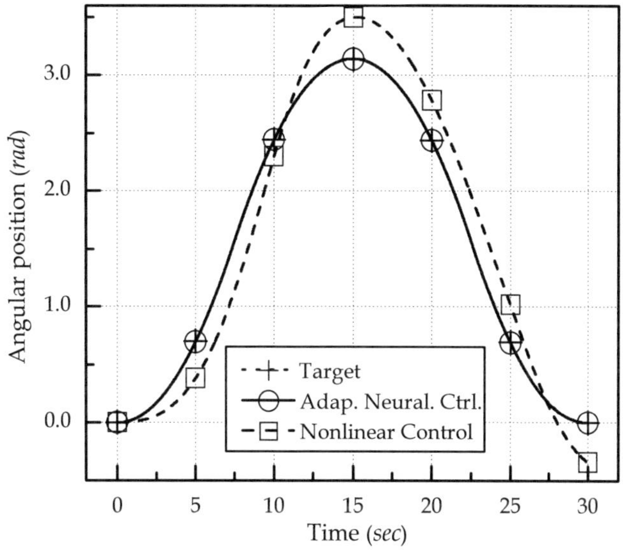

Fig. 3. Evolution of the angular position θ_1 (*rad*)

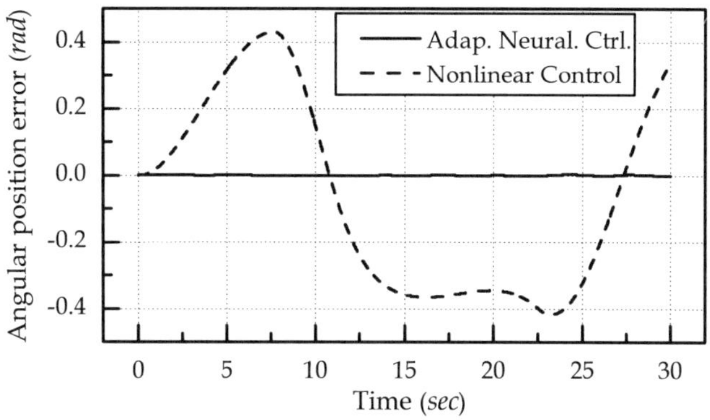

Fig. 4. Evolution of the angular position error $\tilde{\theta}_1$ (*rad*)

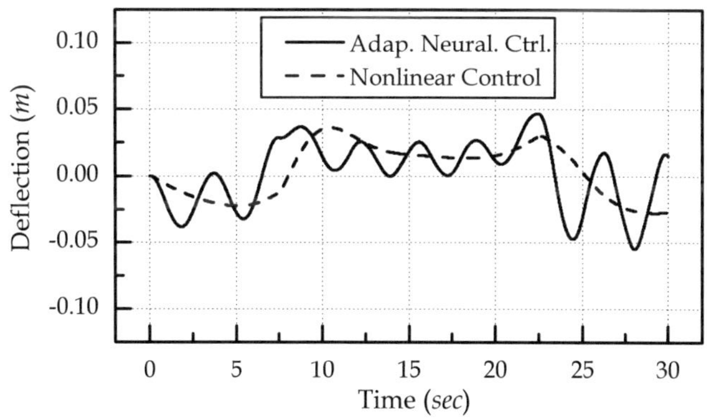

Fig. 5. Evolution of the deflection f_1 (*m*)

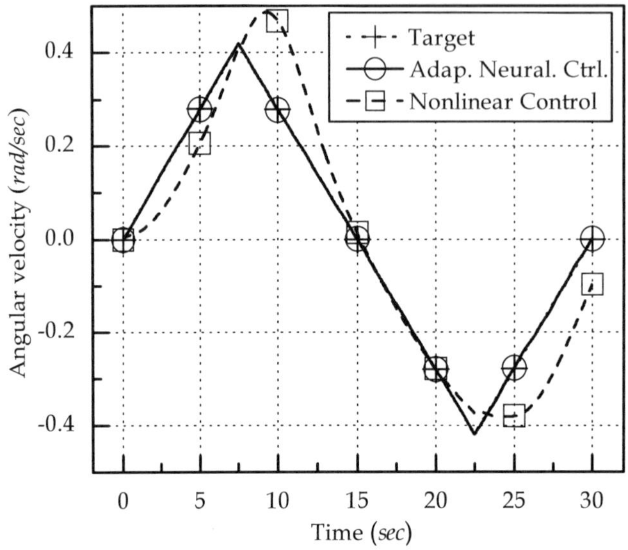

Fig. 6. Evolution of the angular velocity $\dot{\theta}_1$ (*rad/sec*)

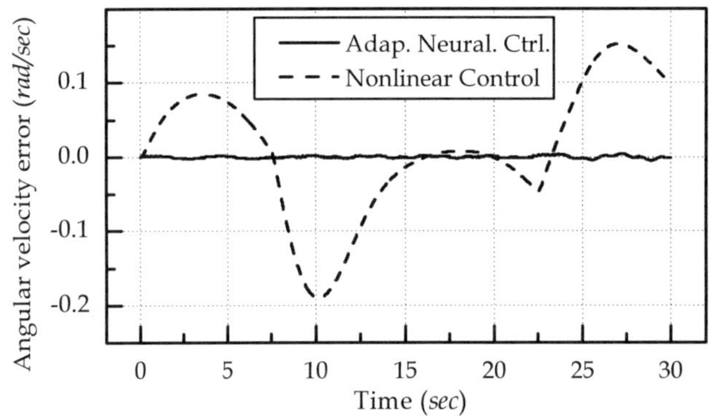

Fig. 7. Evolution of the angular velocity error $\dot{\tilde{\theta}}_1$ (*rad/sec*)

	Maximum Error		Root Mean Square Error	
Variable	θ_1 (*rad*)	$\dot{\theta}_1$ (*rad/sec*)	θ_1 (*rad*)	$\dot{\theta}_1$ (*rad/sec*)
Adap. Neural Control	3.07×10^{-3}	5.19×10^{-3}	1.14×10^{-3}	1.60×10^{-3}
Nonlinear Control	4.31×10^{-1}	1.90×10^{-1}	2.98×10^{-1}	8.92×10^{-2}

Table 2. Error on joint 1

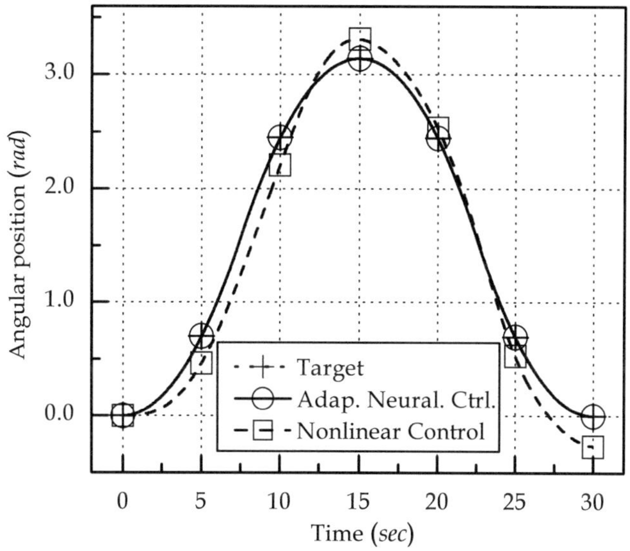

Fig. 8. Evolution of the angular position $\theta_2\,(rad)$

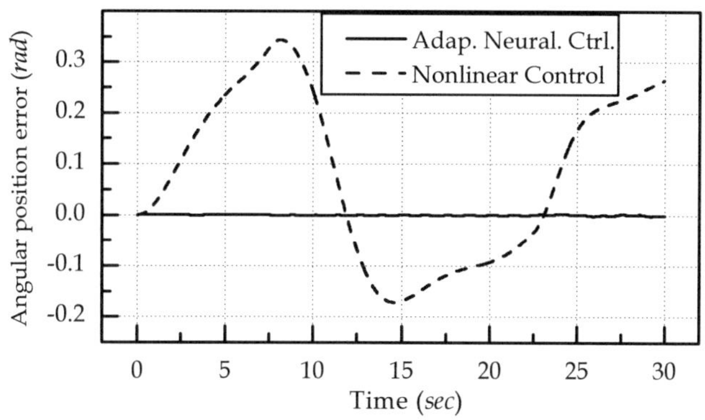

Fig. 9. Evolution of the angular position error $\tilde{\theta}_2\,(rad)$

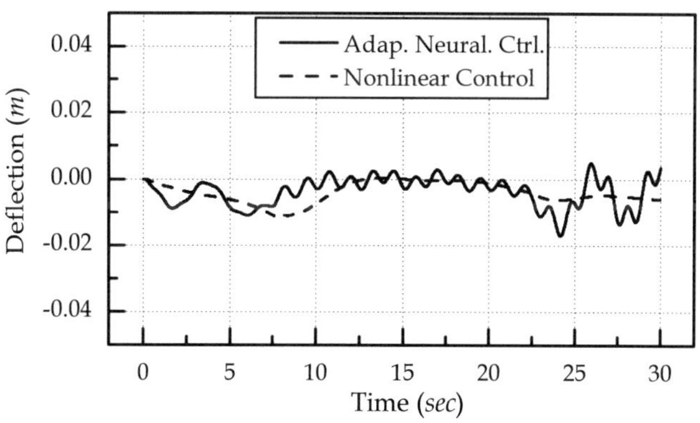

Fig. 10. Evolution of the deflection $f_2\,(m)$

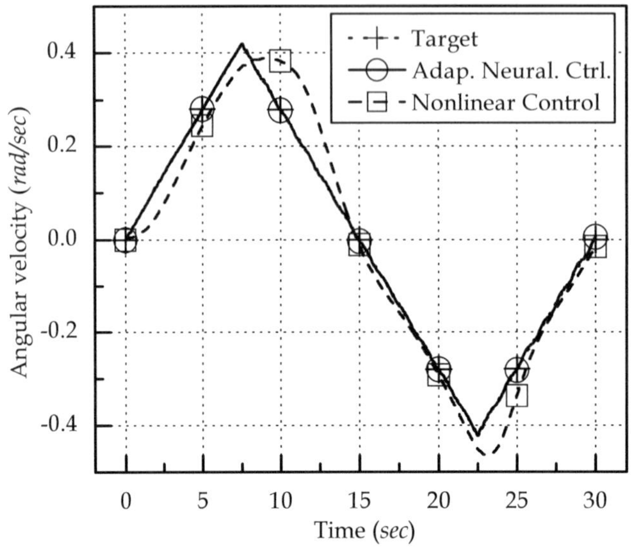

Fig. 11. Evolution of the angular velocity $\dot{\theta}_2$ (*rad/sec*)

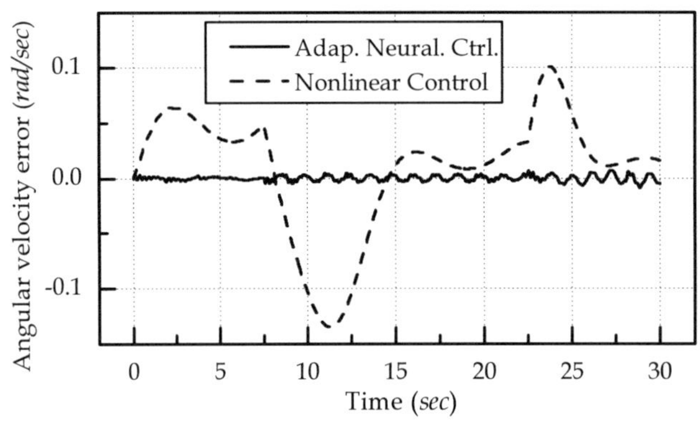

Fig. 12. Evolution of the angular velocity error $\dot{\tilde{\theta}}_2$ (*rad/sec*)

	Maximum Error		Root Mean Square Error	
Variable	θ_2 (*rad*)	$\dot{\theta}_2$ (*rad/sec*)	θ_2 (*rad*)	$\dot{\theta}_2$ (*rad/sec*)
Adap. Neural Control	2.85×10^{-3}	8.67×10^{-3}	8.43×10^{-4}	2.55×10^{-3}
Nonlinear Control	3.44×10^{-1}	1.35×10^{-1}	1.83×10^{-1}	5.53×10^{-2}

Table 3. Error on joint 2

7. Conclusion

Our goal is to search for intelligent control techniques that improve the performance of the controller and reduce the computation burden. The main idea here is to combine two control techniques, nonlinear control and neural network control.
The new adaptive neural control strategy presented is a combination of two controllers.
The first controller is based on the approximation with neural networks of the robot dynamic equation of the motion. Its aim is to provide a stable and fast control based on the dynamic model of the system. Using artificial neural network in the place of the nonlinear model allow to simplify the structure of the controller reducing its computation time and enhancing its reactivity.
The second controller is based on neural networks that are trained online. Its objective is to ensure that the actual trajectory matches the desired one by compensating errors due to structured and unstructured uncertainty, increasing the precision of the control.
Simulation results on a robot manipulator with two flexible arms have shown the robustness in performance of this control design scheme against adverse effects such as model parameter variations.
In summary, this article provides a novel control structure, to overcome the robotic manipulator control difficulties faced by conventional control schemes when uncertainties (e.g., friction, changing payload, time-varying friction, disturbances) cannot be ignored.

8. Appendix A: Stability analysis

By subtracting (12) from (11a), we obtain the error equation:

$$\mathbf{A_r}\ddot{\tilde{\mathbf{q}}}_\mathbf{r} + \mathbf{A_{re}}\ddot{\tilde{\mathbf{q}}}_\mathbf{e} + \mathbf{h_r}\dot{\tilde{\mathbf{q}}}_\mathbf{r} + \mathbf{h_{re}}\dot{\tilde{\mathbf{q}}}_\mathbf{e} + \mathbf{K_{pr}}\tilde{\mathbf{q}}_\mathbf{r} + \mathbf{K_{vr}}\dot{\tilde{\mathbf{q}}}_\mathbf{r} = 0 \qquad \text{(A.1)}$$

with, $\tilde{\mathbf{q}}_\mathbf{e} = 0 - \mathbf{q_e} = -\mathbf{q_e}$ and $\dot{\tilde{\mathbf{q}}}_\mathbf{e} = 0 - \dot{\mathbf{q}}_\mathbf{e} = -\dot{\mathbf{q}}_\mathbf{e}$ representing the elastic stabilization errors. In addition, rewriting the coupling equation (11b) according to the trajectory and the elastic stabilization error variables ($\tilde{\mathbf{q}}_\mathbf{r}$ and $\tilde{\mathbf{q}}_\mathbf{e}$) gives:

$$\mathbf{A_{er}}\ddot{\tilde{\mathbf{q}}}_\mathbf{r} + \mathbf{A_e}\ddot{\tilde{\mathbf{q}}}_\mathbf{e} + \mathbf{h_{er}}\dot{\tilde{\mathbf{q}}}_\mathbf{r} + \mathbf{h_e}\dot{\tilde{\mathbf{q}}}_\mathbf{e} + \mathbf{K_e}\tilde{\mathbf{q}}_\mathbf{e} = \mathbf{A_{er}}\ddot{\mathbf{q}}_\mathbf{r}^\mathbf{d} + \mathbf{h_{er}}\dot{\mathbf{q}}_\mathbf{r}^\mathbf{d} \qquad \text{(A.2)}$$

Using (A.1) and (A.2), the global error equation becomes:

$$\mathbf{A}\ddot{\tilde{\mathbf{q}}} + \mathbf{h}\dot{\tilde{\mathbf{q}}} + \mathbf{K_p}\tilde{\mathbf{q}} + \mathbf{K_v}\dot{\tilde{\mathbf{q}}} + \mathbf{s_1} = 0 \qquad \text{(A.3)}$$

where the positive constant matrices $\mathbf{K_p}$, $\mathbf{K_v}$ are respectively $\begin{bmatrix} \mathbf{K_{pr}} & 0 \\ 0 & \mathbf{K_e} \end{bmatrix}$, $\begin{bmatrix} \mathbf{K_{vr}} & 0 \\ 0 & 0 \end{bmatrix}$, and

$$\mathbf{s_1} = -\begin{bmatrix} 0 \\ \mathbf{A_{er}}\ddot{\mathbf{q}}_\mathbf{rd} + \mathbf{h_{er}}\dot{\mathbf{q}}_\mathbf{rd} \end{bmatrix}.$$

To study the stability of the global system, the following Lyapunov function is considered:

$$\mathbf{V} = (1/2)\,\dot{\tilde{\mathbf{q}}}^T \mathbf{A}\dot{\tilde{\mathbf{q}}} + (1/2)\,\tilde{\mathbf{q}}^T \mathbf{K_p}\tilde{\mathbf{q}} \qquad \text{(A.4)}$$

by differentiating V and using (A.3), and the fact that $\mathbf{A}$ is a symmetric positive-definite matrix (Kurfess, 2005), we obtain:

$$\dot{\mathbf{V}} = \dot{\tilde{\mathbf{q}}}^{T}((1/2)\dot{\mathbf{A}}-\mathbf{h})\dot{\tilde{\mathbf{q}}} - \dot{\tilde{\mathbf{q}}}^{T}(\mathbf{K_v}\dot{\tilde{\mathbf{q}}}+\mathbf{s_1}) \tag{A.5}$$

The property of passivity of the flexible manipulator implies that $(1/2)\dot{\mathbf{A}}-\mathbf{h}$ is skew symmetric (Lewis, 1999), finally we have:

$$\dot{\mathbf{V}} = -\dot{\tilde{\mathbf{q}}}_{\mathbf{r}}^{T}\mathbf{K_{vr}}\dot{\tilde{\mathbf{q}}}_{\mathbf{r}} + \dot{\tilde{\mathbf{q}}}_{\mathbf{e}}^{T}(\mathbf{A_{er}}\ddot{\mathbf{q}}_{\mathbf{rd}} + \mathbf{h_{er}}\dot{\mathbf{q}}_{\mathbf{rd}}) \tag{A.6}$$

The Lyapunov second method provides that the asymptotic stability of the control is assured if the following conditions are met. $\mathbf{V}$ is strictly positive everywhere except in $\tilde{\mathbf{q}}=\mathbf{0}$ where it is 0 and $\dot{\mathbf{V}}$ is strictly negative everywhere except in $\dot{\tilde{\mathbf{q}}}=\mathbf{0}$ where it is 0.
These conditions are always met if the desired angular velocities and accelerations are not too significant for a given tuning of $\mathbf{K_{vr}}$, so that $\dot{\mathbf{V}}$ remains essentially negative to ensure the control stability.

9. References

Azad, A. K. M.; Tokhi, M. O. & Anand, N. (2003). Teaching of control for complex systems trough simulation. *Proceedings of the 2003 ASEE/WFEO International Colloquium*, American Society for Engineering Education.

Bishop, C. M. (1995). *Neural Networks for Pattern Recognition*. Oxford University Press, New York.

Book, W. J. (1990). Modeling, design, and control of flexible manipulator arms: A tutorial review. *IEEE Proceedings of the 29th Conference on Decision and Control*, pp. 500-506, Honolulu, Haway.

Book, W. J.; Maizza-Neto, O. & Whitney, D. E. (1975). Feedback control of two beam, two joint systems with distributed flexibility. *Journal of Dynamic Systems, Measurement, and Control*, Vol. 97, No. 4, pp. 424-431.

Chedmail, P. & Khalil, W. (1989). Non linear decoupling control of flexible robots. *4th International Conference on Advanced Robotics*, pp. 138-145, Columbus.

Cheng, X. P. & Patel, R. V. (2003). Neural network based tracking control of a flexible macro-micro manipulator system. *Neural Networks*, Vol. 16, pp. 271-286.

De Shutter, J.; Van Brussels, H.; Adam, M.; Froment, A. & Faillot, J. L. (1988). Control of flexible robots using generalized non-linear decoupling. *Proceedings of IFAC*, pp. 113-118.

Dreyfus, G. (2005). *Neural Networks Methodology and Applications*. Springer, ISBN-10 3-540-22980-9, Germany.

Elahinia, M. H. & Ashrafiuon, H. (2001). Nonlinear control of a shape memory alloy actuated manipulator. *ASME Journal Of Vibration Acoustic*, Vol. 123, No. 4, pp. 487-495.

Er, M. J. & Zhou, Y. (2009). *Theory and Novel Applications of Machine Learning*. In-Tech, ISBN: 978-3-902613-55-4.

Giurgiutiu, V.; Jichi, F.; Berman, J. & Kamphaus, J. M. (2001). Theoretical and experimental investigation of magnetostrictive composite beams. *Smart Material Structure*, Vol. 10, No 5, pp. 934-935.

Gupta, M. M.; Jin, L. & Homma, N. (2003). *Static and Dynamic Neural Networks*. Wiley-Interscience, ISBN: 0-471-21948-7.

Hassan, M.; Dubay, R.; Li, C. & Wang, R. (2007). Active vibration control of a flexible one-link manipulator using a multivariable predictive. *Mechatronics*, Vol. 17, pp. 311-323.

Hillsley, K. L. & Yurkovitch, S. (1991). Vibration control of two link flexible robot arm. *Proceedings of the IEEE Conference on Robotics and Automation*, pp. 2121-2126.

Kecman, V. (2001). *Learning and soft computing: support vector machines, neural networks and fuzzy logic models*. The MIT press, ISBN: 0-262-11255-8.

Khorrami, F.; Jain, S. & Tzes, A. (1995). Experimental results on adaptive nonlinear control and input preshaping for multi-link flexible manipulators. *Automatica*, Vol. 31, pp. 83-97.

Krose, B. & Smagt, P. (1996). *An Introduction to Neural Networks*. Amsterdam University Press.

Kuo, C. F. J. & Lee C. J. (2001). Neural network control of a rotating elastic manipulator. *Computers and Mathematics with applications*, Vol. 42, pp. 1009-1023.

Kurfess, T. R. (2005). *Robotics and Automation Handbook*. p. 81, CRC Press, ISBN: 0-8493-1804-1.

Leng, J. & Asundi, A. (1999). Active vibration control system of smart structures based on FOS and ER actuator. Smart Material Structure, Vol. 8, No. 2, pp. 252-256.

Lewis, F. L.; Fitzgerald, J. M.; Walker, I. D.; Cutkosky, M. R.; Lee, K. M.; Bailey, R.; Zhou, C.; Priest, J. W.; Stevens, G. T. Jr. & Liu, K. (1999). *Robotics. Mechanical engineering handbook*. p. 105, CRC Press LLC.

Lin, Y. J. & Lee, T. S. (1992). Comprehensive dynamic modeling and motion/force control of flexible manipulators. *Mechatronics*, Vol. 2, pp. 129-148.

Maouche, A. R. & Attari, M. (2008a). Neural network-based adaptive control of a two-link flexible manipulator. *The Mediterranean Journal of Measurement and Control*, Vol. 4, No 2, pp. 66-75, SoftMotor Ltd., United Kingdom. ISSN: 1743-9310.

Maouche, A. R. & Attari, M. (2008b). Adaptive neural controller of a rotating flexible manipulator. *19th IEEE Symposium on Power Electronics, Electrical Drives, Automation and Motion (SPEEDAM 2008)*, pp. 517-522, June 2008, Ischia, Italy, ISBN: 978-1-4244-1663-9.

Maouche, A. R. & Attari, M. (2007). Hybrid control strategy for flexible manipulators. *17th IEEE International Symposium on Industrial Electronics (ISIE 2007)*, pp. 50-55, June 2007, Vigo, Spain, ISBN: 978-1-4244-0755-2.

Mohamed, Z.; Martins, J. M.; Tokhi, M. O.; Da Costa, J. & Botto, M. A. (2005). Vibration control of a very flexible manipulator system. *Control Engineering Practice*, Vol. 13, pp. 267-277.

Ower, J. C. & Van de Vegt, J. (1987). Classical control design for a flexible manipulator: modeling and control system design. *IEEE Journal of Robotics and Automation*, Vol. RA-3, No. 5, pp. 485-489.

Park, N. C.; Yang, H. S.; Park, H. W. & Park, Y. P. (2002). Position/vibration Control of two-degree-of-freedom arms having one flexible link with artificial pneumatic muscle actuators. *Robotic and Autonomous Systems*, Vol. 40, pp. 239-253.

Pham, C. M. (1992). Identification and control of flexible robots. *Doctorate Thesis of University of Nantes*, France.

Pham, C. M.; Chedmail, P. & Gaultier, M. (1991). Determination of base parameters of flexible links manipulators. *Proc. IMACS Symposium on Modeling and Control of Technological Systems*, pp. 524-529.

Ryu, J. H.; Kwon, D. S. & Hannaford, B. (2004). Control of a flexible manipulator with non collocated feedback: Time-domain passivity approach. *IEEE Transaction on Robotics*, Vol. 20, No. 4, pp. 776-780.

Shan, J.; Liu, H. T. & Sun, D. (2005). Slewing and vibration control of a single-link flexible manipulator by positive feedback (PPF). *Mechatronics*, Vol. 15, pp. 487-503.

Shin, H. C. & Choi, S. B. (2001). Position control of a two-link flexible manipulator featuring piezoelectric actuators and sensors. *Mechatronics*, Vol. 11, No. 6, pp. 707-729.

Slotine, J. J. E. & Lie, W. (1987). On the adaptive control of robot manipulators. *International Newspaper of Robotics Research*, Vol.6, pp. 49-59.

Siciliano, B. & Book, W. J. (1988). A singular perturbation approach to control of lightweight flexible manipulators. *The International Journal of Robotics Research*, Vol. 7, pp. 79-90.

Spong, M. W. (1995). Adaptive Control of flexible joint manipulators: comments on two papers. *Automatica*, Vol. 31, pp. 585-590.

Sun, D.; Mills, J. K.; Shan, J. J. & Tso, S. K. (2004). A PZT actuator control of a single-link flexible manipulator based on linear velocity feedback and actuator placement. *Mechatronics*, Vol. 14, No. 4, pp. 381-401.

Tang, Y.; Sun, F. & Sun, F. (2006). Neural network control of flexible-link manipulators using sliding mode. *Neurocomputing*, Vol. 70, pp. 288-295.

Tian, L. & Collins, C. (2004). A dynamic recurrent neural network-based controller for a rigid-flexible manipulator system. *Mechatronics*, Vol. 14, pp. 471-490.

Ushiyama, M. & Konno, A. (1991). Computed acceleration control for the vibration supression of flexible robotic manipulators. *5th International Conference on Advanced Robotics*, pp. 126-131, Pisa, Italy.

3

Vision-Based Hierarchical Fuzzy Controller and Real Time Results for a Wheeled Autonomous Robot

Pourya Shahmaleki, Mojtaba Mahzoon,
Alireza Kazemi and Mohammad Basiri
Shiraz University
Iran

1. Introduction

One of the most important problems in robotics is motion planning problem, which its basic controversy is to plan a collision-free path between initial and target configurations for a robot. In the framework of motion planning for nonholonomic systems, the wheeled robots have attracted a significant amount of interest. The path planner of a wheeled autonomous robot has to meet nonholonomic constraints and then the movement direction must always be tangent to its trajectory (Paromtichk et. al., 1998; Latombe, 1991, Murray & Sastry, 1993; Lamiraux & Laumond, 2001; Scheuer & Fraichard, 1996). If no obstacles exist on path of the robot, then the robot task is finding the shortest path connecting two given initial and final configurations. The shortest paths for a car like vehicle consist of a finite sequence of two elementary components: arcs of circle (with minimum turning radii) and straight line segments. In any case, the problem is that the curvature is discontinuous between two elementary components, so that these shortest paths cannot be followed precisely without stopping at each discontinuity point to reorient the front wheels. To avoid these stops, several authors have proposed continuous-curvature path planners using differential geometric methods. These planners generate clothoids, cubic spirals, β-splines, quintic polynomials, etc., which are then followed by using a path tracking technique based on, for example, pure-pursuit or predictive control methods (Lamiraux & Laumond, 2001; Scheuer & Fraichard, 1996). Stabilization issues of path-tracking methods for car-like vehicles using the Lyapunov method have been reported in (Walsh et. al., 1994; Tayebi & Rachid, 1996). One of the key technologies of future automobiles is the parking assist or automatic parking control. Control problems of a car-like vehicle are not easy because of the nonholonomic velocity constraints. The truck backer-upper control is a typical nonlinear control problem that cannot be solved by the conventional control techniques.
The goal of controller is to back up a truck to a loading dock from any initial position as quickly and precisely as possible. Backing a truck to the loading dock or parking spot is a difficult task even for a skilled truck driver. The research in parking problem is derived from the study of general motion planning for autonomous robots. In the past few decades, many algorithms have been developed for robot parking planning (Jiang & Seneviratne, 1999; Gomez-Bravo et. al., 2001; Cuesta et. al., 2004; Reeds & Shepp, 1990). The attempts to

solve the truck backer-upper problem, rooted in computational intelligence, can be divided into two groups. The first group of methods seeks the solution through self tuning using neural networks, genetic algorithms or a combination of both. The second group of solutions, based on fuzzy logic, regards the controller as an emulator of human operator. The problem has become an acknowledged benchmark in non-linear control and as an example of a self-learning system in neural networks was proposed by Nguyen and Widrow in 1990 (Nguyen & Widrow, 1989). Careful experiments of their approach showed that the computational effort is very high (Kong & Kosko, 1990). Thousands (about 20000) of back-up cycles are needed before the network learns. Moreover the backpropagation algorithm does not converge for some sets of training samples. Numerous other techniques have been used, including genetic programming (Koza, 1992) Neuro-genetic controller (Schoenauer, & Ronald, 1994) and simplified neural network solution through problem decomposition (Jenkins & Yuhas, 1993). Very interesting contribution is (Tanaka et. al, 1998), where up to ten trailers can be controlled representing those as Takagi-Sugeno models and applying linear matrix inequalities method. A simplified version of the control problem has been extensively investigated in the field of fuzzy control (Ramamoorthy & Huang, 1991; Wang & Mendel, 1992; Ismail & Abu-Khousa, 1996; Kim, 1998; Dumitrache & Buiu, 1999). Also parking problem has been investigated by many researchers in the field of computational intelligence (for example; chang in (Chang et. al, 1995), Schoenauer in (Schoenauer & Ronald, 1994), Wang in (Wang & Mendel, 1992) and Li in (Li & Li, 2007)).

Fuzzy controllers, formulated on the basis of human understanding of the process or identified from measured control actions, can be regarded as emulators of human operators. Fuzzy logic control has more advantages because it can compensate the bad influence by nonlinearity and uncertainties based on advanced human expertise experience, also because it has strong robustness independent of a mathematical model. The other advantages of Fuzzy controllers are that their design is simple, fast, inexpensive, and easily maintained because the rules can be linguistically interpreted by the human experts. Riid & Rustern (Riid & Rustern, 2001) presented a fuzzy supervisory control system over the PID controller to reduce the complexity of the control problem and enhance the control performance. Riid & Rustern in (Riid & Rustern, 2002) demonstrate that problem decomposition leads to more effective knowledge acquisition and improved control performance in fuzzy control. The methodology allows solving complex control problems (truck backer-upper) without loss of functionality that is very difficult with all-in-one approaches and saves design expenses. Li & Chang in (Li & Chang, 2003) addressed the parking problem of a mobile robot by tracking feasible reference trajectories via a fuzzy sliding-mode control. Chen and Zhang in (Chen & Zhang, 1997) have reported a fuzzy controller to park a truck with suboptimal distance trajectories. They chose arcs of circle of minimum turning radii connected with parabolic curves as the optimal trajectories, but the desired parabolic curve to follow has to be given to the controller. More recently Li & Li in (Li & Li, 2007) have presented the fuzzy control system based on a hybrid clustering method and neural network. Sugeno & Murakami (Sugeno & Murakami, 1985) propose an experimental study on parking problem using model car, which is equipped with on-board microprocessor and two supersonic sensors for the measurements of the relative distance and direction. They derive fuzzy control rules by utilizing Sugeno-type fuzzy implications to model the parking experience of a skilled driver. Sugeno et al. (Sugeno et al, 1989) adopt the similar hardware arrangement as that in (Sugeno & Murakami, 1985) to execute the garage parking by employing fourteen fuzzy oral instructions such as "go straight", "slow down", "enter garage" and "speed up". Yasunobu & Murai (Yasunobu & Murai, 1994) exploit the state evaluation fuzzy control and the predictive fuzzy control to achieve the drive knowledge. Only the computer simulations are given to show the effectiveness of the proposed parking control. A skilled-based visual parking control using neural networks and fuzzy is discussed in (Daxwanger & Schmidt, 1995), where two control architectures, the direct neural control and the fuzzy hybrid control, are used to generate the automatic parking commands. The environment information is measured by a video sensor. The control architectures are validated by experiments with an autonomous mobile robot. Tayebi & Rachid (Tayebi & Rachid, 1996) deal with the parking problem of wheeled robot by using time-varying state feedback control law via the Lyapunov direct method. The control law is robust to ensure a global boundedness of the system states under measurement perturbations. The development of a near-optimal fuzzy controller for manoeuvring a car in a parking lot is described in (Leu & Kim, 1998). A cell mapping based method is proposed to systematically group near-optimal trajectories for all possible initial states in the parking lot.

The rules and membership functions of the fuzzy controller are generated using the statistical properties of the individual trajectory groups. An et al. (An et al., 1999) develop an online path-planning algorithm that guides an autonomous mobile robot to a goal with avoiding obstacles in an uncertain world. The established autonomous mobile robot cannot move omni-direction and run on two wheels equipped with a CCD camera. The path-planning algorithm is constructed by three modes: straight mode, spin mode, and avoidance mode. The simulation program and experimental results are developed to check this algorithm by using the garage parking motion. Shirazi & Yih (Shirazi & Yih, 1989) propose an expert's knowledge including symbolic form and nonsymbolic form, where the former can be obtained from expert directly and the latter can be obtained only through an evolutionary process. The evolutionary process consists of three stages: novice, competency, and expert. The developed intelligent control system performs parallel parking to show validity and ability. The fuzzy traveling control of an autonomous mobile robot with six supersonic sensors has been provided in (Ohkita et al, 1993), where the flush problem is considered. The well-known fuzzy theory (Laumond et al, 1994) is now in widespread use such as system identification, function approximation, image compression, prediction, classification, and control. The general characteristic of the fuzzy control is that the IF-THEN fuzzy rules are on the basis of the conventional control strategy and the expert knowledge. It is shown in (Shahmaleki & Mahzoon, 2008; Shahmaleki et al, 2008) that hierarchical control system significantly improves control performance and reduces the design load compared to all-in-one approaches investigated by other researchers. Here, we recommend three approaches to solve parking problem. Finally we select and extend the hierarchical fuzzy control approach to the full truck backer-upper problem.

The path planners described in this research combine two fuzzy modules that provide desired angle value for front wheel so as to generate short paths with continuous curvatures. Approximated trajectories are composed of circular arcs of minimum turning radii and straight line segments. This chapter is structured as follows. Section 2 illustrates the parking problem addressed and its geometric and kinematical constraints. In Sections 3, Fuzzy control is described. In section 4 three control approaches described to solve parking problem. This section shows comparison between recommended approaches and the hierarchical structure of the fuzzy system is selected. Computer simulation results are given to show the validity of the proposed fuzzy logic control algorithms. We have proposed a vision based approach in section 5 for estimating the robot position and direction. Some measures of the car-like robot is extracted from images that are captured using a ceiling mounted camera. These measurements together with the kinematic equations of the robot are used for estimating its position and direction using an extended Kalman filter. The control system has been made and tested on a mobile robot containing kinematics constraints. In Section 6, several experimental results of diagonal parking maneuvers are included to illustrate the efficiency and robustness of the designed controller. Finally, conclusions are given in Section 7.

2. The truck backing up problem

The problem addressed in this research is the diagonal parking of a truck in a constrained domain. The initial state of the truck position is represented by three state variables x, y and φ in Fig. 1. The truck kinematics model is based on the following system of equations (Li & Li, 2007):

$$\begin{cases} x(t+1) = x(t) - \cos(\varphi(t) + \theta((t)) - \sin\big(\theta(t)\big)\sin\big(\varphi(t)\big), \\ y(t+1) = y(t) - \sin(\varphi(t) + \theta\big((t)\big) + \cos\big(\varphi(t)\big)\sin(\theta(t)), \\ \varphi(t+1) = \varphi(t) - arcsin(\frac{2\sin(\theta(t))}{b}) \end{cases} \tag{1}$$

where (x,y) are the coordinates of the vehicle rear axle midpoint, φ is the truck orientation with respect to the horizontal line, b is the length of the truck and the control variable is the steering angle θ, that is the angle of the front wheel with respect to the truck. The truck only moves backward with fixed speed.

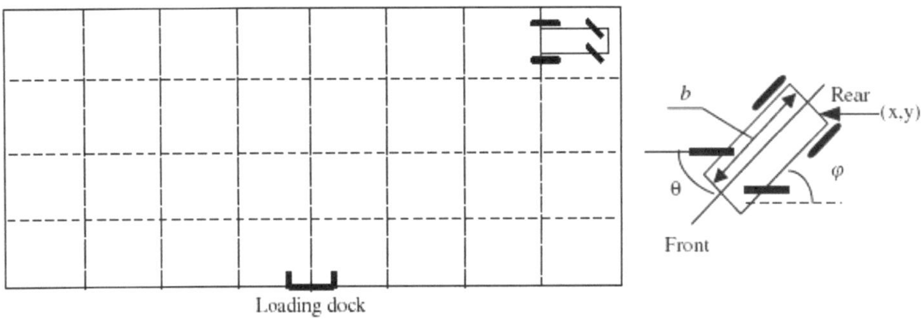

Fig. 1. Diagram of truck and loading dock

3. Fuzzy logic

A typical fuzzy control system consists of four components and the descriptions are stated as follows:

1. Fuzzification Interface: The fuzzification interface performs a conversion from a crisp point into a fuzzy set. The shapes of the membership functions of the linguistic sets are determined according to the expert experience.
2. Knowledge Base: The knowledge base commonly consists two sections: a database and a rule-base. The database contains the membership functions of the fuzzy sets used in

the fuzzy rules and the rule-base contains a number of fuzzy IF-THEN rules. The typical form of fuzzy rules can be expressed as IF precondition, THEN consequence. The canonical fuzzy IF-THEN rules are usually made from the following conditions: (a) Obtaining by the expert knowledge and/or operators experiences. (b) According to the control behavior of the users. (c) According to the characteristic of the plant. (d) Obtaining by self-learning.

3. Inference Engine: The inference engine that performs the fuzzy reasoning upon the fuzzy control rules is the main component of the fuzzy controller. There are varieties of compositional methods in fuzzy inference, such as max-min compositional operation and max-product compositional operation etc.
4. Defuzzification Interface: The defuzzification interface converts the fuzzy output of the rule-base into a non-fuzzy value. The center of area (COA) is the often used method in defuzzification. Suppose $\widetilde{B}$ is a discrete set as $\widetilde{B} = \{y_1, y_2, \ldots, y_n\}$ then COA method can be described as:

$$y^* = \sum_{k=1}^{n} y_k \cdot \mu_{\widetilde{B}}(y_k) \Big/ \sum_{k=1}^{n} \mu_{\widetilde{B}}(y_k) \tag{2}$$

where y* is the crisp value defuzzified from COA.
In this section we illustrate a brief comparison between Classic control and Fuzzy control. Classic control is based on a detailed I/O function OUTPUT= F (INPUT) which maps each high-resolution quantization interval of the input domain into a high-resolution quantization interval of the output domain. Finding a mathematical expression for this detailed mapping relationship F may be difficult, if not impossible, in many applications (Fig 2(a)). But, Fuzzy control is based on an I/O function that maps each very low-resolution quantization interval of the input domain into a very low-low resolution quantization interval of the output domain. As there are only 7 or 9 fuzzy quantization intervals covering the input and output domains the mapping relationship can be very easily expressed using the “if-then” formalism. (In many applications, this leads to a simpler solution in less design time.) The overlapping of these fuzzy domains and their linear membership functions will eventually allow achieving a rather high-resolution I/O function between crisp input and output variables (Fig 2(b)).

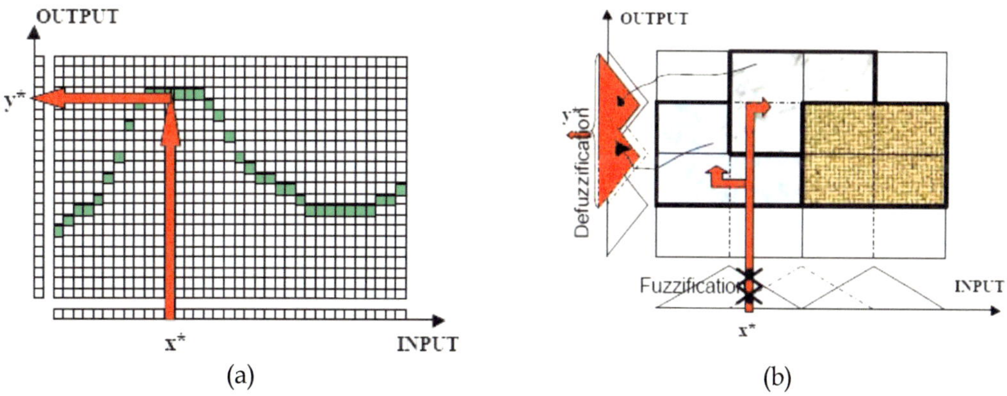

Fig. 2. Comparison between Classic and Fuzzy controls

In Fig 3 structure of fuzzy control is shown.

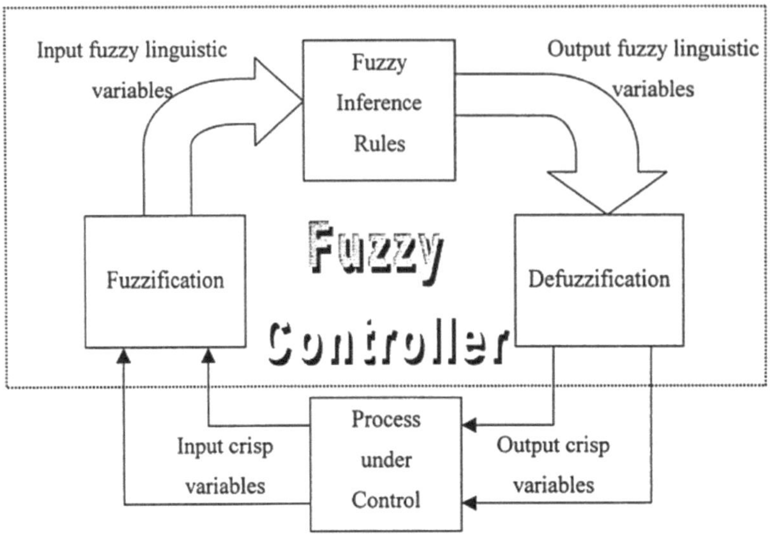

Fig. 3. Structure of Fuzzy control system

4. Designing fuzzy control system

The truck and the loading zone are shown in Fig. 4. The truck has two front steering wheels and two rear driven wheels that cannot move sideways. The coordinate pair (x,y) specifies the rear center position of the truck in the plane. The angle φ increases from -90 toward 270 in a clockwise direction and the steering angle θ is taken as positive if the steering wheel is turned to the right and negative, otherwise. The loading zone is the plane x:[-25,25], y:[0,25]. The goal of this research is to design a Fuzzy Logic Controller (FLC) able to back up the truck into a docking situation from any initial position that has enough clearance from the docking station. The controller should produce the appropriate steering angle θ = [-40°, 40°] at every stage to make the truck back up to a configuration with x=0, y=0, φ=90 (that is the desired parking space) from any initial position (x, y and φ) and to stop there. Thus controller is a function of state variables:

$$\theta = f(x,y,\varphi), \tag{3}$$

The y coordinate is not used because the straight segments of approximated trajectories are always horizontal. Also typically it is assumed that enough clearance between the truck and the loading dock exists so that the truck y-position coordinate can be ignored, simplifying the controller function to:

$$\theta = f(x,\varphi), \tag{4}$$

Hence only x-position and truck orientation angle φ are inputs of the fuzzy controller and the steering angle θ is the output.
As shown in Fig. 4, the suboptimal goal is that the backward driving involves short trajectories made up of arcs of circle of minimum turning radii and straight line segments, which meet the kinematic constraints in (1).

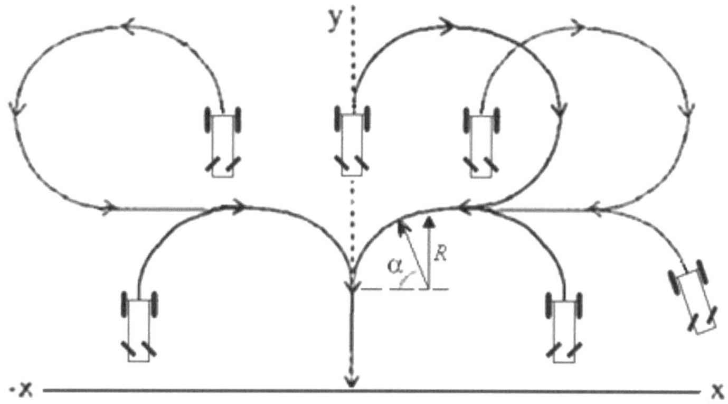

Fig. 4. Ideal trajectories

Analyzing the shortest paths geometrically, a mathematical expression for the steering angle θ which produces curvature of these short paths can be found:

$$\theta = \begin{cases} 40^\circ & if \;\; \varphi > \alpha \\ 0^\circ & if \;\; \varphi = \alpha \\ -40^\circ & if \;\; \varphi < \alpha \end{cases} \tag{5}$$

where the angle α (depends on x)associated with the switching in the steering angle(θ) sign can be calculated as follows:

$$\alpha = \begin{cases} sign(x) \cdot \cos^{-1}(\frac{R - |x|}{R}) & if \;\; |x| < R \\ sign(x) \cdot \frac{\pi}{2} & if \;\; |x| \geq R \end{cases} \tag{6}$$

R being the minimum turning radius corresponding to the maximum curvature (γ) which has a constant value (γ =1/R).

4.1 Integrated approach

Structure of this approach consists of a single module with two inputs (x and φ) and one output (θ) (Fig. 5). It contains five linguistic labels to cover the input variable x and seven labels for the vehicle angle φ (Fig. 6).

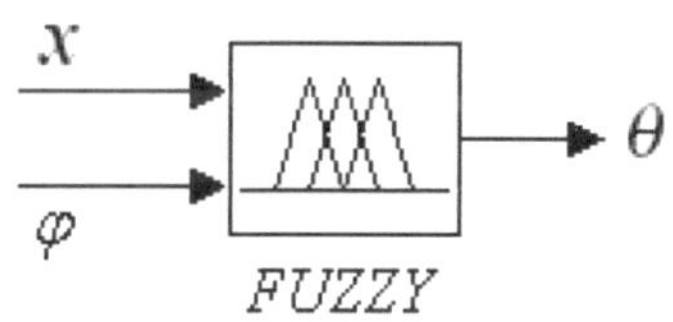

Fig. 5. Structure of fuzzy controller

One triangular and four trapezoidal membership functions (LE, LC, CE, RC, RI) are selected to cover the x variable. Also five triangular and two trapezoidal membership functions (LB, LU, LV, VE, RB, RU, RV) are selected to cover the φ variable.

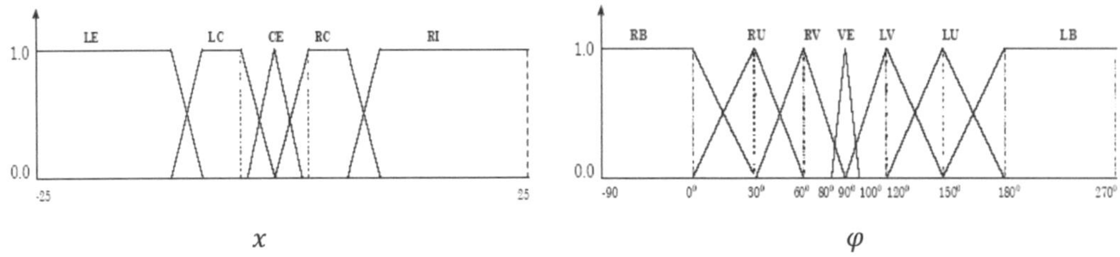

Fig. 6. Mebership functions for x and φ variables (Integrated approach)

The rules are shown in Table 1. Also the "center of gravity" method is used as defuzzification method. The rule base implements a zero-order Takagi-Sugeno inference method.

X / φ	LE	LC	CE	RC	RI
RL	NL	NL	NM	NM	NS
RU	NL	NL	NM	NS	PS
RV	NL	NM	NS	PS	PM
VE	NM	NM	ZE	PM	PM
LV	NM	NS	PS	PM	PL
LU	NS	PS	PM	PL	PL
LL	PS	PM	PM	PL	PL

Table 1. The learned rules for the x and φ variables

The consequents of the rules are the following: ZE=0°, NL=-40°, NM=-30°, NS=-20°, PL=40°, PM=30°, PS=20°.

4.2 Combined approach

As shown in Fig. 7, structure of the controller consists of a fuzzy module and three blocks (Dis, Controller1 and Controller2).

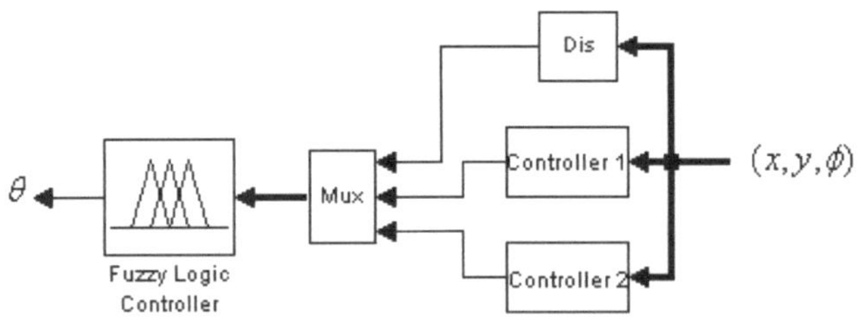

Fig. 7. Structure of the control system

In fact, if we try to find a mathematical expression for the steering angle θ which produces curvature of these short paths, we can recommend equations (7),(8). The angle of wheels (θ1, θ2) is computed in Controller1 and Controller2 based on the equation (7), (8).

$$
\begin{aligned}
&\gamma = ar\cos(x\sqrt{x^2+y^2}) \\
&rel = \phi - \gamma \\
&\beta = rel - round(\frac{rel}{2\pi}) \times 2\pi \\
&\theta 1 = a\gamma + b\beta + c
\end{aligned}
\tag{7}
$$

$$\theta_1 = \begin{cases} \theta_1 & if \quad -\pi/4 < \theta_1 < \pi/4 \\ \pi/4 & if \quad \theta_1 \ge \pi/4 \\ -\pi/4 & if \quad \theta_1 \le -\pi/4 \end{cases}$$

$$\gamma = x$$

$$rel = \phi - \frac{\pi}{2}$$

$$\beta = rel - round(\frac{rel}{2\pi}) \times 2\pi \qquad (8)$$

$$\theta 2 = m\gamma + n\beta + p$$

$$\theta_2 = \begin{cases} \theta_2 & if \quad -\pi/4 < \theta_2 < \pi/4 \\ \pi/4 & if \quad \theta_2 \ge \pi/4 \\ -\pi/4 & if \quad \theta_2 \le -\pi/4 \end{cases}$$

where a, b, c, m, n and p are constant values.
The distance between the vehicle rear axle midpoint and constrained domain is computed by:

$$\text{distance} = \sqrt{(x - x_0)^2 + (y - y_0)^2}^2 \qquad (9)$$

Two S-shaped and Z-shaped membership functions (far, near) are selected to cover the distance universe of discourse (Fig. 8). The consequent of the rules (out1, out2) are singletons. The rules are the following:

1. If distance = near → θ = out1
2. If distance = far → θ = out2

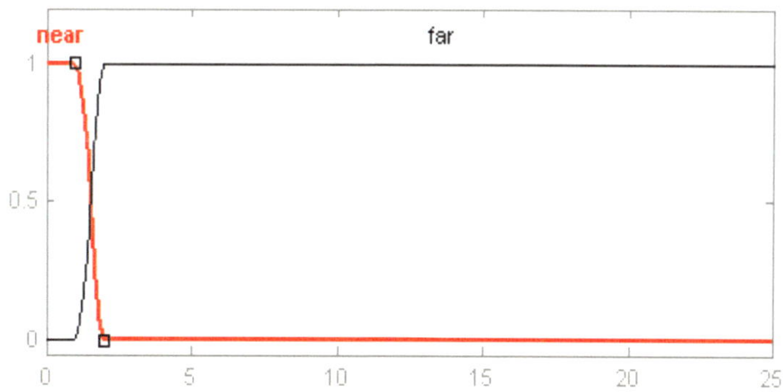

Fig. 8. Membership functions for distance variables

Fig. 9 shows three examples of the generated paths by combined approach.

4.3 Integrated approach

In this section the hierarchical structure is introduced .The scheme is basically made up of two rule bases (Fig. 10).

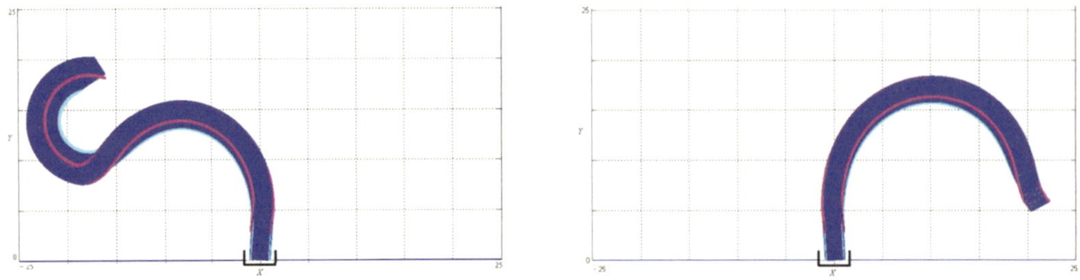

(a) (b)

Fig. 9. Simulated results of the parking maneuvers corresponding to the initial configurations (a) x=-15, y=18, φ =180. (b) x=20, y=9, φ =228

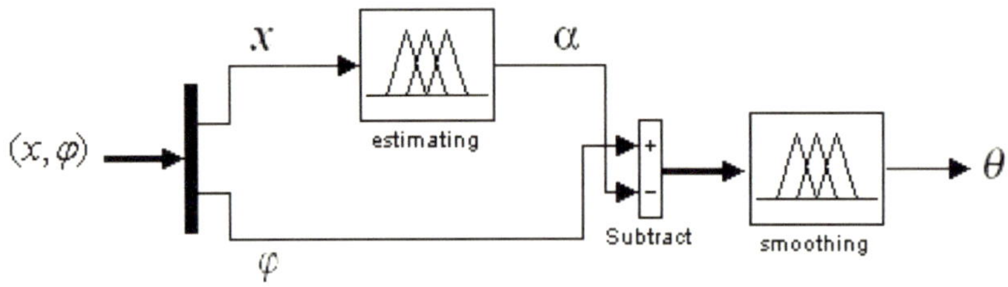

Fig. 10. Structure of hierarchical approach

The first one ("estimating") provides approximately the value of the angle α depending on the input variables x. The second one ("smoothing") provides the desired value for steering angle (θ) depending on the value of difference φ-α. Two triangular and two trapezoidal membership functions (LB, LS, RS and RB) are selected to cover the x universe of discourse and four rules are included in the rule base "estimating". The consequents of the rules (mf1, mf2, mf3, and mf4) are singletons. Also the "center of gravity" method is used for defuzzification.

The rule bases (estimating and smoothing) implements a zero-order Takagi-Sugeno inference method.

The rules are:

1. if (x = LB) → α= mf1
2. if (x = LS) → α= mf2
3. if (x = RS) → α= mf3
4. if (x = RB) → α= mf4,

The rule base "smoothing" also contains two triangular and two trapezoidal membership functions and four rules. The rules are:

1. if (diff = MZ) → θ= nf1
2. if (diff = NZ) → θ= nf2
3. if (diff = PZ) → θ= nf3
4. if (diff = RZ) → θ= nf4,

where MZ, NZ, PZ and RZ are fuzzy sets represented by triangular and trapezoidal membership functions (they cause the smooth switching in the steering angle θ when φ is around α). nf1, nf2, nf3, nf4 are singleton values associated with the angle front wheels. The membership functions for the variables x and diff are shown in Fig. 11.

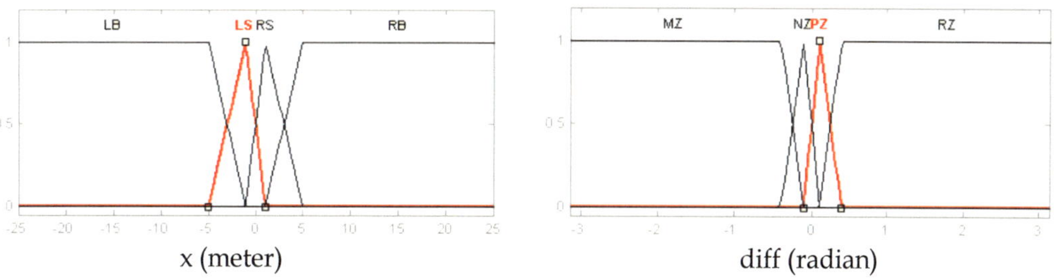

Fig. 11. Membership functions for x and diff variables

The "estimating" module for the hierarchical approach provides a fuzzy approximation for the angle α. The advantage of using this module instead of giving α analytically is that the required computational cost is reduced. Using normalized triangular and trapezoidal membership functions for the antecedents of the rules and a zero-order Takagi-Sugeno inference engine makes this approximation piecewise linear, which means that only several additions and products need to be implemented. The computational cost of additions and products is less than that of a nonlinear function such as Arcos (.) in (6).

Fig. 12 shows the variations of θ versus x and φ corresponding to (5) and (6). These equations are associated with an on-off control because the θ value presents abrupt changes, and would require stopping the robot to perform this switching.

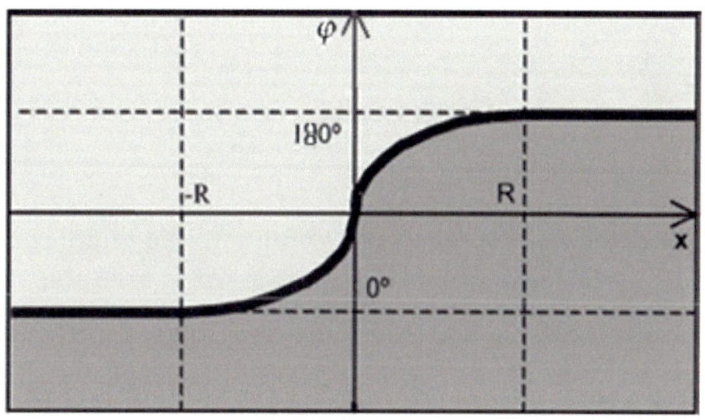

Fig. 12. Steering angle θ versus x and φ for short paths. The dark color presents the θ=-40°and the light one presents θ=40°

Three fuzzy modules (integrated, combined and hierarchical) described previously are zero-order Takagi-Sugeno systems whose input membership functions always overlap each other. Hence, the subgoal of providing continuous-curvature and short paths is achieved.

Comparing the three approaches for designing the controller the hierarchical one is more efficient since it generates paths but with small number of rules. Besides it provides the higher smoothness near the target configuration (x=0). As a result, the hierarchical module was selected as the control system.

Simulated results using the present hierarchical scheme for the different initial positions are shown in Fig. 13. In this figure, t indicates the parking duration. It can be seen how the generated paths (Fig. 13) are very close to the ideal paths (Fig. 4) made up of circular arcs and straight lines.

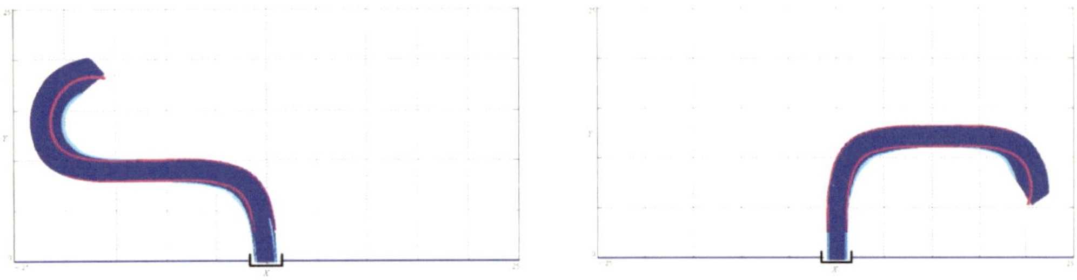

(a) (b)

Fig. 13. Results of the parking maneuver corresponding to the initial configurations (a) x=-20, y=18.4, φ=120°, t=78 steps, (b) x=17.5, y=8, φ=252°, t=72 steps

Further, according to the robot kinematics equations, the work of Li and Li (Li & Li, 2007) has been used for comparison. Fig.14 shows simulated results of Li and Li (Li & Li, 2007) for the same initial conditions of Fig.13.

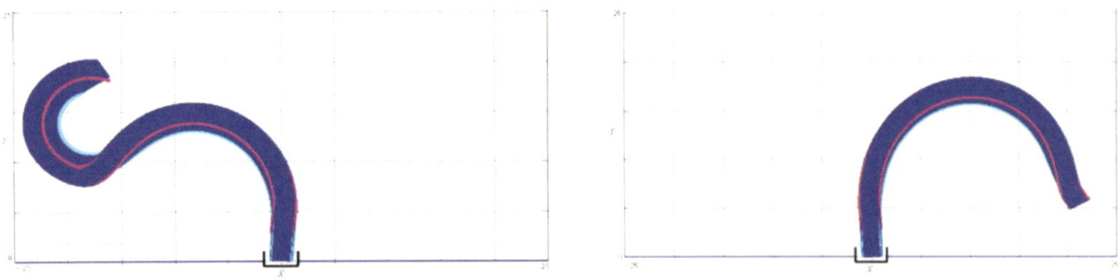

(a) (b)

Fig. 14. Results of the parking maneuver corresponding to the initial configurations
(a) x=-20, y=18.4, φ=120°, t=93 steps, (b) x=17.5, y=8, φ=252°, t=86 steps, (Li & Li, 2007)

An advantage of this approach is that the rules are linguistically interpretable and the controller generates paths with 8 rules compared with 35 used by (Riid & Rustern, 2002). Besides it provides the higher smoothness near the target configuration (x=0). Also, parking durations are shorter than those obtained by (Li & Li, 2007) under the same initial conditions. In this work, trajectories are composed of circular arcs and straight segments but in other methods, trajectories are composed of circular arcs.

5. Real time experimental studies

As shown in Fig. 15(a), the designed mobile robot has a 30cm×20cm×10cm, aluminium body with four 7cm diameter tires. It contains an AVR-ATMGEA64 micro controller, running at 16 MHz clock. The robot is equipped with three 0.9 degree stepper motors, two for the back wheels and one guides the steering through a gear box. The control of the mobile robot motion is performed on two levels, as demonstrated in Fig. 15(b). This two-layer architecture is very common in practice because most mobile robots and manipulators usually do not allow the user to impose accelerations or torques at the inputs. It can also be viewed as a simplification to the problem as well as a more modular design approach. The high level control (Hierarchical Fuzzy Controller) determines the steering angle θ of the robot considering the position (x,y) and angle (φ) of the robot which is received from the vision system. While the low level controller receives the output of high level control and determines steering angle of the front wheel and the speed of two rear wheels differentially.

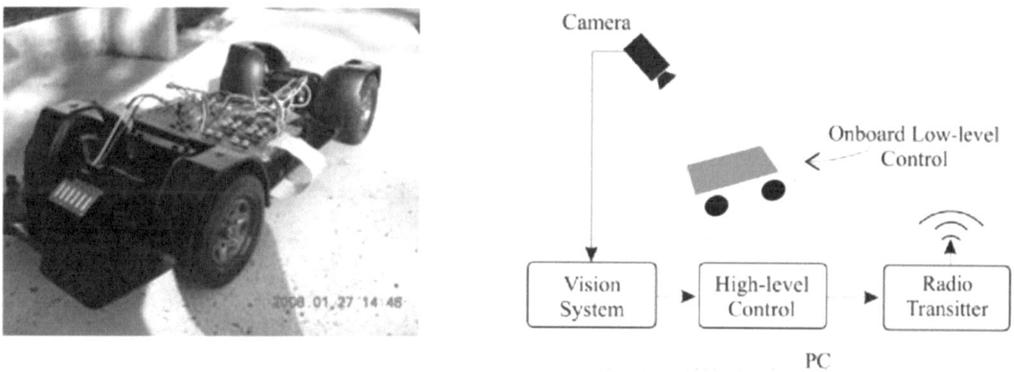

(a) (b)

Fig. 15. (a) Designed mobile robot. (b) The control architecture of the mobile robot

The structure of real control system is shown in Fig. 16.

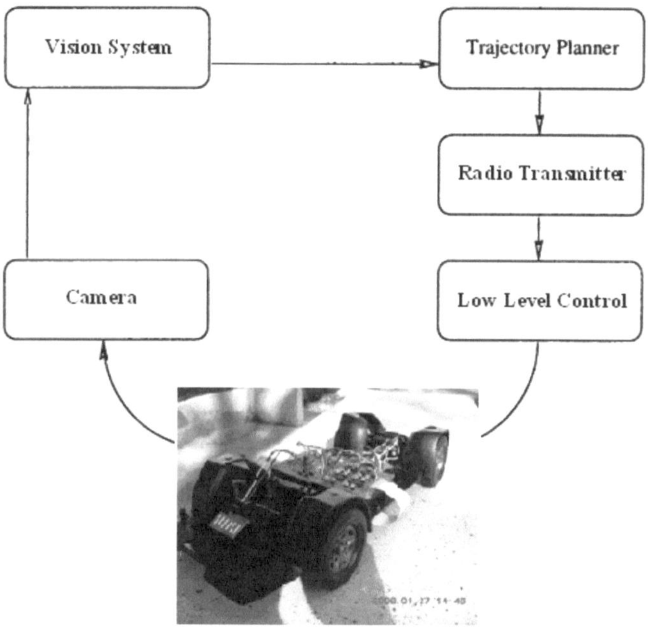

Fig. 16. The structure of real control system

5.1 Vision subsystem

For the backer-upper system to work in a real environment it is necessary to obtain the car position and orientation parameters. For this task different sensing and measuring instruments have been used in the literature. Some authors (Demilri & Turksen, 2000) have used sonar to identify the location of the mobile robots in a global map. This is achieved by using fuzzy sets to model the sonar data and by using the fuzzy triangulation to identify the robots position and orientation. Other authors have used analogue features of RFID tags system (Miah & Gueaieb, 2007) to locate the car-like mobile robot. Vision based position estimation has been also used for this task. In (Chen & Feng, 2009) a hardware implemented vision based method is used to estimate the robot position and direction. They use a camera mounted on the mobile robot and estimate the car-like robot position and direction using profiles of wavelet coefficients of the captured images and using of a self organizing map neural network. Each neuron categorizes measurements of a location and direction bin. This method is limited in that it works based on recognizing the part of parking that is in the view field of robot's camera. This parking view classification based approach, requires new training if the parking space is changed. Also it has not the potential for localizing free parking lots and other robots or obstacles which may be required in real applications.

A ceiling mounted camera can provide a holistic view to the location. Using a CCD camera as measuring device to capture images from parking area, and using image processing and tracking algorithms, we can estimate position and direction of the object of interest. This approach can be used in multi-agent environments to localize other objects and obstacles and even free parking lot positions. Here we assume just one robot and no obstacles. Also, we assume that the camera has been installed on the ceiling in the center of parking zone and at a proper height such that we can ignore perspective effects at corners of the captured images. Thus a linear calibration can be used for conversion between the (i, j) pixel indices in the image and the (x, y) coordinates of the parking zone. This assumption can introduce some approximation errors. As will be described here, using a prior knowledge of the car kinematic in an extended Kalman filtering framework can correct these measurement errors. With this configuration and assumptions a simple non realistic solution for position and direction estimation can be used as follows. Set two different color marks on top of the car in middle front and rear wheels position. Then from the captured image extract the two colored marks and find their center. Assume (x_r, y_r) and (x_f, y_f) be coordinates of middle rear and front points then (x, y) input variables of the fuzzy controller can be estimated from (x_r, y_r) after some calibration. The direction φ of the car-like robot relative to x-axis can also be determined using:

$$\varphi = \tan^{-1}\frac{(y_f - y_r)}{(x_f - x_r)} \tag{10}$$

Note that the $tan^{-1}(.)$ function used here should consider signs of y_f-y_r and x_f-x_r terms so that it can calculate the direction in the range $[0,2\pi]$ or equivalently $[-\pi, \pi]$. Such a function in most programming environments is commonly named *atan2(.,.)* which perceives $y_f - y_r$ and $x_f - x_r$ separately and calculates the true direction accordingly.

This is a simple solution for non-realistic experimental conditions. However it is necessary to consider more realistic applications of the backer-upper system. So we should eliminate strong non-realistic constraints like hand marking the car with two different color marks.

Here we propose a method based on Hough transform for extracting measurements to estimate car position and orientation parameters. Using Hough transform we can just extract the orientation from the border lines of the car, but the controller subsystem needs the direction φ in range $[-\pi, \pi]$ to calculate correct steering angle. To find the true direction we use a simple pattern classification based method to discriminate between front and rear sides of the car-like robot from its pixel gray values. This classifier trains the robots image and is independent of the parking background. Also it can be trained to work for different moving objects.

We can use extracted measurements of each frame to directly estimate (x, y, φ) state variables. But since extracted measurements are not accurate enough, we use these measurement parameters together with kinematic equations (1) of the plant as a state transition model in an extended Kalman filter to estimate the state variables (x, y, φ) of the robot more accurately.

5.2 Car position extraction using Hough transform

Hough transform (HT) first proposed by Hough (Hough, 1962) and improved by Duda & Hart (Duda & Hart, 1972) is a feature extraction method which is widely used in computer vision and image processing. It converts edge map of an image into a parametric space of a given geometric shape. Edge map can be extracted using edge extraction methods which filter the image to extract high frequency parts (edges) and then apply a threshold to get a binary matrix. HT tries to find noisy and imperfect examples for a given shape class within an image. There exists HTs for lines, circles and ellipses.

For example classic Hough transform, finds lines in a given image. A line can be parameterized in the Cartesian coordinate by slope (m) and interception (b) parameters (Hough, 1962). Each point (x, y) of the line can be constrained by the equation $y = mx + \mathrm{b}$. However this representation is not well-formed for computational reasons. The slope of near vertical lines, go to infinity hence it is not a good representation for all possible lines. The classic Hough transform proposed by Duda and Haart (Duda & Hart, 1972) uses a polar representation in which lines are shown by two parameters r and θ in the polar coordinate. Parameter r is length of the vector started from origin and perpendicularly connected to the line (distance of line to the origin) and θ is the angle between that vector and x axis.

Classic Hough transform calculates a 2D parameter map matrix for quantized values of (r,θ) parameters. An algorithm determines lines with (r,θ) values that pass through each edge point of the image and increases votes of those (r,θ) bins in the matrix. For each edge point this accumulation is carried out. Finally the peaks in the parameter map show the most perfect lines that exist in the image. The following equation relates the (x,y) Cartesian coordinate of line points with the r,θ polar line parameters, as previously defined.

$$\begin{cases} y = -\dfrac{\cos\theta}{\sin\theta}x + \dfrac{r}{\sin\theta} \\ r(\theta) = x\cos\theta + y\sin\theta \end{cases} \tag{11}$$

For any edge point (x_i,y_i), equation (11) provides a sinusoidal curve in terms of r and θ parameters. Points on this curve determine all lines (r_j,θ_j) that pass through the edge point (x_i,y_i). For each edge point votes of all cells of the parameter matrix that fall on the corresponding sinusoidal curve are increased.

The external boundary of the car-like robot is approximated by a rectangle. To extract four lines of this rectangle in each input image frame, first calculate the edge map of the image using an edge extraction algorithm. Then apply Hough transform and extract dominant peaks of the parameter map. Then among these peaks we search to select four lines that satisfy the constraints of being edges of a rectangle corresponding to car-like robot size. Four selected lines should approximately form a a×b rectangle where a and b are width and length of the car-like robot.

Let the four selected lines have parameters (r_i,θ_i), $i = 1,2,3,4$. In order to extract the rectangle formed by these four lines, four intersection points (x_j,y_j), $j = 1,2,3,4$ of perpendicular pairs should be calculated. Solving for the linear system in equation (12), intersection point (x_0,y_0) of two sample lines (r_1,θ_1) and (r_2,θ_2) can be determined.

$$\begin{cases} x_0 \cos\theta_1 + y_0 \sin\theta_1 = r_1 \\ x_0 \cos\theta_2 + y_0 \sin\theta_2 = r_2 \end{cases} \tag{12}$$

If the lines are not parallel, the unique solution is given by equation (13).

$$\begin{cases} x_0 = \dfrac{r_1 \sin\theta_2 - r_2 \sin\theta_1}{\sin(\theta_2 - \theta_1)} \\ y_0 = \dfrac{-r_1 \cos\theta_2 + r_2 \cos\theta_1}{\sin(\theta_2 - \theta_1)} \end{cases} \tag{13}$$

A problem with HT is that it is computationally expensive. However its complexity can be reduced since position and orientation of the robot is approximately known in the tracking procedure. Thus HT just should be calculated for a part of the image and a range of (r,θ) around current point. Also the level of quantization of (r,θ) can be set as large as possible to reduce the time complexity. Relative coarse bin sizes for (r,θ) also help to cope with little curvatures in the border lines of the car-like robot. This is at the expense of reducing the estimated position and direction resolution. The relative degraded resolution of (r,θ) due to coarse bin sizes can be restored by the correction and denoising property of Kalman filter. Note that the computation complexity of Kalman filter is very low relative to HT, since the former manipulates very low dimensional extracted measurements while the latter manipulates high dimensional image data.

5.3 Determining car direction using classification

Using equation (13), four corners of the approximately rectangular car border can be estimated. Now it is necessary to specify which pair of these four points belongs to the rear and which pair belongs to the front side of the car. We can not extract any information from Hough transform about the rear-front points assignment. But this assignment is required to determine middle rear wheels points (x_r,y_r) and also the signed direction φ of the car.

To solve this problem we adopt a classification-based approach. For each frame, using the four estimated corner points of the car, a rectangular area of $n_a \times n_b$ pixels of the car-like object is extracted. Then extracted pixels are stacked in a predefined order to get a $n_a \times n_b$ feature vector. A classifier that is trained using training data, is used to determine the direction using these feature vectors. However, due to large number of features, it is necessary to apply a feature reduction transformation like principle component analysis

(PCA) or linear discriminant analysis (LDA) before the classification (Duda et al, 2000). These linear feature transforms reduce the size of feature vectors by selecting most informative or discriminative linear combinations of all features. Feature reduction, reduces the classifier complexity hence the amount of labeled data that is required for training the classifier. Different feature reduction and classifier structures can be adopted for this binary classification task. Here we apply PCA for feature reduction and a linear support vector machine for classification task. Supprot Vector Machine (SVM) proposed by Vapnik (Vapnik, 1995) is a large margin classifier based on the concept of structural risk minimization. SVM provides good generalization capability. Its training, using large number of data, is time consuming to some extent, but for classification it is as fast as a simple linear transform. Here we use SVM because we want to create a classifier with good generalization and accuracy, using small number of training data.

LDA is a supervised feature transform and provides more discriminative features relative to PCA hence it is commonly preferred to PCA. But the simple LDA reduces the number of features to at most $C-1$ features where C is number of classes. Since our task is a binary classification, hence using LDA we just would get one feature that is not enough for accurate direction classification. Thus we use PCA to have enough features after feature reduction. To create our binary direction sign classifier, first we train the PCA transform. To calculate principle components, mean and covariance of feature vectors are estimated then eigen value decomposition is applied on the covariance matrix. Finally N eigen vectors with greater corresponding eigen values, are selected to form the transformation matrix W. This linear transformation reduces dimension of feature vectors from $n_a \times n_b$ to N elements. Here in experiments $N = 10$ eigen values provides good results.

To train a binary SVM, reduced feature vectors with their corresponding labels are first normalized along each feature by subtracting the mean and dividing by the standard deviation of that feature. About 100 training images are sufficient. These examples should be captured in different points and directions in the view field of the camera. The car pixels extracted from each training image, can be resorted in two feature vectors one from front to rear which takes the label -1 and one from rear to front which takes the label +1. In the training examples position of the car and its pixel values are extracted automatically using Hough transform method described in previous section. But the rear-front labeling should be assigned by a human operator. This binary classification approach provides accuracy higher than 97% which is completely reliable. Because the car motion is continuous, we can correct possible wrong classified frames using previous frames history.

Using this classification method the front-rear assignment of the four corner points of the car is determined. Now Corner points are sorted in the following defined order to form an 8 dimensional measurement vector $Y^I = [x_{r_1}, y_{r_1}, x_{r_2}, y_{r_2}, x_{f_1}, y_{f_1}, x_{f_2}, y_{f_2}]^T$. The r_1, r_2, f_1, f_2 subscripts denote in order, the rear-left, rear-right, front-left and the front-right corners of the car.

From the four ordered corner points in the measurement vector Y^I, we can also directly calculate an estimate of the car position state vector to form another measurement vector $Y^D = [x_r, y_r, \varphi_{rf}]^T$ where (x_r, y_r) is the middle rear point coordinate and φ_{rf} is the signed direction of rear to front vector of the car-like robot relative to the x-axis. The superscripts D and I in these two measurement vectors show that they are directly or indirectly related to the state variables of the car-like robot that is required in the fuzzy controller. The measurement vector Y^D can be determined from measurement vector Y^I using equation (14).

$$Y^D = \begin{pmatrix} x_r \\ y_r \\ \varphi_{rf} \end{pmatrix} = \begin{pmatrix} (x_{r_1} + x_{r_2})/2 \\ (y_{r_1} + y_{r_2})/2 \\ \tan^{-1} \dfrac{(y_{f_1} + y_{f_2}) - (y_{r_1} + y_{r_2})}{(x_{f_1} + x_{f_2}) - (x_{r_1} + x_{r_2})} \end{pmatrix} \tag{14}$$

In the next section we will illustrate a method for more accurate estimation of state parameters by filtering these inaccurate measurements in an extended Kalman filtering framework.

5.4 Tracking the car state parameters with extended Kalman filter

Here we illustrate the simple and extended Kalman filters and their terminology and then describe our problem formulation in terms of an extended Kalman filtering framework.

5.4.1 Kalman filter

The Kalman filter (Kalman, 1960) is an efficient Bayesian optimal recursive linear filter that estimates the state of a time discrete linear dynamic system from a sequence of measurements which are perturbed by Gaussian noise. It is mostly used for tracking objects in computer vision and for identification and regulation of linear dynamic systems in control theory. Kalman filter considers a linear relation between measurements Y and state variables X of the system that is commonly named as the observation model of the system. Another linear relation is considered for state transition, between state variables in time step t, X_t and in time step t-1, X_{t-1} and the control inputs u_t of the system. These linear models are formulated as follows:

$$\begin{cases} X_t = F_t X_{t-1} + B_t u_t + w_t\,, & w_t \sim N(0,Q) \\ Y_t = H_t X_t + v_t\,, & v_t \sim N(0,R) \end{cases} \tag{15}$$

In equation (15), F_t is the dynamic model, B_t is the control model, w_t is the stochastic process noise model, H_t is the observation model, v_t is the stochastic observation noise model and u_t is the control input of the system. Kalman filter considers the estimated state $\hat{X}$ as a random vector with Gaussian distribution and a covariance matrix P. In following equations the notation $\hat{X}_{i|j}$ is used for the estimated state vector in time step i by using measurement vectors up to time step j.

The prediction estimates of state are given in equation (16), where $\hat{X}_{t|t-1}$ is the predicted state and $P_{t|t-1}$ is the predicted state covariance matrix. Note that in the prediction step just the dynamic model of the system is used to predict what would be the next state of the system. The prediction result is a random vector so it has its covariance matrix with itself.

$$\begin{cases} \hat{X}_{t|t-1} = F_t \hat{X}_{t-1|t-1} + B_{t-1} u_{t-1} \\ P_{t|t-1} = F_t P_{t-1|t-1} F_t^T + Q_{t-1} \end{cases} \tag{16}$$

In each time step before the current measurement is prepared we can estimate the predicted state then we use the acquired measurements from the sensors to update our predicted belief according to the error. The updated estimates using the measurements are given in equation (17). In this equation, Z_t is the innovation or prediction error, S_t is the innovation covariance, K_t is the optimal Kalman gain, $\hat{X}_{t|t}$ is the updated estimate of system state and $P_{t|t}$ is the updated or posterior covariance of the state estimation in time step t. The Kalman gain balances the amount of contribution of dynamic model and the measurement to the state estimation, according to their accuracy and confidence.

$$\begin{cases} Z_t = Y_t - H_t \hat{X}_{t|t-1} \\ S_t = H_t P_{t|t-1} H_t^T + R_t \\ K_t = P_{t|t-1} H_t^T S_t^{-1} \\ \hat{X}_{t|t} = \hat{X}_{t|t-1} + K_t Z_t \\ P_{t|t} = (I - K_t H_t)\, P_{t|t-1} \end{cases} \tag{17}$$

In order to use Kalman filter in a recursive estimation task we should specify dynamic and observation models F_t, H_t and some times the control model B_t. Also we should set initial state $\hat{X}_{0|0}$ and its covariance $P_{0|0}$ and prior process noise and measurement noise covariance matrices Q_0, R_0.

5.4.2 Extended Kalman filter

Kalman filter proposed in (Kalman, 1960) has been derived for linear state transition and observation models. These linear functions can be time variant that result in different F_t and H_t matrices in different time steps t. In extended Kalman filter (Bar-Shalom & Fortmann, 1988), the dynamic and observation models are not required to be linear necessarily. The models just should be differentiable functions.

$$\begin{cases} X_t = f(X_{t-1}, u_t) + w_t\,, & w_t \sim N(0, Q) \\ Y_t = h(X_t) + v_t\,, & v_t \sim N(0, R) \end{cases} \tag{18}$$

Again w_t and v_t are process and measurement noises which are Gaussian distributions with zero mean and Q, R covariance matrices.

In extended Kalman filter functions f (.) and h (.) can be used to perform prediction step for state vector $\hat{X}_{t|t-1}$ but for prediction of covariance matrix $P_{t|t-1}$and also in the update step for updating state and covariance matrix we can not use this non-linear functions. However, we can use a linear approximation of these non linear functions using the first partial derivatives around the predicted point $\hat{X}_{t|t-1}$. So for each time step t, Jacobian matrices of functions f (.) and h (.), should be calculated and used as linear approximations for dynamic and observation models in that time step.

5.5 Applying extended Kalman filter for car position estimation

Now we illustrate the dynamic and observation models to be used in the extended Kalman filtering framework. The dynamic model should predict the state vector $X_t = [x_t, y_t, \varphi_t]^T$ from existing state vector $X_{t-1} = [x_{t-1}, y_{t-1}, \varphi_{t-1}]^T$ and the control input to the car-like robot which is the steering angle θ_{t-1}. This is just the kinematic equations of the car-like robot that is given in equation (1). This equation considers unit transition velocity between time steps. This should be replaced with a translation velocity parameter V that is unknown. It can be embedded as an extra state variable to X to form the new state vector $X^v = [X; V]$ or may be left as a constant. The state transition function for the new state vector used here is given in equation (19).

$$X_t^V = \begin{bmatrix} x_t \\ y_t \\ \varphi_t \\ V_t \end{bmatrix} = f(X_{t-1}^V, \theta_{t-1}) = \begin{bmatrix} x_{t-1} - V_{t-1}[\cos(\varphi_{t-1} + \theta_{t-1}) - sin(\theta_{t-1})\cos(\varphi_{t-1})] \\ y_{t-1} - V_{t-1}[\sin(\varphi_{t-1} + \theta_{t-1}) + cos(\varphi_{t-1})\sin(\theta_{t-1})] \\ \varphi_{t-1} - V_{t-1} \arcsin(2sin(\theta_{t-1})/b) \\ V_{t-1} \end{bmatrix} \quad (19)$$

The observation model should calculate measurements from current state vector. As we have considered two measurements $Y^I = [x_{r_1}, y_{r_1}, x_{r_2}, y_{r_2}, x_{f_1}, y_{f_1}, x_{f_2}, y_{f_2}]^T$ and $Y^D = [x_r, y_r, \varphi]^T$, we would have two observation models correspondingly. First observation model is a nonlinear function $Y^I = h_I\,(X_t^V)$ since its calculation of it requires some $cos(\varphi)$ and $sin(\varphi)$ terms. The second observation model is an identity function $Y_t^D = h_D\,(X_t^V) = X_t$ that is $H_t = I_{3\times4}$. To prevent complexity we used the direct measurement vector hence identity observation model. Now the extended Kalman filter can be set up. Initial state vector can be determined from Y_0^D that is extracted from first frame the velocity can be set to 1 for initial step. Update steps of the filtering will correct the speed. The Initial state covariance matrix and process and measurement noise covariance matrices are initialized with diagonal matrices that contain estimations of variance of corresponding variables.

For each input frame first the predicted state is calculated using prediction equations and state transition function (19), then HT is computed around current position and direction and best border rectangle is determined from extracted lines, then signed direction is determined using the classification. Then measurement Y_t^D is calculated. Finally we use this measurement vector to update the state according to extended Kalman filter update equations. Then x_t, y_t, φ_t values of the updated state parameters are passed to the high level fuzzy control to calculate the steering angle θ which is passed to the robot and also is used in the state transition equation (19) in the next step.

6. Results

In order to test the designed controller, the truck is backed to the loading dock from two different initial positions (Fig. 17). Hierarchical control system is very suitable for the implementation of the multi-level control principle and bringing it back together into one functional block. Experimental and simulation results using the present hierarchical scheme for different initial positions are shown in Fig. 17. In this figures, t indicates the parking

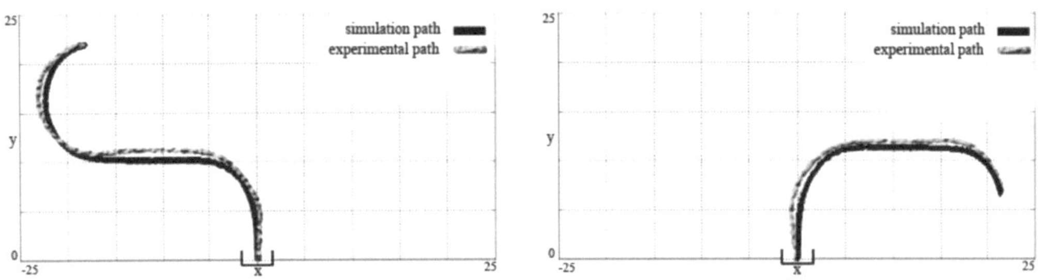

(a) (b)

Fig. 17. Experimental and simulation results of the parking maneuver corresponding to the initial configurations (a) x=-20, y=18.4, φ =60, t=78 steps, (b) X=17.5, y=4, φ =162, t=69 steps duration. It can be seen how the generated paths (Fig. 17) are very close to the ideal paths (Fig. 4) made up of circular arcs and straight lines.

Fig.18 illustrates how the steering angle "given by the hierarchical fuzzy controller" in short paths of Fig.17 is continuous, so the robot can move continuously without stopping.

The difference between generated paths (Fig. 17) is attributed to error of the vision subsystem, in estimating x,y,φ position variables. This error is propagated to the output of the controller and finally to the position of robot in the real environment.

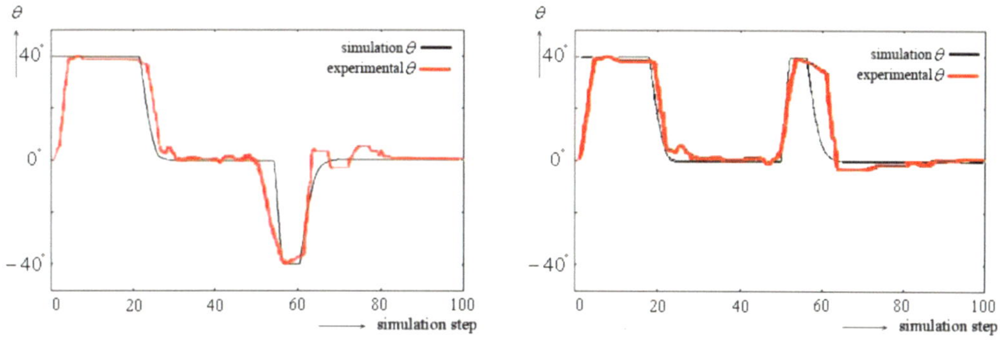

(a) (b)

Fig. 18. (a) Experimental and simulation steering angle transitions for the paths in Fig. 17(a), (b) Experimental and simulation steering angle transitions for the paths in Fig. 17(b)

7. Conclusion

A fuzzy control system has been described to solve the truck backer-upper problem which is a typical problem in motion planning of nonholonomic systems. As hierarchy is an indispensable part of human reasoning, its reflection in the control structure can be expected to improve the performance of the overall control system. The main benefit from problem decomposition is that it allows dealing with problems serially rather than in parallel. This is especially important in fuzzy logic where large number of system variables leads to exponential explosion of rules (curse of dimensionality) that makes controller design extremely difficult or even impossible. The "divide and rule" principle implemented through hierarchical control system makes it possible to deal with complex problems without loss of functionality. It has also been shown that problem decomposition is vital for successful implementation of linguistic analysis and synthesis techniques in fuzzy modelling and controlling because a hierarchy of fuzzy logic controllers simulates an existing hierarchy in the human decision process and keeps the linguistic analysis less complicated so that it is manageable. In this work the proposed controller has a hierarchical structure composed of two modules which adjust the proper steering angle of front wheels similar to what a professional driver does. The computational cost is also less because we don't have to work with nonlinear function such as "Arccos (.)". Compared with traditional controller, this fuzzy controller demonstrates advantages on the control performance, robustness, smoothness, rapid design, convenience and feasibility. Trajectories are composed of circular arcs and straight segments and as a result the hierarchical approach produces shorter trajectories in comparison with other methods. The control system has been simulated with a model of a mobile robot containing kinematics constraints. The experimental results obtained confirm that the designed control system meets its specifications: the robot is stopped at the parking target with the adequate orientation and short paths with continuous-curvature are generated during backward maneuver. The vision system utilizes measurements extracted from a ceiling mounted camera and estimates the mobile robot position using an extended Kalman filtering scheme. This results in correction and denoising of the measured position by exploiting the kinematic equations of the robot's motion.

8. References

Paromtichk, I.; Laugier, C.; Gusev, S. V. & Sekhavat, S. (1998). Motion control for parking an autonomous vehicle, *Proc. Int. Conf. Control Automation, Robotics and Vision*, vol. 1, pp. 136–140

Latombe, J. C. (1991). Robot motion planning, Norwell, MA: Kluwer Murray, R. M. & Sastry, S. S. (1993). Nonholonomic motion planning: Steering using sinusoids, *IEEE Trans. Automat. Contr.*, vol. 38, pp. 700–715

Lamiraux, F. & Laumond, J.-P. (2001). Smooth motion planning for car-like vehicles, *IEEE Trans. Robot. Automat.*, vol. 17, pp. 498–502

Scheuer, A. & Fraichard, T. (1996). Planning continuous-curvature paths for car-like robots, *Proc. IEEE Int. Conf. Intelligent Robots and Systems*, vol. 3, Osaka, Japan, pp. 1304–1311

Walsh, G.; Tylbury, D.; Sastry, S.; Murray, R. & Laumond, J. P. (1994). Stabilization of trajectories for systems with nonholonomic constraints, *IEEE Trans. Automat. Contr.*, vol. 39, pp. 216–222

Tayebi, A. & Rachid, A. (1996). A time-varying-based robust control for the parking problem of a wheeled mobile robot, *Proc. IEEE Int. Conf. Robotics and Automation*, pp. 3099–3104

Jiang, K. & Seneviratne, L. D. (1999). A sensor guided autonomous parking System for nonholonomic mobile robots, *Proc. IEEE Int. Conf. Robotics and Automation*, pp. 311–316

Gomez-Bravo, F.; Cuesta, F. & Ollero, A. (2001). Parallel and diagonal parking in nonholonomic autonomous vehicles, *Engineering Applications of Artificial Intelligence*, New York: Pergamon, vol. 14, pp. 419–434

Cuesta, F.; Bravo, F. G. & Ollero, A. (2004). Parking maneuvers of industrial-like electrical vehicles with and without trailer, *IEEE Trans. Ind. Electron.*, vol. 51, pp. 257–269

Reeds, J. A. & Shepp, R. A. (1990). Optimal path for a car that goes both forward and backward, *Pacific J. Math.*, vol. 145, no. 2, pp. 367–393.

Nguyen, D. & Widrow, B. (1989). The truck backer-upper: An example of self learning in neural network, *Proc. of the International Joint Conference on Neural Networks*, Washington DC, pp. 357-363

Kong, S. & Kosko, B. (1990). Comparison of fuzzy and neural truck backer-upper control systems, *Proc. IJCNN*, vol. 3, pp. 349-358

Koza, J.R. (1992). A genetic approach to the truck backer upper problem and the inter-twined spirals problem, *Proc. Int. Joint Conf. Neural Networks*, Piscataway, NJ, vol. 4, pp. 310-318

Schoenauer, M. & Ronald, E. (1994). Neuro-genetic truck backer-upper controller, *Proc. First Int. Conf. Evolutionary Comp.*, pages 720-723. Orlando, FL, USA

Jenkins, R.E. & Yuhas, B.P. (1993). A Simplified Neural Network Solution Through Problem Decomposition: The Case of the Truck Backer-Upper, *IEEE Trans. Neural Networks*, vol. 4, no. 4, pp. 718-720

Tanaka, K.; Kosaki, T. & Wang, H.O. (1998). Backing Control Problem of a Mobile Robot with Multiple Trailers: Fuzzy Modelling and LMI-Based Design, IEEE *Trans. Syst., Man, Cybern.*, Part C, vol. 28, no. 3, pp. 329-337

Ramamoorthy, P.A. & Huang, S. (1991). Fuzzy expert systems vs. neural networks - truck backer-upper control revisited, *Proc. IEEE Int. Conf. Systems Engineering*, pp. 221-224

Wang, L.-X. & Mendel, J.M. (1992). Generating fuzzy rules by learning from examples. IEEE *Trans. Systems, Man, and Cybernetics*, vol. 22, no. 6, pp. 1414-1427

Ismail, A. & Abu-Khousa, E.A.G. (1996). A Comparative Study of Fuzzy Logic and Neural Network Control of the Truck Backer-Upper System, *Proc. IEEE Int. Symp. Intelligent Control*, pp. 520-523

Kim, D. (1998). Improving the fuzzy system performance by fuzzy system ensemble, *Fuzzy Sets and Systems*, vol. 98, pp. 43-56

Dumitrache, I. & Buiu, C. (1999). Genetic learning of fuzzy controllers, *Mathematics and Computers in Simulation*, vol. 49, pp. 13-26

Chang, J.S.; Lin, J.H. & Chiueh, T.D. (1995). Neural networks for truck backer upper control system, *Proc. International IEEE/IAS Conference on Industrial Automation and Control*, Taipei, pp. 328–334.

Schoenauer, M. & Ronald, E. (1994). Neuro-genetic truck backer-upper controller, *Proc. of the First IEEE Conference on Evolutionary Computation*, Part 2(of 2), Orlando, pp. 720–723

Wang, L.X. & Mendel, M. (1992). Fuzzy basis function, universal approximation, and orthogonal least-squares learning, *IEEE Trans. Neural Networks*, 3 (5), 807–814

Li,Y. & Li, Y. (2007). Neural-fuzzy control of truck backer-upper system using a clustering method, *NeuroComputing*, 70, 680-688

Riid, A. & Rustern, E. (2001). Fuzzy logic in control: truck backer-upper problem revisited, *The 10th IEEE International Conference on Fuzzy Systems*, Melbourne, pp. 513–516.

Riid, A. & Rustern, E. (2002). Fuzzy hierarchical control of truck and trailer, *The 8th Biennal Baltic Electronic Conference*, dcc,ttu,ee, Tallinn, pp. 343–375

Li, T. - H. S. & Chang, S.-J. (2003). Autonomous fuzzy parking control of a car-like mobile robot, *IEEE Trans. Syst., Man,Cybern.*, A, vol.3, pp. 451–465

Chen G. & Zhang, D. (1997). Back-driving a truck with suboptimal distance trajectories: A fuzzy logic control approach, *IEEE Trans. Fuzzy Syst.*, vol. 5, pp. 369–380

Shahmaleki, P. & Mahzoon, M. (2008). Designing a Hierarchical Fuzzy Controller for Backing-up a Four Wheel Autonomous Robot, *American Control Conference*, Seattle, 10.1109/ACC.2008.4587269

Shahmaleki, P.; Mahzoon, M. & Ranjbar, B. (2008). Real time experimental study of truck backer upper problem with fuzzy controller, *World Automation Congress, WAC 2008*, Page(s):1 - 7

Sugeno, M. & Murakami, K. (1985). An experimental study on fuzzy parking control using a model car, *Industrial Applications of Fuzzy Control*, M. Sugeno, Ed. North-Holland, The Netherlands, pp. 105–124

Sugeno, M.; Murofushi, T.; Mori, T.; Tatematsu, T. & Tanaka, J. (1989). Fuzzy algorithmic control of a model car by oral instructions, *Fuzzy Sets Syst.*, vol. 32, pp. 207–219

Yasunobu, S. & Murai, Y. (1994). Parking control based on predictive fuzzy control, *Proc. IEEE Int. Conf. Fuzzy Systems*, vol. 2, pp. 1338–1341

Daxwanger, W. A. & Schmidt, G. K. (1995). Skill-based visual parking control using neural and fuzzy networks, Proc. IEEE Int. Conf. System, Man, Cybernetics, vol. 2, pp. 1659–1664

Tayebi, A. & Rachid, A. (1996). A time-varying-based robust control for the parking problem of a wheeled mobile robot, *Proc. IEEE Int. Conf. Robotics and Automation*, pp. 3099–3104

Leu, M. C. & Kim, T. Q. (1998). Cell mapping based fuzzy control of car parking, *Proc. IEEE Int. Conf. Robotics Automation*, pp. 2494–2499

An, H.; Yoshino, T.; Kashimoto, D.; Okubo, M.; Sakai, Y. & Hamamoto, T. (1999). Improvement of convergence to goal for wheeled mobile robot using parking motion, *Proc. IEEE Int. Conf. Intelligent Robots Systems*, pp. 1693–1698

Shirazi, B. & Yih, S. (1989). Learning to control: a heterogeneous approach, *Proc. IEEE Int. Symp. Intelligent Control*, pp. 320–325 O

hkita, M.; Mitita, H.; Miura, M. & Kuono, H. (1993). Traveling experiment of an autonomous mobile robot for a flush parking, *Proc. 2nd IEEE Conf. Fuzzy System*, vol. 2, Francisco, CA, pp. 327–332

Laumond, J. P.; Jacobs, P. E.; Taix, M. & Murray, R. M. (1994). A motion planner for nonholonomic mobile robots, *IEEE Trans. Robot. Automat.*, vol. 10, pp. 577–593

Demilri, K. & Turksen, I.B. (2000). Sonar based mobile robot localization by using fuzzy triangulation, *Robotics and Autonomous Systems*, vol. 33, pp. 109-123

Miah, S. & Gueaieb, W. (2007). Intelligent Parallel Parking of a Car-like Mobile Robot Using RFID Technology, *Robotics and Sensor Environments*, IEEE International Workshop On, pp. 1-6

Chen, Ch.Y. & Feng, H.M. (2009). Hybrid intelligent vision-based car-like vehicle backing systems design, *Expert Systems with Applications*, vol. 36, issue 4, pp. 7500-7509

Hough, P. (1962). Methods and means for recognizing complex patterns, U.S. Patent 3069654

Duda, R. & Hart, P. (1972). Use of the Hough Transformation to Detect Lines and Curves in Pictures, *Comm. ACM*, vol. 15, pp. 11–15

Duda, R.; Hart, P. & Stork, D. (2000). *Pattern Classification* (2nd ed.), Wiley Interscience

Vapnik, V. N. (1995). *The Nature of Statistical Learning Theory*, Springer-Verlag, New York

Kalman, R. (1960). A new approach to linear filtering and prediction problems, *Transactions of ASME – Journal of Basic Engineering*, 82, pp. 32-45

Bar-Shalom, Y. & Fortmann, T. E. (1988). *Tracking and data association*, San Diego, California: Academic Press, Inc.

4

Smooth Path Generation for Wheeled Mobile Robots Using η^3-Splines

Aurelio Piazzi, Corrado Guarino Lo Bianco and Massimo Romano
University of Parma, Department of Informatics Engineering
Italy

1. Introduction

The widespread diffusion of wheeled mobile robots (WMRs) in research and application environments has emphasized the importance of both intelligent autonomous behaviors and the methods and techniques of motion control applied to these robot vehicles (Choset et al., 2005; Morin & Samson, 2008). In particular, the motion control of WMRs can be improved by planning smooth paths with the aim to achieve swift and precise vehicle movements. Indeed, smooth paths in conjunction with a suitable or optimal velocity planning lead to high-performance trajectories that can be useful in a variety of applications (Kant & Zucker, 1986; Labakhua et al., 2006; Suzuki et al., 2009).

At the end of the eighties Nelson (Nelson, 1989) pointed out that Cartesian smooth paths for WMRs should possess continuous curvature. He proposed two path primitives, quintic curves for lane change maneuvers and polar splines for symmetric turns, to smoothly connect line segments. In the same period, also Kanayama and Hartman (Kanayama & Hartman, 1989) proposed the planning with continuous curvature paths. They devised the so-called cubic spiral, a path primitive that minimizes the integral of the squared curvature variation measured along the curve. Subsequently, Delingette *et al.* (Delingette et al., 1991) proposed the "intrinsic spline", a curve primitive that makes it possible to achieve overall continuous curvature and whose curvature profile is a polynomial function of the arc length.

A line of research starting with Boissonnat *et al.* (Boissonnat et al., 1994) and continued in (Scheuer & Laugier, 1998; Kito et al., 2003) evidenced the advisability to plan paths not only with continuous curvature, but also with a constraint on the derivative of the curvature. In particular, Fraichard and Scheuer (Fraichard & Scheuer, 2004) presented a steering method, called *CC Steer*, leading to paths composed of line segments, circular arcs, and clothoids where the overall path has continuous bounded curvature and bounded curvature derivative. On this topic, Reuter (Reuter, 1998) went further. On the ground of avoiding jerky motions, he presented a smoothing approach to obtain trajectories with continuously differentiable curvature, i.e. both curvature and curvature derivative are continuous along the robot path.

Reuter's viewpoint was enforced in (Guarino Lo Bianco et al., 2004b) where it was shown that in order to generate velocity commands with continuous accelerations for a unicycle robot, the planned path must be a G^3-path, i.e., a path with third order geometric continuity

(continuity along the curve of the tangent vector, curvature, and derivative of the curvature with respect to the arc length). More specifically, considering the classic kinematic model of the unicycle (cf. 1) we have that the Cartesian path generated with linear and angular continuous accelerations is a G^3-path and, conversely, given any G^3-path there exist initial conditions and continuous-acceleration commands that drive the robot on the given path. A related path-inversion algorithm was then presented to obtain a feedforward (open-loop) smooth motion generation that permits the independent planning of both the path and the linear velocity. For mobile robots engaged in autonomous and event-driven navigation it emerged the necessity to perform iterative path replanning in order to comply with changing guidance tasks. The resulting composite path must retain G^3-continuity of the whole path in order to avoid breaks of motion smoothness. In this context, it is useful a G^3-path planning tool that permits, on one hand, interpolating an arbitrary sequence of Cartesian points with associated arbitrary tangent directions, curvatures, and curvature derivatives, and on the other hand, shaping the path between two consecutive interpolating points according to the current navigation task.

An answer to this necessity emerging from G^3-path replanning is a Cartesian primitive, called η^3-spline, succinctly presented in (Piazzi et al., 2007). It is a seventh order polynomial spline that allows the interpolation of two arbitrary Cartesian points with associated arbitrary G^3-data (unit tangent vector, curvature, and curvature derivative at the path endpoint) and depends on a vector (η) of six parameter components that can be used to finely shape the path. The η^3-spline, a generalization of the η^2-spline presented in (Piazzi&Guarino Lo Bianco, 2000; Piazzi et al., 2002), can generate or approximate, in a unified framework, a variety of simpler curve primitives such as circular arcs, clothoids, spirals, etc.

This chapter exposes the motivation and the complete deduction of the η^3-splines for the smooth path generation od WMRs. Sections are organized as follows. Section 2 introduces the concept of third order geometric continuity for Cartesian curves and paths. A brief summary of the path inversion-based control of WMRs (Guarino Lo Bianco et al., 2004b) is reported in Section 3. Section 4 proposes the polynomial G^3-interpolating problem and exposes its solution, the η^3-spline, defined by explicit closed-form expressions (cf. (4)-(19) and Proposition 2). This curve primitive enjoys relevant and useful properties such as completeness, minimality, and symmetry (Properties 1-3). Section 5 presents a variety of path generation examples. A note on the generalization of η^3-splines is reported in Section 6. Conclusions are made in Section 7.

2. G^3-continuity of Cartesian curves and paths

A curve on the $\{x, y\}$-plane can be described by the map

$$\mathbf{p} : [u_0, u_1] \to \mathbb{R}^2 \ , \ u \to \mathbf{p}(u) = [\alpha(u)\ \beta(u)]^T$$

where $[u_0, u_1]$ is a real closed interval. The associated "path" is the image of $[u_0, u_1]$ under the vectorial function $\mathbf{p}(u)$, i.e., $\mathbf{p}([u_0, u_1])$. We say that curve $\mathbf{p}(u)$ is regular if $\dot{\mathbf{p}}(u) \in C_p([u_0, u_1])$ and $\dot{\mathbf{p}}(u) \neq 0\ \forall\ u \in [u_0, u_1]$ (C_p denotes the class of piecewise continuous functions). The arc length measured along $\mathbf{p}(u)$, denoted by s, can be evaluated with the function

$$f : [u_0, u_1] \to [0, s_f] \ , \ u \to s = \int_{u_0}^{u} \left\|\dot{\mathbf{p}}(\xi)\right\| d\xi$$

where $\|\cdot\|$ denotes the Euclidean norm and s_f is the total curve length, so that $s_f = f(u_1)$. Given a regular curve $\mathbf{p}(u)$, the arc length function $f(\cdot)$ is continuous over $[u_0, u_1]$ and bijective; hence its inverse is continuous too and is denoted by

$$f^{-1} : [0, s_f] \to [u_0, u_1] \ , \ s \to u = f^{-1}(s) \ .$$

Associated with every point of a regular curve $\mathbf{p}(u)$ there is the orthonormal moving frame, referred in the following as $\{\boldsymbol{\tau}(u), \boldsymbol{\nu}(u)\}$, that is congruent with the axes of the $\{x, y\}$-plane and where $\boldsymbol{\tau}(u) = \dot{\mathbf{p}}(u)/\left\|\dot{\mathbf{p}}(u)\right\|$ denotes the unit tangent vector of $\mathbf{p}(u)$. For any regular curve such that $\ddot{\mathbf{p}}(u) \in C_p([u_0, u_1])$, the scalar curvature $\kappa_c(u)$ and the unit vector $\boldsymbol{\nu}(u)$ are well defined according to the Frenet formula $\frac{d\boldsymbol{\tau}}{ds}(u) = \kappa_c(u)\boldsymbol{\nu}(u)$ (see for example (Hsiung, 1997, p. 109)). The resulting curvature function can be then defined as

$$\kappa_c : [u_0, u_1] \to \mathbb{R} \ , \ u \to \kappa_c(u) \ .$$

The scalar curvature can be also expressed as a function of the arc length s according to the notation:

$$\kappa : [0, s_f] \to \mathbb{R} \ , \ s \to \kappa(s) \ .$$

Hence, this function can be evaluated as $\kappa(s) = \kappa_c(f^{-1}(s))$. In the following, "dotted" terms indicate the derivative of a function made with respect to its argument, so that $\dot{\kappa}_c := \frac{d\kappa_c}{du}$ whereas $\dot{\kappa} := \frac{d\kappa}{ds}$.

Definition 1 (G^1-,G^2- and G^3-curves) *A parametric curve* $\mathbf{p}(u)$ *has first order geometric continuity, and we say* $\mathbf{p}(u)$ *is a* G^1*-curve, if* $\mathbf{p}(u)$ *is regular and its unit tangent vector is a continuous function along the curve, i.e.,* $\boldsymbol{\tau}(\cdot) \in C^0([u_0, u_1])$. *Curve* $\mathbf{p}(u)$ *has second order geometric continuity, and we say* $\mathbf{p}(u)$ *is a* G^2*-curve, if* $\mathbf{p}(u)$ *is a* G^1*-curve,* $\ddot{\mathbf{p}}(\cdot) \in C_p([u_0, u_1])$, *and its scalar curvature is continuous along the curve, i.e.,* $\kappa_c(\cdot) \in C^0([u_0, u_1])$ *or, equivalently,* $\kappa(\cdot) \in C^0([0, s_f])$. *Curve* $\mathbf{p}(u)$ *has third order geometric continuity, and we say* $\mathbf{p}(u)$ *is a* G^3*-curve, if* $\mathbf{p}(u)$ *is a* G^2*-curve,* $\dddot{\mathbf{p}}(\cdot) \in C_p([u_0, u_1])$, *and the derivative with respect to the arc length s of the scalar curvature is continuous along the curve, i.e.,* $\dot{\kappa}(\cdot) \in C^0([0, s_f])$.

Barsky and Beatty (Barsky&Beatty, 1983) introduced G^1- and G^2- curves in computer graphics. G^3-curves have been proposed in (Guarino Lo Bianco et al., 2004b) for the inversion-based control of WMRs. The related definition of G^i-paths is straightforwardly introduced as follows.

Definition 2 (G^1-, G^2- and G^3-paths) *A path of a Cartesian plane, i.e., a set of points in this plane, is a* G^i*-path* ($i = 1, 2, 3$) *or a path with i-th order geometric continuity if there exists a parametric* G^i*-curve whose image is the given path.*

Hence, G^3-paths are paths with continuously differentiable curvature. The usefulness of planning with such paths was advocated by Reuter (Reuter, 1998) on the grounds of avoiding slippage in the motion control of wheeled mobile robots.

3. Inversion-based smooth motion control of WMRs

Consider a WMR whose nonholonomic motion model is given by

$$\begin{cases} \dot{x}(t) &= v(t)\cos\theta(t) \\ \dot{y}(t) &= v(t)\sin\theta(t) \\ \dot{\theta}(t) &= \omega(t) \end{cases} \tag{1}$$

As usual, x and y indicate the robot position with respect to a stationary frame, θ is the robot heading angle, and v and ω are its linear and angular velocities to be considered as the control inputs.

In order to achieve a smooth control, inputs $v(t)$ and $\omega(t)$ must be C^1-functions, i.e., linear and angular accelerations have to be continuous signals. It is useful to define an "extended state" of model (1) that also comprises the inputs and their first derivatives:

$$\{x(t), y(t), \theta(t), v(t), \dot{v}(t), \omega(t), \dot{\omega}(t)\}.$$

Then, the following local Smooth Motion Planning Problem (SMPP) can be posed (Guarino Lo Bianco et al., 2004b).

SMPP: Given any assigned traveling time $t_f > 0$, find control inputs $v(\cdot), \omega(\cdot) \in C^1([0, t_f])$ such that the WMR, starting from any arbitrary initial extended state

$$\mathbf{p}_A = [x_A \; y_A]^T = [x(0) \; y(0)]^T, \; \theta_A = \theta(0),$$

$$v_A = v(0), \; \dot{v}_A = \dot{v}(0), \; \omega_A = \omega(0), \; \dot{\omega}_A = \dot{\omega}(0),$$

reaches any final, arbitrarily assigned, extended state

$$\mathbf{p}_B = [x_B \; y_B]^T = [x(t_f) \; y(t_f)]^T, \; \theta_B = \theta(t_f),$$

$$v_B = v(t_f), \; \dot{v}_B = \dot{v}(t_f), \; \omega_B = \omega(t_f), \; \dot{\omega}_B = \dot{\omega}(t_f).$$

The solution of the above problem, exposed in (Guarino Lo Bianco et al., 2004b), can be used in a motion control architecture based on the iterative steering approach (Lucibello & Oriolo, 1996). In such a way, a swift high-performance motion of the WMR can be achieved, while intelligent or elaborate behaviors are performed. The solution to SMPP is based on a path dynamic inversion procedure that needs the planning of a G^3-path connecting $\mathbf{p}_A$ with $\mathbf{p}_B$. This relies on the following result.

Proposition 1 (Guarino Lo Bianco et al., 2004b) *Assign any $t_f > 0$. If a Cartesian path is generated by model (1) with inputs $v(t), \omega(t) \in C^1([0, t_f])$ and $v(t) \neq 0 \; \forall t \in [0, t_f]$ then it is a G^3-path. Conversely, given any G^3-path there exist inputs $v(t), \omega(t) \in C^1([0, t_f])$ with $v(t) \neq 0 \; \forall t \in [0, t_f]$ and initial conditions such that the path generated by model (1) coincides with the given G^3-path.*

The G^3-path connecting $\mathbf{p}_A$ with $\mathbf{p}_B$ must satisfy interpolating conditions at the endpoints that depend on the initial and final extended states of the WMR. Consider, for example, the case $v_A > 0$ and $v_B > 0$. Then, angles θ_A and θ_B between the x-axis and the endpoint unit tangent vectors must coincide with the heading angles of the WMR at the initial and final poses (see Fig. 1). Moreover, curvatures and their derivatives with respect to the arc length can be determined at the endpoints according to the expressions:

$$\begin{aligned} \kappa_A &= \frac{\omega_A}{v_A}, \; \dot{\kappa}_A = \frac{\dot{\omega}_A v_A - \omega_A \dot{v}_A}{v_A^3} \\ \kappa_B &= \frac{\omega_B}{v_B}, \; \dot{\kappa}_B = \frac{\dot{\omega}_B v_B - \omega_B \dot{v}_B}{v_B^3} \end{aligned}$$

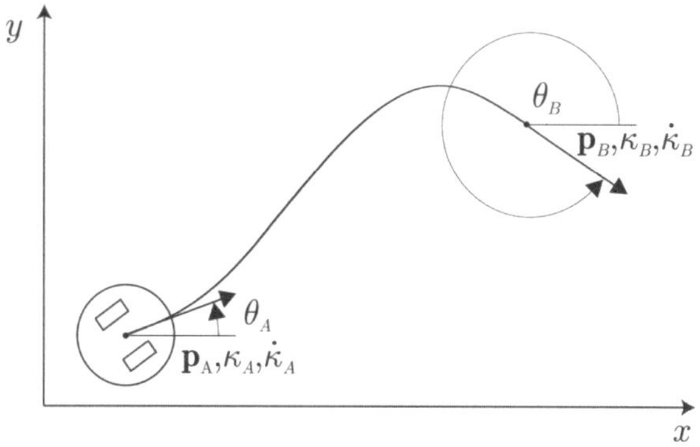

Fig. 1. A G^3-path connecting $\mathbf{p}_A$ with $\mathbf{p}_B$ for the SMPP.

The critical cases $v_A = 0$ and/or $v_B = 0$, as well as other cases, are discussed in (Guarino Lo Bianco et al., 2004b).

Proposition 1 makes clear that continuous curvature paths, or G^2-paths, may be insufficiently smooth for a motion planning with fast and swift robot maneuvers. Indeed, with G^2-paths the command accelerations can be discontinuous causing the possible slippage of the WMR. On the contrary, G^3-paths are well suited for smooth motion planning because they can be followed with continuous accelerations commands.

4. The η^3-splines

In the context of smooth iterative steering for WMRs, the previous section has recalled the necessity of planning G^3-paths having arbitrary interpolating conditions at the endpoints. Then, a natural approach is to find a polynomial curve for the associated interpolation problem. This justifies the introduction of the following formal problem.

The polynomial G^3-interpolating problem: *Determine the minimal order polynomial curve which interpolates two given endpoints $\mathbf{p}_A = [x_A\ y_A]^T$ and $\mathbf{p}_B = [x_B\ y_B]^T$ with associated unit tangent vectors defined by angles θ_A and θ_B, scalar curvatures κ_A and κ_B, and curvature derivatives with respect to the arc length $\dot{\kappa}_A$ and $\dot{\kappa}_B$ (see Fig. 1). Assume that interpolating data $\mathbf{p}_A$, $\mathbf{p}_B \in \mathbb{R}^2$, θ_A, $\theta_B \in [0, 2\pi)$, κ_A, $\kappa_B \in \mathbb{R}$ and $\dot{\kappa}_A$, $\dot{\kappa}_B \in \mathbb{R}$ can be arbitrarily assigned.*

The solution proposed for the above interpolating problem is given by a seventh order polynomial curve $\mathbf{p}(u) = [\alpha(u)\ \beta(u)]^T$, $u \in [0, 1]$ defined as follows

$$\alpha(u) := \alpha_0 + \alpha_1 u + \alpha_2 u^2 + \alpha_3 u^3 + \alpha_4 u^4 + \alpha_5 u^5 + \alpha_6 u^6 + \alpha_7 u^7 \tag{2}$$

$$\beta(u) := \beta_0 + \beta_1 u + \beta_2 u^2 + \beta_3 u^3 + \beta_4 u^4 + \beta_5 u^5 + \beta_6 u^6 + \beta_7 u^7 \tag{3}$$

$$\alpha_0 = x_A \tag{4}$$

$$\alpha_1 = \eta_1 \cos\theta_A \tag{5}$$

$$\alpha_2 = \frac{1}{2}\eta_3 \cos\theta_A - \frac{1}{2}\eta_1^2 \kappa_A \sin\theta_A \tag{6}$$

$$\alpha_3 = \frac{1}{6}\eta_5 \cos\theta_A - \frac{1}{6}\left(\eta_1^3 \dot{\kappa}_A + 3\eta_1\eta_3\kappa_A\right)\sin\theta_A \tag{7}$$

$$\alpha_4 = 35(x_B - x_A) - \left(20\eta_1 + 5\eta_3 + \frac{2}{3}\eta_5\right)\cos\theta_A + \left(5\eta_1^2\kappa_A + \frac{2}{3}\eta_1^3\dot{\kappa}_A + 2\eta_1\eta_3\kappa_A\right)\sin\theta_A - \left(15\eta_2 - \frac{5}{2}\eta_4 + \frac{1}{6}\eta_6\right)\cos\theta_B - \left(\frac{5}{2}\eta_2^2\kappa_B - \frac{1}{6}\eta_2^3\dot{\kappa}_B - \frac{1}{2}\eta_2\eta_4\kappa_B\right)\sin\theta_B \quad (8)$$

$$\alpha_5 = -84(x_B - x_A) + (45\eta_1 + 10\eta_3 + \eta_5)\cos\theta_A - \left(10\eta_1^2\kappa_A + \eta_1^3\dot{\kappa}_A + 3\eta_1\eta_3\kappa_A\right)\sin\theta_A + \left(39\eta_2 - 7\eta_4 + \frac{1}{2}\eta_6\right)\cos\theta_B + \left(7\eta_2^2\kappa_B - \frac{1}{2}\eta_2^3\dot{\kappa}_B - \frac{3}{2}\eta_2\eta_4\kappa_B\right)\sin\theta_B \quad (9)$$

$$\alpha_6 = 70(x_B - x_A) - \left(36\eta_1 + \frac{15}{2}\eta_3 + \frac{2}{3}\eta_5\right)\cos\theta_A + \left(\frac{15}{2}\eta_1^2\kappa_A + \frac{2}{3}\eta_1^3\dot{\kappa}_A + 2\eta_1\eta_3\kappa_A\right)\sin\theta_A - \left(34\eta_2 - \frac{13}{2}\eta_4 + \frac{1}{2}\eta_6\right)\cos\theta_B - \left(\frac{13}{2}\eta_2^2\kappa_B - \frac{1}{2}\eta_2^3\dot{\kappa}_B - \frac{3}{2}\eta_2\eta_4\kappa_B\right)\sin\theta_B \quad (10)$$

$$\alpha_7 = -20(x_B - x_A) + \left(10\eta_1 + 2\eta_3 + \frac{1}{6}\eta_5\right)\cos\theta_A - \left(2\eta_1^2\kappa_A + \frac{1}{6}\eta_1^3\dot{\kappa}_A + \frac{1}{2}\eta_1\eta_3\kappa_A\right)\sin\theta_A + \left(10\eta_2 - 2\eta_4 + \frac{1}{6}\eta_6\right)\cos\theta_B + \left(2\eta_2^2\kappa_B - \frac{1}{6}\eta_2^3\dot{\kappa}_B - \frac{1}{2}\eta_2\eta_4\kappa_B\right)\sin\theta_B \quad (11)$$

$$\beta_0 = y_A \quad (12)$$

$$\beta_1 = \eta_1\sin\theta_A \quad (13)$$

$$\beta_2 = \frac{1}{2}\eta_3\sin\theta_A + \frac{1}{2}\eta_1^2\kappa_A\cos\theta_A \quad (14)$$

$$\beta_3 = \frac{1}{6}\eta_5\sin\theta_A + \frac{1}{6}\left(\eta_1^3\dot{\kappa}_A + 3\eta_1\eta_3\kappa_A\right)\cos\theta_A \quad (15)$$

$$\beta_4 = 35(y_B - y_A) - \left(20\eta_1 + 5\eta_3 + \frac{2}{3}\eta_5\right)\sin\theta_A - \left(5\eta_1^2\kappa_A + \frac{2}{3}\eta_1^3\dot{\kappa}_A + 2\eta_1\eta_3\kappa_A\right)\cos\theta_A - \left(15\eta_2 - \frac{5}{2}\eta_4 + \frac{1}{6}\eta_6\right)\sin\theta_B + \left(\frac{5}{2}\eta_2^2\kappa_B - \frac{1}{6}\eta_2^3\dot{\kappa}_B - \frac{1}{2}\eta_2\eta_4\kappa_B\right)\cos\theta_B \quad (16)$$

$$\beta_5 = -84(y_B - y_A) + (45\eta_1 + 10\eta_3 + \eta_5)\sin\theta_A + \left(10\eta_1^2\kappa_A + \eta_1^3\dot{\kappa}_A + 3\eta_1\eta_3\kappa_A\right)\cos\theta_A + \left(39\eta_2 - 7\eta_4 + \frac{1}{2}\eta_6\right)\sin\theta_B - \left(7\eta_2^2\kappa_B - \frac{1}{2}\eta_2^3\dot{\kappa}_B - \frac{3}{2}\eta_2\eta_4\kappa_B\right)\cos\theta_B \quad (17)$$

$$\beta_6 = 70(y_B - y_A) - \left(36\eta_1 + \frac{15}{2}\eta_3 + \frac{2}{3}\eta_5\right)\sin\theta_A - \left(\frac{15}{2}\eta_1^2\kappa_A + \frac{2}{3}\eta_1^3\dot{\kappa}_A + 2\eta_1\eta_3\kappa_A\right)\cos\theta_A - \left(34\eta_2 - \frac{13}{2}\eta_4 + \frac{1}{2}\eta_6\right)\sin\theta_B + \left(\frac{13}{2}\eta_2^2\kappa_B - \frac{1}{2}\eta_2^3\dot{\kappa}_B - \frac{3}{2}\eta_2\eta_4\kappa_B\right)\cos\theta_B \quad (18)$$

$$\begin{aligned}\beta_7 \;=\;& -20\,(y_B - y_A) + \left(10\eta_1 + 2\eta_3 + \frac{1}{6}\eta_5\right)\sin\theta_A \\ & + \left(2\eta_1^2\kappa_A + \frac{1}{6}\eta_1^3\dot{\kappa}_A + \frac{1}{2}\eta_1\eta_3\kappa_A\right)\cos\theta_A + \left(10\eta_2 - 2\eta_4 + \frac{1}{6}\eta_6\right)\sin\theta_B \\ & - \left(2\eta_2^2\kappa_B - \frac{1}{6}\eta_2^3\dot{\kappa}_B - \frac{1}{2}\eta_2\eta_4\kappa_B\right)\cos\theta_B\end{aligned} \tag{19}$$

The real parameters η_i, $i = 1,\dots, 6$ which appear in (4)-(19), can be freely selected and influence the path shape without violating the endpoint interpolating conditions. They can be packed together to form a six-dimensional vector $\boldsymbol{\eta} := [\eta_1\ \eta_2\ \eta_3\ \eta_4\ \eta_5\ \eta_6]^T$, and the parametric curve (2)-(3) will be concisely denoted in the following as $\mathbf{p}(u, \boldsymbol{\eta})$ or, informally, $\boldsymbol{\eta}^3$*-spline*. Vector $\boldsymbol{\eta}$ spans in $\mathcal{H} := \mathbb{R}_+^2 \times \mathbb{R}^4$ ($\mathbb{R}_+$ denotes the set of positive real numbers).
Coefficient expressions (4)-(19) were deduced by solving a nonlinear equation system associated to the endpoint interpolation conditions. The correctness of the provided expressions is formally stated by the following proposition.

Proposition 2 *The parametric curve* $\mathbf{p}(u; \boldsymbol{\eta})$ *satisfies any given set of interpolating data* $\mathbf{p}_A$, θ_A, κ_A, $\dot{\kappa}_A$ *and* $\mathbf{p}_B$, θ_B, κ_B, $\dot{\kappa}_B$, *for all* $\boldsymbol{\eta} \in \mathcal{H}$.

Proof – Basically, *Proposition 2* asserts that curve $\mathbf{p}(u; \boldsymbol{\eta})$ fulfills any set of interpolating conditions independently from the choice of $\boldsymbol{\eta}$. The proof can be then established by direct computation. Owing to definitions given for $\kappa(s)$ and $\kappa_c(u)$, it is possible to write

$$\frac{d\kappa_c}{du} = \frac{d\kappa}{ds}\frac{ds}{du} = \frac{d\kappa}{ds}\left\|\dot{\mathbf{p}}(u)\right\|,$$

or, briefly,

$$\dot{\kappa}_c(u) = \dot{\kappa}(s)\left\|\dot{\mathbf{p}}(u)\right\|.$$

Bearing in mind this result and the definition of the unit tangent vector $\boldsymbol{\tau}(u)$, curve $\mathbf{p}(u; \boldsymbol{\eta})$ satisfies the assigned boundary conditions if the following equalities hold for all $\boldsymbol{\eta} \in \mathcal{H}$.

$$\mathbf{p}(0;\boldsymbol{\eta}) = \mathbf{p}_A, \tag{20}$$

$$\mathbf{p}(1;\boldsymbol{\eta}) = \mathbf{p}_B, \tag{21}$$

$$\boldsymbol{\tau}(0;\boldsymbol{\eta}) = \frac{\dot{\mathbf{p}}(0;\boldsymbol{\eta})}{\left\|\dot{\mathbf{p}}(0;\boldsymbol{\eta})\right\|} = \begin{bmatrix}\cos\theta_A \\ \sin\theta_A\end{bmatrix}, \tag{22}$$

$$\boldsymbol{\tau}(1;\boldsymbol{\eta}) = \frac{\dot{\mathbf{p}}(1;\boldsymbol{\eta})}{\left\|\dot{\mathbf{p}}(1;\boldsymbol{\eta})\right\|} = \begin{bmatrix}\cos\theta_B \\ \sin\theta_B\end{bmatrix}, \tag{23}$$

$$\kappa_c(0;\boldsymbol{\eta}) = \kappa_A, \tag{24}$$

$$\kappa_c(1;\boldsymbol{\eta}) = \kappa_B, \tag{25}$$

$$\dot{\kappa}_c(0;\boldsymbol{\eta}) = \dot{\kappa}_A\left\|\dot{\mathbf{p}}(0;\boldsymbol{\eta})\right\|, \tag{26}$$

$$\dot{\kappa}_c(1;\boldsymbol{\eta}) = \dot{\kappa}_B\left\|\dot{\mathbf{p}}(1;\boldsymbol{\eta})\right\|. \tag{27}$$

First consider conditions (20) and (21). Taking into account that the parametric curve $\mathbf{p}(u;\,\boldsymbol{\eta})$ is described by means of (2) and (3), and its coefficients are defined according to (4)–(19), it is immediate to verify that, as required, $\mathbf{p}(0;\,\boldsymbol{\eta}) = \mathbf{p}_A$ and $\mathbf{p}(1;\,\boldsymbol{\eta}) = \mathbf{p}_B\ \forall\boldsymbol{\eta} \in \mathcal{H}$.
Further, differentiating $\mathbf{p}(u;\,\boldsymbol{\eta})$ with respect to u, we obtain $\dot{\mathbf{p}}(u;\boldsymbol{\eta}) = [\dot{\alpha}(u;\boldsymbol{\eta})\ \dot{\beta}(u;\boldsymbol{\eta})]^T$ where

$$\dot{\alpha}(u) = \alpha_1 + 2\alpha_2 u + 3\alpha_3 u^2 + 4\alpha_4 u^3 + 5\alpha_5 u^4 + 6\alpha_6 u^5 + 7\alpha_7 u^6, \tag{28}$$

$$\dot{\beta}(u) = \beta_1 + 2\beta_2 u + 3\beta_3 u^2 + 4\beta_4 u^3 + 5\beta_5 u^4 + 6\beta_6 u^5 + 7\beta_7 u^6. \tag{29}$$

Evaluating $\dot{\mathbf{p}}\,(u;\,\boldsymbol{\eta})$ for $u = 0$ and $u = 1$, and considering (4)–(19), it is easy to verify that, for all $\boldsymbol{\eta} \in \mathcal{H}$,

$$\dot{\mathbf{p}}\,(0;\boldsymbol{\eta}) = \eta_1[\cos\theta_A \quad \sin\theta_A]^T,$$
$$\dot{\mathbf{p}}\,(1;\boldsymbol{\eta}) = \eta_2[\cos\theta_B \quad \sin\theta_B]^T.$$

Bearing in mind that $\eta_1,\ \eta_2 \in \mathbb{R}_+$, it follows that

$$\left\|\dot{\mathbf{p}}(0;\boldsymbol{\eta})\right\| = \eta_1 \tag{30}$$

$$\left\|\dot{\mathbf{p}}(1;\boldsymbol{\eta})\right\| = \eta_2 \tag{31}$$

and, consequently, as desired

$$\boldsymbol{\tau}(0;\boldsymbol{\eta}) = \dot{\mathbf{p}}(0;\boldsymbol{\eta})/\left\|\dot{\mathbf{p}}(0;\boldsymbol{\eta})\right\| = [\cos\theta_A \ \ \sin\theta_A]^T\ ,$$
$$\boldsymbol{\tau}(1;\boldsymbol{\eta}) = \dot{\mathbf{p}}(1;\boldsymbol{\eta})/\left\|\dot{\mathbf{p}}(1;\boldsymbol{\eta})\right\| = [\cos\theta_B \ \ \sin\theta_B]^T\ .$$

According to the theory of planar curves, the scalar curvature can be evaluated by means of the formula

$$\kappa_c = \frac{\dot{\alpha}\ddot{\beta} - \ddot{\alpha}\dot{\beta}}{(\dot{\alpha}^2 + \dot{\beta}^2)^{3/2}}\ , \tag{32}$$

where $\ddot{\alpha}(u)$ and $\ddot{\beta}(u)$ can be obtained by differentiating (28) and (29), i.e.,

$$\ddot{\alpha}(u) = 2\alpha_2 + 6\alpha_3 u + 12\alpha_4 u^2 + 20\alpha_5 u^3 + 30\alpha_6 u^4 + 42\alpha_7 u^5, \tag{33}$$

$$\ddot{\beta}(u) = 2\beta_2 + 6\beta_3 u + 12\beta_4 u^2 + 20\beta_5 u^3 + 30\beta_6 u^4 + 42\beta_7 u^5. \tag{34}$$

Applying (4)–(19) to (28)–(34) and evaluating $\kappa_c(u)$ for $u = 0$ and $u = 1$ we verify that, for all $\boldsymbol{\eta} \in \mathcal{H}$,

$$\kappa_c(0;\boldsymbol{\eta}) = \kappa_A, \qquad \text{and} \qquad \kappa_c(1;\boldsymbol{\eta}) = \kappa_B\ .$$

The first derivative of $\kappa_c(u)$ with respect to u is given by

$$\frac{\mathrm{d}\kappa_c}{\mathrm{d}u} = \frac{(\dot{\alpha}\dddot{\beta} - \dddot{\alpha}\dot{\beta})(\dot{\alpha}^2 + \dot{\beta}^2) - 3(\dot{\alpha}\ddot{\beta} - \ddot{\alpha}\dot{\beta})(\dot{\alpha}\ddot{\alpha} + \dot{\beta}\ddot{\beta})}{(\dot{\alpha}^2 + \dot{\beta}^2)^{5/2}}\ . \tag{35}$$

An explicit differentiation of (33) and (34) makes it possible to write

$$\ddot{\alpha}(u) = 6\alpha_3 + 24\alpha_4 u + 60\alpha_5 u^2 + 120\alpha_6 u^3 + 210\alpha_7 u^4 \,, \quad (36)$$

$$\dddot{\beta}(u) = 6\beta_3 + 24\beta_4 u + 60\beta_5 u^2 + 120\beta_6 u^3 + 210\beta_7 u^4 \,. \quad (37)$$

The initial and final values of $\dot{\kappa}_c$ can be then obtained from (35) by applying (4)–(19) to (28)–(37):

$$\dot{\kappa}_c(0;\boldsymbol{\eta}) = \dot{\kappa}_A\eta_1 = \dot{\kappa}_A \left\|\dot{\mathbf{p}}(0;\boldsymbol{\eta})\right\| \,,$$
$$\dot{\kappa}_c(1;\boldsymbol{\eta}) = \dot{\kappa}_B\eta_2 = \dot{\kappa}_B \left\|\dot{\mathbf{p}}(1;\boldsymbol{\eta})\right\| \,.$$

Both equalities hold for all $\boldsymbol{\eta} \in \mathcal{H}$. ■

The next result shows how the introduced η^3-spline is a complete parameterization of all the seventh order polynomial curves interpolating given endpoint data.

Property 1 (Completeness) *Given any seventh order polynomial curve* $\mathbf{q}(u)$, $u \in [0; 1]$ *with* $\dot{\mathbf{q}}(0) \neq 0$ *and* $\dot{\mathbf{q}}(1) \neq 0$ *which satisfies a given set of interpolating conditions* $\mathbf{p}_A$, θ_A, κ_A, $\dot{\kappa}_A$ *and* $\mathbf{p}_B$, θ_B, κ_B, $\dot{\kappa}_B$, *there exists a parameter vector* $\boldsymbol{\eta} \in \mathcal{H}$ *such that* $\mathbf{p}(u; \boldsymbol{\eta})$ *coincides with* $\mathbf{q}(u)$.

Proof - Consider a seventh order polynomial curve $\mathbf{q}(u)$ defined as follows

$$\mathbf{q}(u) := \begin{bmatrix} \gamma(u) \\ \delta(u) \end{bmatrix}$$

where

$$\gamma(u) = \gamma_0 + \gamma_1 u + \gamma_2 u^2 + \gamma_3 u^3 + \gamma_4 u^4 + \gamma_5 u^5 + \gamma_6 u^6 + \gamma_7 u^7 \,,$$

$$\delta(u) = \delta_0 + \delta_1 u + \delta_2 u^2 + \delta_3 u^3 + \delta_4 u^4 + \delta_5 u^5 + \delta_6 u^6 + \delta_7 u^7 \,.$$

We assume that $\dot{\mathbf{q}}(0) \neq 0$, $\dot{\mathbf{q}}(1) \neq 0$, and all the interpolating conditions at the path endpoints are satisfied.

Considering the zero and first order boundary conditions we have

$$\mathbf{q}(0) = \begin{bmatrix} \gamma(0) \\ \delta(0) \end{bmatrix} = \begin{bmatrix} x_A \\ y_A \end{bmatrix} = \mathbf{p}_A,$$
$$\mathbf{q}(1) = \begin{bmatrix} \gamma(1) \\ \delta(1) \end{bmatrix} = \begin{bmatrix} x_B \\ y_B \end{bmatrix} = \mathbf{p}_B,$$

$$\boldsymbol{\tau}(0) = \frac{\dot{\mathbf{q}}(0)}{\left\|\dot{\mathbf{q}}(0)\right\|} = \frac{\begin{bmatrix} \dot{\gamma}(0) \\ \dot{\delta}(0) \end{bmatrix}}{\left(\dot{\gamma}(0)^2 + \dot{\delta}(0)^2\right)^{1/2}} = \begin{bmatrix} \cos\theta_A \\ \sin\theta_A \end{bmatrix},$$

$$\boldsymbol{\tau}(1) = \frac{\dot{\mathbf{q}}(1)}{\left\|\dot{\mathbf{q}}(1)\right\|} = \frac{\begin{bmatrix} \dot{\gamma}(1) \\ \dot{\delta}(1) \end{bmatrix}}{\left(\dot{\gamma}(1)^2 + \dot{\delta}(1)^2\right)^{1/2}} = \begin{bmatrix} \cos\theta_B \\ \sin\theta_B \end{bmatrix},$$

which can be rewritten as follows

$$\gamma(0) = x_A \,, \quad (38)$$

$$\delta(0) = y_A \,, \quad (39)$$

$$\gamma(1) = x_B\,, \tag{40}$$

$$\delta(1) = y_B\,, \tag{41}$$

$$\dot\gamma(0) = [\dot\gamma(0)^2 + \dot\delta(0)^2]^{\frac{1}{2}} \cos\theta_A\,, \tag{42}$$

$$\dot\delta(0) = [\dot\gamma(0)^2 + \dot\delta(0)^2]^{\frac{1}{2}} \sin\theta_A\,, \tag{43}$$

$$\dot\gamma(1) = [\dot\gamma(1)^2 + \dot\delta(1)^2]^{\frac{1}{2}} \cos\theta_B\,, \tag{44}$$

$$\dot\delta(1) = [\dot\gamma(1)^2 + \dot\delta(1)^2]^{\frac{1}{2}} \sin\theta_B\,. \tag{45}$$

Further, taking into account that the scalar curvature $\kappa_c(u)$ of $\mathbf{q}(u)$ and its first derivative $\dot\kappa_c(u)$ can be evaluated with expressions similar to (32) and (35), the second and third order boundary conditions can be explicitly written as follows

$$\frac{\dot\gamma(0)\ddot\delta(0) - \ddot\gamma(0)\dot\delta(0)}{\left(\dot\gamma(0)^2 + \dot\delta(0)^2\right)^{3/2}} = \kappa_A\,, \tag{46}$$

$$\frac{\dot\gamma(1)\ddot\delta(1) - \ddot\gamma(1)\dot\delta(1)}{\left(\dot\gamma(1)^2 + \dot\delta(1)^2\right)^{3/2}} = \kappa_B\,, \tag{47}$$

$$\frac{[\dot\gamma(0)\dddot\delta(0) - \dddot\gamma(0)\dot\delta(0)][\dot\gamma(0)^2 + \dot\delta(0)^2] - 3[\dot\gamma(0)\ddot\delta(0) - \ddot\gamma(0)\dot\delta(0)][\dot\gamma(0)\ddot\gamma(0) + \dot\delta(0)\ddot\delta(0)]}{[\dot\gamma(0)^2 + \dot\delta(0)^2]^{\frac{5}{2}}} = \dot\kappa_A[\dot\gamma(0)^2 + \dot\delta(0)^2]^{\frac{1}{2}}\,, \tag{48}$$

$$\frac{[\dot\gamma(1)\dddot\delta(1) - \dddot\gamma(1)\dot\delta(1)][\dot\gamma(1)^2 + \dot\delta(1)^2] - 3[\dot\gamma(1)\ddot\delta(1) - \ddot\gamma(1)\dot\delta(1)][\dot\gamma(1)\ddot\gamma(1) + \dot\delta(1)\ddot\delta(1)]}{[\dot\gamma(1)^2 + \dot\delta(1)^2]^{\frac{5}{2}}} = \dot\kappa_B[\dot\gamma(1)^2 + \dot\delta(1)^2]^{\frac{1}{2}}\,. \tag{49}$$

The proof requires to show that there exists a parameter vector $\boldsymbol{\eta} \in \boldsymbol{\mathcal{H}}$ such that $\mathbf{p}(u; \boldsymbol{\eta})$ coincides with $\mathbf{q}(u)$, i.e., such that ($i = 0, 1, \ldots, 7$)

$$\begin{array}{l} \alpha_i = \gamma_i \\ \beta_i = \delta_i \end{array}\,.$$

To this aim, select the η_i parameters as follows

$$\eta_1 := [\dot\gamma(0)^2 + \dot\delta(0)^2]^{\frac{1}{2}}\,, \tag{50}$$

$$\eta_2 := [\dot\gamma(1)^2 + \dot\delta(1)^2]^{\frac{1}{2}}\,, \tag{51}$$

$$\eta_3 := \ddot\delta(0)\sin\theta_A + \ddot\gamma(0)\cos\theta_A\,, \tag{52}$$

$$\eta_4 := \ddot\delta(1)\sin\theta_B + \ddot\gamma(1)\cos\theta_B\,, \tag{53}$$

$$\eta_5 := \dddot{\delta}(0)\sin\theta_A + \dddot{\gamma}(0)\cos\theta_A\,, \tag{54}$$

$$\eta_6 := \dddot{\delta}(1)\sin\theta_B + \dddot{\gamma}(1)\cos\theta_B\,. \tag{55}$$

Owing to the above assignments, the interpolating conditions (38)–(49) on the polynomial curve $\mathbf{q}(u)$ can be rewritten as follows

$$\gamma(0) = x_A\,, \tag{56}$$

$$\delta(0) = y_A\,, \tag{57}$$

$$\gamma(1) = x_B\,, \tag{58}$$

$$\delta(1) = y_B\,, \tag{59}$$

$$\dot{\gamma}(0) = \eta_1\cos\theta_A\,, \tag{60}$$

$$\dot{\delta}(0) = \eta_1\sin\theta_A\,, \tag{61}$$

$$\dot{\gamma}(1) = \eta_2\cos\theta_B\,, \tag{62}$$

$$\dot{\delta}(1) = \eta_2\sin\theta_B\,, \tag{63}$$

$$\cos\theta_A\ddot{\delta}(0) - \sin\theta_A\ddot{\gamma}(0) = \eta_1^2\kappa_A\,, \tag{64}$$

$$\cos\theta_B\ddot{\delta}(1) - \sin\theta_B\ddot{\gamma}(1) = \eta_2^2\kappa_B\,, \tag{65}$$

$$\cos\theta_A\dddot{\delta}(0) - \sin\theta_A\dddot{\gamma}(0) = \eta_1^3\dot{\kappa}_A + 3\eta_1\eta_3\kappa_A\,, \tag{66}$$

$$\cos\theta_B\dddot{\delta}(1) - \sin\theta_B\dddot{\gamma}(1) = \eta_2^3\dot{\kappa}_B + 3\eta_2\eta_4\kappa_B\,. \tag{67}$$

From definitions (4) and (12) and relations (56) and (57) obviously we get

$$\alpha_0 = x_A = \gamma(0) = \gamma_0\,,$$
$$\beta_0 = y_A = \delta(0) = \delta_0\,.$$

Analogously, from (5) and (13), and taking into account (60) and (61), it is possible to infer

$$\alpha_1 = \eta_1\cos\theta_A = \dot{\gamma}(0) = \gamma_1\,,$$
$$\beta_1 = \eta_1\sin\theta_A = \dot{\delta}(0) = \delta_1\,.$$

Owing to definition (52) and condition (64), we obtain from (6) and (14)

$$\begin{aligned}\alpha_2 &= \frac{1}{2}[\ddot{\delta}(0)\sin\theta_A + \ddot{\gamma}(0)\cos\theta_A]\cos\theta_A - \frac{1}{2}[\ddot{\delta}(0)\cos\theta_A - \ddot{\gamma}(0)\sin\theta_A]\sin\theta_A\\ &= \frac{1}{2}\ddot{\gamma}(0) = \gamma_2\,,\end{aligned}$$

$$\begin{aligned}\beta_2 &= \frac{1}{2}[\ddot{\delta}(0)\sin\theta_A + \ddot{\gamma}(0)\cos\theta_A]\sin\theta_A + \frac{1}{2}[\ddot{\delta}(0)\cos\theta_A - \ddot{\gamma}(0)\sin\theta_A]\cos\theta_A \\ &= \frac{1}{2}\ddot{\delta}(0) = \delta_2\,.\end{aligned}$$

In the same way, from (7) and (15), and taking into account (54) and (66), it is possible to verify

$$\begin{aligned}\alpha_3 &= \frac{1}{6}[\dddot{\delta}(0)\sin\theta_A + \dddot{\gamma}(0)\cos\theta_A]\cos\theta_A - \frac{1}{6}[\dddot{\delta}(0)\cos\theta_A - \dddot{\gamma}(0)\sin\theta_A]\sin\theta_A \\ &= \frac{1}{6}\dddot{\gamma}(0) = \gamma_3\,,\end{aligned}$$

$$\begin{aligned}\beta_3 &= \frac{1}{6}[\dddot{\delta}(0)\cos\theta_A + \dddot{\gamma}(0)\sin\theta_A]\cos\theta_A + \frac{1}{6}[\dddot{\delta}(0)\sin\theta_A - \dddot{\gamma}(0)\cos\theta_A]\sin\theta_A \\ &= \frac{1}{6}\dddot{\delta}(0) = \delta_3\,.\end{aligned}$$

Focusing on the α_i expressions given by (8)–(11), the η_i definitions (52)–(55), and the interpolating equations (56), (58), (60), (62), (64)–(67) we obtain, after some algebraic manipulations, to express α_4, α_5, α_6, and α_7 as a linear combinations of $\gamma(0)$, $\dot{\gamma}(0)$, $\ddot{\gamma}(0)$, $\dddot{\gamma}(0)$, $\gamma(1)$, $\dot{\gamma}(1)$, $\ddot{\gamma}(1)$, and $\dddot{\gamma}(1)$:

$$\alpha_4 = 35[\gamma(1)-\gamma(0)] - 15\dot{\gamma}(1) - 20\dot{\gamma}(0) + \frac{5}{2}\ddot{\gamma}(1) - 5\ddot{\gamma}(0) - \frac{1}{6}\dddot{\gamma}(1) - \frac{2}{3}\dddot{\gamma}(0)\,,$$

$$\alpha_5 = 84[\gamma(0)-\gamma(1)] + 45\dot{\gamma}(0) + 39\dot{\gamma}(1) + 10\ddot{\gamma}(0) - 7\ddot{\gamma}(1) + \dddot{\gamma}(0) + \frac{1}{2}\dddot{\gamma}(1)\,,$$

$$\alpha_6 = 70[\gamma(1)-\gamma(0)] - 34\dot{\gamma}(1) - 36\dot{\gamma}(0) + \frac{13}{2}\ddot{\gamma}(1) - \frac{15}{2}\ddot{\gamma}(0) - \frac{1}{2}\dddot{\gamma}(1) - \frac{2}{3}\dddot{\gamma}(0)\,,$$

$$\alpha_7 = 20[\gamma(0)-\gamma(1)] + 10[\dot{\gamma}(0)+\dot{\gamma}(1)] + 2[\ddot{\gamma}(0)-\ddot{\gamma}(1)] + \frac{1}{6}[\dddot{\gamma}(0)+\dddot{\gamma}(1)]\,.$$

Then, by virtue of the definition given for $\gamma(u)$, it is easy to verify that $\alpha_i = \gamma_i$ for $i = 4, 5, 6, 7$. A similar procedure can be also adopted for the remaining β_i coefficients by manipulating (52)–(55), (57), (59), (61), (63), (64)–(67) and replacing the resulting expressions in (16)–(19). These final passages are left to the interested reader.

The minimality of the η^3-spline is the focus of the next statement. ■

Property 2 (Minimality) *The curve* $\mathbf{p}(u;\,\eta)$ *is the minimal order polynomial curve interpolating any arbitrarily given set of data* $\mathbf{p}_A$, $\mathbf{p}_B \in \mathbb{R}^2$, θ_A, $\theta_B \in [0, 2\pi)$, κ_A, $\kappa_B \in \mathbb{R}$ *and* $\dot{\kappa}_A$, $\dot{\kappa}_B \in \mathbb{R}$.

Proof – Proposition 2 and Property 1 have shown that the η^3-spline $\mathbf{p}(u;\,\eta)$ is the family of all polynomial curves, till to the seventh order, interpolating any given endpoint data. Hence, if a sixth or lower order polynomial curve exists interpolating any assigned set of boundary conditions, it must coincide with $\mathbf{p}(u;\,\eta)$ for some appropriate $\eta \in \mathcal{H}$. Consider the following boundary conditions (leading to a so-called lane-change path):

$$\mathbf{p}_A = [0\ 0]^T\,,\ \mathbf{p}_B = [2\ 1]^T\,,\ \theta_A = \theta_B = 0,\ \kappa_A = \kappa_B = 0,\ \dot{\kappa}_A = \dot{\kappa}_B = 0\,,$$

and evaluate the η^3-spline curve using (4)–(19)

$$
\begin{aligned}
\alpha(u;\boldsymbol{\eta}) \;=\;& \eta_1 u + \tfrac{1}{2}\eta_3 u^2 + \tfrac{1}{6}\eta_5 u^3 \\
&+ \left[70 - 20\eta_1 - 5\eta_3 - \tfrac{2}{3}\eta_5 - 15\eta_2 + \tfrac{5}{2}\eta_4 - \tfrac{1}{6}\eta_6\right] u^4 \\
&+ \left[-168 + 45\eta_1 + 10\eta_3 + \eta_5 + 39\eta_2 - 7\eta_4 + \tfrac{1}{2}\eta_6\right] u^5 \\
&+ \left[140 - 36\eta_1 - \tfrac{15}{2}\eta_3 - \tfrac{2}{3}\eta_5 - 34\eta_2 + \tfrac{13}{2}\eta_4 - \tfrac{1}{2}\eta_6\right] u^6 \\
&+ \left[-40 + 10\eta_1 + 2\eta_3 + \tfrac{1}{6}\eta_5 + 10\eta_2 - 2\eta_4 + \tfrac{1}{6}\eta_6\right] u^7 \\
\beta(u;\boldsymbol{\eta}) \;=\;& 35u^4 - 84u^5 + 70u^6 - 20u^7
\end{aligned}
$$

Evidently, $\beta(u; \boldsymbol{\eta})$ is a strict seventh order polynomial that does not depend on $\boldsymbol{\eta}$. Thus, it is not possible to interpolate the given data with a sixth or lower order polynomial curve. ∎

Proposition 2 and Property 2 show that the η^3-spline is the solution to the introduced G^3-interpolating problem. Moreover, the η^3-spline represents a family of curves that depends on a symmetric parameterization induced by the chosen $\boldsymbol{\eta}$ vector. This property, presented below formally, may be useful in shaping the η^3-spline by varying the η_i components.

Property 3 (Symmetry) *Assume* $\eta_1 = \eta_2 = v \in \mathbb{R}_+$, $\eta_3 = -\eta_4 = w \in \mathbb{R}$, $\eta_5 = \eta_6 = z \in \mathbb{R}$ *and define* $\boldsymbol{\eta} = [v\ v\ w\ -w\ z\ z]^T$. *Moreover, consider* $\theta_A = \theta_B = \theta \in [0, 2\pi)$, $\kappa_A = \kappa_B = 0$, $\dot{\kappa}_A = \dot{\kappa}_B = 0$. *Then, for any* $\mathbf{p}_A$ *and* $\mathbf{p}_B$, *curve* $\mathbf{p}(u; \boldsymbol{\eta})$ *satisfies the following symmetry relation*

$$\mathbf{p}(u;\boldsymbol{\eta}) = \mathbf{p}_A + \mathbf{p}_B - \mathbf{p}(1-u;\boldsymbol{\eta}) \tag{68}$$

$\forall u \in [0, 1]$, $\forall v \in \mathbb{R}_+$, $\forall w, z \in \mathbb{R}$.

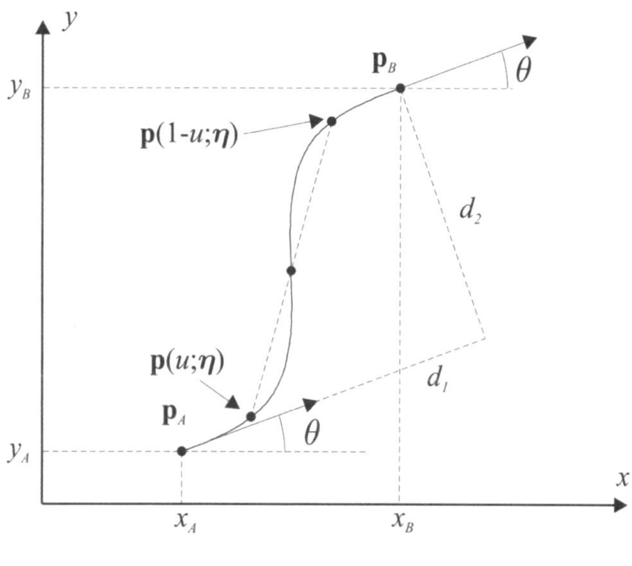

Fig. 2. A graphical interpretation of symmetry relation (68).

Proof – It is always possible to find $d_1, d_2 \in \mathbb{R}$ such that (cf. Fig. 2)

$$\mathbf{p}_B = \mathbf{p}_A + d_1 \begin{bmatrix} \cos\theta \\ \sin\theta \end{bmatrix} + d_2 \begin{bmatrix} -\sin\theta \\ \cos\theta \end{bmatrix}$$

Curve $\mathbf{p}(u; \boldsymbol{\eta})$, evaluated by means of (4)–(19) and the assigned interpolating conditions, can be expressed as

$$\mathbf{p}(u;\boldsymbol{\eta}) = \begin{bmatrix} x_A \\ y_A \end{bmatrix} + v\begin{bmatrix} \cos\theta \\ \sin\theta \end{bmatrix} u + \frac{1}{2}w\begin{bmatrix} \cos\theta \\ \sin\theta \end{bmatrix} u^2 + \frac{1}{6}z\begin{bmatrix} \cos\theta \\ \sin\theta \end{bmatrix} u^3$$
$$+\begin{bmatrix} \cos\theta & -\sin\theta \\ \sin\theta & \cos\theta \end{bmatrix}\begin{bmatrix} 35d_1 - 35v - \frac{15}{2}w - \frac{5}{6}z \\ 35d_2 \end{bmatrix} u^4 +$$
$$+\begin{bmatrix} \cos\theta & -\sin\theta \\ \sin\theta & \cos\theta \end{bmatrix}\begin{bmatrix} -84d_1 + 84v + 17w + \frac{3}{2}z \\ -84d_2 \end{bmatrix} u^5 + \tag{69}$$
$$+\begin{bmatrix} \cos\theta & -\sin\theta \\ \sin\theta & \cos\theta \end{bmatrix}\begin{bmatrix} 70d_1 - 70v - 14w - \frac{7}{6}z \\ 70d_2 \end{bmatrix} u^6 +$$
$$+\begin{bmatrix} \cos\theta & -\sin\theta \\ \sin\theta & \cos\theta \end{bmatrix}\begin{bmatrix} -20d_1 + 20v + 4w + \frac{1}{3}z \\ -20d_2 \end{bmatrix} u^7$$

Now, use (69) to evaluate $\mathbf{p}(u;\boldsymbol{\eta}) + \mathbf{p}(1-u;\boldsymbol{\eta})$. Some algebraic manipulations are required to obtain

$$\mathbf{p}(u;\boldsymbol{\eta}) + \mathbf{p}(1-u;\boldsymbol{\eta}) = 2\begin{bmatrix} x_A \\ y_A \end{bmatrix} + \begin{bmatrix} \cos\theta & -\sin\theta \\ \sin\theta & \cos\theta \end{bmatrix}\begin{bmatrix} d_1 \\ d_2 \end{bmatrix} = \mathbf{p}_A + \mathbf{p}_B$$

and conclude that, evidently, (68) holds $\forall u \in [0, 1]$, $\forall v \in \mathbb{R}_+$, $\forall w, z \in \mathbb{R}$. ■

The $\boldsymbol{\eta}^3$-spline can be used to generate or approximate a variety of path primitives, for example among many others, clothoids, spirals, circular arcs, etc. (see the next section). The most fundamental primitive, i.e. the line segment, can be obtained with appropriate interpolating conditions regardless of the shaping vector $\boldsymbol{\eta}$.

Property 4 (**Line segments generation**) *Define* $d = \|\mathbf{p}_B - \mathbf{p}_A\|$ *and assume* $x_B = x_A + d\cos\theta$, $y_B = y_A + d\sin\theta$, $\theta_A = \theta_B = \theta \in [0, 2\pi)$, $\kappa_A = \kappa_B = 0$, $\dot{\kappa}_A = \dot{\kappa}_B = 0$. *Then,* $\mathbf{p}(u;\boldsymbol{\eta})$ *is a line segment* $\forall \boldsymbol{\eta} \in \mathcal{H}$.

Proof – Take the assigned interpolating conditions and evaluate the $\boldsymbol{\eta}^3$-spline coefficients by means of (4)–(19). Few algebraic manipulations lead to the following expression for the parametric curve $\mathbf{p}(u;\boldsymbol{\eta})$

$$\mathbf{p}(u;\boldsymbol{\eta}) = \begin{bmatrix} x_A \\ y_A \end{bmatrix} + f(u;\boldsymbol{\eta})\begin{bmatrix} \cos\theta \\ \sin\theta \end{bmatrix}, \tag{70}$$

where $f(u;\boldsymbol{\eta})$ is the following scalar function

$$\begin{aligned} f(u;\boldsymbol{\eta}) &= \eta_1 u + \tfrac{1}{2}\eta_3 u^2 + \tfrac{1}{6}\eta_5 u^3 \\ &+ \left[35d - \left(20\eta_1 + 5\eta_3 + \tfrac{2}{3}\eta_5\right) - \left(15\eta_2 - \tfrac{5}{2}\eta_4 + \tfrac{1}{6}\eta_6\right)\right] u^4 \\ &+ \left[-84d + (45\eta_1 + 10\eta_3 + \eta_5) + \left(39\eta_2 - 7\eta_4 + \tfrac{1}{2}\eta_6\right)\right] u^5 \\ &+ \left[70d - \left(36\eta_1 + \tfrac{15}{2}\eta_3 + \tfrac{2}{3}\eta_5\right) - \left(34\eta_2 - \tfrac{13}{2}\eta_4 + \tfrac{1}{2}\eta_6\right)\right] u^6 \\ &+ \left[-20d + \left(10\eta_1 + 2\eta_3 + \tfrac{1}{6}\eta_5\right) + \left(10\eta_2 - 2\eta_4 + \tfrac{1}{6}\eta_6\right)\right] u^7. \end{aligned}$$

It is easy to verify that $f(0;\boldsymbol{\eta}) = 0$ and $f(1;\boldsymbol{\eta}) = d$. Thus, equation (70) proves that $\mathbf{p}(u;\boldsymbol{\eta})$ belongs to the segment line joining $\mathbf{p}_A$ with $\mathbf{p}_B$ $\forall u \in [0, 1]$ and $\forall \boldsymbol{\eta} \in \mathcal{H}$. ■

5. Path generation with η^3-splines

As shown in the previous section, the η^3-spline depends on a vector $\boldsymbol{\eta}$ of parameters that can be freely selected to shape the spline while preserving the interpolating conditions at the path endpoints. Specifically, parameters η_1, η_3, and η_5 influence the curve at its beginning whereas η_2, η_4, and η_6 affect the curve ending. Parameters η_1 and η_2 can be interpreted as "velocity" parameters. Parameters η_3, η_4 and η_5, η_6 are "twist" parameters that depend on the curve accelerations and curve jerks at the path endpoints respectively(cf. definitions given in (50)-(55)).
Some examples illustrate the path shaping by varying the η_i parameters. Consider the following interpolating conditions that lead to symmetric lane-change curves: $\mathbf{p}_A = [0\ 0]^T$, $\mathbf{p}_B = [4\ 3]^T$, $\theta_A = \theta_B = 0$, $\kappa_A = \kappa_B = 0$, $\dot{\kappa}_A = \dot{\kappa}_B = 0$. Fig. 3 shows the influence of the velocity parameters on the curve shape by plotting ten splines with $\eta_1 = \eta_2 = 1, 2, \ldots 10$ while maintaining $\eta_3 = \eta_4 = \eta_5 = \eta_6 = 0$ for all the curves. Curves of Fig. 4 are drawn by assuming $\eta_1 = \eta_2 = 5$, $\eta_5 = \eta_6 = 0$ and $\eta_3 = -\eta_4 = -50, -40, \ldots, 40, 50$. They depict the effect of varying the twist acceleration parameters η_3 and η_4. Fig. 5 considers the case $\eta_1 = \eta_2 = 5$, $\eta_3 = -\eta_4 = 10$ and plots curves with twist jerk parameters assuming the values $\eta_5 = \eta_6 = -1000, -900, \ldots, 900, 1000$.
The proposed examples make evident that, acting on the shaping parameter vector $\boldsymbol{\eta}$, a wide variety of curves satisfying the boundary conditions can be obtained. This suggests choosing $\boldsymbol{\eta}$ to generate optimal curves. Different optimality criteria may be chosen depending on the desired WMR motion smoothness. For instance, consider a robot motion with constant linear velocity. According to the inversion-based control proposed in (Guarino Lo Bianco et al., 2004b), the path arc length and the angular velocity are given by $s = vt$ and $\omega(t) = v\kappa(vt)$ with $t \in [0, t_f]$. Hence, to minimize the maximum absolute value of the angular velocity we are requested to minimize the maximum absolute value of the path curvature, i.e. to solve the problem $\min_{\boldsymbol{\eta} \in \mathcal{H}} \max_{s \in [0, s_f]} |\kappa(s; \boldsymbol{\eta})|$. Alternatively, to avoid rough movements of the WMR, we

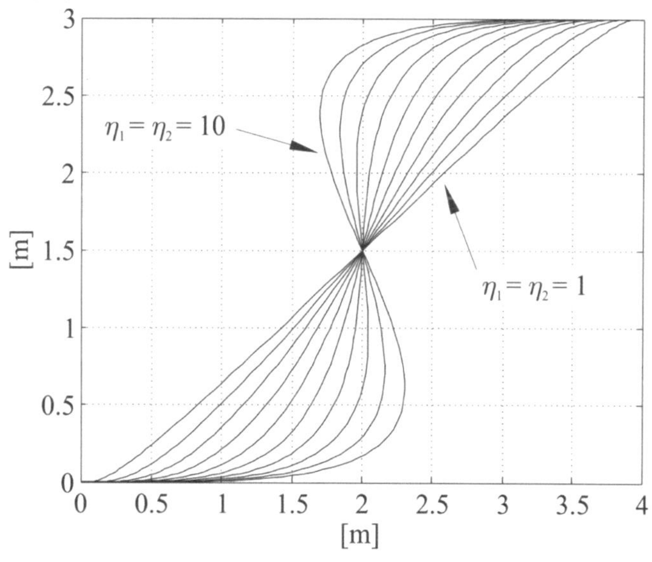

Fig. 3. Lane-change curves with $\eta_3 = \eta_4 = \eta_5 = \eta_6 = 0$.

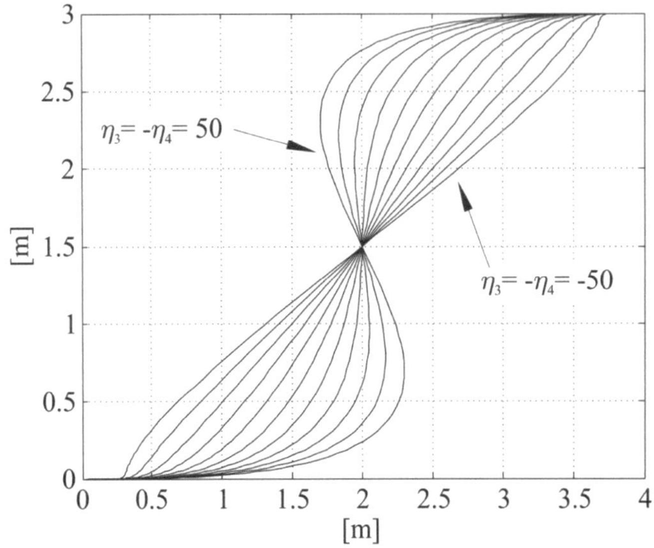

Fig. 4. Lane-change curves with $\eta_1 = \eta_2 = 5$ and $\eta_5 = \eta_6 = 0$.

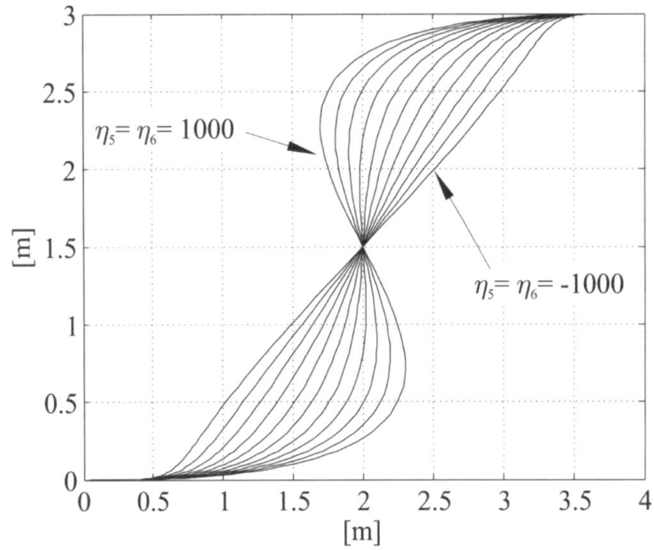

Fig. 5. Lane-change curves with $\eta_1 = \eta_2 = 5$ and $\eta_3 = -\eta_4 = 10$.

may desire to minimize the maximum absolute value of the angular acceleration along the planned path that leads to the problem

$$\min_{\boldsymbol{\eta}\in\mathcal{H}} \max_{s\in[0,s_f]} \left|\dot{\kappa}(s;\boldsymbol{\eta})\right| . \tag{71}$$

In many planning cases, an acceptable sub-optimal solution to (71) can be obtained by a rough heuristic rule: $\eta_1 = \eta_2 = \|\mathbf{p}_A - \mathbf{p}_B\|$ and $\eta_3 = \eta_4 = \eta_5 = \eta_6 = 0$. This rule can be viewed as the straightforward extension of an analogous rule proposed in (Piazzi et al., 2002) for the path planning of quintic η^2-splines (see also (Guarino Lo Bianco & Piazzi, 2000) where computational results of optimal path planning were reported). In Fig. 6 this rule has been applied to a lane-change curve with variations on the curvature derivative at the initial path point. The chosen interpolating conditions are: $\mathbf{p}_A = [0\ 0]^T$, $\mathbf{p}_A = [4\ 3.5]^T$, $\theta_A = \theta_B = \pi/2$, $\kappa_a = \kappa_B = 0$, and $\dot{\kappa}_B = 0$, while the curvature derivative in $\mathbf{p}_A$ takes the values $\dot{\kappa}_A = -5, 0, 5$. According to the heuristic rule, we have chosen $\boldsymbol{\eta} = [5.3151\ 5.3151\ 0\ 0\ 0\ 0]^T$. The curves of Fig. 6 show that variations of the boundary curvature derivative may have a neat impact on the shape of the η^3-spline.

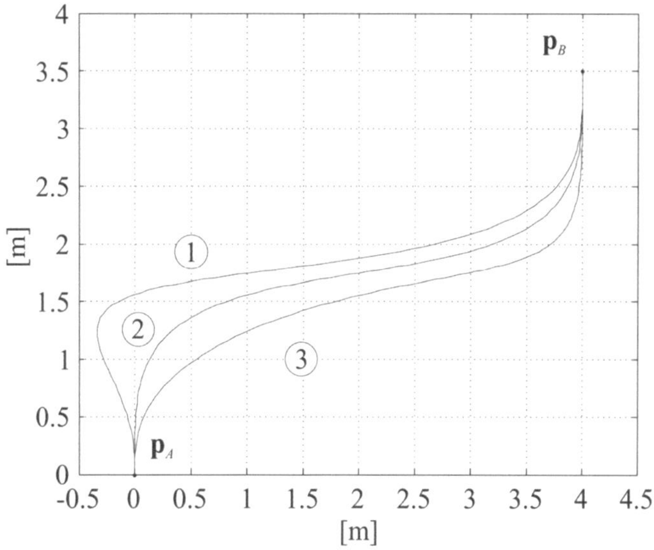

Fig. 6. Modifying a lane-change curve by perturbing $\dot{\kappa}_A$.

When appropriate endpoint interpolating conditions are chosen, the η^3-spline can approximate a variety of primitive curves. Fig. 7 shows the plot of an η^3-spline that is a good approximation of a clothoid (curve with label 2): it has $\dot{\kappa}(s) \simeq 0.04\ \forall\ s \in [0, s_f]$. The interpolating conditions are $\mathbf{p}_A = [0\ 0]^T$, $\mathbf{p}_B = [2.0666\ 1.0568]^T$, $\theta_A = 0$, $\theta_B = 1.4450$, $\kappa_A = 0$, $\kappa_B = 1.1333$, $\dot{\kappa}_A = \dot{\kappa}_B = 0.04$ and the shaping parameters have been fixed as $\eta_1 = \eta_2 = 2.37$ and $\eta_3 = \eta_4 = \eta_5 = \eta_6 = 0$. The other two curves plotted in Fig. 7 departs from the clothoid by modifying the curvature derivative in $\mathbf{p}_B$, with variation ±0.4, so that to obtain $\dot{\kappa}_B = -0.36$ (curve 1) and $\dot{\kappa}_B = 0.44$ (curve 3).

The last example depicted in Fig. 8 reports a composite G^3-path completely generated with η^3-splines. It is made of five curves: a lane-change curve, a line segment, a cubic spiral (i.e. a curve whose tangent direction is a cubic function of the arc length, cf. (Kanayama & Hartman, 1989)), a generic twirl arc, and a circular arc. The interpolating and shaping parameters are reported in Table 1.

6. A note on η^k-splines

The concept of geometric continuity of planar curves and paths, introduced in Section 2, can be generalized as follows (D^k denotes the k-th order derivative operator).

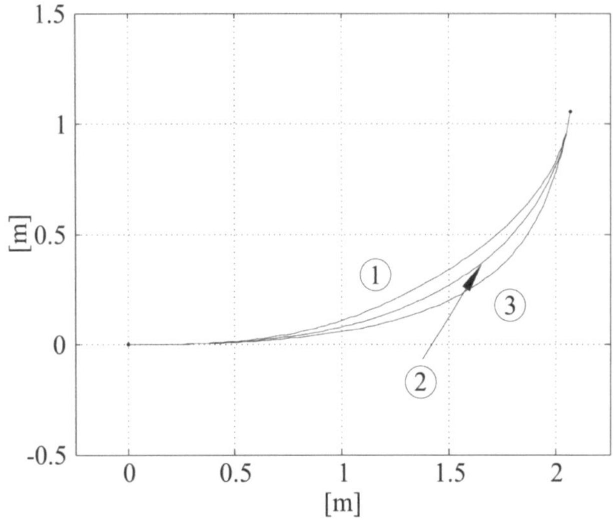

Fig. 7. Modifying a clothoid by perturbing $\dot{\kappa}_B$.

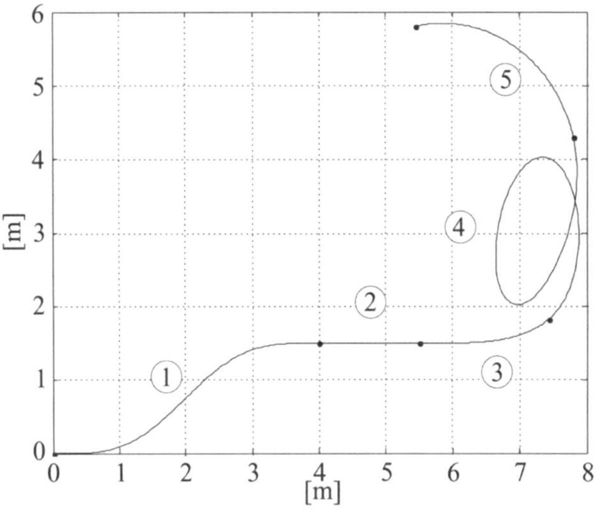

Fig. 8. A composite G^3-path made by $\boldsymbol{\eta}^3$-splines.

curve	$\mathbf{p}_A$	$\mathbf{p}_B$	θ_A	θ_B	κ_A	κ_B	$\dot{\kappa}_A$	$\dot{\kappa}_B$	η_1	η_2	η_3	η_4	η_5	η_6
lane-change curve	0 0	4 1.5	0	0	0	0	0	0	4.27	4.27	0	0	0	0
line segment	4 1.5	5.5 1.5	0	0	0	0	0	0	0	0	0	0	0	0
cubic spiral	5.5 1.5	7.4377 1.8235	0	0.6667	0	1	0	1	1.88	1.88	0	0	0	0
generic twirl arc	7.4377 1.8235	7.8 4.3	0.6667	1.8	1	0.5	1	0	7	10	10	-10	4	4
circular arc	7.8 4.3	5.4581 5.8064	1.8	3.3416	0.5	0.5	0	0	2.98	2.98	0	0	0	0

Table 1. The interpolating parameters and the η_i coefficients used to generate the composite G^3-path of Fig. 8.

Definition 3 (G^k-curves; $k \geq 2$) *A parametric curve* $\mathbf{p}(u)$ *has k-th order geometric continuity and we say* $\mathbf{p}(u)$ *is a* G^k*-curve if* $\mathbf{p}(u)$ *is* G^{k-1}*-curve,* $D^k\mathbf{p}(\cdot) \in C^p([u_0, u_1])$*, and* $D^{k-2}\kappa(\cdot) \in C^0([0, s_f])$.

Definition 4 (G^k-paths; $k \geq 2$) *A set of points of a Cartesian plane is a* G^k*-path if there exists a parametric* G^k*-curve whose image is the given path.*

Roughly speaking, the k-th order geometric continuity of curves amounts to the continuity of the curvature function up to the $(k - 2)$-nd derivative. In a more general setting, geometric continuity is treated in (Peters, 2002). Now we can naturally state the polynomial G^k-interpolating problem as the generalization of the G^3-problem of Section 4.

The polynomial G^k-interpolating problem: *Determine the minimal order polynomial curve which interpolates two given endpoints* $\mathbf{p}_A = [x_A\ y_A]^T$ *and* $\mathbf{p}_B = [x_B\ y_B]^T$ *with associated unit tangent vectors defined by angles* θ_A *and* θ_B*, and curvature derivatives* $D^i\kappa_A$ *and* $D^i\kappa_B$ *for* $i = 0, 1, \ldots, k - 2$. *All the endpoint interpolating data can be arbitrarily assigned.*

Then, following the approach proposed in Section 4, we could derive the η^k-spline as the solution of the above problem characterized by minimality, completeness, and symmetry. For $k = 2$ this has been done in (Piazzi & Guarino Lo Bianco, 2000) and the deduced η^2 splines are quintic polynomial curves that depend on a four-dimensional η vector ($\eta \in \mathbb{R}_+^2 \times \mathbb{R}^2$ is the vector of the shaping parameters). The η^2-splines have been proposed for autonomous guidance of cars (Piazzi et al., 2002) and of wheeled omnidirectional robots (Guarino Lo Bianco et al., 2004a). The remaining elementary cases k = 1 and k = 0 leading to the η_1-spline and the η_0-spline are reported in the Appendix.

On the grounds of the already found η^k-splines (k = 0, 1, 2, 3) we infer that the general η^k-spline is a polynomial curve with order equal to $2k + 1$ and whose parameterization depends on a shaping vector η with $2k$ components ($\eta \in \mathbb{R}_+^2 \times \mathbb{R}^{2(k-1)}$). Closed-form expressions of the η^k-spline could be generated by suitably devised computer algebra procedures.

Formally define η^k**-Paths** as the set of all the paths given by the η^k-splines for all $\eta \in \mathbb{R}_+^2 \times \mathbb{R}^{2(k-1)}$ and all the endpoint interpolation data. Then, the following property holds.

Property 5 η^k**-Paths** $\subset \eta^{k+1}$**-Paths** *for all* $k \in \mathbb{N}$.

This property helps to explain why the η^3-splines are quite good in approximating standard curve primitives. Indeed, it was already shown in (Guarino Lo Bianco & Piazzi, 2000) that η^2-splines can very well approximate circular arcs or clothoids so that η^3-splines can only better the approximations with further curve inclusions such as, for example, cubic spirals (see previous Section 5). Following (Guarino Lo Bianco et al., 2004b) we can foresee the use of η^4-splines to achieve for WMRs the generation of velocity commands with continuous jerk signals (signals with continuous acceleration derivatives).

7. Conclusions

This chapter has presented the η^3-spline, a seventh order polynomial curve that interpolates between two Cartesian points with arbitrary assigned tangent vectors, curvatures and curvature derivatives. This curve primitive, given with explicit closed-form expressions, depends on a shaping parameter vector that can be freely chosen to shape or optimize the path. An advantage of the new spline over other curve primitives, such as clothoids or polynomial spirals, is the avoidance of any numerical integration/procedure to evaluate the curve coordinates. Properties of the η^3-spline such as completeness, minimality, and symmetry have been also reported. Investigations on optimal η^3-splines have been reported in (Guarino Lo Bianco & Gerelli, 2007; Gerelli, 2009). How to achieve high-performance motion control of WMRs with the η^3-spline in conjunction with obstacle avoidance capabilities has been addressed in (Villagra & Mounier, 2005; Chang & Liu, 2009).

8. Appendix

Case k = 1 (solution to the polynomial G^1-interpolating problem). The η^1-spline is a third-order polynomial curve $\mathbf{p}(u;\,\eta)$ $u \in [0, 1]$, $\eta = [\eta_1 \eta_2]^T \in \mathbb{R}^2_+$ with coefficients defined as follows.

$$
\begin{aligned}
\alpha_0 &= x_A \\
\alpha_1 &= \eta_1 \cos\theta_A \\
\alpha_2 &= 3\,(x_B - x_A) - 2\eta_1 \cos\theta_A - \eta_2 \cos\theta_B \\
\alpha_3 &= -2\,(x_B - x_A) + \eta_1 \cos\theta_A + \eta_2 \cos\theta_B \\
\beta_0 &= y_A \\
\beta_1 &= \eta_1 \sin\theta_A \\
\beta_2 &= 3\,(y_B - y_A) - 2\eta_1 \sin\theta_A - \eta_2 \sin\theta_B \\
\beta_3 &= -2\,(y_B - y_A) + \eta_1 \sin\theta_A + \eta_2 \sin\theta_B
\end{aligned}
$$

Case k = 0 (solution to the polynomial G^0-interpolating problem). For completeness we also give the η^0-spline which is simply expressed by the first-order curve $\mathbf{p}(u) = \mathbf{p}_A + \mathbf{p}_B \cdot u$, $u \in [0, 1]$. There are no shaping parameters as the curve is just the line segment connecting $\mathbf{p}_A$ with $\mathbf{p}_B$. Note that zero-order geometric continuity coincides with the standard notion of function continuity (C^0-continuity).

9. References

Barsky, B. A. & Beatty, J. C. (1983). Local control of bias and tension in beta-spline. *Computer Graphics*, Vol. 17, No. 3, pp. 193–218.

Boissonnat, J.-D., Cérézo, A. & Leblond, J. (1994). A note on shortest paths in the plane subject to a constraint on the derivative of the curvature. Tech. Rep. 2160, INRIA, Rocquencourt, France.

Chang, H.-C. & Liu, J.-S. (2009). High-quality path planning for autonomous mobile robots with η^3-splines and parallel genetic algorithms. *Robotics and Biomimetics, 2008. ROBIO 2008. IEEE International Conference on*, pp. 1671–1677. doi: 10.1109/ROBIO.2009.4913252.

Choset, H., Lynch, K., Hutchinson, S., Kantor, G., Burgard, W., Kavraki, L. & Thrun, S. (2005). *Principles of Robot Motion: Theory, Algorithms, and Implementations*. The MIT Press, Cambridge, MA.

Delingette, H., Hébert, M.&Ikeuchi, K. (1991). Trajectory generation with curvature constraint based on energy minimization. *Proc. of the IEEE-RSJ Int. Conf. on Intelligent Robots and Systems*, pp. 206–211. Osaka, Japan.

Fraichard, T. & Scheuer, A. (2004). From Reeds and Shepp's to continuous-curvature paths. *IEEE Trans. on Robotics*, Vol. 20, No. 6, pp. 1025–1035.

Gerelli, O. (2009). *Optimal Constrained Planning for Complex Mechatronic Systems*. Phd thesis, Universitá di Parma, Dipartimento di Ingegneria dell'Informazione, Parma, Italy.

Guarino Lo Bianco, C. & Gerelli, O. (2007). Optimal path generation for wheeled mobile robots with η^3-splines. *Proceedings of the 13th IEEE IFAC Int. Conf. on Methods and Models in Automation and Robotics*, pp. 1049–1054. Szczecin, Poland.

Guarino Lo Bianco, C.&Piazzi, A. (2000). Optimal trajectory planning with quintic G^2-splines. *Procs of the IEEE Intelligent Vehicles Symposium, IV2000*, pp. 620–625. Dearborn (MI), USA.

Guarino Lo Bianco, C., Piazzi, A. & Romano, M. (2004a). Smooth control of a wheeled omnidirectional robot. *Proc. of the IFAC Symp. on Intelligent Autonomous Vehicles, IAV2004*. Lisbon, Portugal.

— — — (2004b). Smooth motion generation for unicycle mobile robots via dynamic path inversion. *IEEE Trans. on Robotics*, Vol. 20, No. 5, pp. 884–891.

Hsiung, C.-C. (1997). *A First Course in Differential Geometry*. International Press, Cambrige,MA.

Kanayama, Y. & Hartman, B. (1989). Smooth local path planning for autonomous vehicles. *Proc. of the IEEE Int. Conf. on Robotics and Automation*, Vol. 3, pp. 1265–1270. Scottsdale, AZ (US).

Kant, K. & Zucker, S. (1986). Toward efficient trajectory planning: the path-velocity decomposition. *Int. J. of Robotics Research*, Vol. 5, No. 3, pp. 72–89.

Kito, T., Ota, J., Katsuki, R., Mizuta, T., Arai, T., Ueyama, T. & Nishiyama, T. (2003). Smooth path planning by using visibility graph-like method. *Proc. of the IEEE Inter. Conf. on Robotics & Automation*, pp. 3770–3775. Taipei, Taiwan.

Labakhua, L., Nunes, U., Rodrigues, R. & Leite, F. (2006). Smooth trajectory planning for fully automated passengers vehicles: Spline and clothoid based methods and its simulation. J. P. J. Andrade-Cetto, J.-L. Ferrier&J. Filipe, (Eds.) *Informatics in Control Automation and Robotics*. Springer, Berlin.

Lucibello, P. & Oriolo, G. (1996). Stabilization via iterative state steering with application to chained-form systems. *Proc. of the 35th IEEE Conf. on Decision and Control*, Vol. 3, pp. 2614–2619. Kobe, Japan.

Morin, P. & Samson, C. (2008). Motion control of wheleed mobile robots. B. Sicilano & O. Khatib, (Eds.) *Springer Handbook of Robotics*, pp. 799–826. Springer, Berlin.

Nelson, W. (1989). Continuous steering-function control of robot carts. *IEEE Trans. on Ind. Electronics*, Vol. 36, No. 3, pp. 330–337.

Peters, J. (2002). Geometric continuity. G. Farin, J. Hoschek & M.-S. Kim, (Eds.) *Handbook of Computer Aided Geometric Design*, pp. 193–229. North-Holland.

Piazzi, A. & Guarino Lo Bianco, C. (2000). Quintic G^2-splines for trajectory planning of autonomous vehicles. *Procs of the IEEE Intelligent Vehicles Symposium*, pp. 198–203. Dearborn (MI), USA.

Piazzi, A., Guarino Lo Bianco, C., Bertozzi, M., Fascioli, A. & Broggi, A. (2002). Quintic G^2-splines for the iterative steering of vision-based autonomous vehicles. *IEEE Trans. On Intelligent Transportations Systems*, Vol. 3, No. 1, pp. 27–36.

Piazzi, A., Guarino Lo Bianco, C. & Romano, M. (2007). η^3-splines for the smooth path generation of wheeled mobile robots. *Robotics, IEEE Transactions on*, Vol. 23, No. 5, pp. 1089–1095. ISSN 1552-3098. doi:10.1109/TRO.2007.903816.

Reuter, J. (1998). Mobile robots trajectories with continuously differentiable curvature: an optimal control approach. *Proc. of the 1998 IEEE/RSJ Int. Conf. on Intelligent Robots and Systems*, Vol. 1, pp. 38–43. Victoria, B.C., Canada.

Scheuer, A. & Laugier, C. (1998). Planning sub-optimal and continuous-curvature paths for car-like robots. *Proceedings of the 1998 IEEE/RSJ International Conference on Intelligent Robots and Systems*, pp. 25–31. Victoria, B.C., Canada.

Suzuki, Y., Thompson, S.&Kagami, S. (2009). Smooth path planning with pedestrian avoidance for wheeled robots: Implementation and evaluation. *Autonomous Robots and Agents, 2009. ICARA 2009. 4th International Conference on*, pp. 657–662. doi:10.1109/ICARA.2000.4803910.

Villagra, J. & Mounier, H. (2005). Obstacle-avoiding path planning for high velocity wheeled mobile robots. *Proceedings of the IFAC World Congress*. Prague, Czech Republic.

5

Motion Synthesis and Coordinated Control in the Multi-Axle-Driving-Vehicle

Yunhua Li and Liman Yang
Beijing University of Aeronautics and Astronautics
China

1. Introduction

Multi-axle-Driving-vehicle is a kind of the mobile robot and it also belongs to the construction machinery which generally has heavy load-capacity and huge size. As shown in Fig.1 and Fig.2, both hoisting-girder transporter and DCY900 transportation vehicle are multi-axle-driving vehicles. In order to decrease the ground-contacting pressure, the ground-clearance of the vehicle chassis and the size of the tyre, and to increase the loading-capacity and the passing performance of the vehicle, the multi-axle driving and independently controlled steer axles are employed, and also distributed electro-hydraulic proportional control is applied to the heavy type of vehicles. Such a large-scale vehicle has to deal with the complex motion control problem. The controlled output motions of multiple axles should meet certain matching condition or corresponding relationship so as to make the whole vehicle to realize the expected contouring motion trace. For example, all the powered steer axles have to be coordinately controlled in real time in order to achieve smooth and accurate steering motion without slipping and sliding. Besides the steering function, the steer axles are also designed to automatically level the vehicle body when it moves in an uneven terrain. It follows that the motion synthesis and coordinated control methods should concurrently cope with the tasks and motions of multiple subsystems.

Conventionally, coordinated control of a simple mechatronics system is realized through a centralized control scheme in which each of the actuators is directly linked to the controller through cable in a point-to-point manner. However, for a complex multi-tasking mechatronic system with a large number of subsystems and actuators, such a control scheme is impractical. This is especially true for a large-scale multi-axle vehicle because it is huge in size and has many distributed subsystems to be arranged anywhere in the vehicle. If a centralized control scheme is employed, it will result in a very messy wiring scheme.

Thanks for the advanced network technologies, which provide us an effective way to realize coordinated control for the multi-axle driving vehicles. In a network environment, all the control devices such as sensors, actuators, and controllers are distributed and simply linked together through network interfaces (e.g., Field-bus, Industrial Ethernet, and mobile net) so as to achieve coordination and resources sharing efficiently. In convention, a network-based mechatronic control system is called an NCS (Networked Control System), which has many advantages over a centralized control system, e.g., low installation cost, ease of system maintenance, simplicity in failure diagnosis, and high flexibility in system management

(Lian et al., 2002). Therefore, the NCS is an ideal solution for the motion synthesis and coordinated motion control of large-scale and complex mechatronic systems.

The conventional approach for motion synthesis and coordinated motion control employs the actuator-level tracking error as the major performance index. A feedback and feedforward controller is then individually designed for each axis to achieve its planned motion profile. Such a control strategy is not appropriate for a complex mechatronics system to accomplish multitasks with distributed and coordinated operations. Apparently, it will be more effective to evaluate the contour-tracking accuracy, i.e., the difference between the actual and targeted motion trajectories in the system level. Besides, an effective feedback and feedforward controller combined with a cross-coupled control law can be developed to significantly improve the contour control accuracy. There are a number of representative works in the related areas. A multi-axis task-coordination approach (Tomizuka & Niu, 2001) is presented to form the first loop of the feedback and feedforward control, in which an accurate plant model is needed. A new variable gain cross-coupled control method based on system-level tracking errors is proposed (Yeh & Hsu, 2003). A kind of task-space nonlinear sliding mode observer is introduced to control a synchronized double-cylinder system. Through theoretical analysis and Lab-based experimental study, the effectiveness of the system-level contour control strategy has been demonstrated (Sun& George, 2002). A multi-axis motion synchronization strategy is developed in which the asymptotic convergence of both tracking and synchronization errors are achieved (Liu, 2005). In order to improve contouring performance of the retrofitted milling machine, a self-tuning adaptive control strategy combined with cross-coupled control of axial motion is designed (Yan & Lee, 2005). For large-scale multi-axle vehicles, NCS-oriented motion synthesis framework and crossed-couple control algorithm are investigated (Li et al., 2007) and the practical engineering applications on Hoisting-girder transporter are explored (Yang et al., 2009).

In this chapter, addressing to the motion synthesis and coordinated control of multi-axle driving vehicles, we shall discuss the basic background knowledge, the operation principle, the kinematical models and coordinated control methodology to be concerned in the traveling and steering systems of the multi-axle driving vehicles. Firstly, the NCS fundamental knowledge and common motion synthesis modes of vehicle steering are outlined, and a kind of networked-based travelling and steering system is proposed for multi-axle construction machinery. Then, the kinematical models of two-axle vehicle and multi-axle vehicle are respectively established. Furthermore, for multi-axle driving vehicle, the travelling and steering hydraulic system design are provided, and the multi-axle coordinated control strategy are developed. Finally, the experimental investigations on the DCY transportation vehicle and track-laying machine for high speed railway are explored.

Fig. 1. Hoisting-girder transporter with 900T load

Fig. 2. DCY900 powered transportation vehicle

2. Fundamental knowledge

2.1 Networked control system

A typical network control system (NCS) is shown in Fig.3. It is a spatially distributed system in which the communication between sensors, actuators, and controllers occurs through a shared band-limited digital communication network (Hespanha et al., 2007). However, in broad sense, NCSs also include many types even covering traditional DCS and remote networked control systems based on internet.

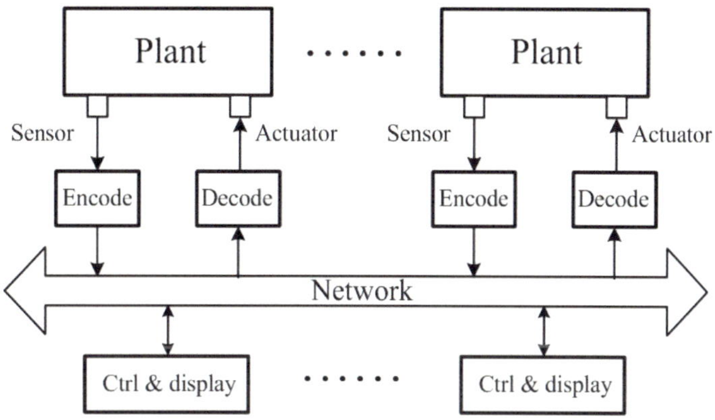

Fig. 3. General NCS architecture

In view of physical realization, the NCS can be classified into different types such as serial-bus configuration, Field-bus configuration, mobile network, and industry Ethernet etc. According to the control node types, the NCS can also be classified into three basic styles: the sensor/actuator node style, the coupling node style, and the controller node style. In the former two styles, the control closed-loops are built by network communication, in which the sensing and controlling data are transmitted by network. While last style is similar to DCS (Distributed Control System) which almost real control tasks are executed in intelligent nodes and only some commands and warning signals are transmitted on network. The mathematical descriptions of the three kinds of NCS are given as follows.

a. Sensor/actuator node style NCS
Considering the ith actuator node, the dynamic equation and control law are respectively as follows:

$$\begin{cases} \dot{\boldsymbol{x}}_i = \boldsymbol{f}_i(\boldsymbol{x}_i, u_i),\ y_i = h_i(\boldsymbol{x}_i) \\ u_i = u_i(r_i, y_i) \end{cases} \tag{1}$$

where $\boldsymbol{x}_i, \boldsymbol{f}_i \in \mathbb{R}^{n_i}$, $\boldsymbol{x}_i$ is the state vector (i=1,2,...,m), m denotes the node number, and n_i denotes the dimension number of the state vector of the plant to be controlled by the ith node. The above equation set consists of the state equations for the actuators and the controlled plants, and the output equations (at the actuator nodes) as well as the control algorithm for master control node. The outputs and control signals of the each node are transmitted through network. Obviously, it can be also view as a kind of generalized centralized-control system connected through Field-bus.

b. Coupling node style NCS
For this case, there are n_i plants to be controlled by the ith node. The dynamic equation and the control law are respectively as follows:

$$\begin{cases} \dot{\boldsymbol{x}}_{il} = \boldsymbol{f}_{il}(\boldsymbol{x}_{il}, u_{il}),\ y_{il} = h_{il}(\boldsymbol{x}_{il}) \\ u_{il} = u_{il}(r_{il}, y_{il}) \end{cases} \tag{2}$$

where $\boldsymbol{x}_{il}, \boldsymbol{f}_{il} \in \mathbb{R}^{n_{il}}$ $(i = 1,\cdots,m;\ l = 1,\cdots,n_i\ )$, $\boldsymbol{x}_i$ is the state vector, m denotes coupling node number, n_i denotes the number of the plants controlled by the ith node, and n_{il} denotes the dimension number of state vector of the lth plant controlled by the ith coupling node. Equation (2) is composed of the state equations and the output equations of plants controlled by the coupling node as well as the control algorithm of the master control node. The outputs and control signals of the nodes are transmitted through network. Style 2 is degenerated into style 1 when l=1.

c. Controller node style NCS

$$\begin{cases} \dot{\boldsymbol{x}}_{il} = \boldsymbol{f}_{il}(\boldsymbol{x}_{il}, u_{il}), y_{il} = h_{il}(\boldsymbol{x}_{il}) \\ u_{il} = u_{il}(r_{il}, y_{il}), r_{il} = r_{il}(t_k) \end{cases} \tag{3}$$

where $\boldsymbol{x}_{il}, \boldsymbol{f}_{il} \in \mathbb{R}^{n_{il}}$ $(i = 1,\cdots,m;\ \ l = 1,\cdots,n_i\ )$, $\boldsymbol{x}_i$ is the state vector, m represents the controller node number, n_i represents the output number of the plants controlled by the ith controller node, and n_{il} is the dimension number of the state vector of the lth plant controlled by the ith node. Equation (3) is composed of the state equations and the output equations of the plants controlled by the controller nodes, the control law determined by the ith node, and the reference control signals produced by motion planning. The outputs and control signals of the nodes are transmitted through network.

In general, an NCS may contain the three basic styles mentioned above or their hybrid styles. For the third style of NCS, its logic and function diagram is shown in Fig.4, which describes the system logic and function arrangement, the relationship of transmitted signals, and the control loops.

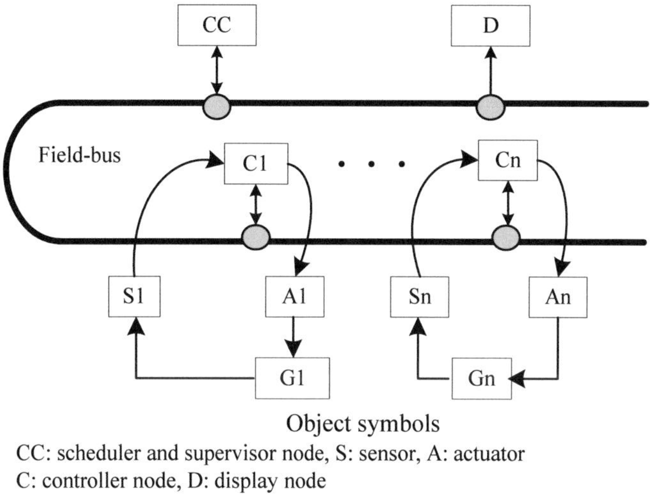

Fig. 4. Block diagram of an NCS

In the third style of NCS, a mechatronic system that consists of multiple distributed subsystems is equivalent to a MIMO system with transmitting delay. The output-motion synthesis depends on a set of tasks performed on the nodes. Each node can control one or several plants with a feedback or feedforward controller. Information exchange among the nodes through field-bus makes all plant outputs be controlled for system-level contour tracking so that motion synthesis and coordinate motion control can be realized.

2.2 Motion synthesis modes of steering system

The conventional motion synthesis modes include mechanical (typically like linkage, gear, and cable), pneumatic, hydraulic and electrical transmissions. However, they are unsuitable for large-scale multi-axle vehicles in which many spatially distributed physical components are needed and the complicated operation functions are required usually. For instance, the mechanical mode is very difficult to realize accurate motion synthesis and multiple manipulation modes. The electrical scheme has to face the problems like as complex wiring, difficult maintenance, high fault ratio and hard expansion. From the preceding introduction of NCS, we can see that the distributed and networked structure of NCS is helpful for information share and integration as well as intelligent decision-making. As result, it provides an ideal framework for the motion synthesis and coordinated motion control of large-scale and distributed construction machinery (Li & Yang, 2005).In this section, the conventional ways of mechanical and full hydraulic motion synthesis are described with example of the construction vehicle's steering control and a new based-networked synthesis scheme is developed.

a. Mechanical steering

The earliest steering scheme is Ackerman's steering trapezium, it is shown in Fig.5. The motion synthesis is undertaken by the linkage mechanism and the wheel system. It has the advantages of exact transmission, reliability, easy fabrication, simple operation and high transmission efficiency. But, it can't usually realize the stepless speed regulation and the transmitting of the power for long distance, and also its structure is also complicated relatively. The collocation of the transmission mechanism is very difficult and the motions among mechanisms are not easy to control and integrate, so that it doesn't realize the flexible multi-mode steering. It also makes against decreasing the gap to ground and improving the passing and smoothing ability. Due to the above disadvantages, this mechanical transmitting mode only works in the special condition, thus it can't fit the agile manipulating demands of modern construction machinery.

The electrical, pneumatic, or hydraulic steering scheme can solve the problem of the force-assistant, which makes it possible to the steer the heavy vehicle. Among of them, because of high power-density and rapid response, the hydraulic power steering is widely used in the construction machinery.

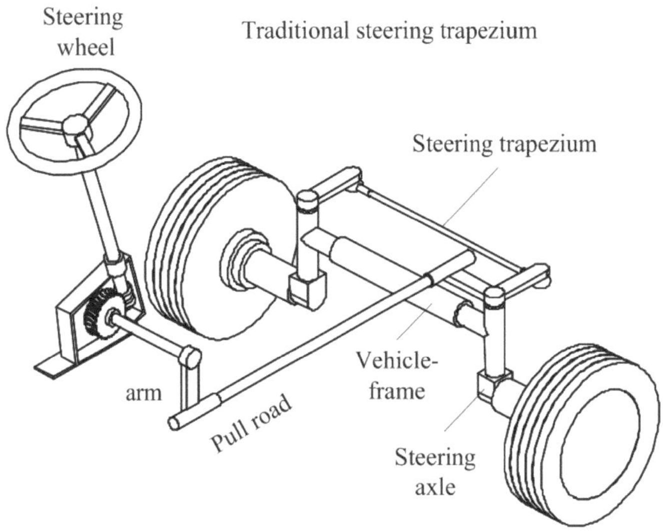

Fig. 5. Ackerman's steering trapezium

b. Full-hydraulic steering

The most common type of hydraulic steering system is full-hydraulic steering system. It is a closed loop control system by using the meter motor to realize the hydraulic-internal-feedback. It can simplify the structure of the steering system and decrease the manipulating force of the steering system, which is a good choice for the vehicles with two axles. Actually, it is still belong to Akerman's mechanical linkage steering, and the only difference is its hydraulic assistant force function. Obviously, it can't also realize the steering of the vehicles with more than two axles. Moreover, another defect of it is low efficiency. But at present, the load-sensitive system has been adopted, which is composed of electro-hydraulic proportional pumps and multi-path electro-hydraulic proportional valves (Kemmetmüller, 2007). This technique can improve the efficiency of construction machine to some extent.

c. Based-networked electro-hydraulic steering

In view of the advantages of hydraulic transmission on power transmission as well as the opportunities of network technique on information share and integration, a distributed and numerical manipulating control scheme based on field-bus network and electro-hydraulic proportional control is proposed for motion synthesis and coordinated control of construction machinery with multi-axle driving vehicles.

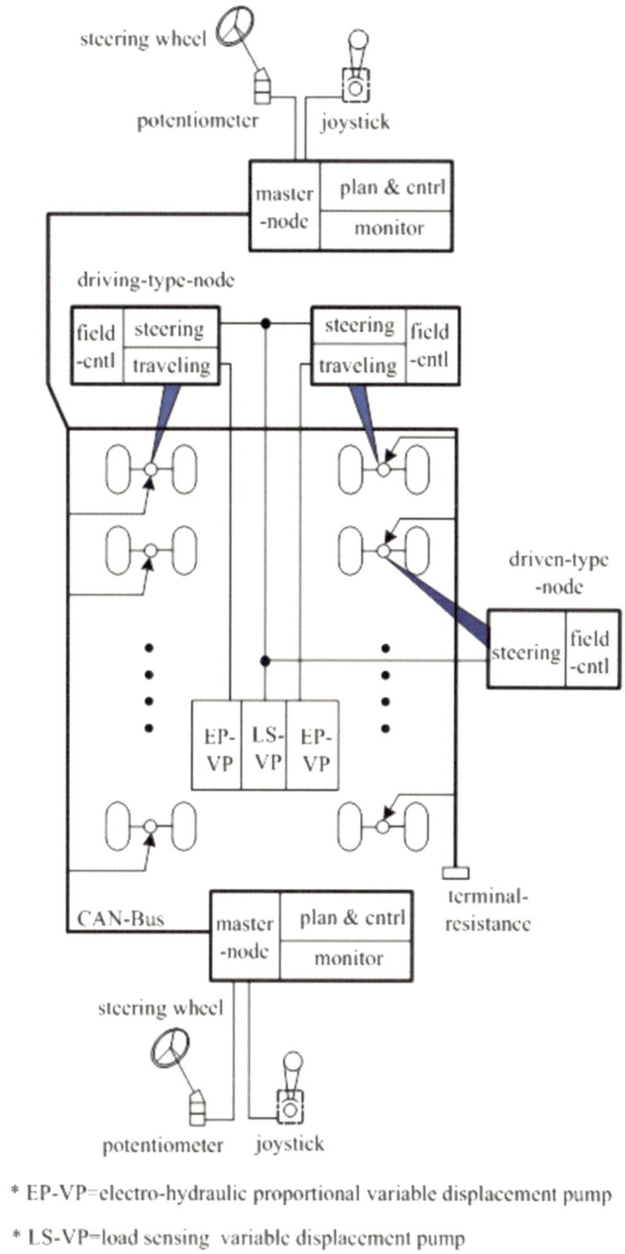

Fig. 6. NCS-based electro-hydraulic steering system

Without loss of generality, this method is applied using the DCY series of transportation vehicle (Li et al., 2007) as the example. The control principle of steering system is shown in Fig. 6. The independent steering mechanism is adopted, i.e. a single axle is driven by a valve-controlled hydraulic cylinder. Each wheel axle can be controlled by intelligent node on the CAN-Open network to turn any angle. In Fig.6, the types of the nodes contain the master controller located in the cab to receive all kinds of operation commands, and the field nodes such as driving-type-node and driven-type-node to be placed at the two sides of the vehicle body. Each of them controls several groups of steering, driving and suspending mechanisms. In virtue of software trapezoid (kinematical resolution), this scheme can achieve multiple steering modes such as diagonal steering, longitudinal steering, front (rear) axle steering, and center steering, etc..

The kinematical models of individual steering mode are established in advance and memorized in the master controller. During the running, the expected turning angle of every wheel is resolved from steering wheel in master node according to kinematical model and transmitted to local controller node by bus data exchange. Thus, motion synthesis is implemented through multiple closed-loop controls of steering mechanisms in the same time. In principle, as long as each individual wheel can turn to its expected angle precisely, the whole vehicle can realize the pure rolling steering, in which all axles turn around the rotation center without slipping and sliding.

3. Kinematics analysis of two-axle driving vehicle

For convenient comprehension, we firstly analyze the two-axle driving vehicle. As shown in Fig.7, the vehicle has two driving wheels and a driven wheel which can turn any angle. The differential speed steering is employed while traveling. In Fig.7, *OXY* denotes global coordination and P*xy* denotes mobile coordination built on the vehicle reference point *P*. Define the state vector of (X,Y,θ), where (X,Y) is position coordinate of point *P* in *OXY* and θ is the driving orientation angle, i.e. the included angle between *x*-axis of *Pxy* and *X*-axis of *OXY*. The axle space of two driving wheel is 2*B* and axle-space length between driving wheels and driven wheel is *W*. Suppose the left and right driving wheels' linear speeds are given as v_l and v_r, thus the resultant speed along the *Px*-axis can be get

$$v_x = \frac{1}{2}(v_l + v_r) \tag{4}$$

The turning speed of vehicle is given as

$$\dot{\theta} = \frac{1}{2B}(v_r - v_l) \tag{5}$$

Suppose O' is turning center and the turning radius $R = \overline{O'P}$, R can be accumulated by differential speed steering relation $v_l / v_r = (R - B)/(R + B)$, thus

$$R = \frac{v_l + v_r}{v_l - v_r} B = \frac{v_x}{\dot{\theta}} \tag{6}$$

The driven wheel is free wheel here. Its rotation angle φ can be obtained.

$$\varphi = \arctan\frac{W}{R} = \arctan\frac{W(v_r - v_l)}{B(v_r + v_l)} \tag{7}$$

The kinematic model of reference point *P* in OXY coordination can be obtained as

$$\begin{cases} \dot{X} = v_x \cos\theta = \dfrac{1}{2}(v_l + v_r)\cos\theta \\ \dot{Y} = v_x \sin\theta = \dfrac{1}{2}(v_l + v_r)\sin\theta \\ \dot{\theta} = \omega_z = \dfrac{1}{2B}(v_r - v_l) \end{cases} \tag{8}$$

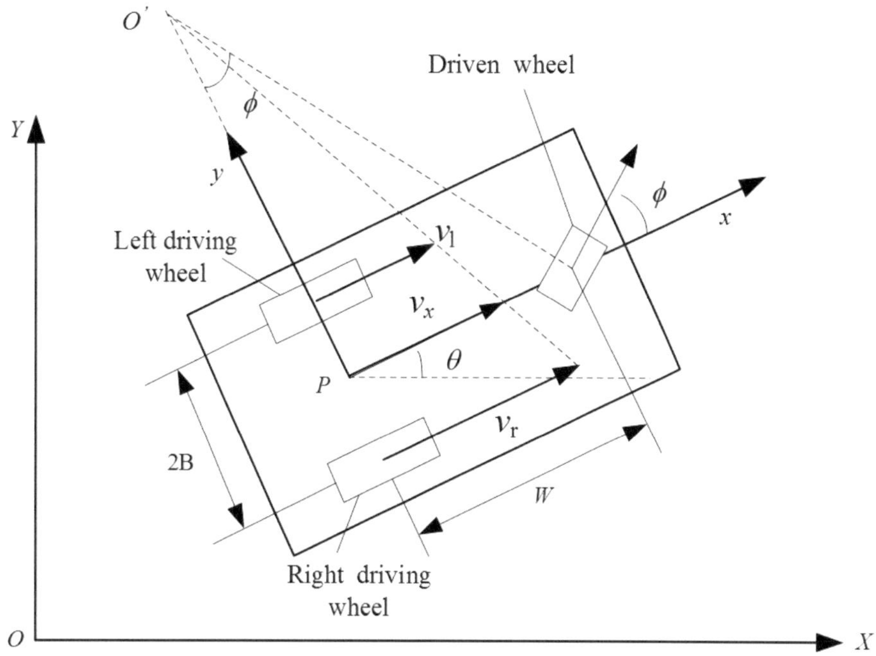

Fig. 7. Kinematic schematic of two-axle driving vehicle

4. Kinematics analysis of multi-axle driving vehicle

For multi-axle driving vehicle, independent steering machines are necessary for all wheels including driving wheels and driven wheels to realize the pure rolling around certain center while steering. Taking an eight-wheel vehicle as an example, kinematic schematic is shown in Fig.8. Assume that the vehicle is rotated around point O' with steering wheel rotates α at a certain moment. The position coordinate of vehicle center P is (X_P, Y_P) and the orientation angle is θ in global coordination OXY. The rotation angles of wheels are represented as φ_i $(i = 1, 2, ..., 8)$ relative to Px-axis. The linear speeds of all wheels are v_i $(i = 1, 2, ..., 8)$ and the traveling speed of vehicle center point P is V_P . Suppose space lengths between adjacent wheel-axles are equal, denoted as L. The left-right direct-axle space is $2B$. Define the whole vehicle turning radius $R = \overline{O'P}$ and each wheel turning radius $R_i (i = 1, 2, ..., 8)$ is the length from O' to wheel-axle center.

In order to achieve the pure-rolling steering without slipping, the rotation angles φ_i and linear speeds v_i of all wheels must match certain geometrical relation. Let left first wheel trace the steering wheel rotation angle, i.e. $\varphi_1 = \alpha$, thus $R_1 = 1.5L/\sin\varphi_1$ and the vehicle turning radiuses can be obtained

$$R = R_1 \cos\varphi_1 + B = 1.5L\cot\varphi_1 + B \tag{9}$$

The rotation angles φ_i should meet the steering trapezium, described as

$$\begin{cases} \varphi_1 = \alpha, \quad \varphi_4 = -\varphi_1 \\ \varphi_2 = \arctan\dfrac{0.5L}{(R-B)}, \quad \varphi_3 = -\varphi_2 \\ \varphi_5 = \arctan\dfrac{1.5L}{(R+B)}, \quad \varphi_8 = -\varphi_5 \\ \varphi_6 = \arctan\dfrac{0.5L}{(R+B)}, \quad \varphi_7 = -\varphi_6 \end{cases} \tag{10}$$

Respective wheel's turning radius is expressed as

$$R_i = \begin{cases} (R-B)/\cos\varphi_i \ , \ i = 1,2,3,4 \\ (R+B)/\cos\varphi_i \ , \ i = 5,6,7,8 \end{cases} \tag{11}$$

In the ideal situation, all wheels roll purely around center O' in which the linear speeds v_i are proportional to relative the turning radiuses R_i . In this case, the vehicle can be looked as a rigid body rotating around fixed-axis, thus turning velocity can be expressed as $\dot{\theta} = \dfrac{V_P}{R} = \dfrac{v_i}{R_i}$. Suppose left first wheel is a driving wheel and its speed v_1 is given, the vehicle center speed can be get $V_P = v_1\dfrac{R}{R_1} = v_1(\cos\varphi_1 + \dfrac{B}{1.5L}\sin\varphi_1)$. Consequently, the kinematical function of vehicle motion can be depicted as

$$\begin{cases} \dot{X}_P = V_P\cos\theta = v_1(\cos\varphi_1 + \dfrac{B}{1.5L}\sin\varphi_1)\cos\theta \\ \dot{Y}_P = V_P\sin\theta = v_1(\cos\varphi_1 + \dfrac{B}{1.5L}\sin\varphi_1)\sin\theta \\ \dot{\theta} = \omega = \dfrac{V_P}{R} = \dfrac{v_1\sin\varphi_1}{1.5L} \end{cases} \tag{12}$$

Other driving wheels' linear speeds should meet $v_i = \omega R_i$. During the entire steering procedure, if the actual wheel (axle) turning angles or driving wheels' speeds can't keep matching with their planned values, the pure-rolling condition will not be satisfied. Consequently, the serious slippage of the wheels relative to the ground will be generated, and also the unbalanced force among steering mechanisms will be induced due to the actuation redundancy.

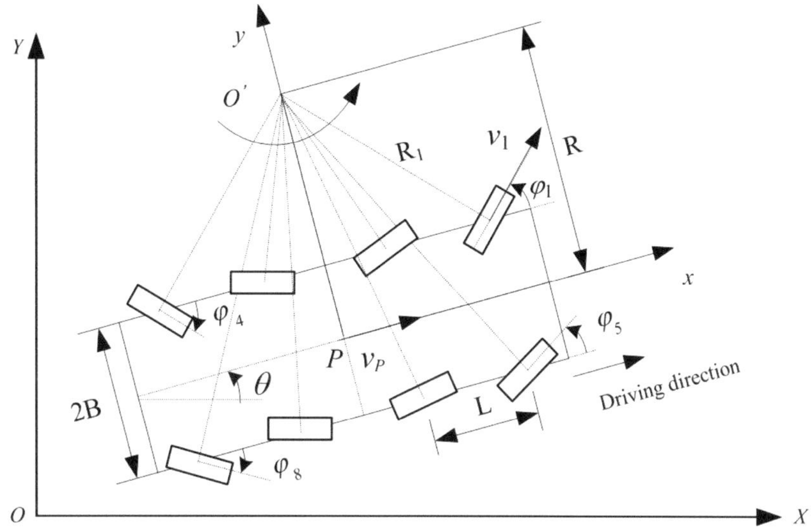

Fig. 8. Kinematic schematic of multi-axle driving vehicle

5. Hydraulic control of travelling and steering of multi-axle vehicle

5.1 Travelling hydraulic system

Travelling hydraulic system is one typical pump-control-motor system as shown in Fig.9. In general, the closed-type hydraulic circuit with one or two proportional variable-displacement pumps is adopted and each pump drives multiple parallel variable-displacement motors to drive the vehicle. This kind of motor can switch between two working conditions of slow speed and large torque as well as high speed and little torque. Through the switch and combination of motors' displacement, three or four speed stages can be formed, and the stepless speed-adjustment in every stage can be achieved by controlling the displacement of pump. As stopping to steer, all wheels can be switched into "free wheel" state. Furthermore, the speed sensors can be installed on the motors to attain closed-loop travelling speed control.

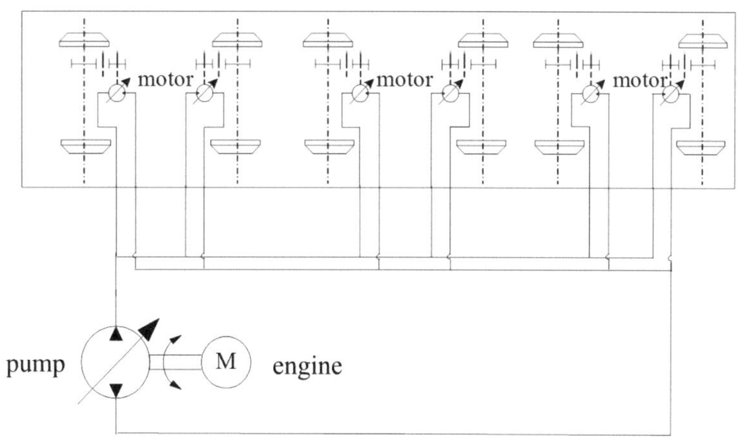

Fig. 9. Travelling hydraulic system schematic

5.2 Steering hydraulic system

According to the dynamic analysis in the steering procedure, when the velocity of the vehicle is low and the lateral slip angle is very small, the steering belongs to natural steering (Hosaka, 2004). In this case, the rotational dynamic influence to the vehicle chassis can be omitted. The whole steering motion is governed by the dynamic equations of each wheel. As shown in Fig. 10, a single wheel's steering hydraulic system contains proportional amplifier components, a valve-controlled cylinder, and the steering linkage mechanism.

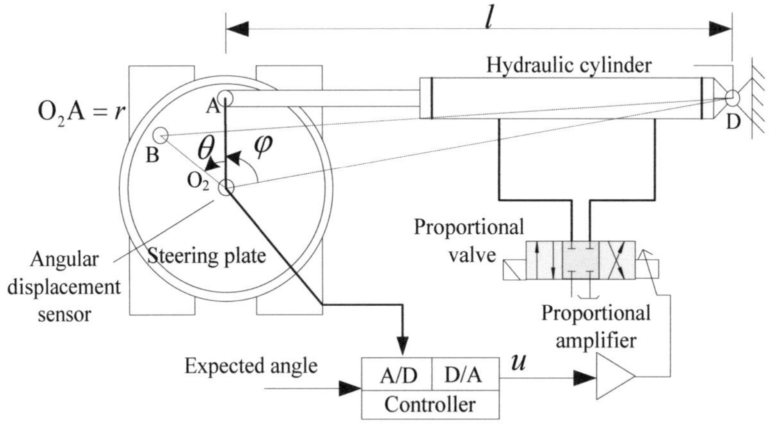

Fig. 10. Steering hydraulic system schematic

In light of the hydraulic work principles and taking u and θ as the control input and output of the independent steering mechanism, we obtained the following governing equation.

$$\begin{cases} (k_{\mathrm{v}}\dfrac{u}{u_{\mathrm{m}}})\sqrt{1-\mathrm{sgn}(\dfrac{u}{u_{\mathrm{m}}})\dfrac{p_{\mathrm{L}}}{p_{\mathrm{s}}}} = A_r r(\theta)\dot{\theta} + \dfrac{V_t}{4\beta_e}\dot{p}_{\mathrm{L}} + C_a p_{\mathrm{L}} \\ p_{\mathrm{L}} A_r r(\theta) = J\ddot{\theta} + B_m\dot{\theta} + T_{\mathrm{L}}\,\mathrm{sgn}(\dot{\theta}) \\ r(\theta) = \dfrac{r\sqrt{l^2+r^2}\sin(\theta+\varphi)}{\sqrt{l^2+2r^2+2rl\sin\theta-2r^2\cos\theta}} \\ T_{\mathrm{L}} = 2Z\xi\sqrt{B_0^2/4+b^2/8} \end{cases} \tag{13}$$

where $r(\theta)$ represents the equivalent arm of the cylinder's thrust force applied onto the turn-plate. This function is deduced from the geometric relationship of turn-plate and the cylinder. Since the movement of the piston within its stroke does not result in significant changes of the moment of the cylinder's thrust force, we consider it as a constant in order to simplify the analysis. The notations of the symbols are listed in Table 1.

6. Multi-axles motion synthesis and coordinated control

For multi-axle driving vehicles, the motion synthesis and coordinated control problem occurs in the steering procedure while travelling. The wheel-axles turning angles are expected to satisfy the pure-rolling condition described by Eqn. (10). However, affected by many factors, the actual wheel (axle) turning angles are very difficult to keep matching with its planned values during the entire steering procedure. Consequently, the system-level contouring error will be generated, which will eventually result in serious slippage of the wheels relative to the ground. As the error increases, the wheel tires will wear off and the unbalanced force among steering mechanisms will be induced due to the actuation redundancy. Therefore, in this section we discuss the cross-coupled control method to solve contouring error.

Symbol —Notation
A_r - effective area of cylinder, B_0 - distance between two tires
b - width of the tire, ξ - steering damp coefficient,
C_a —equivalent leak coefficient, J - equivalent rotating inertia,
k_v - proportional amplification coefficient , Q_L —load flow,
p_l - load pressure, p_s - supplying pressure,
u - input of the proportional amplifier, u_m - maximum input,
θ - wheel turning angle, T_l - torque to resist steering
V_t - equivalent volume, β_e - equivalent-volume elastic modulus
Z - vertical load acting on steering mechanism

Table 1. Symbol notations of the hydraulic steering mechanism

The contouring error is the shortest distance from current position to expected trajectory. At first let's see a simple two-axis output example as shown in Fig.11. Considering a linear contour with angle θ between the expected line and the X-axis, the contouring error e_c can be depicted by the following equation

$$e_c = -e_x \sin\theta + e_y \cos\theta \tag{14}$$

where e_x and e_y are the tracking errors of X and Y axes respectively.

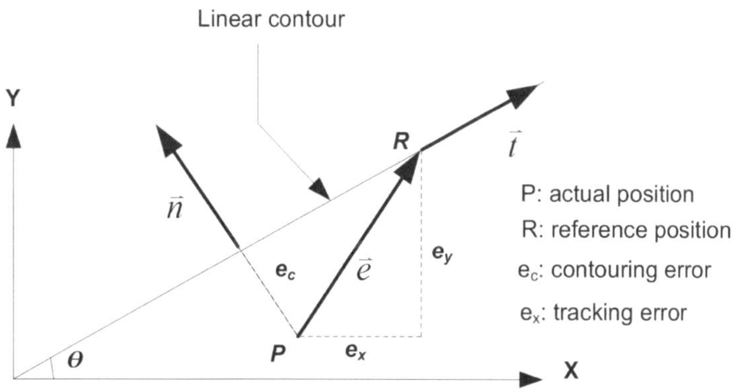

Fig. 11. Linear contour illustration

Cross-coupling control technology provides advantages and opportunities to improve synchronization performance of multi-axle outputs. Over past decade, the cross-coupling concept has been developed and widely used in multi-axle motion synthesis such as reducing contouring error of CNC machines (Zhong,2002) and contour tracking control of mobile robot (Sun,2002; Rodriguez & Nijmeijer, 2004). Here, a simple PID cross-coupled controller based on real-time feedback and information sharing is presented to resolve the synchronization problem of multi-axle vehicle while steering. The basic idea is to select one wheel's actual turning angle as reference and the relative angles of the other wheels are solved from the motion equations. The tracking and contouring errors of each wheel can be obtained by comparing its actual feedback value with its two expected ones from steering wheel and fiducial wheel. Then, two closed-loop PID control laws are designed respectively for the contouring error and the tracking error. As a result, the controller can satisfy coherence demand of contour tracking in the process of steering. This control method with contour error tracking of single-axle is illustrated in Fig.12.

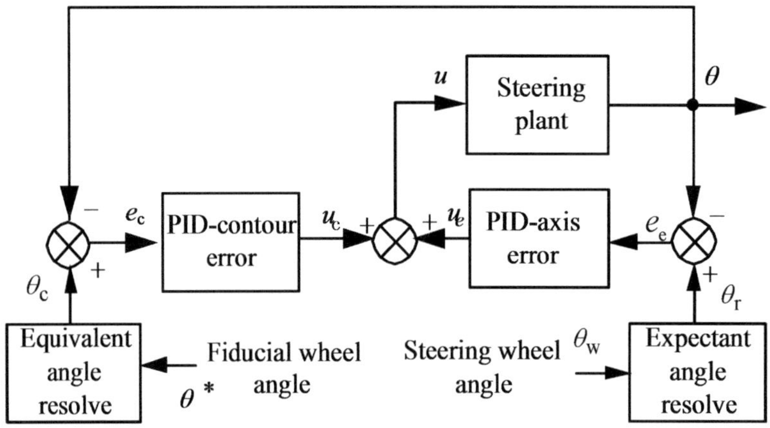

Fig. 12. Block diagram of the steering controller with contouring error

In Fig.12, the output of each axle needs to meet two requirements. One is the expected angle θ_r resolved from the steering wheel command, and the other is the theoretically equivalent angle θ_c resolved from the current feedback value of the angle of fiducial wheel. As a result, the tracking error and contouring error are expressed as $e_e = \theta_r - \theta$ and $e_c = \theta_c - \theta$. The corresponding control law is composed of two parts, i.e., $u = u_e + u_c$. The piecewise PID control law is adopted in u_e and u_c. The expressions of the control law u_e and u_c are respectively given by:

$$u(k) = u_e(k) + u_c(k) \tag{15}$$

$$u_e(k) = u_e(k-1) + P_e(e_e(k))\Delta e_e(k) + I_e(e_e(k))e_e(k) + D_e(e_e(k))\Delta^2 e_e(k) \tag{16}$$

$$u_c(k) = u_c(k-1) + P_c(e_c(k))\Delta e_c(k) + I_c(e_c(k))e_c(k) + D_c(e_c(k))\Delta^2 e_c(k) \tag{17}$$

where $P_i(e_i(k))$, $I_i(e_i(k))$, and $D_i(e_i(k))$ ($i = \text{e,c}$) respectively denote the function of proportional gain, integral gain, and differential gain of u_e and u_c with $\Delta e_i(k) = e_i(k) - e_i(k-1)$, and $\Delta^2 e_i(k) = \Delta e_i(k) - \Delta e_i(k-1)$.

Note that the method to solve the contouring error is based on the transformation to the reference angle. Hence, we can change the contouring tracking problem to a synchronized tracking problem. The mathematical relationship between the contouring error and the tracking error can be readily depicted from Fig. 13. Here, x_1 is the equivalent angle based on the steering kinematics formula from fiducial wheel's angle, x_2 denotes the real steering angle of the discussed axle, and $\overline{PR}$ and $\overline{PC}$ denote the tracking error and the contouring error respectively. Because $e_c = \overline{PC} = \overline{BP}\cos(\pi/4)$, $\overline{BP}$ can be used to express the contouring error. Obviously, through this transformation, the contouring tracking problem can be converted to a synchronized tracking problem so that the computation of the contouring error is simplified.

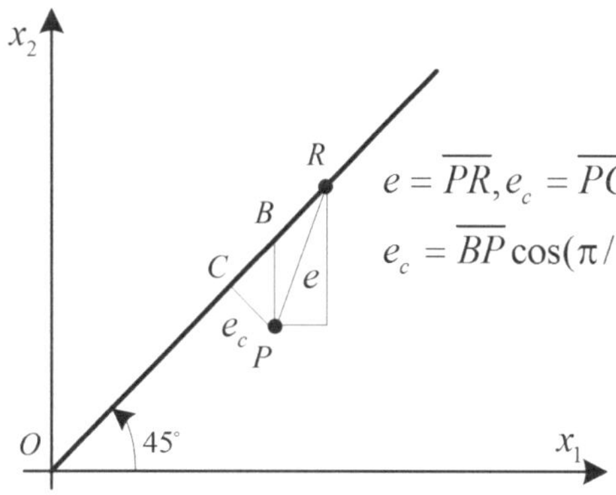

Fig. 13. Simplification of contouring error

7. Manipulating control of the multi-axle-electrohydralic-control-transpoter

The DCY series of transportation vehicles are designed and commissioned to transfer huge and heavy objects. Such a vehicle, integrated with electro-hydraulic proportional control and networked control, has multiple functions such as travelling, steering, leveling, and lifting. In order to reduce the supporting load on each of the vehicle wheels, a number of supporting and driving wheels are employed. Without loss of generality, the following study is carried out using the DCY270 model as the example.

7.1 Overall framework design of control system

As shown in Fig. 14, the DCY270 transportation vehicle is designed to carry 270 tons load for building industry and ship manufacturing. It has 10 axles and 20 wheels that can steer in multiple modes. An integrated solution approach based on the NCS (through the field-bus with CAN-Open protocols) and distributed electro-hydraulic proportional control is proposed to perform various functions. The control principle of steering system is shown in Fig.15. There are three intelligent nodes on the CAN-Open network. The CC node is the master controller located in the cab to receive all kinds of operation commands. C1 and C2 nodes represent the two controllers placed at the two sides of the vehicle body. Each of them controls four groups of steering, driving and suspending mechanisms close to it. Fig.15 (a) shows the steering mechanism of a single axle driven by a valve-controlled hydraulic cylinder, while Fig.15 (b) shows the configuration of the CC.

Fig. 14. DCY 270 powered transportation vehicle

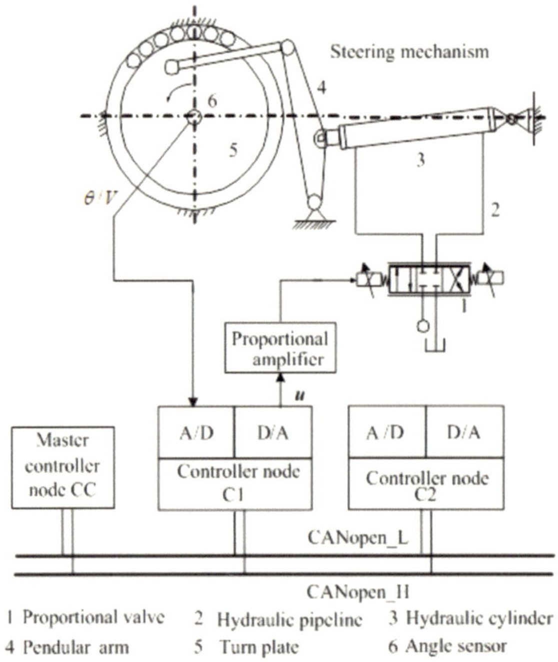

(a). Diagram of the steering system based on CAN-bus

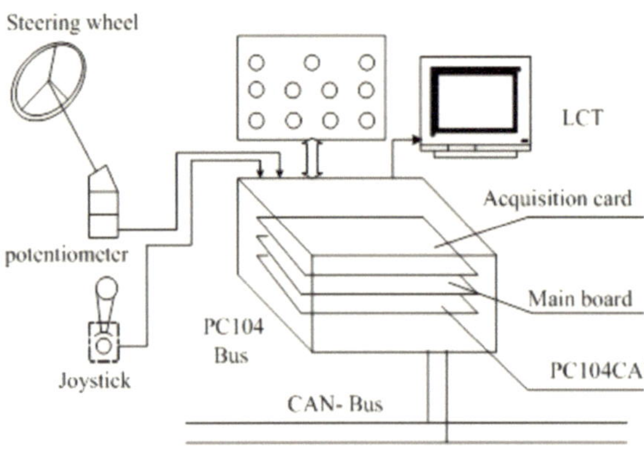

(b). Configuration of the master/slave controller node

Fig. 15. NCS-based electro-hydraulic control diagram for steering system

7.2 Multimode steering system based on NCS

In order to perform different operation tasks, the transportation vehicle is required to steer in multiple steering modes such as diagonal steering, longitudinal steering, front (rear) axle steering, and center steering (as shown in Fig.16). According to this figure, the steering kinematics models under different steering modes can be formulated. In principle, as long as each individual wheel can turn to its expected angle precisely, the whole vehicle can realize different steering modes, in which all axles turn around the rotation center without slipping and sliding, i.e., a pure rolling steering.

Obviously, the traditional Ackerman's steering trapezium can only be employed for the simple steering. Moreover, the design and manufacture of the steering linkage mechanism are difficult and costly. However, the networked-control provides an effective way to realize multimode steering.

Here, we give out the kinematic model of the normal steering (longitudinal steering) mode as shown in the up-right corner of Fig. 16. The rotation angles of left and right five axles are set as α_i and $\beta_i\,(i=1,2,\cdots,5)$ respectively. The notations of B, L_1, L_2, L_3 and R are shown in Fig. 16. In terms of forward traveling direction, positive and negative turnings are identified with counterclockwise and clockwise directions, respectively. α_1 is controlled by the rotational angle of the steering wheel θ_w, while other axles' expected angles are calculated from the given motion relationship. The expected angle of each axle is as follows.

$$\begin{cases} \alpha_1 = i_w \theta_w, \\ R = 0.5(L_1 \cot\alpha_1 + B), \quad \tan\alpha_2 = 0.5L_2/(R-0.5B), \\ \tan\alpha_3 = 0.5L_3/(R-0.5B), \quad \tan\beta_1 = 0.5L_1/(R+0.5B), \\ \tan\beta_2 = 0.5L_2/(R+0.5B), \quad \tan\beta_3 = 0.5L_3/(R+0.5B), \\ \alpha_2 = -\alpha_4, \ \alpha_1 = -\alpha_5, \ \beta_2 = -\beta_4, \ \beta_1 = -\beta_5 \end{cases} \tag{18}$$

where i_w is the transmission ratio between α_1 and θ_w.

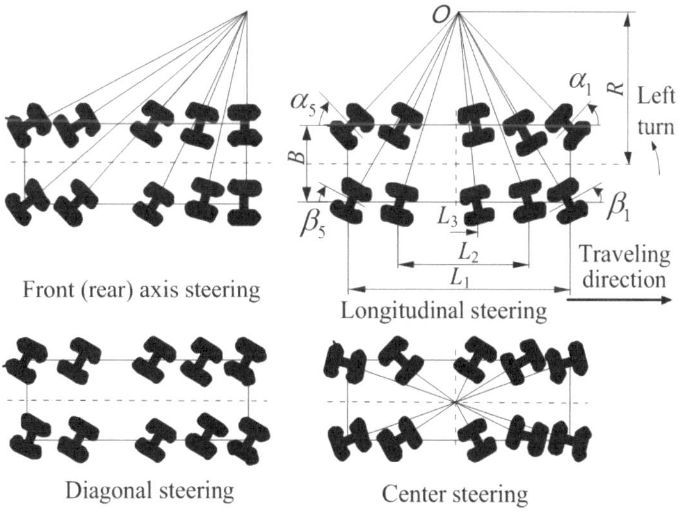

Fig. 16. Kinematics diagrams of different steering modes

Using the same method, the kinematics models of the other three kinds of steering modes, i.e., front (rear) axle steering, diagonal steering, and center steering, are given in (19), (20) and (21), respectively as follows.

$$\begin{cases} \alpha_5 = i_w \theta_w, R = L_1 \cot \alpha_5 + 0.5B \ , \\ \tan \alpha_4 = 0.5(L_1 + L_2)/(R - 0.5B), \\ \tan \alpha_3 = 0.5(L_1 - L_3)/(R - 0.5B), \\ \tan \alpha_2 = 0.5(L_1 - L_2)/(R - 0.5B), \\ \alpha_1 = 0, \ \beta_1 = 0, \cot \beta_5 - \cot \alpha_5 = 2B / L_1, \\ \cot \beta_4 - \cot \alpha_4 = 2B/(L_1 + L_2), \\ \cot \beta_3 - \cot \alpha_3 = 2B/(L_1 - L_3), \\ \cot \beta_2 - \cot \alpha_2 = 2B/(L_1 - L_2). \end{cases} \tag{19}$$

$$\alpha_i = \beta_i = i_w \theta_w, i = 1,2,3,4,5 \tag{20}$$

$$\begin{cases} \tan \alpha_1 = -L_1 / B, \tan \alpha_2 = -L_2 / B, \tan \alpha_3 = -L_3 / B \\ \tan \beta_1 = L_1 / B, \tan \beta_2 = L_2 / B, \tan \beta_3 = L_3 / B \\ \alpha_1 = -\alpha_5, \quad \alpha_2 = -\alpha_4, \quad \beta_2 = -\beta_4, \quad \beta_1 = -\beta_5 \end{cases} \tag{21}$$

Note that for the centre steering mode, the driving motors in the left and right sides are controlled by two pumps respectively. As a result, the center steering in the same direction can be achieved.

7.3 Motion synthesis strategy for contouring control

For the steering system of the DCY270 transportation vehicle, above cross-coupling PID control algorithm is adopted and the detailed resolution is given as follows. The ten wheels (axles) are divided into two groups: $\alpha_1 \sim \alpha_5$ and $\beta_1 \sim \beta_5$ controlled by C1 and C2, respectively. One wheel is selected as the norm of each group, i.e. α_1 and β_5. At the same time α_1 is sent to C2 as β_5's norm through CAN-bus from C1. Taking longitudinal steering as an example, the task of resolving the expected angles can be carried out by the master control node CC according to (18), such that $\bar{\alpha}_1 = i_w \theta_w$, $\bar{\alpha}_1 \rightarrow \bar{\alpha}_2 \sim \bar{\alpha}_5$, and $\bar{\alpha}_1 \rightarrow \bar{\beta}_1 \sim \bar{\beta}_5$. Here, we use the operator $\rightarrow$ to indicate that the right terms are determined from the left known term. Equivalent angles are computed from real feedback α_1 according to (18), which can be expressed as $\alpha_1 \rightarrow \hat{\alpha}_2 \sim \hat{\alpha}_5, \alpha_1 \rightarrow \hat{\beta}_5$, and $\beta_5 \rightarrow \hat{\beta}_1 \sim \hat{\beta}_4$. Then we can obtain the tracking errors: $e_i = \bar{\alpha}_i - \alpha_i$, or $e_i = \bar{\beta}_i - \beta_i$ (i=1,2,...,5) with $e_{ei} = e_i$, and the contouring error: $e_{ci} = \hat{\alpha}_i - \alpha_i$ or $e_{ci} = \hat{\beta}_i - \beta_i$ (i=1,2,...,5). Each node's closed-loop control is formed by two control components u_e and u_c. The tracking error and contouring error can be restrained by using the piecewise PID control law described in (15)-(17). In the practical application, the magnitude of the control signal is restricted in a suitable range according to the specification of the driving equipments and control targets. The expressions of the control signal and the PID parameters are given.

$$u(k)=\begin{cases}u_{\max}\,\mathrm{sgn}(e(k)) & |e(k)|>\delta_1\\ u(k) & \xi_1<|e(k)|\le\delta_1\\ u(k-1) & |e(k)|\le\xi_1\end{cases}\tag{22}$$

$$\{P_e,I_e,D_e\}=\begin{cases}\{p_{e1},i_{e1},d_{e1}\} & \xi_1<|e(k)|\le r_1\\ \{p_{e2},i_{e2},d_{e2}\} & r_1<|e(k)|\le\delta_1\end{cases}\tag{23}$$

$$\{P_c,I_c,D_c\}=\begin{cases}\{p_{c1},i_{c1},d_{c1}\} & \xi_2<|e_c(k)|\le r_2\\ \{p_{c2},i_{c2},d_{c2}\} & r_2<|e_c(k)|\le\delta_2\end{cases}\tag{24}$$

where, $0<\xi_1<r_1<\delta_1$, $0<\xi_2<r_2<\delta_2$. The saturation values and dead zone of control signal are decided according to the actual case. In view point of the practical requests of the system, the rapidity is not important but the static error and overshoot should be avoided. Relying on such control requirements, the piecewise PID parameters are estimated by empirical values and further optimized in practical tests.

Combining the kinematical model and steering control law with the dynamic model of electro-hydraulic steering mechanism, the block diagram of the overall vehicle steering system is shown in Fig.17.

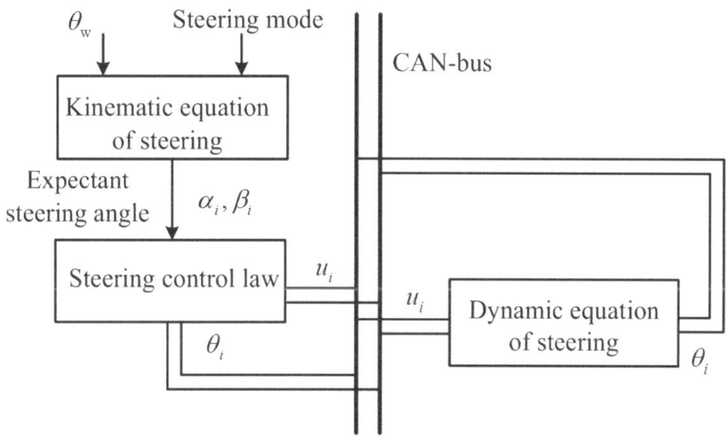

Fig. 17. Block diagram of the overall steering system

7.4 Experiments

The test of center steering, diagonal steering and longitudinal steering are performed and actual effects are shown in Fig. 18 In the longitudinal steering control, the cross-coupled control law is adopted and the contouring errors of left-right symmetrical wheel 2, 4, 7, 9 are measured and drawn as in Fig.19. The result indicates whole vehicle's contouring precise satisfies the manipulation performance demand.

Fig. 18. Test of multimode steering

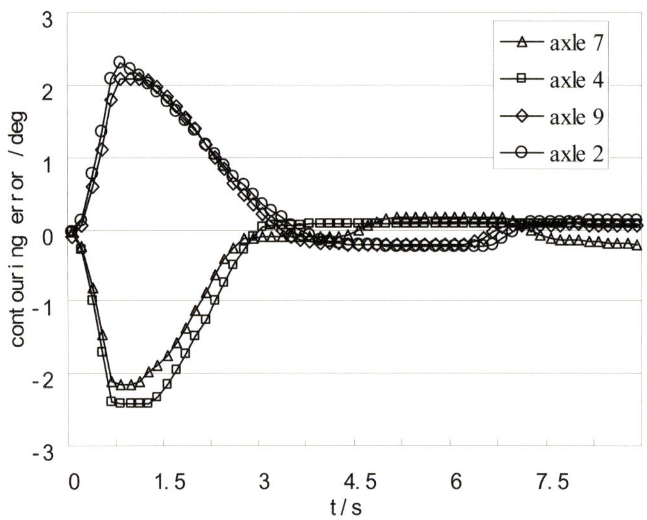

Fig. 19. Experimental result of cross-coupled control law

8. Manipulating control of track-laying-machinery for high speed railway

The high-speed railway rail-laying vehicles is one kind of large special automatic machine developed to pave the sleepers and long rails on the new lines so as to satisfy the demands of high speed railway with seamless lines (Zhao et al., 2006). Automatic drive function of rail-laying vehicles can make it follow the scheduled route and control running speed automatically. It is very important to ensure the precision of paving railway lines.

8.1 Framework of automatic drive system

Shown as in Fig.20, the rail-laying vehicle consists of guided pedrails, bodywork and power bogies. It is one special type of multi-axle driving vehicle. Guided pedrails and power bogies provide traction, bodywork connects guided pedrails by swing assembly, and the steering structure is articulated. Vehicles' automatic drive includes speed control and steering control.

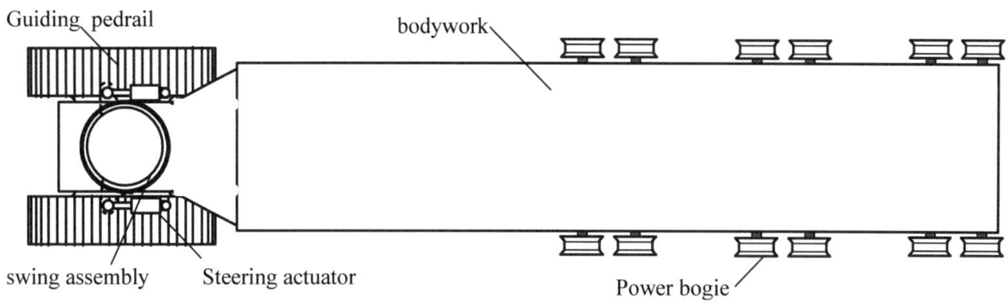

Fig. 20. Structure schematic of rail-laying vehicle

The travelling hydraulic system of rail-laying vehicles is composed of the hydraulic system of guided pedrails and power bogie. They all employ hydrostatic pressure transmission in closed circuit and high-speed driving scheme containing variable stroke pump, variable motor and reduction box. The design works of travelling control focus on the speed synchronism and allocating fraction between guided pedrails and power bogies.

The steering hydraulic system is mounted on guided pedrails. Its main components include proportional valve, two steering hydraulic-cylinders and swing assembly. Automatic steering control system consists of controller, ultrasonic sensors, angular coder, hydraulic steering mechanism.

In consideration of the structure characteristic and control demands, the control system adopts the scheme of electro-hydraulic proportional control system based on network combining CAN-bus network to control separate actuators such as valves, pumps and motors. The configuration and interfaces of control systems are showed in Fig 21. Hardware platform contains integrative IPCs, Fieldbus controllers and sensors. EPEC controller of Finland Company is chosen as Field bus controller. It provides digital input/output, analog input/output and PWM output, which can replace proportional amplifier to drive valve directly. Four EPEC control modules and one IPC are networked by CAN protocol.

8.2 Steering control system

The steering mechanism is articulated on the guided pedrail by swing hydraulic-cylinders. Swing cylinders keep or change the comparative angle between padrail and bodywork vehicle to make rail layer move on straight line or curve with different radius. Fig.22 shows the hydraulic steering working principle.

Rail-laying vehicle follows straight line along the indicating route during laying rail to realize automatic guide. Considering the disturbing affection of light, fog, and rain in field, shoe ultrasonic sensor is selected to measure the distance between the line and pedrail. If vehicle deviate the datum line, the shoe ultrasonic sensor can reflect the deviation. At same time, one rotation coder installed in the pedrail panel can measure the current steering angle. These two signals are feedbacked to controller through CANbus, and then controller executes the corresponding algorithm to drive cylinders to correct the travelling direction of the pedrail. Fig.23 is automatic steering control block diagram.

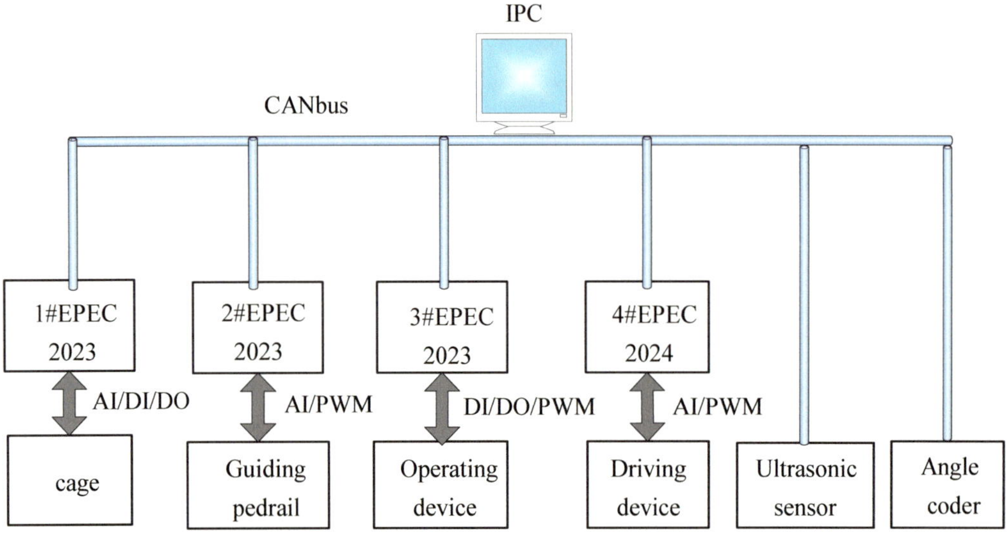

Fig. 21. Configuration of automatic control system

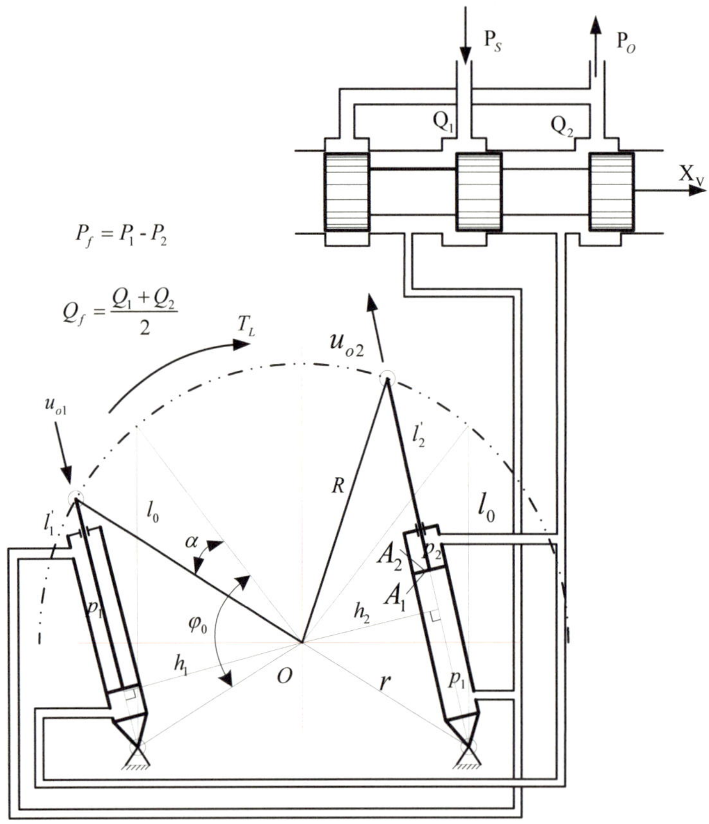

Fig. 22. Hydraulic steering system schemes

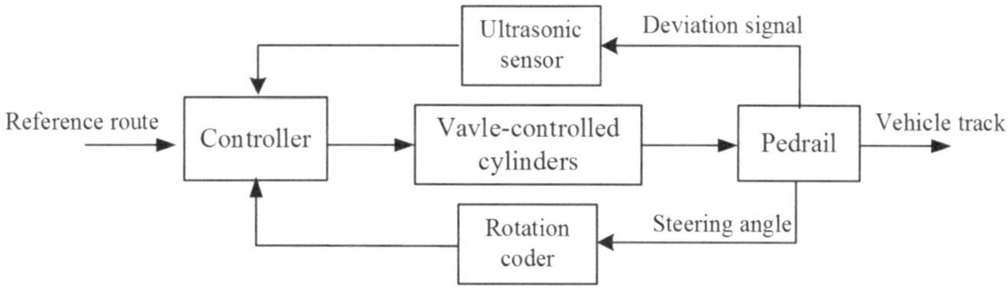

Fig. 23. Automatic steering control block diagram

8.3 Travelling control system

The traction is generated by both pedrail vehicle and power bogie. Their travelling hydraulic systems employ hydrostatic pressure transmission in closed circuit and high-speed drive scheme. Two variable pumps respectively drive six motors of three power bogies and two motors of pedrail. The hydraulic system working pressure is 27 MPa and peak pressure is 30 MPa. Four speed stages can be switched by controlling the number of working motors and displacement of motors, and the stepless speed-adjustment in every stage can be realized by adjusting the displacement of pump. Of course, the speed sensors are adopted to achieve speed closed-loop control shown as in Fig.24.

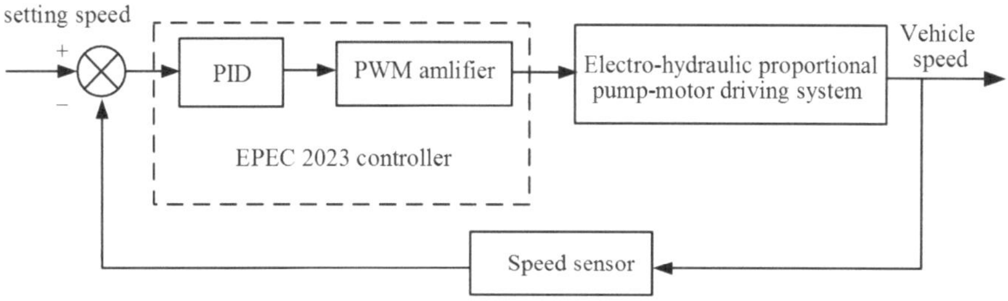

Fig. 24. Schematic diagram of closed-cycle speed control

Because pedrail and power bogies have independent drive systems, the synchronism of pedrail and three power bogies is very important while travelling. The synchronous solution depends on the principle "Power bogie is priority and pedrail is assistant". The controller observes the system pressure and motor rotation speed, and adjust the displacements of the pumps and motors to ensure the travelling synchronization between pedrail chassis and power bogies referring to the speed of power bogies wheel and load change. Six motors of power bogies are synchronous naturally for cohering effect between wheels and rails.

Allocating traction appropriately is essential to ensure system efficiency. Both work condition and load may change while travelling, but those will all reflect in load change. As a result, the displacements of pedrail motors and bogies motors are adjusted relying on load condition to ensure the best allocation of traction between pedrail and bogies.

9. Conclusion

Focusing on the problems on the motion synthesis and coordinated control of multi-axle driving vehicles, we have proposed the realization framework based on networked control, and have established the mathematic model of multi-mode-steering kinematics. Then, a kind of the coordinated control method for multi-axle driving vehicles has been studied.

This study has resulted in the successful development of three kinds of typical heavy construction field robots, i.e. DCY900 transporter, Hoisting-girder transporter with 900t load, and DPG500 high-speed railway rail-laying vehicles.

10. References

Hespanha, J.; Naghshtabrizi, P. & Xu, Y. (2007) A survey of recent results in networked control systems, *Proceedings of the IEEE*, Vol.95 No.1 pp.138-162

Kemmetmüller, W. (2007). Mathematical modelling and nonlinear controller design for a novel electro-hydraulic power steering system, *IEEE/ASME Trans. Mechatron.*, Vol. 12, No.1, pp. 85–97

Lian, F.; Moyne, J. & Tibury, D. (2002). Network design consideration for distributed control systems, *IEEE Trans. Contr. Syst. Technol.*, Vol.10, No.2, pp. 297-307

Li, Y.; Yang, G. (2005). Investigation on network control for control and manipulating system of the vehicle and construction machinery, *Engineering Science*, Vol. 7, No.9, pp. 25-29

Li, Y.;Yang, L., & Yang, G. (2007). Network-based coordinated motion control of large-scale transportation vehicles. *IEEE/ASME Trans. Mechatron.*, Vol. 12, No.2, pp208-215

Niu, W.; Tomizuka, M. (2001). A new approach of coordinated motion control subjected to actuator saturation, *ASME J. Dyn. Syst., Meas., Control*, Vol.123, No.3, pp. 496-504

Rodriguez, A.; Nijmeijer, H. (2004). Mutual synchronization of robots via estimated state feedback: a cooperative approach, *IEEE Trans. Contr. Syst. Technol.*, Vol. 12, No. 4, pp. 542-554

Sun, H.; George T. (2002). Motion synchronization for dual-cylinder electrohydraulic lift systems, *IEEE/ASME Trans. on Mechatronics*, Vol.7, No.2, pp.171-181

Sun, D.; Mills (2002). Adaptive synchronized control for coordination of multi-robot assembly tasks, *IEEE Trans. Robot. Autom.*, Vol. 18, No.4, pp. 498–510

Tomizuka, M.; Hu, J. & Chiu, T. (1992). Synchronization of two motion control axes under adaptive feedforward control, *ASME J. Dyn. Syst.,Meas., Contr.*, Vol. 114, No. 2, pp. 196–203

Yang, L.; Guo, Z. & Li, Y. (2009). Posture measurement and coordinated control of twin hoisting-girder transporters based on hybrid network and RTK-GPS, *IEEE/ASME Transactions on Mechatronics*, Vol.14, No.2, pp.141-150

Yeh, S.; Hsu, P. (2002). Estimation of the contouring error vector for the cross-coupled control design, *IEEE/ASME Trans. Mechatronics*, Vol. 7, No. 1, pp. 44–51

Yeh, S.; Hsu, P. (2003). Analysis and design of integrated control for multi-axis motion systems, *IEEE Transactions on Control Systems Technology*, Vol.11, No.3, pp.375-382

Zhong, Q.; Shi, Y. & Mo, J. (2002). A linear cross coupled control system for high speed machining, *Int. J. Adv. Manuf. Technol.*, Vol. 19, pp. 558–563

Zhao, Y.; Li, Y. & Yang, L. (2006). Research on automatic drive technology of high-speed railway rail-laying vehicles, *IEEE International Conference on Robotics, Automation and Mechatronics*, Thailand, June , 2006

Hosaka, M. (2004). Yaw rate control of electric vehicle using steering-by wire system, *IEEE AMC2004*, pp.31-34, Kawasaki, 2004

Liu, H.; Dong S. (2005). Uniform synchronization in multi-axis motion control , *Proceedings of the American Control Conference* , pp.4537-4542, 2005

Yan, M.; Lee, M. & Yen, P. (2005). Theory and application of a combined self-tuning adaptive control and cross-coupling control in a retrofit milling machine, *Mechatronics*, Vol.15, No.2, pp. 193-211

6

A Novel Traction Control for Electric Vehicle without Chassis Velocity

Dejun Yin and Yoichi Hori
University of Tokyo
Japan

1. Introduction

Due to the growing concern about global environmental problems and shrinking non-renewable energy sources, research on electric vehicles and hybrid electric vehicles is once again attracting significant attention. Meanwhile, significant improvements in power electronics, energy storage and control technology have made electric vehicles fully feasible, preparing the state of the art for their return to the market (Chau et al., 2008; Affanni et al., 2005; Nagai, 2007).

Beside the advantages for the environment, manufacture and maintenance, from the viewpoint of control technology, the most distinct advantages of electric vehicles have not been well recognized. Since electric vehicles and some specially configured hybrid electric vehicles are driven by electric motors, the advantages provided to these electric vehicles can be summarized as follows (Hori, 2004):

1. Quick torque generation
2. Easy torque measurement
3. Possibility of independently equipped motors for each wheel

On the other hand, considering the different regions of the world, the increase of the mobility shows a clear correlation to the gross domestic product. With further economic growth, we can predict an even greater increase in mobility and in traffic density throughout the world. For this reason, vehicle motion control systems have been developed to provide active safety control, and have made significant technological progress over the last decade to enhance vehicle stability and handling performance in critical dynamic situations by introducing computer control technology.

From the development history of vehicle motion control, it can also be found that, effective operation of any vehicle control system is based on some basic assumptions, for example, the output torque being able to accurately work on the vehicle. For this purpose, traction control, as the most primary active safety control for vehicles, is developed to ensure the effectiveness of the torque output. The key to traction control is antislip control, when the vehicle is driven or brakes on a slippery road, especially for light vehicles because they are more inclined to skid on slippery roads. Traction control must not only guarantee the effectiveness of the torque output to maintain vehicle stability, but also provide some information about tire-road conditions to other vehicle control systems. Moreover, in electric vehicles a well-managed traction control system can cover the functions of ABS, because electric motors can generate deceleration torque as easily as acceleration torque (Mutoh et al., 2007). Based on the core traction control, more complicated two-degree-of-freedom motion control for vehicles can be synthesized by introduction of some additional information on steering angle, yaw rate, lateral acceleration and so on (Saito et al., 2002; Fujimoto et al., 2004). Moreover, from the viewpoint of the relation between safety and cost, a more advanced traction control synthesis also means lower energy consumption.

However, actual vehicles present challenges to research on traction control. For example, the real chassis velocity is not available, and the friction force that it is to drives the vehicle is immeasurable in real time. Conventional traction control in internal combustion engine vehicles, according to nonlinear tire models such as the Magic Formula (Pacejka & Bakker, 1992), has to calculate the chassis velocity and the slip ratio to acquire maximum friction force from the road. For this purpose, due to physical and economic reasons, the non-driven wheels are usually utilized to provide an approximate vehicle velocity. However, this method is not applicable when the vehicle is accelerated by 4WD systems or decelerated by brakes equipped in these wheels. For this reason, the accelerometer measurement is also used to calculate the velocity value, but it cannot avoid offset and error problems. Other sensors, e.g., optical sensors (Turner & Austin, 2000), sensors of magnetic markers (Lee & Tomizuka, 2003; Suryanarayanan & Tomizuka, 2007), etc., can also obtain the chassis velocity. However, they are too sensitive and reliant on the driving environment or too expensive to be applied in actual vehicles.

In order to overcome the obstacles of calculating chassis velocity, some controllers in electric vehicles, for example Model Following Control (MFC), do not need information on chassis velocity or even acceleration sensors. In these systems, the controllers only make use of torque and wheel rotation as input variables for calculation. Fewer sensors contribute not only to lower cost, but also higher reliability and greater independence from driving conditions, which are the most outstanding merits of this class of control systems. Accordingly, research on more practical and more sophisticated antislip control based on MFC continues until now. Sakai et al. proposed a primary MFC system for antislip control (Sakai & Hori, 2001). Saito et al. modified it and proposed a novel stability analysis to determine the maximum feedback gain, and furthermore, used the antislip control as a core subsystem and extended it to two-degree-of-freedom motion control (Saito et al., 2002; Fujimoto et al., 2004). Akiba et al. improved the control performance by introduction of back electromotive force, and added a conditional limiter to avoid some of its inherent drawbacks (Akiba et al., 2007). Nevertheless, these control designs based on compensation have to consider the worst stability case to decide the compensation gain, which impairs the performance of antislip control. Furthermore, gain tuning for some specific tire-road conditions also limits the practicability of this method.

Therefore, this work, making use of the advantages of electric vehicles, focuses on development of a core traction control system based on Maximum Transmissible Torque Estimation (MTTE) which requires neither chassis velocity nor information about tire-road conditions. In this system, use is made of only the torque reference and the wheel rotation to estimate the maximum transmissible torque to the road surface, then the estimated torque is applied for antislip control implementation. Moreover, the proposed control method is expected to provide a general approach for traction control, as well as a basis for more complicated and advanced motion control in electric vehicles.

The rest of the paper is structured as follows. A novel topology of traction control, based on the proposed maximum transmissible torque estimation, is presented in Section 2, and an equivalent model is used to provide the stability analysis of the control system. Section 3 discusses the experimental electric vehicle and presents the experimental results, as well as additional simulation results. These comparative results are followed by a detailed discussion including stability, control performance and robustness in Section 4.

2. Maximum transmissible torque estimation for antislip control

2.1 Longitudinal model and dynamic analysis

Because only longitudinal motion is discussed in this paper, the dynamic longitudinal model of the vehicle can be described as in Fig. 1 and the parameter definition is listed in Table 1. Generally, the dynamic differential equations for the calculation of longitudinal motion of the vehicle are described as follows:

$$J_w \dot{\omega} = T - rF_d \tag{1}$$

$$M\dot{V} = F_d - F_{dr} \tag{2}$$

$$V_w = r\omega \tag{3}$$

$$F_d(\lambda) = \mu N \tag{4}$$

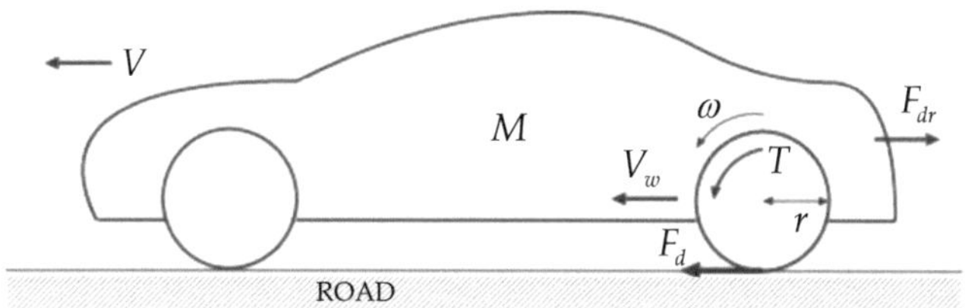

Fig. 1. Dynamic longitudinal model of the vehicle

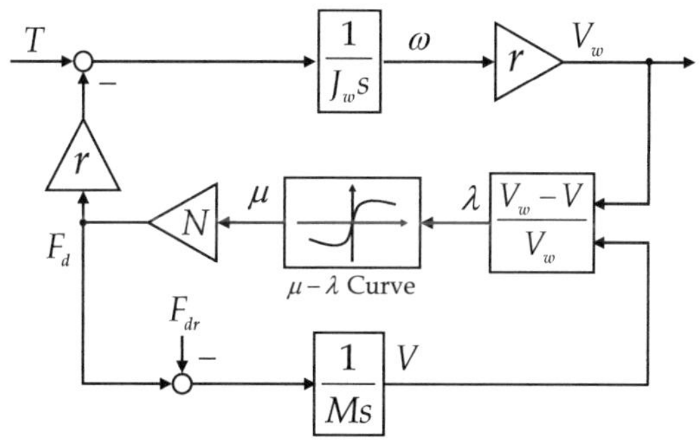

Fig. 2. One-wheel vehicle model with Magic Formula.

The interrelationships between the slip ratio and friction coefficient can be described by various formulas. Here, as shown in Fig. 2, the widely adopted Magic Formula is applied to build a vehicle model for the following simulations. The slip ratio in (4) is defined as (5). Fig. 3 describes a typical relationship between slip ratio and friction coefficient.

$$\lambda = \frac{V_w - V_x}{\max(V_w, V_x)} \tag{5}$$

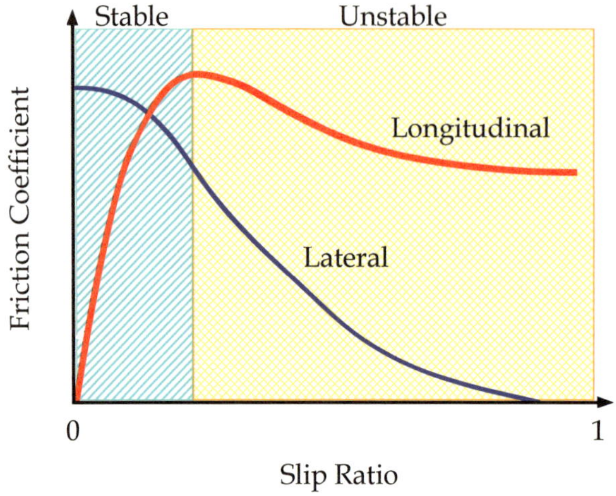

Fig. 3. A typical μ - λ curve

2.2 Maximum transmissible torque estimation

In this work, in order to avoid the complicated μ - λ relation, only the dynamic relation between tire and chassis is considered based on the following considerations, which transform the antislip control into maximum transmissible torque control.

1. Whatever kind of tire-road condition the vehicle is driven in, the kinematic relationship between the wheel and the chassis is always fixed and known. Although this relationship also exists in the dynamics of an internal combustion engine vehicle, it is very difficult to realize this consideration in these vehicles. However, the rigidity of an electric motor makes this possible.
2. During the acceleration phase, considering stability and tire abrasion, well-managed control of the velocity difference between wheel and chassis is more important than the mere pursuit of absolute maximum acceleration.
3. If the wheel and the chassis accelerations are well controlled, the difference between the wheel and the chassis velocities, i.e. the slip, is also well controlled.

According to (1) and (3), the driving force, i.e. the friction force between the tire and the road surface, can be calculated as (6). Assuming T is constant, it can be found that the higher V_w, the lower F_d. In normal road conditions, F_d is less than the maximum friction force from the road and increases as T goes up. However, when slip occurs, F_d will become smaller than the maximum friction force that the tire-road relation can provide and cannot increase with T. Here, there are only two parameters, r and J_w, so F_d is easily calculated in most tire-road conditions.

$$F_d = \frac{T - J_w \dot{\omega}}{r} \tag{6}$$

When slip starts to occur, the difference between the velocities of the wheel and the chassis becomes larger and larger, i.e. the acceleration of the wheel is larger than that of the chassis. Furthermore, according to the Magic Formula, the difference between the accelerations will make the slip more severe.

Symbol	Definition
J_w	Wheel Inertia
V_w	Wheel Velocity (Circumferential Velocity)
ω	Wheel Rotation
T	Driving Torque
r	Wheel Radius
F_d	Friction Force (Driving Force)
M	Vehicle Mass
N	Vehicle Weight
V	Chassis Velocity (Vehicle Velocity)
F_{dr}	Driving Resistance
λ	Slip Ratio
μ	Friction Coefficient

Table 1. Parameter list

Therefore, the condition that the slip does not start or become more severe is that the acceleration of the wheel is close to that of the chassis. Moreover, considering the μ - λ relation described in the Magic Formula, an appropriate difference between chassis velocity and wheel velocity is necessary to provide the friction force. Accordingly, (7) defines α as a relaxation factor to describe the approximation between the accelerations of the chassis and the wheel. In order to satisfy the condition that slip does not occur or become larger, α should be close to one.

$$\alpha = \frac{\dot{V}^*}{\dot{V}_w^*}, \text{ i.e. } \alpha = \frac{(F_d - F_{dr})/M}{(T_{\max} - rF_d)r/J_w} \tag{7}$$

With a designed α, when the vehicle enters a slippery road, $T_{\max}$ must be reduced adaptively following the decrease of F_d to satisfy (7), the no-slip condition.

Since the friction force from the road is available from (6), the maximum transmissible torque, $T_{\max}$ can be calculated as (8). This formula indicates that a given F_d allows a certain maximum torque output from the wheel so as not to increase the slip. Here, it must be pointed out that driving resistance, F_{dr}, is assumed to be zero, which will result in an over evaluation of $T_{\max}$ and consequently impair the anti-slip performance. However, F_{dr} is a variable related with the chassis velocity and the vehicle shape, and can be calculated or estimated in real time if higher anti-slip performance is required or if the vehicle runs at high speed (Sakai et al., 1999; Cao et al., 2006). Although the vehicle mass, M can also be estimated online (Ikeda et al., 1992; Winstead & Kolmanovsky, 2005; Phornsuk et al., 2006), in this paper it is assumed to be constant. Due to these assumptions, the following sections will evaluate and analyze the robustness to disturbances in driving resistance and the perturbation in vehicle mass.

$$T_{\max} = \left(\frac{J_w}{\alpha M r^2} + 1 \right) r F_d \tag{8}$$

Finally, the proposed controller can use $T_{\max}$ to constrain the torque reference if necessary.

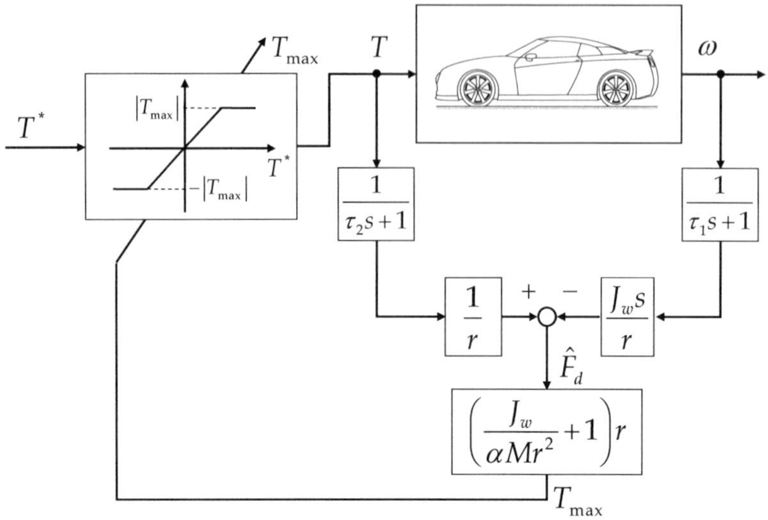

Fig. 4. Primary control system based on MTTE

2.3 Controller design

The torque controller is designed as in Fig. 4, in which the limiter with a variable saturation value is expected to realize the control of torque output according to the dynamic situation. Under normal conditions, the torque reference is expected to pass through the controller without any effect. On the other hand, when on a slippery road, the controller can constrain the torque output to be close to T_{max}.

Firstly, the estimator uses the driving torque generated by the motor and the rotation speed of the wheel to calculate the friction force, and then estimates the maximum transmissible torque according to (8). Finally, the controller utilizes the estimated torque value as a saturation value to limit the commanded torque, T^*. In essence, the estimation shown in Fig. 4 is a disturbance observer.

Here, although it will cause some phase shift, duo to the low resolution of the shaft encoder installed in the wheel, a low pass filter (LPF) with a time constant of τ_1 is introduced to smooth the digital signal, ω, for the differentiator which follows. In order to keep the filtered signals in phase, another LPF with a time constant of τ_2 is added for T.

In the actual controllers, the wheel radius, r, in the blocks can be combined to reduce the calculations.

2.4 Stability analysis

Considering that the Magic Formula included in the vehicle model shown in Fig. 2 is non-linear, this work makes use of an equivalent model for stability analysis to decide on parameters.

Slip occurs when part of the outputted torque cannot be transmitted to the chassis by the tire-road interaction, resulting in lower chassis acceleration than that of wheel. Here, (9) uses Δ to describe the ratio of the under-transmitted torque.

In addition, taking into account the ideal state and the worst-case slip in which the wheel spins completely idly, that is, the inertia of the whole system is equal to the inertia of wheel, J_w, the variation range of Δ, as (10), is available too.

$$\dot{V} - \dot{V}_w = -\frac{\Delta T}{Mr} \tag{9}$$

$$\Delta \in [0,\ Mr^2/J_w] \tag{10}$$

According to (1), (2) and (9), the dynamic longitudinal model of the vehicle can be simplified as in (11), a Single-Input-Single-Output system which masks the complicated interaction among tire, chassis and road, which contributes to the stability analysis. That is, the unwanted wheel acceleration that causes slip can be regarded as the result of a decrease in system inertia. And, Δ can also be treated as a description of variation in system inertia.

$$J\dot{\omega} = T \tag{11}$$

Here, as shown in (12), J is the equivalent inertia of the whole vehicle system from the viewpoint of the driving wheel, and J_n the nominal inertia where no slip occurs.

$$J = \frac{J_n}{1+\Delta},\ J_n = J_w + Mr^2 \tag{12}$$

Consequently, use is made of (12) to take place of the vehicle model shown in Fig. 4 for stability analysis. When the vehicle rapidly accelerates on a slippery road, the estimated T_{max} will constrain T^* and take its place to be treated as the input value to the motor. In this case, the whole system will automatically transform into a closed feedback system, as shown in Fig. 5.

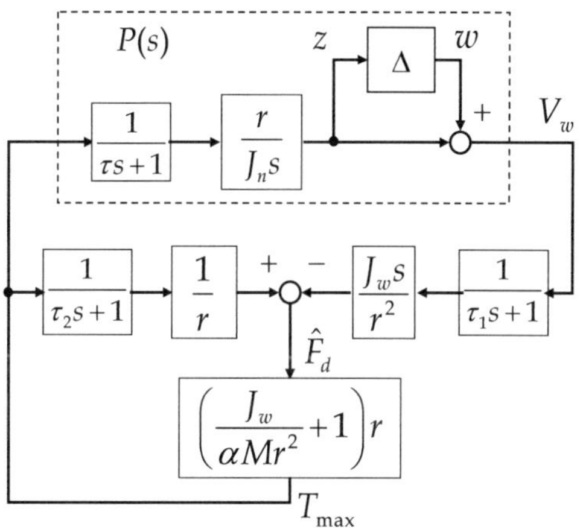

Fig. 5. Equivalent closed-loop control system

Here, in order to analyze the stability easily, the delay of the electro-mechanical system is simplified as a LPF with a time constant of τ. However, in the following simulation, according to the behaviour of real vehicle, the system delay is simulated by a pure delay and a LPF.

The system of Fig. 6, which is the equivalent block diagram of Fig. 5, is used for the analysis of the closed-loop stability against Δ, the model variation. T_{zw}, the transfer function from w to z in Fig. 6, is described in (13).

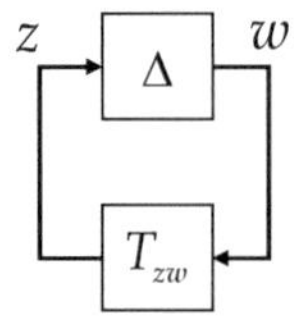

Fig. 6. Equivalent block diagram for stability analysis

$$T_{zw} = \frac{-J_w K}{J_n r \tau \tau_1 s^2 + J_n((r-K)\tau + r\tau_1)s + J_n r - Mr^2 K} \tag{13}$$

Here,

$$K = \left(\frac{J_w}{\alpha M r^2} + 1 \right) r \tag{14}$$

As a result, the following conditions in (16) must be satisfied to ensure the closed-loop stability, that is, ensure the real part of the roots of the characteristic equation, (15), to be negative (Ioannou & Sun, 1995). Here, τ_2 is assumed equal to τ_1 to simplify the solution.

$$1 - T_{zw}\Delta = 0 \tag{15}$$

$$\begin{cases} \alpha > \dfrac{1 - J_w \Delta / Mr^2}{1+\Delta} \\ \tau_1 > \dfrac{J_w \tau}{\alpha M r^2} \end{cases} \tag{16}$$

It can be found in (16) that if there is no limiter, when the vehicle runs in a normal state, a must be larger than 1 to fulfil the requirement for stability. However, considering (8), when a is larger than one, T_{max} will be always restrained to be smaller than the torque that the tire-road interface can provide, which will impair the acceleration performance. Therefore, in this work, a is designed to be slightly smaller than one to ensure acceleration performance while improving the antislip performance.

2.5 Compensation for acceleration performance

In real experiments, even in normal road conditions, T_{max} may be smaller than T^* due to system delay at the acceleration start, which will cause suddenly commanded acceleration to be temporarily constrained by T_{max} during the acceleration phase.

In order to avoid this problem, the increasing rate of T^* is amplified as a stimulation to force the under-evaluated T_{max} to the acceleration reference. T_{max} is used instead of T_{max} as the input to the controller, whose relation is described by (17). Here, G is a compensation gain. Additionally, the over-expanded T_{max} can be automatically constrained by the following controller.

$$T'_{max} = T_{max} + \dot{T}^* G \; (\dot{T}^* > 0) \tag{17}$$

The value of G should be decided corresponding to the total delay in the system and the maximum acceleration rate that the vehicle permits. Generally, larger delay or a higher acceleration rate requires a larger value for G.

Fig. 7. COMS3 - a new experimental electric vehicle

3. Experiments and simulation

3.1 Experimental electric vehicle

In order to implement and verify the proposed control system, a commercial electric vehicle, COMS, which is made by TOYOTA AUTO BODY Co. Ltd., shown in Fig. 7 was modified to fulfil the experiments' requirements. Each rear wheel is equipped with an Interior Permanent Magnet Synchronous Motor (IPMSM) and can be controlled independently after modification.

As illustrated in Fig. 8, a control computer is added to take the place of the previous ECU to operate the motion control. The computer receives the acceleration reference signal from the acceleration pedal sensor, the forward/backward signal from the shift switch and the wheel rotation from the inverter. Then, the calculated torque reference of the left and the right rear wheel are independently sent to the inverter by two analog signal lines. Table 2 lists the main specifications of the experimental electric vehicle.

3.2 Experiments

Based on the algorithm shown in Fig. 4, the controllers were synthesized and discretized, and then implemented to COMS3 for experiments.

In these experiments, the slippery road was simulated by an acrylic sheet with a length of 1.2 m and lubricated with water. However, in all of the experiments, the friction coefficient between the tire and road surface was not measured and was unknown.

The initial velocity of the vehicle was set higher than 1 m/s to avoid the immeasurable zone of the shaft sensors installed in the wheels. However, the results of some additional experiments which were not included in this dissertation proved that the proposed control algorithm can also work when the vehicle starts off on a slippery road. In the comparative experiments, the initial velocities were kept to the same value as much as possible.

Here, it must be pointed out that in order to detect the chassis velocity and avoid mounting additional sensors for the experiment, only the left rear wheel is driven by the motor, while the right rear wheel rolls freely to provide a reference value of the chassis velocity for comparison. Due to the short track and small torque, the yaw moment resulting from the difference of torque between the left and the right wheels was too small to affect the

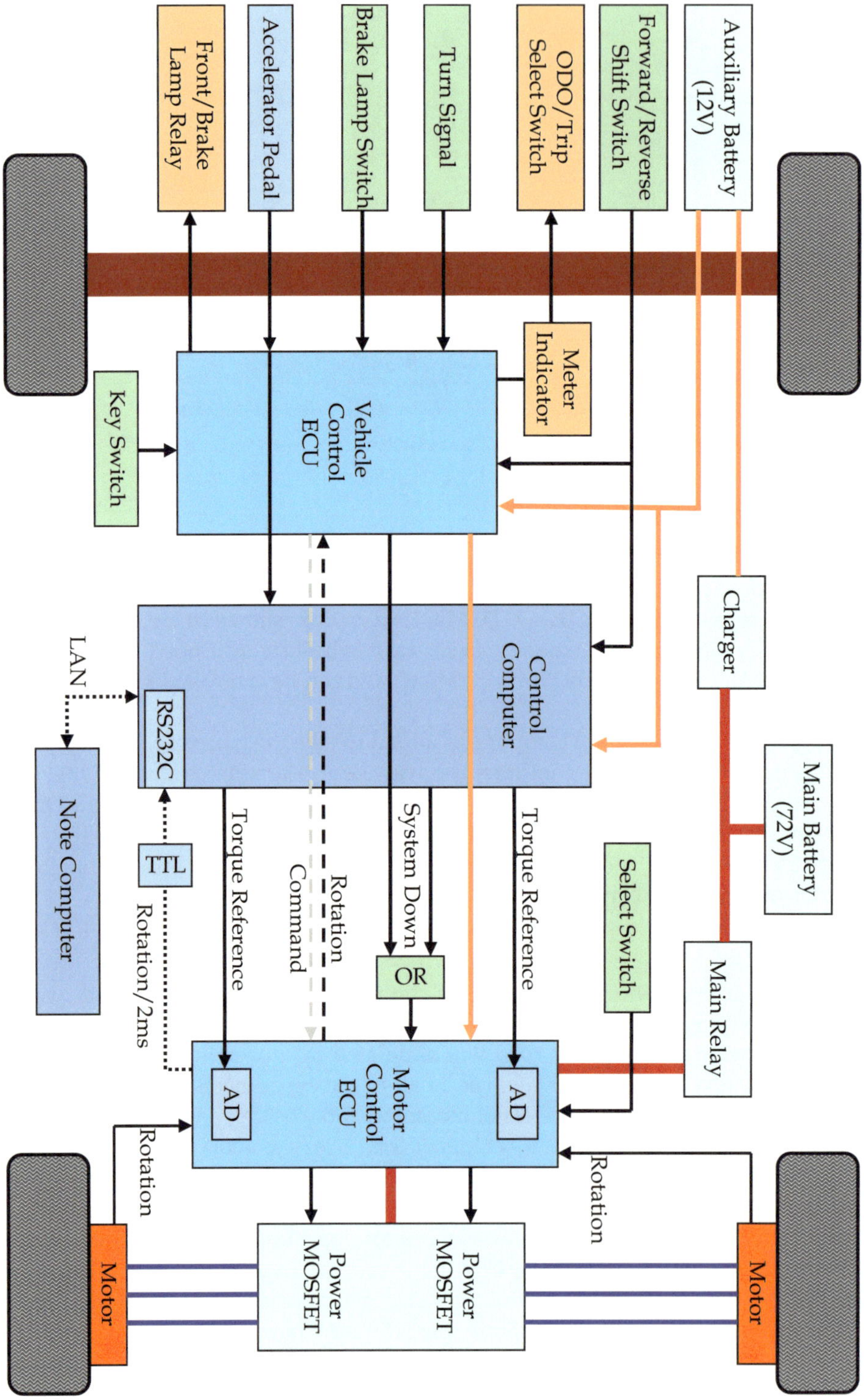

Fig. 8. Schematic of the electrical system of COMS3

Total Weight	360 kg
Max. Power	2000 W × 2
Max. Torque	100 Nm × 2
Wheel Inertia	0.5 kgm2 × 2
Wheel Radius	0.22 m
Sampling Time	0.01 s
Controller	PentiumM 1.8G, 1 GB RAM
A/D and D/A	12 bit
Shaft Encoder	36 pulse/round

Table 2. Specification of COMS3

experiments. Furthermore, this experimental method makes the vehicle skid more easily on the simulated slippery road.

In the following experiments where the robustness to disturbances in driving resistance was evaluated, the vehicle was attached to concrete blocks that were connected in a chain and dragged along the road to simulate the driving resistance due to wind or an uphill slope. The simulated driving resistance was measured by a dynamometer.

Fig. 9 describes the comparison of control performance between the control based on MTTE and MFC, as well as the non-control case.

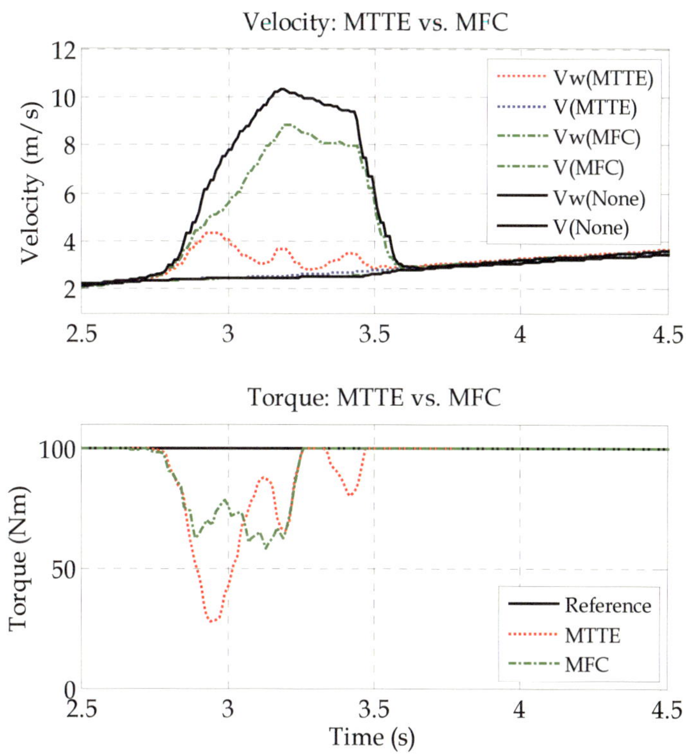

Fig. 9. Comparison of experimental results of two control designs

Because the vehicle mass varies significantly in a real driving environment, In order to evaluate the robustness to variation in vehicle mass, some comparative experiments were performed with different nominal mass in the proposed controller while keeping the real vehicle mass fixed at 360 kg. Fig. 10 provides these comparative results, in which the nominal mass varied from 180 kg to 360 kg.

Fig. 11 describes the results of the experiment with driving resistance. In these experiments, the driving resistance was simulated at 230 N, corresponding to the air resistance of a BMW 8-series running at a speed of 86 km/h.

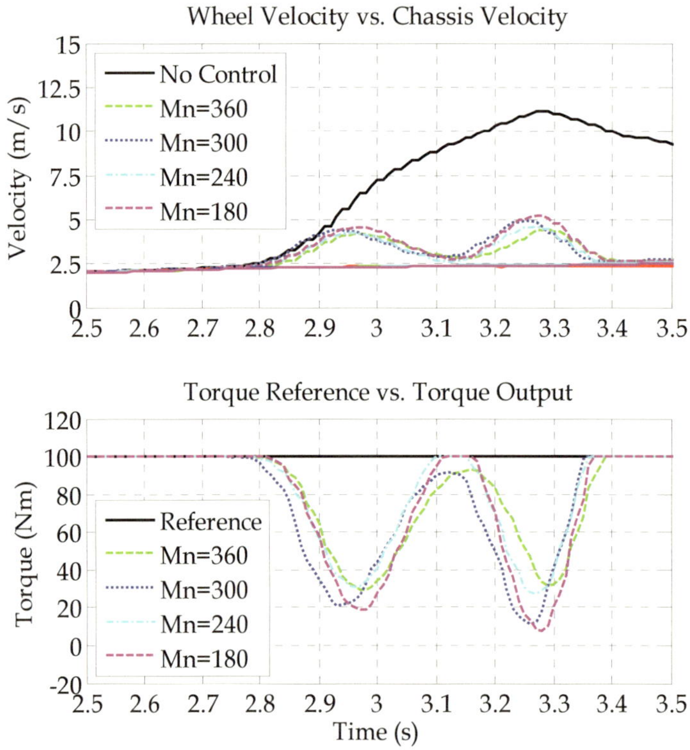

Fig. 10. Comparison of experimental results with variation in vehicle mass

3.2 Simulation

Due to the limitation of experimental equipment and facilities, the existing experimental results could not provide a comprehensive comparative demonstration to sufficiently reflect the features and essence of the proposed control topology. Therefore, numerical simulations were performed to provide more detailed comparisons and analysis, in which parameters could be set more precisely, providing finer insight into the controller behaviour than is possible through experiments alone. The following analysis and discussion will based on the experimental results, as well as on the simulation results.

Simulation systems were synthesized based on the algorithm in Fig. 4 and the vehicle model in Fig. 2. Here, it is must pointed that, in the vehicle model, a LPF with a time constant of 1 ms is placed after the friction coefficient to simulate the dynamics of the tire. The time constant is linearly related to the softness of the tire.

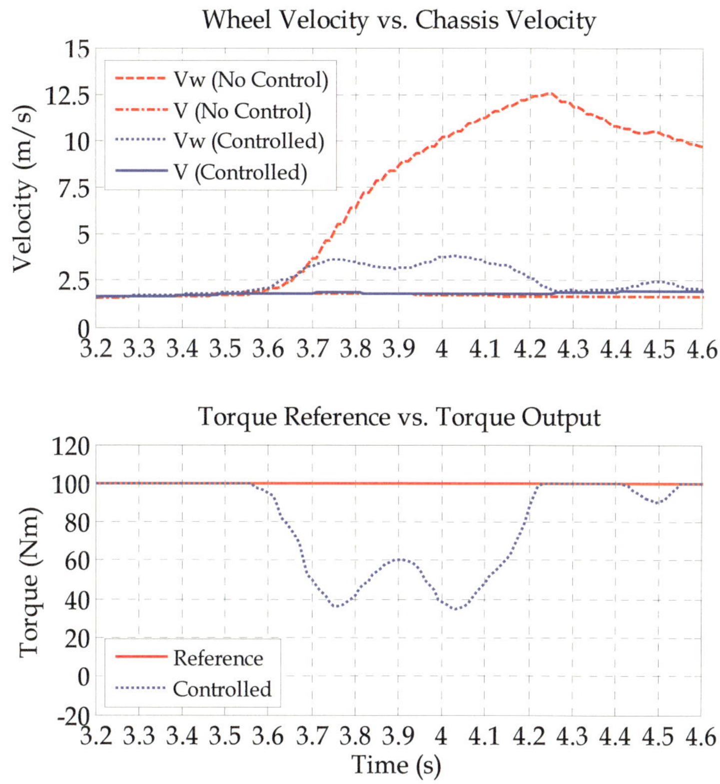

Fig. 11. Comparison of experimental results with driving resistance of 230 N

Fig. 12 illustrates the stability of the control system in which a is designed to be 0.9 for two different slip states. In these simulations, the system delay and time constants in LPFs are shortened to make the primary tendency clear. Here, system delay was set to 0, and time constants in the LPFs were set to 20 ms. In the simulation of severe slip, the maximum friction coefficient is set to 0.3, and a slight slip of 0.6. Here, the commonly used λ and Δ values are utilized to describe the extent of slip.

4. Discussion

4.1 Stability analysis

When a is smaller than one, according to (1) and (8), it can be found that when the vehicle runs in no-slip conditions, as described as (18), $T_{\max}$ will be larger than T^*, and the unwanted torque will be eliminated by the limiter, which keeps the system stable and responsive to the driver's torque reference. In this case, the torque reference from the driver passes through the controller and enters the motor, which makes the whole system operate like an open-loop control system.

$$T_{\max} = T + \frac{J_w}{r}\left(\frac{1}{\alpha}\dot{V} - \dot{V}_w\right) \tag{18}$$

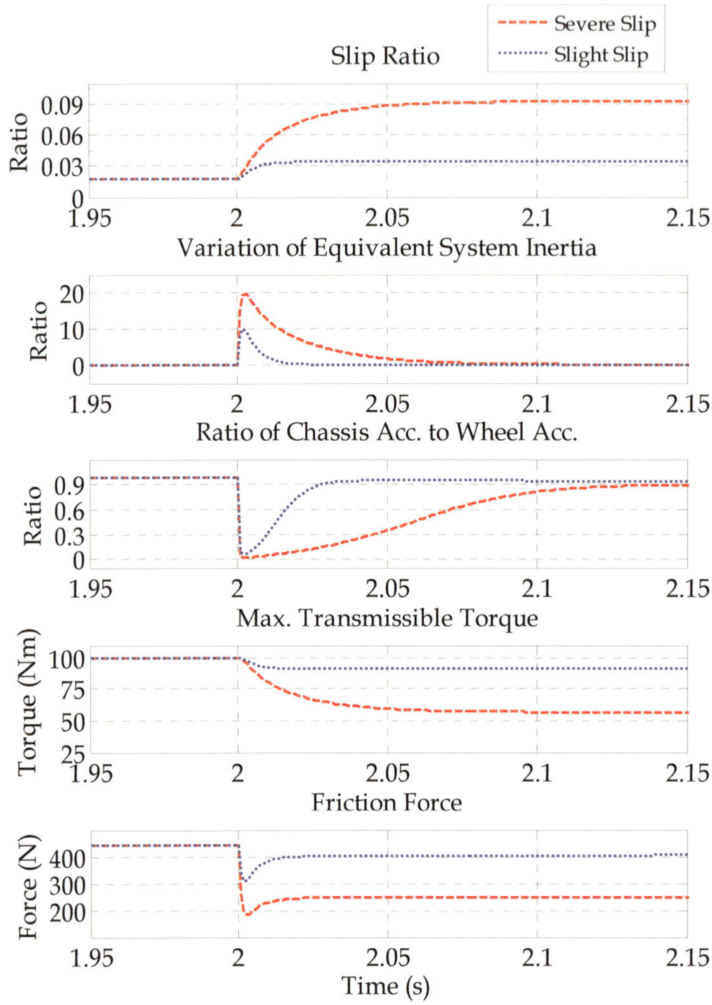

Fig. 12. System stability in different slippery conditions

On the other hand, when the vehicle enters a slippery road, as described in Fig. 12, due to the system delay, a sudden slip will occur at the first, and then, the whole system will work in two different states:

1. Slight slip that makes (16) valid, that is, the system is theoretically unstable. Therefore, at this time, the difference between the accelerations of the wheel and the chassis is in an expanding state, and this state will last until the ratio of the acceleration of the chassis to that of the wheel arrives at the designed value, a. However, a well designed a will allow T_{max} to rise to increase the slip properly, according to the Magic Formula, so as to provide an increased friction force, as expected.
2. Severe slip that satisfies (16) occurs. The system is stable, that is, T_{max} will become smaller and smaller to restrain the slip. The ratio of the acceleration of the chassis to that of the wheel will become larger and larger to meet the designed a.

In conclusion, the simulations and experiments indicate that, a relaxation factor a which is smaller than 1 makes the system work in a critical state, which results in the best antislip performance while keeping the system stable.

4.2 Control performance

Fig. 9 also shows that compared to the no-control case, the difference between the wheel velocity and the chassis velocity caused mainly by the delay in the control system does not increase. The estimated maximum transmissible torque is close to the input reference torque in the normal road, and corresponds to the maximum friction force allowed by the slippery tire-road surface. Moreover, the comparison with the control based on MFC demonstrated this control performance furthermore (Yin et al., 2009).

Perturbations in vehicle mass and the disturbances in driving resistance are usually the most uncertain factors in real driving environments. Fig. 10 and Fig. 11 show that the proposed control system has high robustness to perturbations in vehicle mass and disturbances in driving resistance.

4.3 Theoretical analysis with partially linearized model

Detailed analysis of control characteristics can be performed with a partially linearized vehicle model. When the control system operates in the closed-loop control state, if only the basic fact that the friction coefficient decreases with the velocity difference between the wheel and chassis is considered, the whole controlled system can be simplified as in Fig. 13.

Note that when the vehicle accelerates in a normal condition, that is, the friction force increases with the slip, considering the nonlinear tire model, in this model K_u should be a varying negative value. However, K_u is partially assumed to be positive, because the following discussion is limited to the condition where the slip ratio is in the unstable area as shown in Fig. 3.

In Fig. 13, F_{do} is the friction force between the tire and road surface when antislip control starts. M_n denotes the nominal vehicle mass and generally it is equal to the mass of the vehicle and the driver. Here, in order to simplify the analysis, the delay in the vehicle system is ignored and τ_2 is assumed to be equal to τ_1. Differentiation of the difference between the velocities is introduced due to the integral relation between the forces and the velocities.

The transfer function from F_{do} to A_D, and the one from F_{dr} to A_D can be defined as T_{Ad} and T_{Ar} respectively, in which J is the equivalent system inertia.

$$J = J_w + Mr^2 \tag{19}$$

$$T_{Ad} = \frac{-J\tau_1 s^2 + \left(\dfrac{M}{\alpha M_n} - 1\right) J_w s}{MJ_w \tau_1 s^2 + \left(MJ_w - JK_u \tau_1\right) s + \left(\dfrac{M}{\alpha M_n} - 1\right) J_w K_u} \tag{20}$$

$$T_{Ar} = \frac{J_w \tau_1 s^2 + J_w s}{MJ_w \tau_1 s^2 + \left(MJ_w - JK_u \tau_1\right) s + \left(\dfrac{M}{\alpha M_n} - 1\right) J_w K_u} \tag{21}$$

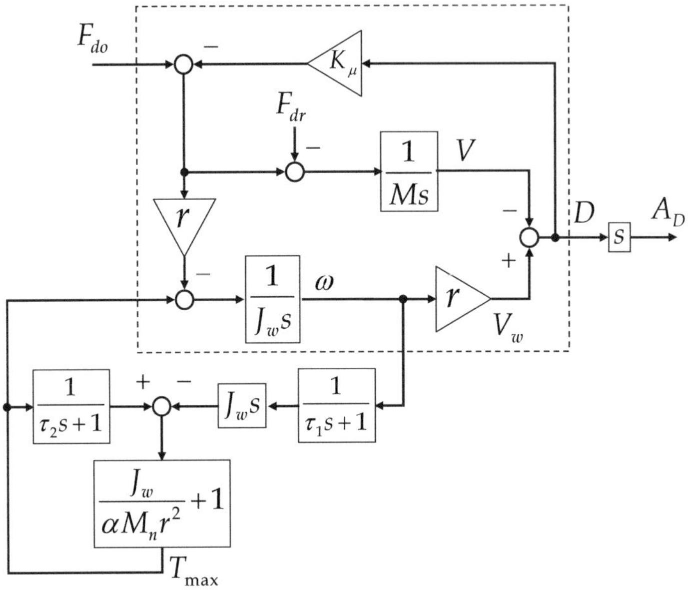

Fig. 13. Partially linearized control system

Here, Equation (20) is used as an example to examine the relation between the friction force and the differential of the velocity difference as follows.

$$\begin{aligned}
\lim_{t\to\infty}\frac{A_D(t)}{F_{do}(t)} &= \lim_{s\to 0} T_{Ad}(s) \\
&< \lim_{s\to 0}\frac{-J\tau_1 s^2+\left(\dfrac{M}{\alpha M_n}-1\right)J_w s}{MJ_w\tau_1 s^2+\left(MJ_w-JK_u\tau_1\right)s} \\
&= \frac{\dfrac{M}{\alpha M_n}-1}{MJ_w-JK_u\tau_1}J_w
\end{aligned} \tag{22}$$

Equation (22) shows that the differential of the slip is proportional to the friction force, which agrees with the additional simulation results shown in Fig. 14, in which small time constants in LPFs and small delay in the electromechanical system were adopted to examine the relation between the friction force and the velocity difference, *a* was set to 0.5, and the velocity difference representing the extent of the slip is equal to the difference between the wheel velocity and chassis velocity.

Equation (22) also indicates that when τ_1 or K_u is considerably larger, that is, the system delay causes the controller to be unable to follow the quickly varying friction force, the slip will become larger. Here, F_{do} is assumed to be a unit step reference.

4.4 Applied for two-degree-of-freedom motion control

From the viewpoint of the object of control, in the traction control based on MFC, the nominal model in the observer is the whole vehicle which varies drastically when the

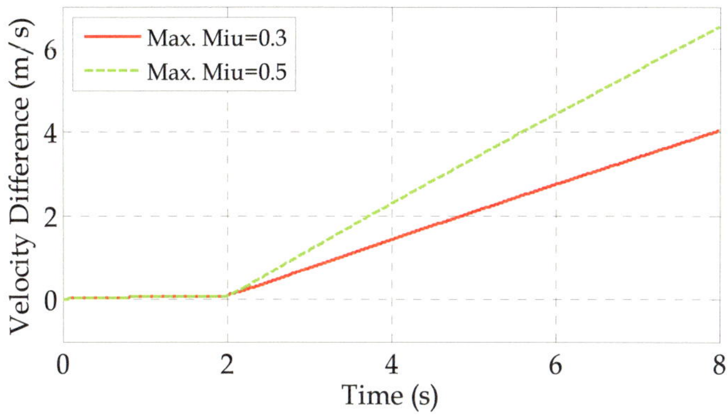

Fig. 14. Comparison of simulation results in different roads with small time constants

vehicle mass changes or load transfer among the driving wheels occurs during acceleration, braking or when the vehicle enters a curve. Therefore, this type of control method can be thought of as a control for the chassis behavior, which introduces the possibility that this control will interfere with other chassis control systems.

In contrast to this, the proposed control method is essentially based on control of wheel behavior rather than control of the whole chassis behavior, which decreases the interference with chassis controls, and also provides a sound independent basis for them.

Here, this work presents an example to demonstrate the application of the proposed traction control to planar motion control of electric vehicles. Fig. 15 shows the architecture of a Direct Yaw Moment Control enhanced by MTTE. Here, the steering-wheel angle, the estimated vehicle linear velocity and friction coefficient are used to calculate the nominal yaw rate, then the difference between the measured yaw velocity and the nominal yaw rate acts as the input to the traction distribution module. Finally, the traction distribution module generates the nominal torque reference for the MTTE.

5. Conclusion

This work proposed an estimator of maximum transmissible torque and applied it to the control of the driving motors in electric vehicles for slip prevention. This estimator, which does not calculate chassis velocity, instead using only the input torque and output rotation of the wheel, provides a good foundation for antislip control. The effectiveness of the estimation demonstrated that motors can act not only as actuators but also as a good platform for state estimation because of their inherently fast and accurate torque response. The experiments and simulations verified the effectiveness of the estimation in antislip control. Additionally, this estimator is also expected to provide the maximum transmissible torque for other vehicle control systems to enhance their control performance when the vehicle runs in slippery conditions.

The controller designed to co-operate with the estimator can provide higher antislip performance while maintaining control stability. When excessive torque is commanded, this controller constrains the control output to follow the actual maximum driving force between the tire and the road surface to prevent slip. In addition, the acceleration compensation resolved the problem of deterioration of pedal response due to system delay.

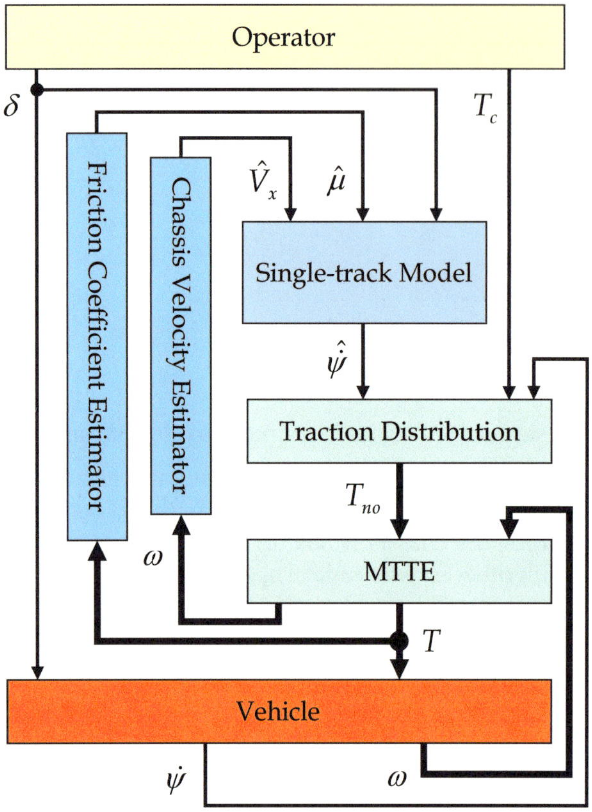

Fig. 15. A DYC system based on MTTE

Comparative experiments and simulations with variation of control variables and road conditions demonstrated the advantages of the proposed control design. The proposed control not only has a better antislip performance and higher adaptability in different tire-road conditions, but also has greater robustness to perturbations in vehicle mass and disturbances in driving resistance. These features enhance the applicability and practicality of this method. Furthermore, the detailed discussion and analysis on the experiments and simulations agreed with above evaluations and provided theoretical support for the proposed control topology. These advantages qualify the proposed control method as a general approach for traction control, as well as a basis for more sophisticated and advanced motion control in electric vehicles.

6. References

Affanni, A; Bellini, A; Franceschini, G; Guglielmi, P; & Tassoni, C. (2005). Battery Choice and Management for New-Generation Electric Vehicles, IEEE Transaction on Industrial Electronics, Vol. 52, No. 5, pp. 1343-1349.

Akiba, T.; Shirato, R.; Fujita, T. & Tamura, J. (2007). A study of Novel Traction Control Model for Electric Motor Driven Vehicle, Proceeding of 4th Power Conversion Conference-NAGOYA, pp. 699-704.

Cao, J. Y.; Cao, B. G. & Liu, Z. (2006). Driving Resistance Estimation Based on Unknown Input Observer, Journal of Applied Sciences, pp. 888-891.

Chau, K. T.; Chan, C. C. & Liu, C. (2008). Overview of Permanent-Magnet Brushless Drives for Electric and Hybrid Electric Vehicles, IEEE Transaction on Industrial Electronics, Vol. 55, No. 6, pp. 2246-2257.

Fujii, K. & Fujimoto, H. (2007). Traction control based on slip ratio estimation without detecting vehicle speed for electric vehicle, Proceeding of Fourth Power Conversion Conference-NAGOYA, pp.688-693.

Fujimoto, H.; Saito, T.; & Noguchi, T. (2004). Motion Stabilization Control of Electric Vehicle under Snowy Conditions Based on Yaw-Moment Observer, Proceeding of IEEE International Workshop on Advanced Motion Control 2004, pp.35-40.

Hori, Y. (2004). Future Vehicle Driven by Electricity and Control -Research on Four-Wheel-Motored "UOT Electric March II", IEEE Transaction on Industrial Electronics, Vol. 51, No. 5 pp. 954-962.

Ikeda M.; Ono T. & Aoki N. (1992). Dynamic mass measurement of moving vehicles, Transactions of the Society of Instrument and Control Engineers, Vol. 28, No. 1, pp. 50-58.

Lee, H. & Tomizuka, M. (2003). Adaptive Vehicle Traction Force Control for Intelligent Vehicle Highway Systems (IVHSs), IEEE Transaction on Industrial Electronics, Vol. 50, No. 1, pp. 37-47.

Mutoh, N.; Hayano, Y.; Yahagi, H. & Takita, K. (2007). Electric Braking Control Methods for Electric Vehicles With Independently Driven Front and Rear Wheels, IEEE Transaction on Industrial Electronics, Vol. 54, No. 2, pp.1168-1176.

Nagai, M. (2007). The perspectives of research for enhancing active safety based on advanced control technology, Vehicle System Dynamics, Vol. 45, No. 5, pp. 413-431.

Pacejka, H. B. & Bakker, E. (1992). The Magic Formula Tyre Model, Vehicle System Dynamics, Vol. 21, No. 1, pp. 1-18.

Ioannou P. A. & Sun J. (1995). Robust Adaptive Control, Prentice Hall, pp. 79-104.

Phornsuk, R. A.; Hiroshi, H.; Masatoshi, A. & Shigeto, O. (2006). Adaptive controller design for anti-slip system of EV, Procceding of 2006 IEEE Conference on Robotics, Automation and Mechatronics, No. 4018749.

Saito, T.; Fujimoto, H. & Noguchi, T. (2002). Yaw-Moment Stabilization Control of Small Electric Vehicle, Procceding of Industrial Instrumentation and Control, IEE Japan, pp. 83-88.

Sakai, S. & Hori, Y. (2001). Advantage of Electric motor for anti skid control of electric vehicle, European Power Electronics Journal, vol. 11, No. 4, pp. 26-32.

Sakai, S.; Sado, H. & Hori, Y. (1999). Motion Control in an Electric Vehicle with 4 Independently Driven In-Wheel Motors, IEEE Transaction on Mechatronics, vol.4, No.1, pp. 9-16.

Suryanarayanan, S. & Tomizuka, M. (2007). Appropriate Sensor Placement for Fault-Tolerant Lane-Keeping Control of Automated Vehicles, IEEE/ASME Transactions on Mechatronics, Vol. 12, No. 4, pp. 465-471.

Turner, J. D. & Austin, L. (2000). Sensors for Automotive Telematics, Measurement Science & Technology, Vol. 11, No. 2, pp. 58-79.

Winstead, V. & Kolmanovsky, I. V. (2005). Estimation of Road Grade and Vehicle Mass via Model PredictiveControl, Proceeding of IEEE Conference on Control Applications 2005, WC2.3.

Yin, D.; Oh, S.; & Hori, Y. (2009). A Novel Traction Control for EV Based on Maximum Transmissible Torque Estimation, IEEE Transaction on Industrial Electronics, Vol. 56, No. 6, pp. 2086-2094.

7

Formal Verification of Hybrid Automotive Systems

Jairam Sukumar[1], Subir K Roy[1], Kusum Lata[2] and Navakanta Bhat[2]
[1]*Texas Instruments*
[2]*Indian Institute of Science*
India

1. Introduction

State of the art automotive systems have become extremely complex in terms of functionality, system architecture and implementation. These systems consist of significant portions of embedded software as well, running in excess of millions of lines of C code. Given such enormous complexities it is never an easy task to verify their functional correctness. Traditional simulation based validation schemes cannot hope to cover the entire specification domain of such a system, as creating test cases to cover all the different corner cases of functional behaviour is extremely difficult. The number of such test cases can be extremely large; it is typically a manual process, at best. Moreover, the time taken to run all the test cases to verify the complete system under all possible scenarios, environmental constraints and input conditions will be inordinately large. This can strongly impact the system design and implementation turn-around-time, thereby reducing the window of opportunity in the market for the product. The amount of complex components being integrated into main stream automotive systems is continuously on the rise, latest being MEMS based sensor components. These MEMS components are becoming a part and parcel of modern automotive electronics systems, ranging from dynamic vehicle performance systems and safety systems to, infotainment and other accessories. To illustrate an example, the heart of any modern automotive engine cruise control (ACC) system includes MEMS based sensors, such as, accelerometers and gyroscopes. The verification of such MEMS components is, in general, hard and difficult, and their integration into a hybrid system exacerbates an already complex hybrid system verification challenge.
Formal analysis of hybrid systems integrating MEMS components, such as sensors, requires their models to be comprehended and integrated into the model of the hybrid system itself. While the cruise control behaviour is primarily discrete, the MEMS sensor behaviour is continuous and highly non-linear. Thus, a system designer will need to validate, both, the discrete time and the continuous time behaviour in any such hybrid system. The single largest challenge in the verification of such hybrid systems lies in appropriately modelling individual components and integrating them within a single verification framework. Another strong motivation in employing formal verification for the validation of hybrid systems, also, is to circumvent the well known drawbacks inherent in a simulation based validation approach. In the domain of discrete time formal verification, a wide range of commercial tools exist which have matured with an increasingly large user base. These

tools, however, are incapable of validating traditional hybrid systems, let alone, hybrid systems integrating MEMS based sensor components. In this chapter, we first present the complexity inherent in hybrid systems and then show how formal verification can be employed to verify them.

2. Advances in automotive systems

Advances in automotive electronics have triggered large scale integration of myriad features for all categories of vehicles. The two fundamental vectors of feature development include safety and passenger comfort. These features can be further classified based on their means of integration, viz., passive and active. While the passive features amount to pre-installed and pre-verified components, each active feature needs a continuous sampling of ambient conditions and based on the condition, feedback actions are taken by the vehicular controller. All these features hence also require processing capabilities much higher than requirements needed in the past. Some of the active mode examples include integrated park and lane assistant, cruise control, climate control, etc.

Advent of these new features also brings to the horizon, the challenge of modelling and analyzing new types of sensors to support these features. We focus on one such type: the MEMS based sensors. The computational complexity of modelling MEMS components, increases manifolds as one needs to solve or model non-linear partial differential equations (PDE) to accurately capture the sensor characteristics.

This chapter attempts to address the challenges posed by integration of these components into mainstream VLSI systems and their verification. The chapter helps the reader to appreciate the nuances of verification methodologies and propose recipes to formally verify computationally complex hybrid automotive systems. A methodology is described, based on transformation techniques that can be deployed to solve these problems. We introduce Simulink-Stateflow based modelling and verification platform [1], as shown in Figure 1. The figure explains how the continuous and discrete time systems interact together in a Matlab based environment. It is shown how this can be used to integrate some of the complex MEMS based components into mainstream VLSI systems. We also introduce the reader to CheckMate, a formal analysis software tool available in public domain. Simulation traces from Simulink-Stateflow framework are used in the formal analysis engine present in CheckMate, instead of being derived implicitly by numerical integration, or explicitly by transformation based approaches.

To illustrate this approach, we choose one representative real life control system, fairly popular, from the automotive hybrid system space. This hybrid system directly interfaces with a MEMS based gyroscope used as a speed sensor. We then deploy our proposed approach to perform a formal analysis of the MEMS integrated ACC system (MIACCS). We model the adaptive cruise control system in a traditional simulation framework, and then proceed to build a formal analysis framework for the same system. The chapter is divided into the following sections. Section 2 introduces the reader to the state of the art in hybrid formal verification methodologies. Section 3 explains about the transformation based techniques for formal analysis. We discuss formal analysis platforms in Section 4 and introduce the reader to one of them in Section 5 viz. CheckMate, which works on Simulink-Stateflow (SSF) based system. SSF based methods are the most widely used platforms across the industry for hybrid and real time system. Finally we illustrate the approach with the case of a MEMS based adaptive cruise control system in Section 6 and through 8.

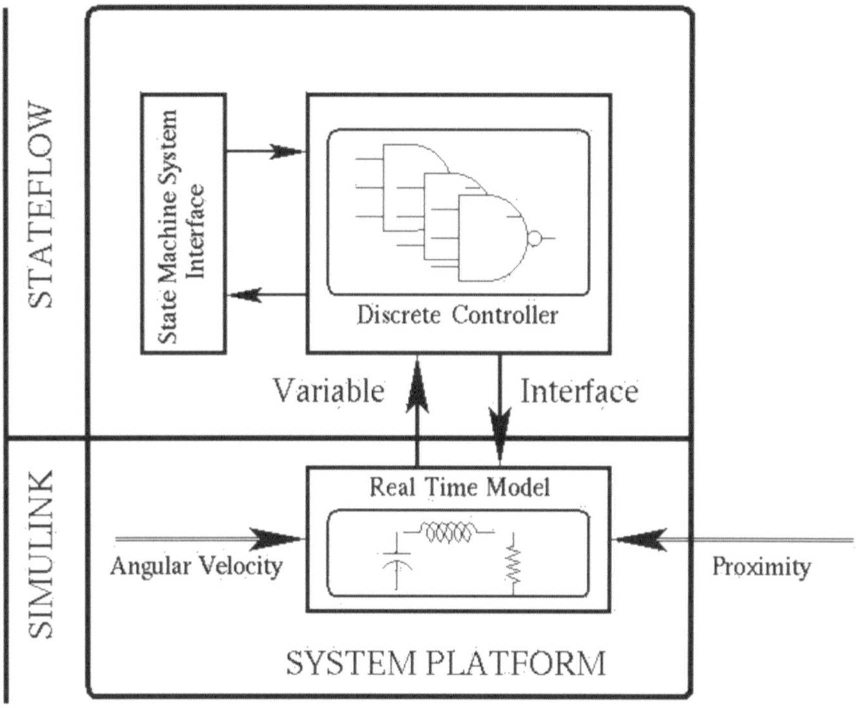

Fig. 1. Simulink Stateflow Platform

3. Hybrid verification methodologies: past, present & future

3.1 Introduction to hybrid formal analysis

Specifically the very low Defective Parts Per Million (DPPM) requirements to meet the stringent automotive fail safety norms and standards, poses new hurdles and therefore newer challenges to overcome them. Thus, it has become a topic of active research in the recent past [2] [3] [4]. Present generation, state of art automobiles, use MEMS based sensors for measurement of different vehicular parameters [5]. Given the fact that, verification of such mainstream automotive control systems need to be based on formal approaches, validation of system behaviour integrating such MEMs components, results in additional complexities. A hybrid system, typically, includes both discrete and continuous time components. To analyze hybrid systems, consisting of, both, discrete behaviour and continuous behaviour components, it becomes necessary to partition these distinct functional behaviours and use specialized analysis engines to target each behaviour domain. A typical hybrid system formulation is shown in Figure 2. First of all, there needs to exist a grammar to formally describe the hybrid system. A given hybrid system is then partitioned into its representative discrete and continuous time systems. Both these partitions are individually solved through their respective solvers. Continuous time solvers include symbolic analysis based solvers, while discrete time solutions can be obtained through traditional graph traversal based methods. Finally both these solutions are integrated by interleaving of variable, at each time step of computation. Decision to change a state is taken based on corresponding conditions being satisfied to trigger these changes.

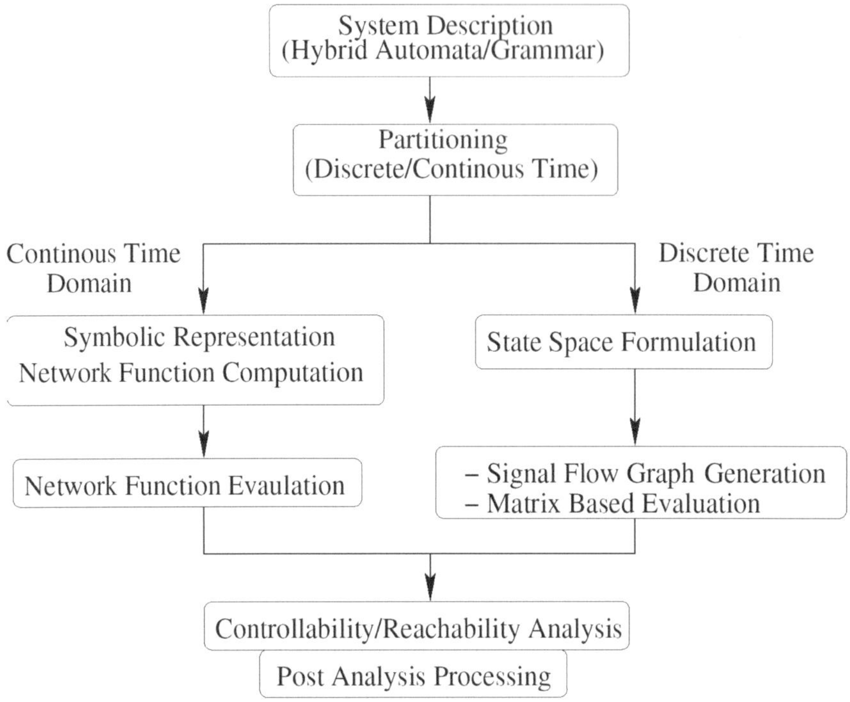

Fig. 2. Formal Verification Flow for Hybrid Systems

For a detailed introduction to the timed and hybrid automata interested readers are referred to [6] and [7]. A discrete time variable needs to be identified and then used to sample the system behaviour in temporal space. After each time step, the system state is determined and the state equations are solved. Changes in the state of the system are then evaluated based on state values and guards corresponding to the transition edges of the automata. The continuous time system formulation is thus, analyzed through time- discretisation and then solved through numerical routines based on Runge-Kutta or Newton-Raphson methods for non-linear dynamic functions [8]. A time discretisation step is followed by linearization of non-integer functions. For each time step, the continuous time domain state variables are concurrently analyzed with the discrete time variables to determine the reachability of the state space of the hybrid system needed to establish truth validity of the safety properties. These terms are explained in later in Section 3.2. Thus, the overall model checking methodology is based on progression in the time domain. Figure 3 illustrates the above methodology.

2.2 Formal analysis of timed automata

The concept of timed automata is illustrated to the reader by a simple example of a temperature controller. A diagram of a temperature controller is shown in Figure 4a. The controller has two states *viz.* ON and OFF. The controller enters into the initial ON state with the temperature variable x being initialized to the room temperature T_{room}. The controller switches the heater which increases the temperature with a linear rate until the temperature attains an upper bound. The heater is cut-off by the controller and the system cools down with a rate given by the differential equation shown in the OFF state (Figure 4b).

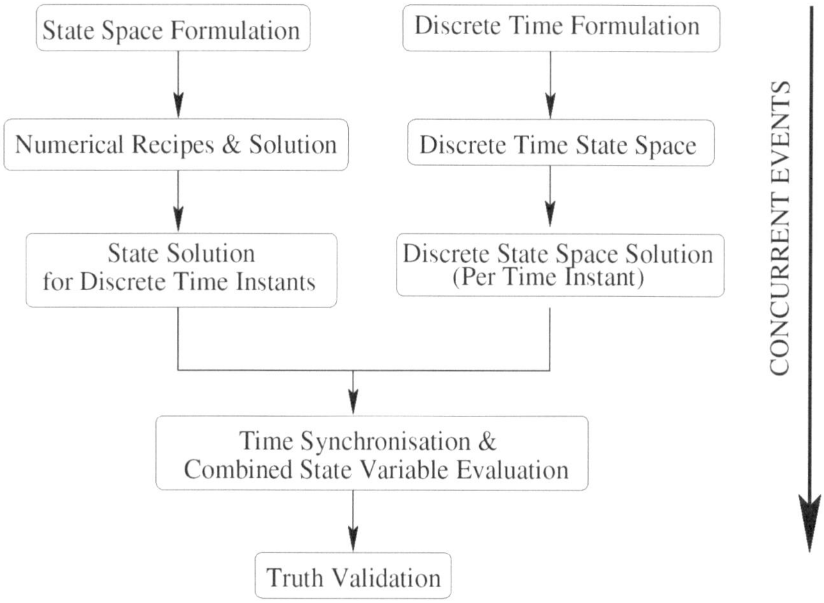

Fig. 3. Formal Verification Flow for Timed Automaton

State equations can be solved using Laplace transforms as follows:
ON STATE:

$$\frac{dx}{dt} = \phi(t) \tag{1}$$

$$X(s) = \frac{1}{s}\phi(s) \tag{2}$$

$$x(t) = \phi(t) + C \tag{3}$$

OFF STATE:

$$\frac{dx}{dt} + \alpha x - f(t) \tag{4}$$

$$sX(s) + \alpha X(s) = F(s) \tag{5}$$

$$X(s) = \frac{F(s)}{s+\alpha} \tag{6}$$

$$x(t) = e^{-\alpha t} f(t) + C \tag{7}$$

It can also be seen that for a class of rational polynomials, there exists a simple partial fraction decomposition which simplifies the Inverse Laplace Transform computation. Equations (3) and (7) are exact functions with respect to the time variable. Thus, exact values of these functions corresponding to each dynamic state variable (for example, temperature)

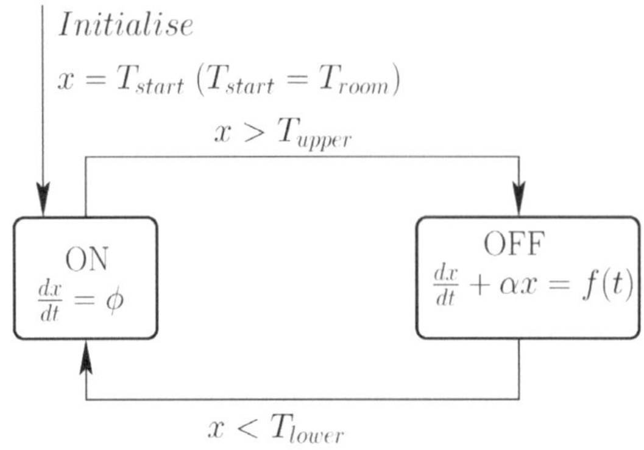

Fig. 4.a State Transition Graph of Temperature Controller

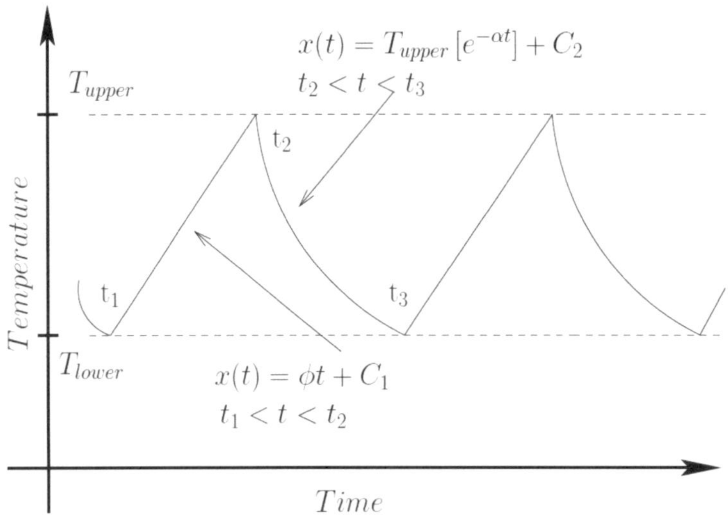

Fig. 4.b System Response

at different time points can be easily and exactly computed to determine the set of reachable states. The steady state solution space, which is also the reachable state space, is shown in Figure 4b. The reachable state space varies with the parameters Φ and α present in the differential equations. The reader should note that, for this parameterized reachable state space corresponding to the temperature variable, constraint solvers need to be deployed to prove the specification correctness.

3. Transformation based formal analysis

State of the art methods of formal analysis are based on timed or hybrid automata.
There also exists a symbolic analytical method, for the solution of system description. If an analytical model can be obtained based on exact modeling, it can provide potential computational benefits. Hence let us explore solving linear differential equations symbolically through Laplace transforms instead of simulation. We can transform the state equations in the time domain and then solve the set of algebraic equations on them through the well known method of pole residue decomposition of transfer functions. The solutions

in the original time domain obtained through inverse Laplace transforms are then bound, based on the constraints imposed on the corresponding differential equations. Constraint solvers can be used to solve for these bounds and establish the truth of the properties.

3.1 Partial decomposition based solution

Let us know look at a simple n^{th} order linear differential equation and assume it to be a part of the continuous time system. Let us also assume all state equations to be defined by linear differential equations. The equations will be of the standard form:

$$A_n \frac{d^n x}{dt^n} + A_{n-1} \frac{d^{n-1} x}{dt^{n-1}} + \ldots + A_1 \frac{dx}{dt} + A_0 x = f(t) \tag{8}$$

The time domain differential equation is re-written as the following:

$$(A_n D_n + A_{n-1} D_{n-1} + \ldots + A_1 D + A_0)x = f(t) \tag{9}$$

Here **D** is the differential operator. Transforming the time domain equation in (8)-(9) into the frequency domain through Laplace transform, we arrive at the following equation:

$$(A_n s^n + A_{n-1} s^{n-1} + \ldots + A_1 s)X(s) = F(s) \tag{10}$$

$$X(s) = \frac{F(s)}{A_n s^n + A_{n-1} s^{n-1} + \ldots + A_1 s} \tag{11}$$

Assuming X(s) can be transformed into a rational polynomial of the type A(s)/B(s), regular polynomial packages like Matlab™ or Octave can be used to convert this rational polynomial into the pole/residue form. The characteristic equation can be re-written as:

$$X(s) = \sum \frac{z_i}{s_i + p_i} + R \tag{12}$$

The time domain solution to (12) can be written as:

$$x(t) = \sum_i z_i e^{p_i t} \tag{13}$$

Simple Matlab/Octave routines can be used to solve a variety of rational polynomials. The reader can try many of such combinations. This approach can be rendered into a simple automated flow which invokes math solvers to solve the standard pole-zero decomposition problems. Using these computed poles and residues it is easy to derive the time domain response of the dynamic components in a hybrid system.

3.2 Constraint formulation and truth validation

Let us now look at the constraint formulation mechanism used in the context of the above example.

We need to formally state the property to be verified. It can be formally described as a constraint. However we need a formal language to describe this constraint. One can imagine a domain space corresponding to the span of a specified property. We call this the constraint space. The constraint space needs to be intersected with the reachable state space (or the solution space) of the state variables obtained by approaches based on Laplace transforms or by numerical integration of differential equations. If the constraint property space subsumes the reachable state space, the correctness of the controller behaviour is established on its formal model. The solution space of the state variable and the property to be checked can be represented as:

$$X(t) = \begin{cases} X_1(t); t \in D_1 \\ X_2(t); t \in D_2 \\ X_3(t); t \in D_3 \end{cases} \tag{14}$$

The set $D1 \cup D2 \cup D3$ provides the complete time domain region of operation of the system. Let, for all specification S_i the constraint space for its corresponding property be P_i. The objective is to obtain the region of intersection of P_i and $X(t)$. For the temperature controller discussed earlier, we explain the constraint analysis methodology. Let us define three properties; that describe three different safety requirements of the controller in example which we mentioned previously:

- Temperature of the system remains between $[T_{upper} + T_{lower}]/2$ and T_{lower}.
- The Temperature of the system is always below $T_{upper.}$
- The Temperature of the system is always between 300C and T_{room}. (Where T_{upper}>300C and T_{room}<T_{lower})

In other words the above three properties can be specified as a part of functional specification of the hybrid system. The functionality of the implemented controller should ensure that these specifications are met. The specifications can be transformed into mathematical inequalities or constraints as shown in the

S/N	Functional Spec	Controller Specifications
1	$T_{lower} < X(t) < [T_{upper} + T_{lower}]/2$	$T_{lower} < x(t) < T_{upper}$
2	$X(t) < T_{upper}$	$T_{lower} < x(t) < T_{upper}$
3	$T_{room} < x(t) < 300; T_{upper} > 300 \,\&\&\, T_{room} < T_{lower}$	$T_{lower} < x(t) < T_{upper}$

TABLE I

A simple method as mentioned above is to formulate the solution surface of the constraints and evaluate the intersection. Solutions that can be incorporated in 2 or 3-dimensional surfaces can be very easily visualized. The solution surface for the properties in Table I can be illustrated as shown in Figure 5. It can be easily noted that the functional specification 1 is not satisfied as the controller specification does not bound the corresponding property. This results in a failure of the property to meat the desired specification.

4. Hybrid analysis platforms

Let us progress to the next level, where we now discuss the complexities involved with ordinary and partial differential equations, and see how we can tackle them. We assume

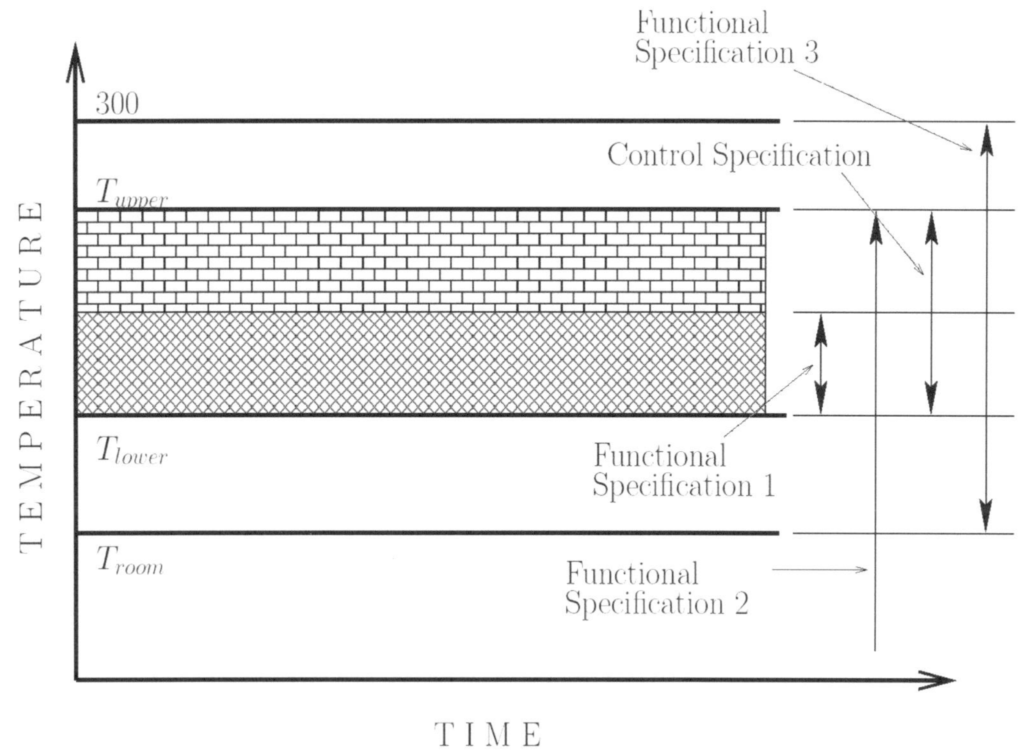

Fig. 5. Steady State System Temperature Response Domains

that the reader is fairly well aware about the basics of differential equations. Analysis of the continuous time domain behaviour modelled by Ordinary/Partial Differential Equations (O/PDE) can be solved by using a Linear Program Solver such as, LP Solver [9] for obtaining the state space solutions and discrete event solvers like SAT solvers [10] for discrete state transitions. In addition, to render the analysis formal, a reasoning engine to interpret the correctness of the behaviour of system implementation against the specified set of formal properties is needed. A fully automated approach would the golden ideal; however, even for the simplest hybrid system with linear behaviour, it has been theoretically proven that establishing the truth value of properties captured in a first order logic on real numbers is un-decidable and NP-hard [11-13] . Therefore, several automated approaches which are inexact or approximate [14,15,16,17] and other non-automated approaches, based on theorem proving, requiring user inputs to drive the formal analysis [18, 19]; have arisen to tackle the intrinsic complexity inherent in hybrid systems. The presence of components like MEMS based sensors renders the formal analysis even harder, with their dynamic behaviour being modelled and described only by higher order non-linear partial differential equations (NL-PDEs). It is well known that behavioural models based on higher order non-linear partial differential equations are not amenable to pure formal analysis [2].

At the same time, a complete system level analysis suite based purely on simulation approaches for hybrid systems with such components leads to unacceptably high run times, in order to cover every possible combination of corner case behaviours due to internal interactions within system components and external interactions with the system environment. While, detailed analysis of MEMs structures can be performed by finite element based approaches, carried out within respective energy domains of corresponding sub-systems [20], integrating such analysis framework for system level formal analysis results in huge computational bottlenecks [2]. Symbolic simulation methods are an alternative and can replace traditional simulation [21]. These are being increasingly used in many domains for system validation, where, an analytical solution of the system is achieved through symbolic analysis using solvers of various types [21]. Truth decidability in these cases becomes a computationally difficult task due to the infinite cardinality of the state spaces of the continuous dynamical systems. The authors of [21] give the family of linear differential equations with a decidable reachability problem, by symbolically computing the reachable state sets by posing it as a quantifier elimination problem in the decidable theory of reals. Public domain quantifier elimination tools such as REDLOG ([22]) and QEPCAD ([23]) implement these approaches and have been used in symbolic verification of hybrid systems. As discussed earlier, real life hybrid systems, however, require complex linear and non-linear differential equations to model their dynamic behaviour, thereby rendering symbolic approaches computationally expensive. Reachable state sets for these systems are computed approximately, using numerical methods based on time step integration of differential equations to contain the complexity in computing the exact reachable state sets. This approximate computation is either based on polyhedra, or level sets, or ellipsoids [24], [25], [26] and [27]. In [3], the authors explore new algorithms for accurate event detection for simulation based reachability analysis and abstraction methods. The advantages of transformation based approaches discussed earlier in Section 3, to alleviate problems arising out of misses in the detection of important system behaviour events can be easily seen. This helps in decreased computational overheads arising from numerical integration, accurate event detection and, therefore, increased robustness in formal analysis of hybrid system behaviour. It is easy to see that the analytical form of the proposed approach also, makes it amenable towards computation of approximate reachable state sets, needed for automated formal analysis. While the technique based on transformation approaches is appealing for analyzing hybrid system components described behaviourally by linear differential equations, it is inappropriate for MEMS based components described by non-linear PDEs, which may not even have an analytical solution describing their behavior. However, it is possible to obtain approximate analytical models for some of the MEMS based components, such as, gyroscopes, which involve a single convolution operation in either the time or frequency domains. Even this is computationally expensive for formal analysis. Some of these simulation complexity aspects of MEMS components have been described in [32].

5. Introduction to CheckMate

In this section we give a very brief introduction to CheckMate [27], a public domain tool from CMU. CheckMate can be employed to verify hybrid systems modeled as hybrid automata, either formally using model checking, or through a simulation mechanism. This solution is built on the very popular Simulink/Stateflow Framework (SSF) from Mathworks, widely accepted both in the academia and the industry. CheckMate supports three important custom SSF blocks, viz., Switched Continuous System Block (SCSB), Polyhedral Threshold Block (PTHB), and Finite State Machine Block (FSMB). A hybrid system is modeled primarily using these three blocks. CheckMate, however, supports a few other blocks present in SSF.

A SCSB is used to define the system continuous dynamics in terms of first order differential equations. A PTHB generates events whenever the system crosses a specified threshold described in terms of a linear constraint. This generated event is used as an input to the FSMB to trigger transitions from one state to another. Based on the sink state of a transition edge that is reached, the SCSB block on reaching that state generates the continuous state trajectory using the dynamics corresponding to that state.

Verification is performed using three distinct phases. In the first phase, it allows verification of the hybrid system using traditional simulation based on numerical integration, as supported in SSF. Thus, CheckMate models can be simulated in a manner similar to any other general SSF model. In the next phase, Explore, beginning with the initial location of the hybrid automata, it checks whether each simulation trajectory, starting with different initial conditions corresponding to the set of corner vertices in the convex polytope initial continuous set, satisfies a given formal property specified as an ACTL formula. It informs the user in case of any violation. In the third and final phase, Verify, CheckMate performs formal verification.

6. Adaptive Cruise Control system: a case study

The concept of adaptive cruise control (ACC) system has been developed to aid vehicular traffic on highways. It is an automatic closed loop system, through which a driver during a long drive, can volitionally transfer control to an intelligent vehicular controller system.

Figure 6 illustrates a speed control mechanism used in automobiles. A MEMS sensor viz. a gyroscope is attached at the wheel base to provide details on the vehicular speed. Speed and proximity sensors are also installed at the front and rear end of the vehicle to continuously monitor the speed and distance of the vehicle ahead and behind. All these details are then sampled by engine controller to take a suitable course of action for the vehicle.

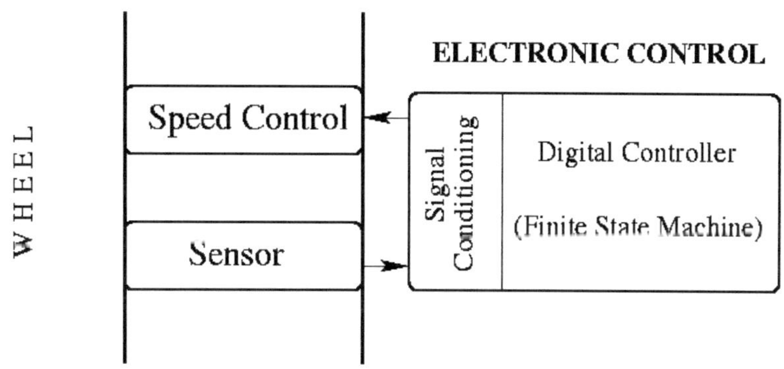

Fig. 6. Closed Loop Vehicular Speed Control

The vehicular control is marked by a closed loop control chain, which is executed in a continuous polling mode. This system is primarily a state machine, which is designed to translate the nature of the traffic conditions into a vehicular and generate an action for each state. The action is again governed by the parameters of highway control viz. speed limits, minimum proximity etc. Figure 7 describes the state transition graph for the ACC system.

The system behaviour consists of four states, *viz,* 'HALT', 'ACCELARATE', 'CRUISE' and 'RETARD'. The variables Xp (for proximity to the front vehicle) and V (for speed) govern the assignments to different states and the transitions between these states. The engine

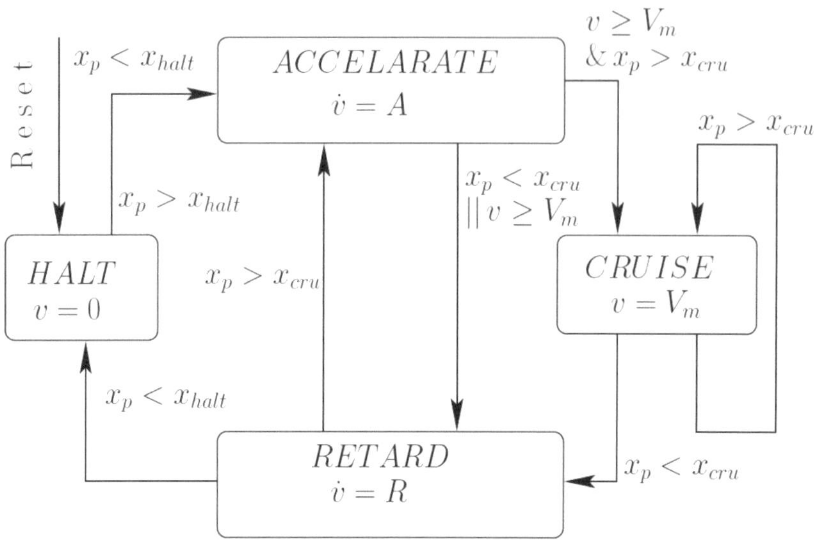

Fig. 7. ACC State Transition Graph

controller samples these states and accordingly sends control signals to various automotive peripheral subsystems to act. These functional tasks include acceleration, retardation or maintaining constant speed of the vehicle. The sensors on the vehicle, keep polling for front and rear vehicular proximity for a continuous update of system state.

7. Introduction to gyroscope model

This section provides an introduction to analysis and modeling of a gyroscope. Gyroscope is a device that is used to measure the angular velocity of a system. It primarily uses the concept of Coriolis force, to translate angular motion to linear motion detection, which is captured in the form of capacitance variation in measurements. The reader is encouraged to look into basic text books [28] [29] applied mechanics and MEMS [30] to understand the concepts of mechanical motion and gyroscope basics. [20] provides a very good summary paper for design and analysis of micromechanical gyroscope. For the clarity of the reader and completeness of the subject, we present a very basic analytical overview of the gyroscope. Figure 8 shows a mathematical model of a gyroscope and its Simulink representation.

A gyroscope can be abstracted as a simple set of coupled differential equations whose output can be equated to a capacitance equivalent to the angular velocity. The reader should note that, simple as it seems, there exists a fundamental difference between this model and most other models based on differential equations. The conventional models would utilize Laplace transform so that the convolution operations in time domain translate to multiplication operation in the frequency domain, thus, resulting in simpler computations. However in the case of gyroscopes, there exist both multiplication and convolution operation in both time and frequency domains. Hence a simplified computation is not possible. The reader is encouraged to explore options to simplify this computationally challenging problem. In this case we resort to a simplified time domain based simulation approach through Simulink.

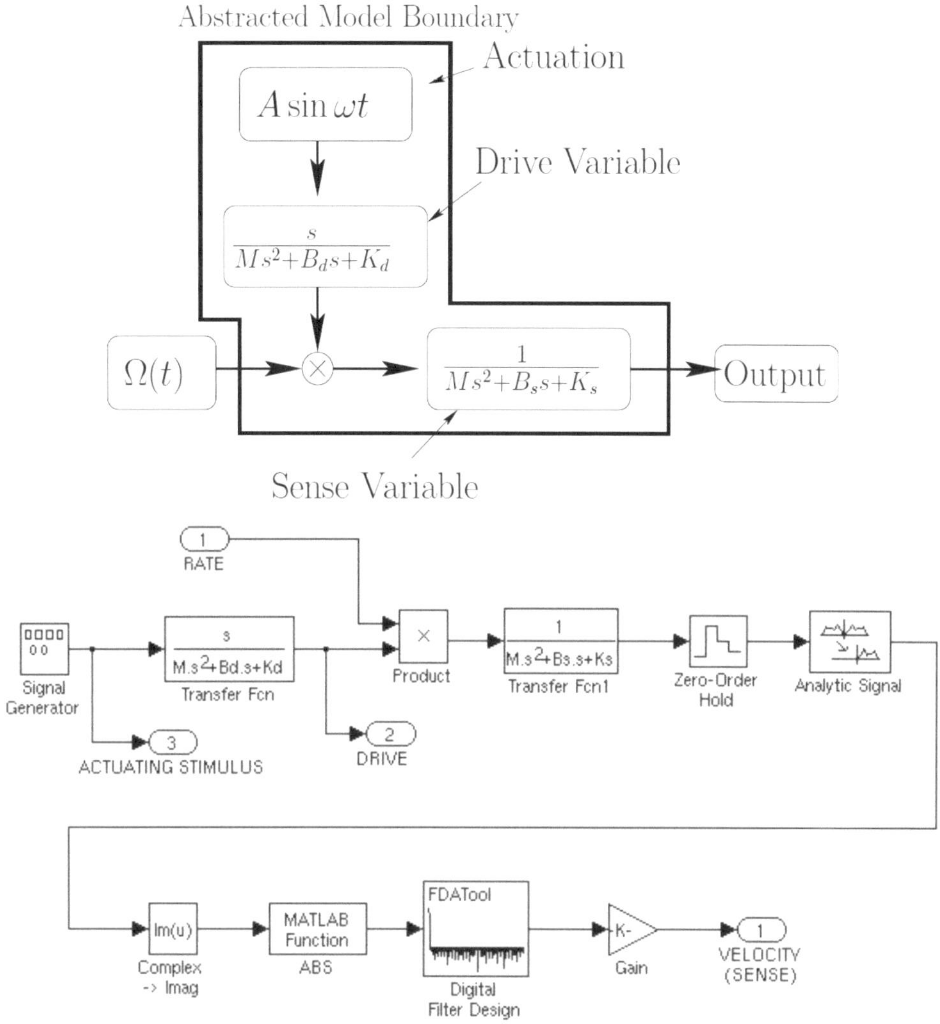

Fig. 8. Gyroscope Model [20] & Simulink Representation

8. Formal verification of ACC system

8.1 Planning the models

Let us first explore how the given system can be casted into the Check Mate framework. It is assumed that the reader is well aware of the Simulink and Stateflow tools. It is a fairly simple task to translate the ACC system requirements into an SSF diagram. These tasks today are routine for practicing system design engineers who program in Simulink/Stateflow in Matlab. The state transition graph of the ACC state machine (FSM) can be added as is, into the Stateflow system of Matlab. For each state in the hybrid automata, there is a need to associate, a real time Simulink model, which continuously modifies the state variables of the system. At each time step during the temporal evaluation of the hybrid system, based on state specific behaviour, conditions associated with each transition edge are evaluated. An illustration of the same is shown in Figure 9. Simulink representation of a real time system involves simple hook up of maths based models, which are either user created or linked Simulink library instantiations. Similarly Stateflow diagrams can also be easily created through templates available in the Matlab framework. Reader can go through standard Matlab help manuals and tutorials to get a better grasp of these basics. The differential equation pertaining to each state can be easily modelled by Simulink modules. Figure 10 illustrates the integrated ACC system, with assumptions upon availability of the sensor inputs through direct means. Now let us look at the ACC system in more detail. Figure 7 represents its state transition graph.

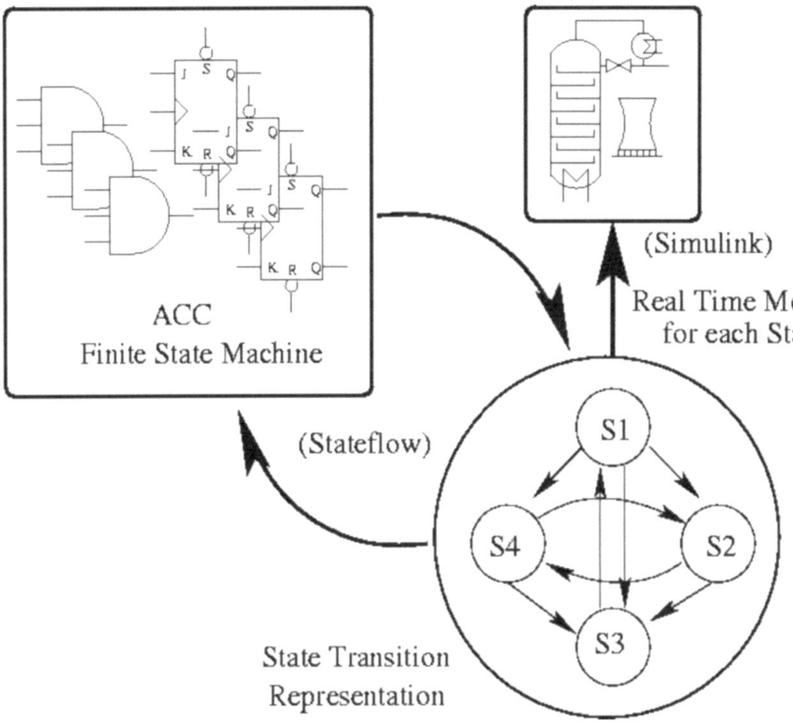

Fig. 9. Representation of a Simulink State flow system

This same representation has been cast into the system shown in Figure 10. Each state shown in the figure, models a real time differential equation defining that state. A switch multiplexes the activation of the state equation in time domain depending on the state decision. The Stateflow sub-box shown in the figure generates the control signal for the switch. This sub-box models the ACC transition graph. This implementation forms one of the simplest representations of an integrated Simulink-Stateflow system. The reader should again keep in mind that this system still does not include the sensor model into the overall model of the ACC system. Notably there are two points to mention. Firstly, there exists a sensor (not shown in the state transition graph), which senses the vehicular velocity, later processed by the controller. The need for the integration of the sensor into the main ACC system arises because due to the fact that, there exists error in measurements taken at the sensor output of the sensor, which may not be accounted into the error-analysis of the formal analysis model. The integration of the sensor helps solve this problem. Secondly, the sensor processing and the ACC system processing time steps are the same in a Matlab based setup. Hence within the framework of a single discrete event solver, one cannot freeze the simulation time step against the other. This problem is illustrated in Figure 11. As the case in our example, we cannot freeze the ACC system time step progression and wait for the sensor to provide an event signalling completion. However the current versions of the Matlab solvers do not address this requirement. Working within the Matlab setup, can pose challenges, which need not even be addressed. A work around to this problem involves translating one of the time variables to static. In other words, if we can characterize the sensor model, not only can we account the model into the framework, but also perform a formal analysis with respect to the correctness of the sensor functionality.

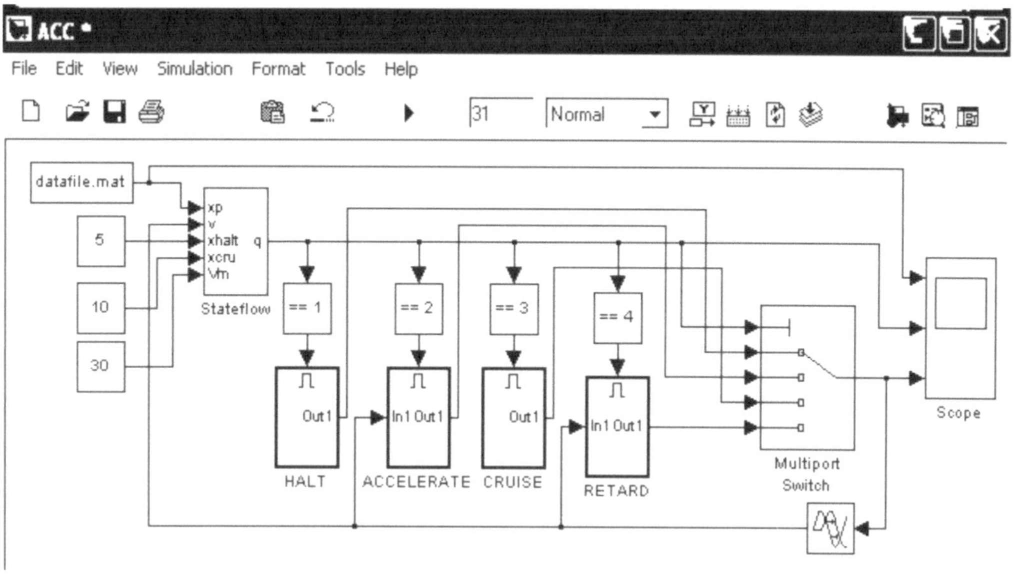

Fig. 10. Representation of an ACC System.

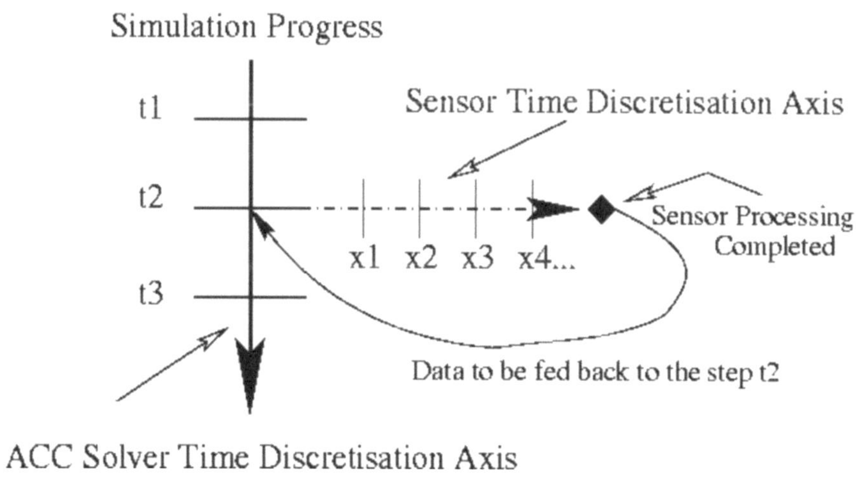

Fig. 11. Multiple Time Discretisation Stack Requirements

8.2 CheckMate model of ACC system

We now introduce the CheckMate modeling environment. A basic introduction has already been provided to the reader. In this section, we illustrate how a Simulink-Stateflow based system can be translated into a CheckMate model. The reader should note that the CheckMate model consists of a restricted set of the larger Simulink-Stateflow system. However the way the same system is represented, is in a different format. The first step to translation lies in transforming the Stateflow part of the SSF model into a combination of a state machine and a SCSB model in the CheckMate system. The SCSB model is a mathematical equivalent to the switch in the real time system. The second step then translates each guard condition on outgoing transition edges corresponding to a state into a PTB block. Figure 12 and 13 show the ACC model in CheckMate. The diagram shows the PTHB modules used in modeling all the state transitions. The reader is encouraged to go through reference [33] that illustrates the formal verification approach to the validation of MEMS based hybrid systems.

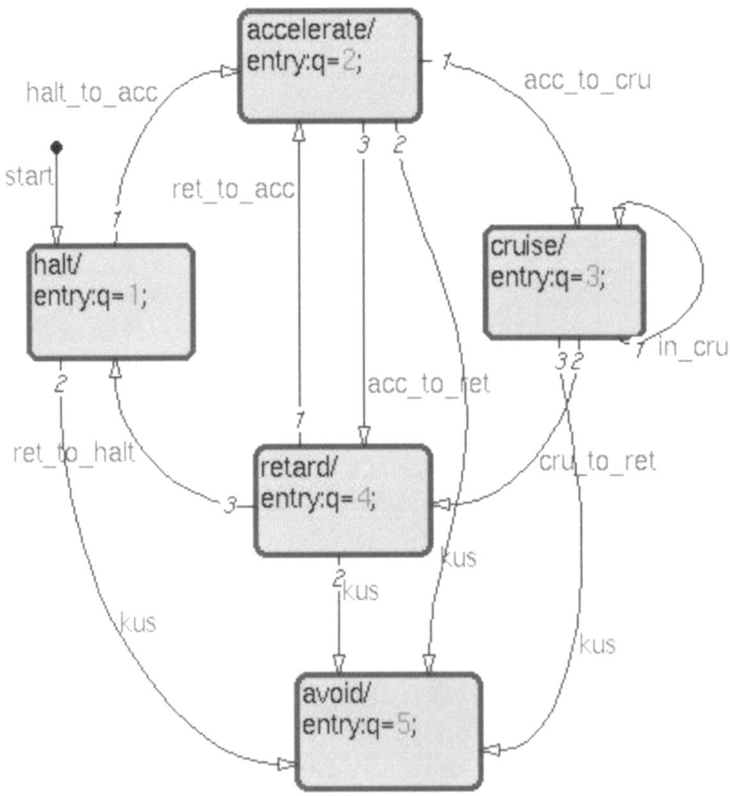

Fig. 12. State Flow Representation of ACC System in Matlab

The specification of the ACC system can then be validated in the SSF model through time domain simulation method. These specifications can be captured as a set of properties, which we then formally verify in CheckMate. The sensor model in SSF uses several continuous time domain dynamic components that do not belong to the set of dynamic components allowed by CheckMate. This along with the multiple time step requirements leads one to explore an alternative method to integrate a MEMS sensor into the hybrid system.

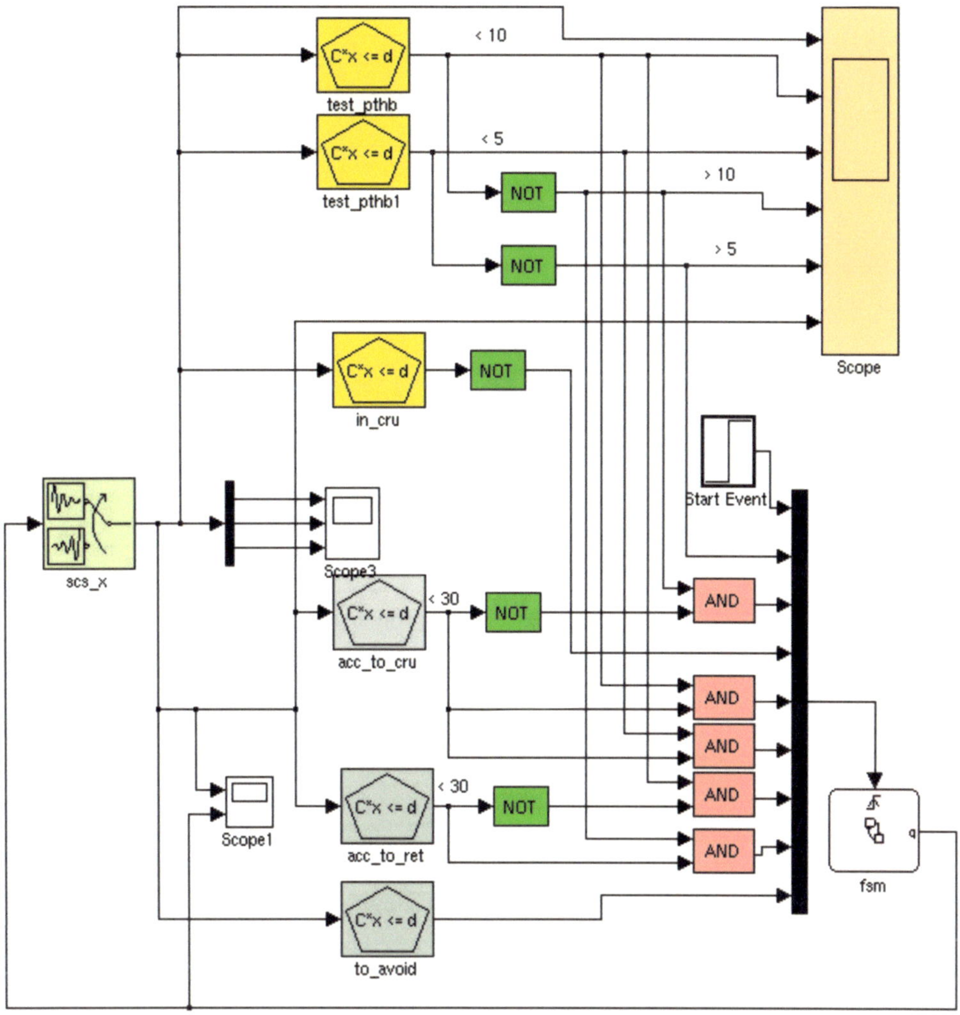

Fig. 13. Check Mate Representation of ACC System

8.3 Integrating sensors

Let us now devote ourselves to the task of integrating a component like a MEMS sensor. We have already discussed in Section 8.1 our reasons for searching for an alternative method to integrate real time sensors. In our case this is the MEMS gyroscope model. The standard SSF model of a gyroscope uses several continuous time domain dynamic components which do not belong to the set allowed by CheckMate. There are two methods possible to approach this problem. First method shown in Figure 14 is to build an independent interrupt driven discrete time solver solution.

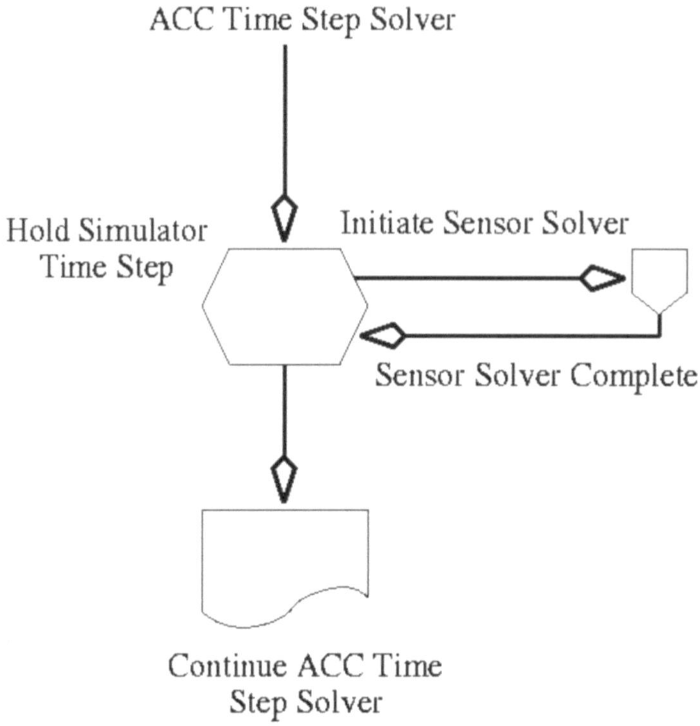

Fig. 14. An Independent Interrupt driven Solver Solution Approach

The second method consists of characterising the time domain behaviour of the sensor and capturing this behaviour as an empirical model. The empirical model can be implemented as a simple look-up-table (LUT) which can be easily interpolated (or extrapolated) as per the input conditions in the field. The data-points for the LUT can be obtained for a range of velocity values by carrying out dynamic simulation on an exact macro-model of the MEMS gyroscope in the SSF framework (Figure 7). To integrate this empirical LUT model of the MEMS gyroscope, it is necessary to make changes in its implementation code. We can access the LUT model through function calls in CheckMate to get the desired outputs and thereby obtain the continuous time trajectories needed for formal analysis. For non-sensor based systems, where there are no multiple time step requirements, we can also use a simulation approach and integrate it with the main system solver.

Readers should note that an LUT based approach will have a limited reach as a solution approach. It is clear that this approach cannot be used for general hybrid systems having dynamic components described with a system of strongly non-linear differential-algebraic equations, as in analog mixed signal design blocks. We can also look at using the exact simulation traces available from the general model in SSF. However they cannot be applied to real time sensor models due to reasons discussed earlier. For other real time components, the exact simulation traces enable the *Explore* and the *Verify* phase to construct accurate flow-pipes, and generate better approximations to these flow-pipes. This is illustrated in Figure 15. The method provides an alternative path to choose between a Simulink or a formal Checkmate model used in the computation. However for simplicity sake, let us restrict ourselves to the LUT based approach.

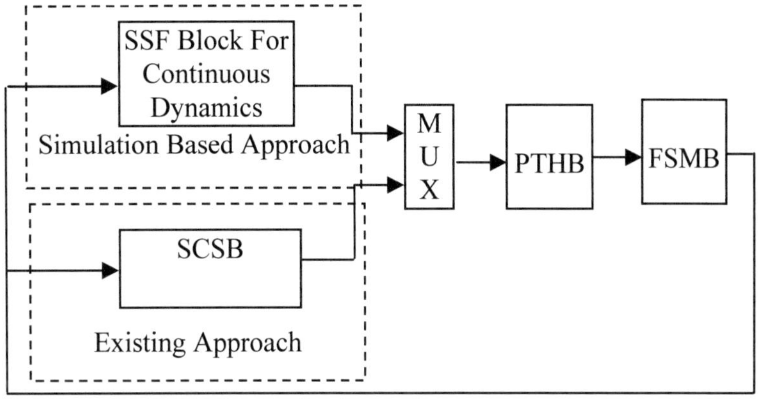

Fig. 15. Hybrid Component Integration Approaches

Figure 15 also includes the LUT macro-model between the switched dynamic block and the PTHB modules. A restriction imposed by CheckMate is that for formal analysis it assumes a hybrid system to be closed. In other words it needs to be a closed loop system. The ACC system model can easily be seen to be open with respect to the velocity of the leading vehicle V_L. This is because the control actions in the hybrid automata of the ACC system, depends on the behaviour of the leading vehicle resulting from changes in its velocity V_L. Regular SSF framework allows modelling of an open hybrid systems. However, it needs some effort to model this in CheckMate. We can model such a scenario by addition of a redundant equation in terms of V_L, V_T and proximity (X_p) in which we render V_L as a parameter (V_T and X_p as the closed system state variables).The equation used is, $R = \int (V_L - V_T)dt = X_p$.

8.4 Verifying the specification properties

Let us now look at some of the controller specifications to be formally analysed:

P1: The tracking vehicle should never retard above X_{cru}.
P2: The tracking vehicle should never accelerate when $X_p < X_{halt}$.
P3: The tracking vehicle should not cruise when $X_p < X_{cru}$.
P4: The value of proximity X_p in all states will be always greater than 0.
P5: For $X_p < X_{halt}$ tracking vehicle is always in the HALT state.
P6: When $X_p > X_{cru}$ tracking vehicle never goes to the HALT state.
P7: When $V > V_m$ and $X_p > X_{cru}$ and tracking vehicle is always in the CRUISE state.

The ACC model without the MEMS gyroscope block can be easily verified using CheckMate. Adding the sensor block causes CheckMate to report non-compliance. In the STG (State Transition Graph) of the hybrid automata, five states have been used. The output values of the state-space variables from a previous state become the initial values for the next state continuous dynamics. To get the velocity value of the tracking vehicle to be zero before entering the halt state, we can add an intermediate state where the velocity is brought down to zero. The properties can be represented through ACTL. The same properties can also be verified by simulation. These results are shown in Figure 16.

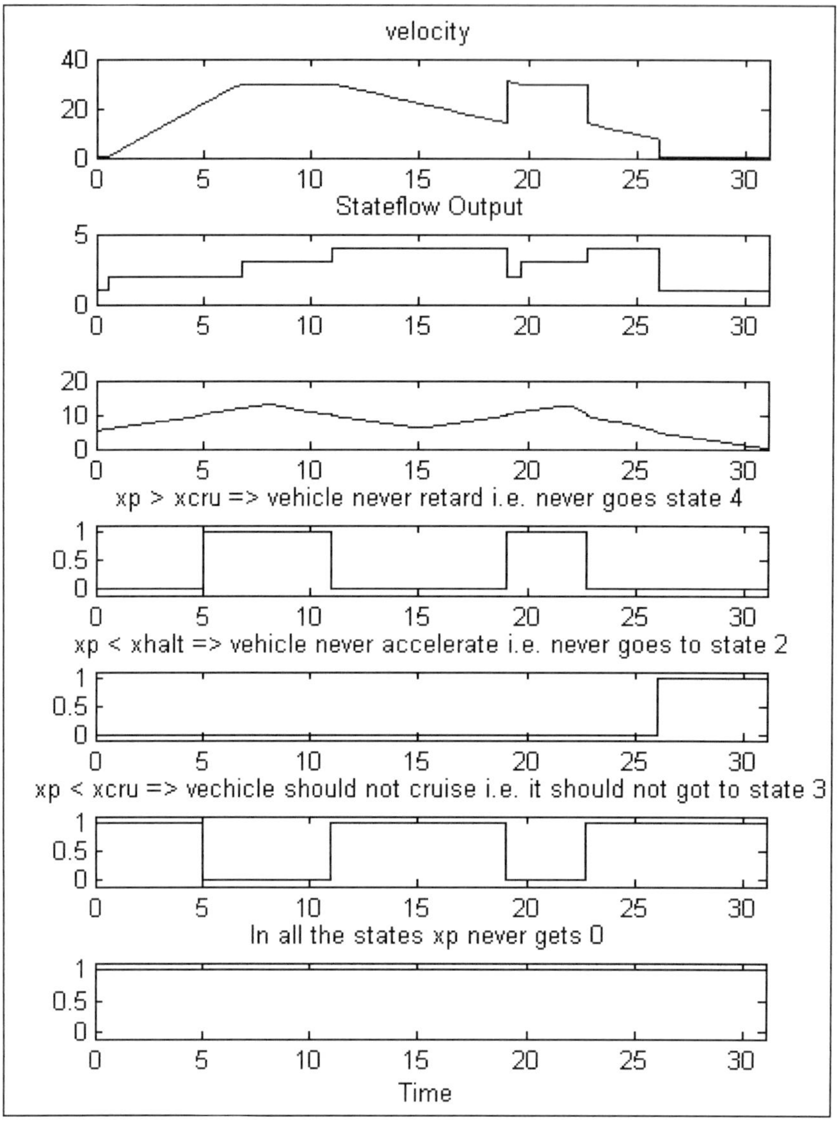

Fig. 16. SSF output of ACC model for different properties

9. Conclusions

In this chapter we give a brief introduction the reader on the basics of formal verification of hybrid systems, which are primarily targeted for automotive systems. CheckMate based approach has been introduced to explore other available hybrid system design and verification frameworks for analysing hybrid system implementation. The reader is then walked through a case study of an adaptive cruise control system, which contains a MEMS based velocity sensor. After reading this chapter, the reader should be able to plan a hybrid system formal analysis platform and apply various methods to integrate complex sensors and other complex real time components.

10. References

[1] Simulink State-Flow, http://www.mathworks.com/products/stateflow/
[2] C. J. Tomlin, I. Mitchell, A. M. Bayen, M. Oishi, Computational techniques for the verification of hybrid systems, Proc. IEEE 91 (7) (2003), 986{1001.
[3] R. Alur. et. al., Hierarchical modeling and analysis of embedded systems, Proc. IEEE 91 (1) (2003) 11{28.}
[4] R. Alur and D. L. Dill, "A theory of timed automata," Theoretical Computer Science, vol. 126, no. 2, pp. 183–235, 1994.
[5] D R Sparks, Application of MEMS technology in automotive sensors and actuators; Micromechatronics and Human Science, 1998. MHS '98. Proceedings of the 1998 International Symposium on 25-28 Nov. 1998 Page(s):9 - 15
[6] T. A. Henzinger, "The theory of hybrid automata," Logic in Computer Science, vol. 96, no. 1, pp. 278–292, 1996.
[7] P. Kopke, T. Henzinger, A. Puri and P. Varaiya, What's decidable about Hybrid Automata?, ACM Symposium on Theory Of Computing, 1995, 372-382.
[8] W. Press, S. Teukolsky, W. Vetterling, and B. Flannery, Numerical Recipes in C, 2nd ed. Cambridge, UK: Cambridge University Press,1992.
[9] E. M. Clarke, A. Biere, R. Raimi, and Y. Zhu. Bounded Model Checking Using Satisfiability Solving, Formal Methods in System Design, 19(1):7-34, 2001.
[10] Bernd Becker, et. al, Bounded Model Checking and Inductive Verification of Hybrid Discrete-continuous Systems,GI/ITG/GMMWorkshop,2002.
[11] André Platzer and Jan-David Quesel. KeYmaera: A Hybrid Theorem Prover for Hybrid Systems, Automated Reasoning, Third International Joint Conference, IJCAR 2008, Sydney, Australia, Proceedings, volume 5195 of LNCS, pages 171-178. Springer-Verlag, 2008.
[12] T. A. Henzinger, The theory of hybrid automata, LICS, IEEE Computer (1996) 278-292.
[13] E. Abraham-Mumm, M. Steffen and U. Hannemann: Verification of hybrid systems: Formalization and proof rules in PVS, ICECCS, IEEE Computer Society (2001) 48-57.
[14] "PHAVer: Polyhedral hybrid automaton verifier", http://www.cs.ru.nl/goranf/, 2006.
[15] Eugene A, Thao D, and Maler, O., "The d/dt Tool for Verification of Hybrid Systems",CAV'02 - Computer Aided Verification, 2002, Copenhagen, Denmark, July 2002, 365-370, LNCS 2404. http://www-verimag.imag.fr/tdang/ddt.html
[16] F. D. Torrisi and A. Bemporad, HYSDEL --- A tool for generating computational hybrid models for analysis and synthesis problems, IEEE Transactions on Control Systems Technology, 2004, Volume 12, 235-249.
[17] CMU CheckMate website, http://www.ece.cmu.edu/~webk/checkmate/
[18] André Platzer and Jan-David Quesel. KeYmaera: A Hybrid Theorem Prover for Hybrid Systems, Automated Reasoning, Third International Joint Conference, IJCAR 2008, Sydney, Australia, Proceedings, volume 5195 of LNCS, pages 171-178. Springer-Verlag, 2008.
[19] E. Abraham-Mumm, M. Steffen and U. Hannemann: Verification of hybrid systems: Formalization and proof rules in PVS, ICECCS, IEEE Computer Society (2001) 48-57.
[20] H. Xie, G. K. Fedder, Integrated microelectromechanical gyroscopes, Journal of Aerospace Engg. 6 (2) (2002) 65{75}

[21] G. Lafferriere, G. J. Pappas, and S. Yovine, "Symbolic reachability computation for families of linear vector fields," Journal of Symbolic Computation, vol. 32, no. 3, pp. 231–253, 2001.

[22] A. Dolzman, T. Sturm, Redlog : Computer algebra meets computer logic, ACM SIGSAM Bulletin 31 (2) (1997) 2{9}

[23] G. Collins, H. Hong, Partial cylindrical algebraic decomposition for quantifier elimination, Journal of Symbolic Computation 12 (1991) 299{328}

[24] A. Kurzhanski, P. Varaiya, Ellipsoidal techniques for Reachability analysis, in: B. Krogh, N. Lynch (Eds.), Hybrid Systems: Computation and Control, Vol. 1790 of LNCS, Springer, 2000, pp. 203{213}.

[25] A. Chutinan, B. Krogh, and Verification of polyhedral-invariant hybrid automata using polygonal Flow pipe approximations, in: F. Vaandrager, J. H. Van Schuppen (Eds.), Hybrid Systems: Computation and Control, Vol. 1569 of LNCS, Springer, 1999.

[26] T. Dang, O. Maler, Reachability analysis via face lifting, in: T. Henzinger, S. Sastry (Eds.), Hybrid Systems: Computation and Control, Vol. 1386 of LNCS, Springer, 1998, pp. 96{109}.

[27] Chutinan, B. Krogh, Computational techniques for hybrid system verification, IEEE Trans. Automatic Control 48 (1) (2003) 64{75.}

[28] Irving Shames, Engineering Mechanics: Dynamics, 4th Edition, Prentice Hall 1997

[29] Jerry Ginsberg , Engineering Dynamics (Hardcover), Cambridge University Press; 3rd edition, 2007

[30] Stephen D. Senturia , Microsystem Design (Hardcover)

[31] André Platzer and Jan-David Quesel. KeYmaera: A Hybrid Theorem Prover for Hybrid Systems, Automated Reasoning, Third International Joint Conference, IJCAR 2008, Sydney, Australia, Proceedings, volume 5195 of LNCS, pages 171-178. Springer-Verlag, 2008.

[32] R. M. Kirby, G. E. Karniadakis, O. Mikulchenko and K. Mayaram. An Integrated Simulator for Coupled Domain Problems in MEMS, IEEE Journal of Micromechanical Systems, Vol. 10, No. 3, September 2001, 379-391

[33] Jairam S., Kusum Lata, Subir K. Roy and Navakanta Bhat, "Formal Verification of a MEMS Based Adaptive Cruise Control System", *Proceedings of Modeling and Simulation of Micro Systems,* June 2008, Boston, USA.

8

Rolling Stability Control of In-wheel Motor Electric Vehicle Based on Disturbance Observer

Kiyotaka Kawashima, Toshiyuki Uchida and Yoichi Hori
University of Tokyo
Japan

1. Introduction

1.1 In-wheel-motor electric vehicle's advantages and application to vehicle motion control

Electric vehicles (EVs) with distributed in-wheel-motor systems attract global attention not only from the environmental point of view, but also from the vehicle motion control. In-wheel-motor EVs can realize high performance vehicle motion control by utilizing advantages of electric motors which internal combustion engines do not have. The EV has the following four remarkable advantages (Hori et al., 2004):

- Motor torque response is 10-100 times faster than internal combustion engine's one. This property enables high performance adhesion control, skid prevention and slip control.
- Motor torque can be measured easily by observing motor current. This property can be used for road condition estimation.
- Since an electric motor is compact and inexpensive, it can be equipped in each wheel. This feature realizes high performance three dimensional vehicle motion control.
- There is no difference between acceleration and deceleration control. This actuator advantage enables high performance braking control.

Slip prevention control is proposed utilizing fast torque response. Road condition and skid detection methods are developed utilizing the advantage that torque can be measured easily (Hori et al., 2004). Yawing stability control, side slip angle estimation and control methods are also proposed by utilizing a distributed in-wheel-motor system (Fujimoto et al., 2006 & 2007; Hori et al., 2007).

1.2 Background and purpose of the research

The purpose of this paper is to propose integrated rolling and yawing stability control (RSC and YSC). Rollover stability is important for all classes of light-vehicles such as light trucks, vans, SUVs and especially, for EVs which have narrow tread and high CG because EV is suitable for relatively small vehicle and human height does not change. According to the data from NHTSA, ratio of rollover accidents of pick ups' and vans' crashes in 2002 was only 3% against whole accidents. However, nearly 33% of all deaths from passenger vehicle crashes are due to rollover accidents (NHTSA, 2007).

Therefore, RSC is very important not only for ride quality but also for safety. The RSC system has been developed by several automotive makers and universities (Liebemann et al., 2004; Tseng et al., 2007; Yi et al., 2008). Rollover detection systems, such as rollover index (*RI*) (Yi et al., 2008) and Time-to- rollover (*TTR*) (Peng et al., 2001) are proposed for mitigating critical rolling motion. Every system controls braking force on each wheel independently and suppresses sudden increase of lateral acceleration or roll angle. However, since braking force is the average value by pulse width modulation control of brake pad, brake system cannot generate precise torque or positive torque. In the case of in-wheel-motor, both traction and braking force can be realized quickly and precisely. In addition to actuator advantages, RSC is designed by utilizing two-degrees-of-freedom (2-DOF) control based on disturbance observer (DOB) (Kawashima et al., 2008). For the vehicle motion control field, DOB is applied to vehicle yaw/pitch rate control (Fujimoto et al., 2006 & 2007) and 2-DOF control is applied to the electric power steering control (Guvenc et al., 2004). There are three reasons to utilize DOB: 1) disturbance suppression, 2) nominalize lateral vehicle model and 3) tracking capability to reference value. DOB loop that suppresses the effect of disturbance is faster than outer loop that achieves tracking capability. Designing DOB for traction force is not applicable for ICEV, because engine torque is not accurately known and long time delay exists. Therefore, DOB is applicable only in case of EVs. The tracking capability and robustness for lateral acceleration disturbance against such as side blast are realized by the proposed method. However, roll and yaw stabilities are incompatible. High rolling stability makes vehicle behavior under steer. On the other hand, high yawing stability to avoid vehicle side slip, vehicle roll stability is not guaranteed. In the next section, electronic stability program (ESP) on EV is introduced using *RI* based on vehicle geometry and dynamics model, which achieves integrated three dimensional vehicle motion control.

2. Electric stability program for electric vehicle

2.1 Introduction of electric stability program

Fig. 1. shows concept of ESP for EV. ESP consists of two systems; vehicle/road state estimation system (S1) and integrated vehicle motion control system (S2). S1 integrates

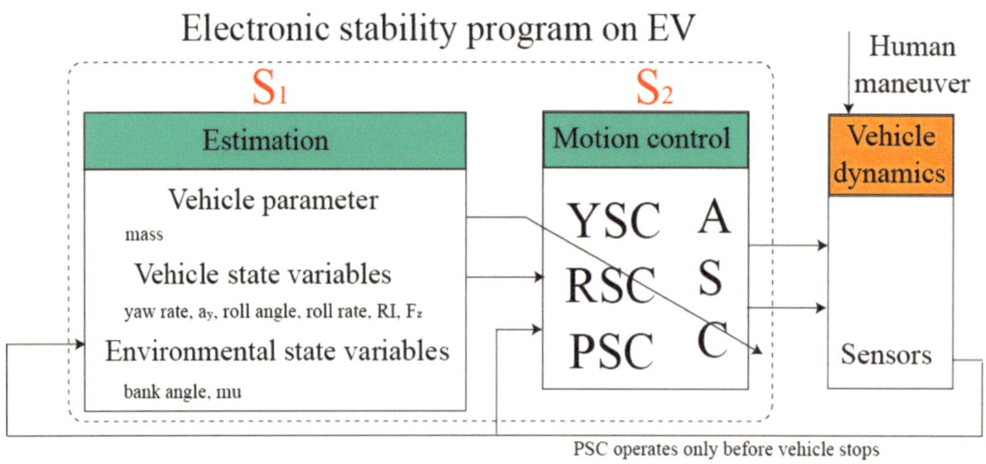

Fig. 1. EPS based on DOB

information from sensors (accelerometer, gyro, GPS, suspension stroke and steering angle sensors) and estimates unknown vehicle parameters (mass), vehicle state variables (yaw rate, lateral acceleration, roll angle, roll rate and normal forces on tires) and environmental state variables (Tseng et al., 2007; Hac et al., 2004). According to the information from S1, S2 controls vehicle dynamics using RSC and YSC, pitching stability control (PSC) and anti-slip control (ASC). According to *RI*, which is calculated by S1, a proper stability control strategy (YSC, RSC or mixed) is determined. RSC is based on DOB and nominal vehicle state is calculated by a controller. If there are errors between calculated and actual dynamics, it is compensated by differential torque.

2.2 A scheme of integrated vehicle motion control

Lateral acceleration is composed of vehicle side slip, yaw rate and longitudinal speed.

$$a_y = V(\dot{\beta} + \gamma) \tag{1}$$

If constant vehicle speed is assumed and lateral acceleration is suppressed, yaw rate is also suppressed as long as differentiation of side slip is not controlled. This physical constraint makes RSC and YSC incompatible. Therefore, rollover detection is necessary for integrated control. In order to detect rollover, Yi proposed *RI* ($0 < RI < 1$) as a rollover detection (Yi et al., 2008). When *RI* is high which means a vehicle is likely to roll over, the weight of RSC is set as high. On the other hand, *RI* is small, which means a vehicle is not likely to roll over, the weight of YSC is set as high. Control algorithm is simple and given by following equation. Fig. 2. shows block diagram of three dimensional integrated vehicle motion control.

$$\begin{aligned} N^* &= f(RI, N_{RSC}, N_{YSC}, N_{DOB}) \\ &= RI*N_{RSC} + (1 - RI) * N_{YSC} + N_{DOB} \end{aligned} \tag{2}$$

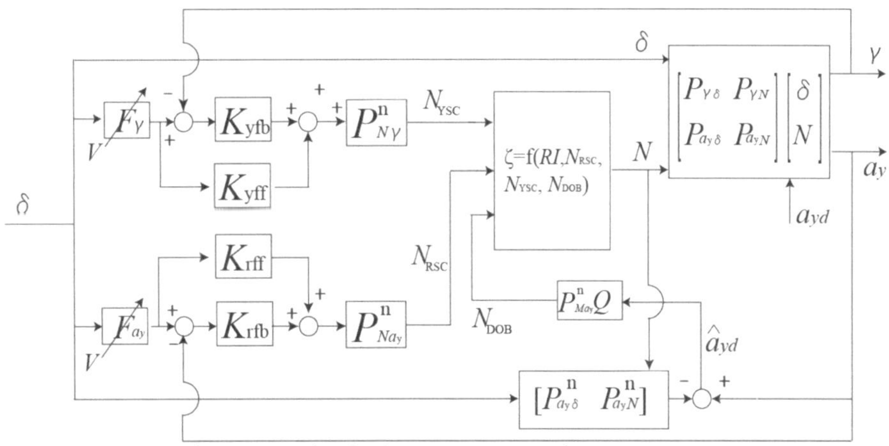

Fig. 2. Block diagram of integrated vehicle motion control

3. Estimation system

S1 is composed of vehicle parameters, state variables and environmental state variables estimation system. In this section, vehicle state variable estimation system is mainly introduced. According to the estimated state variables, *RI*, a distribution ratio of RSC and YSC is determined.

3.1 Lateral acceleration and roll angle observer

Fig. 3. and fig. 4. show four wheel model and rolling model of electric vehicle.
Vehicle motion is expressed as the following three linear equations.

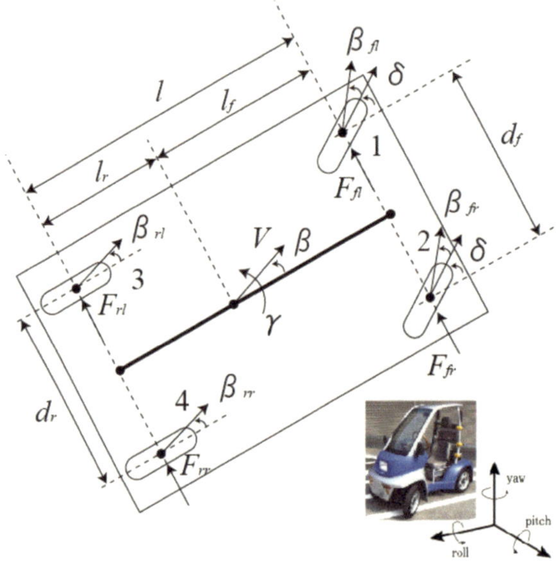

Fig. 3. Four wheel model

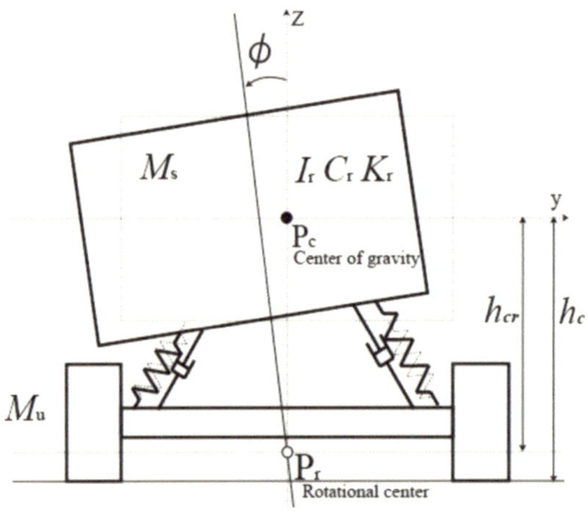

Fig. 4. Rolling model

Lateral motion:

$$
\begin{aligned}
MV(s\beta+\gamma) &= F_{yfl}+F_{yfr}+F_{yrl}+F_{yrr} \\
&= -2c_f(\beta+\frac{l_f}{V}\gamma-\delta)-2c_r(\beta-\frac{l_r}{V}\gamma)
\end{aligned}
\tag{3}
$$

Yawing motion:

$$I_{yaw}s\gamma = (F_{yfl} + F_{yfr})l_f - (F_{yrl} + F_{yrr})l_r$$
$$= -2c_f(\beta + \frac{l_f}{V}\gamma - \delta)l_f + 2c_r(\beta - \frac{l_r}{V}\gamma)l_r + N \quad (4)$$

Rolling motion:

$$M_s h_{cr} a_y = I_r\ddot{\phi} + C_r\dot{\phi} + K_r\phi - M_s g h_{cr}\sin\phi \quad (\phi < \phi_{wheel-lift-off}) \quad (5a)$$

$$M_s h_{cr} a_y = I_{r2}\ddot{\phi} - M_{sgh_{cr}}\sin\phi + M_s g\frac{d}{2}\cos\phi \quad (\phi > \phi_{wheel-lift-off}) \quad (5b)$$

Here, these motion equations need to be expressed as state equations to design observer. Observer gain matrix, however, becomes 2 * 4 matrix if whole equations are combined. To reduce redundancy of designing gain matrix, tire dynamics and rolling dynamics are separated. A matrix, A_{rt} connects two state equations. From eq.(3) and eq.(4), state equation is expressed as,

$$\dot{x}_t = A_t x_t + B_t u, \quad (6)$$

$$y_t = C_t x_t + D_t u. \quad (7)$$

It is noted that there is feedforward term in the transfer function from u to y_t. Therefore, to eliminate feedforward term and design stable observer, x_t vector is defined using differential torque and steering angle as the following equations,

where, $x_t = \begin{bmatrix} a_y - c_2\delta & \dot{a}_y - c_2\dot{\delta} - b_1 N - c_1\delta \end{bmatrix}^T$,

$y_t = a_y, u = \begin{bmatrix} N & \delta \end{bmatrix}$,

$A_t = \begin{bmatrix} 0 & 1 \\ -a_0 & -a_1 \end{bmatrix}, B_t = \begin{bmatrix} b_1 & c_1 \\ a_1 b_1 + b_0 & a_1 c_1 + c_0 \end{bmatrix}$,

$C_t = \begin{bmatrix} 1 & 0 \end{bmatrix}, D_t = \begin{bmatrix} 0 & c_2 \end{bmatrix}$.

$a_0 = \dfrac{4c_f c_r l^2}{MI_y V^2} - \dfrac{2(c_f l_f - c_r l_r)}{I_y}, a_1 = \dfrac{2M(c_f l_f^2 + c_r l_l^2) + 2I_y(c_f + c_r)}{MI_y V}$,

$b_0 = \dfrac{2(c_f + c_r)}{MI_y}, b_1 = -\dfrac{2(c_f l_f - c_r l_r)}{MI_y V}, c_0^{'} = \dfrac{4c_f c_r l}{MI_y}, c_1^{'} = \dfrac{4c_f c_r l_r l}{MI_y V}$,

$c_0 = c_0^{'} - a_0 c_2, c_1 = c_1^{'} - a_1 c_2, c_2 = \dfrac{2c_f}{N}$.

From eq.(5a), state space equation is,

$$\dot{x}_r = A_r x_r + A_{rt} y_t, \quad (8)$$

$$y_r = C_r x_r, \tag{9}$$

where, $x_t = \begin{bmatrix}\phi & \dot{\phi}\end{bmatrix}^T, y_r = \dot{\phi},$

$$A_r = \begin{bmatrix} 0 & 1 \\ -\dfrac{K_r - M_s g h_{cr}}{I_r} & -\dfrac{C_r}{I_r} \end{bmatrix}, A_{rt} = \begin{bmatrix} 0 & 0 \\ \dfrac{M_s h_{cr}}{I_r} & 0 \end{bmatrix},$$

$$C_r = \begin{bmatrix}0 & 1\end{bmatrix}.$$

It should be noted that lateral acceleration dynamics expressed as eq.(6) is a linear time varying system depending on vehicle speed. The states are observable at various longitudinal speed except for a very low speed. In the following sections, for repeatability reason, experiment has been done under constant speed control. Observer gains are defined by pole assignment.

These parameters are based on the experiment vehicle"Capacitor-COMS1" developed in our research group. The method to evaluate the values of c_f, c_r are referred to the paper (Takahashi et al., 2006). Since rolling dynamics was unknown, model identification is conducted to derive roll model. Constant trace method is applied to the rolling model parameters identification. From equation (5a), lateral acceleration $\hat{a}_y$ is written as

$$\hat{a}_y(k \mid \theta) = \hat{\theta}^T \xi(k), \tag{10}$$

where, $\theta = [I_{roll} \quad C_{roll} \quad K_{roll}]^T$, $\xi = [\ddot{\phi} \quad \dot{\phi} \quad \phi]^T$.

The algorithm of the constant trace method is to update forgetting factor λ, such that trace of gain matrix P, is maintained as constant.

Due to the forgetting factor, when ξ is big, θ can be identified with good precision, and when ξ is small and little information, θ is seldom updated. With constant trace method, stable parameter estimation is achieved. Update equation is written by the following equation.

$$\varepsilon(k) = a_y(k) - \hat{\theta}^T(k-1)\xi(k) \tag{11}$$

$$\hat{\theta}(k) = \hat{\theta}(k-1) + \frac{P(k-1)\xi(k)}{1 + \xi^T(k)P(k-1)\xi(k)}\varepsilon(k) \tag{12}$$

$$P(k) = \frac{1}{\lambda(k)}\left\{P(k-1) - \frac{P(k-1)\xi(k)\xi^T(k)P(k-1)}{1 + \xi^T(k)P(k-1)\xi(k)}\right\} \tag{13}$$

$$\lambda(k) = 1 - \frac{|P(k-1)\xi(k)|}{1 + \xi^T(k)P(k)\xi(k)} \frac{1}{tr[P(0)]} \tag{14}$$

where, ε is output error.

Utilizing constant trace method to the experimental result, angular frequency $\sqrt{K_r / I_r}$ = 17.2 (rad/sec) and damping coefficient $\sqrt{1/(2I_r K_r)}C_r$ = 0.234 (1/sec). Fig. 5. shows detected acceleration information by sensor and calculated acceleration with estimated parameter $\hat{\theta}$ and ξ. From the figure, the two lines merge and parameter identification is succeeded.

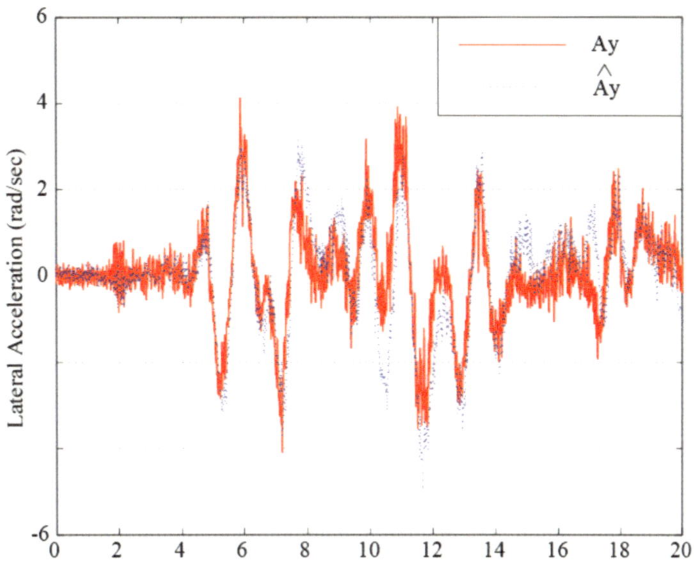

Fig. 5. Title of figure, left justified

3.2 Rollover index

RI is a dimensionless number which indicates a danger of vehicle rollover. *RI* is defined using the following three vehicle rolling state variables; 1)present state of roll angle and roll rate of the vehicle, 2)present lateral acceleration of the vehicle and 3)time-to-wheel lift. *RI* is expressed as eq. (15),

$$RI = C_1\left(\frac{|\phi|\dot{\phi}_{th} + \phi_{th}|\dot{\phi}|}{\phi_{th}\dot{\phi}_{th}}\right) + C_2\left(\frac{|a_y|}{a_y}\right) + (1 - C_1 - C_2)\left(\frac{|\phi|}{\sqrt{\phi^2 + \dot{\phi}^2}}\right), \quad if \quad \phi(\dot{\phi} - k_1\phi) > 0 \tag{15}$$

$$RI = 0, \qquad else \ if \quad \phi(\dot{\phi} - k_1\phi) < 0$$

where, *C*1,*C*2 and *k*1 are positive constants (0 < *C*1, *C*2 < 1). *ayth* is defined by vehicle geometry.

Fig. 6. shows equilibrium lateral acceleration in rollover of a suspended vehicle. It shows the relation between vehicle geometry such as h,d and K_r and vehicle states such as ϕ and a_y. From the static rollover analysis, critical lateral acceleration a_{yth} which induces rollover is defined. Phase plane analysis is conducted using a_{yth} and roll dynamics.

Fig. 7. shows phase plane plot under several initial condition $(\phi, \dot{\phi})$ at critical lateral acceleration. Consequently, ϕ_{th} and $\dot{\phi}_{th}$ are defined by the analysis.

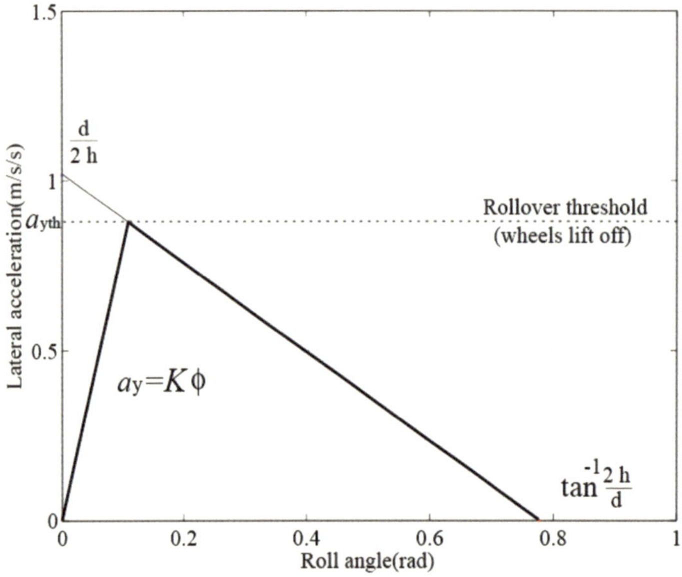

Fig. 6. Equilibrium lateral acceleration in rollover of a suspended vehicle

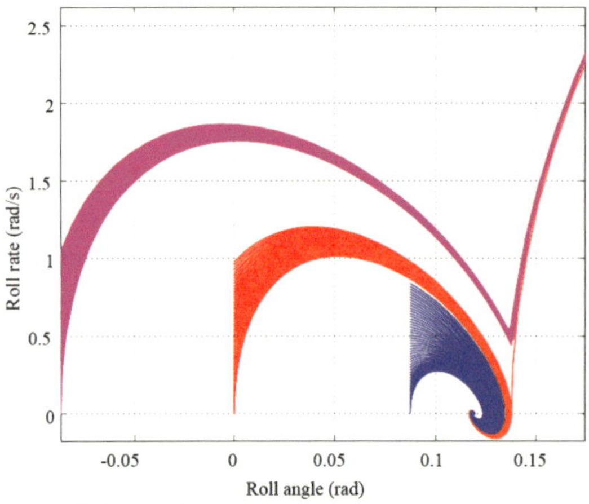

Fig. 7. Phase plane plot of roll dynamics

4. Integrated motion control system

4.1 Rolling stability control based on two-degree-of-freedom control

In this section, RSC based on 2-DOF control which achieves tracking capability to reference value and disturbance suppression is introduced. For RSC, lateral acceleration is selected as controlling parameter because roll angle information is relatively slow due to roll dynamics (about 100ms).

(a) Lateral acceleration disturbance observer

Based on fig. 8., transfer function from reference lateral acceleration u, δ and a_{yth} to a_y is expressed as the following equation. Roll moment is applied by differential torque N^* by right and left in-wheel-motors. Reference value of lateral acceleration is given by steering angle and vehicle speed.

$$a_y = \frac{P_{a_yN}P^n_{Na_y}(K_{ff}+K_{fb})}{1+P_{a_yN}P^n_{Na_y}K_{fb}}u + \frac{P_{a_y\delta}}{1+P_{a_yN}P^n_{Na_y}K_{fb}}\delta + \frac{1}{1+P_{a_yN}P^n_{Na_y}K_{fb}}a_{yd}. \tag{16}$$

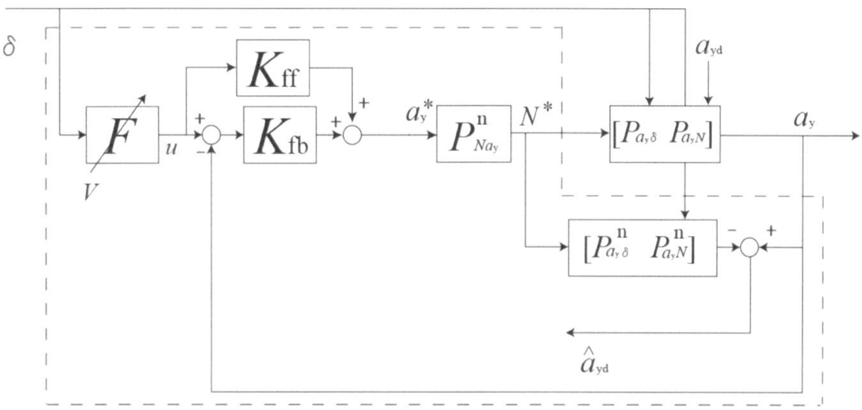

Fig. 8. Block diagram of lateral acceleration DOB

Tracking capability and disturbance suppression are two important performances in dynamics system control and can be controlled independently. On the other hand, one-degree-of-freedom (1-DOF) control such as PID controller loses important information at subtracting actual value from reference one. In the control, there is only one way to se feedback gain as high to improve disturbance suppression performance, however the gain makes the system unstable. Hence 2-DOF control in terms of tracking capability and disturbance suppression is applied to RSC. Proposed lateral acceleration DOB estimates external disturbance to the system using information; V, δ, N and a_y.

Fig. 8. also shows the block diagram of lateral acceleration DOB.
Estimated lateral acceleration disturbance $\hat{a}_{yth}$ and a_y are expressed as

$$\hat{a}_{yd} = a_y - P^n_{a_yN}N^* - P^n_{a_y\delta}\delta, \tag{17}$$

$$a_y = P_{a_yN}N^* + P_{a_y\delta}\delta + a_{yd}. \tag{18}$$

$$\hat{a}_{yd} = \frac{P^n_{Na_y}}{P_{Na_y}}\left(\left(\frac{P_{Na_y}}{P^n_{Na_y}} - 1\right)a_y + (P_{a_y\delta} - P^n_{a_y\delta})\delta + a_{yd}\right). \tag{19}$$

In eq. (19), the first and the second terms are modeling errors and the third term is lateral disturbance. If modeling error is small enough, $\hat{a}_{yth}$ is approximately equal to actual lateral acceleration disturbance.

(b) Disturbance suppression and normalize of roll model

Fig. 9. shows the proposed 2-DOF control for RSC.

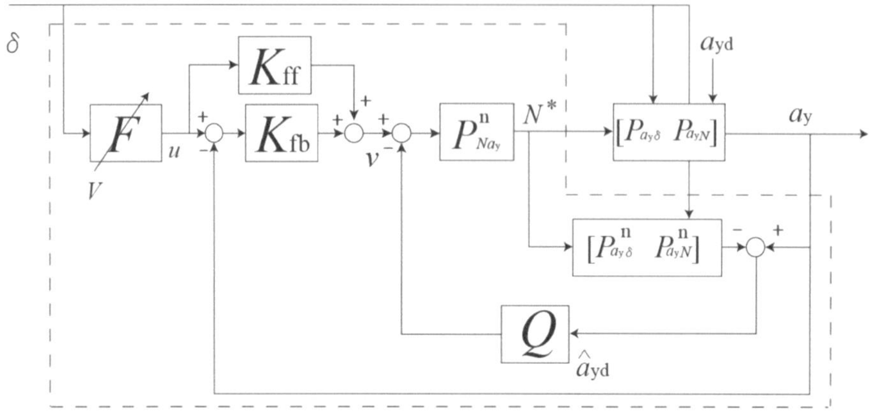

Fig. 9. Block diagram of 2-DOF for RSC based on DOB

Estimated lateral acceleration disturbance is fedback to lateral acceleration reference multiplied by filter Q.

$$a_y^* = v - Q\hat{a}_{yd.} \tag{20}$$

Filter Q is low pass filter and expressed as the following equation (Umeno et al., 1991). In this study, the cut-off frequency is set as 63 rad/s.

$$Q = \frac{1+\sum_{k=1}^{N-r} a_k(\tau s)^k}{1+\sum_{k=1}^{N} a_k(\tau s)^k}, \tag{21}$$

where, r must be equal or greater than relative order of the transfer function of the nominal plant. Substituting eq. (19) to eq. (17) and (20), the following equation is defined.

$$a_y = v + P^n_{a_y\delta}\delta + (1-Q)\hat{a}_{yd}. \tag{22}$$

Disturbance, which is lower than the cut-off frequency of Q and vehicle dynamics, is suppressed by DOB. In addition to the function of disturbance rejection, the plant is nearly equal to nominal model in lower frequency region than the cut-off frequency. Therefore the proposed RSC has the function of model following control.

4.2 Yawing stability control

As fig. 2. shows, YSC is yaw rate control. Yaw rate reference value is defined by steering angle and longitudinal vehicle speed. Transfer function from yaw rate reference and steering angle is expressed as the following equation.

$$\gamma = \frac{P_{\gamma M}P^n_{N\gamma}(K_{ff}+K_{fb})}{1+P_{\gamma N}P^n_{N\gamma}K_{fb}}u + \frac{P_{\gamma\delta}}{1+P_{\gamma N}P^n_{N\gamma}K_{fb}}\delta. \tag{23}$$

5. Simulation results

Three dimensional vehicle motion simulations have been conducted with combination software of CarSim 7.1.1 and MATLAB R2006b/Simulink. At first, the effectiveness of RSC is verified. Lateral acceleration disturbance is generated by differential torque for repeatability reason of experiments. In the simulation, lateral blast is generated at straight and curve road driving, the proposed DOB suppresses the disturbance effectively. To show the effectiveness of ESP, lateral acceleration response and trajectory at curving are compared. It is shown that lateral acceleration is unnecessarily suppressed only with RSC, however, tracking capability to yaw rate reference is achieved by ESP.

5.1 Effectiveness of RSC

(a)Vehicle Stability under Crosswind Disturbance

Vehicle stability of RSC under crosswind disturbance is demonstrated. At first, the vehicle goes straight and a driver holds steering angle (holding steering wheel as 0 deg). Under 20 km/h vehicle speed control, crosswind is applied during 3-6 sec. Fig. 10. shows the simulation results.

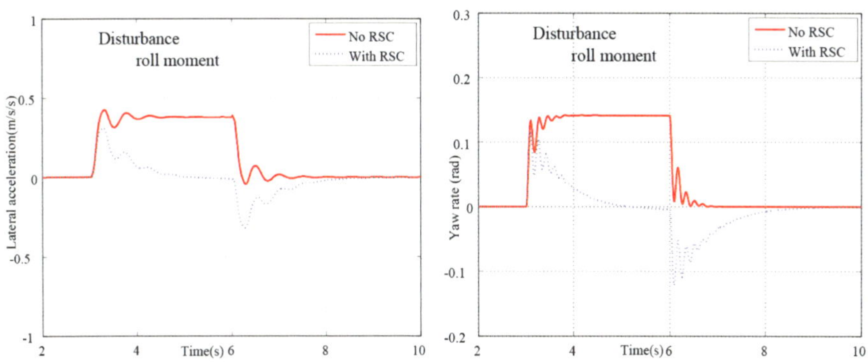

(a) Lateral acceleration (b) Yaw rate

Fig. 10. Simulation result of RSC: Disturbance suppression at straight road driving

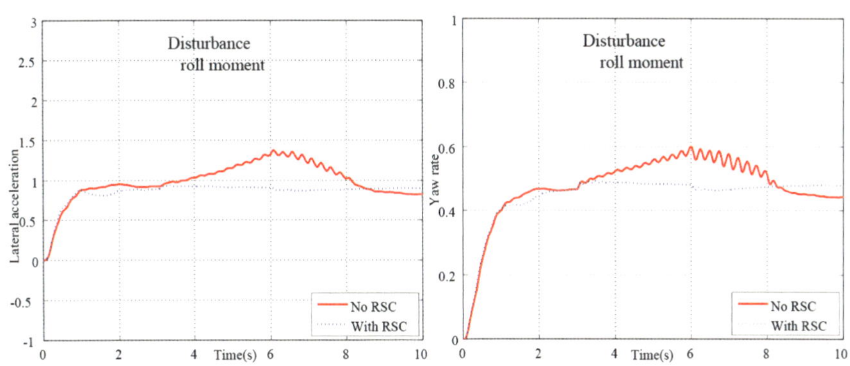

(a) Lateral acceleration (b) Yaw rate

Fig. 11. Simulation result of RSC: Disturbance suppression at curve road driving

When proposed RSC is activated, the proposed lateral acceleration DOB detects the lateral acceleration disturbance and suppresses it. Then, disturbance is applied at curve road driving. Under 20km/h constant speed control as well, 180 deg step steering is applied with roll moment disturbance during 3-6 sec. Fig. 11. shows decrease of lateral acceleration since disturbance is rejected perfectly by differential torque with RSC. The robustness of RSC is verified with simulation results.

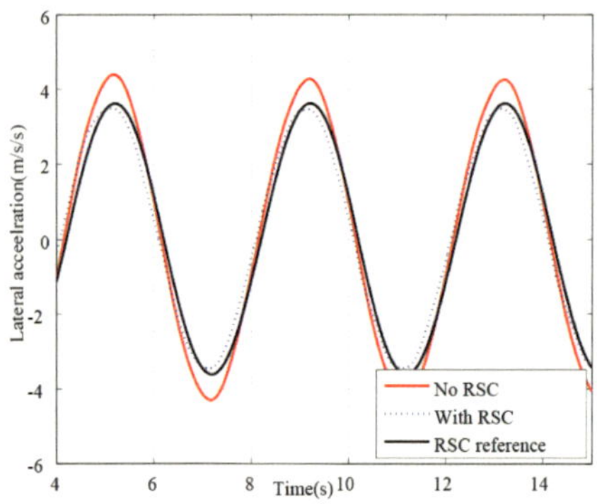

Fig. 12. Simulation result of RSC: Tracking capability to reference value

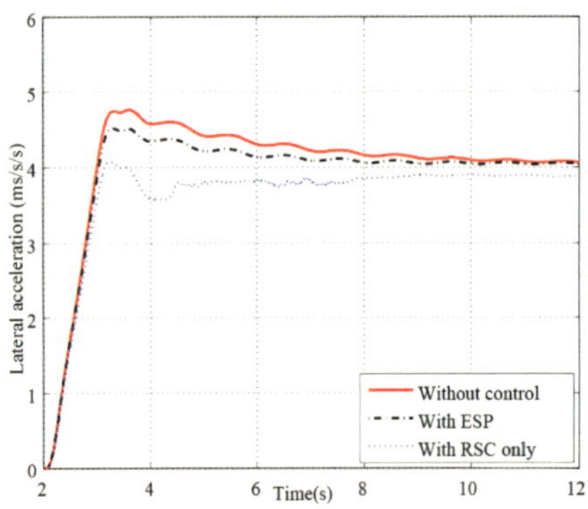

(a) Lateral acceleration

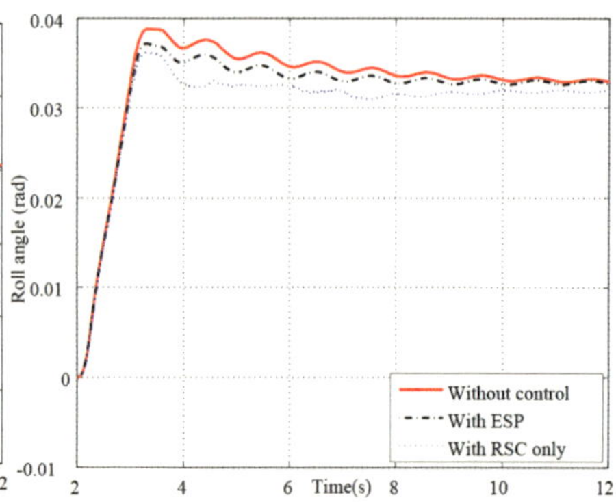

(b) Roll angle

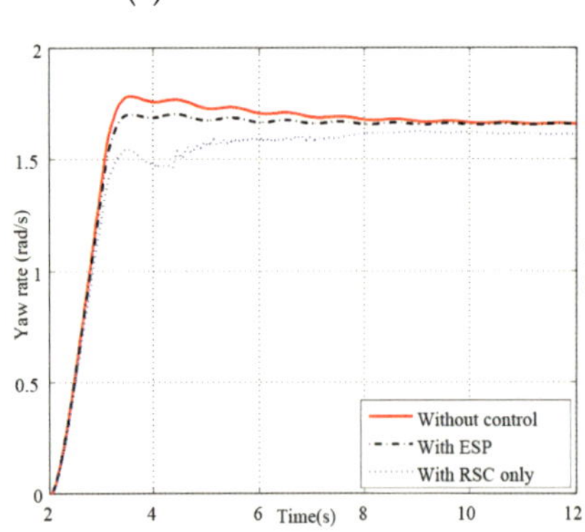

(c) Yaw rate

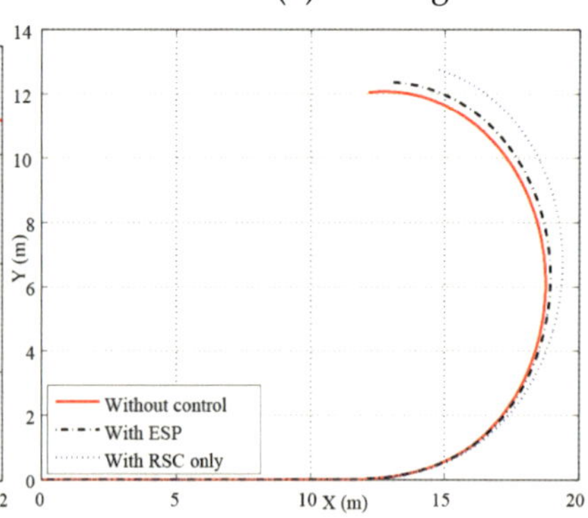

(d) Trajectory

Fig. 13. Simulation results of ESP: Step steering maneuver

(b)Tracking capability to reference value

In this section, tracking capability of RSC to reference value is verified with simulation results. Under 20km/h vehicle speed control, 180 deg sinusoidal steering is applied and reference value of lateral acceleration is 80% of nominal value. Fig. 12. shows that lateral acceleration follows reference value with RSC.

5.2 Effectiveness of EPS

Rollover experiment can not be achieved because of safety reason. Under 20km/h constant speed control, 240 deg step steering is applied. From fig. 13., with only RSC case, even though the danger of rollover is not so high, lateral acceleration is strongly suppressed and trajectory of the vehicle is far off the road. On the other hand, with ESP case, the rise of lateral acceleration is recovered and steady state yaw rate is controlled so that it becomes close to no control case.

6. Experimental results

6.1 Experimental setup

A novel one seater micro EV named "Capacitor COMS1" is developed for vehicle motion control experiments. The vehicle equips two in-wheel motors in the rear tires, a steering sensor, an acceleration sensor and gyro sensors to detect roll and yaw motion. An upper micro controller collects sensor information with A/D converters, calculates reference torques and outputs to the inverter with DA converter. In this system, sampling time is 1 (msec). Fig. 14. shows the vehicle control system and Table 1. shows the specifications of the experimental vehicle.

At first, disturbance suppression performance and tracking capability to reference value are verified with experimental results. Then, effectiveness of ESP is demonstrated. In the experiment, since vehicle rollover experiment is not possible due to safety reason, step response of lateral acceleration and yaw rate are evaluated.

6.2 Effectiveness of RSC

(a)Vehicle Stability under Crosswind Disturbance

For repeatability reason, roll moment disturbance is generated by differential torque. Under 20 km/h constant speed control, roll moment disturbance is applied from 1 sec. The disturbance is detected by DOB and compensated by differential torque of right and left inwheel motors. Here, the cut-off frequency of the low pass filter is 63 rad/s.

Fig. 15. shows disturbance suppression during straight road driving. Step disturbance roll moment (equivalent to $0.5\,m/s^2 * h_{cr}$) is applied around 1 sec. In the case without any control and only with FB control of RSC, lateral acceleration is not eliminated and vehicle trajectory is shifted in a wide range. On the other hand, in the case with DOB, disturbance is suppressed and vehicle trajectory is maintained.

Fig. 16. shows the experimental results of disturbance suppression at curve road driving. Under 20 km/h constant speed control, 240 deg steering is applied and disturbance is applied at around 2.5 sec. In this case, data is normalized by maximum lateral acceleration. In the case with RSC DOB, whole effect of disturbance is suppressed as no disturbance case. In the case without RSC, lateral acceleration decreases about 25% and vehicle behavior becomes unstable.

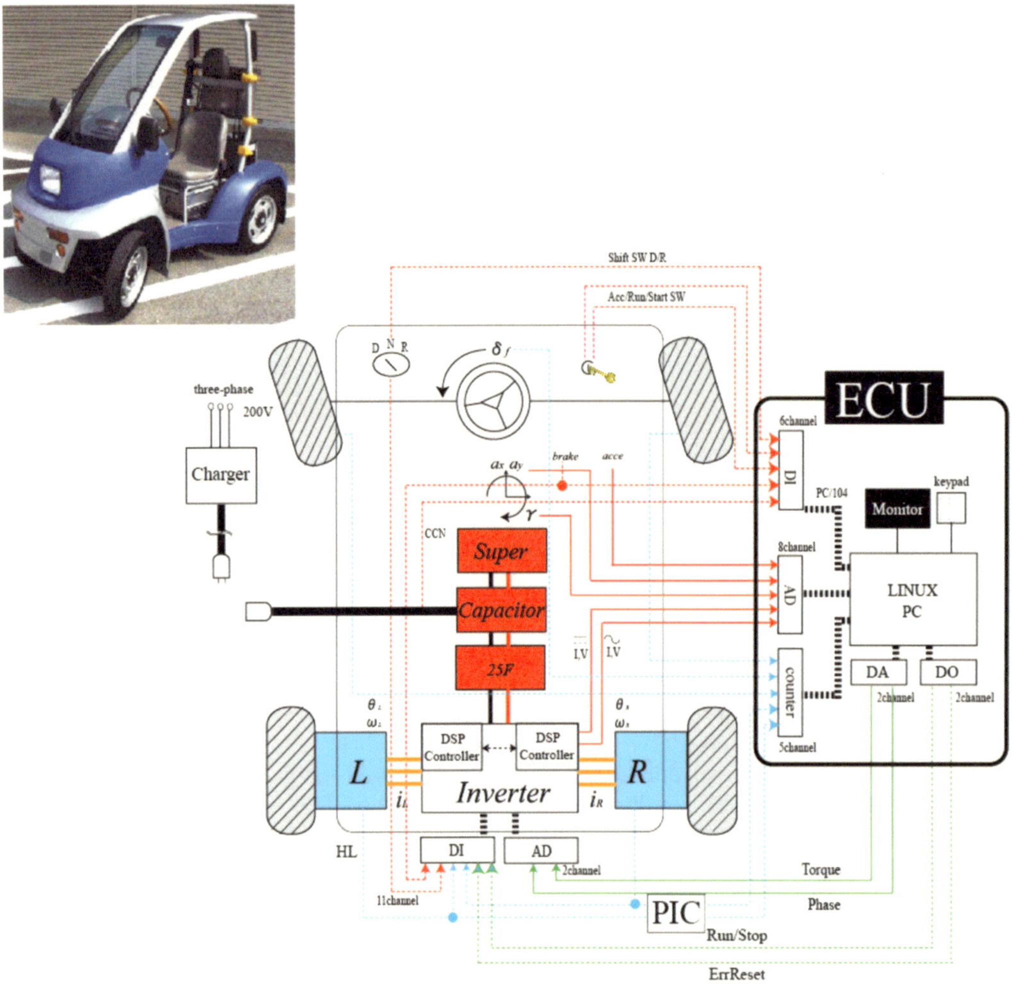

Fig. 14. Control system of experimental vehicle

Motor	
Category	IPMSM
Phase/Pole	3/12
Rating power/Max	0.29kW/2kW
Max torque	100Nm
Max velocity	50km/h
Inverter	
Switching Hardware	MOS FET
Control method	PWM vector control

Table 1. Drive train specification of experimental vehicle

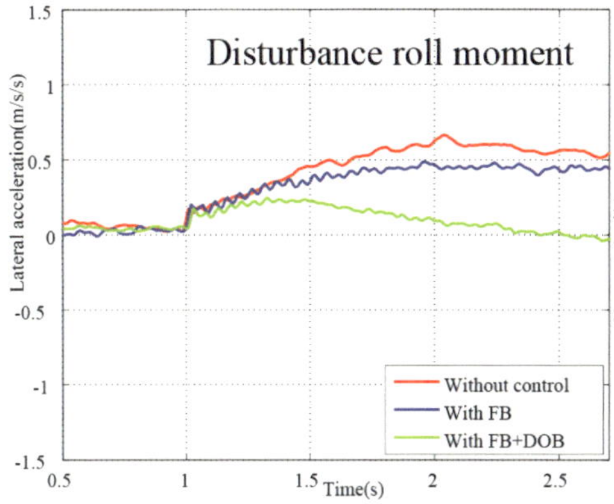

(a) Lateral acceleration

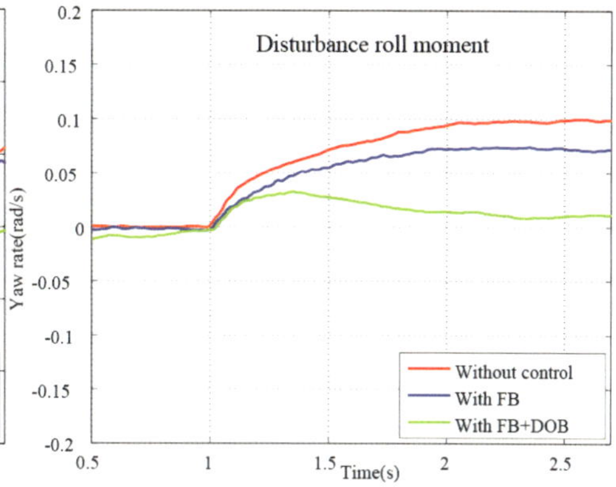

(b) Yaw rate

Fig. 15. Experimental result of RSC: Disturbance suppression at straight road driving

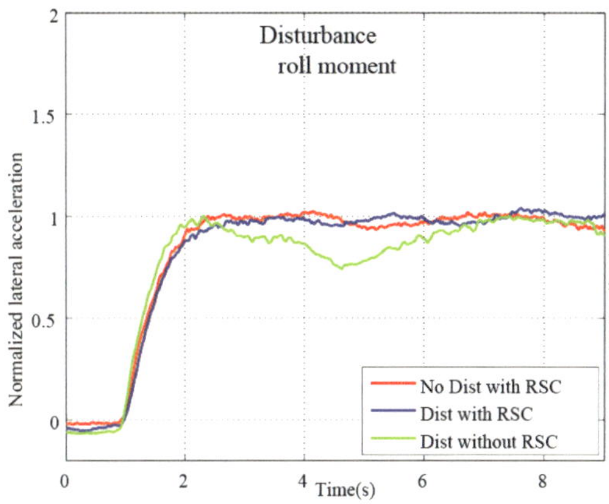

(a) Lateral acceleration

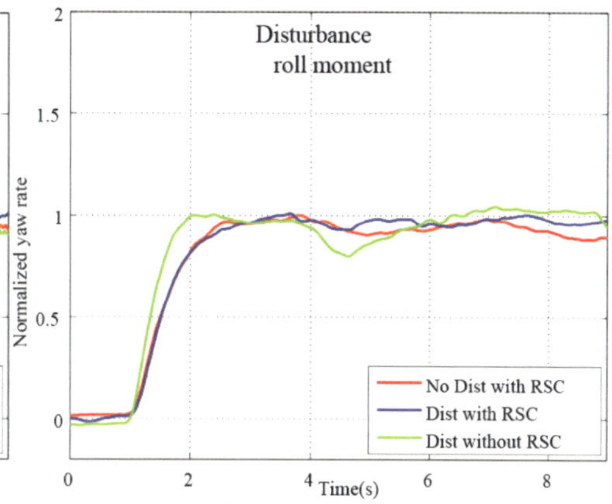

(b) Yaw rate

Fig. 16. Experimental result of RSC: Disturbance suppression at curve road driving

(b)Tracking capability to reference value

In the previous section, since it was assured that the inner DOB loop is designed properly, tracking capability to reference value is verified with experimental results. 180 deg sinusoidal steering is applied and reference lateral acceleration is 80% of nominal value. The outer loop is designed with pole root loci method. Fig. 17. shows that in the case with RSC, tracking capability to reference value is achieved.

6.3 Effectiveness of EPS

Effectiveness of ESP is demonstrated by experiments. For safety reason, rollover experiment is impossible. Therefore, experimental condition is the same as 5.2. Under 20km/h constant speed control, 180 deg step steering is applied.

Fig. 18. shows that in the case with only RSC, lateral acceleration and yaw rate are strongly suppressed. On the other hand, in the case with ESP, yaw rate is recovered close to reference value. In addition, the rise of lateral acceleration is also recovered and stable cornering is achieved with ESP.

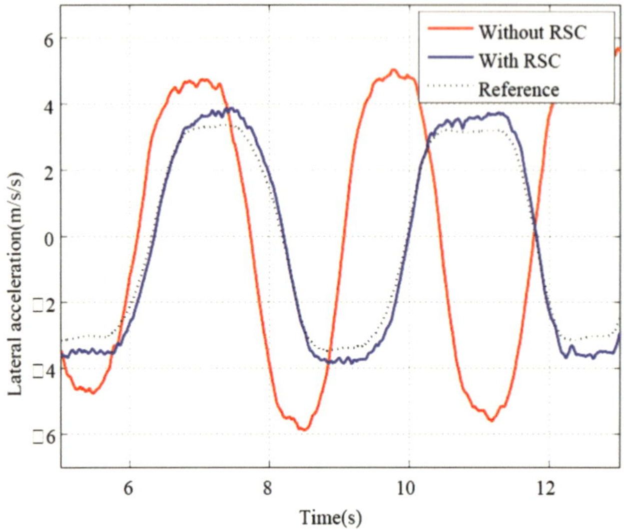

Fig. 17. Experimental result of RSC: Tracking capability to reference value

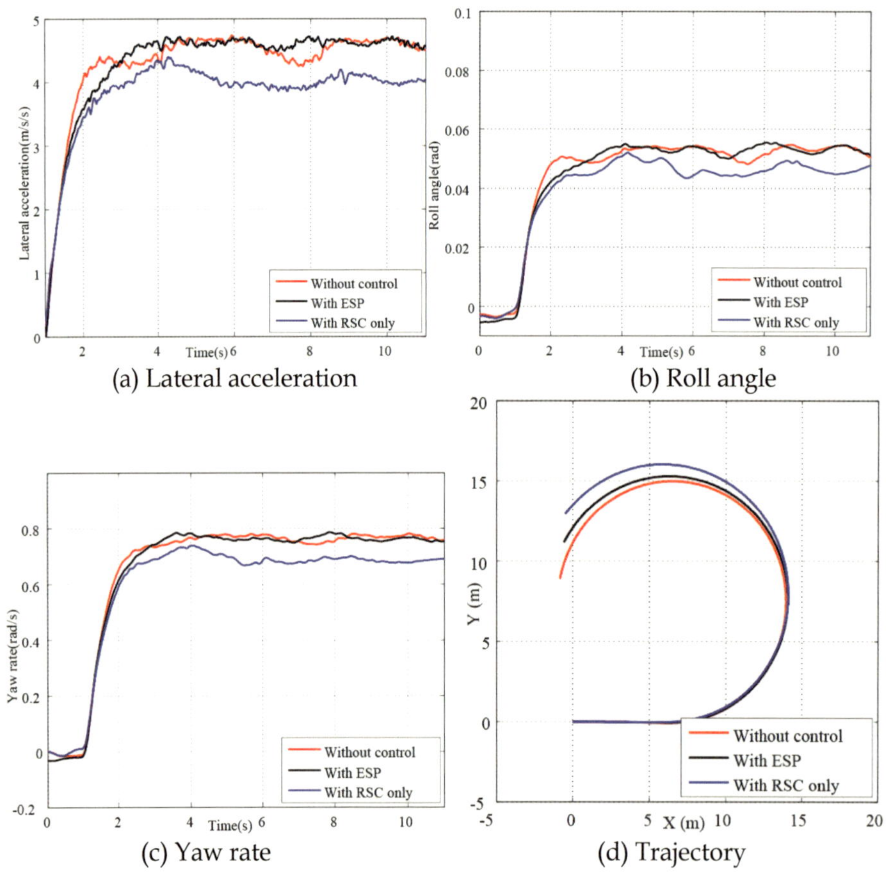

Fig. 18. Simulation results of ESP: Step steering maneuver

7. Conclusion

In this paper, a novel RSC based on ESP utilizing differential torque of in-wheel-motor EV is proposed. Effectiveness of novel RSC designed by 2-DOF control is verified with simulation and experimental results. Then incompatibility of RSC and YSC is described and ESP is proposed to solve the problem utilizing *RI* which is calculated using estimated value of estimation system of ESP. Experimental results validates the proposed ESP.

8. Acknowledgement

The author and the work are supported by Japan Society for the Promotion of Science.

9. References

Yoichi Hori, "Future Vehicle driven by Electricity and Control-Research on Four Wheel Motored UOT Electric March II", IEEE Transaction on Industrial Electronics, Vol.51, No.5, pp.954-962, 2004.10

Hiroshi Fujimoto, Akio Tsumasaka, Toshihiko Noguchi, "Vehicle Stability Control of Small Electric Vehicle on Snowy Road", JSAE Review of Automotive Engineers, Vol. 27, No. 2, pp. 279-286, 2006.04

Shinsuke Satou, Hiroshi Fujimoto, "Proposal of Pitching Control for Electric Vehicle with In-Wheel Motor", IIC-07- 81 IEE Japan, pp.65-70, 2007.03 (in Japanese).

Peng He, Yoichi Hori, "Improvement of EV Maneuverability and Safety by Dynamic Force Distribution with Disturbance Observer", WEVA-Journal, Vol.1, pp.258-263, 2007.05

National highway traffic safety administration, Safercar program, http://www.nhtsa.gov/

E. K. Liebemann, "Safety and Performance Enhancement: The Bosch Electronic Stability Control(ESP)", SAE Technical Paper Series, 2004-21-0060, 2004.10

Hongtei E. Tseng, et al, "Estimation of land vehicle roll and pitch angles", Vehicle System Dynamics, Vol.45, No.5, pp.433-443, 2007.05

Kyongsu Yi, et al, "Unified Chassis Control for Rollover Prevention, Maneuverability and Lateral Stability", AVEC2008, pp.708-713, 2008.10

Bo-Chiuan Chen, Huei Peng, "Differential-Braking-Based Rollover Prevention for Sport Utility Vehicles with Human-in- the-loop Evaluations", Vehicle System Dynamics, Vol.36, No.4-5, pp359-389, 2001.

Kiyotaka Kawashima, Toshiyuki Uchida, Yoichi Hori, "Rolling Stability Control of In-wheel Electric Vehicle Based on Two-Degree-of-Freedom Control", The 10th International Workshop on Advanced Motion Control, pp. 751-756, Trento Italy, 2008.03

Bilin Aksun Guvenc, Tilman Bunte, Dirk Odenthal and Levent Guvenc, "Robust Two Degree-of-Freedom Vehicle Steering Controller Design", IEEE Transaction on Control Systems Technology, Vol. 12, No. 4, pp.627-636, 2004.07

A. Hac, et. al, "Detection of Vehicle Rollover", SAE Technical Paper Series, 2004-01-1757, SAEWorld Congress, 2004.

N. Takahashi, et. al, "Consideration on Yaw Rate Control for Electric Vehicle Based on Cornering Stiffness and Body Slip Angle Estimation", IEE Japan, IIC-06-04, pp.17-22, 2006

Takaji Umeno, Yoichi Hori, "Robust Speed Control of DC Servomotors Using Modern Two Degrees-of-Freedom Controller Design", IEEE Transaction Industrial Electronics, Vol.38, No. 5, pp.363-368, 1991.10

Nomenclature

a_x, a_y : Longitudinal and lateral acceleration

a_{yd} : Lateral acceleration disturbance

a_{yth} : Critical lateral acceleration

c_f, c_r : Front and rear tire cornering stiffness

C_r : Combined roll damping coefficient

d, d_f, d_r : Tread at CG, front and rear axle

$F_{yfl}, F_{yfr}, F_{yrl}, F_{yrr}$: Tire lateral forces

g : Gravity acceleration

h_c, h_{cr} : Hight of CG and distance from CG to roll center

I_r, I_{r2} : Moment of inertia about roll axis (before and after wheel-lift-off)

I_y : Moment of inertia about yaw axis

K_r : Combined roll stiffness coefficient

l, l_f, l_r : Wheelbase and distance from CG to front and rear axle

M, M_s, M_u : Vehicle, sprung and unsprung mass

N : Yaw moment by differential torque

V, V_w : Vehicle and wheel speed

β, γ, δ : Body slip angle, yaw rate and tire steering angle

$\phi, \dot{\phi}, \phi_{th}, \dot{\phi}_{th}$: Roll angle, roll rate, threshold of roll angle and roll rate

9

Terrestrial and Underwater Locomotion Control for a Biomimetic Amphibious Robot Capable of Multimode Motion

Junzhi Yu, Qinghai Yang, Rui Ding and Min Tan
Laboratory of Complex Systems and Intelligence Science, Institute of Automation
Chinese Academy of Sciences
China

1. Introduction

The advancement of mechatronic devices and computer science has provided an impulse to fast-moving robotic technology in last decades. Taking the category of robots as an example, besides the industrial robots for manufacturing, the list of emerging robots for spaceflight, navigation, medical nursing, service, military purposes and so on, are growing (Yang et al., 2007). Further, there are many application-specific robots being developed and used today across a wide variety of domains. An accompanying drawback is that conventional robots can only work in a single working condition. For instance, the terrestrial mobile robots are functionally unable to propel in water owing to lacking necessary aquatic propelling units or waterproof treatment, while the underwater robots mostly have not sufficient locomotion ability on land since the locomotion will undergo stronger friction than it encounters in viscosity medium. Developing versatile robots adapting to changing environments faces significant challenge. Amphibious robots, with dual locomotion for mixed water-land environments, draw great attention and interest from academics and engineers all over the world (Ijspeert et al., 2005, 2007; Healy & Bishop, 2009). No doubt, they are very important tools when executing terrestrial and/or underwater related operations in complex surroundings (e.g. in the combat zone). In particular, military robots are currently being applied to many missions in Iraq and Afghanistan ranging from mine detection, surveillance, as well as logistics to rescue operations. Besides military applications, the well-developed amphibious robots that are highly maneuverable and adaptable to changeable terrains will cover more complex real-world missions, including ecological monitoring, amphibious reconnaissance, safety check, search and rescue, etc.

Compared with other single-function robots, the existing amphibious robots capable of operating both on land and under water are relatively rare. Generally speaking, they tend to fall into two primary categories: legged and snake-like. Since irregular and uneven terrain is the salient feature of water-land environment, many amphibious robots conventionally utilized leg-like locomotion on rough terrains. Some examples include the lobster robot constructed by J. Ayers group in Northeastern University of US (Ayers, 2004), the ALUV with six legs to duplicate crab by IS Robotics and Rockwell for the purpose of sensing or mine detection (Greiner et al., 1996), as well as the robotic crab built by Harbin Engineering

University in China (Wang et al., 2005). Although these legged robots with waterproofing treatment can operate on land and underwater, the aquatic locomotion is restricted to the ocean floor, which greatly reduces their workspace. Moreover, the mechanical configuration and the control algorithms related to these robots are highly complicated. Some other robots use improved legged structures as leading driving devices, such as the simplified wheel-leg propellers of Wheg IV built by Case Western Reserve University (CWRU) and the Naval Postgraduate School (NPS) to mimic cockroach's outstanding locomotion ability (Boxerbaum et al., 2005; Harkins et al. 2005), driving fins of robot turtle called Madeleine in Nekton Research (Kemp et al. 2005), and the paddles and semicircular legs applied to a series of legged amphibious robots developed by McGill University and its cooperative universities (Prahacs et al. 2005; Georgiades et al. 2009). The modified legged amphibious robots exhibit faster locomotion speed and better mobility, whilst maintaining a strong adaptability.

Aside from leg-like mode, snake-like locomotion is also utilized to achieve amphibious movements in a biomimetic manner. Some snakes in nature possess unique biological properties making them survive in various geographical environments, offering design inspiration in creating novel robots. Typically, ACM-R5 and AmphiBot are two robotic prototypes with different design philosophies. The ACM-R5 composed of multiple joints with 2 DOFs is built by robotics lab in Tokyo Institute of Technology and is the latest version in their research on snake-like robot since 1970s (Yamada et al. 2005). While the AmphiBot is constructed by Swiss Federal Institute of Technology and can crawl on land like snake and swim in water like lamprey (Ijspeert et al., 2005, 2007).

At present, most studies on amphibious robots mainly concentrate on locomotion mechanisms, control algorithms as well as their implementation. There is still a big gap between the actual performance of the existing robots and that of the biological counterpart in terms of speed, maneuverability and terrain adaptability. At the same time, the amphibious operation capabilities both on land and under water can hardly be guaranteed. One of the key causes is the difficulty posed by multifunctional driving mechanisms and steady control methods. This problem is further complicated by the fact that effective mechanism for direct control over the robot's position and orientation is unavailable. Based on our previous research on the mechatronic design and motion control of biomimetic robotic fish/dolphin (Yu et al. 2004, 2007), this chapter presents the preliminary results of our attempts to create an amphibious robot, "AmphiRobot", which is capable of multimode motion. The AmphiRobot takes the carangiform swimming as the primary locomotion pattern under water and the wheel-like motion as the basic way on land. Considering slender body structure of the robot, a body deformation steering approach is proposed for the locomotion on land, which employs the propelling units' departure from the longitudinal centerline of the whole body. Meanwhile, a chainlike network model of Central Pattern Generator (CPG) based on the nonlinear oscillator has been established for the underwater locomotion, which comprises the tail fin CPG and pectoral fin CPG. Benefitting from the reasonable mass distribution, the promethean swiveling body device, which can revolve all of the propelling-units in ±90°, executes the smooth transition of fish-like motion and dolphin-like swimming without additional counterweight. Compared with the existing amphibious robots, the multi-purpose, amphibious propulsive mechanism that combines carangiform or dolphin-like swimming with wheel-like motions achieves efficient movements both under water and on land possibly, which endows the robot with more substantial terrain adaptability.

The rest of the chapter is organized as follows. The bio-inspired mechanisms, mechanical design as well as system implementation are outlined in Section 2. A body deformation steering approach to locomotion control on land is offered in Section 3. The CPG based swimming control is presented in Section 4. Finally, Section 5 concludes the chapter with the outline of future work.

2. Mechatronic design of a biomimetic amphibious robot

2.1 Biological inspiration

In past millions of years, the fact that amphibians can survive in complex and changeful environments reveals that every amphibian possesses unique traits and dexterous structures suitable for the current living conditions. Replicating such biological morphologies, structures, functions, working principles, controlling mechanisms, etc., in the context of biorobotics, will greatly promote the insight of researchers on amphibians and accelerate the investigation on amphibious robots (Bandyopadhyay, 2004, 2005). Since bio-inspired design is the blending of biology, mechanics, mechanical engineering, electronics and computer control into an integrated system, it is never an easy task to copy nature exactly and essentially. For the convenience of engineering practice, a partial biomimetic approach is commonly employed. That is, only part of the biomimetic robot, which may be the morphology, mechanical structure, function, locomotion or control principle, is similar to the biological counterpart, whereas other parts are the same as different prototypes or are not bio-inspired at all. The AmphiRobot in this work combines the locomotion features of carangiform fish and dolphin together and also integrates the characters of wheeled devices. Compared with the propellers-driven mechanisms, on the one hand, fish takes advantage of the coordinated motion of its body, fins and tail to achieve efficient and agile swimming performance (Sfakiotakis et al. 1999). The AmphiRobot therefore takes fish-like swimming as the main motion mode by using a set of modular propelling units and caudal peduncle. On the other hand, dolphin relies on the coordination of oscillating tail fluke and pectoral flippers to perform fast and efficient propulsion. Its fluke oscillation in the vertical plane, rather than the oscillation of fish tail in the horizontal plane, endows dolphin with better maneuverability while pitching (Fish & Rohr, 1999). A promethean swiveling body device is further introduced to unite the fish-like and dolphin-like swimming into the AmphiRobot, enabling the robot to convert the motion between these two modes agilely.

2.2 Design specifications

The design of the AmphiRobot is directed by the following guidelines:

- To be modular: The module-oriented design allows us to quickly alter the length of the robot by adding or removing modules, as well as to replace failed module;
- To be waterproof: Each module, plus head, is individually watertight. Even leakage occurred in one module will merely damage a single joint, which will not affect overall function of the robot;
- To be transparent: The side panels of head and each module are made of Perspex, facilitating the monitor of operation and trouble shooting;
- To be slightly buoyant: When inactive, most of the robot body should stay under the surface of water with the longitudinal axis parallel with the surface and the robot at a pre-set depth under water should revert to the surface passively;

- To have distributed actuators and power: Each module and head carry batteries for their own DC motors or servomotors, which not only prolong the working period of robots, but also strengthen the modularity;
- To be stable in every locomotion mode: The robot should be stable in both fish-like motion and dolphin-like one when inactive, so the centre of mass of module should be placed just at the geometric centre so that the module can float stably while the centre of mass of head should be placed below the geometric centre to ensure the stability of the whole robot.

2.3 Mechanical design

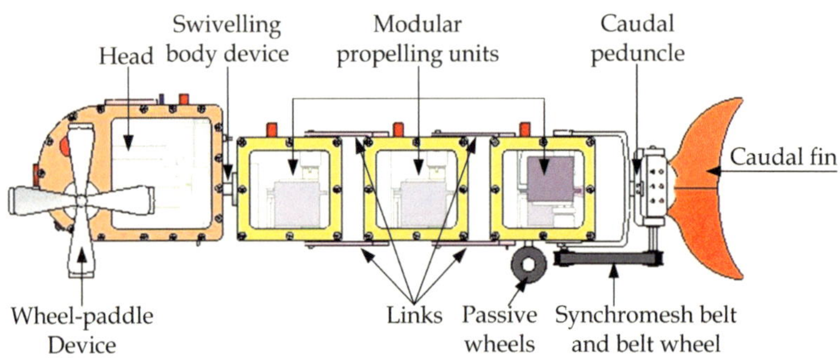

Fig. 1. The overall structure of the AmphiRobot-I with multimode motion

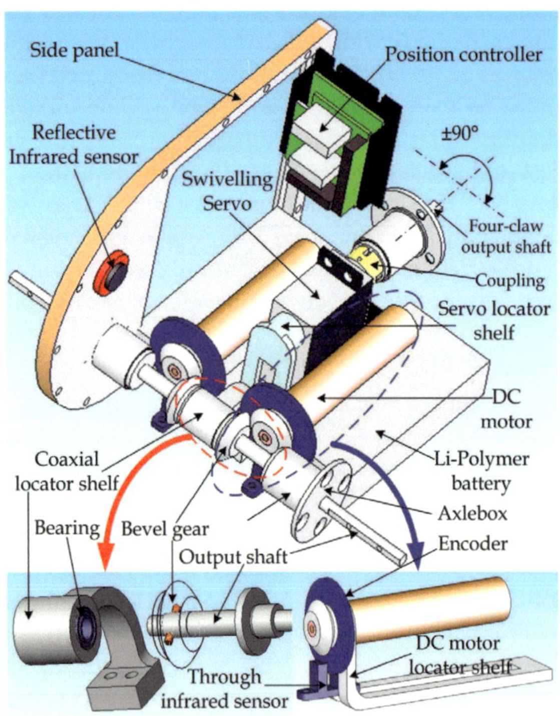

Fig. 2. Schematic illustration of the head structure in the AmphiRobot-I

As shown in Fig. 1, the AmphiRobot is composed of a head, alternative wheel-paddle and flipper, swivelling body device, modular propelling units and caudal peduncle. The framework of the robot parts is manufactured by the alloy aluminium fabrication, with the use of transparent plexiglass on both sides, which will aid in making impervious to surrounding water and inspecting the running state of built-in components. The head of the AmphiRobot serves as the control center, including a pair of DC motors, control circuits and other core components, as illustrated in Fig. 2. Two DC motors and their controllers are located in the head symmetrically, with their rotation output transmitted by a pair of mutual-engaged bevel wheels. By altering the output direction 90°, the output shafts rotate to obtain the power in the head. As depicted in Fig. 3, the alternative wheel-paddle and flipper can be assembled to the output shafts of DC motors respectively, to achieve various amphibious motion modes.

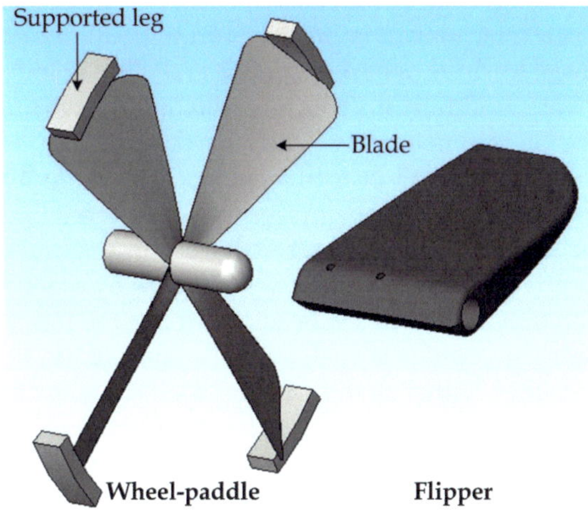

Fig. 3. Interchangeable wheel-paddle and flipper

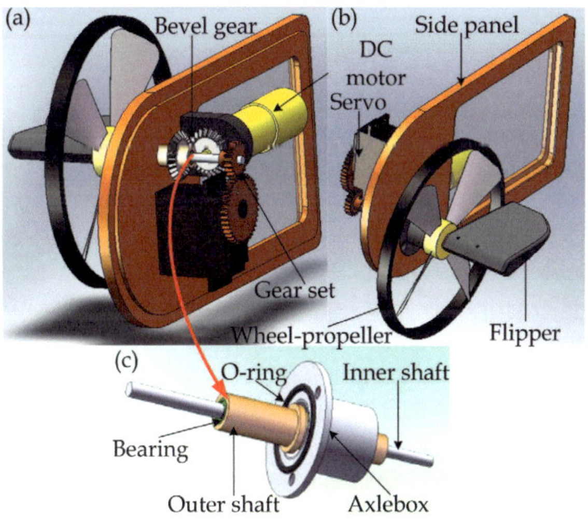

Fig. 4. An integrated wheel-propeller-fin mechanism. (a) Back view. (b) Front view. (c) Mechanical configuration of a composite coaxial shaft

In water, with the flipper mounted to the output shaft, the jigging motion along with lateral oscillations of modular propelling units can implement moving forward or backward, turning, and pitching. In the meantime, continuous rotating of the flipper on land will drive the robot to crawl forward. The crawling exhibited in the robot, in particular, allows better obstacle-negotiation capability, but this movement is relatively slow and insufficient. It will readily lead to the head vibration, which implicates the system unsteadiness. However, a wheel-paddle with four feet at the end of four spokes is employed and assembled in each side of the robot's head, which covers the shortage of flipper. The similarity with wheel on structure can improve moving speed dramatically while its tetrapod design also takes into account the climbing obstacle capability. By utilizing special shaped design of wheel-paddle blades, its continuous rotating can drive the robot by thrust vertical to robot's profile plane which can help turn under water.

Furthermore, to make separate wheel-paddle and flipper more compact and flexible, a hybrid wheel-propeller-fin mechanism is proposed in the AmphiRobot-II, where a unique coaxial shaft is employed to drive wheel-propeller and flipper individually. With such a mechanism, the underwater and terrestrial locomotion might be simultaneously guaranteed in performance and be autonomously switched in control. Fig. 4 shows the schematic representation of the integrated wheel-propeller-fin mechanism. As a crucial component, the composite coaxial shaft shown in Fig. 4c comprises two independent outputs: the inner shaft and the outer shaft. The former actuated by servo drives the artificial flipper, while the latter actuated by DC motor drives the wheel-propeller. Because symmetrically fixed drive shaft will laterally take up too much room in the head unit (only 150 mm in width), an extension fixture has to be utilized. The composite shaft joins the outer of the side panel via axlebox of the custom-built sealing unit. Rather than fastening the DC motor and servo firmly to the bottom of the head, they are vertically arranged on the inner of side panel. Specifically, the servo drives the inner shaft via a gear set with a reduction ratio of 1:2, which allows the flipper to forward flapping or reversing flapping in a range of 0–360° (discontinuous). Notice that the flipper in this fashion is capable of forward and backward swimming, as well as pitch motions by adjusting the angle of attack of the fin.

2.4 Hardware configuration

The AmphiRobot possesses many DOFs for flexible locomotion. The control system is therefore required to manage a multitude of servomotors and DC motors, plus sensors, communication module, etc., bringing forward high demands. Fig. 5 illustrates the overall structure of the control system specially developed for the robot, whose kernel is the master board based on the ARM AT91RM9200 microcontroller produced by Atmel Corporation.

The control of DC motors depends on the matching position controllers which link DC motors and AT91RM9200 together and communicate with AT91TM9200 via RS-232 ports. The controllers can realize multiple motion modes of DC motors, such as (profile) velocity mode, (profile) position mode, homing mode, and so on. When the switch is on, the controllers enter homing mode to find the initial position of wheel-paddle device or flipper, and then stay still waiting for the orders from AT91RM9200. If the robot is on land, the controllers are in velocity mode and make motors rotate continuously; otherwise, if in water, the controllers will switch to position mode and realize the motors' jigging motion around a middle position. The operating modes via controller and corresponding motion of motors always depend on the orders from upper control platform.

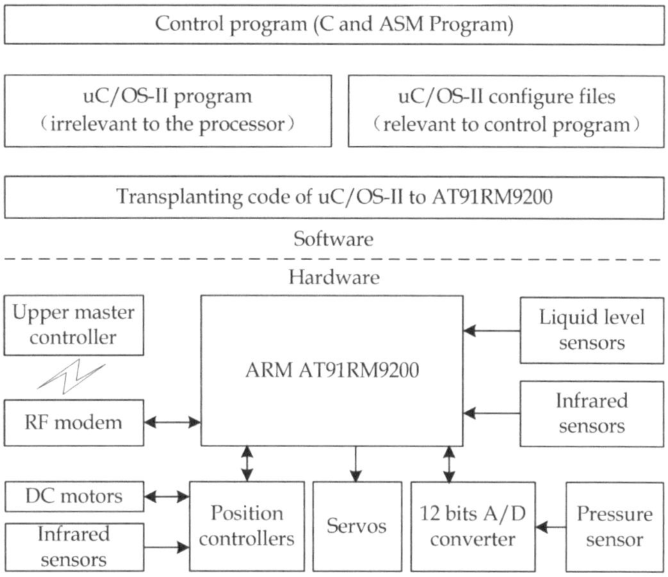

Fig. 5. Hardware and software structure in the AmphiRobot

Pressure sensor, with an analog voltage output, is used for depth detection in water. There are two kinds of infrared sensors which are reflective and through infrared sensors as shown in Fig. 2. The former are installed in the right front and bilateral position of robot head serving as robot's eyes for obstacle detection, and the detection range can be adjusted through variable resistors. The latter are mounted together with the motors' output shafts and output signals are connected to position controller for location of the absolute position of DC motors when power on. The two liquid level sensors, located in the head and the last propelling-unit respectively, output switching signals to judge whether the robot is on land or in water. These sensors with different functions make up a sensing system to provide ambient information for robot and lay the foundation for multimode locomotion control.

2.5 Software configuration

As described in Fig. 5, the control program for the AmphiRobot is based on the real-time kernel uC/OS-II, a preemptive kernel, guaranteeing fast response to the changes of control parameters and external disturbance. The overall control system consists of an operator, a processing centre, sensing inputs, and actuating mechanisms. Firstly, commands from the operator are transmitted to AT91RM9200 through radio waves and sensing signals are input as interrupts. Secondly, the central processing unit interprets the commands and interrupts, verifies their validity, and then sends corresponding signals to motor controllers in the head and servomotors in propelling-units. Finally, the actuators will receive the signals and move in a desired fashion.

The critical requirement for the control program is the stability and swift response. Each motion mode of AmphiRobot corresponds to an identical user task in uC/OS-II, which largely facilitates and accelerates implementation of motion modes, and also ensures the stability of program. Due to the very short time with interrupt off, the tests show that AmphiRobot can always answer for the orders and the changing conditions. More details can be referred to Ding et al., 2009.

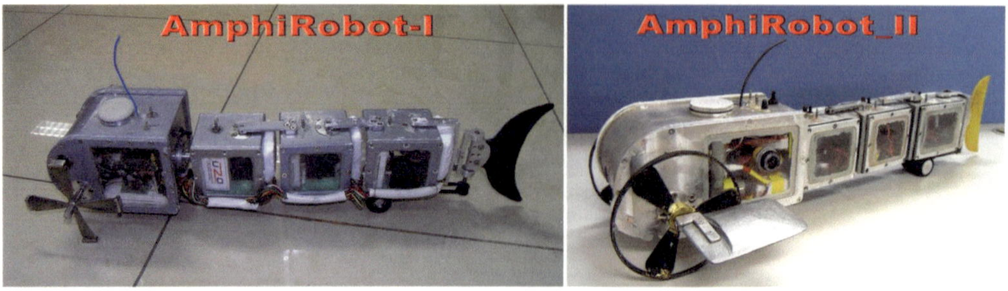

Fig. 6. Developed robotic prototype: AmphiRobot-I and AmphiRobot-II

2.6 Experimental setup

Base on the above hardware and software design, as shown in Fig. 6, two robotic prototypes have been successfully fabricated in our laboratory. The dimensions for these two prototypes are about 640mm × 190mm × 110mm and 700mm × 320mm × 150mm, respectively.

At present, two control modes have been applied to the amphibious prototype: the manual mode and the automatic control mode. For the former, a custom-built remote controller has been developed. Since radio frequency (RF) waves are severely attenuated under water, once the depth underwater exceeds 300 mm, the RF link will become unreliable and even unconnected. A relatively autonomous mode with the aid of onboard sensors, therefore, is further employed to obstacle-avoidance, diving and surface, aquatic-terrestrial transition, etc. In actual test, the prototype can easily implement the switch between fish-swimming mode and dolphin-swimming mode by regulating the swiveling body mechanism, and can also perform efficient propulsion in each mode. Detailed control methods to terrestrial and underwater locomotion will be elaborated in Sections 3 and 4, respectively.

3. Terrestrial locomotion control

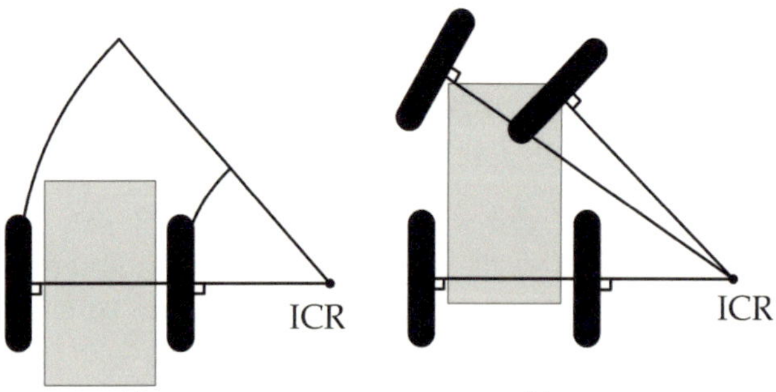

(a) (b)

Fig. 7. Two cases for formation of instantaneous center of rotation (ICR) in two locomotion modes on land. (a) The differential drive case. (b) The ackerman steering case.

As a rule, the mobile robot locomotion on land involves differential drive, steered wheel drive, synchronous drive, omni-directional drive and ackerman steering. For the AmphiRobot equipped with wheel-paddles (AmphiRobot-I) or wheel-propeller-fin mechanism (AmphiRobot-II), wheeled locomotion is the basic mode on land. It seems that

the differential drive in Fig. 7a is a reasonable locomotion form. But the slender body of the robot and the lateral friction from rear passive wheels have somewhat negative influence on steering. The poor actual steering performance verifies the incongruity of differential steering. Through careful analysis on the basic configuration, the drive of the AmphiRobot is more similar to that of car drive shown in Fig. 7b. The perpendiculars of two mutually independent wheels and the fixed wheels of a car form an instantaneous center of rotation (ICR) and the orientation of the car will be altered. When the robot body remains straight, the fore driving wheel-like part and rear passive wheels are parallel, no ICR is formed, and the robot moves forward. Benefiting moderately from the carangiform swimming mode in water, the robot's body shape can be varied when the modular propelling units departure from their central positions, and then the perpendiculars of wheel-paddles and passive wheels intersect and an ICR is formed which makes the robot maneuver on land. Such a maneuvering procedure is hereinafter referred to as "body-deformation steering."
For our robot with three fish-like propelling units, the rotations of the second or third unit independently and the coordinated oscillations of the two units will make the body shape change and meet the requirements of forming an ICR. So there exist three ways available to steer the robot agilely on land. These three methods form different ICRs corresponding to different turning radii, and the following parts will deal with geometry-based analyses and optimization.

3.1 Coordinated deflection via the last two propelling units

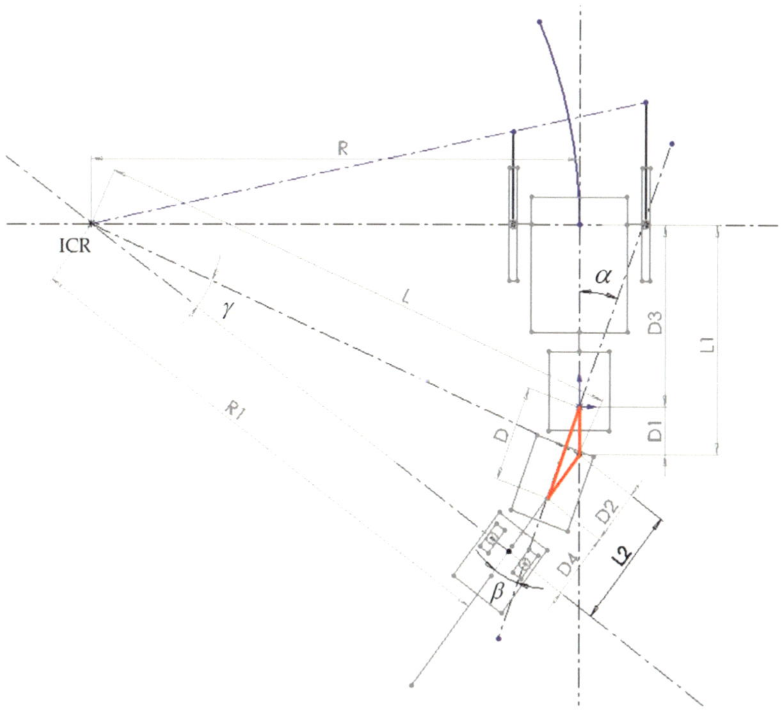

Fig. 8. Illustration of changing the body shape via the coordinated deflections of the second and third propelling units, wherein the perpendiculars of wheel-like part and passive wheels form an ICR

As shown in the Fig. 8, both the second unit and the third one departure from their middle positions with offset of α and β, respectively. The body shape turns from a straight line into an approximate arc shape so that an ICR is coming into being.

Once the deflection angles of propelling units are given, $D1$ and $D2$ can be calculated according to the below relation of side length and interior angle of triangle:

$$\frac{D1}{\sin\beta}=\frac{D}{\sin(\pi-\alpha-\beta)} \qquad \frac{D2}{\sin\alpha}=\frac{D}{\sin(\pi-\alpha-\beta)} \tag{1}$$

where D is a constant. The two right triangles share the same hypotenuse, having

$$\begin{cases}\sin(\alpha+\beta-\gamma)=\dfrac{L1}{L}\\ \sin\gamma=\dfrac{L2}{L}\end{cases} \Rightarrow \frac{L1}{\sin(\alpha+\beta-\gamma)}=\frac{L2}{\sin\gamma} \tag{2}$$

where $L1 = D1 + D3$, $L2 = D2 + D4$, specially $D3$ and $D4$ are known variables.

Combing (1) and (2), γ can be solved, and then the turning radius associated with the specific deflection can be derived through $R = L1 \times \cot(\alpha + \beta - \gamma)$:

$$R=\frac{D(\sin\alpha+\sin\beta*\cos(\alpha+\beta))+D3*\sin(\alpha+\beta)*\cos(\alpha+\beta)+D4*\sin(\alpha+\beta)}{\sin^2(\alpha+\beta)} \tag{3}$$

3.2 Deflection of the second or third propelling unit separately

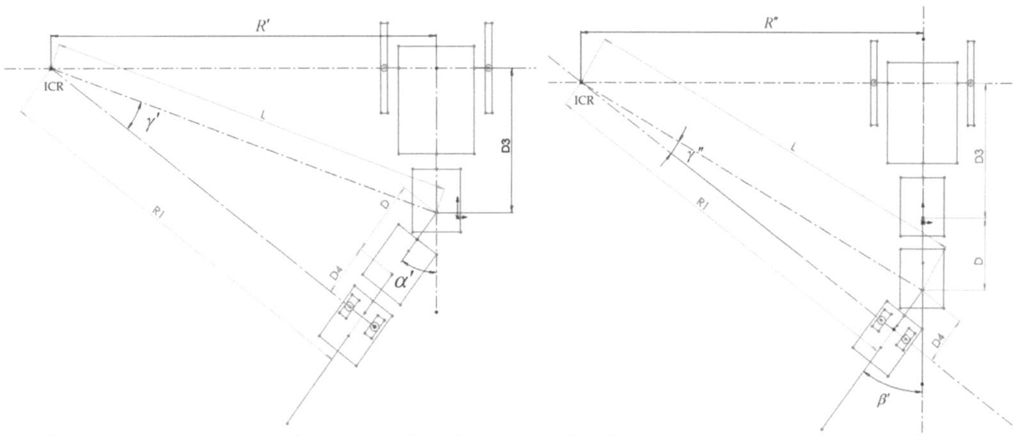

(a) Separate rotating of the second unit (b) Separate rotating of the third unit

Fig. 9. Illustration of deflecting the second or third propelling unit separately to change the body shape and form new ICRs

As illustrated in Fig. 9, the separate rotating of the second or third unit will also form ICR. According to the above calculation method, the following relationship can be yielded:

$$R'=\frac{D3*\cos\alpha+D+D4}{\sin\alpha} \qquad R''=\frac{(D+D3)\cos\beta+D4}{\sin\beta} \tag{4}$$

3.3 Comparison of three methods

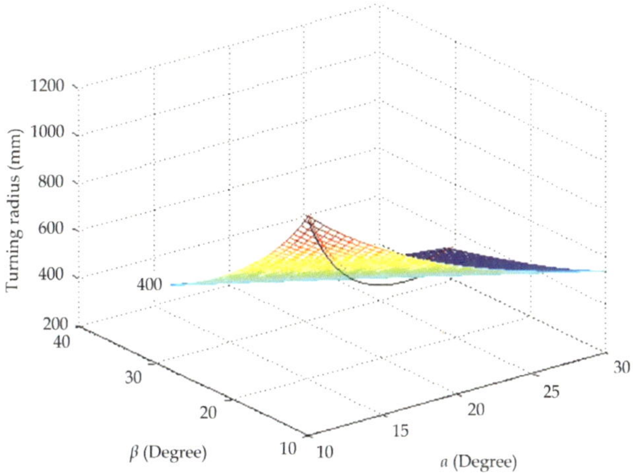

Fig. 10. Comparative result for three methods to form ICR

To further acquire the optimal turning mode, the radii of three turning cases with the same deflection angels are analytically compared. That is, the following relation holds:

$$\alpha + \beta = \alpha' = \beta' \tag{5}$$

Fig. 10 indicates the comparative result, where the hood facet is the turning radius related to coordinated deflection of the second and third units. The red and black curves denote the turning radii corresponding to the independent deflection of the second and third units. As can be seen, the following inequality is satisfied:

$$R'' < R < R' \tag{6}$$

With the equal deflections, the case employing the third propelling unit to rotate around its spin axis generates the minimal radius, in which the robot is more maneuverable on land. Furthermore, the individual rotating of the third unit brings about less deviation of passive wheels from the longitudinal centerline, which is beneficial to the stability of the robot.

3.4 Differential velocity

Suppose that the revolutions, angular velocities and linear velocities of the left and right wheel-like part are n_L, n_R, ω_L, , ω_R and V_L,, V_R, respectively, the angular velocity of the robot is ω. As can be observed in Fig. 8, the left and right wheel-paddles hold the same angular velocity, i.e.,

$$\begin{cases} V_R = r\omega_R = r\dfrac{2\pi n_R}{60} = \omega * \left(R'' + \dfrac{d}{2}\right) \\ V_L = r\omega_L = r\dfrac{2\pi n_L}{60} = \omega * \left(R'' - \dfrac{d}{2}\right) \end{cases} \tag{7}$$

where r is the radius of wheel-like part and d is the distance between the two wheel-like parts. Another equation expressed in the revolutions can be further written as

$$\frac{n_R}{n_L} = \frac{R'' + \frac{d}{2}}{R'' - \frac{d}{2}}. \tag{8}$$

3.5 Kinematic models

Base on the optimal body-deformation steering by the deflection of the third propelling unit, the two robot frames are established in the midpoints of the rear wheels and forward wheel-like part, taking the heading of the third propelling unit and head module as the orientations of X axis, and the orientations from right wheel to left wheel as that of Y axis, as shown in Fig. 11.

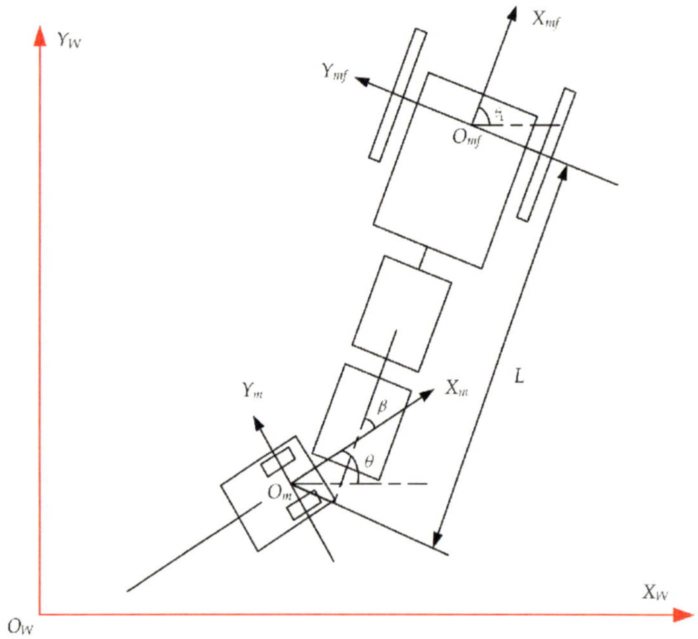

Fig. 11. Schematic of kinematic model based on the body-deformation steering

For the convenience of theoretical analysis, some suppositions are made below:

- Robot moving on a horizontal plane;
- Point contact of the wheels with ground which are not deformable;
- Pure rolling and no slipping, skidding or sliding;
- Steering axis orthogonal to the surface.

It can be easily obtain the following equation:

$$\begin{cases} \dot{x} = v\cos(\theta + \beta) \\ \dot{y} = v\sin(\theta + \beta) \end{cases} \tag{9}$$

where v is the motion speed of the midpoint of driving wheel-like part; θ is the direction angle, i.e. the angle between X_m and X_W; β is the steering angle of guide wheels; $\dot{x}$ and $\dot{y}$ are the translational velocities of O_{mf}.

Via rotating of the third unit causes deviation of passive wheels from the longitudinal central line, and the offset in such a situation is very little compared with the distance between fore and rear wheels. The following equation can therefore be approximated as:

$$\dot{\theta} = (v / L)\tan\beta \tag{10}$$

where L is the distance between the two midpoints, and $\dot{\theta}$ is the rotary velocity.
The equations (10) and (11) compose one of the kinematic models of the robot. Notice that wheel can not move in the direction perpendicular to the wheel plane. To be specific, O_{mf} can not translate along the Y_{mf} axis for the fore wheel and O_m can not move along Y_m axis. So the following non-holonomic constraint for the amphibious robot holds:

$$\begin{cases} \dot{x}\sin\theta - \dot{y}\cos\theta + L\dot{\theta}\cos\beta = 0 \\ \dot{x}\sin(\theta+\beta) - \dot{y}\cos(\theta+\beta) = 0 \end{cases} \tag{11}$$

Due to the special steering method, the rear wheels have different radius with that of fore wheel, another kinematic model can be described below:

$$\begin{cases} \dot{x}' = v\cos(\theta+\beta) + v\tan\beta\sin(\theta+\beta) \\ \dot{y}' = v\sin(\theta+\beta) + v\tan\beta\cos(\theta+\beta) \\ \dot{\theta} = (v / L)\tan\beta \end{cases} \tag{12}$$

where $\dot{x}'$ and $\dot{y}'$ are the translational velocities of O_m in the world frame.

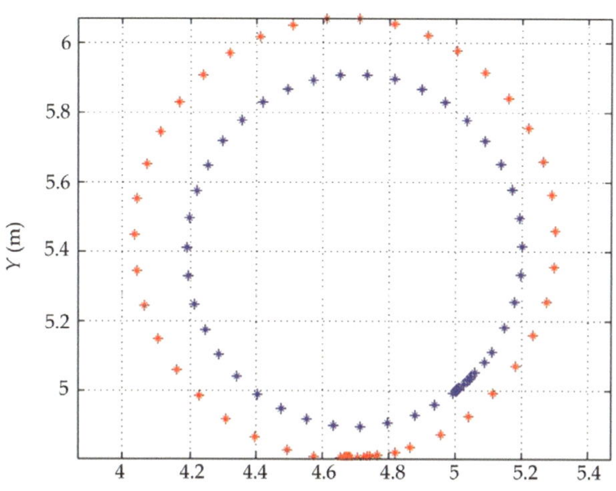

Fig. 12. The simulation result of two kinematic models

Based on the kinematic equations (9), (10) and (12), the locomotion trajectories of the robot can be simulated. As shown in Fig. 12, the circle in blue asterisk is the trajectory of fore wheel, and the red one is that of rear passive wheels. The error between these two trajectories is largely caused by the approximation of (10), which is the departure of midpoint of rear wheels from the longitudinal centerline.

Additionally, combining (7) and (10), the relationship between the steering angle and the turning radius can further be derived:

$$\tilde{R} = \frac{(D3 + D + D4 * \cos\beta)}{\tan\beta} \tag{13}$$

Notice that this is only an approximate result, corresponding to the accuracy radius R'' in (4). A comparative result for this relation is plotted in Fig. 13, where the blue curve denotes the turning radius calculated through the geometric analysis, while the red one is based on the kinematics model. Though the error increases with rising steering angle, the absolute error is relatively small as opposed to the turning radius, which is acceptable in practice. When the steering angle is about 45°, the turning radius approaches a body length of the robot. Notice also that a body length means the length between the fore wheel-like part and rear wheels as the body maintains straight.

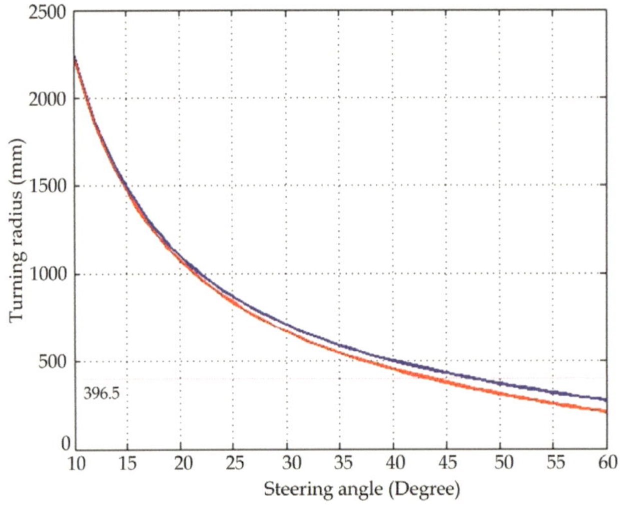

Fig. 13. The comparison of turning radii

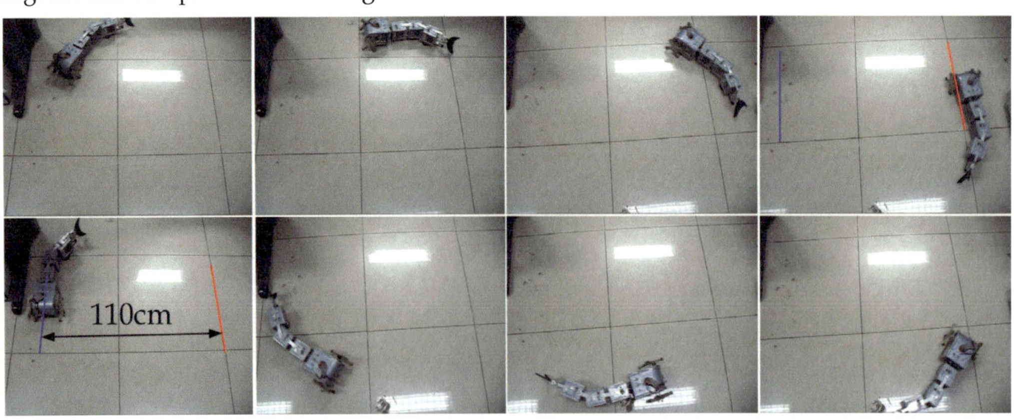

Fig. 14. Image sequence (anticlockwise) of performing circular motion via the body-deformation steering

3.6 Experimental results

To verify the body-deformation steering method and the analysis about the turning radius, experiments have been carried out on the ceramic tile-paved floor in our lab. During testing, the head and the first two propelling units keep in a straight line, the third unit is rotated by a specific angle of 37° which remains the same all the while, with the velocities of the left and right wheel-paddles according with the relationship in (8), and a circular motion is executed, as shown in Fig. 14. Note that the blue and red lines in two snapshots stand for the same locations in the actual ground. The measured steering radius is about 550 mm, and the calculated radii from (4) and (13) are 552 mm and 506 mm, respectively. The negligible error between the theoretical and experimental performance demonstrates the validity of the utilized body-deformation steering approach.

4. Underwater locomotion control

Neurobiology studies have shown that the locomotion of animals is governed hierarchically by the central nervous system, from the cerebral cortex level, the brainstem level, to the spinal cord level (Delcomyn, 1980). Fundamental rhythmic movements in locomotion, such as swimming, walking, running, and flying, are produced by central pattern generators (CPGs) at the spinal cord level. A CPG is a neuronal circuit capable of producing rhythmic patterns of neural activity automatically and unconsciously. The rhythmic pattern activates motor neurons that control the muscles generating the rhythmic movements. At the higher level, CPGs are networks of neurons that can produce coordinated oscillatory signals without oscillatory inputs. The sensory input or descending input from higher elements can regulate the frequency and phase of the rhythmic patterns by altering the intrinsic properties of the neurons and the synaptic strengths and connectivity among them (Delcomyn, 1980; Ijspeert, 2008).

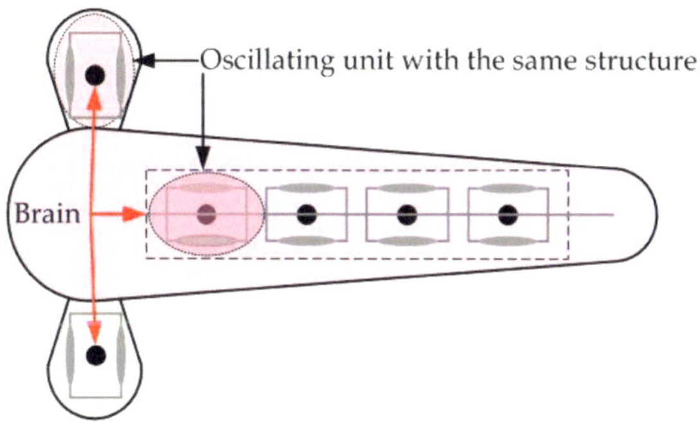

Fig. 15. The simplified CPG model for the AmphiRobot

Typically, there are two kinds of CPG models: chain structure and network structure. The former is commonly used by multi-limbs animals, especially higher vertebrates, e.g. cat, dog, etc. The latter is usually found in aquatic animals such as lamprey and salamander. In the chain CPG, the oscillators are arranged as a line and rhythmic signal is transmitted from one side to another with phases imposed to oscillators. Since the basic locomotion mode of the AmphiRobot in water is carangiform swimming, a bio-inspired controller, i.e. CPG-based control model is employed to generate steady fish-like swimming. From the perspective of biology, fish, as a vertebrate, possesses the similar neural structure similar to chain-like CPG. One motion freedom generally corresponds to an oscillating unit and several oscillating units constitute different and complete topology of CPGs. Therefore, the general structure of the AmphiRobot is functionally depicted in Fig. 15, including pectoral CPG and tail CPG. The oscillating units have the identical structure which comprises two nonlinear oscillators denoting the extensor and flexor, separately.

4.1 Model of nonlinear oscillator

For steady anguilliform and carangiform swimming, a propulsive wave form (hereafter referred to as body wave) that results from the progression of muscular contraction from head to tail is exhibited (Barrett, 1996). The replication of traveling body wave allows the robot to propel itself forward in water. This body wave is usually fitted by a sinusoidal function in bio-inspired engineering. Similar to lamprey model, the nonlinear oscillator is described as a phase oscillator with controlled amplitude (Ijspeert et al, 2007):

$$\begin{cases} \dot{\theta}_i = 2\pi f_i + \sum_j a_j w_{ij} \sin\left(\theta_j - \theta_i - \phi_{ij}\right) \\ \ddot{a}_i = \tau_i \left\{ \frac{\tau_i}{4}\left(A_i - a_i\right) - \dot{a}_i \right\} \end{cases} \tag{14}$$

where θ_i and a_i denote state variables representing the phase and the amplitude of oscillator, f_i and A_i indicate the intrinsic frequency and amplitude, and τ_i is a positive constant determining the convergence speed from a_i to A_i. The couplings between oscillators are defined by the weight w_{ij} and phase biases ϕ_{ij} .

The output x_i is a positive oscillatory signal expressed by (15) and it is the actual controlling parameters of the AmphiRobot.

$$x_i = a_i \left\{1 + \cos\left(\theta_i\right)\right\} \tag{15}$$

To analyze the converging character of amplitude, the second equation in (14) can be rewritten as

$$\ddot{a} + \tau\dot{a} + \frac{\tau^2}{4}a = \frac{\tau^2}{4}A \tag{16}$$

So the general solution of preceding equation can be derived below:

$$a = \left(C_1 + C_2 t\right) \cdot e^{-\frac{\tau}{2}t} + A \tag{17}$$

Because t is a positive constant, a will approach A when $t \to \infty$. That is, the state variable a will ultimately converge to A. The convergence speed is determined by τ. The bigger τ is, the faster the convergence will be.

4.2 CPG network model

Two intrinsic cellular reaction mechanisms influence the formation of CPG. One is the rhythmic excitation of single pacemaker cell and another is the synaptic connection between components of neural network. The rhythmic excitation of pacemaker is generated by the oscillation of membrane potential which is driven by several ionic mechanisms. However, the implementation of membrane oscillation is difficult in engineering practice. In our method, the external stimulus is utilized to excite the oscillation of nonlinear oscillators. The rhythmic excitation supplied by pacemaker cell appears at different occasions in different neurons. Between the two identical pacemaker cells which activate extensor and flexor respectively, there exists the synaptic inhibition. Namely, when a pacemaker cell is active, another does not fire. The synaptic inhibition guarantees the interactive activity of CPG. Oscillating condition and network mechanism including weight and phase are generally involved in CPG model. The network mechanism in the AmphiRobot includes the intrinsic connections of pectoral CPG and tail CPG and mutual connections between them.

The network model of AmphiRobot is schematically shown in Fig. 16. The input is the drive which is divided into left part and right one, driving the left and right side of robot body respectively. The pitch arcs represent the connections between oscillators and the arrows are the connection directions. In particular, the oscillating units in the tail CPG possess the same structure and their coordinated oscillations generate thrust for efficient and flexible swimming. We remark that the couplings between different units may be downwards, upwards and contralateral connections. Specifically, the couplings between the tail CPG and the pectoral CPG are defined as the unidirectional connections from oscillators of tail to those of pectoral fins. Some activation limitation is further imposed on the unidirectional couplings, which will be addressed later.

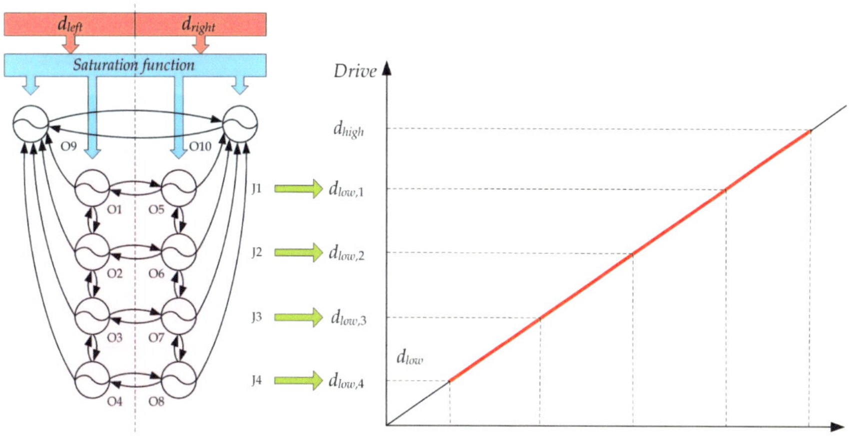

(a) The network with coupling (b) The input drive

Fig. 16. The CPG network for the AmphiRobot

4.3 The design of saturation function

To bridge the input drive and oscillators, a special saturation function is imported, which has a direct impact on the output of CPG. Considering the motion mechanism of biological opponent and the energy saving during slow swimming, the AmphiRobot adopts the oscillating length-based control strategy. Each joint is assigned a different threshold which is ascending upwards except that pectoral fins have the same threshold as Joint 4 (J4), as shown in Fig. 16. When the input is below the minimum threshold d_{low} (the same as $d_{low,4}$), all the oscillators can not oscillate and the robot remains still. Once the input is large than $d_{low,4}$, the O4, O8 and O9, O10 starts to oscillate, leading to the swing of J4 and pectoral fins, where the AmphiRobot is in the slow-swimming mode. As the input keeps increasing, the J3–J1 will participate in the propulsion. At the same time, oscillating frequency and amplitude are in proportion to the input drive and keep increasing. When the input drive reaches the maximum threshold d_{high}, the frequency and amplitude achieve the maximum values, respectively, and the robot therefore obtains the maximum velocity. Once the input is larger than d_{high}, d_{high} will be adopted as the input:

$$\text{if } d_{left,right} \geq d_{high} \quad d_{left,right} = d_{high}$$

The amplitude of oscillator has the same functional form in (18). Oscillators in each unit have the same coefficients while different parameters in different units.

$$A = \begin{cases} k_A d + b_A & \text{if } d \geq d_{low} \\ A_{sat} & \text{otherwise} \end{cases} \tag{18}$$

The frequency of oscillator has different saturation function for the tail CPG and the pectoral CPG. When the left and right input drives are different, the left and right oscillators will oscillate with distinct frequency and output waves possess the variant periods, the subtraction of which generates a new wave with modified period. Although the non-synchronous frequency is feasible in engineering, the coupling of excitation-inhibit on left and right side will be broken. All the oscillators of the tail CPG are consequently allocated the same frequency:

$$f_{tail} = \begin{cases} k_{f,tail} \dfrac{d_L + d_R}{2} + b_{f,tail} & \text{if } \min(d_L, d_R) \geq d_{low} \\ k_{f,tail} \cdot \max(d_L, d_R) + b_{f,tail} & \text{if } \max(d_L, d_R) \geq d_{low} \geq \min(d_L, d_R) \\ f_{sat} & \text{otherwise} \end{cases} \tag{19}$$

When the input drive satisfies the oscillating conditions, oscillators will work. In such a case, the drive is called "effective drive" expressed as d_{effect}. On the other hand, the motion of pectoral fins does not possess the symmetry and each pectoral fin can work independently. Thus, the pectoral fins have the same functional form with different parameters:

$$f_{pec} = \begin{cases} k_{f,pec} d + b_{f,pec} & \text{if } \min(d_L, d_R) \geq d_{low} \\ f_{sat} & \text{otherwise} \end{cases} \tag{20}$$

4.4 Parametric configuration

At present, the trial-and-error method based on the simulating techniques of computer is employed to obtain the characteristic parameters associated with the built CPG model. Tables 1 and 2 summarize the related parameters for saturation function and for couplings between oscillators to achieve steady swimming in water. For simplicity, the phase lag between the left and right pectoral fins is defined as 0. The phase lag between oscillators in the same unit of tail CPG is defined as π according to the flexor-extensor structure. Since the traveling wave propagates from head to tail, the downward phase lag is defined as $\pi/4$, accordingly the upward phase as $-\pi/4$. The pectoral fins and J4 oscillate in phase and thereby the phase relationship between pectoral and tail oscillators can be derived.

Variables	Symbol	The tail CPG	The pectoral CPG
Number of oscillators	N	8	2
Time constant	τ	20	
Minimum threshold	d_{low}	[2.5, 2, 1.5, 1] (J1–J4)	1
Maximum threshold	d_{high}	5	
Frequency coefficient	$[k_f, b_f]$	[0.45, 0.3]	[0.5, 0.3]
Amplitude coefficient	$[k_A, b_A]$	[0.030, 0.10] (J1) [0.035, 0.12] (J2) [0.045, 0.14] (J3) [0.055, 0.16] (J4)	[0.08, 0.06]
Saturation frequency	f_{sat}	0	
Saturation amplitude	A_{sat}	0	

Table 1. The parameters in saturation function

Variables	Symbol	The tail CPG	The pectoral CPG
Coupling (in the tail CPG)	(downwards) $[w_{ij}, \phi_{ij}]$ (upwards) $[w_{ij}, \phi_{ij}]$ (contralateral) $[w_{ij}, \phi_{ij}]$	$[10, \pi/4]$ $[10, -\pi/4]$ $[10, \pi]$	
Coupling (the tail CPG to the pectoral CPG)	$[w_{19}, \phi_{19}]$ $[w_{29}, \phi_{29}]$ $[w_{39}, \phi_{39}]$ $[w_{49}, \phi_{49}]$ $[w_{5,10}, \phi_{5,10}]$ $[w_{6,10}, \phi_{6,10}]$ $[w_{7,10}, \phi_{7,10}]$ $[w_{8,10}, \phi_{8,10}]$	$[0/30, 3\pi/4]$ $[0/30, 2\pi/4]$ $[0/30, \pi/4]$ $[0/30, 0]$ $[0/30, -\pi/4]$ $[0/30, -2\pi/4]$ $[0/30, -3\pi/4]$ $[0/30, -\pi]$	
Coupling (in the pectoral CPG)			[10, 0]

Table 2. The parameters for couplings between oscillators

Notice that two kinds of weight from the tail CPG to the pectoral CPG are pre-set. When the robot moves slowly, the weight is 0. While in fast swimming, the weight is chosen as 30. The critical condition for weight switch is triggered by the value of effective drive. The weight is 0 when $d_{effect} < 3$ and 30 when $d_{effect} \geq 3$.

4.5 On-line generation of control parameters

In the conventional body wave-based control method, the oscillating positions in a period is fixed. The speed control mainly depends on the time delay between two positions (Yu et al., 2004). The less the delay is, the faster the robot swims. The robot obtains the maximum speed in case of the minimum time delay. Due to the fixed but limited control points, the fish-like oscillations is relatively stiff. In addition, the robot is unable to adjust the characteristics according to the external environment and internal state. Considering these factors, CPG-based model is integrated to produce control parameters in real time.

In practice, the subtraction of output of two oscillators in one tail unit is used to actuate the corresponding servomotor, as shown in (21), whereas the outputs of pectoral oscillators are directly adopted as control signal.

$$\varphi_i = x_i - x_{i+4} \tag{21}$$

where i=1, 2, 3, 4. An example of the input drive for straight swimming is plotted in Fig. 17, and corresponding output signal for oscillating units in Fig. 18.

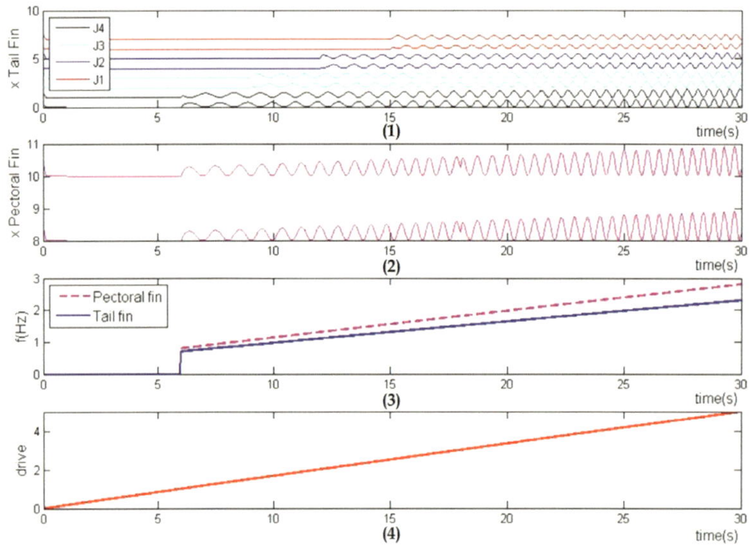

Fig. 17. The activity of CPG in straight swimming

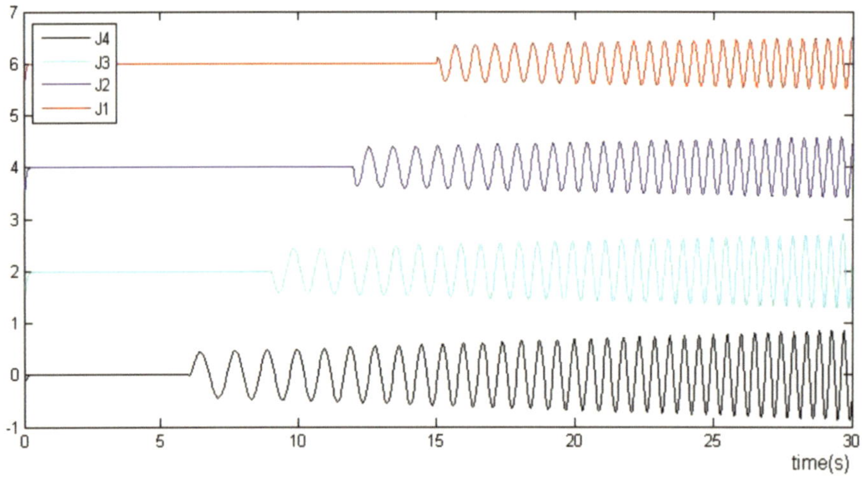

Fig. 18. The applied control signals in straight swimming

Further, assuming $\dot{a}_i = y_i$ and $\dot{y}_i = \ddot{a}_i$, a set of reduced-order equation is employed to yield control data:

$$\begin{cases} \dot{\theta}_i = 2\pi f_i + \sum_j a_j w_{ij} \sin\left(\theta_j - \theta_i - \phi_{ij}\right) \\ \dot{a}_i = y_i \\ \dot{y}_i = \tau_i \left(\frac{\tau_i}{4}\left(A_i - a_i\right) - y_i \right) \end{cases} \tag{22}$$

The 10 oscillators of the CPG network are made up of 30 reduced-order equations, which means considerable computing burden and an efficient numerical method is obligatory. With the processing frequency of 180 MHz in the ARM AT91RM9200 microcontroller, the 4-order and three-order Runge-Kutta in tests consumed about 43 ms and 33 ms, respectively, both of which are higher than 20 ms, the intrinsic control period of servomotors. To mitigate the computation consumption, a simplified Euler method is utilized to solving the equation (22). The iterative method of Euler is outlined as below:

$$y_{n+1} = y_n + hf\left(x_n, y_n\right) \tag{23}$$

The computing amount of the Euler method is only a quarter of that of four-order Runge-Kutta. The time consumption is reduced to about 13–14 ms, which satisfies the requirement of real-time computation.

4.6 Experimental results

During testing, different drives with bilateral symmetry and varied strength are applied to the AmphiRobot. All the joints from J1 to J4 are involved in the propulsion sequentially. A comparison of actual swimming wave and the simulating wave are depicted in Fig. 19. Notice that the snapshots of swimming are clipped from the experimental video and the straight lines in the simulating results denote joint J1–J4. In Fig. 19a, pectoral fins and joint J4 initially started to oscillate simultaneously when the input drive reached the threshold of these oscillators, and others joints remained still in their middle positions. By analogy, the oscillators corresponding to the J3–J1 from Figs. 19b–d participated in the driving, and the oscillating frequency and amplitude increased in proportion to the input drive. In the comparisons, the actual moving pattern agrees with simulation result, which verifies the feasibility of the CPG network. Note also that an irregular jerk emerged while powered on and it was caused by the fact that random numbers were assigned to the initial values of (22). This jerk disappeared quickly when the random initial values converged to the regular oscillating wave. The jerk demonstrates a potential advantage of CPG model that the controller can switch to stable locomotion from an arbitrary state and eliminate disturbance effectively.

Furthermore, the relationship between speed and symmetrical input drives in straight swimming is summarized in Fig. 20. When the inputs exceeded the minimum threshold, the AmphiRobot began to swim. Because only pectoral fins and J4 were involved in propulsion, the swimming speed was very low. The profile of moving velocity was similar with an exponential curve and the acceleration increased constantly when the drives ranged from $d_{low,4}$ to $d_{low,1}$. Though the swimming speed continued to augment when the inputs were above $d_{low,1}$, the acceleration decreased remarkably. The swimming attained the maximum speed when inputs reached maximum acceptable drives. The whole profile was clearly divided into two distinct phases by the point $d_{low,1}$. In the first stage, besides the speed increased with increasing frequency and amplitude, the active body length involved in the swimming increased continually because of the participation of more joints. Nevertheless, all the joints functioned and the active body length remained invariant in the second stage. The two-phase profile demonstrated that the oscillating body length plays an important role in the swimming speed of the AmphiRobot.

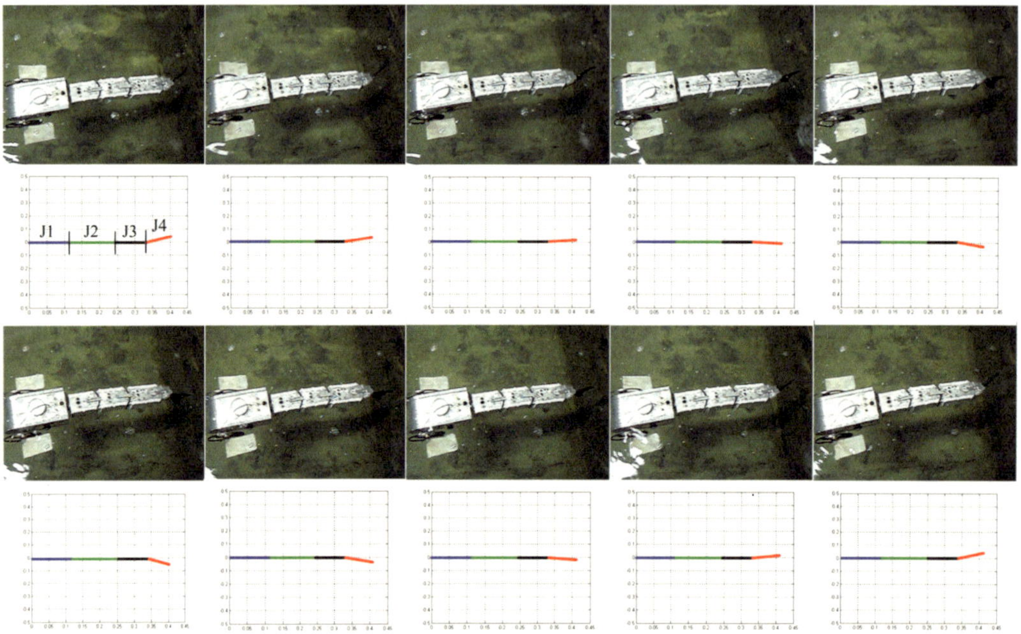

(a) drive=1

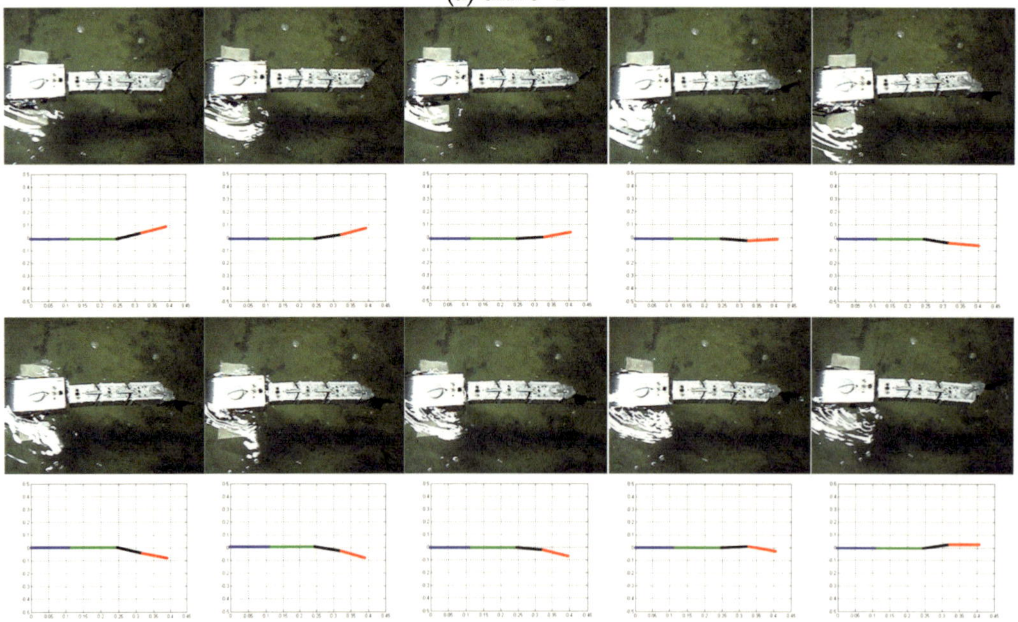

(b) drive=1.5

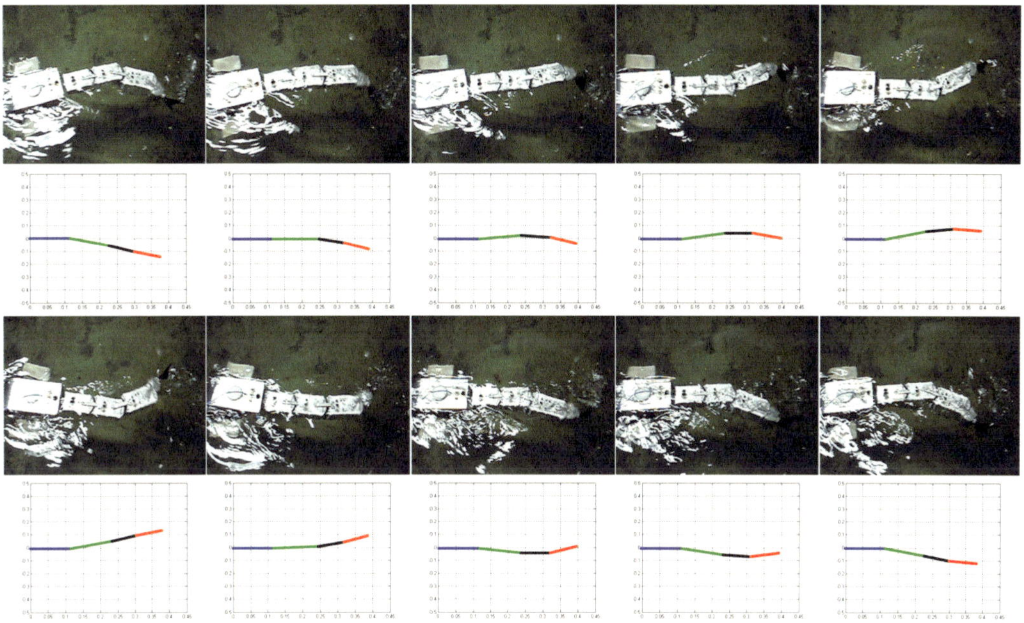

(c) drive=2

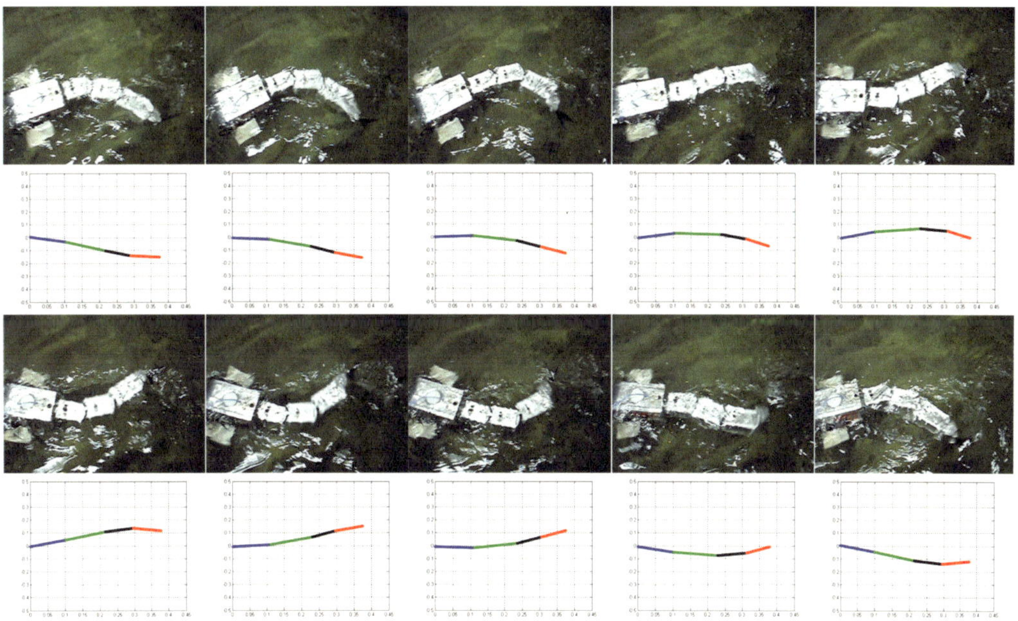

(d) drive=2.5

Fig. 19. A comparison of actual swimming and simulation results

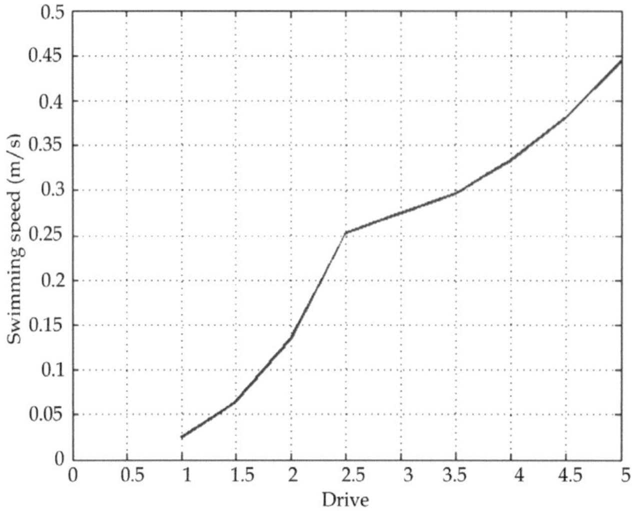

Fig. 20. The relationship of swimming speed and drive difference

5. Conclusion

This chapter has reviewed some of the issues involved in creating a multimode amphibious robot, especially its mechanical design and motion control, in a biomimetic manner. Based on the body structure, motion characteristics of amphibians, two generations of multimode biomimetic amphibious robots, named "AmphiRobot", have been developed. For terrestrial movements, a geometry based steering method called body-deformation steering has been proposed and optimized, taking advantage of the wheel-like mechanisms attached to the robot. At the same time, a chainlike CPG network responsible for coordinated swimming between multi-joint tail and artificial pectoral fins has been built. The aquatic control parameters mainly involve the length of undulation part, oscillating frequency and amplitude cooperatively regulated by the threshold values of the saturation function for each propelling unit. The real-time online calculation of controlling parameters has been also implemented. Preliminary testing results, both on land and in water, have demonstrated the effectiveness of the proposed control scheme. However, the amphibious locomotion performance of the AmphiRobot is still far behind that of animals in terms of speed and agility, especially in complex unstructured environments. More cooperative efforts from materials, actuators, sensors, control as well as learning aspects will be needed to improve the robot locomotor skills in unstructured and even unknown surroundings.
The ongoing and future work will focus on the analysis and optimization of locomotion control for autonomous movements as well as flexible water-land transitions. Hydrodynamic experiments based hybrid mechanical/electrical optimization, of course, is a plus for real-world applications.

6. Acknowledgement

The authors would like to thank Prof. Weibing Wang in the Machine and Electricity Engineering College, Shihezi University, for his contribution to mechanical design and fabrication of the AmphiRobot.

This work was supported in part by the National Natural Science Foundation of China under Grants 60775053 and 60505015, in part by the Municipal Natural Science Foundation of Beijing under Grant 4082031, in part by the National 863 Program under Grant 2007AA04Z202, and in part by the Beijing Nova Programme (2006A80).

7. References

Ayers, J. (2004). Underwater walking, *Arthropod Structure and Development*, Vol. 33, pp. 347-360

Bandyopadhyay, P.R. (2004). Guest editorial: biology-inspired science and technology for autonomous underwater vehicles, *IEEE J. Ocean. Eng.*, Vol. 29, No. 3, pp. 542-546

Bandyopadhyay, P.R. (2005). Trends in biorobotic autonomous undersea vehicles, *IEEE J. Ocean. Eng.*, Vol. 30, No. 1, pp. 109-139

Barrett, D.S. (1996). Propulsive efficiency of a flexible hull underwater vehicle, *Dissertation for the Doctoral Degree*, Cambridge, MA: Massachusetts Institute of Technology

Boxerbaum, A.; Werk, P.; Quinn, R. & Vaidyanathan, R. (2005). Design of an autonomous amphibious robot for surf zone operation: Part I mechanical design for multi-mode mobility, *Proc. of the 2005 IEEE/ASME International Conference on Advanced Intelligent Mechatronics*, pp. 1460-1464

Delcomyn, F. (1980). Neural basis for rhythmic behaviour in animals, *Science*, Vol. 210, pp. 492-498

Ding, R.; Yu, J.; Yang, Q.; Hu, X. & Tan, M. (2009). Platform-level design for a biomimetic amphibious robot, *Proc. of IEEE International Conference on Robotics and Biomimetics*, Bangkok, Thailand, pp. 977-982

Fish, F.E. & Rohr, J.J. (1999). Review of dolphin hydrodynamics and swimming performance, *Technical Report 1801*, SPAWARS System Center San Diego, CA

Georgiades, C.; Nahon, M. & Buehler, M. (2009). Simulation of an underwater hexapod robot, *Ocean Engineering*, Vol. 36, pp. 39-47

Greiner, H.; Shectman, A.; Won, C.; Elsley, D. & Beith, P. (1996). Autonomous legged underwater vehicles for near land warfare, *Proc. Symp. Autonom. Underwater Vehicle Tech.*, pp. 41-48

Harkins, R.; Ward, J.; Vaidyanathan, R.; Boxerbaum, A. & Quinn, R. (2005). Design of an autonomous amphibious robot for surf zone operation: Part II - hardware, control implementation and simulation, *Proc. of the 2005 IEEE/ASME International Conference on Advanced Intelligent Mechatronics*, pp. 1465-1470

Healy, P.D. & Bishop, B.E. (2009). Sea-dragon: an amphibious robot for operation in the littorals, *Proc. of 41st Southeastern Symp. on System Theory*, University of tennessee Space Institute, Tullahome, TN, USA, pp. 266-270

Ijspeert, A.J.; Crespi, A. & Cabelguen, J.-M. (2005). Simulation and robotic studies of salamander locomotion: applying neurobiological principles to the control of locomotion in robots, *Neuroinformatics*, Vol. 3, pp. 171-196

Ijspeert, A.J.; Crespi, A.; Ryczko, D. & Cabelguen, J.-M. (2007). From swimming to walking with a salamander robot driven by a spinal cord model, *Science*, Vol. 315, No. 5817, pp. 1416-1420

Ijspeert, A.J. (2008). Central pattern generators for locomotion in animals and robots: a review, *Neural Networks*, Vol. 21, No. 4, pp. 642-653

Kemp, M.; Hobson, B. & Long, J. (2005). Madeleine: an agile auv propelled by flexible fins, *Proc. of the 14th International Symposium on Unmanned Untethered Submersible Technology (UUST)*, pp. 1-6

Lauder, G.V.; Anderson, E.J.; Tangorra, T.J. & Madden, P.G.A. (2007a). Fish biorobotics: kinematics and hydrodynamics of self-propulsion, *The Journal of Experimental Biology*, Vol. 210, pp. 2767-2780

Lauder, G.V. & Madden, P.G.A. (2007b). Fish locomotion: kinematics and hydrodynamics of flexible foil-like fins, *Exp. Fluids*, Vol. 43, pp. 641-653

Prahacs, C.; Saunders, A.; Smith, M.; McMordie, D. & Buehler, M. (2005). Towards legged amphibious mobile robotics, *J. Eng. Design and Innovation* (online), vol. 1, part. 01P3, Available: www.cden.ca/JEDI/index.html

Sfakiotakis, M.; Lane, D.M. & Davies, J.B.C. (1999). Review of fish swimming modes for aquatic locomotion, *IEEE J. Oceanic Eng.*, Vol. 24, No. 2, pp. 237-252

Yamada, H.; Chigisaki, S.; Mori, M.; Takita, K.; Ogami, K. & Hirose, S. (2005). Development of amphibious snake-like robot ACM-R5, *Proc. of 36th Int. Symposium on Robotics*, pp. 433-440

Yang, Q.; Yu, J.; Tan, M. & Wang, S. (2007). Amphibious biomimetic robots: a review, *Robot*, Vol. 29, No. 6, pp. 601-608

Yu, J.; Tan, M.; Wang, S. & Chen, E. (2004). Development of a biomimetic robotic fish and its control algorithm, *IEEE Trans. Syst., Man, Cybern. B, Cybern.*, Vol. 34, No. 4, pp. 1798-1810

Yu, J.; Hu, Y.; Fan, R.; Wang, L. & Huo, J. (2007). Mechanical design and motion control of a biomimetic robotic dolphin, *Advanced Robotics*, Vol. 21, No. 3-4, pp. 499-513

Wang, L.; Sun, L.; Chen, D.; Zhang, D. & Meng, Q. (2005). A bionic crab-like robot prototype, *Journal of Harbin Engineering University*, Vol. 26, No. 5, pp. 591-595

10

Autonomous Underwater Vehicle Motion Control during Investigation of Bottom Objects and Hard-to-Reach Areas

Alexander Inzartsev, Lev Kiselyov,
Andrey Medvedev and Alexander Pavin
Institute of Marine Technology Problems (IMTP FEB RAS)
Far East Branch of the Russian Academy of Sciences
Russia

1. Introduction

Modern Autonomous Underwater Vehicles (AUVs) can solve different tasks on sea bosom research, objects search and investigation on the seabed, mapping, water area protection, and environment monitoring. In order to solve problems of bottom objects survey AUV has to move among the obstacles in a small distance from the seabed. Such motion is connected with active manoeuvring, changes in speed and direction of the movement, switch and adaptive correction of modes and control parameters. This can be exemplified by using AUV for geologic exploration and raw materials reserves estimation in the area of seamounts, which are guyots with rugged topography. Such problems arise during vehicle manoeuvring near artificial underwater point or extended objects (for example, dock stations or underwater communications). The problems of ocean physical fields' survey are of a particular interest. These are the problems of bathymetry and seabed mapping as well as signature areas of search objects.

To perform these tasks AUV must be equipped with the systems that can define the positions of the vehicle body against the obstacles and search objects. As a rule, acoustic distance-measuring systems (multibeam and scanning sonars, and also groups of sonars with the fixed directional diagram) and other vision systems are used for these purposes. AUV path planning is carried out with the use of current sensory data due to the lack of a priori information. On the basis of measured distances the current environment model and the position of the vehicle are defined. Then taking into account vehicle dynamic features the direction of probable movement and usable motion modes are evaluated. At each control phase a motion replanning is carried out taking into account new data received from sensors and changed surrounding.

The paper presents the results of research and working outs based on the many years of experience of the Institute of Marine Technology Problems (IMTP) FEB RAS (Ageev et al., 2005). It also gives examples of realization of the offered solutions in the structure and algorithms of motion control of certain autonomous underwater vehicles-robots.

2. Control system peculiarities of AUV capable to work at severe environment

The use of AUV to perform different operations under conditions of difficult informative uncertain or extreme surrounding requires a developed complex of positioning, control, and computer vision systems onboard the vehicle. In the overall structure of control system one can mark such basic systems providing AUV functioning as an equipment carrier, and information and searching functions.

The basis of control system is a local area network composed of several computers. It provides motion control and emergency and search functions. To organize AUV's local area network high speed channels (Ethernet) and quite slow exchange serial channels are used. To form the control navigation and sensors' data are used. Emergency sensors are used for AUV safety. Remote change of AUV mission can be carried out with the help of acoustic link. Positioning system plays an important role. Positioning accuracy is acquired by using on-board autonomous navigation system including inertial positioning system, angular and position measuring devices, and acoustic Doppler log. An accumulating dead-reckoning error can be decreased by means of integration of hydroacoustic and stand-alone data by operating AUV with hydroacoustic navigation facilities with long or ultra-short base.

Search systems incorporated computer vision systems differ on physical principles and methods of data acquisition. Acoustic systems include high-frequency and low-frequency side-scan and sector-scan sonars as well as subbottom profiler. Current-conducting objects can be found with the use of electromagnetic locator (EML). A video system carries out imaging and object recognition. It includes photo and video cameras.

The information from sensors and measuring systems are usually stored for the following mapping of researched area (ecological, geophysical, etc.) If necessary, this information can be used in real time, for example, for contouring the areas with abnormal characteristics of measured fields.

System architecture of programmed control has hierarchic three-level organization (strategic, tactic and executive levels). Program-task (mission) for the vehicle is programmed on the highest level and in general it contains the description of desired motion path and operation modes of onboard equipment. Tactic level contains a set of vehicle behavior models (function library) and a scheduler that coordinates their work. The lowest level carries out tactical commands. To do this it contains a set of servocontrollers. Control algorithms providing "reflex" motion among the obstacles work on the lowest level.

The propulsion system is used for spatial motion, positioning, and obstacle avoiding. It provides free motion modes (motion in wide speed range, hovering, and free trim motion). There are stern and bow propulsion sections. Control forces and moments are created with the help of four stern mid-flight and several stern and fore lateral thrusting propulsions. Multi-beam echoranging system (ERS) with the range up to 75 meters is used for working out corresponding controls and obstacles detection. ERS sonars are oriented on the front aspect under different angles to vehicle fore-and-aft axis (forward, down, sideway, up).

3. Motion modes and AUV dynamics peculiarities

Trajectories of arbitrary forms are required for bottom objects survey, constructions inspection, docking with mooring facilities or homing beacons. Not only basic motion modes but more difficult modes of dynamic positioning at variable speed and circular change of thrust vector direction (start-stop, reverse, transversal, etc.) must be performed. Among typical practical tasks of this class are:

- maneuvering in specified area near the target at variable speed and heading correction, pointing to the target (signal source), approaching to the target and point positioning,
- lengthy objects search and survey,
- path selection in the rugged bottom relief.

In many cases the said missions are interconnected and can correspond to different phases of a particular vehicle mission. So we shall consider them as components of single scenery for rather complex missions' performance.

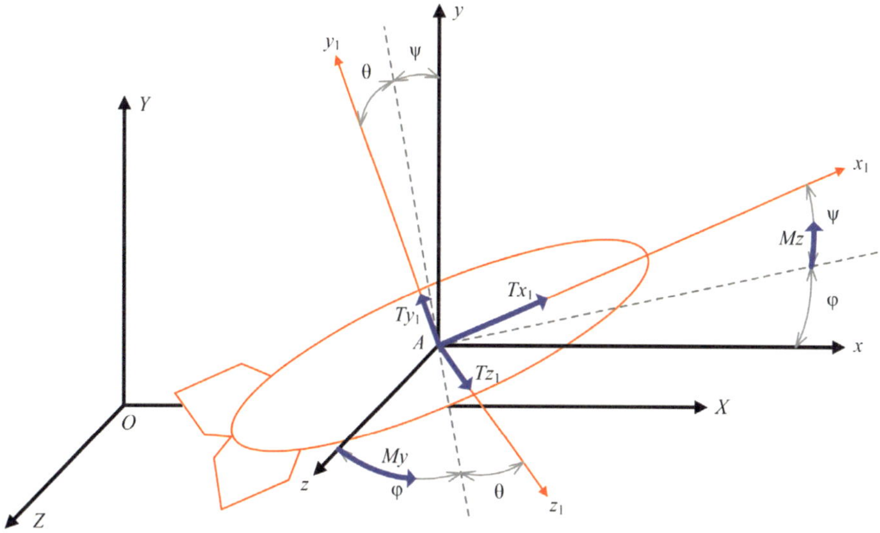

Fig. 1. A system of coordinates and flow pattern of force in trimetric projection

Let's equate the model of AUV spatial motion as (fig. 1):

$$
\begin{aligned}
& m_x\dot{\upsilon} = -R_x + P\cdot\sin\vartheta + T_x\cdot\cos\alpha\cdot\cos\beta - T_y\cdot\cos\alpha - T_z\cdot\sin\beta,\\
& m_y\upsilon\dot{\vartheta} = R_y + P\cdot\cos\vartheta + T_x\cdot\cos\alpha + T_y\cdot\cos\alpha,\\
& J_z\ddot{\psi} = M_z + M_0\cdot\sin\psi + M_z^{ctrl},\\
& J_y\ddot{\varphi} = M_y + M_y^{ctrl},\\
& \dot{X} = \upsilon\cdot\cos\vartheta\cdot\cos\varphi + \upsilon_{Tx},\\
& \dot{Y} = \upsilon\cdot\sin\vartheta + \upsilon_{Ty},\\
& \dot{Z} = \upsilon\cdot\sin\varphi\cos\vartheta + \upsilon_{Tz}.\\
& m_x = M+\lambda_{11},\ m_y = M+\lambda_{22},\ m_z = M+\lambda_{33},\ I_y = I_{yy}+\lambda_{55},\ I_z = I_{zz}+\lambda_{66},
\end{aligned}
\tag{1}
$$

where λ_{11}, λ_{22}, λ_{33}, λ_{55}, λ_{66}– added masses and liquid inertia moment, T_{x1}, T_{y1}, T_{z1}, M_y^{ctrl}, M_z^{ctrl} – projection of control forces and moments in a system of coordinates dependent on the vehicle, υ - speed against the flow, φ, ψ – heading and vehicle pitch correspondingly, ϑ, χ - angles of ascent and motion swing, R_x, R_y, R_z, M_y, M_z – hydrodynamic forces and moments, M_0 – moment of stability, υ_{Tx}, υ_{Ty}, υ_{Tz} – current velocity vector components which have constant, variable or random character, P – variable buoyancy depending, in particular, on the depth of the vehicle descent.

According to the general formulation area survey is performed with the help of maneuvering piecewise-constant speed and path program near the target (object) and start-stop control mode at point dynamic positioning or along the contour.

Horizontal motion area (X, Z) can be defined by one of the following methods:

- coordinates of the target point $\{X_T, Z_T\}$, local area radius r_T and distance to the target d_T, distance d_B to the signal source (transponder) and bearings ϑ_B,
- optional close circuit g(X,Z)=0, against which the vehicle displacement d_i is defined in directions dependent of the vehicle,
- linear zone $|aX+bZ+c| \leq \Delta_l$ width Δ_l against extended object and relative linear $\{\Delta X_l, \Delta Z_l\}$ and angular $\Delta\varphi_l$ vehicle motions.

Control responses created by the stern and bow propulsions in the trimetric projection connected with the vehicle are given by (Ageev et al., 2005; Kiselev & Medvedev, 2009):

$$
\begin{aligned}
&T_x = (T_{SU} + T_{SR} + T_{SB} + T_{SL}) \cdot \cos\delta\ , \\
&T_y = (T_{SU} - T_{SB}) \cdot \sin\delta + T_{BH} = T_y^S + T_{BV}\ , \\
&T_z = (T_{SR} - T_{SL}) \cdot \sin\delta + T_{BV} = T_z^S + T_{BH}\ , \\
&M_z^{ctrl} = (T_{SU} - T_{SB})(x_{TS} \sin\delta + y_{TS} \cos\delta) + T_{BV} \cdot x_{TB} = T_y^S \cdot d + T_{BV} \cdot x_{TB}\ , \\
&M_y^{ctrl} = (T_{SR} - T_{SL})(x_{TS} \sin\delta + z_{TS} \cos\delta) + T_{BH} \cdot x_{TB} = T_z^S \cdot d + T_{BH} \cdot x_{TB}, \\
&d = (x_{TS} \cdot \sin\delta + y_{TS} \cdot \cos\delta) / \sin\delta = x_{TS} + y_{TS} \cdot ctg\delta\ , \\
&T_S = T_S^{\max} \cdot sat\left(\frac{U_T^S}{T_S^{\max}}\right),\ T_B = T_B^{\max} \cdot sat\left(\frac{U_T^B}{T_B^{\max}}\right).
\end{aligned}
\qquad (2)
$$

where T_{SU}, T_{SB} – vertical channel stern mid-flight propulsions thrusts (upper and bottom correspondingly), T_{SR}, T_{SL} - horizontal channel stern mid-flight propulsions thrusts (right and left correspondingly), T_{BH}, T_{BV} - horizontal and vertical bow maneuvering thrusts, x_{TS}, y_{TS}, δ - coordinates and pitch angle of stern mid-flight propulsions, x_{TB} - axial coordinate of bow maneuvering propulsion, $U^S{}_T$, $U^B{}_T$ - control functions for stern and bow propulsion sections.

As is clear from set of equations one and the same control responses can be created by means of applying different work patterns of stern and bow propulsions. A practical application has the following modes:

- cruising motion;
- low speed motion.

The first mode is characterized by the fact that vehicle spatial motion is carried out by means of changing of attack angle with the help of variables T_x, $M_Y{}^{CTRL}$, $M_Z{}^{CTRL}$. At the same time only stern mid-flight propulsions form the mentioned forces and moments. This mode is used for vehicle control only at cruising speed.

In the second mode all propulsion sections are used to form vehicle motion, and control is performed according to five degrees of freedom with the help of variables T_x, T_Y, T_Z, M_Y^{CTRL}, M_Z^{CTRL}. This mode is used for vehicle control at low speed and during hovering.

Complex AUV motions are carried out by means of combination of these two modes. Let's analyze several possible control methods. They differ by the logic of program algorithm and by system dynamics in performing complex spatial motions.

4. Motion control in the rugged bottom relief

AUV usage for seabed layer survey is connected with the organization of equidistant motion (motion at equally distance from the seabed) and bypassing or bending around the obstacles. Equidistant motion control assumes the formation of an equidistant model on the basis of echoranging data and data on vehicle relative motion. In this case control can be organized as an adjusted program that can forecast spatial equidistant path and direct the vehicle along it. In a plane case the task is simplified and consists in stabilization of positioning and angular error formed with the help of several range sensors. Such control method was implemented in different versions of the majority of the vehicles designed by IPMT FEB RAS (Ageev et al., 2005).

Characteristic features of the task can be illustrated on the example of control organization during seamounts (guyots) survey. They are distinguished by sharp changeable microrelief and different obstacle along the motion path (Ageev et al., 2000; Smoot, 1989). AUV use for seamounts survey is mainly connected with the geologic exploration and raw materials reserves estimation (for example, the resources of ferro-manganese nodules in the Pacific Ocean created on the guyots tilted areas). Common characteristics of the guyot macrorelief are:

- cone form with side angle up to 30°- 40° near the top;
- flat top covered with the fall-outs, the edge can have barriers;
- nodules are created in the guyot upper vein systems;
- sides and top can have picks and gorges;
- side can have terraces (width up to several kilometers), edges can have peaks and barriers.

The main objective of the survey is the estimation of amount of minerals in the given area and the conditions for the following exploitation. The second task in using AUV is reduced to SSS survey. The first task can be partly solved by using photo and TV survey. It is a rather complicated task because it is quite difficult due to the necessity to approach to the surface up to 3-5 meters.

Let's consider potential obstacles in more detail.

Peaks are rather large underwater mounts with pike. The vehicle must pass such obstacles sideways.

Barriers and peaks can be found on terrace edges and guyot top edge. Fault ridge height can be up to dozen of meters. Bypassing of low barrier is rather simple. Peak bypassing during moving from below is a more complicated task. In this case the vehicle has to stop forward motion and emerge staying at an allowable distance from the obstacle.

Breaks and gorges are not the survey objects and the vehicle must go above them. The major problem is to recognize this land shape.

Let's consider several motion peculiarities in typical mode taking into account dynamic features of the vehicle and power requirements. Broadly speaking, the selection of motion modes is rather optional. The following variants are possible:

- motion along the side with zero pitch or with the pitch that corresponds the side angle;
- obstacle bending "without a pause" at permanent or variable speed;
- deceleration or back motion with transfer to hovering when the obstacle that cannot be bypassed "without a pause" is found;
- body scanning without forward motion for obstacle heighting and maximal visual angle;
- complex obstacle avoiding (peak, high hurdle) with the use of backward motion at big pitch and attack angle.

It is necessary to choose the most energetically efficient motion modes. So the modes with absolute minimum resistance are more preferable. It is connected with providing an "optimal" angle of attack that corresponds preset current speed. Not all of the abovementioned modes meet this requirement. So, moving along the side with zero pitch cannot be considered appropriate as the basic mode, as in this case there can be high angles of attack, and energy consumption can be reduced only at the expense of speed decreasing. In just the same way it is difficult to provide energetically efficient mode at complicated maneuvering near the obstacle as vehicle security prioritizes. As a consequence there appears additional energy consumption for motion performing. One more peculiarity is incomplete, unreliable, fuzzy information about the bottom configuration. It leads to the suitability of construction of hybrid control structure with fuzzy-logic elements (Ageev et al., 2000; Kiselev & Medvedev, 2009). Let us cite as an example the results of motion modeling in vertical plane for such typical control modes as obstacle bending "without a pause" at equidistant curve with regard to relief, bending around high and rapid obstacles, maneuvers on tracking another more complicated bottom forms.

For descriptive reasons sonar beams are depicted at several points of motion path. The length of each beam corresponds to ERS radius of action. In most cases complicated obstacles bending is carried out with the use of deceleration and back motion modes. In all considered cases control system keeps equidistant motion at preset distance of 3 meters. When the obstacle is found it performs maneuver on its avoiding. The use of fuzzy-logic elements with failures in ERS work allows leveling equidistant motion path, especially at unreliable information intensification.

4.1 Motion along the side with preset pitch

The mostly widespread case during guyots' research is moving along the smooth slope. Creation of control forces and moments with the help of four stern and one bow propulsions gives a chance for free selection of propulsion thrust values betweenness. In particular, if vertical force T_Y and moment M_z are defined, then it is possible to find the equations for all thrust components at presence of all additional kinematical connections from static equations. At the same time with the purpose of energy minimization it's possible to let that depth stabilization and motion along smooth lope is carried out at cruising mode (with the use of mid-flight propulsions), and during maneuvering and moving along steep slope stern and fore propulsions work simultaneously.

Actual angle of attack is defined by correlation of vertical thrust components, buoyancy, and uplift hydrodynamic force. As the last one nonlinearly depends on speed and angle of attack, it is obvious that power spent for motion is also in nonlinear dependence of angle of attack. This can be approximately evaluated on the basis of empirical data.

4.2 Various obstacle bending

For single obstacles avoiding such as "bench", "hurdle", "den", "cutting", and so on, typical control based on the echoranging data can be used.

When the obstacle is found a pitch that is corresponding its height is created. Vehicle dynamic features and control character are analogous to the previous case. Besides, safety distance is under control. It allows obstacle bending "without a pause" and without deceleration as well as bow propulsion switching on.

The motion mode is chosen depending on slope gradient Δ calculated on the basis of ERS data at each points of motion path.

Let d_i (i = 1...4) – is rangers' distances directed correspondingly down, at an angle, forward, and at an angle upward; α_i, - angle between i and i +1 ranger, i φ_i – angle between vehicle fore-and-aft axis and i- th ranger, ψ - pitch. Then slope gradient is calculated as follows:

$$\Delta = \psi + \max_{i=1}^{3}\left(\varphi_i - arctg\left(\frac{d_{i+1}\cdot\sin\alpha_i}{d_i - d_{i+1}\cdot\sin\alpha_i}\right)\right),\quad \varphi_i = (90^o - \alpha_i) + \sum_{k=1}^{i}\alpha_k \tag{3}$$

Median filtering is used to eliminate influence of rangers' noise on calculated slope.

Let δ_1 and δ_2 are slope gradient threshold values at which the switch from cruising motion mode to the deceleration or stern propulsion mode takes place. At small slope gradient ($|\Delta|<\delta_1$) a cruising motion mode at the speed of $\upsilon \approx 1$ m/s is used. This mode is provided by the work of vehicle stern propulsions.

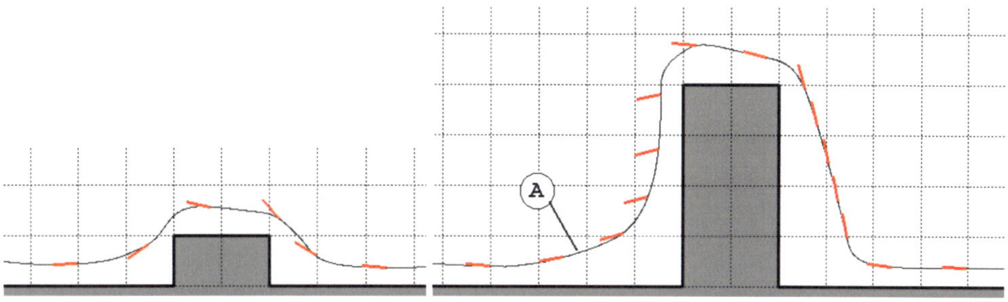

Fig. 2. Obstacle bending: "without a pause" (left) and high "hurdle" (right)

To form control M_z positional error $dY = Y_T - \min(d_1, d_2\cos\alpha)$ (here Y_T - preset height over the bottom), program pitch in the shape of calculated surface steep Δ, angular speed on pitch ψ', and safety distance on the front ranger d_3 are used:

$$M_z = K_1\cdot dY + K_2\cdot(\Delta-\psi) + K_3\cdot\dot{\psi} + K_4\cdot\left(1/d_3\right) \tag{4}$$

At the same time small obstacles are bended "without a pause". Fig. 2 (left) shows the example of obstacle bending when its height is compared with the height of vehicle motion. It depicts motion path and positions of the casing at every 20 seconds of simulation time. The width of coordinate grid square side is 10 m.

At large slope gradient ($\delta_1<|\Delta|<\delta_2$) vehicle speed reduces up to 0.5 m/s. To create control moments the collaboration of stern and fore propulsions is used. It allows creating much larger trim angles.

To avoid high and rapid obstacles ($|\Delta|>\delta_2$) vehicle stops its forward motion and keeps allowable distance to the obstacle. Upon that the distance to the obstacle is controlled according to the upper and front rangers. Simultaneously upward motion on the slope with pitch stabilization takes place. Fig. 2 (right) shows the example of "hurdle" obstacle bending.
Control thrusts and moments are formed as follows:

$$T_z = K_1 \cdot (\min(d_3, d_4) - D) + K_2 \cdot \dot{x} ,$$
$$T_y = \begin{cases} T_{y\max}, & \Delta \le 90^o \\ K_1 \cdot (d_4 - 2Y_T) + K_2 \cdot \dot{\psi}, & \Delta > 90^o \end{cases}, \quad D = \begin{cases} 2Y_T, & \Delta \le 90^o \\ 30, & \Delta > 90^o \end{cases}, \tag{5}$$
$$M_z = K_1 \cdot (\psi_T - \psi) + K_2 \cdot \dot{\psi}$$

where K_1, K_2 – are control parameters, ψ_T - target pitch.
As distinct from previous case AUV cannot perform bending of such an obstacle "without a pause" due to its great height.
At point "A" vehicle decelerates and starts stabilizing preset distance to the obstacle. This distance is in proportion to preset motion height. Vehicle deceleration is performed smoothly due to timely obstacle detection.

4.3 Complicated obstacles avoiding

A slope with gradient of 90° and more is considered to be "peak" or "cave" obstacle. The stabilized distance to the obstacle increases up to 30÷40 m. It brings to backward motion under the "peak", or to the fact that the vehicle doesn't enter the "cave". Fig. 3 shows the example of vehicle motion in the area of such an obstacle. The size of the obstacle is so that it is fully in the field of ERS vision. On the basis of this data vehicle emerges without entering into the "cave".

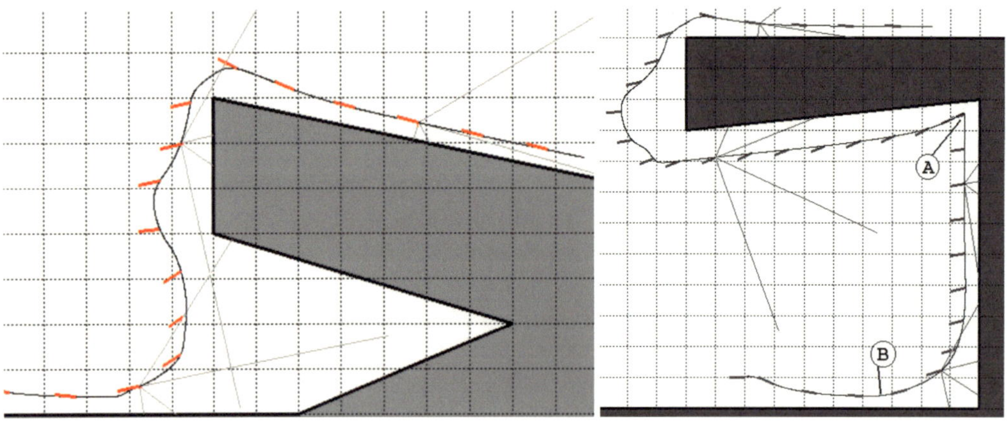

Fig. 3. "Cave" obstacle bending (left figure) and "peak" obstacle bending (right figure)

Obstacle bending of such type is characterized by the use of deceleration and backward motion modes. The angle of attack can vary up to 180°. Fig. 3 (right) shows the example of vehicle motion under the "peak". This case is similar to the one described above. The only difference is that AUV cannot beforehand estimate the character of the obstacle due to its

great size. As a result only at point "A" a vehicle can define that it is under the "peak" and starts backward motion.
This motion algorithm allows avoiding getting into the "gorges" if their width is compared to the ERS radius of action.
The example of AUV upward motion on the slope with rather rugged relief is depicted in fig. 4.

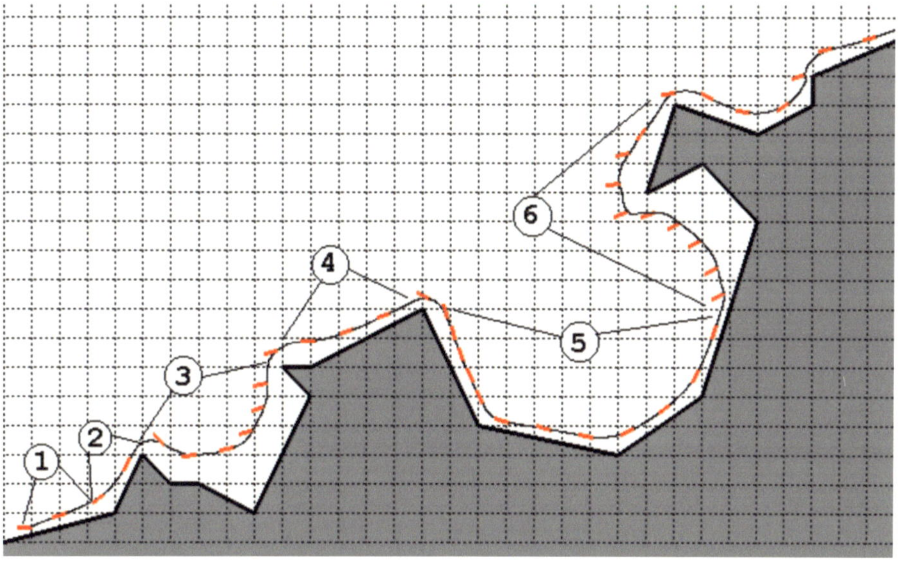

Fig. 4. Rugged relief motion

The whole motion path consists of several areas. Each area has its own motion mode. AUV preset motion height over the bottom is 3 m. Areas 1,2,4,5 are represented as the motion modes along the slope that has both positive and negative steep from 70° to -65°. At area 3 the vehicle performs "peak" bending. Motion at great distance from the slope at this area is explained by the fact that "peak" in the motion direction was found beforehand. At area 6 the vehicle performs exit from under the "peak" of a greater size in the same way as it was described earlier.

5. Extended objects search and tracking

Characteristic feature of the task is in organization of extended line search according to the AUV systems signals for the following object tracking in the given survey zone (Ageev et al., 2005; Inzartsev & Pavin, 2009). Practical approaches to perform such task with the use of echosounder (Inzartsev & Pavin, 2006; Pavin, 2006), magnetometric, electromagnetic (Kukarskih & Pavin, 2008) and video (Scherbatyuk et al., 2000) systems are known. Such decisions were used in AUVs "AE-2", "XP-21", and "R-1". Motion control is formed by means of choosing general direction and its correction according to the contact with the object. In fuzzy situations search motions are performed in limited area.
When the object is found vehicle linear and angular motion parameters with regard to extended line are defined. These parameters are an input data for AUV control system. The task is to make so that the AUV trajectory "in average" to be as close to the tracked object as possible in the presence of positioning and dynamic errors.

If location of extended object is defined beforehand with preciseness enough for coming of the vehicle into the point of contact establishing with the detecting devices, then the vehicle mission includes:

- arrival to the object area and contact search with the object;
- maneuvering near the object and detecting of extended line orientation;
- extended line tracking at given "zone" that corresponds the area of steady state contact;
- return to the search program at occasional loss of contact with the object.

Acquisition system can include different devices that allow finding the object according to the short-range signals and identify it against the background of false signals. To solve this problem the computer video system must include high-resolution survey sonars, video system and magnetometric or electromagnetic detecting devices.

Let's illustrate general provisions on the example of underwater cable inspection with the use of video system and electromagnetic locator designed by IPMT FEB RAS. Fig. 5 shows the layout of devices used by AUV.

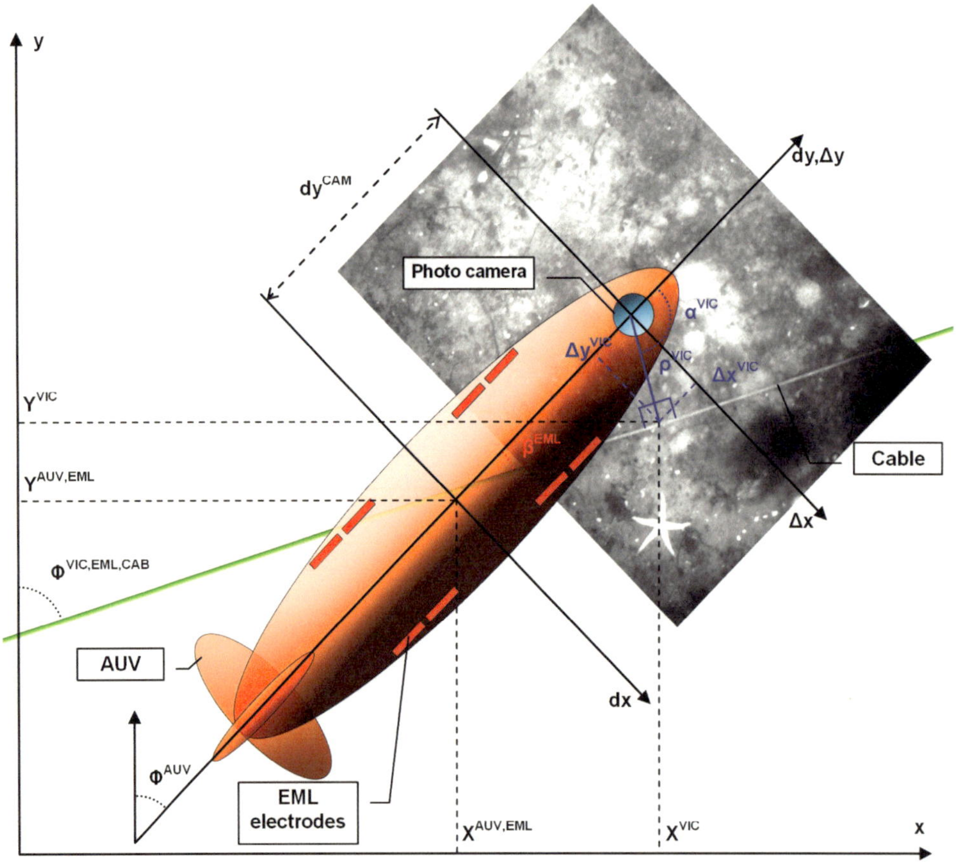

Fig. 5. AUV coordinates and devices layout

When the object is found with the help of video system at the output of recognition system for each frame the following set of values is pointed out:

- direction of recognized extended object with regard to image fore-and-aft axis;

- distance from the center of the frame to the linear object;
- length of visible part of the object.

Received parameters are used to define the position of the object in inertial system of coordinates.

In the mode of object tracking the control data is formed as:

$$\begin{aligned}\varphi_{tag} &= \varphi_{AUV} + \Delta\varphi_{line}\ ,\\ \alpha_{tag} &= \Delta\varphi_{line} \cdot sat\left(K_p \cdot \Delta d_Y + K_d \cdot \dot{d}_Y\right),\\ v_{tag} &= f\,(h_{tag}, t_{VIC}) + K_v \cdot |\sin(\alpha_{tag})|\ .\end{aligned} \qquad (6)$$

Where:

$\Delta\varphi_{line}$ - object orientation in the system of coordinates connected to the camera;
K_p, K_d - amplification constants for positional and differential components;
K_v - constant of proportionality;
α_{tag} - target attack angel;
Δd_Y - preset position stabilization error in diametral plane,
$\dot{d}_Y$ - AUV motion speed in cross direction;
$f(h_{tag}, t_{VIC})$ - function evaluating dependence of preset AUV speed from the motion height and operational period of video image processing system.

Extended object position according to the data of electromagnetic detecting system is defined at the moment of maximum potentials on receiving electrodes. Estimated

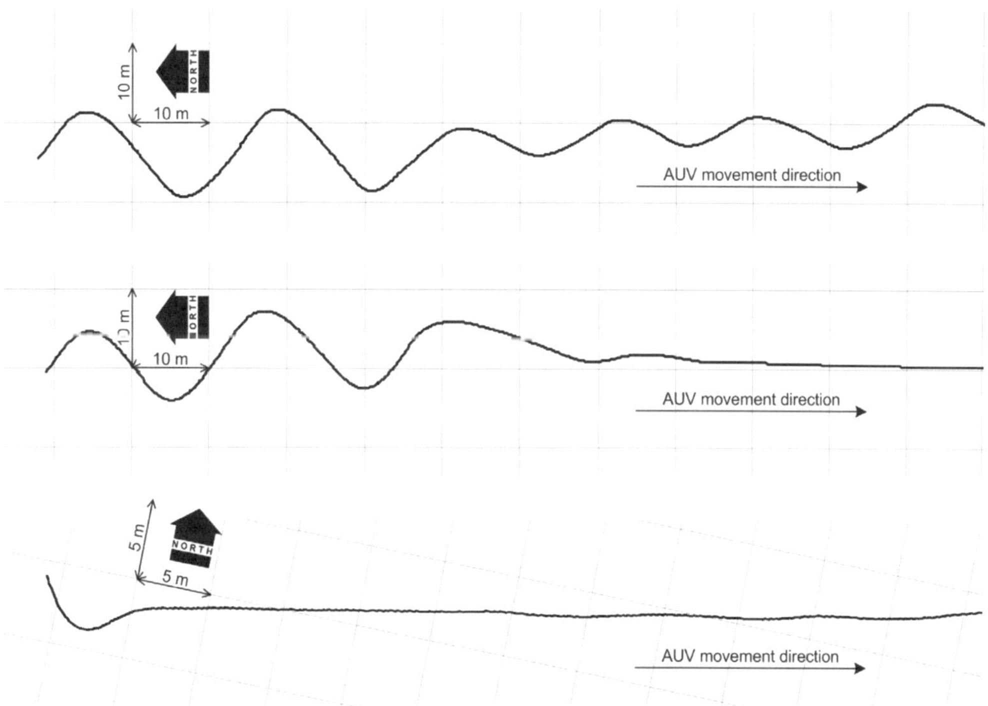

Fig. 6. Cable tracking with the use of on EML (upper figure), EML and video system (central figure), only video system (lower figure)

probability of the search object existence is defined on the basis of the values of potential and speed of its increasing for the period of time preceding the maximum. Particular dependence is chosen empirically on the basis of device characteristics, AUV velocity (and height), and inspected object characteristics (electromagnetic characteristics, diameter, shell thickness and so on).

At coprocessing of video, electromagnetic and navigation signals the control of the contacts with the object is performed, and the data for AUV control system is worked out (Inzartsev & Pavin, 2008). Fig. 6 shows the graphs illustrating the process of cable tracking at separate and joint use of electromagnetic locator and video system.

6. Guiding to the given target in the action of disturbances

Let's analyze the task of AUV motion control in conditions of incomplete or unreliable information about environment and action of shift current. To be more specific let's dwell upon the analysis of vehicle motion in guiding to the given target from random starting position. This task can be considered as a component of the closing-in scenario and coupling with underwater target.

In this case it is necessary to provide approach and hovering of the vehicle above the object (at the target) in the dynamic positioning mode. Guiding algorithm can use information received from navigation-piloting sensors and positioning system data on the vehicle coordinates, pointing to the target and distance to it. AUV motion control is conducted by transformation of control forces and moments into the thrusts, its calculated by the program and created by thrust-steering complex.

In general the algorithm must provide:

- classification of incalculable attack moments to work out adequate behavior;
- current vector autodetection with sufficient accuracy to compensate crabbing in stabilization tasks;
- behavior shaping based on the history of AUV and control algorithm conditions.

Let's call R_1 - area radius where guiding to the target is performed, R_0 - maneuvering area radius, R - distance from vehicle to target.

Under conditions of constant current the following mode of vehicle approaching to the target is logical:

- motion on bearings (azimuth) of the target on preset constant speed at $R>R_1$,
- speed reduction at signing on the area $R_0 \leq R \leq R_1$ and turn into the direction corresponding to the bearings sign (or value),
- position stabilization at $R<R_0$.

While moving into given small area program algorithm logic forms speed change mode, relative bearing φ_k and turn directions:

$$\varphi_{k+1} = \begin{cases} \varphi_k, & r_k \leq r_g, \\ \varphi_k + \Delta\varphi_k \cdot sign(\varepsilon_k), & r_k > r_g, \end{cases} \tag{7}$$

$$\varepsilon_k = \varphi_k - arctg\,(Z_k - Z_g)/(X_k - X_g).$$

It's obvious that current influence leads to shifting of the program path against the accepted coordinate system, and to solve the problem it's necessary to form the control possessing with "robustness" property to external resistance.

At coordinate path points definition this task resolves into choosing control providing minimum "miss" Δ in guiding to the target coordinates and path length minimum S, preassigned by points of intersection $P_k = \{X_k, Z_k\}$:

$$\Delta = \sum \left[(X(t_k) - X_k)^2 + (Z(t_k) - Z_k)^2 \right]^{1/2}, \\ S = \sum \left[(X(t_k) - X(t_{k-1}))^2 + (Z(t_k) - Z(t_{k-1}))^2 \right]^{1/2}, \tag{8}$$

where t_k – is path discrete intervals.

At the final stage of the area path investigation the task of approaching to the target and target positioning is being solved. Control possessing (U_x, U_z) provides approaching to the target and positioning near it in the basis of PID control. It's possible at known relative position of the vehicle and target. PID-control with limitations for the value of control responses is described:

$$U_x = K_1 \cdot \Delta X + K_2 \cdot \dot{X} + K_3 \cdot \int_{t_0}^{t} \Delta X dt, \; U_z = K_1 \cdot \Delta Z + K_2 \cdot \dot{Z} + K_3 \cdot \int_{t_0}^{t} \Delta Z dt, \\ T_{x_1} = U_x \cdot \cos\varphi + U_z \cdot \sin\varphi, \; T_{z_1} = -U_x \cdot \sin\varphi + U_z \cdot \cos\varphi, \\ \sqrt{(T_{x1}^2 + T_{z1}^2)} < T_{\max} \tag{9}$$

In many cases positioning data required for control, has imprecise or failure character and it leads to the conclusion that a fuzzy logic vehicle can be used. The advantages of such an approach are:

- description of the vehicle behavior on the formalized language;
- availability of a priori information about the system for improvement of adjustment quality;
- "robustness" with regard to changeable conditions;
- a priori adjustment of membership function (MF) parameters to reduce the time required for identification of these parameters;
- possibility to perform nonlinear transfer between motion modes.

In program algorithm based on fuzzy regulator a standard scheme is used: fuzzification (making fuzzy) - production deduction rules- defuzzification (making logic). MF are represented as piecewise-linear forms. Such a choice provides simplicity of software implementation and computation speed.

Parameters of input and output variables of the regulator are adjusted for every certain AUV model in accordance with its hydrodynamic features.

Let's consider input linguistic variable as an example: "Distance-to-go to the target" (fig. 7). Each variable term (near- R_0, middle - R_1, far - R_2 guiding zone) has its own velocity mode. Parameter values of MF are chosen in accordance with spatial restrictions at defined velocity mode. The restrictions are known from calculating hydrodynamic studies.

For output variables: output consists of thrusts/moments corresponding certain discrete velocity.

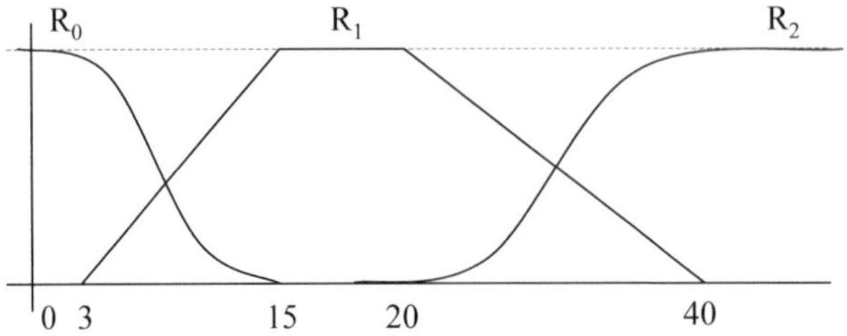

Fig. 7. Linguistic variable "Distance to the target" with three terms (range of values in meters)

From here the reactions of vehicle motion system on possible situations come. Situation is one of probable positions of vehicle casing against the target. In fuzzy version the situation looks as follows: "target to the right" and "vehicle in the near radius R_0", "target behind" and "the target is far from R_2" and so on. Response to the situation can be described as follows: "if a situation is", then "speed-up", "no sideway motion" and "turn to the right". Deduction rules are represented in the form of fuzzy association map (FAM) (fig. 8).

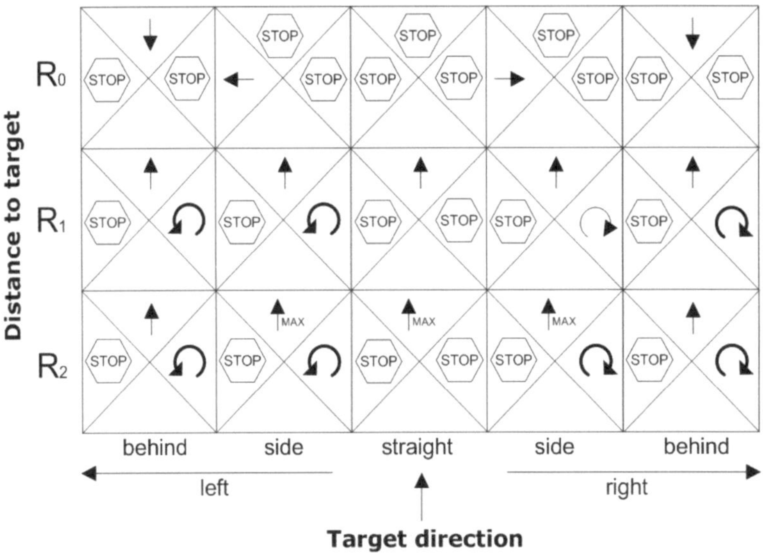

Fig. 8. Fuzzy controller deduction rules for motion to the target

Fig. 9 shows the example of motion paths derived in algorithm modeling. The example illustrates the process of vehicle guiding to the target in two cases: in constant and variable currents. Change of current vector projection for the second case is preassigned by periodic function: $V_{Tx} = V_{Ty} = A\sin(\omega t)$, where $A = 0.5$ m/s; $\omega = 0.008$ Hz.

The motion is built from three linked modes: guiding along the distance at preset velocity in far zone, approaching to the target at variable velocity on entering into specified area, dynamic positioning with holding the vehicle head to stream near the target.

Two parameters are used in controlling: target direction in relation to vehicle motion direction and distance to the target.

At known current course, vehicle and target coordinates the distance D and direction φ_t.

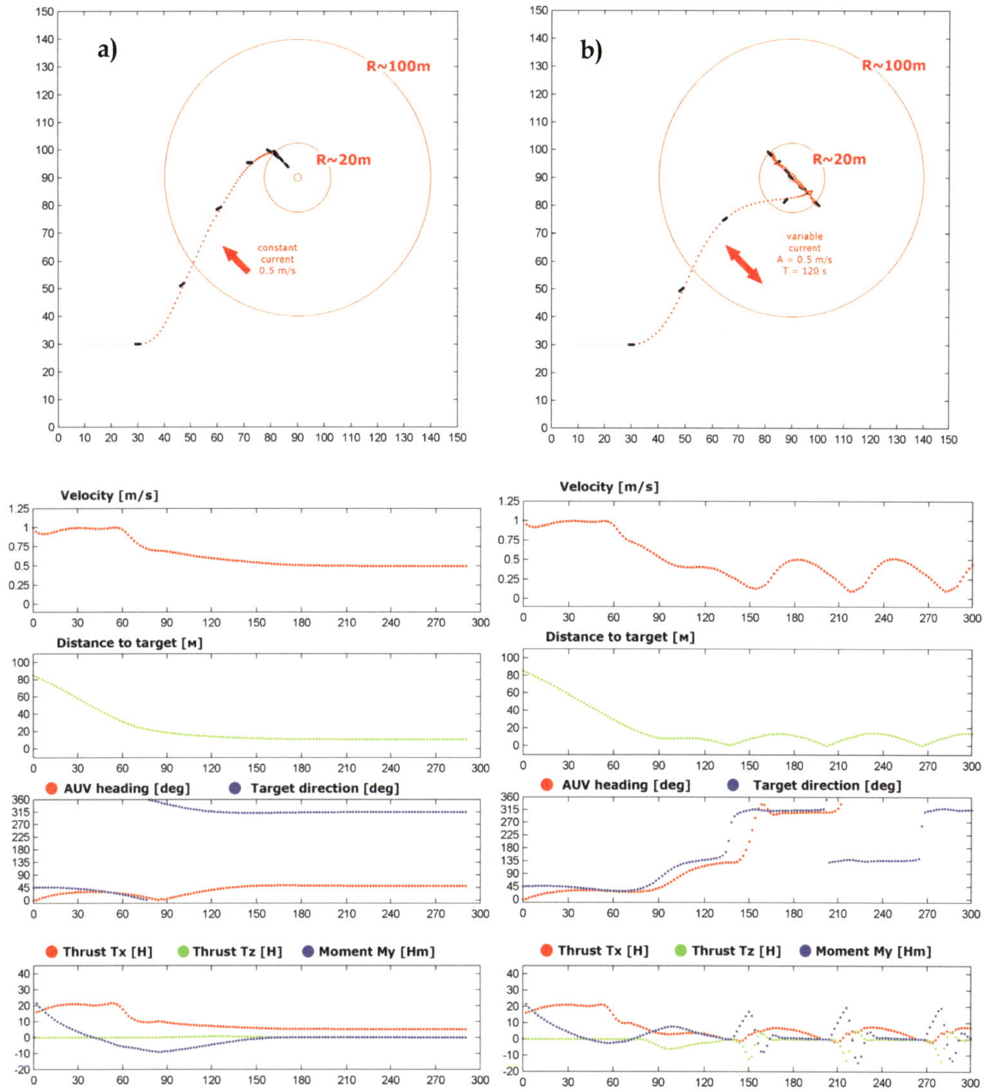

Fig. 9. The results of modeling, a) - motion path and parameters at constant current, b) - motion path and parameters at variable current (time in seconds)

The use of fuzzy regulator brings to implicit switching of motion modes, so switching zonal boundaries (fig. 8) conditionally divide the underwater space into the zones. That zones correspond different conditions at given motion modes.

Under conditions of constant current the vehicle approaches the target and hover near it positioning itself head to stream. Under conditions of shifting current the vehicle positions close to the target, hauls and changes the speed so that to stay at the mean near the target.

Another way of solving analogous task is connected with the control organization in the case of motion to guiding sonar transponder, and to approach it discrete distance measurement R and direction (bearing) measurements θ are used. At that the task is to provide decreasing of d_m upon the average, keeping $d_m = 0$ and vehicle angle orientation on target bearing.

7. AUV motion control during investigation of ocean physical fields

The tasks of AUV motion control are usually connected with bottom survey, object search, and inspection in near-bottom space as well as physical fields' measurements (Ageev et al., 2005; Kiselev & Medvedev, 2009). Manifold AUV uses for search-inspection tasks, seabed survey, bottom mapping, and aquatic medium monitoring can be considered on common grounds at the root of which the idea of ocean physical fields lies. So, for example, the task of route selection in the rugged bottom topography is a special case of a more common task of bottom path investigation and navigation according to the bathymetric map. More generally similar task arises in organization of any physical field path investigation. Spatial structure of such physical fields possesses the following features: changeability, abnormal level, and correlation in field geometry, etc.

In hands-on experience physical fields measurements in water columns and near the bottom are based on creation of survey network bound to base points, horizons or bottom points. According to the whole ocean scale observations held in different periods of time and in different places, the scientists receive average information about structure of sudden (random) fields. Though, as a rule, they are considered to be static, homogeneous, and isotropic. Small-scale phenomena research is carried out by means of establishing stations on the oceanographic grounds with the following machine processing of received information. Depth and area measurement network creates a system of transverse sections characterizing spatial structure of the field. Data received in the result of substitution of continuous field by network of point measurements is used in future for field mapping, i.e., for its reconstruction in any point by means of basis measurements profiling in network node. Nowadays along with traditional methods of oceanographic measurements the methods of path measurements with the help of autonomous, remotely operated, and towed vehicles are used. So the use of AUV has a number of advantages, especially during complex measurements at great depth and in extreme surrounding.

Fig. 10 shows the examples of bathymetric mapping and temperature field mapping with the help of data received by AUV "Klavesin" during Lomonosov Ridge research in polar expedition "Arctic Zone-2007". During this experiment bathymetric, hydrographical, and other measurements were conducted while following along a programmed path near the bottom by geographical referencing the measurements with preciseness that AUV navigation facilities can provide.

Field mapping according to the measurement data is a common task, though rather labor-intensive. In simplified version one can confine to building separate field realization, isolines or other sections in particular. Two interconnected tasks have an independent meaning: navigation according to known map elements and motion organization according to field isolines (sections).

Let's dwell upon probable variants of problem description.

Let $\xi(X,Z)$ be a variable characterizing flat field section that can be defined as isoline map $\xi(X,Z)$=const. Field measuring device during path motion $\{X(t), Z(t)\}$ gives field measurement ξ (X(t), Z(t)) with an accidental error. Let's suppose that vehicle (measuring device) geographical coordinates and speed are defined by on-board navigation system with

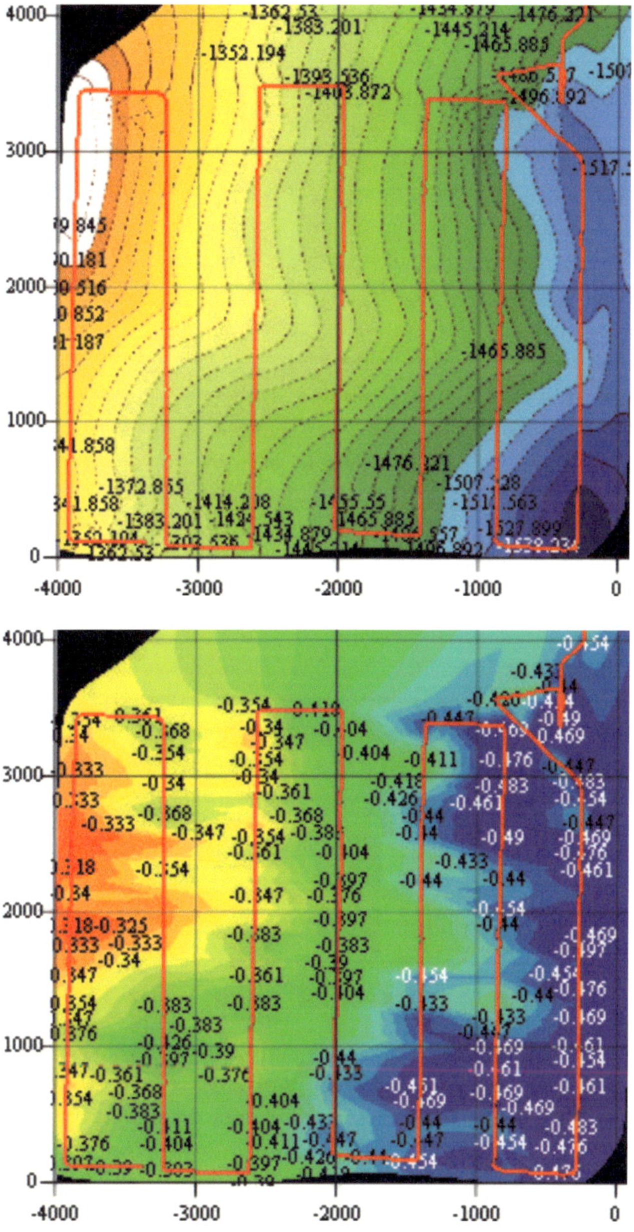

Fig. 10. Fragments of bathymetric mapping (upper) and temperature field mapping (bottom) during research over Lomonosov Ridge in Arctic zone with the help of AUV "Klavesin"

errors $\{\Delta X_a, \Delta Z_a\}$, $\{\Delta v_x, \Delta v_z\}$. Let's define field variability along trajectory by field gradient value $|\Delta\xi|$ or change $\Delta\xi = |\nabla\xi| v\Delta t$ on time interval Δt.

Control $U(v, X, Z, \xi, \Delta\xi)$ must be organized so that:

a. $\xi(X(t), Z(t)) = \xi^0 = const$ or $v\nabla\xi = 0$;

b. trajectory goes through all points $\{X_k, Z_k\}$, fulfilling condition $\xi=\xi_{max}$ or $\xi = \xi_{min}$;

c. trajectory goes through all points fulfilling condition $|\nabla\xi\ (X(t),Z(t)))|=|\nabla\xi|max$ or $|\Delta\xi\ (X(t), Z(t))| = |\Delta\xi|\,max$.

The first case corresponds to motion along preset isoline; the second one is edging of an area according to the points with extreme values of field level. This can be of interest during edging of signatures created by search objects. The last case corresponds to motion along the point with maximum rating of field gradient (change).

In all three control variants vehicle coordinates must be known. Though in some cases (for example, during isoline tracking "at the mean") it is enough to orient velocity vector in accordance with flection of reproduced trajectory.

During integrated performing of control task and state vector evaluating knowledge of field map and its elements with preciseness better than accuracy dead reckoning allows clarifying the position of the vehicle. In this case common problem description consists in construction of computational procedure which algorithm depends on estimation covariance matrix, measured field values, gradient, and spectrum patterns of sensors noises. An alternative variant of the task performing is given below.

7.1 Path control with field isolines search and tracking

Isoline motion joins the tasks of field mapping and motion control along isoline paths, motion along which contains constant value. Program algorithm in this case must contain conditions controlling ordered transfer from one curve to another as well as angular motion control law that displays isoline flection. To choose correctly the direction of transfer between field isolines it is necessary to gain information about direction of gradient vector. This information can be gained by means of measuring gradient components with the help of differential sensors or scheme imitating gradient calculation in search motions. In this case parameters of search trajectory and radius of isoline flection as well as period of search motions and time of data updating must be coordinated.

The use of data on field for motion control is equal to inclusion of field parameter $\xi(X,Z)$ into extended vector of system condition (1) with additional equation:

$$\xi = \nabla\xi_x(X,Z)\cdot V\cos\chi + \nabla\xi_z(X,Z)\cdot V\sin\chi, \tag{10}$$

where path angle is χ, drift angle and route are connected by the equation $\chi = \varphi - \beta$.

During motion along isoline $\xi(X,Z) = \xi^0 = const$ velocity vector $v_x=v\cos\chi$, $v_z=v\sin\chi$ must obey "at the mean" to kinematic condition:

$$\nabla\xi_x V_x + \nabla\xi_z V_z = 0, \tag{11}$$

which can be expressed in terms of gradient projection on axis connected with vehicle: $\beta = arctg(\nabla\xi_{x1}/\nabla\xi_{z1})$or in the form:

$$tg\varphi = (\nabla\xi_{x_1}V_x + \nabla\xi_{z_1}V_z)/(\nabla\xi_{x_1}V_z - \nabla\xi_{z_1}V_x), \tag{12}$$

Equation (12) gives values for programmed course if another motion parameters are known. Let's consider the task of motion control during search and tracking of given isoline $\xi = \xi_0$, considering that control vector consists of two components - one of them for position control, the other one for orientation control.

Let's define "distance" D_ξ from point with current field value to preset isoline by equation

$$D_\xi = |\xi - \xi_0| / |\nabla \xi|, \tag{13}$$

and projection of gradient vector on motion direction as

$$p = p_v \nabla \xi = |\nabla \xi| \cos \gamma,\ \gamma = \beta + arctg(\nabla \xi_{z_1} / \nabla \xi_{x_1}) \tag{14}$$

The choice of motion direction at yield of isoline in accordance with gradient direction must obey condition: $(\xi - \xi_0)\ p < 0$. If not, it's necessary to perform search motion that properly orients velocity vector.
Let's motion control to the point with coordinates

$$X_\xi = D_\xi \cdot \cos \varphi_s,\ Z_\xi = D_\xi \cdot \sin \varphi_s, \tag{15}$$

define as:

$$\begin{aligned} U_x &= K_1(X - X_\xi) + K_2 \cdot \dot{X} + K_3 \cdot \int_0^t (X - X_\xi) dt, \\ U_z &= K_1(Z - Z_\xi) + K_2 \cdot \dot{Z} + K_3 \cdot \int_0^t (Z - Z_\xi) dt, \\ \dot{X} &= v \cdot \cos \varphi,\ \ \dot{Z} = v \cdot \sin \varphi, \end{aligned} \tag{16}$$

where K_1, K_2, K_3 are control parameters.
In projections on vehicle axis we'll receive:

$$\begin{aligned} U_{x_1} &= U_x \cdot \cos \varphi + U_z \cdot \sin \varphi, \\ U_{z_1} &= -U_x \cdot \sin \varphi + U_z \cdot \cos \varphi, \end{aligned} \tag{17}$$

where above-mentioned restrictions on a control take place.
Generally control law (24, 25) providing yield of isoline and motion in a given vicinity Δ_0 can be written as:

$$U_{\varphi,\xi} = \begin{cases} (U_x, U_z), & |\Delta \xi| > \Delta_0, \\ U_\varphi = K_\varphi(\varphi - \varphi_s) + K_{\dot{\varphi}} \dot{\varphi} + K_\xi D_\xi sign(p \Delta_\xi), & |\Delta \xi| \le \Delta_0, \end{cases} \tag{18}$$

Let's consider an example where several features of the task are evident.
Let's define field $\xi(X,Z)$ by isoline class approximated by cubic parabolas like $Z = aX^3+bX^2+cX+d$.
Let's define values for control and orientation:

$$\begin{aligned} &\nabla_\xi = (3aX^2 + 2bX + c, 1), |\nabla_\xi| = \left[1 + (3aX^2 + 2bX + c)^2\right]^{1/2}\ \acute{a} \\ &D_\xi = |Z - (aX^3 + bX^2 + cX + d)| / |\nabla_\xi|,\ \varphi_s = arctg(3aX^2 + 2bX + c), \\ &X_\xi = D_\xi(3aX^2 + 2bX + c),\ Z_\xi = 1 \cdot D_\xi \end{aligned} \tag{19}$$

Motion modeling results for three isoline tracking variants are represented in fig. 11.

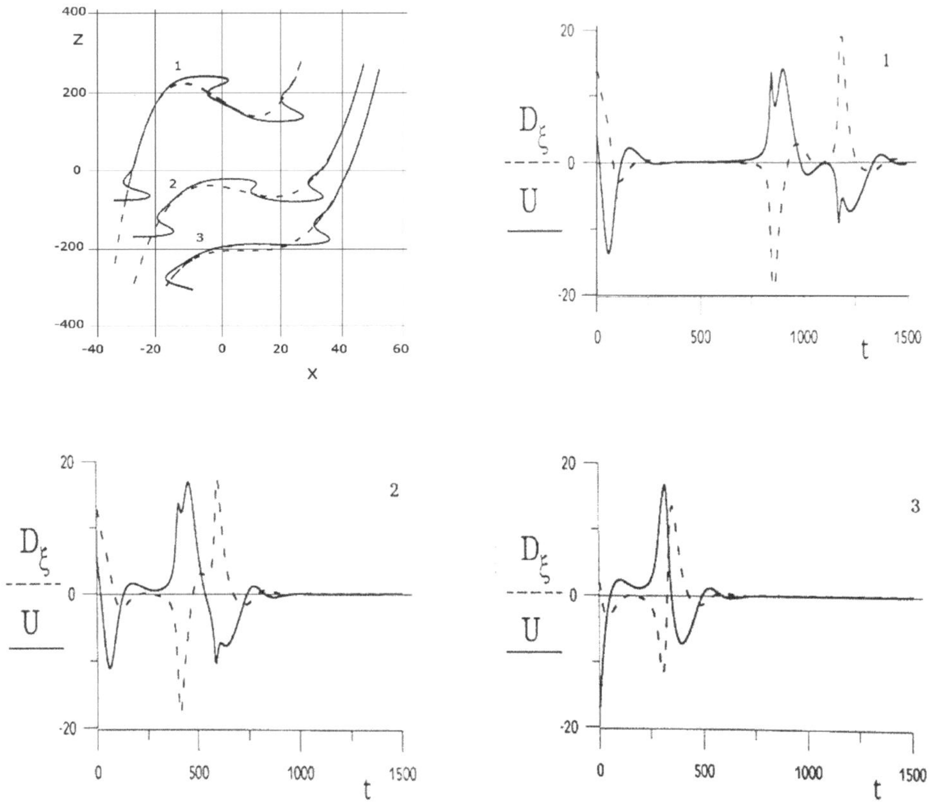

Fig. 11. An example of isoline tracking defined by cubic parabolas in vicinity Δ_0=25 (1. a = 0.01, b=0, c=-5, d=200; 2. a = 0.005, b=0, c=-2, d=50; 3. a = 0.005, b=0.04, c=0, d=-150)

Expanded scale on X-coordinate is chosen to show greater clearness of process dynamics during isoline tracking.

In this example the most characteristic fact is that in points with greatest isoline flection the switch of control takes place depending on value of given vicinity of isoline tracking, and, consequently radius of vehicle circulation on curved trajectory.

8. Conclusion

The results of the research on development of flexible AUV control systems presented in the paper are based on practical experience of the IMTP FEB RAS. Relevance of the problems addressed in the paper stems mainly from the need for performing new complex tasks under conditions of uncertain and extreme surrounding when the AUV autonomous operation time is extended. It is crucial to develop control systems with integrated processing of search and navigation information. Control and navigation systems developed by IMTP FEB RAS enable performance of many search-and-survey operations in the sea. Navigation-control facilities are the basis for generating complex AUV behavior missions and for performing "intelligent" control scenarios. This is particularly true for inspection of underwater objects and structures, bottom relief and physical fields. The report presents the results of the major collective work, so the authors would like to thank the following colleagues whose materials were used during preparation of the paper: N.I. Rylov, Yu.V. Matvienko, O.Yu. Lvov, Yu.V. Vaulin, A.A. Boreyko, A.K. Kukarskih.

9. References

Ageev, M.; Kiselyov, L. & Inzartsev, A. (2000). Some Aspects of AUV Control in Sea Mounts Survey, *Marine Technologies*. #3, Vladivostok, Dalnauka. - P. 46-55. (in Russian)

Ageev, M.; Kiselyov L.; Matviyenko Yu. et al. (2005). *Autonomous Underwater Vehicles. Systems and Technologies*, (in Russian), Ed. Acad. Ageev M. D, Nauka, Moscow, Russia

Inzartsev, A.; Kamorniy, A.; Lvov, O.; Matviyenko, Yu. & Rylov, N. (2007) AUV Application For Scientific Research In Arctic Zone, *Underwater Research and Robotics*. #2. pp. 5-14. (in Russian)

Inzartsev, A. & Pavin, A. (2006). AUV Behavior Algorithm While Inspecting of Partly Visible Pipeline, *Proceedings of the OCEANS 2006 MTS/IEEE Conference*, ISBN 1-4244-0115-1, September 2006, Boston, MA, USA

Inzartsev, A. & Pavin, A. (2008). AUV Cable Tracking System Based on Electromagnetic and Video Data, *Proceedings of OCEANS'08 MTS/IEEE Kobe-Techno-Ocean'08 (OTO'08)*, ISBN: 978-1-4244-2126-8, April 2008, Kobe, Japan

Inzartsev, A. & Pavin, A. (2009) AUV Application for Inspection of Underwater Communications, *Underwater Vehicles* edited by Inzartsev, A., In-Tech Publishers, Vienna, pp. 215-234, ISBN 978-953-7619-49-7, open access: www.intechweb.org

Kiselev, L.; Inzartsev, A.; Matviyenko, Yu. & Rylov, N. (2007) Current issues of navigation and control in autonomous underwater vehicle development, *Proc. of International Conference on Subsea Technologies SubSeaTech'2007*, June 25-28, St.Petersburg, Russia, ISBN 5-88303-409-8

Kiselyov, L. & Medvedev A. (2009). Typology Models of Dynamic AUV Spatial Motion, *Proceedings of 6th International Symposium on Underwater Technology*, April 21-24, Wuxi, China, pp. 120-125.

Kukarskih, A. & Pavin, A. (2008). Using of Electromagnetic Searcher for Inspection of Metallic Cable by Means of AUV, *Proceedings of OCEANS'08 MTS/IEEE Kobe-Techno-Ocean'08 (OTO'08)*, ISBN: 978-1-4244-2126-8, April 2008, Kobe, Japan

Pavin, A. (2006). The Pipeline Identification Method Basing on AUV's Echo-Sounder Data, *Proceedings of the OCEANS 2006 MTS/IEEE Conference*, ISBN 1-4244-0115-1, September 2006, Boston, MA USA

Scherbatyuk, A.; Boreyko, A. & Vaulin, Yu. (2000). AUV Operation Based on Video Data Processing: Some IMTP Experience, *Workshop on Sensors and Sensing Technology for Autonomous Ocean Systems*, Oct -Nov 2000, Hawaii

Smoot, C. (1989). The Marcus-Wake Seamounts and Guyots as Paleofracture Indicators and their Relation to the Dutton Ridge, *Marine Geology*, pp. 117-131.

11

Integrated Positioning System of Autonomous Underwater Robot and Its Application in High Latitudes of Arctic Zone

Alexander Inzartsev, Alexander Kamorniy, Lev Kiselyov,
Yury Matviyenko, Nicolay Rylov, Roman Rylov and Yury Vaulin
Institute of Marine Technology Problems (IMTP FEB RAS)
Far East Branch of the Russian Academy of Sciences
Russia

1. Introduction

Nowadays various robots are developed and used for deep-water operations and oceanic research including autonomous unmanned underwater vehicles (AUV), which are particularly effective for operations at great depths, under ice, and in other extreme surrounding. A considerable number of such vehicles have been developed in several countries. They are designed for object search, bottom configuration survey, geological survey, scientific research and a wide range of military missions.

Modern underwater vehicles generally comprise positioning systems that include onboard autonomous, acoustic and satellite positioning systems (Romero & Lester, 2000; Theseus AUV; Maridan AUV). For example, the autonomous vehicle *Hugin* by *C&C* is equipped with an inertial positioning system (IPS) based on fiber-optic gyros, which is integrated with an *RD Instruments* Doppler log, a depth sensor, a height sensor and an ultra-short base acoustic system (USBL) by *Konsberg Simrad.* The *Maridan* company jointly with the Technical University of Denmark and *Kearfott Guidance and Navigation Corporation*, USA, have developed the *Marpos* system to be installed onboard an AUV. *Marpos* is an integrated Doppler inertial positioning system with the high-precision strap-down inertial positioning system *KN5053* as its core equipped with laser gyros, which was developed by the *Kearfott* company. The IPS is adjusted using the data of the *RDI* Doppler log, which measures the vehicle speed over the bottom or through the water while a *DGPS* receiver is used for surface positioning.

The similar method was applied for development of the *Oracle* vehicle by *Thales-Bluefin.* Its positioning is based on the *Litton LN-250 MIMU* system consisting of three fiber-optic gyros and three accelerometers installed on the inertial unit. In addition the positioning system incorporates the following: a digital quartz pressure (depth) sensor, an ultra-short base acoustic positioning system, and a 600 kHz Doppler log.

The positioning system of *Boeing/Fugro/ Oceaneering* vehicles is based on total integration of all the available sensors including an *IPS*, a Doppler log, a height sensor, a depth sensor, long-base systems and ultra-short base systems.

Russian Institute of Marine Technology Problems (IMTP) FEB RAS has many years of experience of developing and using AUVs for solving practical tasks at depths up to 6,000 m

(Ageev et al., 2005). This also includes experience of development and operation of various positioning devices. The positioning devices that had been developed over the past years had various operating ranges, various precision rates and differed considerably in the system integration approach but in general they were designed to solve the tasks of enabling secure and reliable positioning of AUVs for various applications.

Judging from the experience of all previous work the list of the said problems includes the following:

- determination and display onboard the carrier ship of the current AUV position in the shallow sea and deep sea,
- secure performance of operation missions near the sea bottom and bottom-level obstructions,
- mission performance control onboard the ship,
- obtaining the AUV system status information onboard the ship,
- current AUV positioning onboard the AUV,
- efficiency improvement and expansion of the positioning system operation range by eliminating the fixed devices (beacons),
- high-speed data exchange between the AUV and the ship via an acoustic communication link,
- transporting the vehicle into the acoustic devices operation range and providing communication for short-range control.

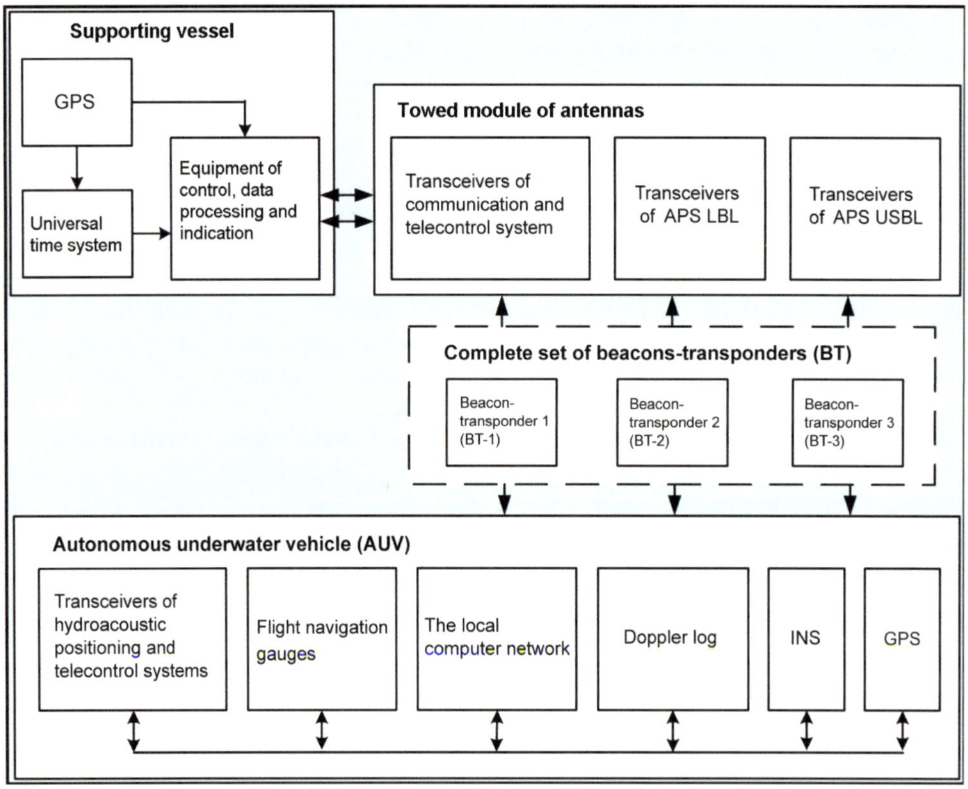

Fig. 1. AUV positioning devices

The positioning equipment of modern vehicles developed by the IMTP FEB RAS comprises units of onboard autonomous, acoustic and satellite positioning systems (Ageev et al., 2005; Kiselyov et al., 2004; Inzartsev et al., 2007a; Inzartsev et al., 2007b). Each system in its turn is a system of devices incorporated in the integral base configuration of AUV systems and shipboard equipment (fig. 1).
Let us consider next some architecture features and characteristics of particular systems incorporated in the navigation complex.

2. Acoustic positioning complex

The core of the AUV navigation system is an acoustic positioning complex which incorporates a long-base (LB) and ultra-short base (USB) acoustic positioning systems (APS). The complex integrates structurally shipboard antennas of the systems, shipboard support equipment, software is unified. The operation mode of the complex is selected based on the current tactics of the AUV application.
The complex enables structuring the configuration of navigation facilities depending on a type of the problems to be solved and operating environment. As a rule the following complex configurations are used:

- LB-APS with the range up to 15 km and relative error of 10^{-3} can be combined with a low-speed remote-control system (Table 1);
- USB-APS with the range up to 10 km and relative error of 10^{-2}; it can be integrated with an LB- APS (Table 2);
- ACS, an acoustic system for the AUV communication with the carrier ship with the range of 6-10 km, 4,000 bit/s speed and error rate of 10^{-2}.

The positioning complex equipment consists of a set of transponders, a set of AUV transceiving equipment and a set of shipboard equipment. As a whole the complex ensures navigation, remote control, telemetry and search operation control by analyzing video images or sonar images transmitted from the AUV. The acoustic system and its modifications, where, besides the positioning system, a telemetry system and a remote-control system were implemented, too, have successfully enabled actual AUV maritime operations during the last three decades.

Operation depth	Up to 6000 m
Range	10-15 km
Range relative measurement error	No more than $0,5*10^{-2}$
Bandwidth	11-14 kHz
Beacon self-sustaining period	Up to 0.25 of a year
System installation time (with transponders positioning)	Up to 6 hours in shallow sea and up to 24 hours in deep sea
Size of shipboard electronic equipment	0.6 x 0.55 x 0.12 m
Size of transponder (diameter x length)	0.14 x 0.85 m
Size of towed antenna module	0.9 x 0.4 x 1.8 m

Table 1. LB-APS performance characteristics

Operation depth	Up to 6000 m
Range	Up to 10 km
Bearing angle error	No more than 1-2°
Range relative measurement error	No more than $0.5*10^{-2}$
Operating frequency	11.75 kHz
System installation time	1 hour
Size of shipboard electronic equipment	0.6 x 0.55 x 0.12 m
Size of towed antenna module	0.9 x 0.4 x 1.8 m

Table 2. USB-APS performance characteristics

Let us cite as an example the results of search operations in the area of the military helicopter crash near the coast of the Peter the Great Bay of the Japan Sea (Ageev et al., 2005). During the first phase of the operations when the site of the helicopter crash was searched for, a large area was surveyed using side scan sonar (SSS). The motion path is shown in fig. 2a.

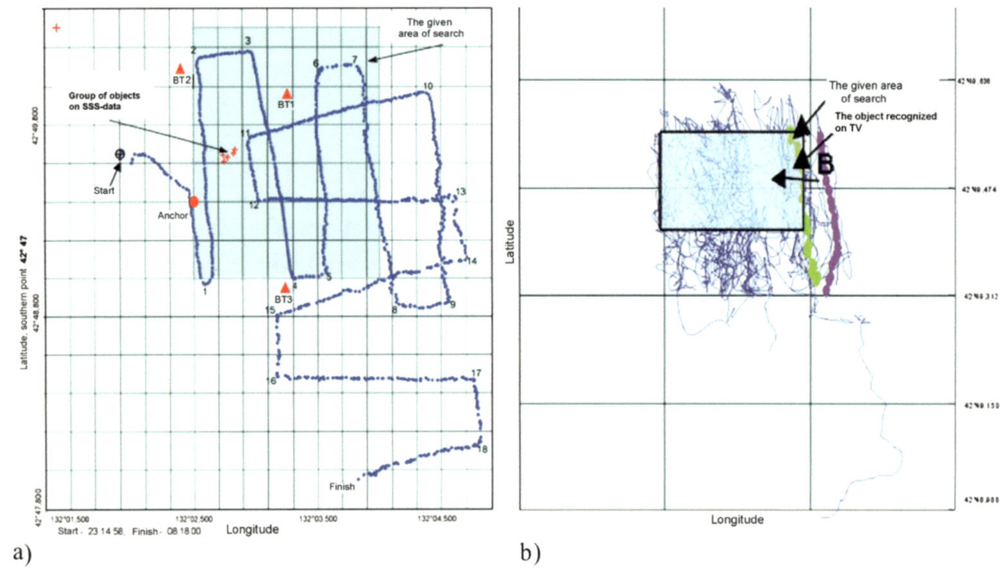

Fig. 2. Area path investigation during helicopter search. a) large-scale SSS survey, S = 22 sq. km, Δ = 20 m; b) detailed video survey, S=0.16 sq. km, Δ = 5 m

Points of beacon locations, the carrier ship anchorage, and a series of targets detected on the SSS screen are marked on the path. The total survey area was equal approximately to 22 sq. km. The depth of the site in the operations area was approximately 70 m, the horizontal range was up to 6 km, and the target positioning accuracy was approximately 20 m. During the second phase close search was carried out for precise positioning of the detected targets. In the area of 400 x 400 m the search was carried out with the use of video system with the vehicle travel height equal to 5 m. In order to avoid gaps in the search square the vehicle motion was remotely controlled in short straight-line tacks at a distance of several meters from the preceding ones. The results of the search operations in the area are given in fig. 2b where the AUV paths for three launches are shown; each launch has 5 to 6-hour duration and they are overlapped in the search square. The total area of video observation was no less than 100,000 sq. m, and tack positioning accuracy was less than 10 m which allowed accomplishing the mission: to detect and to position the targets.

3. Onboard autonomous positioning complex

The onboard autonomous navigation system (BANS) generally encompasses positioning and steering sensors (a depth gauge, a magnetic compass, a gyroscopic compass, a heel sensor, a pitch sensor, a specific velocity meter, an absolute velocity meter, i.e. an electromagnetic log and a Doppler log (EL, DL), angular-rate sensors, an inertial positioning system (IPS), an echoranging system (ERS), a GPS receiver). Depending on the BANS configuration the data provided by the measuring devices is used to enhance reliability and accuracy of the system operation.

The BANS itself represents a stratified module formed by a dead reckoning system, the IPS and the GPS receiver operated by a local area network. Autonomous navigation facilities of that configuration can enable performance of complex programmed missions and gathering data on the vehicle condition during its motion. Solving the problems of objects investigation and work of the vehicle under conditions of autonomy brings to necessity of development of positioning complex, which allows perform autonomous motion correction by the current vehicle coordinates and the given coordinates of targets or references.

When the AUV autonomous operation time is extended it is important to provide positioning with as much accuracy as possible using integrated BANS. In general navigation error during determination of vehicle coordinates by means of dead-reckoning depends on a number of factors: instrumental errors of measuring devices, in particular, gyros drift, initial data input errors, velocity measurement errors, especially at unaccounted stream, and IPS initial alignment errors.

It's obvious that when there is no correction from external measuring devices the cumulative positioning error accumulates in time, and when the vehicle operation time is extended it becomes intolerably large. When the vehicle operates in shallow area error can be corrected due to the possibility to make corrections with the help of GPS during episodic vehicle surfacing.

The following variants of integration and correction of navigation information onboard the AUV are of the greatest practical interest at present.

- Correction of IPS (in the complete configuration or in a gyros mode) with regard to DL (near the bottom) and GPS (on the sea surface);
- Reciprocal correction of gyroscopic and magnetic compasses in various AUV operation modes;
- Integrated processing of the information from BANS and LB (USB) APS onboard receiver;
- correction of BANS via an acoustic communication link and remote control using APS and GPS data.

Let's note some peculiarities of navigational calculations, and cite as an example integrated positioning system of AUV "Klavesin".

AUV onboard positioning complex is realized in two configurations. They differ in the type of used IPS. In one of them a fiber-optic gyroscopic compass "Octans III" by French company iXSEA is used as IPS, in the other one the IPS is mechanical without dynamic tuning gyros. In this system speed correction is programmatically provided by the use of measurements of acoustic Doppler log. Kalman filter is used for combined data processing from both internal (gyros, accelerometers) and external sources (DL, GPS receiver).

As it was mentioned, positioning (local and absolute) in autonomous system is performed by means of dead-reckoning. Velocity vector data received with the help of water speed log or bottom speed log is used for it. In these cases velocity vector components $V=(V_x,V_y)$ against are given by:

$$V_x = V^{rel}\cos\psi\,\cos\varphi\,,\;\; V_y = V^{rel}\cos\psi\,\sin\varphi\,, \tag{1}$$

$$V_x = V_x^{abs}\cos\varphi + V_y^{abs}\sin\varphi\,,\;\; V_y = V_x^{abs}\sin\varphi + V_y^{abs}\cos\varphi, \tag{2}$$

where V^{rel} – speed against the flow, V_x^{abs}, V_y^{abs} - average values of longitudinal and cross stream velocity components, measured by Doppler log, ψ, φ - path and vehicle trim difference, measured by IPS.

When the vehicle operates in a limited area or in the carrier-ship tracking mode, autonomous coordinates can be corrected through combined processing of the BANS and APS data or through positioning data transmission to the vehicle via acoustic communication link together with telemetry data and remote control commands.

BANS task consists in path and speed measuring, measurements of speed projections on coordinate axis according to these data, and measurements integration for positioning. If water speed log is used then available current data (V_{Tx} , V_{Ty}) is taken into account during measuring. System operation can be described by formula:

$$\begin{aligned} X_c &= X(t_0) + \int_{t0}^{t} (V^{rel}\cos\psi\,\cos\varphi + V_{Tx})\,d\tau\,, \\ Y_c &= Y(t_0) + \int_{t0}^{t} (V^{rel}\cos\psi\sin\varphi + V_{Ty})\,d\tau\,, \end{aligned} \tag{3}$$

or

$$\begin{aligned} X_c &= X(t_0) + \int_{t0}^{t} V_X^{abs}\cos\varphi\,d\tau\,, \\ Y_c &= Y(t_0) + \int_{t0}^{t} V_y^{abs}\sin\varphi\,d\tau\,, \end{aligned}$$

where X_c, Y_c – coordinated reckoned with the help of BANS, $X(t_0)$, $Y(t_0)$ - reference coordinates entered before start.

Range evaluations r_i (i =1...n) from vehicle to transponders with coordinates (X_i, Y_i) and evaluations of the vehicle coordinates (X_{abs}, Y_{abs}) which can be received by APS are connected by equations:

$$\left(X_{abs}-X_i\right)^2+\left(Y_{abs}-Y_i\right)^2=r_i^2,\ i=1,..n\ . \tag{4}$$

The task of integration and correction (Romero & Lester, 2000) consists of elimination of large initial dead-reckoning errors of current positioning and current vector evaluation. To do it data of different systems is used at each step of navigational calculations. Integrated positioning algorithm performs: positioning on the basis of navigation-piloting sensors, response selection from hydroacoustic transponders, hydroacoustic positioning on the basis of distance-measurement information, coordinates correction and current speed evaluation. Reckoning system input parameters are the path of gyroscopic and magnetic compasses, pitch, relative or absolute velocity, and preliminary evaluation of current velocity components (when water-speed log is used). Output parameters are course made per one correction cycle, and coordinates. The process of signals selection from APS transponders provides false signals filtration, which was caused by refraction effects and multipath condition, and formation of distances for the following positioning.

The variants of integration and system accuracy evaluation given above were tested during repeated field tests in the deep and shallow sea. Some results of positioning characteristics experimental tests during AUV performing different operations are given below.

3.1 Positioning process in AUV control system

Positioning process takes an important place in AUV control system. To a large extent system architecture of control is defined by requirements of organization of navigational calculations. Let's mention some peculiarities of this process.

1. Navigational calculations are made in distributed environment of AUV's local area network (LAN) (fig. 3). Positioning process is organized as a combination of informative interrelated tasks performed on separate computers and LAN microcontrollers. All navigation facilities (IPS, positioning and steering sensors, APS signal transducers, etc.) are the abonents of several segments of this network, and two computers perform navigational calculations simultaneously.
2. In many cases navigational calculations have "interactive" character. In other words positioning process can require from AUV control system performance of several additional actions (not foreseen by program-task) for accuracy improvement or disambiguation in positioning. Suchlike situations can happen during operations at great depths (for vehicle reference coordinates positioning after long descent.
3. "Interactivity" of positioning process is developed in set of commands of program-task (mission) used for required motions specifying.

Navigational calculations are organized on two AUV control system hierarchy levels. The basic calculation process is performed on the execution level. Positioning process besides sensors data receiving and processing specifies data and commands exchange and enquiry messages to AUV control system nucleus (by means of distributed data base).

The basic operation process consists of integrated calculations for integration of LB APS transducers responses or USB APS data and BANS reckoned data. As a rule, such mode doesn't affect performing AUV major mission. In disadvantageous conditions when there is a large positioning error according to data of different systems, a necessity to perform test motions (for disambiguation) arises. This situation becomes obvious during positioning process and is based on the rate of convergence of iterative loop calculations, value of closure error, etc. At the same time a corresponding enquire is formed to the highest (coordinating) hierarchy level for performing such motions.

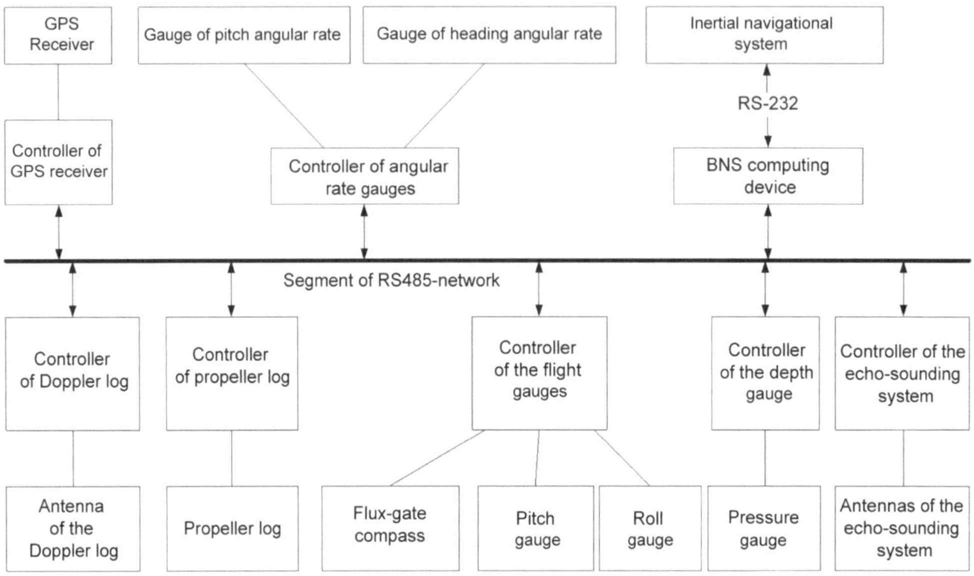

Fig. 3. Elements of local area network

Depending on the status of the scheduler this requirement can be denied, queued or immediately executed. In the last case performing of the program-task is discontinued, and the processing procedure starts (standard or described in mission). As a rule, standard processing procedures are performed during hovering mode or motion along the special search path. Certain parameters of the said procedure performance can be defined by positioning process itself. From mission support of such AUV behavior consists in the use of several operation control functions (Ageev et al., 2005):

- description of desired path configuration: TACK_…(…), POSITION_…(…);
- defining of conditions of navigational devices (for example, IPS): GET_…();
- loading out-of-order inquiry processing of positioning process SIGNAL(…).

The functions of the first two categories are also used in standard libraries of coordinating level. A corresponding notification is sent to the carrier-ship through acoustic communication link in the beginning of unplanned part of the mission performance.

4. Results of sea trials and experimental operation

4.1 AUV positioning complex accuracy test in shallow sea

The tests of positioning complex were carried out on AUV "MT-98" during trajectory measurements under ground test conditions in one of the Peter the Great Bay creeks.

It follows from the comparison of the results of the multiply conducted experiments that average runout rate for BANS positioning is minimal when an IPS (in a gyroscopic compass mode) is integrated with a Doppler log. Relative error averaged according to several vehicle launches comprised 28.5 m/h, which corresponds to the positioning error of approximately 1% for the whole distance covered.

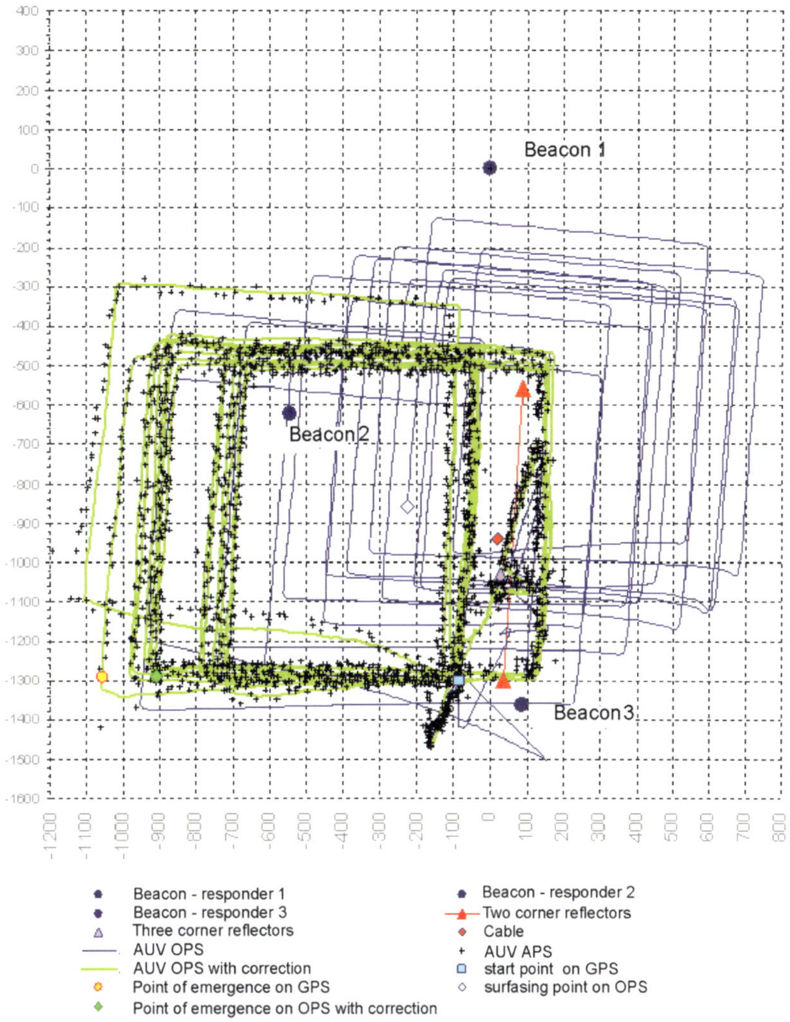

Fig. 4. AUV motion path during positioning complex tests

Further system modification was connected with increasing the AUV autonomous operation time. Fig 4 shows the AUV motion paths obtained during a 17-hour vehicle launch while it was positioned using APS, BANS and GPS devices. The programmed path was set in the form of repeated squares within the range of the three APS transponders. Various types of the AUV motion paths shown in figure correspond to the following test conditions. Positioning process using the BANS was carried out according to the gyroscopic compass and electromagnetic (impeller) log data. The obtained BANS coordinates were simultaneously corrected based on the LB-APS distance-measurement information. To compare the obtained results and to determine a BANS accumulating error the LB-APS and GPS measurement data was used (during surfacing). In addition it was assumed that the LB-APS internal "point" error with respect to the actual position of an object and a similar observation error according to GPS data do not exceed 10 – 15 m, which allows adopting these systems as a "reference". Under these conditions the BANS reckoning error (without corrections based on APS data) accumulated upon completion of the program with regard to the surfacing location coordinates based on GPS data was 933 m which corresponds to the runout rate of 54 m/h. The BANS and LB-APS integration virtually enables reducing the BANS reckoning error down to the corresponding LB-APS level (15 m).

4.2 AUV preparation for operating in the polar latitudes

In different periods of North development and research to perform operations, always hot and difficult, the most perfect technologies were used. Nowadays in Arctic Zone underwater vehicle, among them vehicles-robots, are used. We all know about the operations performed by underwater vehicles under the ice on evaluation of bottom configuration in places where cables and pipelines are installed and about the operations on fiber-optic cable installation. The importance of the operations is specified by growing interest to the resources within seabed covered by solid ice. Until now the Arctic Ocean seabed has been explored using individual sounding carried out by icebreakers or drifting polar stations. Though modern atomic icebreakers can bring scientific expeditions to any part of Arctic Zone, they cannot provide all range of necessary polar research. Application of underwater robots operated onboard icebreakers appears to be the most appropriate method to investigate bathymetric, physical, and geomorphologic characteristics of the Arctic seabed in the area of widespread ice cover. The first operational experience in high latitudes of the Arctic zone using underwater robots was received in August 2007 in the Arctic Ocean near Lomonosov Ridge (Inzartsev et al., 2007a). The expedition of the atomic icebreaker "Russia" investigated the geological characteristics of the seabed at depths of 1,500-1,600 meters in the area of over 50 sq. km.

Fig. 5. AUV "Klavesin" onboard the atomic icebreaker "Russia"

Preliminary integrated checkout of the vehicle efficiency in high latitudes was carried out earlier onboard the atomic icebreaker "Russia" during the expedition to the North Pole in summer 2007.

Further the paper discusses the stages of preparation to the research in Arctic, gives some scientific data received during the deep-water descents, and evaluates research results.

The operations were carried out with the help of AUV "Klavesin" (fig. 5). This vehicle developed by IMTP FEB RAS is designed for supervisory and searching tasks under conditions of open water at depth up to 6000 m.

However, normal work of the vehicle under conditions of polar latitudes and solid ice cover demanded serious changes in organization of its operation, navigation and communication facilities, descent and ascent technologies.

These adjustments were dictated by the extreme operating environment which includes:

- AUV descent and ascent operations through the ice opening, which size is comparable to the size of the carrier,
- ice drift in the exploration area;
- latitude dependence of accuracy of the magnetic sensors and gyroscopes.

During preparations for the high-latitude expedition the ways of solving a number of problems due to these factors became basic. Several details of the preparation work are given below.

4.3 Accommodation of AUV onboard control system to the operations in high latitudes

AUV "Klavesin" is a multi-purpose system equipped with sophisticated facilities for autonomous and acoustic positioning and communication, a configurable control system enabling search operations in an autonomous mode or using acoustic remote control equipment. To fit the polar conditions the AUV standard equipment was supplemented with a series of special function modules and base units:

- the system performing the procedure of AUV automatic transporting to the onboard antenna was developed; the procedure is initiated when the mission is accomplished;
- for AUV precision control during AUV ascent in the ice opening a standard set of remote-control commands transmitted via acoustic communication link was changed;
- to make the vehicle stay on the surface when the mission is accomplished to decrease its floatability under conditions of desalination of surface layer of water a special mode of stabilization with the help of vertical thrusting propulsions was introduced.

The most important task was to develop and debug a system of the AUV homing to the carrier ship. After completing simulation research and full-scale experiments a sequence of operations was determined for the AUV to implement the homing algorithm.

At the first stage the AUV performs search motion along the path in the shape of circle, forms array of range value to the shipboard antenna depending on the current path, and takes a bearing corresponding to maximum speed of range attention. Having found required direction the vehicle moves to the carrier-ship along the fixed route. When AUV approaches to the shipboard antenna not closer than 100 m it starts moving in the "figure-of-eight" (in the center of which the shipboard antenna stays), and waits for the commands from the shipboard. The final stage of the control during AUV homing to the ice opening is carried out by the operator in the acoustic remote control mode.

The elements of the described algorithm were tested in the sea during AUV preparation. Figure 6 shows one of AUV motion paths during its homing to the shipboard antenna in the open-water at the initial distance of 750 m.

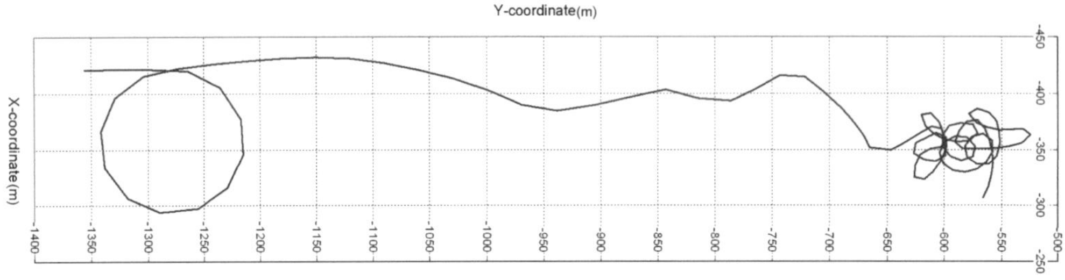

Fig. 6. An example of AUV autonomous homing path to the shipboard antenna

These trials proved that homing to the carrier is performed rather quickly. In the mentioned experiment at cruising speed 1 m/s an average speed of approaching to the antenna comprised taking into account all movements of the vehicle 0.57 m/s, and not taking into account initial search movement - 0.8 m/s.

The AUV's operation under ice is necessary not only for accurate positioning of the current operations, but also for monitoring and ensuring the AUV return to the carrier ship. When the vehicle moves 10-15 km away from the launch ice opening it is important to ensure reliable acoustic link with the shipboard antenna module. At the same time, it is imperative to provide ongoing monitoring of the communication link condition in order to avoid risks of losing acoustic contact. In this case antenna module becomes a towed acoustic beacon to which AUV is homed when the mission is accomplished. For open-water and mid-latitude operations AUV "Klavesin" is equipped with hydroacoustic navigation and control facilities the application of which on the North Pole in normal operations mode is limited by a number of circumstances.

Operation of USB APS requires no transponders. It is usually equipped with a magnetic course sensor, which has low accuracy in polar latitudes. Installing bottom acoustic transponders, both return and single-use, and the LB APS on the site are ineffective due to drift of ice floe. If the homing acoustic antenna drifting with the ship moves too far off, conditions for acoustic monitoring and operation control onboard the ship deteriorate dramatically. Installing surface LB APS transponders also has its shortcomings. Firstly, each transponder current positioning requires integrating with the regular navigation receiver, as well as with coordinates transmitter in the control desk; then the data input is required. Secondly, it is necessary to install the transponders at the depths of at least 250-300 m to provide their proper work taking into account the peculiarities of vertical distribution of sound speed in Arctic latitudes. The sizes of the ice opening, and, thus, the transponders' measuring base are limited. At the same time unpredictable drift of transponders, installed on flexible umbilicals, brings to considerable errors and failures of navigation system operation.

Basic elements of the positioning complex include an inertial positioning system (IPS) and an acoustic Doppler log. During preparations for the high-latitude expedition operation procedure was worked out for the gyrocompass "Octans-III" by French company iXSEA and the Doppler log developed by the IMTP FEB RAS. As a result the following scheme was implemented for AUV positioning guidance.

Three maximum-spaced LB APS transponders were installed around the ice opening chosen for AUV descending – ascending. The transponders' coordinates were determined at the time of their installation and directly before AUV starting. Then they were entered into the positioning program as persistent data. Transponders' location was measured at regular intervals, and updated measurements were entered into the positioning program. The current ship's position and its antenna position were determined by regular satellite positioning receiver. Taking into account received data the ice floe drift and the location of transponders' measuring base were evaluated.

AUV starting point coordinates on the surface were recorded. Then the AUV mission starting point at the bottom and the starting point of the onboard positioning system operation respectively were determined based on the LB APS data. During the mission the AUV current path was reckoned based on readings of the absolute velocity meter, the course indicator, the depth gauge, the heel sensor and the pitch sensor installed onboard.

Based on the telemetry data transmitted from the AUV via acoustic communication link the AUV motion path was monitored onboard the carrier ship in real time. Navigational plot simultaneously displayed the drift path of the carrier ship with the base of transponders and AUV motion path in respect of drifting transponders' base (fig. 7). The reckoning system's resultant error was corrected by a series of discrete points where the AUV position was calculated based upon the LB APS data using the refined coordinates of the transponders.

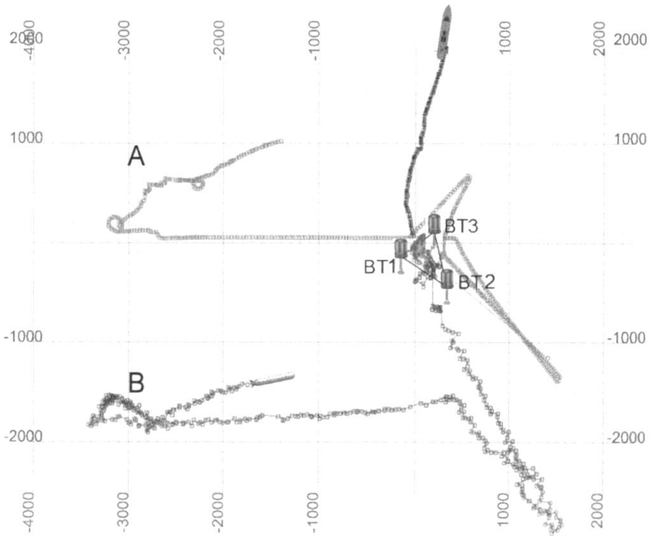

Fig. 7. AUV motion path displayed on the navigational plot: A - in respect of the bottom according to the onboard positioning system data; B - in respect of drifting transponders' base according to the LB APS data.

After its mission had been accomplished, the AUV performed automatic location of the shipboard acoustic antenna module. At the final homing stage before its ascent the AUV position in the ice opening was controlled using the vehicle distance from the antenna module and each transponder. Commands for the last procedures of ascent (ascent from the depth of 20 m and then 5 m) were sent when the AUV was in the nearest position to the shipboard antenna (no more than 20-25 m) and in the center of the ice opening (determined according to the distance from AUV to the transponders).

4.4 The results of the research

The expedition performed operations on the Lomonosov Ridge in the area of the point with the coordinates 84°40' N and 149°10' E at the condition of ice cover approximately 9.5 points (solid ice cover with single rare ice openings sized up to 100 m.) and with the speed of drift of ice floe up to 0.5 knots. Firstly, a trial AUV descent at the depth up to 100 m for ballasting and checking system operation was carried out. The received results allowed coming to a decision about deep-water launches.

Two operational descents with echo-ranging survey of the bottom, environment measurements, acoustic profiling and photographing of separate bottom areas were performed. During operational launches AUV position was controlled onboard the carrier-ship with AUV motion path displayed in real time and presenting AUV current condition parameters - coordinates, speed, course, depth, height, and direct distance from shipboard antenna.

The navigation scheme and technique described above enabled AUV positioning and control, monitoring its mission accomplishment, and ensuring the vehicle's precise arrival at the ice opening for ascent. At the final stage of AUV mission - ascent after 22 hour of self-contained operation - the control of vehicle direct distance from carrier ship antenna and installed transponders was provided. Range measurement error didn't exceed 10 m at that moment, and when appearing on the surface AUV was in 10-15 m from the board of the carrier ship and in 20-30 m from its antenna.

Analyzing available data positioning accuracy can be approximately evaluated. During the 22-hour long launch cumulative uncorrected error of the onboard positioning system, which was defined as deviation between the ascent point coordinates determined by the onboard positioning system and the coordinates obtained during GPS observation, was equal to 1,370 m or approximately 60 m/hour. This error had been accumulated and formed from the following sources:

- error of geographical coordinates for the mission starting point at the bottom. AUV starting point coordinates on the surface were determined rather accurately, but during descent (approximately 50 min) AUV moved along complicated path, and its location was monitored by APS using drifting transponders' base. Estimated position of the starting point according to APS was corrected by compensation of transponders' base drifting with error approximately 50 m.
- dead-reckoning error of the onboard positioning system. According to the results of the experiments carried out during system debugging the cumulative reckoning error was less than 1% of traversed path. It comprises less than 50 m/h at speed 1 m/s.
- dead-reckoning error during AUV ascent and homing at depths excluding Doppler log efficiency. Vehicle speed data was worked out by water speed log, and its accuracy is essentially lower than that of a Doppler log. Total operation time of the reckoning system in a homing mode was at least 3 hours, it also influenced cumulative error.

Evaluation listed above is not final, as accepted configuration of navigation facilities has additional possibilities of correcting reckoned coordinates and considerably reducing positioning error. Reduction of error is achieved by means of positioning separate points of reckoned path to the points calculated at this period of time according to LB APS with the use of drifting transponders relocation. Error in determination of coordinates calculated according to the LB APS data can be compared to the relative range measurement error (no more than one percent for the disadvantageous working conditions) and comprises 60 m at the range of 6000 m. Then, as it was mentioned, onboard the carrier ship besides speed and course data necessary for reckoning, the telemetry data on depth and height are received, and direct AUV range from the homing antenna with precise coordinates are continuously controlled. If the vehicle performs rectilinear equal tacks, then drift parameters of the carrier ship and abovementioned basic data allow positioning the vehicle according to changes of range data from the homing antenna using simple mathematical models. These coordinates positioning error comprises about 2% from the current range (for the conditions of carried out operations - about 100 m).

Good positioning facilities of AUV "Klavesin" allowed efficiently perform a number of research operations during deep-water descents under ice in High Latitudes. During abovementioned expedition the following operations were carried out with the help of underwater vehicle:

- bathymetric survey of seabed area equal 50 sq. km,
- echo-ranging survey of seabed surface,
- acoustic profiling,
- strip survey of some seabed areas,
- sea water temperature and electric conductivity measurements.

Let's mention some results of the performed operations.

Bathymetric survey was carried out by means of direct measurements of vehicle descent with the use of depth sensor, and measurements of AUV distance to the bottom with the use of echoranging system. At AUV speed 1 m/s discreteness of the data received comprises 1 m. Bathymetric cumulative error doesn't exceed 3 m. All measurements are made in international reference coordinate system WGS-84. A bathymetric map of the area is made according to the measurement data.

Echo-ranging survey of seabed area was carried out with the help of low-frequency and high-frequency side-scan sonars (LF SSS and HF SSS). A combined SSS-image (plot) of operation area and separate high-resolution fragments of bottom and biological payloads were received. The results of SSS-survey illustrate the character of seabed and bottom objects of different nature.

Seabed acoustic profiling was performed during vehicle motion at 30 m from the bottom. The swath was approximately 30 m, profiling depth 30-50 m. Geological structure of deep-sea and sediment layers were explored. It allowed evaluating morphological characteristics of the bottom structure.

Hydrologic research included sea water temperature and electric conductivity measurements. This data was used for sound velocity calculation. The character of temperature dependences on depth and formation of vertical distribution of sound velocity is detected. Vertical temperature profiles, electric conductivity and sound velocity profiles as well as map of near bottom temperature field were made on the basis of these measurements.

Seabed photo survey was carried out at 0.75÷5.1 m. Photos of many biological payloads sheltered in silt with exit openings are of a great interest.

5. Conclusion

1. An autonomous unmanned underwater vehicle for scientific research was used for the first time in the world history under ice in the Arctic polar latitudes. The possibility of its use for bottom characteristics research was practically proved.

2. As a result of the research the unique information about the seabed characteristics, which cannot be accessed using any other equipment was obtained. Based on the obtained data a bathymetric map and a sonar image plot of the explored seabed area were composed. Acoustic sounding bottom profiles, vertical temperature, electric conductivity and sound velocity profiles were generated.
3. The materials gained during the expedition can be of a scientific interest for the maritime law, marine biology, geology, and marine science specialists.

6. Acknowledgments

The authors thanks IMTP FEB RAS members - all those who took part in development and trials of AUV positioning complex as well as the colleagues from the institutions who took part in organization and complex testing of AUV systems. Especially authors would like to thank A. Pavin whose materials were used during preparation of the paper.

7. References

Ageev, M.; Kiselyov, L.; Matviyenko, Yu. et al. (2005). *Autonomous Underwater Vehicles. Systems and Technologies* (in Russian), Ed. Acad. Ageev M. D. (Moscow: Nauka, 2005). - 398 p.

Inzartsev, A.; Kamorniy, A.; Lvov, O.; Matviyenko, Yu. & Rylov N. (2007a). AUV Application For Scientific Research In Arctic Zone (in Russian). *Underwater Research and Robotics*, 2007, № 2. - P. 5-14.

Inzartsev, A.; Kiselyov, L.; Matviyenko, Yu. & Rylov, N. (2007b). Actual Problems of Navigation and Control at Creation of Autonomous Underwater Vehicles. *Proceedings of International Conference on Subsea Technologies* (SubSeaTech'2007), June 25-28, 2007, St.Petersburg, Russia, ISBN 5-88303-409-8.

Kiselyov, L.; Inzartsev, A.; Matviyenko, Yu.; Vaulin, Yu. et al. (2004). Underwater Navigation, Control and Orientation (in Russian). *Mechatronics, Automation, Control*, 2004, № 5. - P. 23-28.

Maridan AUV, web-site: www.maridan.atlas-elektronik.com

Romeo, J. & Lester, G. (2001). Navigation is Key to AUV Missions. *Sea Technology,* 2001, Vol. 42, № 12, P.24-29.

Theseus AUV, web-site: www.ise.bc.ca/theseus.html

12

Intelligent Flight Control of an Autonomous Quadrotor

Syed Ali Raza and Wail Gueaieb
University of Ottawa, Canada

1. Introduction

This chapter describes the different steps of designing, building, simulating, and testing an intelligent flight control module for an increasingly popular unmanned aerial vehicle (UAV), known as a quadrotor. It presents an in-depth view of the modeling of the kinematics, dynamics, and control of such an interesting UAV. A quadrotor offers a challenging control problem due to its highly unstable nature. An effective control methodology is therefore needed for such a unique airborne vehicle.

The chapter starts with a brief overview on the quadrotor's background and its applications, in light of its advantages. Comparisons with other UAVs are made to emphasize the versatile capabilities of this special design. For a better understanding of the vehicle's behavior, the quadrotor's kinematics and dynamics are then detailed. This yields the equations of motion, which are used later as a guideline for developing the proposed intelligent flight control scheme.

In this chapter, fuzzy logic is adopted for building the flight controller of the quadrotor. It has been witnessed that fuzzy logic control offers several advantages over certain types of conventional control methods, specifically in dealing with highly nonlinear systems and modeling uncertainties. Two types of fuzzy inference engines are employed in the design of the flight controller, each of which is explained and evaluated.

For testing the designed intelligent flight controller, a simulation environment was first developed. The simulations were made as realistic as possible by incorporating environmental disturbances such as wind gust and the ever-present sensor noise. The proposed controller was then tested on a real test-bed built specifically for this project. Both the simulator and the real quadrotor were later used for conducting different attitude stabilization experiments to evaluate the performance of the proposed control strategy. The controller's performance was also benchmarked against conventional control techniques such as input-output linearization, backstepping and sliding mode control strategies. Conclusions were then drawn based on the conducted experiments and their results.

1.1 Quadrotor background

Louis Bréguet and Jacques Bréguet, two brothers working under the guidance of Professor Charles Richet, were the first to construct a quadrotor, which they named Bréguet Richet Gyroplane No. 1 Breguet-Richet-1907. The first flight demonstration of Gyroplane No. 1

with no control surfaces was achieved on 29 September 1907. Figure 1 shows the huge quadrotor with double layered propellers being prepared for its first manned flight.

Fig. 1. Bréguet Richet Gyroplane No. 1 Rumerman (2002).

Later, two additional designs were developed and experimental flights were conducted. The first, by Georges de Bothezat and Ivan Jerome in 1922, had six-bladed rotors placed at each end of an X-shaped truss structure, as shown in Figure 2.

Fig. 2. Quadrotor designed by George De Bothezat, February 21, 1923 Rumerman (2002).

The second, shown in Figure 3, was built by Étienne Œ hmichen in 1924, and set distance records, including achieving the first kilometer long helicopter flight.

Fig. 3. Œ hmichen quadrotor designed in 1924 Rumerman (2002).

At present, apart from military endeavours, UAVs are also being employed in various commercial and industrial applications. In particular, these include the use of unmanned helicopters for crop dusting or precision farming Sugiura et al. (2003), and microwave autonomous copter systems for geological remote sensing Archer et al. (2004). STARMAC Waslander et al. (2005) is a multi-agent autonomous rotorcraft, which has potential in security-related tasks, such as remote inspections and surveillance. The commercially available quadrotor kit called DraganFlyer Inc. (2008) has become a popular choice for aerial mapping and cinematography.

UAVs are subdivided into two general categories, fixed wing UAVs and rotary wing UAVs. Rotary winged crafts are superior to their fixed wing counterparts in terms of achieving higher degree of freedom, low speed flying, stationary flights, and for indoor usage. A quadrotor, as depicted in Figure 4, is a rotary wing UAV, consisting of four rotors located at the ends of a cross structure. By varying the speeds of each rotor, the flight of the quadrotor is controlled. Quadrotor vehicles possess certain essential characteristics, which highlight their potential for use in search and rescue applications. Characteristics that provide a clear advantage over other flying UAVs include their Vertical Take Off and Landing (VTOL) and hovering capability, as well as their ability to make slow precise movements. There are also definite advantages to having a four rotor based propulsion system, such as a higher payload capacity, and impressive maneuverability, particularly in traversing through an environment with many obstacles, or landing in small areas.

As illustrated by the conceptual diagram in Figure 4, the quadrotor attitude is controlled by varying the rotation speed of each motor. The front rotor (M_f) and back rotor (M_b) pair rotates in a clockwise direction, while the right rotor (M_r) and left rotor (M_l) pair rotates in a counter-clockwise direction. This configuration is devised in order to balance the drag created by each of the spinning rotor pairs. Figure 5 shows the basic four maneuvers that can be accomplished by changing the speeds of the four rotors. By changing the relative speed of the right and left rotors, the roll angle of the quadrotor is controlled. Similarly, the pitch angle is controlled by varying the relative speeds of the front and back rotors, and the yaw angle by varying the speeds of clockwise rotating pair and counter-clockwise rotating pair. Increasing or decreasing the speeds of all four rotors simultaneously controls the collective thrust generated by the robot. A roll motion can be achieved while hovering by increasing the speed of the right rotor, while decreasing the speed of the left rotor by the same amount. Hence, the overall thrust is kept constant.

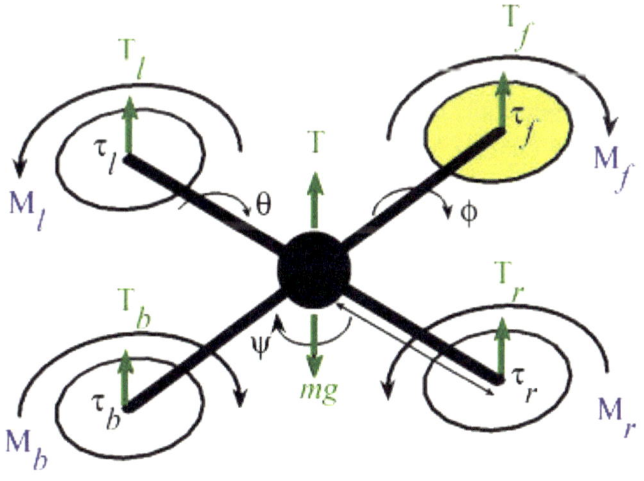

Fig. 4. Conceptual diagram of a quadrotor.

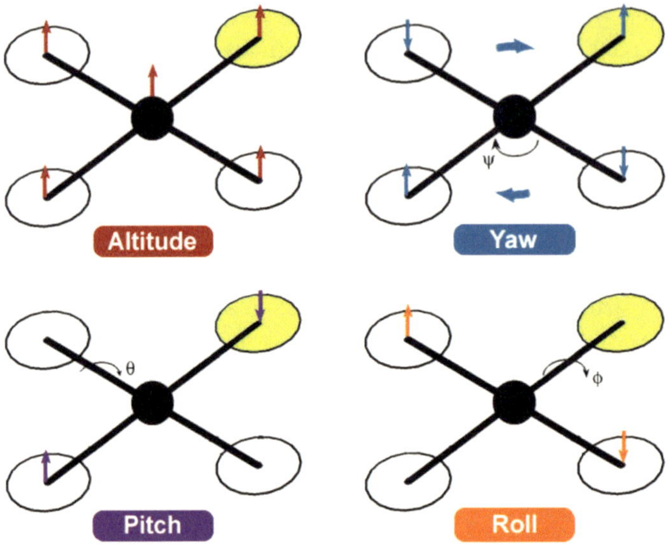

Fig. 5. Quadrotor dynamics.

In the past few years, much research has already been conducted on the modeling and control of quadrotors. Many control techniques, as summarized in Table 1, are proposed in the literature, however, excluding STARMAC, their primary focus is mostly for indoor flight control and therefore do not account for uncertainties and external disturbances. Lyapunov stability theory is used for stabilization and control of the quadrotor in Bouabdallah et al. (2004a) and Dzul et al. (2004). Conventional PD^2 feedback and PID structures are used for simpler implementation of control laws, and comparison with LQR based optimal control theory is presented in Tayebi and McGilvray (2006) and Bouabdallah et al. (2004b). Backstepping control is also proposed with the drawback of higher computational loads in Guenard et al. (2005). Visual feedback is applied in many cases, using onboard or offboard cameras for pose estimation by Altug et al. (2002) and Guenard et al. (2008). Fuzzy logic control techniques have also been proposed Coza and Macnab (2006), along with neural networks Tarbouchi et al. (2004) and reinforcement learning Waslander et al. (2005).

Many quadrotor test-beds have been constructed in different research projects, where simulators are also developed for testing the control laws beforehand. In Kivrak (2006), LQR is used for attitude stabilization of a commercially available Draganflyer Vti quadrotor model in MATLAB Simulink. In another project, the modeling, design, and control of a Miniature Flying Robot (MFR), named OS4 was accomplished Bouabdallah (2007), where a mathematical model was developed for the simulation and control of a mini quadrotor using linear and nonlinear control methods.

2. Quadrotor's kinematics and dynamics

Mathematical modelling provides a description of the behaviour of a system. The flight behaviour of a quadrotor is determined by the speeds of each of the four motors, as they vary in concert, or in opposition with each other. Hence, based on its inputs, a mathematical representation of the system can be used to predict the position and orientation of the quadrotor. The same can further be used to develop a control strategy, whereby manipulating the speeds of individual motors results in achieving the desired motion.

Project	Control Technique	Picture
STARMAC, Stanford University, 2005 Waslander et al. (2005)	Reinforcement Lrn.	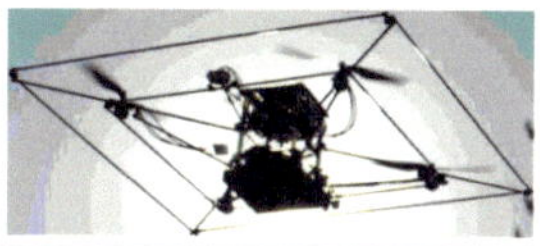
OS4, EPFL, December 2006 Bouabdallah (2007)	Backstepping	
Pennsylvania State University, Hanford, 2005 Hanford (2005)	PI	
Helio-copter, Brigham Young University, Fowers, 2008 Fowers (2008)	Visual feedback	
HMX-4, Pennsylvania State University, 2002 Altug et al. (2002)	Feedback Lin.	
Quad-Rotor UAV, University of British Columbia Chen and Huzmezan (2003)	MBPC and $H\infty$	
Quad-Rotor Flying Robot, Universiti Teknologi Malaysia Weng and Shukri (2006)	PID	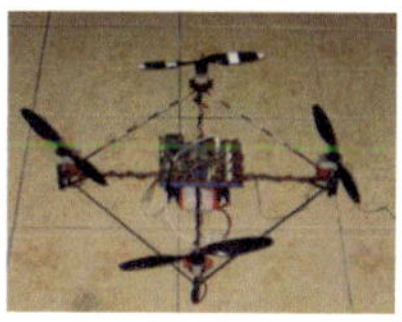

Table 1. Quadrotor flight control techniques used in various projects.

To derive the full mathematical model of the quadrotor, we need to define its kinematics and dynamics first. The kinematic equations provide a relation between the vehicle's position and velocity, whereas the dynamic model defines the relation governing the applied forces and the resulting accelerations.

2.1 Reference frames

Before getting into the equations of kinematics and dynamics of the quadrotor, it is necessary to specify the adopted coordinate systems and frames of reference, as well as how transformations between the different coordinate systems are carried out.

The use of different coordinate frames is essential for identifying the location and attitude of the quadrotor in six degrees of freedom (6 DOF). For example, in order to evaluate the equations of motion, a coordinate frame attached to the quadrotor is required. However, the forces and moments acting on the quadrotor, along with the inertial measurement unit (IMU) sensor values, are evaluated with reference to the body frame. Finally, the position and speed of the quadrotor are evaluated using GPS measurements with respect to an inertial frame located at the base station.

Thus, three main frames of reference are adopted, as shown in Figure 6:

1. The inertial frame, $\mathcal{F}_i = (\vec{x_i}, \vec{y_i}, \vec{z_i})$, is an earth-fixed coordinate system with the origin located on the ground, for example, at the base station. By convention, the x-axis points towards the north, the y-axis points towards the east, and the z-axis points towards the center of the earth.
2. The body frame $\mathcal{F}_b = (\vec{x_b}, \vec{y_b}, \vec{z_b})$, with its origin located at the center of gravity (COG) of the quadrotor, and its axes aligned with the quadrotor structure such that the x-axis $\vec{x_b}$ is along the arm with front motor, the y-axis $\vec{y_b}$ is along the arm with right motor, and the z-axis $\vec{z_b} = \vec{x_b} \times \vec{y_b}$, where 'x ' denotes the cross product.
3. The vehicle frame, $\mathcal{F}_v = (\vec{x_v}, \vec{y_v}, \vec{z_v})$, is the inertial frame with the origin located at the COG of the quadrotor. The vehicle frame has two variations, $\mathcal{F}_\phi$ and $\mathcal{F}_\theta$. $\mathcal{F}_\phi$ is the vehicle frame, $\mathcal{F}_v$, rotated about its z-axis $\vec{z_v}$ by an angle ψ so that $\vec{x_v}$ and $\vec{y_v}$ are aligned with $\vec{x_b}$ and $\vec{y_b}$, respectively. $\mathcal{F}_\theta$ is frame $\mathcal{F}_\phi$ rotated about its y-axis, $\vec{y_\phi}$, by a pitching angle, θ, such that $\vec{x_\phi}$ and $\vec{z_\phi}$ are aligned with $\vec{x_b}$ and $\vec{z_b}$, respectively.

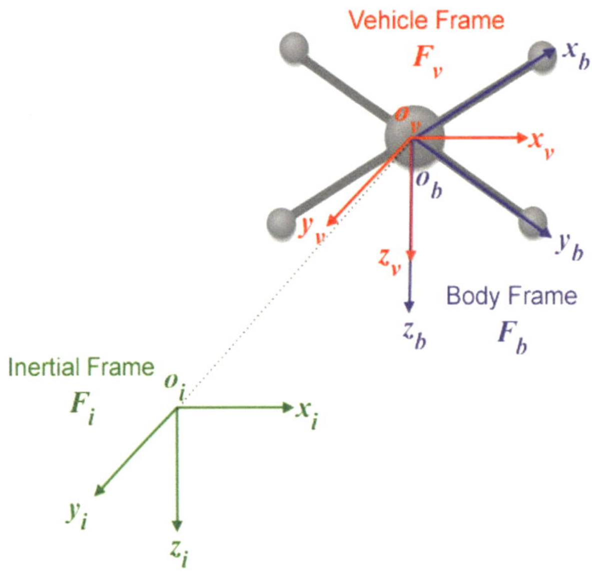

Fig. 6. The inertial, body and vehicle frames of reference.

Translation and rotation matrices are used to transform one coordinate reference frame into another desired frame of reference. For example, the transformation from $\mathcal{F}_i$ to $\mathcal{F}_v$ provides the displacement vector from the origin of the inertial frame to the center of gravity (COG) of the quadrotor. Also, the transformation from $\mathcal{F}_v$ to $\mathcal{F}_b$ is rotational in nature, therefore yielding the roll, pitch and yaw angles.

2.2 Quadrotor's kinematics

Let $P_{\mathcal{F}}^T = [p_x, p_y, -p_z]$ and $\Omega_{\mathcal{F}}^T = [\phi, \theta, \psi]$ denote the quadrotor's position and orientation within a given frame $\mathcal{F}$. The relation between the quadrotor's speed in the three predefined frames is expressed as

$$\begin{bmatrix} \dot{p}_x \\ \dot{p}_y \\ -\dot{p}_z \end{bmatrix}_{\mathcal{F}_i} = \begin{bmatrix} \dot{p}_x \\ \dot{p}_y \\ \dot{p}_z \end{bmatrix}_{\mathcal{F}_v} = \left[R_{\mathcal{F}_v}^{\mathcal{F}_b}\right]^T \begin{bmatrix} \dot{p}_x \\ \dot{p}_y \\ \dot{p}_z \end{bmatrix}_{\mathcal{F}_b} \tag{1}$$

Where $\left[R_{\mathcal{F}_v}^{\mathcal{F}_b}\right]^T \in \mathbb{R}^{3\times 3}$ is the rotation matrix that maps frame $\mathcal{F}_b$ to frame $\mathcal{F}_v$ and is defined by

$$\left[R_{\mathcal{F}_v}^{\mathcal{F}_b}\right]^T = \begin{bmatrix} c\theta c\psi & s\phi s\theta c\psi - c\phi s\psi & c\phi s\theta c\psi + s\phi s\psi \\ c\theta s\psi & s\phi s\theta s\psi + c\phi c\psi & c\phi s\theta s\psi - s\phi c\psi \\ -s\theta & s\phi c\theta & c\phi c\theta \end{bmatrix}$$

with $s\theta = \sin\theta$ and $c\theta = \cos\theta$. The same notation applies for $s\phi$, $c\phi$, $s\psi$, and $c\psi$.

The rotational motion relationship can therefore be derived using the appropriate state variables, such as the vehicle frame angles (ϕ, θ, and ψ) and the body frame angular rate ($\dot{\phi}$, $\dot{\theta}$, and $\dot{\psi}$). However, in order to do so, these variables need to be brought into one common frame of reference. Using rotation matrices to transform vehicle frames $\mathcal{F}_\phi$, $\mathcal{F}_\theta$, and $\mathcal{F}_v$ into the body frame of reference $\mathcal{F}_b$, we get

$$\begin{bmatrix} \dot{\phi} \\ \dot{\theta} \\ \dot{\psi} \end{bmatrix}_{\mathcal{F}_b} = R_{\mathcal{F}_\phi}^{\mathcal{F}_b}(\phi) \begin{bmatrix} \dot{\phi} \\ 0 \\ 0 \end{bmatrix} + R_{\mathcal{F}_\phi}^{\mathcal{F}_b}(\phi)\, R_{\mathcal{F}_\theta}^{\mathcal{F}_\phi}(\theta) \begin{bmatrix} 0 \\ \dot{\theta} \\ 0 \end{bmatrix} + R_{\mathcal{F}_\phi}^{\mathcal{F}_b}(\phi)\, R_{\mathcal{F}_\theta}^{\mathcal{F}_\phi}(\theta)\, R_{\mathcal{F}_v}^{\mathcal{F}_\theta}(\psi) \begin{bmatrix} 0 \\ 0 \\ \dot{\psi} \end{bmatrix}$$

where $R_{\mathcal{F}_\theta}^{\mathcal{F}_\phi} = \begin{bmatrix} 1 & 0 & 0 \\ 0 & c\phi & s\phi \\ 0 & -s\phi & c\phi \end{bmatrix}$, $R_{\mathcal{F}_v}^{\mathcal{F}_\theta} = \begin{bmatrix} c\theta & 0 & -s\theta \\ 0 & 1 & 0 \\ s\theta & 0 & c\theta \end{bmatrix}$ and $R_{\mathcal{F}_\phi}^{\mathcal{F}_b}(\phi) = R_{\mathcal{F}_\theta}^{\mathcal{F}_b}(\theta) = R_{\mathcal{F}_\psi}^{\mathcal{F}_b}(\psi) = I$.

Therefore,

$$\begin{bmatrix} \dot{\phi} \\ \dot{\theta} \\ \dot{\psi} \end{bmatrix}_{\mathcal{F}_b} = \begin{bmatrix} 1 & 0 & -s\theta \\ 0 & c\phi & s\phi c\theta \\ 0 & -s\phi & c\phi c\theta \end{bmatrix} \begin{bmatrix} \dot{\phi} \\ \dot{\theta} \\ \dot{\psi} \end{bmatrix}_{\mathcal{F}_v}$$

It follows that,

$$\begin{bmatrix} \dot{\phi} \\ \dot{\theta} \\ \dot{\psi} \end{bmatrix}_{\mathcal{F}_v} = \begin{bmatrix} 1 & s\phi\tan\theta & c\phi\tan\theta \\ 0 & c\phi & -s\phi \\ 0 & s\phi\sec\theta & c\phi\sec\theta \end{bmatrix} \begin{bmatrix} \dot{\phi} \\ \dot{\theta} \\ \dot{\psi} \end{bmatrix}_{\mathcal{F}_b} \tag{2}$$

Equations (1) and (2) represent the quadrotor's equations of motion.

2.3 Quadrotor's dynamics

To build the dynamic model of the quadrotor we will use Newton-Euler formalism, while adopting the following assumptions:

1. The quadrotor structure is a rigid body.
2. The quadrotor frame is symmetrical.

3. The COG of the quadrotor coincides with the center of the rigid frame.

The moment of inertia is calculated by assuming the quadrotor as a central sphere of radius r and mass M_o surrounded by four point masses representing the motors. Each motor is supposed to have a mass m and attached to the central sphere through an arm of length l, as shown in Figure 7.

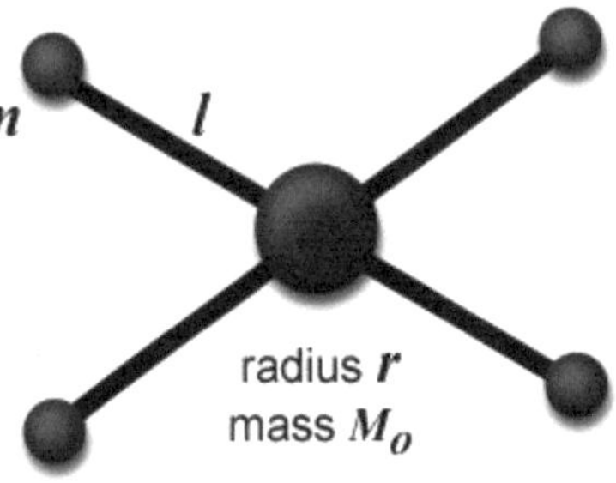

Fig. 7. Moment of inertia.

Due to the symmetry of the quadrotor about all three axes, its inertial matrix becomes symmetrical and is defined by

$$J = \begin{bmatrix} j_x & 0 & 0 \\ 0 & j_y & 0 \\ 0 & 0 & j_z \end{bmatrix}$$

where $j_x = j_y = j_z = \frac{2M_0 r^2}{5} + 2l^2 m$.

The dynamics of the quadrotor under external forces applied to its COG and expressed in the body frame is derived by applying Newton-Euler formulation Beard (2008)

$$\begin{bmatrix} MI_{3\times 3} & 0 \\ 0 & I_{3\times 3} \end{bmatrix} \begin{bmatrix} \ddot{P}_{\mathcal{F}_b} \\ \ddot{\Omega}_{\mathcal{F}_b} \end{bmatrix} + \begin{bmatrix} \dot{\Omega}_{\mathcal{F}_b} \times M\dot{P}_{\mathcal{F}_b} \\ \dot{\Omega}_{\mathcal{F}_b} \times J\dot{\Omega}_{\mathcal{F}_b} \end{bmatrix} = \begin{bmatrix} F_{\mathcal{F}_b} \\ \tau_{\mathcal{F}_b} \end{bmatrix}$$

where M is the quadrotor's total mass, and $F^T = [f_x\ f_y\ f_z]$ and $\tau^T = [\tau_\phi\ \tau_\theta\ \tau_\psi]$ are the external force and torque vectors applied on the quadrotor's COG. The terms τ_ϕ, τ_θ, and τ_ψ are the roll, pitch and yaw torques respectively.

Thus, the translational dynamic model can be written as

$$\begin{bmatrix} \ddot{p}_x \\ \ddot{p}_y \\ \ddot{p}_z \end{bmatrix}_{\mathcal{F}_b} = \begin{bmatrix} \dot{\psi}\dot{p}_y - \dot{\theta}\dot{p}_z \\ \dot{\phi}\dot{p}_z - \dot{\psi}\dot{p}_x \\ \dot{\theta}\dot{p}_x - \dot{\phi}\dot{p}_y \end{bmatrix}_{\mathcal{F}_b} + \frac{1}{M} \begin{bmatrix} f_x \\ f_y \\ f_z \end{bmatrix}_{\mathcal{F}_b}$$

while the rotational model is

$$\begin{bmatrix} \ddot{\phi} \\ \ddot{\theta} \\ \ddot{\psi} \end{bmatrix} = J^{-1} \left\{ \begin{bmatrix} 0 & \dot{\psi} & -\dot{\theta} \\ -\dot{\psi} & 0 & \dot{\phi} \\ \dot{\theta} & -\dot{\phi} & 0 \end{bmatrix} J \begin{bmatrix} \dot{\phi} \\ \dot{\theta} \\ \dot{\psi} \end{bmatrix} + \begin{bmatrix} \tau_\phi \\ \tau_\theta \\ \tau_\psi \end{bmatrix} \right\}$$

$$= \begin{bmatrix} \frac{j_y - j_z}{j_x}\dot{\theta}\dot{\psi} \\ \frac{j_z - j_x}{j_y}\dot{\phi}\dot{\psi} \\ \frac{j_x - j_y}{j_z}\dot{\phi}\dot{\theta} \end{bmatrix}_{\mathcal{F}_b} + \begin{bmatrix} \frac{1}{j_x}\tau_\phi \\ \frac{1}{j_y}\tau_\theta \\ \frac{1}{j_z}\tau_\psi \end{bmatrix}_{\mathcal{F}_b}$$

2.4 Aerodynamic forces and torques

With the derived kinematic and dynamic model, we will now define the forces and torques acting on the quadrotor. The forces include the aerodynamic lift generated by each rotor, and the gravitational pull acting in counter to the total lift generated. The moments are the torques generated in order to achieve the roll, pitch and yaw movements. The following forces and torques are produced:

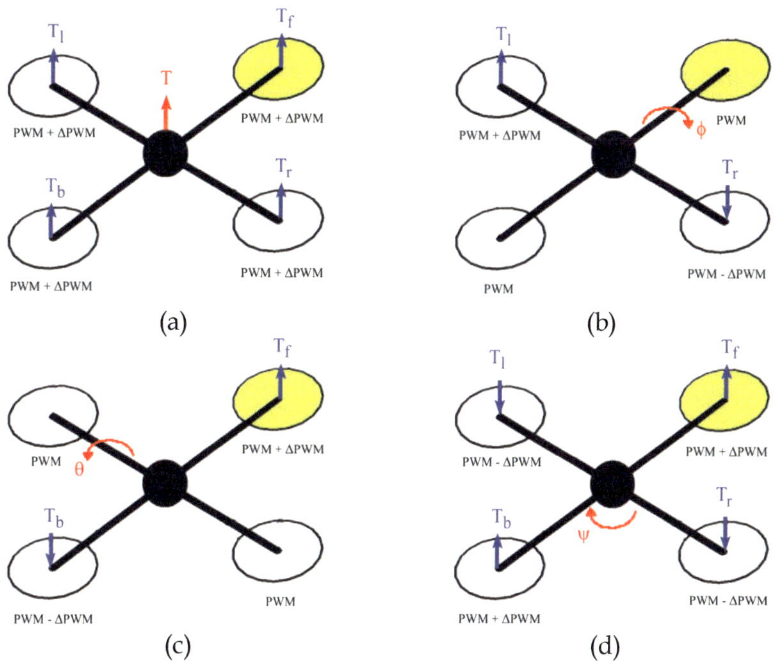

Fig. 8. Forces and moments acting on the quadrotor: (a) Quadrotor thrust; (b) Rolling torque; (c) Pitching torque; and (d) Yawing torque.

Upward Force (Thrust): The total quadrotor thrust is the sum of the thrust produced by each propeller, as depicted in Figure 8(a):

$$T = T_f + T_r + T_b + T_l$$

Rolling Torque: This is the torque produced by increasing the left rotor's thrust while decreasing that of the right rotor, or vice versa, as shown in Figure 8(b):

$$\tau_\phi = l(T_l - T_r)$$

Pitching Torque: The pitching torque in Figure 8(c) is produced by increasing the front rotor's thrust while decreasing that of the back rotor, or vice versa:

$$\tau_\theta = l(T_f - T_b)$$

Yawing Torque: The yawing torque is the result of all four individual torques generated due to the spinning rotors. The front and back rotors spin in the clockwise direction, while the left and right rotors spin in the counterclockwise direction. As shown in Figure 8(d), an imbalance between these two pairs results in a yawing torque causing the quadrotor to rotate about its z-axis:

$$\tau_\psi = \tau_f + \tau_b - \tau_r - \tau_l$$

Gravitational Force (weight): Along with the other forces, the gravitational force acts on the COG of the quadrotor. In the vehicle frame this force is expressed as

$$W_{\mathcal{F}_v} = \begin{bmatrix} 0 \\ 0 \\ Mg \end{bmatrix}$$

with g being the gravitational constant. Therefore, in the body frame, the weight can be written as

$$W_{\mathcal{F}_b} = R_{\mathcal{F}_v}^{\mathcal{F}_b} \begin{bmatrix} 0 \\ 0 \\ Mg \end{bmatrix} = \begin{bmatrix} -Mgs\theta \\ Mgc\theta s\phi \\ Mgc\theta c\phi \end{bmatrix}$$

Including the forces and torques acting on the system, the equations of motion become as defined below.

$$\begin{bmatrix} \ddot{p}_x \\ \ddot{p}_y \\ \ddot{p}_z \end{bmatrix}_{\mathcal{F}_b} = \begin{bmatrix} \dot{\psi}\dot{p}_y - \dot{\theta}\dot{p}_z \\ \dot{\phi}\dot{p}_z - \dot{\psi}\dot{p}_x \\ \dot{\theta}\dot{p}_x - \dot{\phi}\dot{p}_y \end{bmatrix} + \begin{bmatrix} -gs\theta \\ gc\theta s\phi \\ gc\theta c\phi \end{bmatrix} + \begin{bmatrix} 0 \\ 0 \\ \frac{-f_z}{M} \end{bmatrix}$$

$$\begin{bmatrix} \ddot{\phi} \\ \ddot{\theta} \\ \ddot{\psi} \end{bmatrix}_{\mathcal{F}_b} = \begin{bmatrix} \frac{j_y - j_z}{j_x}\dot{\theta}\dot{\psi} \\ \frac{j_z - j_x}{j_y}\dot{\phi}\dot{\psi} \\ \frac{j_x - j_y}{j_z}\dot{\phi}\dot{\theta} \end{bmatrix} + \begin{bmatrix} \frac{1}{j_x}\tau_\phi \\ \frac{1}{j_y}\tau_\theta \\ \frac{1}{j_z}\tau_\psi \end{bmatrix}$$

3. Flight controller design

This section details the development of a fuzzy logic flight controller for the quadrotor. A generalized overview of fuzzy logic control and the advantages it offers for nonlinear control applications are presented. Based on the dynamics and kinematics derived in the previous section, the autonomous flight control strategy is thereby introduced. The proposed fuzzy logic controller is implemented with two types of inference engines for comparison.

3.1 Fuzzy logic control

Since its inception in Zadeh (1965), fuzzy logic has been applied to various fields of engineering, manufacturing, business, and medicine, among others. Within the area of engineering, control systems offer significant applications for fuzzy logic, designated as fuzzy logic control. Before getting into details with regards to fuzzy logic control, we would first like to provide some basic facts about fuzzy systems.

Fuzzy logic control offers a great advantage over some conventional control methods which heavily depend on the exact mathematical model of the control system, specifically in dealing with nonlinear systems subjected to various types of uncertainties. Being independent of the plant's parameters sets fuzzy controllers apart from their conventional counterparts. Fuzzy controllers in general can be designed intuitively in light of the knowledge acquired on the behavior of the system in hand. This knowledge is often gained through experience and common sense, regardless of the mathematical model of the dynamics governing this behavior. For example, in learning how to ride a bike, humans try to build a set of common sense rules and learn from their failures without paying any attention to the dynamic model of the bike. Fuzzy logic control tries to mimic this type of human-like reasoning and embrace it within a pre-defined mathematical model to automate the control of complex systems characterized by ill-defined mathematical models, for example.

3.2 Flight control algorithm

The quadrotor is an under-actuated system with four actuators controlling its six degrees-of-freedom position/orientation. The flight controller is responsible for achieving two challenging goals simultaneously: (i) controlling the quadrotor's position, while (ii) stabilizing its attitude, i.e., orientation (roll, pitch and yaw angles). More specifically, given a desired position (p_x, p_y, p_z) and yaw angle ψ, the goal is to design a controller to force these control states to converge to their respective desired values, while maintaining the pitch and roll angles as close to zero as possible.

Let PWM_{mot} denote the PWM value of motor $mot \in \{f, r, b, l\}$ for the front, right, back, and left motors, respectively. Then, the thrust and torque applied on the quadrotor by motor mot can be expressed as

$$T_{mot} = K_T \times \text{PWM}_{mot}$$
$$\tau_{mot} = K_\tau \times \text{PWM}_{mot}$$

where K_T and K_τ are motor-dependent parameters. This yields

$$\begin{bmatrix} \text{PWM}_f \\ \text{PWM}_r \\ \text{PWM}_b \\ \text{PWM}_l \end{bmatrix} = G \times \begin{bmatrix} T \\ \tau_\phi \\ \tau_\theta \\ \tau_\psi \end{bmatrix}$$

with

$$G = \begin{bmatrix} K_T & K_T & K_T & K_T \\ 0 & -l \times K_T & 0 & l \times K_T \\ l \times K_T & 0 & -l \times K_T & 0 \\ -K_\tau & K_\tau & -K_\tau & K_\tau \end{bmatrix}^{-1}.$$

The above equations provide a basic understanding of how the angular speed of each motor contribute to the overall thrust and torques exerted on the quadrotor. This knowledge will serve as a guideline in developing a rule base of the direct fuzzy logic controller, as depicted in Figure 9.

Three fuzzy controllers are designed to control the quadrotor's roll (ϕ), pitch (θ) and yaw (ψ) angles, denoted by FLC_ϕ, FLC_θ, and FLC_ψ, respectively, with the former two serving as attitude stabilizers. Three fuzzy controllers, FLC_x, FLC_y and FLC_z, are further designed to control the quadrotor's position. All six fuzzy controllers have identical inputs, (i) the error $e = \tilde{(.)} = (.)_d - (.)$, which is the difference between the desired signal $(.)_d$ and its actual value (.), and (ii) the error rate $\dot{e}$. The first input (error) is normalized to the interval [−1,+1], while the second (error rate) is normalized to the interval [−3,+3].

$$U_x = \mathrm{FLC}_x(\tilde{p}_x, \dot{\tilde{p}}_x) \qquad U_\theta = \mathrm{FLC}_\theta(\tilde{\theta}, \dot{\tilde{\theta}})$$
$$U_y = \mathrm{FLC}_y(\tilde{p}_y, \dot{\tilde{p}}_y) \qquad U_\phi = \mathrm{FLC}_\phi(\tilde{\phi}, \dot{\tilde{\phi}})$$
$$U_z = \mathrm{FLC}_z(\tilde{p}_z, \dot{\tilde{p}}_z) \qquad U_\psi = \mathrm{FLC}_\psi(\tilde{\psi}, \dot{\tilde{\psi}})$$

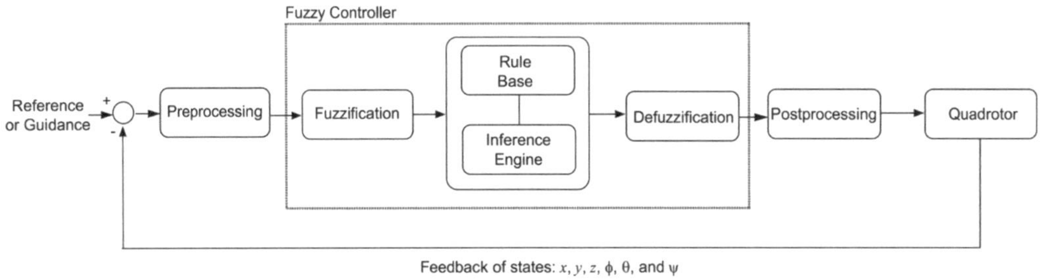

Fig. 9. Control scheme.

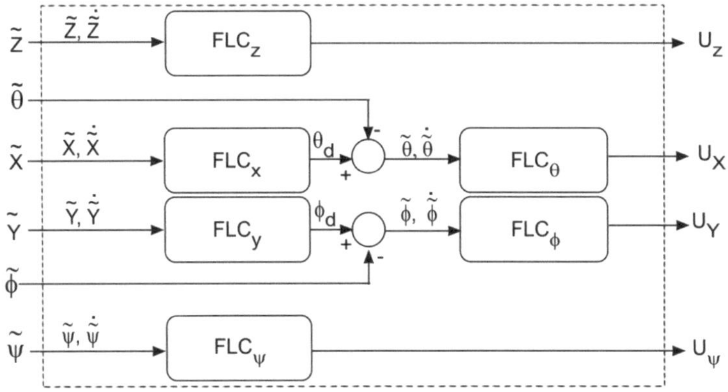

Fig. 10. Flight controller block diagram.

In this control strategy, the desired pitch and roll angles, θ_d and ϕ_d, are not explicitly provided to the controller. Instead, they are continuously anticipated by controllers FLC_x FLC_y in such a way that they stabilize the quadrotor's attitude. The input and output membership functions (Figure 11) of each FLC are tuned empirically and are finalized as follows:

$$\mu_N(e) = \mathrm{trapezoid}(-1, -0.15, 0) \qquad \mu_Z(e) = \mathrm{triangle}(-0.15, 0, 0.15)$$
$$\mu_P(e) = \mathrm{trapezoid}(0, 0.15, 1) \qquad \mu_N(\dot{e}) = \mathrm{trapezoid}(-3, -1.5, 0)$$
$$\mu_Z(\dot{e}) = \mathrm{triangle}(-1.5, 0, 1.5) \qquad \mu_P(\dot{e}) = \mathrm{trapezoid}(0, 1.5, 3)$$
$$\mu_N(U) = \mathrm{trapezoid}(-1, -0.85, 0) \qquad \mu_Z(U) = \mathrm{triangle}(-0.1, 0, 0.1)$$
$$\mu_P(U) = \mathrm{trapezoid}(0, 0.85, 1)$$

A unified rule base comprising nine IF-THEN rules is developed and is presented in Table 2.

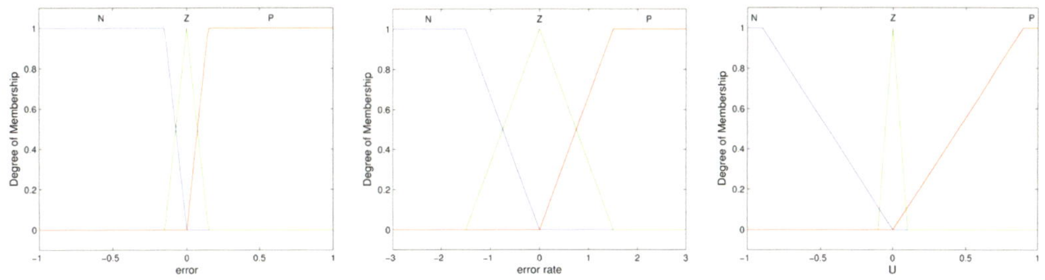

(a) Input variable error e. (b) Input variable error rate $\dot{e}$. (c) Output variable U.

Fig. 11. Input and output membership functions.

$\dot{e}$ \ e	N	Z	P
N	N	N	Z
Z	N	Z	P
P	Z	P	P

Table 2. The rule base of the fuzzy controller.

For it to be modular and independent of the quadrotor's parameters, the fuzzy logic controllers are bounded by pre-processing and post-processing blocks (Figure 9). The pre-processing module calculates the error e and error rate $\dot{e}$ and normalizes them to the intervals [−1,+1] and [−3,+3], respectively. The post-processing block uses the controllers output signals to calculate the PWM value of each motor as follows:

$$\begin{aligned}
\text{PWM}_f &= Sat(U_Z + U_X - U_\psi + \text{Offset}) \\
\text{PWM}_r &= Sat(U_Z + U_Y + U_\psi + \text{Offset}) \\
\text{PWM}_b &= Sat(U_Z - U_X - U_\psi + \text{Offset}) \\
\text{PWM}_l &= Sat(U_Z - U_Y + U_\psi + \text{Offset})
\end{aligned}$$

where 'Offset' is a priori-defined bias to counter balance the weight of the quadrotor. The resultant PWM values are saturated to a maximum threshold that depends on the maximum possible speed of the motors used.

It is important to note that this control scheme does not depend on the kinematic and dynamic equations derived in section 2. Those equations are only used to build the quadrotor's model in the simulator, which would be unnecessary with a real quadrotor. Being independent of the plant's parameters sets the fuzzy controllers apart from conventional control systems, which depend in one way or the other on the plant's mathematical model. The fuzzy controllers are designed in light of the knowledge acquired on the quadrotor's behavior and from its dynamic model. Therefore, changing the quadrotor or some of its physical parameters like the mass and inertia does not require redesigning the fuzzy logic controller. Instead, the postprocessing module may need to be fine-tuned to optimize the controller's performance, such as to calibrate the offset, for instance.

Two different fuzzy inference engines are implemented: (i) a Mamdani, and (ii) a Takagi-Sugeno-Kang (TSK) fuzzy model. The Mamdani fuzzy inference method uses a min-max operator for the aggregation and the centroid of area method for defuzzification. One known problem with this type of controller is the high computational burden associated to it, especially when implemented on an embedded system. To alleviate this problem, a zero order TSK fuzzy inference engine is implemented for comparison. In this model, the output membership functions of the Mamdani fuzzy controller are replaced with three fuzzy singletons $N = -1$, $Z = 0$ and $P = +1$.

4. Numerical and experimental results

To test the proposed fuzzy logic flight controller with both inference engines and study their performances, a simulation environment is first developed. After satisfactory performance results are attained, the controller is implemented on a quadrotor testbed. The details of the quadrotor simulator and the real-world test-bed are presented in the following subsections.

4.1 Simulation results

The quadrotor simulator is implemented in MATLAB Simulink, as shown in Figure 12. The equations of motion derived earlier are used to model the quadrotor. The inputs to the quadrotor are taken as the four PWM speed values of the motors. To make the simulations more realistic, sensory noise and environmental disturbances such as wind are also taken into account. Medium wind gust speeds are generated based on real data from Canada Weather Statistics, Statistics (2009). The wind disturbance is incorporated as two further inputs representing the north and east wind condition. The quadrotor model outputs are the linear and angular accelerations that are integrated twice to obtain the position and orientation vectors. The angular accelerations $\ddot{\phi}$, $\ddot{\theta}$, and $\ddot{\psi}$ are degraded with a white noise and then used as a feedback to the fuzzy controller.

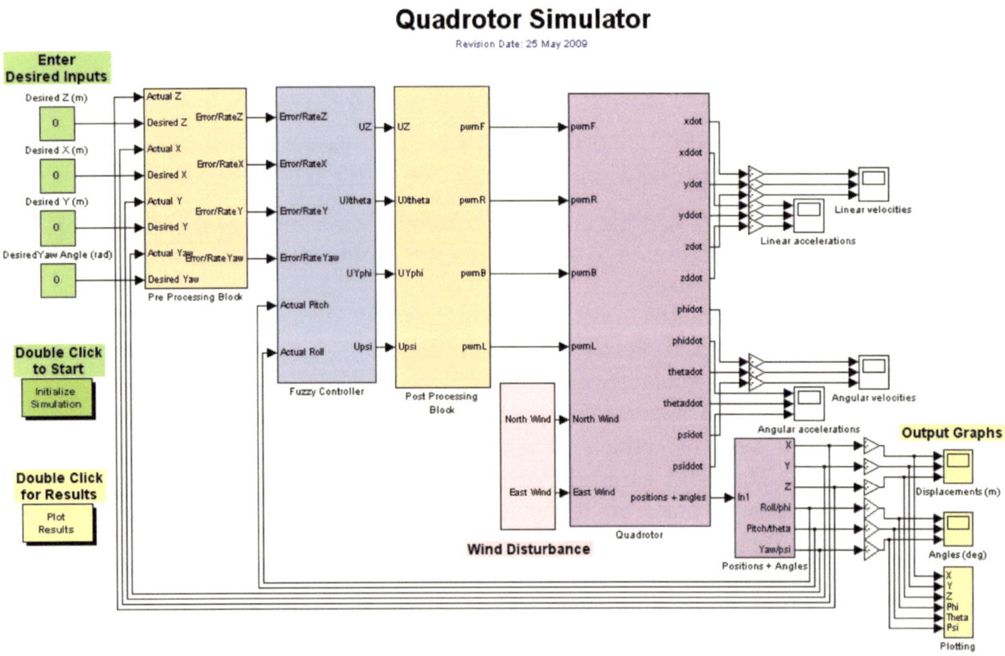

Fig. 12. MATLAB Simulink block diagram of the quadrotor simulator.

The values used for the quadrotor's dynamic parameters are: $M = 0.765$ Kg, $l = 0.61$ m, $J_x = J_y = 0.08615$ Kg·m2, $J_z = 0.1712$ Kg·m2, $K_T = 5.45$, $K_{tau} = 0.0549$, Offset = $Mg/(4K_T) = 0.344$.

The six identical fuzzy controllers are developed using the MATLAB Fuzzy Logic Toolbox. The input and output variables with membership functions described earlier are set accordingly. The fuzzy controllers are used in a configuration as shown in Figure 10. The inputs and outputs of the flight controller are pre- and post-processed, respectively.

The environmental disturbances are introduced such that a white noise is added to the angular accelerations $\ddot{\phi}$, $\ddot{\theta}$, and $\ddot{\psi}$ for emulating the inertial measurement unit (IMU) sensor. The IMU signals are further processed through rate transitions to incorporate the ADC sampling rate.

The user-defined inputs are the desired translatory coordinates $P_{\mathcal{F}_i}$ with respect to the inertial frame, and the desired yaw angle ψ. The desired pitch and roll angles are implicitly set to zero to achieve attitude stabilization.

Three simulations are conducted to test the performance of the proposed fuzzy logic flight controllers with both inference engines. The system's initial states are set to zero, while the desired quadrotor's position and orientation are set to $P^T_{\mathcal{F}_i}$ = [10, 10, 25] m and $\Omega^T_{\mathcal{F}_i}$= [0, 0, 30] degrees in all three simulations. The purpose of the simulation is to assess the performance of the fuzzy logic controller and compare the accuracy of the two fuzzy inference engines under different disturbance conditions. The first simulation is run without any disturbances. In the second simulation, the controller is subjected to sensor noise. In the third simulation, it is subjected to sensor noise and medium north-east wind gust of 10 m/s.

The simulation results presented in Figures 13 and 14, demonstrate the satisfactory performance of the proposed controller despite the presence of sensor noise and wind disturbances. The Mamdani fuzzy controller converges to the desired states relatively faster than its TSK counterpart. The yaw angle drift under wind disturbance is clearly visible with the TSK controller.

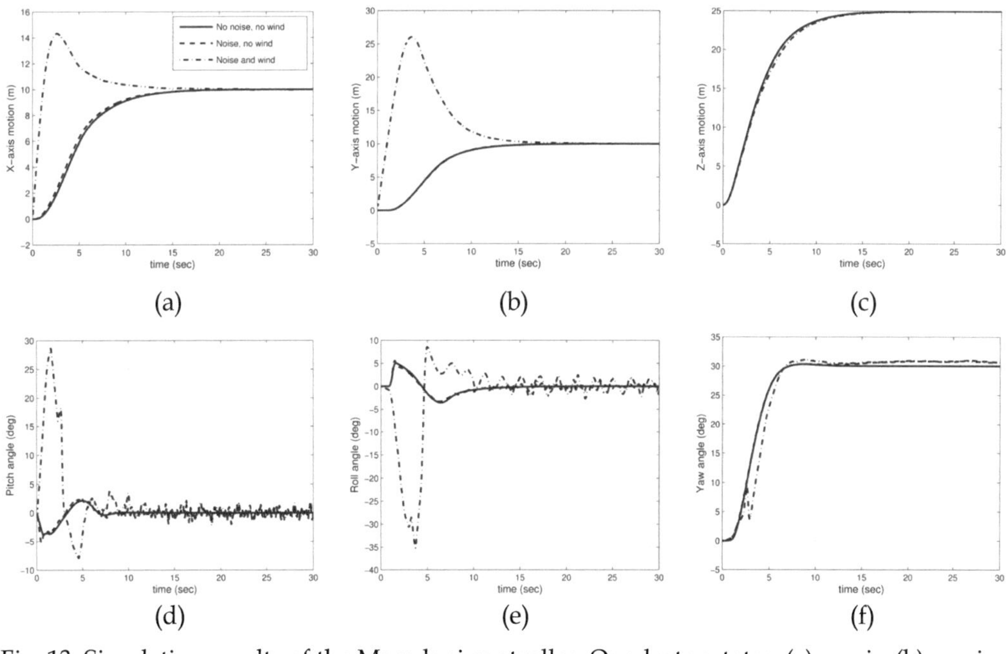

Fig. 13. Simulation results of the Mamdani controller. Quadrotor states: (a) x-axis; (b) y-axis; (c) z-axis (altitude); (d) pitch (θ); (e) roll (ϕ); and (f) yaw (ψ).

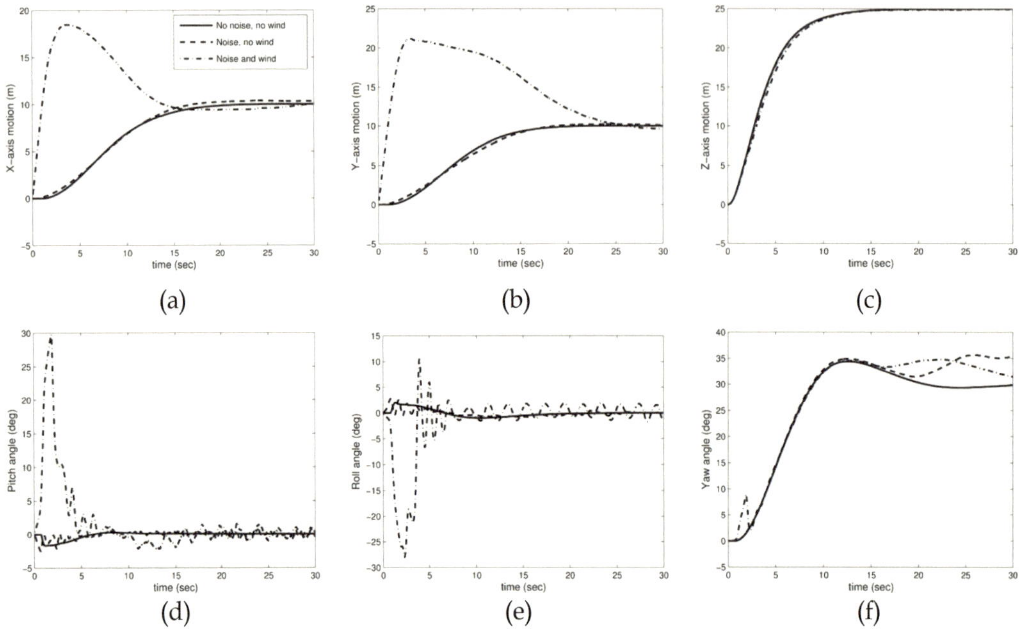

Fig. 14. Simulation results of the TSK controller. Quadrotor states: (a) x-axis; (b) y-axis; (c) z-axis (altitude); (d) pitch (θ); (e) roll (ϕ); and (f) yaw (ψ).

4.2 Quadrotor test-bed and experimental results

The quadrotor test-bed, as shown in Figure 15, comprises light weight carbon fiber rods of length 0.61 m connected to a center piece forming the desired cross structure for the frame. The four propulsion units are made of brushless DC motors connected to electronic speed controllers to provide actuation. Different propellers are also tested to find the optimal thrust to power consumption ratio. The system is powered by a 2100 mAh, 11.1-Volt lithium polymer battery. The quadrotor's specifications are summarized in Table 3.

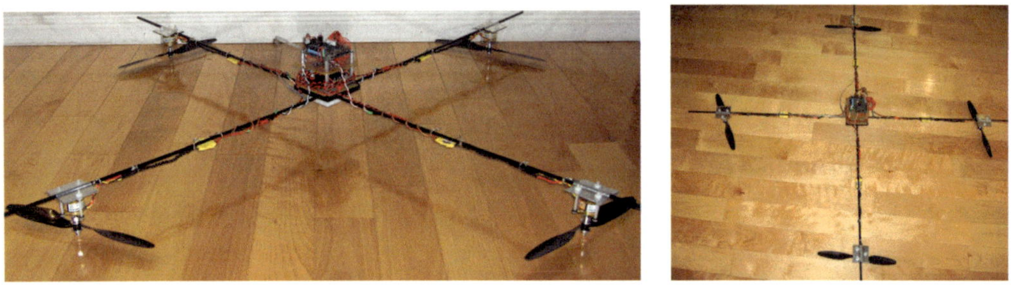

Fig. 15. The quadrotor test-bed.

The IMU is designed using a triple axis 3g accelerometer ADXL330, a dual axis 500 degree-per-sec rate gyroscope IDG300 for pitch and roll angle rates. The yaw angle acceleration is measured using a single axis 300 degree-per-sec ADXRS300 gyroscope. The bandwidths of all the sensors are trimmed to 10 Hz. The MEMS inertial sensors are combined with complementary filtering techniques for the online estimation of the attitude angles.

Component	Specifications	Weight (g)
4 Carbon fiber rods	Frame arms of 0.61m length	52
Center piece	Frame center joint	58
4 Motor mounts	Aircraft grade aluminum sheets	40
4 Brushless DC motors	RIMFIRE 1000kV, 135Watts	164
4 Electronic Speed Controllers	ElectriFly 12A ESC 150Watts	56
4 Composite propellers	Counter rotating sets L=10", P=4.5"	24
4 Propeller adapters	Collet type aluminum 3-5mm	20
Lithium polymer battery	11.1V 2100mAh	150
Power supply board	Hybrid PSU	70
IMU sensor board	ADXL330, IDG300, ADXRS300	44
Axon board	ATmega640	34
Standoffs and screws	Aluminum frame assembly	53
	Total Weight	765

Table 3. The quadrotor test-bed component summary.

The processing is accomplished with a 34-gram Axon board based on ATmega640 microcontroller running at 16 MHz clock. The IMU data is sampled using an onboard 10-bit ADC. The proposed flight control is implemented using the TSK fuzzy controllers converted to a look-up table for a higher computational efficiency. The flight controller, elaborated in Algorithm 1, is set to operate at a bandwidth of 100 Hz.

Algorithm 1 Quadrotor Flight Control Algorithm

Initialize:

UART 0,1,2,3

Analog and Digital Ports

Timer 0,1,2,3,4

ADC clock = 125 KHz

PWM Timers 3,4

Body:

IF (button pressed) THEN Arm the motors.

  Desired Coordinates

  GPS Coordinates

  ADC Gyro and Accelerometer Signals

  IMU Measurement Calibrations

  Complementary Filtering for Attitude Estimates

  Transmit the Attitude Estimates

  Pre-processing: Error and Error Rate Calculation

CALL the 6 Fuzzy Controllers

  Calculate $Ux, Uy, Uz, U\phi, U\theta, U\psi$

  Post-processing: Calculate PWM_f, PWM_b, PWM_r, PWM_l

  Set PWM timers

Repeat

END

The first experiment is designed to test the quadrotor's hovering and attitude stabilization capabilities. So, the desired position $P_{\mathcal{F}_i}$ is pre-defined as the quadrotor's current position. Both, experimental and simulation results are reported in Figure 16. In simulation, the controller managed to keep the pitch and roll angles within the interval of [−3,+4] degrees. However, in reality, the errors of these angles were fluctuating between −8 and +7 degrees for the pitch, and −6 and +12 degrees for the roll. The main difference between the simulation and experimental results stems from the vibration of the frame from which the test-bed was made. In addition, the 0.61-m long carbon fiber arms were bending and twisting when excited by the motors. This put the fuzzy logic controller to a higher challenge than what was originally anticipated. Yet, it was successful in forcing the pitch and roll angles to within an acceptable range.

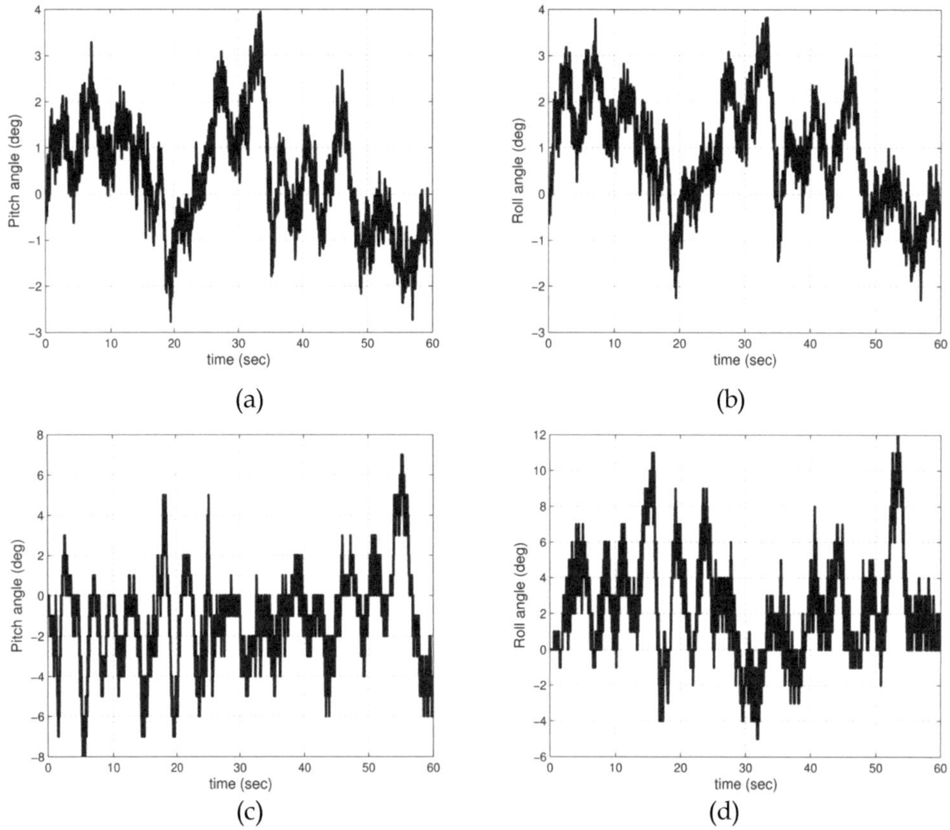

Fig. 16. Simulation and experimental attitude stabilization results: (a) simulator pitch (θ); (b) simulator roll (ϕ); (c) test-bed pitch (θ); and (d) test-bed roll (ϕ).

In the second experiment, the controller was tested under harsher conditions so as to evaluate its behavior if the quadrotor collides into an obstacle, or if it is faced with other types of disturbances. Hence, the previous experiment was repeated, but this time one of the quadrotor's arms is abruptly tapped in the middle of the flight. The results are shown in Figure 17. As can be seen, the controller was able to quickly bring the pitch and roll angle errors back to within a safe range.

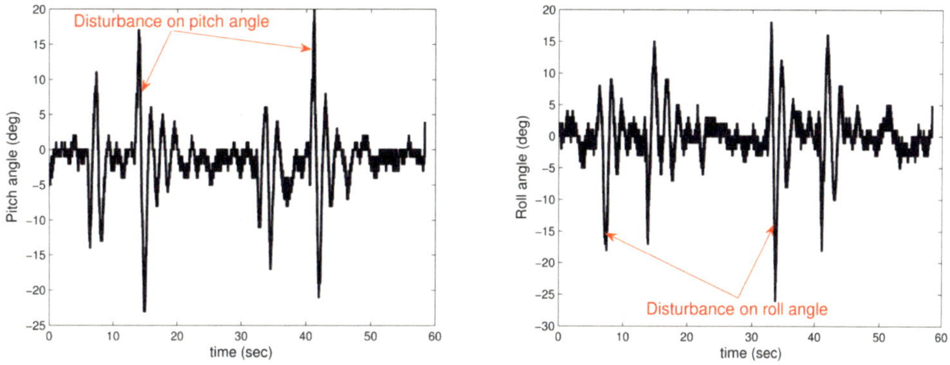

Fig. 17. Experimental results of attitude stabilization under external disturbances: (a) test-bed pitch θ); and (b) test-bed roll (ϕ).

5. Conclusion and future directions

This chapter addressed the problem of autonomous flight control of a quadrotor UAV. It was conducted as a research project at the School of Information Technology and Engineering (S.I.T.E.), University of Ottawa. Detailed mathematical modeling of the quadrotor's kinematics and dynamics was provided. A modular fuzzy logic approach was proposed for the autonomous control of quadrotors in general, without the need for a precise mathematical model of their complex and ill-defined dynamics. The fuzzy technique was implemented through Mamdani and TSK inference engines for comparison purposes. The controller comprises six individual fuzzy logic modules designated for the control of the quadrotor's position and orientation. The investigation on the two types of control methodologies was conducted in a simulator environment, where disturbances such as wind conditions and sensor noise were incorporated for a more realistic simulation. The fuzzy flight controller was eventually implemented on a quadrotor test-bed specifically designed and built for the project.

The experiments were first conducted on the simulator before being validated on the test-bed. The results demonstrated a successful control performance especially with the Mamdani inference engine. When compared to other conventional techniques applied for a similar purpose Altug et al. (2002), Bouabdallah et al. (2004b), the proposed methodology showed a higher robustness despite the induced disturbances.

The future work is directed towards achieving fully autonomous flight in outdoor environments. Furthermore, adaptive fuzzy control techniques will be investigated to automatically tune some of the controller's parameters online, to further optimize its performance.

6. References

E. Altug, J.P. Ostrowski, and R. Mahony. Control of a quadrotor helicopter using visual feedback. In *Proceedings of the 2002 IEEE International Conference on Robotics and Automation*, pages 72–77, 2002.

F. Archer, A. Shutko, T. Coleman, A. Haldin, E. Novichikhin, and I. Sidorov. Introduction, overview, and status of the microwave autonomous copter system MACS. *Proceedings of the 2004 IEEE International Geoscience and Remote Sensing Symposium*, 5:3574–3576, Sep 2004.

Randal W. Beard. Quadrotor dynamics and control. Brigham Young University, February 19 2008. URL http://www.et.byu.edu/groups/ece490quad/control/quadrotor.pdf. lecture notes.

S. Bouabdallah. Design and control of quadrotors with application to autonomous flying. Master's thesis, Swiss Federal Institute of Technology, 2007.

S. Bouabdallah, P. Murrieri, and R. Siegwart. Design and control of an indoor micro quadrotor. In *Proceedings of the International Conference on Robotics and Automation*, 2004a.

S. Bouabdallah, A. Noth, and R. Siegwart. PID vs LQ control techniques applied to an indoor micro quadrotor. In *International Conference on Intelligent Robots and Systems*, 2004b.

M. Chen and M. Huzmezan. A combined MBPC/ 2 DOF H_∞ controller for a quad rotor UAV. In *AIAA Guidance, Navigation, and Control Conference and Exhibit*, 2003.

C. Coza and C.J.B. Macnab. A new robust adaptive-fuzzy control method applied to quadrotor helicopter stabilization. In *Annual meeting of the North American Fuzzy Information Processing Society*, pages 454–458, 2006.

A. Dzul, P. Castillo, and R. Lozano. Real-time stabilization and tracking of a four rotor mini rotorcraft. *IEEE Transaction on Control System Technology*, 12(4): 510–516, 2004.

S. G. Fowers. Stabilization and control of a quad-rotor micro-UAV using vision sensors. Master's thesis, Brigham Young University, August 2008.

N. Guenard, T. Hamel, and V. Moreau. Dynamic modeling and intuitive control strategy for an X4-flyer. In *International Conference on Control and Automation*, pages 141–146, 2005.

N. Guenard, T. Hamel, and R. Mahony. A practical visual servo control for an unmanned aerial vehicle. *IEEE Transactions on Robotics*, 24(2):331–340, 2008.

S. D. Hanford. A small semi-autonomous rotary-wing unmanned air vehicle. Master's thesis, Pennsylvania State University, December 2005.

Draganfly Innovations Inc. Industrial aerial video systems and UAVs, October 2008. URL http://www.draganfly.com/.

A.Ő. Kivrak. Design of control systems for a quadrotor flight vehicle equipped with inertial sensors. Master's thesis, The Graduate School of Natural and Applied Sciences of Atilim University, December 2006.

J. Rumerman. Helicopter development in the early twentieth century. [Online] Available, December 2002. URL www.centennialofflight.gov.

Canada Weather Statistics. Courtesy of Environment Canada. [Online] Available, February 2009. http://www.weatherstats.ca/.

R. Sugiura, T. Fukagawa, N. Noguchi, K. Ishii, Y. Shibata, and K. Toriyama. Field information system using an agricultural helicopter towards precision farming. *Proceedings of the 2003 IEEE/ASME International Conference on Advanced Intelligent Mechatronics*, 2:1073–1078, July 2003.

M. Tarbouchi, J. Dunfied, and G. Labonte. Neural network based control of a four rotor helicopter. In *International Conference on Industrial Technology*, pages 1543– 1548, 2004.

A. Tayebi and S. McGilvray. Attitude stabilization of a VTOL quadrotor aircraft. *IEEE Transaction on Control System Technology*, 14(3):562–571, May 2006.

S. L. Waslander, G. M. Hoffmann, J. S. Jang, and C. J. Tomlin. Multi-agent quadrotor testbed control design: integral sliding mode vs. reinforcement learning. In *International Conference on Intelligent Robots and Systems*, pages 468–473, 2005.

K. W. Weng and M. Shukri. Design and control of a quad-rotor flying robot for aerial surveillance. *4th Student Conference on Research and Development (SCOReD 2006)*, pages 173–177, 2006.

L.A. Zadeh. Fuzzy sets. *Information and Control*, 8:338–353, 1965.

13

Microgravity Experiment for Attitude Control of A Tethered Body by Arm Link Motion

Masahiro Nohmi
Kagawa University
Japan

1. Introduction

Tethered satellite systems (TSS) offer various attractive potential applications in space, and therefore, they have been an object of study for over two decades (Rupp & Laue, 1978). In these studies, the tether extension strategy utilizes the gravity force and/or the centrifugal force associated with the orbital motion. Typically, it is assumed that the tether is to be extended for about 20-100km along the local vertical. Mass, elasticity, and deflection of the tether cannot be neglected in such a case. The TSS is then characterized by various dynamical effects including three dimensional rigid body dynamics, swinging and vibration, in-plane and out-of-plane motions of the tether, transverse vibrations of the main satellite, and so on. Due to these diverse dynamical effects, the TSS can become unstable. Therefore, at an early stage, active tether tension control has been recognized to be indispensable. Since then, various control approaches have been proposed, including both linear regulators (Bainum & Kumar, 1980) and nonlinear control methods (Pines, D. J. et al, 1990), control strategies employing thrusters, tension in the tether line or motion of the offset of the tether extending point have been investigated (Modi, V. J. et al., 1992).

Tethered Space Robot (TSR), which is connected to a mother spacecraft through a piece of tether, is a new type of space robot system proposed in the previous work (Nohmi et al., 2001). TSR differs significantly from the TSS studied so far, mainly in three aspects. First, we assume that the tether is to be extended for a relatively short distance, 10-100m. Second, we do not envision gravity force and/or centrifugal force influencing tether extension. Rather, we will employ tether extension strategy assisted by an initial translation momentum of the subsystem. Third, we envision the tethered subsatellite to be a multi-body system. Major consequence of the multi-body nature is that its attitude can be controlled under tether tension by its own link motion, instead of reaction wheels, jet, or thrust. When attitude control of the subsystem is possible, robotic motion can be achieved by employing methods borrowed from free-flying space robots studies: the generalized Jacobian matrix based approach (Umetani & Yoshida, 1989), or the virtual manipulator approach (Vafa & Dubowsky, 1990).

This paper discusses attitude control of a tethered space robot based on experimental results under microgravity condition by capsule dropping (at Microgravity Laboratory of Japan; http://www.mglab.co.jp/index.html) and by parabolic flight (at Diamond Air Service corporation; http://www.das.co.jp/new_html/index-static.html). Section 2 describes experimental device for a tethered space robot. Section 3 explains algorithm for attitude control of experimental device of a tethered space robot. Section 4 and section 5 describe microgravity experiment and experimental results, respectively.

2. Attitude control for a tethered space robot

2.1 Design for a tethered space robot

Attitude control of a tethered space robot is based on torque caused by tether tension. Therefore, it is impossible to control attitude around the axis along tether tension. The vertical axes with respect to tether tension can be possible. Figure 1 shows a simple model, which consists of one rigid arm link and one rigid base, and those are connected by one spherical joint. Since tether tension caused no torque, the tethered subsystem is in equilibrium when the mass center of the base, the tether attachment point, and the tether are located on a straight line. O-*xyz* denotes an inertial frame fixed to a tether extension point. O-012 denotes the body fixed frame, whose origin is fixed to the center of the spherical joint, and the 2 axis is along tether extension line at the equilibrium condition. Link motion can obtain torque acting on the base around the 0 axis and the 1 axis, then attitude control around those axes is possible.

2.2 Experimental device for a tethered space robot

Experimental device for a tethered space robot is shown in figure 2, which is developed based on the design as shown in figure 1. Tether is attached to the end of the arm link. All equipments including batteries are mounted on the base without external cables. The parameters are: total mass of the robot is about 4.5kg, scale of the base is 261x261x82mm, and length of the arm link is 100mm, respectively.

Electrical circuit on the experimental device are shown in figure 3. Angular velocities (ω_1 and ω_2) of the base around the 0 axis and the 1 axis are measured by two gyro sensors, respectively. The arm link is actuated by two motors around the 0 axis and 1 axis, respectively. Arm link angle around the 0 axis and 1 axis (ϕ_1 and ϕ_2) measured by potentiometers, respectively.

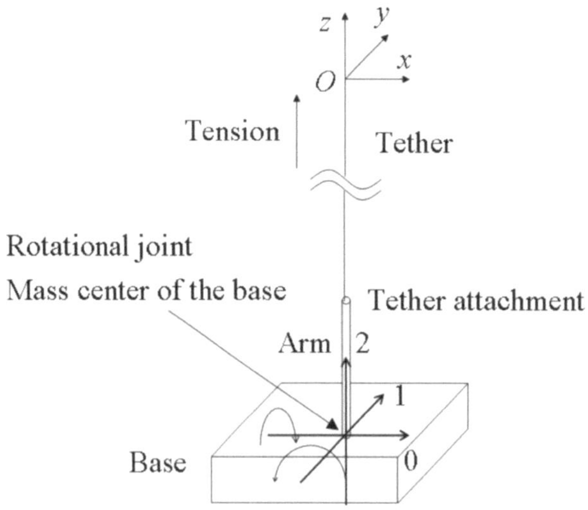

Fig. 1. Simple model for a tethered space robot

Fig. 2. Experimental device for a tethered space robot

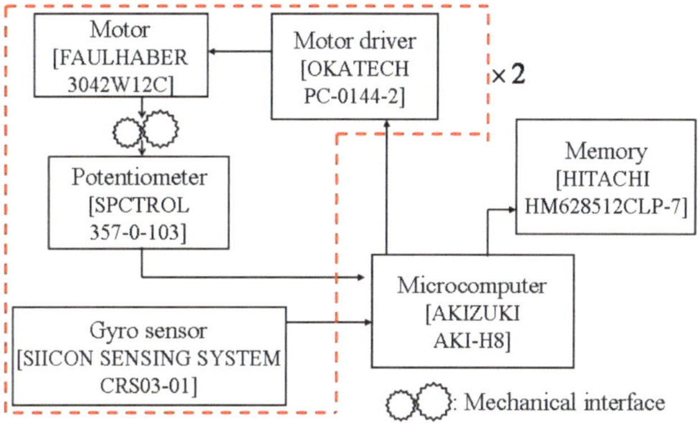

Fig. 3. Electrical circuit on the experimental device

2.3 Joint mechanism for arm link

Joint mechanism consisting of 5 rotational joints as shown in figure 4 was employed for the experimental device. Joint 0 and joint 1 are actuated by motors, respectively. Joint a, joint b, and joint c are passive joints. Then, two actuators along the 0 axis and the 1 axis can control arm link motion around the 0 axis and the 1 axis, respectively.

3. Control algorithm

3.1 Tether tension torque

Figure 5 shows rotational motion of the simple model for a tethered space robot, in case of deviating from the equilibrium. Here, $\boldsymbol{p}$ denotes position vector of the tether attachment point with respect to the origin of O-012. $\boldsymbol{p}_r$ and $\boldsymbol{p}_i$ denote vectors along the 2 axis and the y axis, respectively, and $\| \boldsymbol{p}_r \| = \| \boldsymbol{p}_i \| = \| \boldsymbol{p} \|$. $\boldsymbol{\omega}$ ($= [\ \omega_0\ \omega_1\ \omega_2\ ]^{\mathrm{T}}$) and $\boldsymbol{n}$ denote vectors of angular velocity and tether tension, respectively.

Since it is difficult to measure tether motion, control algorithm is designed under assumption that tether length is long enough compared to scale of a tethered subsystem, and also tether tension is constant, that is, $\boldsymbol{n}$ is constant. Also, assuming that arm link mass can be neglected compared to the base mass, tether tension torque τ_r can be written as:

$$\boldsymbol{\tau}_r = \boldsymbol{p} \times \boldsymbol{n} = -\boldsymbol{n} \times \boldsymbol{p}_r - \boldsymbol{n} \times (\boldsymbol{p} - \boldsymbol{p}_r). \tag{1}$$

The second term in the right hand of equation (1) can be changed by arm link motion.

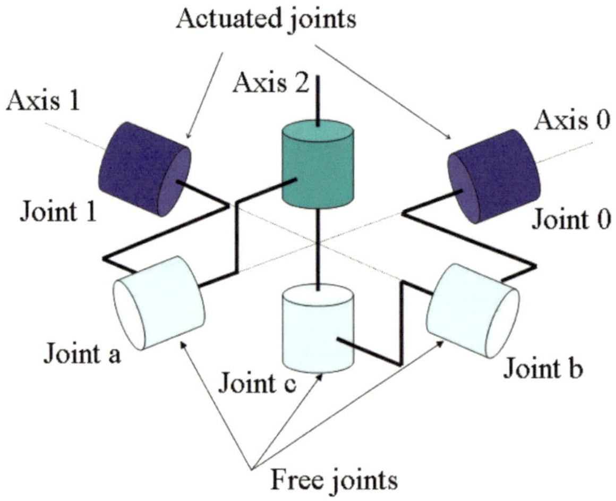

Fig. 4. Joint mechanism for the experimental device

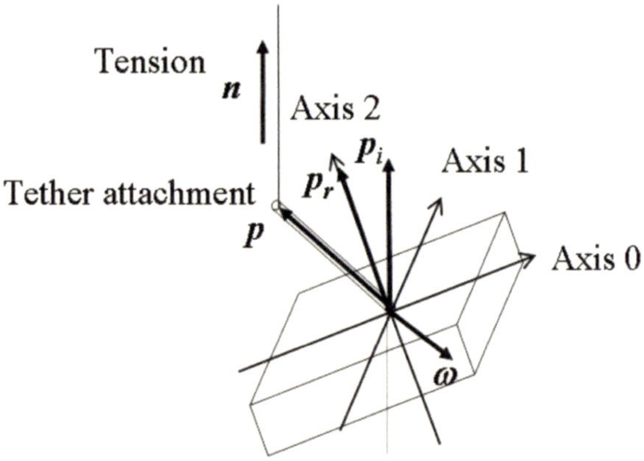

Fig. 5. Tether tension torque acting on the simple model

3.2 Arm link motion control

The tethered subsystem is in the equilibrium when $\boldsymbol{p}_r = \boldsymbol{p}_i$. When arm link motion generates (A) torque acting in direction to stable attitude; and (B) torque damping rotational motion, the PD control for rotational motion of the subsystem is possible. In order to employ the PD control, feedback information of attitude and angular velocity of the base are needed.
Gyro sensors of the experimental device measures angular velocity, however it is difficult to obtain accurate attitude. Then, attitude angle (θ_1 and θ_2) around the 0 axis and the 1 axis are assumed:

$$\theta_i = \int \omega_i dt\ ,\ (i = 0, 1\ ). \tag{2}$$

Also, assuming θ_1, θ_2, ϕ_1, ϕ_2 << 1,

$$\boldsymbol{p} - \boldsymbol{p}_r = l\left[\ \phi_1 \quad -\phi_0 \quad 1\ \right]^{\mathrm{T}}. \tag{3}$$

where l denotes arm link length (= $\|\ \boldsymbol{p}\ \|$). From equations (2) and (3), equation (1) can be rewritten as:

$$\boldsymbol{\tau}_r = -\boldsymbol{n} \times \boldsymbol{p}_r - n\,l\left[\ \phi_0 \quad \phi_1 \quad 0\ \right]^{\mathrm{T}}. \tag{4}$$

Employing the following control equation of arm link motion:

$$\begin{bmatrix} \phi_0 \\ \phi_1 \end{bmatrix} = k_p \begin{bmatrix} \theta_0 \\ \theta_1 \end{bmatrix} + k_d \begin{bmatrix} \omega_0 \\ \omega_1 \end{bmatrix}. \tag{5}$$

where k_p and k_d are control gains, equation (1) can be expressed as:

$$\boldsymbol{\tau}_r = -\boldsymbol{n} \times \boldsymbol{p}_r - n\,l\,k_p \begin{bmatrix} \theta_0 \\ \theta_1 \\ 0 \end{bmatrix} - n\,l\,k_d \begin{bmatrix} \omega_0 \\ \omega_1 \\ 0 \end{bmatrix}. \tag{6}$$

In the right hand of equation (6), the first term is torque by base motion; the second term is (A) torque acting in direction to stable attitude; the third term is (B) torque damping rotational motion, respectively. Then, basic characteristics of the control approach can be summarized as follows:

a. without control (k_p = k_d = 0): the base performs rotational vibration motion due to tether tension torque;
b. P control (k_d = 0): rotational vibration frequency is high, since tether tension torque is large compared to that in case (a);
c. D control (k_p = 0): rotational vibration can be damped;
d. PD control: damping time is short compared to that in case (c), since frequency is high as well as that in case (b).

4. Microgravity experiment by capsule dropping

4.1 Experimental facility

Micro Gravity Laboratory of Japan (MGLAB) provides a high-quality and stable microgravity environment, which is 10^{-5}G order, equivalent to that in space. The experiment capsule is dropped, and a high-quality microgravity environment is created for approximately 4.5 seconds with minimal turbulence. The capsule measures 900mm in outer diameter and 2,280mm in height, and the payload space available for users is 720mm (radius) × 885mm (height), with a maximum weight of 400kg (max. weight loaded by users).

4.2 Experimental device for deployment

In order to perform microgravity experiments for a tethered space robot at MGLAB, experimental device for tether extending mechanism as shown in figure 6 was developed.

The torque transmission device (permanent wave torque) of tether reel can apply constant tension on tether during deployment and retrieval. The load cell measures tension by pushing tether between pulleys, and the encoder measures tether length by rotation of tether reel, respectively.

Deployment mechanism for the robot in MGLAB was designed under consideration of the microgravity experiment sequence of capsule dropping, which is transition from condition under gravitational field to that under the microgravity. Figure 6-(a) shows the initial condition under gravity, and the robot is set on the plate supported by springs. These springs are pressed by the robot mass due to the gravity. When it becomes the microgravity condition, these springs are extended because the gravity force due to the robot mass becomes to be zero, and then the robot is deployed as shown in figure 6-(b). During deployment, tether tension reduces deployable velocity of the robot, then the robot stops to be deployed, and begins to be retrieved.

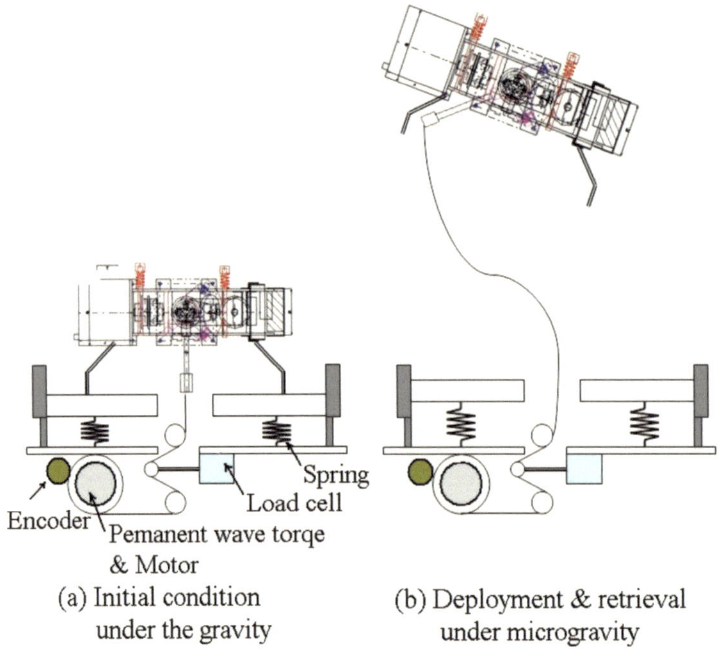

Fig. 6. Deployment mechanism for capsule dropping experiment

4.3 Experimental results and discussion

The experimental results in following three cases are shown in figure 7.

i. P control: $k_p = 2$ and $k_d = 0$, $n = 100$ [gf],
ii. D control: $k_p = 0$ and $k_d = 2$, $n = 60$ [gf],
iii. PD control: $k_p = 3$ and $k_d = 3$, $n = 60$ [gf].

The desired attitude of the base was set as $\theta_0 = \theta_1 = 0$. Figures show time histories, where L denotes tether length, and n denotes load cell data, respectively. Rotation angles around the 0 axis and 1 axis were measured by gyro sensors in case (i), and angular velocities were measured in cases (ii) and (iii), respectively. In graphs of gyro sensors ω_0 and ω_1 or θ_1 and θ_2, and potentiometers ϕ_1 and ϕ_2, dark solid line shows average of 100ms for filtering noise. Here, load cell data in case (ii) was failed to be recorded. It was microgravity condition from t = 0s to t = 4.5s, and large motion at t ~ 4s was due to collision of the robot against to the plate after retrieval. The following characteristics of the attitude control have been confirmed. It is noted that rotational vibration motion of the base attitude could not be suppressed in case of the P control because of k_d = 0, and that its amplitude was increasing.

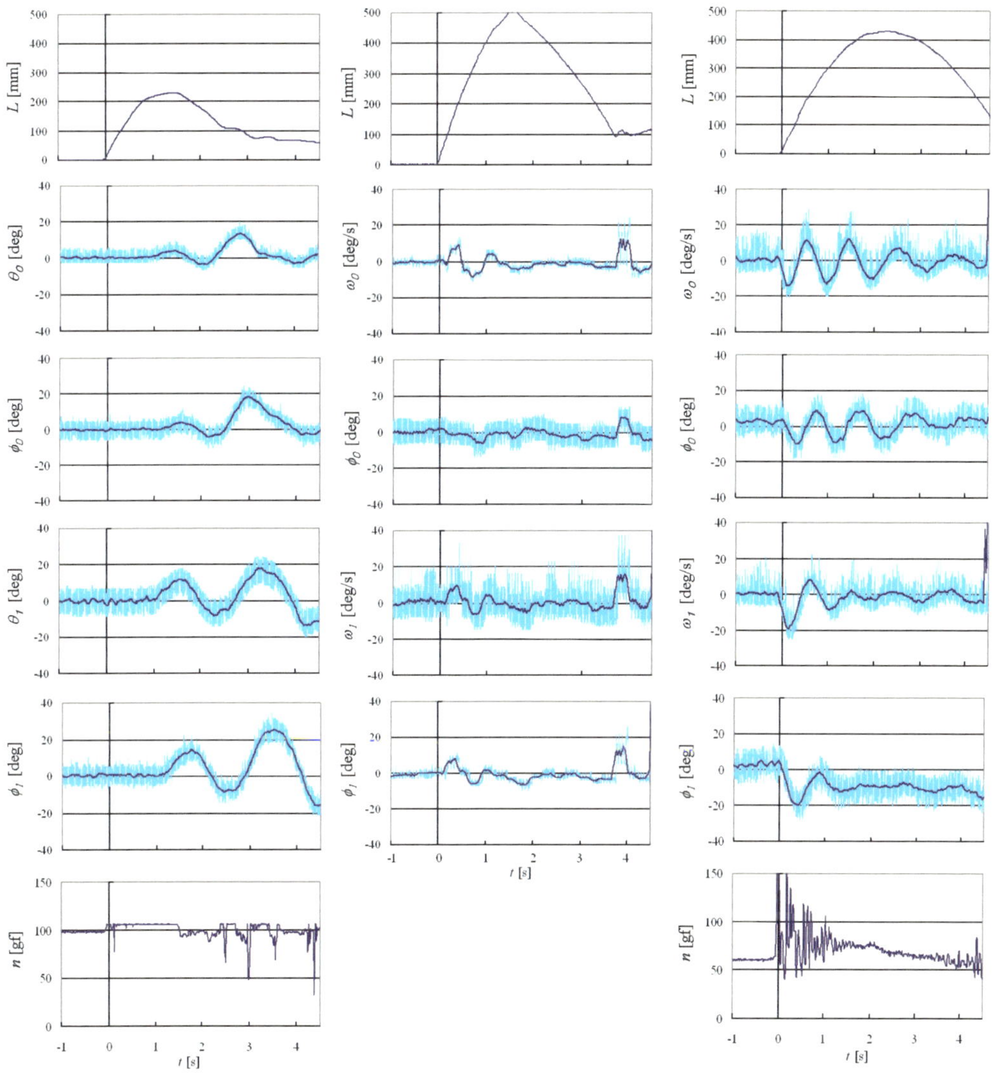

(i) P control (ii) D control (iii) PD control

Fig. 7. Microgravity experimental results by dropping capsule

In case of the PD control, rotational vibration motion was damped. Also, rotational vibration motion could not be observed in case of the D control. Hence, it can be considered from these results that the P control excites rotational vibration motion of the base.
In order to examine the characteristic of the P control, figure 8 shows control input and arm link angle, and also tether tension in case (i). It is noted that tether tension changed large discontinuously when arm link angle was peak value. And, it is also noted that arm link angle was behind the control input. Then, it can be said that the P control excites rotational vibration motion of the base due to these reasons.

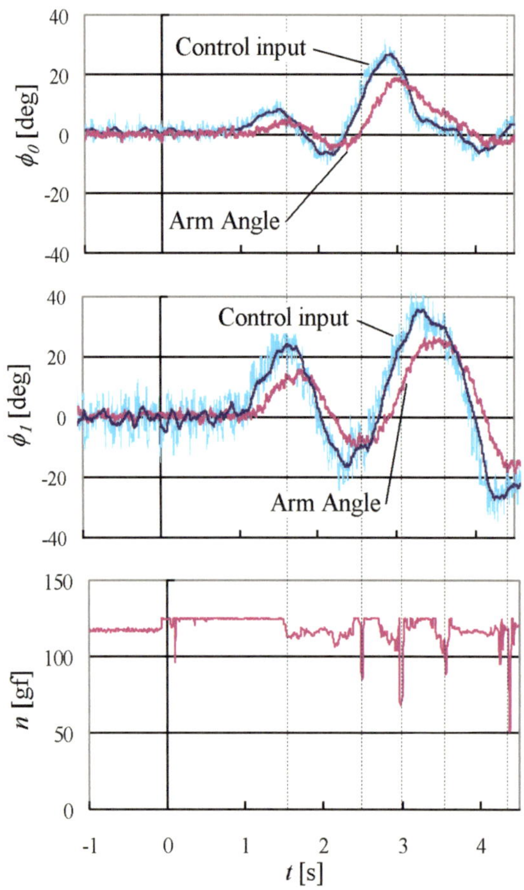

Fig. 8. Arm link motion in case (i) P control

In order to confirm that the D control is suitable for the developed experimental device than the PD control, experiment was performed in case that the P control and the D control are employed in the 0 axis and in the 1 axis, respectively. Figure 9 shows its result. It is noted that rotational vibration motion could be observed in the 0 axis with the P control, and rotational vibration motion could not be observed in the 1 axis with the D control.

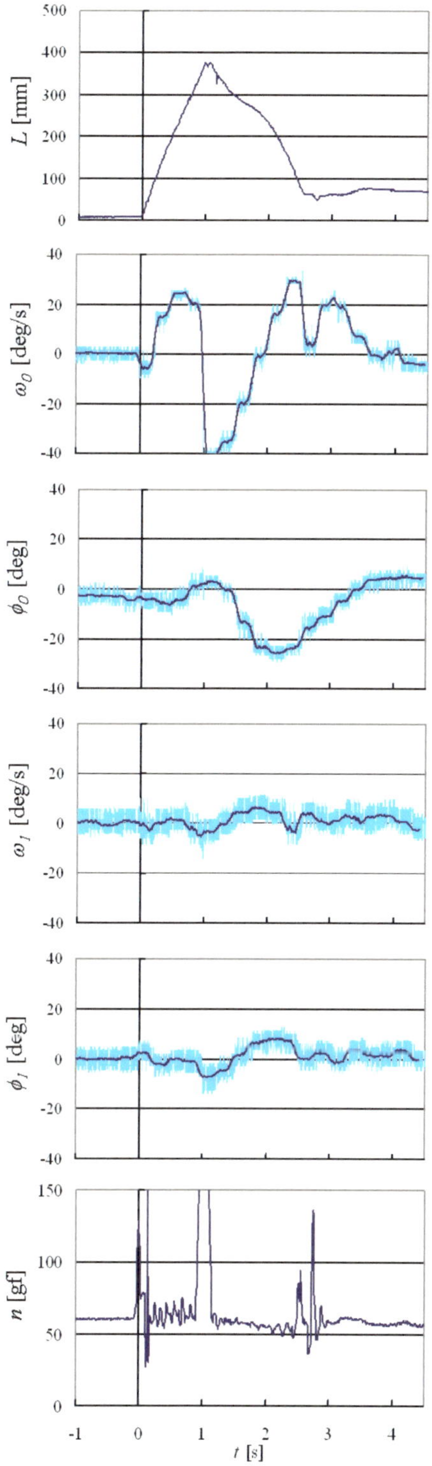

Fig. 9. Compare the D control and the Pcontrl

5. Microgravity experiment by parabolic flight

5.1 Experimental facility

Parabolic fright by the airplane provides microgravity condition, which is performed by Diamond Air Service Corporation in Japan. Microgravity condition is provided in the cabin of the airplane for approximately 20 seconds. Maximum acceleration error is 0.05 G in the vertical direction, 0.01 G in the back and forth, also from side to side, respectively. Scale of the cabin used for the experiment is about 1.5 x 1.5 x 4.8 [m].

5.2 Experimental setting

At the parabolic flight experiment, the experimental device was deployed in the vertical direction in the cabin by constant length tether. Then, a pilot operates the airplane for creating microgravity condition within 0–0.05G in the vertical direction, that is, tether tension was applied at random.

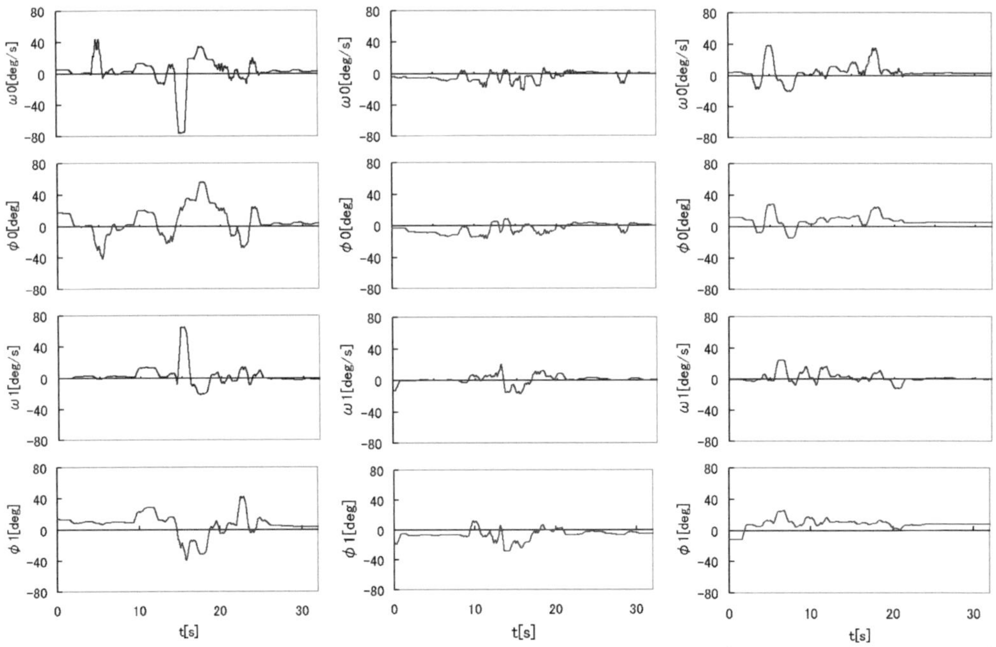

(i) P control (ii) D control (iii) PD control

Fig. 10. Microgravity experimental results by parabolic flight

5.3 Experimental results and discussion

The experimental results in following three cases are shown in figure 10.

i. P control: $k_p = 1$ and $k_d = 0$,
ii. D control: $k_p = 0$ and $k_d = 1$,
iii. PD control: $k_p = 1$ and $k_d = 1$.

The desired attitude of the base was set as $\theta_0 = \theta_1 = 0$. Figure shows time histories of the angular velocities of the base ω_0, ω_1, and ϕ_0, ϕ_1, respectively. The following characteristics of the attitude control have been confirmed. Large rotational motion of the base attitude was observed in case of the P control because it could not be suppressed due to $k_d = 0$. Rotational motion was the smallest in case of the D control, which is better result than that in case of the PD control. Hence, it can be considered that the P control excited rotational motion of the base, since arm link motion changes tether tension, and tether tension excites motion of the tethered subsystem.

6. Conclusion

This paper has described the microgravity experiment of the attitude control for a tethered space robot. A simple model model, which is suitable for attitude control by arm link motion of the multi-body subsystem, and its dynamics have been explained, and also the control equation has been derived. Microgravity experiment was performed in order to confirm and evaluate the attitude control. Experimental device for a tethered space robot is perfectly autonomous and has no external cables, and connected to only tether.

For the microgravity experiment by dropping capsule, tether extending mechanism was designed to apply constant tension on tether, and deployment device was designed under consideration of the experimental sequence. In the microgravity experiment by parabolic flight, tether length was kept constant, and small random tension was applied utilizing characteristics of parabolic flight. The two kinds of microgravity experiment were successfully performed.

It is noted from the microgravity experimental results that the D control can suppress rotational motion of the base, and the P control has possibility to excite rotational motion. The PD control also has such possibility. However it can suppress rotational vibration.

7. Acknowledgement

This work is partially supported by the Japan Space Forum and Grant-in-Aid for Scientific Research and New Energy and Industrial Technology Development Organization

8. References

Rupp, C. C. & Laue, J. H. (1978). Shuttle/Tethered Satellite System, *The Journal of the Astronautical Sciences*, Vol. 26, No. 1, 1978, pp. 1-17.

Bainum, P. M. & Kumar, V. K. (1980). Optimal Control of the Shuttle-Tethered-Subsatellite System, Acta Astronautica, Vol. 7, No. 6, 1980, pp. 1333-1348.

Umetani, Y. & Yoshida, K. (1989). Resolved Motion Rate Control of Space Manipulators with Generalized Jacobian Matrix, *IEEE Transaction on Robotics and Automation,* Vol. 5, No. 3, 1989, pp. 303-314.

Vafa, Z. & Dubowsky, S. (1990). On the Dynamics of Space Manipulators Using the Virtual Manipulator, with Applications to Path Planning, *The Journal of the Astronautical Sciences*, Vol. 38, No. 4, 1990, pp. 441-472.

Pines, D. J.; Flotow, A. H. & Redding, D. C. (1990). Tow Nonlinear Control Approaches for Retrieval of a Thrusting Tethered Subsatellite, *Journal of Guidance, Control, and Dynamics*, Vol. 13, No. 4, 1990, pp. 651-658.

Modi, V. J.; Lakshmanan, P. K. & Misra, A. K. (1992). On the Control of Tethered Satellite Systems, *Acta Astronautica*, Vol. 26, No. 6, 1992, pp. 411-423.

Nohmi, M.; Nenchev, D. N. & Uchiyama, M. (2001). Tethered Robot Casting Using a Spacecraft mounted Manipulator, *AIAA Journal of Guidance, Control, and Dynamics*, Vol. 24, No. 4, 2001, pp. 827-833.

14

Distributed Control of Multi-Robot Deployment Motion

Yu Zhou
State University of New York at Stony Brook
The United States

1. Introduction

A multi-robot system is a collection of mobile robots, each of which is equipped with onboard processors, sensors and actuators and is capable of independent operation and individual autonomous behaviours, collaborating with one another through wireless communications or other forms of interactions to fulfil global goals of the system. The mobile robots bring mobility, sensing capability and processing capability to the system; while a communication network is established among the robots to support data delivery and facilitate collaboration.

Multi-robot systems have higher flexibility, efficiency and reliability than single robots: a team of collaborative robots can accomplish a single task much faster, execute tasks beyond the limits of single robots, perform a complex task with multiple specialized simple robots rather than a super robot, and provide distributed, parallel mobile sensing and processing; a group of robots with heterogeneous capabilities can be organized to handle different tasks; the fusion of information from multiple mobile sensors helps to reduce sensing uncertainty and improve estimation accuracy; and the system function is less influenced by the failure of any individual robot.

Multi-robot systems have numerous applications, from regular civilian tasks, such as surveillance and environment monitoring, to emergency handling, such as disaster rescue and risky material removal, from scientific activities, such as space and deep sea exploration, to military operations, such as de-mining and battle field support, and to largescale agricultural and construction activities. Many applications require a multi-robot system to rapidly deploy into a target environment to provide sensor coverage and execute tasks while maintaining communication connections, and promptly adapt to the changes in the system, environment and task. This imposes significant requirements and challenges on the deployment control of the involved multi-robot systems.

Multi-robot deployment has become a fundamental research topic in the field of multi-robot systems. Both centralized and distributed schemes have been proposed in the literature. In general, centralized control depends on a leading robot to collect the state information of all the member robots, tasks and environment and to determine the appropriate motion of each individual robot. It helps to achieve globally optimal deployment and can be very effective in stable environments. However, centralized processing imposes high computational complexity on the leading robot and makes the multi-robot system vulnerable to the failure of the leader. Moreover, real-time centralized control of multiple robots requires very high communication throughput which is difficult to achieve with the current wireless communication technology. As a result, centralized control has difficulties in adapting to dynamic environments and scaling to large multi-robot systems. Alternatively, distributed control allows each member robot to determine its motion according to the states of itself, its local environment and its interactions with nearby robots and other objects. Distributed processing largely reduces the computational and communication complexities. As a result, distributed control is highly scalable to large multi-robot systems and adaptive to unknown and dynamic environments and changes in multi-robot systems. With properly designed distributed control laws, the desired global goal of a multi-robot system can be achieved as the combined outcome of the self-deployment motion of individual robots.

Recognizing its advantages, we focus our discussions in this chapter on distributed control of multi-robot deployment motion, with the objective to form and maintain sensor coverage and communication connections in a target environment. Section 2 will provide a review of some representative existing distributed multi-robot deployment control schemes. Although distributed multi-robot deployment has received a substantial amount of attention, there has not been enough effort made to address their implementation on realistic robot systems, in particular to explicitly take the kinematic and dynamic constraints into account when determining the deployment motion of individual robots. This disconnection between the control algorithm and physical implementation may degrade the operational effectiveness and robustness of these multi-robot deployment schemes. We will introduce a novel distributed multi-robot deployment control algorithm in Section 3, which takes into account the limited ranges of robot sensing and communication, and naturally incorporates the nonholonomic constraint arising among wheeled robots into individual robots' equation of deployment motion. Simulation results will be reported in Section 4, which proves the effectiveness of the proposed scheme. Section 5 will summarize the proposed scheme and discuss future work.

2. Review of distributed multi-robot deployment schemes

Due to its distributiveness, adaptability and scalability, distributed multi-robot deployment control has attracted a substantial amount of research effort. Here we review some related works on this topic.

One major category of distributed multi-robot deployment control schemes are based on artificial potential or force fields. Parker developed a two-level approach to deploy a homogeneous multi-robot system into an uncluttered environment to observe multiple moving targets (Parker, 1999; Parker, 2002). The low-level control is described in terms of force fields attractive for nearby targets and repulsive for nearby robots. The high-level control is described in terms of the probability of target existence and the probability of a target not being observed by other robots. The summation of the force vectors weighted by the high-level information yields the desired instantaneous location of the robot. The robot's speed and steering commands, which are the functions of the angle between the robot's current orientation and the direction of the desired location, are computed to move the robot in the direction of the desired location. Reif and Wang proposed a "social potential field" method for deploying very large scale multi-robot systems containing hundreds even thousands of mobile robots (Reif & Wang, 1999). Inverse-power force laws between pairs of robots or robot groups were defined, incorporating both attraction and repulsion, to reflect the social relations among robots, e.g. staying close or apart. An individual robot's motion is controlled by the resultant artificial force imposed by other robots and other components of the system. The resulting system displays social behaviors such as clustering, guarding, escorting, patrolling and so on. Howard et al. presented an algorithm for deploying a mobile sensor network in an unknown environment from a compact initial configuration, based on an artificial potential field in which each node is repelled by both obstacles and other nodes (Howard et al., 2002). Poduri and Suktame presented a deployment algorithm for mobile sensor networks to maximize the collective sensor coverage while constraining the degree of the network nodes so that each node maintains a number of connected neighbors, where the interaction between nodes is governed by the repulsive forces among nodes to improve their coverage and the attractive forces to prevent the nodes from losing connectivity (Poduri & Suktame, 2004). Popa et al. proposed a potential field framework to control the behavior of the mobile sensor nodes by combining navigation, attracted by goals and repulsed by obstacles and other nodes, and communication, attracted by maximum communication capacity and avoiding exceeding communication range (Popa et al., 2004). Fan et al. presented a potential field method to ensure the communication among the robots belonging to a formation by adding to each robot one attractive communication force generated by topologically nearby robots (Fan et al., 2005). Ji and Egerstedt presented a collection of graph-based control laws for controlling multi-agent rendezvous and formation while maintaining communication connections, based on weighted graph Laplacians and the edge-tension function (Ji & Egerstedt, 2007).

Closely related, Lam and Liu presented an algorithm for deploying mobile sensor networks such that the network graph approximates the layout of an isometric grid, under the force field defined by the difference between current and ideal local configurations (Lam & Liu, 2006). Jenkin and Dudek presented a distributed method to deploy multiple mobile robots to provide sensor coverage of a target robot (Jenkin & Dudek, 2000). It is formulated as a global energy minimization task over the entire collective in which each robot broadcasts its current position in the target-based coordinate system and moves in the gradient descent direction of its local estimate of the global energy. Butler and Rus presented two event-driven schemes to deploy mobile sensors toward the distribution of the sensed events (Butler & Rus, 2003). In one method, the sensors do not maintain any history of the events, and the robot position is determined by the positions of events like a potential field. In the other method, event history is maintained as a cumulative distribution of events by the sensors for more informed decisions about where to go at each step. With the intention to reduce the communication complexity, Tan presented a distributed self-deployment algorithm for multi-robot systems by combining potential field method with the Delaunay triangulation, which defines the potential field for each robot based on only the one-hop neighbors defined by the Delaunay triangulation (Tan, 2005).

Other than potential/force field methods, Cortes et al. defined the coverage problem as a locational optimization problem, and showed that the optimal coverage is provided by the centroidal Voronoi partitions where each sensor is located at the centroid of its Voronoi cell (Cortes et al., 2004). A gradient decent algorithm is presented to lead the sensor locations converge to the centroidal Voronoi configurations. A similar centroidal Voronoi diagram-based deployment was presented in (Tan et al., 2004). Jiang presented a slightly different method based on the r-limited Voronoi partition (Jiang, 2006). Schwager et al. proposed an adaptive, decentralized controller to drive a network of robots to the estimated centroids of their Voronoi regions while improving sensory distribution over time (Schwager et al., 2007). For this category of methods, local minimum is a potential problem. That is, the robots may be stuck at some Voronoi centroids determined by local configuration and cannot achieve the desired configuration.

Diffusion-based multi-robot deployment schemes were also proposed. Winfield presented a distributed method that deploys a group of mobile robots into a physically bounded region by random diffusion (Winfield, 2000). Kerr et al. presented two physics-based approaches for multi-robot dynamic search through a bounded region while avoiding multiple large obstacles, one based on artificial forces, and the other based on the kinetic theory of gases (Kerr et al., 2005). By mimicking gas flow, the agents will be able to distribute themselves throughout the volume and navigate around the obstacles. Along the same line, Pac et al. proposed a deployment method of mobile sensor networks in unknown environments based on fluid dynamics, by modeling the sensor network as a fluid body and each sensor node as a fluid element (Pac et al., 2006). These methods are designed for continuous sweeping-like coverage, but not suitable for converging multi-robot deployment.

Besides, Bishop presented a method which distributes the functional capability of a swarm of robots to a number of objectives (Bishop, 2007). His method is based on the definition of the capability function of each robot. The primary task (functional coverage) controller is defined based on the definition of the swarm-level objective function. The secondary task (e.g. obstacle avoidance, maintaining of line of sight) is carried out in the null space of the primary task. The potential problems with this method include local minima of the secondary functions and possible incompatibility of the secondary task with the null space of the primary task.

In addition, Jung and Sukhatme addressed the problem of tracking multiple targets using a network of communicating robots and stationary sensors (Jung & Sukhatme, 2002). Their region-based approach controls robot deployment at two levels. They divided a bounded environment into topologically simple convex regions. A coarse deployment controller distributes robots across regions based on the urgency estimates for each region. A target-following controller attempts to maximize the number of tracked targets within a region.

Existing works on distributed multi-robot deployment mostly focus on general schemes. There has not been sufficient attention paid to their implementation on realistic robot systems, e.g. most of existing methods assume reliable information broadcasting among robots to facilitate self-deployment control and multi-robot coordination, which is in fact communication intensive and has reduced robustness in large multi-robot systems. In particular, various kinematic and dynamic constraints must be taken into account in order to determine physically-realizable deployment motion of individual robots. However, there is a lack of a natural framework to incorporate them into deployment control. A very limited number of works have considered kinematic and dynamic constraints explicitly. In general, the dynamic constraints of maximum velocity and maximum acceleration are accommodated by enforcing the computed above-limit acceleration and velocity into the desired ranges (Howard et al., 2002; Jiang, 2006), and the nonholonomic kinematic constraints are ignored by assuming that the robots have holonomic drive mechanisms, i.e. they can move equally well in any direction (Howard et al., 2002; Bishop, 2007). This disconnection between the control algorithm and physical implementation may cause the computed deployment motion unrealizable, and therefore degrades the effectiveness and robustness of the deployment of realistic multi-robot systems.

As an important step towards solving this problem, we propose a novel distributed multirobot deployment control algorithm for deploying a team of mobile robots to establish sensor coverage while maintaining communication connections over a target environment. It takes into account the limited ranges of robot sensing and communication, and in particular naturally incorporates the nonholonomic constraint which arises among wheeled robots. Each member robot self-deploys based on the state of its neighborhood and approaches the desired neighborhood configuration. The resulting local coverage in the neighborhoods of all the robots altogether forms a global coverage of the multi-robot system over the targeted environment.

3. Proposed distributed control algorithm for multi-robot deployment

Targeting to develop a robust distributed multi-robot deployment control scheme which can be reliably implemented in realistic multi-robot systems, we have recently initiated an alternative scheme for distributed multi-robot deployment (Zhou & Tan, 2008; Zhou, 2008). In the following sections, we will provide a extended and detailed description of our original distributed multi-robot deployment algorithm for establishing sensor coverage while maintaining communication connections over targeted environments, based on a lumped dynamics model of involved robots, accommodating the limited robot sensing and communication ranges, and incorporating the nonholonomic kinematic constraint which arises in wheeled robots.

3.1 Objective and assumptions

The objective of the proposed distributed multi-robot deployment algorithm is to deploy a multi-robot system into a targeted environment to form and maintain reliable sensor coverage and communication connections.

To achieve the goal, it is required that

1. The team of robots, each of which has limited independent sensing capability, must collaborate with one another to form and maintain sufficient sensor coverage at the multi-robot system level.
2. The team of robots must maintain globally networked communications at the system level at any time in order to guarantee the information delivery and data sharing in the multi-robot system.

In order to deploy the involved multi-robot system to form and maintain reliable sensor coverage and communication connections in the targeted environment, we propose an overall control strategy as guiding the multi-robot system to approach a desired global deployment configuration which is defined as that each member robot maintains desired distances with nearby objects, including other robots, obstacles and the boundary of the environment.

To facilitate the discussion, we assume that (Fig.1)

1. A group of N mobile robots (nonholonomic or holonomic) are to be deployed into a 2D environment in order to provide sensor coverage while maintaining communication connections, where R_i denotes the ith mobile robot.
2. The underlying environment is a general 2D region which contains objects other than mobile robots, known in general as obstacles (stationary or moving) where O_i denotes the ith obstacle, and can be confined with a boundary denoted by B.

3. Each robot R_i has limited sensing and communication capabilities, represented approximately by a limited circular sensing range with a radius of rs_i and a limited circular communication range with a radius of rc_i respectively. Though the robot sensing range and can be determined by off-line sensor calibration, multiple factors can affect the robot communication range, e.g. the obstacles and humidity. Therefore, varying rc_i is admitted.
4. Each robot R_i updates and maintains a record of its pose (position and orientation) with respect to a global reference frame defined in the environment. The robot self-localization can be accomplished using either relative localization techniques, e.g. odometry and inertial navigation, which is based on the integration of incremental motion, or absolute localization techniques, e.g. GPS, which is based on the measurement of external references, or a fusion of both (Borenstein et al., 1997).

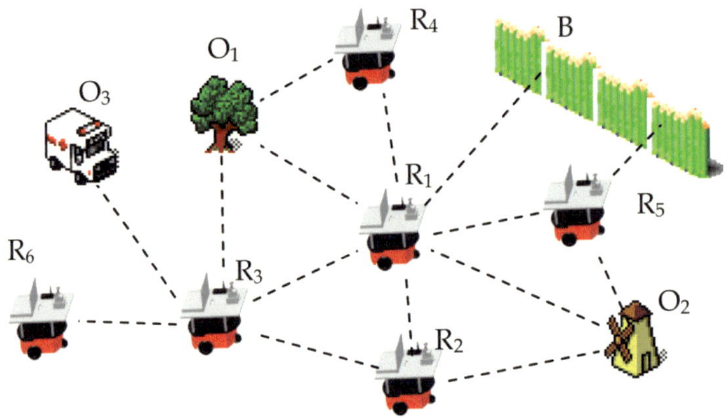

Fig. 1. Multi-robot deployment in a general 2D environment

3.2 Desired deployment configuration

In our scheme, a desired global deployment configuration is defined as such that each member robot maintains desired distances with nearby objects, including other robots, obstacles and the boundary of the environment.

Considering the limited ranges of robot sensing and communication, we define a desired distance a_{ij} between robots R_i and R_j as a designated distance over which a sufficiently large but reliable in-between sensor coverage can be established and reliable in-between wireless communication can be maintained. Since R_i and R_j have circular sensing ranges with radii of rs_i and rs_j respectively, in order to achieve sensor coverage, one should choose $a_{ij} \le rs_i + rs_j$. Meanwhile, in order to achieve communication coverage, one should choose $a_{ij} \le \min(rc_i, rc_j)$ such that R_i and R_j are inside each other's communication range. As a result, in order to achieve both the sensor and communication coverage, one should choose $a_{ij} \le \min(rs_i + rs_j, \min(rc_i, rc_j))$. In practice, the robot sensing range is relatively stable, and can be determined through off-line sensor calibration. The determination of the robot communication range is more complicated, because it is easily affected by various environmental factors in real time, such as surrounding objects and atmospheric conditions. However, a communication calibration and a conservative estimation are helpful. Moreover, the communication range is usually much longer than the sensing range, which means that an estimate of the desired robot-to-robot distance based on the sensing ranges of the involved robots is often reliable.

Besides the desired distance between a pair of robots, one can also specify the desired distance between a robot and an obstacle and the desired distance between a robot and the boundary. In fact, when detecting an obstacle (either stationary or moving), the robot should keep a distance from it to avoid collision while probably maintaining an observation of it. Similarly, in a confined environment, e.g. surrounded by a wall, when detecting the boundary (either stationary or evolving), the robot should also keep a distance from it to avoid collision while maintaining a sufficient sensor coverage of it. Denoting the desired distance between a robot R_i and an obstacle O_j inside its sensing range as b_{ij}, denoting the desired distance between R_i and the boundary B inside its sensing range as c_{ij}, denoting the safety margin of R_i as s_i and noticing $s_i \leq r_i$ in general, one should choose $b_{ij} \in [s_i, r_i]$ and $c_{ij} \in [s_i, r_i]$.

In order to achieve the distributed control of the multi-robot deployment, we decompose the desired global deployment configuration into the desired local deployment configuration around each robot. That is, each robot only needs to approach and maintain desired distances with nearby objects, including other robots, obstacles and the boundary of the environment, in its neighborhood. It results in largely reduced computational and communication complexities. The combined effect of approaching desired local deployment configuration in the neighborhood of each member robot will lead to the desired global deployment configuration at the system level.

To unify the representations, we denote the jth object, which can be another robot, an obstacle or the boundary, in the neighborhood of R_i as T_{ij}, and the desired distance between R_i and T_{ij} as d_{ij} which can be a_{ij}, b_{ij} or c_{ij}, as defined above, corresponding to the actual type of object. In general, a more conservative choice of d_{ij} tends to result in a more reliable but smaller coverage.

3.3 Concept of neighborhood

In order to reduce the computational and communication complexities and achieve distributed control of multi-robot deployment, we propose that each member robot determines its desired motion based on only the state information of other robots and objects in its neighborhood with the intention to approach the desired local deployment configuration in its neighborhood. Therefore, defining the neighborhood for each robot is of high importance to our distributed multi-robot deployment control algorithm. The proposed scheme applies to the following two different definitions of neighborhood:

1. Physical neighborhood: The physical neighborhood of a robot R_i is defined by the robot sensing and communication ranges. If another robot R_j is inside the communication range of R_i, R_i can obtain the state of R_j through communication and retrieve the geometric relationship between R_i and R_j. If R_j is inside the sensing range of R_i, R_i may even sense the state of R_j directly, if appropriate onboard sensors are available. In these cases, we consider that R_j belongs to the neighborhood of R_i. If R_j is outside the sensing and communication ranges of R_i, R_i cannot obtain the state of R_j or retrieve the geometric relationship between R_i and R_j from either direct sensing or communication between them. In these case, we consider that R_j does not belong to the neighborhood of R_i. The concept of physical neighborhood provides a complete count of those physically nearby robots. However, intensive communications among multiple robots may arise in a dense robot gathering, such as the initial stage of the multi-robot deployment process.

2. Topological neighborhood: The topological neighborhood of R_i is defined as the set of its one-hop neighbors on a topological graph representation of the multi-robot system, such as the Delaunay triangulation and the Gabriel graph (Tan, 2005; Sander et al., 2002; Preparata & Shamos, 1985). Since, at any time, there are only a very limited number of one-hop topological neighbors, both the computational and communication complexities are relatively low. However, if the topological relationship among the multiple robots changes, in order to redefine the global topology, multi-hop communications are often necessary and may become intensive. Moreover, depending on the specific type of underlying graph, such as the Delaunay triangulation, we notice that the one-hop neighbors of a boundary node may include those at physically long distances.

Besides robots, an obstacle or the boundary is considered belonging to the neighborhood of a robot R_i only when it is inside the sensing range of R_i.

3.4 Mathematical formulation

The deployment motion of each member robot is governed by its equation of deployment motion. According to the Hamilton's principle (Goldstein, 1980), the optimal deployment motion of a mobile robot R_i during the time period $[t_1, t_2]$ should minimize the total action of R_i during this period, i.e.

$$\mathbf{q}_i(t) = \arg\min_{\mathbf{q}'_i(t)} \int_{t_1}^{t_2} L_i(\mathbf{q}'_i(t))dt \, , \tag{1}$$

where $\mathbf{q}_i(t)$ and $\mathbf{q}'_i(t)$ denote respectively the optimal trajectory and a candidate trajectory of R_i in its configuration space which is spanned by the set of variables uniquely defining the state of R_i, and L_i denotes the Lagrangian of R_i which is defined based on the states of R_i and its neighborhood and will guide R_i to approach the desired neighborhood configuration. In principle, using the method of the variational calculus, one can obtain the following Lagrange's equation for R_i

$$\frac{d}{dt}\left(\frac{\partial L_i}{\partial \dot{\mathbf{q}}_i}\right) - \frac{\partial L_i}{\partial \mathbf{q}_i} = \mathbf{0} \, , \tag{2}$$

which is the equation governing the deployment motion of R_i. Following (2), R_i will self-deploy. The combined effect of the self-deployment motion of all the member robots will lead to the desired global deployment.

Strictly speaking, the dynamics, represented by (2), of a realistic mobile robot can be substantially complicated, with $\mathbf{q}_i$ composed of various motion parameters for the wheels, links and body of the robot. In practice, commercial robot systems mostly provide a transparent lower-level control for the motion of the components, such as the wheels, and users only need to define the motion parameters at the robot level, such as the position, orientation and speed of the whole robot. This is equivalent to an upper-level control of the robot motion, which is based on a lumped model of the robot. Following this practice, we further our discussion and derivation of the distributed multi-robot deployment algorithm with a lumped robot model.

A lumped model for a mobile robot R_i moving in a 2D environment can be defined as following:

1. The position of the robot is represented by the coordinates of its center of mass, (x_i, y_i), in the global frame;
2. The orientation of the robot is represented by the angle θ_i between the longitudinal direction of the robot and the x axis of the global frame;
3. The robot is considered having a point mass m_i at its center of mass;
4. The robot is considered having a moment of inertia I_i about the vertical axis passing through its center of mass.

Based on this 2D lumped robot model, we have

$$\mathbf{q}_i = [x_i \quad y_i \quad \theta_i]^T. \tag{3}$$

We also define the Lagrangian of R_i in the 2D environment as

$$L_i = T_i - U_i, \tag{4}$$

where T_i denotes the kinetic energy of R_i

$$T_i = \frac{1}{2} m_i (\dot{x}_i^2 + \dot{y}_i^2) + \frac{1}{2} I_i \dot{\theta}_i^2, \tag{5}$$

and U_i denotes an artificial potential energy which drives R_i.

The artificial potential energy U_i is defined to move R_i towards its desired neighborhood configuration, based on the definitions of the desired distances between R_i and nearby objects

$$\begin{aligned} U_i = {} & \frac{1}{2} \sum_{j=1}^{n_i} w_{ij} \left(\sqrt{(x_{ij} - x_i)^2 + (y_{ij} - y_i)^2} - d_{ij} \right)^2 \\ & + \chi \frac{1}{2} \left(\operatorname{atan2}\left(\frac{1}{n_i} \sum_{j=1}^{n_i} w_{ij} x_{ij} - x_i, \frac{1}{n_i} \sum_{j=1}^{n_i} w_{ij} y_{ij} - y_i \right) - \theta_i \right)^2 \end{aligned} . \tag{6}$$

Here, n_i denotes the number of other objects in the neighborhood of R_i. With (x_{ij}, y_{ij}) denoting the position of the jth object T_{ij} (in fact the position of the nearest point on T_{ij} relative to R_i) in the neighborhood of R_i, which can be another robot, an obstacle or the boundary, $\sqrt{(x_{ij} - x_i)^2 + (y_{ij} - y_i)^2}$ gives the actual instantaneous distance between R_i and T_{ij}. With d_{ij} denoting the desired distance between R_i and T_{ij}, the term $\frac{1}{2}(\sqrt{(x_{ij} - x_i)^2 + (y_{ij} - y_i)^2} - d_{ij})^2$ defines a potential energy component based on the difference between the desired and actual distances between R_i and T_{ij}, which generates an actuating force to drive R_i towards the desired distance between R_i and T_{ij}. If the actual distance is shorter than d_{ij}, it tends to push R_i away from T_{ij}; if the actual distance is longer than d_{ij}, it tends to pull R_i towards T_{ij}. In addition, w_{ij} is a coefficient weighting the effect among the objects in the neighborhood of R_i. A larger w_{ij} means a bigger influence of T_{ij} on R_i. Meanwhile, $(\frac{1}{n_i}\sum_{j=1}^{n_i} w_{ij} x_{ij}, \frac{1}{n_i}\sum_{j=1}^{n_i} w_{ij} y_{ij})$ defines C_i – the center of mass of the neighborhood of R_i, and $\operatorname{atan2}(\frac{1}{n_i}\sum_{j=1}^{n_i} w_{ij} x_{ij} - x_i, \frac{1}{n_i}\sum_{j=1}^{n_i} w_{ij} y_{ij} - y_i)$ gives the direction angle of the vector pointing from R_i to C_i (where atan2(x, y) gives the arc tangent of y/x, taking into account which quadrant the point (x, y) is in). Therefore, the term $\frac{1}{2}(\text{atan2}(\frac{1}{n_i}\sum_{j=1}^{n_i} w_{ij}x_{ij} - x_i, \frac{1}{n_i}\sum_{j=1}^{n_i} w_{ij}y_{ij} - y_i) - \theta_i)^2$ defines a potential energy component based on the difference between the current orientation of R_i and the direction of C_i relative to R_i, which generates an actuating force to turn R_i towards C_i. This will help to drive R_i towards C_i, and therefore establish a balance in the local deployment. In addition, χ is a coefficient making the translational and rotational terms in (6) compatible. Altogether, (6) defines an artificial potential energy for a member robot according to the difference between the actual and desired configurations in the neighborhood of the robot, which generates the actuating force to drive the robot towards the desired local coverage distances.

Substituting (3)-(6), which define the lumped dynamics model of a mobile robot moving in a 2D environment, into (2), we obtain the basic equation of deployment motion for R_i

$$m_i\ddot{x}_i = -\frac{\partial U_i}{\partial x_i}, \quad m_i\ddot{y}_i = -\frac{\partial U_i}{\partial y_i}, \quad I_i\ddot{\theta}_i = -\frac{\partial U_i}{\partial \theta_i}, \tag{7}$$

Where

$$\frac{\partial U_i}{\partial x_i} = \sum_{j=1}^{n_i} -\frac{w_{ij}(x_{ij}-x_i)(\sqrt{(x_{ij}-x_i)^2+(y_{ij}-y_i)^2}-d_{ij})}{\sqrt{(x_{ij}-x_i)^2+(y_{ij}-y_i)^2}} + \frac{\chi(\frac{1}{n_i}\sum_{j=1}^{n_i} w_{ij}y_{ij} - y_i)(\text{atan2}(\frac{1}{n_i}\sum_{j=1}^{n_i} w_{ij}x_{ij} - x_i, \frac{1}{n_i}\sum_{j=1}^{n_i} w_{ij}y_{ij} - y_i) - \theta_i)}{(\frac{1}{n_i}\sum_{j=1}^{n_i} w_{ij}x_{ij} - x_i)^2 + (\frac{1}{n_i}\sum_{j=1}^{n_i} w_{ij}y_{ij} - y_i)^2},$$

$$\frac{\partial U_i}{\partial y_i} = \sum_{j=1}^{n_i} -\frac{w_{ij}(y_{ij}-y_i)(\sqrt{(x_{ij}-x_i)^2+(y_{ij}-y_i)^2}-d_{ij})}{\sqrt{(x_{ij}-x_i)^2+(y_{ij}-y_i)^2}} - \frac{\chi(\frac{1}{n_i}\sum_{j=1}^{n_i} w_{ij}x_{ij} - x_i)(\text{atan2}(\frac{1}{n_i}\sum_{j=1}^{n_i} w_{ij}x_{ij} - x_i, \frac{1}{n_i}\sum_{j=1}^{n_i} w_{ij}y_{ij} - y_i) - \theta_i)}{(\frac{1}{n_i}\sum_{j=1}^{n_i} w_{ij}x_{ij} - x_i)^2 + (\frac{1}{n_i}\sum_{j=1}^{n_i} w_{ij}y_{ij} - y_i)^2},$$

$$\frac{\partial U_i}{\partial \theta_i} = -\chi(\text{atan2}(\frac{1}{n_i}\sum_{j=1}^{n_i} w_{ij}x_{ij} - x_i, \frac{1}{n_i}\sum_{j=1}^{n_i} w_{ij}y_{ij} - y_i) - \theta_i) \cdot$$

Equation (7) does not include any kinematic constraint. However, in practice, a wheeled robot is under the nonholonomic constraint. That is, at any time it can only have a non-zero speed in its longitudinal direction (i.e. along its orientation) while its side speed is zero. For the lumped robot model, the nonholonomic constraint is defined as

$$\dot{x}_i \sin\theta_i - \dot{y}_i \cos\theta_i = 0 \cdot \tag{8}$$

Then, instead of using (2), the optimal deployment motion of a wheeled robot R_i during the time period $[t_1, t_2]$ should minimize the total action of R_i during this period and satisfy the nonholonomic constraint, i.e.

$$[x_i(t), y_i(t), \theta_i(t)] = \arg \min_{x_i(t), y_i(t), \theta_i(t)} \int_{t_1}^{t_2} L_i dt \text{ subject to } \dot{x}_i \sin\theta_i - \dot{y}_i \cos\theta_i = 0 \cdot \tag{9}$$

Using the method of the variational calculus (Goldstein, 1980), we can obtain the following Lagrange's equation for R_i

$$\begin{aligned} & m_i \ddot{x}_i - \lambda_i \sin\theta_i = -\frac{\partial U_i}{\partial x_i}, \quad m_i \ddot{y}_i + \lambda_i \cos\theta_i = -\frac{\partial U_i}{\partial y_i}, \\ & I_i \ddot{\theta}_i = -\frac{\partial U_i}{\partial \theta_i}, \quad \dot{x}_i \sin\theta_i - \dot{y}_i \cos\theta_i = 0 \end{aligned} \tag{10}$$

where λ_i denotes the Lagrange undetermined multiplier.

Furthermore, it is important for each member robot to converge towards the desired neighborhood configuration. In order to stabilize the deployment motion of each robot around its equilibrium position, a virtual Rayleigh's dissipation function is adopted to provide the necessary damping mechanism

$$F_i = \frac{1}{2}(k_{xi}\dot{x}_i^2 + k_{yi}\dot{y}_i^2 + k_{\theta i}\dot{\theta}_i^2), \tag{11}$$

where k_{xi}, k_{yi} and $k_{\theta i}$ are the viscous damping coefficients associated with the linear and angular velocities of R_i respectively. F_i defines the damping force for each velocity component as

$$\frac{\partial F_i}{\partial \dot{x}_i} = k_{xi}\dot{x}_i, \quad \frac{\partial F_i}{\partial \dot{y}_i} = k_{yi}\dot{y}_i, \quad \frac{\partial F_i}{\partial \dot{\theta}_i} = k_{\theta i}\dot{\theta}_i \cdot \tag{12}$$

Incorporating (12) into (7), we obtain the equation of deployment motion for a holonomic mobile robot with dissipation as

$$m_i \ddot{x}_i = -\frac{\partial U_i}{\partial x_i} - k_{xi}\dot{x}_i, \quad m_i \ddot{y}_i = -\frac{\partial U_i}{\partial y_i} - k_{yi}\dot{y}_i, \quad I_i \ddot{\theta}_i = -\frac{\partial U_i}{\partial \theta_i} - k_{\theta i}\dot{\theta}_i, \cdot \tag{13}$$

Incorporating (12) into (10), we obtain the equation of deployment motion for a nonholonomic mobile robot with dissipation as

$$\begin{aligned} & m_i \ddot{x}_i - \lambda_i \sin\theta_i = -\frac{\partial U_i}{\partial x_i} - k_{xi}\dot{x}_i, \quad m_i \ddot{y}_i + \lambda_i \cos\theta_i = -\frac{\partial U_i}{\partial y_i} - k_{yi}\dot{y}_i, \\ & I_i \ddot{\theta}_i = -\frac{\partial U_i}{\partial \theta_i} - k_{\theta i}\dot{\theta}_i, \quad \dot{x}_i \sin\theta_i - \dot{y}_i \cos\theta_i = 0 \end{aligned} \tag{14}$$

Equations (13) and (14) are the final equations governing the self-deployment motion of a member robot to approach the desired local coverage. In practice, each member robot can online calculate its desired instantaneous acceleration for the deployment motion by substituting its current pose (position and orientation) and the positions of other objects in its neighborhood into (13) (holonomic) or (14) (nonholonomic), and command the lower-level controller to move the robot at the resulting acceleration. By moving each mobile robot in the way defined by its equation of deployment motion, eventually the resulting local coverage in the neighborhoods of all the robots altogether forms a global coverage of the multi-robot system to the targeted environment.

4. Simulation results

The effectiveness of the proposed distributed multi-robot deployment control algorithm has been verified by simulations programmed in Matlab.

4.1 Settings

In the following reported simulations, we assume that 120 mobile robots, each with unit mass and unit moment of inertia, are deployed into a 2D environment. The deployments of both holonomic and nonholonomic mobile robots in both open and corridor environments based on both physical and topological neighborhood consideration have been simulated respectively. We assume that the initially the group of robots are uniformly distributed in a small square region of $x_i \in [-1, 1]$ and $y_i \in [-1, 1]$ with uniformly distributed orientation $\theta_i \in [-\pi, \pi]$. We set the desired distance between a pair of nearby robots as $d_{ij}=5$, the viscous damping coefficients $k_{xi}=k_{yi}=k_{\theta i}=2$, and the relevant weighting coefficients $\chi=w_{ij}=1$. We also discretize the time into a equally-divided sequence $\{t_0, t_1, t_2, \ldots\}$ with the common interval $\Delta t=0.2$. At each time t_k, we calculate the desired instantaneous acceleration of each robot R_i from its equation of deployment motion (13) (if R_i is holonomic) or (14) (if R_i is nonholonomic). The velocity and the pose of R_i at t_k are updated iteratively by numerical integration.

4.2 Deployment in an open environment

Considering 120 holonomic mobile robots being deployed in an open 2D environment, at first we define the self-deployment motion of each mobile robot based on the state of its physical neighborhood. A representative deployment of the robots at t=30 is shown as Fig.2. Then we define the self-deployment motion of each mobile robot based on the state of its topological neighborhood which is generated using the Delaunay triangulation. A representative deployment of the robots at t=30 is shown as Fig.3.

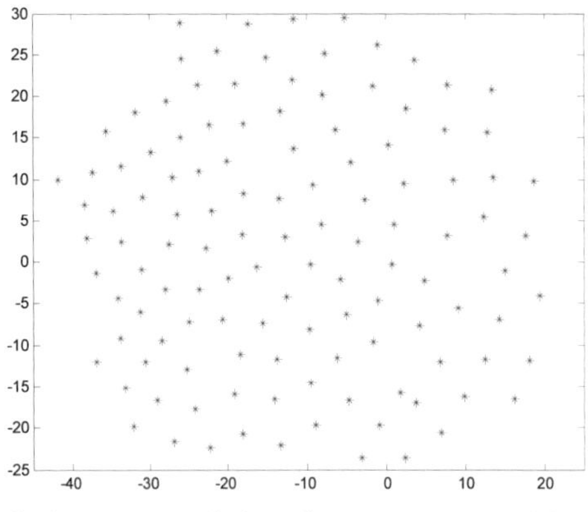

Fig. 2. Deployment of 120 holonomic mobile robots in an open 2D environment based on the state of the physical neighborhood

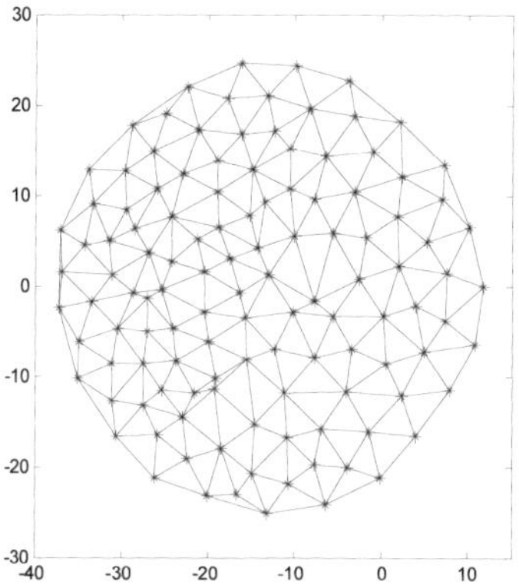

Fig. 3. Deployment of 120 holonomic mobile robots in an open 2D environment based on the state of the topological neighborhood

Considering 120 nonholonomic mobile robots being deployed in an open 2D environment, at first we define the self-deployment motion of each mobile robot based on the state of its physical neighborhood. A representative deployment of the robots at t=30 is shown as Fig.4. Then we define the self-deployment motion of each mobile robot based on the state of its topological neighborhood which is generated using the Delaunay triangulation. A representative deployment of the robots at t=30 is shown as Fig.5.

Figures 2-5 indicates that using the proposed distributed multi-robot deployment control algorithm, both holonomic and nonholonomic mobile robots spread out effectively from their initial gathering to cover the environment, based on the local driving forces defined on either the physical neighborhood or the topological neighborhood. However, while a holonomic mobile robot is capable of moving in any direction at any time (Fig.6), a nonholonomic mobile robot, under the nonholonomic constraint, has to move along its longitudinal direction at any time (Fig.7). For the convenience of display, we show the representative deployment paths of both holonomic (Fig.6) and nonholonomic (Fig.7) mobile robots generated from 12 robots instead of 120 robots, with the arrow head representing the robot orientation. In particular, in the nonholonomic case, backward movement along the longitudinal direction, which satisfies the nonholonomic constraint, is allowed as long as it moves a robot towards the desired neighborhood configuration.

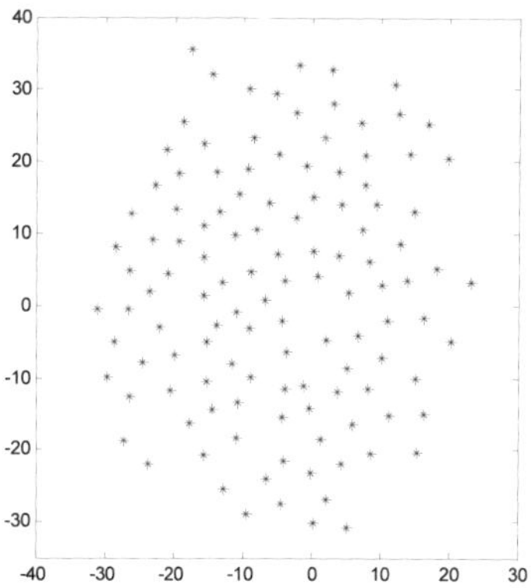

Fig. 4. Deployment of 120 nonholonomic mobile robots in an open 2D environment based on the state of the physical neighborhood

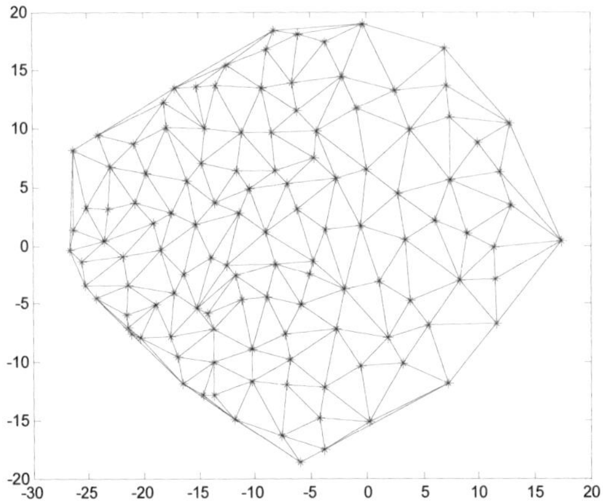

Fig. 5. Deployment of 120 nonholonomic mobile robots in an open 2D environment based on the state of the topological neighborhood

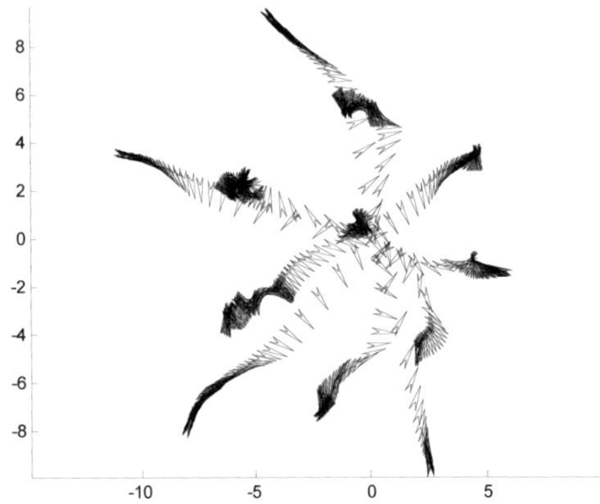

Fig. 6. Deployment paths for 12 holonomic mobile robots

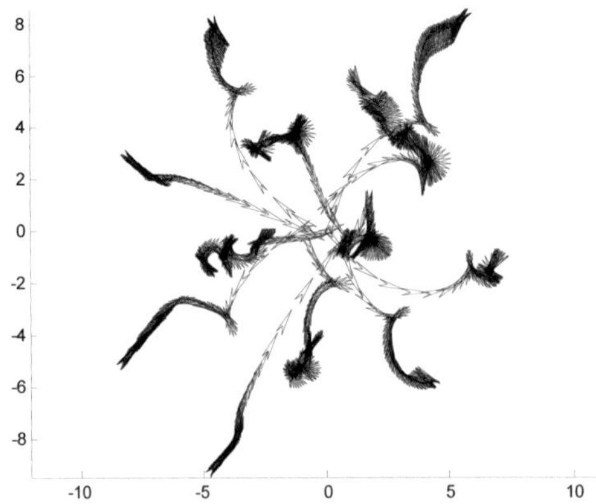

Fig. 7. Deployment paths for 12 nonholonomic mobile robots

4.3 Deployment in a corridor environment

Considering 120 holonomic mobile robots being deployed into a corridor environment with straight walls at y=±15, at first we define the self-deployment motion of each mobile robot based on the state of its physical neighborhood. A representative deployment of the robots at t=30 is shown as Fig.8. Then we define the self-deployment motion of each mobile robot based on the state of its topological neighborhood which is generated using the Delaunay triangulation. A representative deployment of the robots at t=30 is shown as Fig.9.

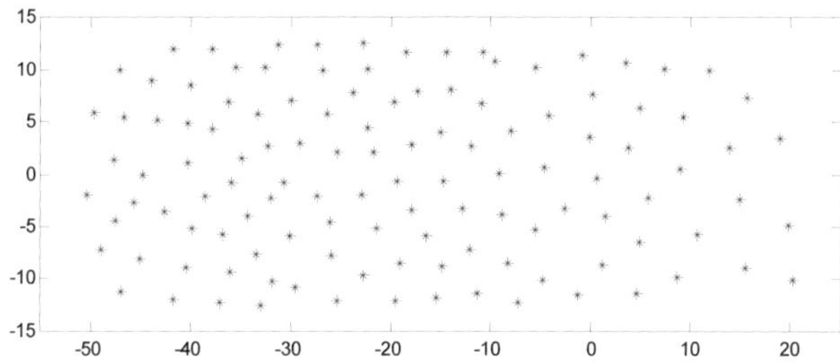

Fig. 8. Deployment of 120 holonomic mobile robots in a corridor environment based on the state of the physical neighborhood

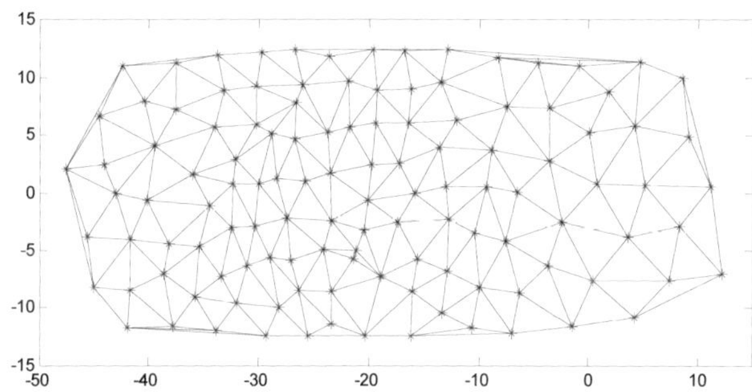

Fig. 9. Deployment of 120 holonomic mobile robots in a corridor environment based on the state of the topological neighborhood

Considering 120 nonholonomic mobile robots being deployed into the same corridor environment, at first we define the self-deployment motion of each mobile robot based on the state of its physical neighborhood. A representative deployment of the robots at t=30 is shown as Fig.10. Then we define the self-deployment motion of each mobile robot based on the state of its topological neighborhood which is generated using the Delaunay triangulation. A representative deployment of the robots at t=30 is shown as Fig.11.

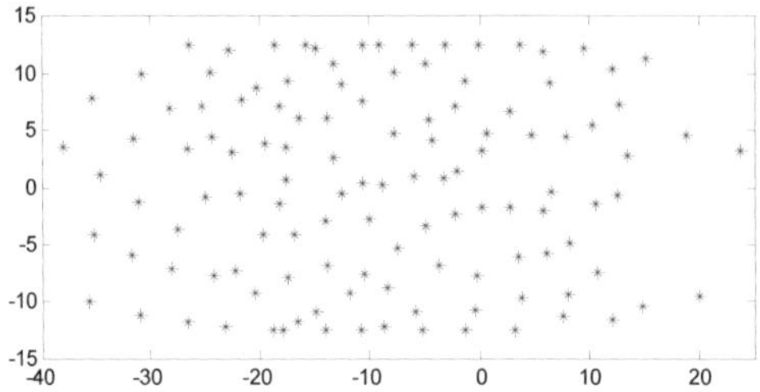

Fig. 10. Deployment of 120 nonholonomic mobile robots in a corridor environment based on the state of the physical neighborhood

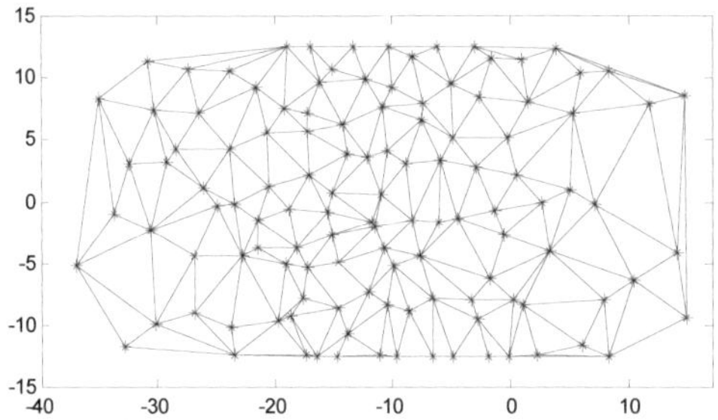

Fig. 11. Deployment of 120 nonholonomic mobile robots in a corridor environment based on the state of the topological neighborhood

Figures 8-11 indicates that using the proposed distributed multi-robot deployment control algorithm, both holonomic and nonholonomic mobile robots spread out effectively from their initial gathering to cover the corridor environment, based on the local driving forces defined on either the geometric neighborhood or the topological neighborhood.

5. Conclusions, discussions and future work

This chapter addresses an important research topic in the field of multi-robot systems, the deployment problem, and introduces a novel distributed multi-robot deployment control algorithm for spreading a team of mobile robots into a targeted environment to form sensor and communication coverage. Each member robot self-deploys according to its equation of deployment motion. The driving force for each mobile robot is defined according to the difference between the actual and desired configurations in the neighborhood of the robot. The Rayleigh's dissipation function is adopted to provide the necessary damping mechanism which maintains the stability of the deployment motion for each robot. Derived from the Hamilton's principle using the method of the variational calculus, the equation of deployment motion naturally incorporates the nonholonomic constraint arising in wheeled robots. Since the equation of deployment motion for each robot depends on only the robot's own kinematic state and its detectable positional relationship with nearby objects, the proposed scheme decentralizes the multi-robot deployment problem into the motion control of individual robots. The combined outcome of the local deployment motion of individual robots leads to the desired global coverage. Simulation results show that the proposed approach can effectively guide the deployment of multi-robot systems.

When deployed into a static environment with fixed obstacles and boundary, the multi-robot system may converge to static sensor coverage. When deployed into a dynamic environment with moving obstacles and evolving boundary, the multi-robot system can change its configuration adaptively. Moreover, the resulting coverage can be a partial instead of complete coverage to the targeted environment, when the environment is larger than the maximum static coverage area of the multi-robot system. After the initial coverage is formed, the multi-robot system can move to provide a mobile coverage of the whole environment. In fact, to take the full advantage of the mobility of the multi-robot system and reduce the operation cost, instead of using a large number of mobile robots to form a static coverage network for a large environment, it is often more efficient to send fewer mobile robots to provide a mobile coverage to the environment. The technique governing multi-robot mobile coverage will be discussed in our future work.

Since the proposed multi-robot deployment control scheme is derived based on a lumped dynamics model of mobile robots and incorporates the nonholonomic kinematic constraint, it helps to result in physically realizable deployment motion in realistic robot systems. However, in order to further improve the implementation robustness of the proposed algorithm to guarantee the physical realizability of the resulting deployment motion, more kinematic and dynamic constraints, such as maximum velocity and acceleration, will be incorporated, and more detailed dynamics models of realistic mobile robots will be studied. Experimental research with physical multi-robot systems will be conducted to verify the results of our algorithm research.

Moreover, we will investigate the convergence property of the whole multi-robot system towards the global optimal coverage as the result of the collection of local coverage. In fact, it is convenient to prove that, if a member robot R_i has a static neighborhood, i.e. other objects in R_i neighborhood are static, the self-deployment motion of R_i, defined by (13) and (14), converges asymptotically to a fixed equilibrium location. We consider the total energy function of each robot R_i as the Lyapunov function candidate

$$E_i = T_i + U_i \cdot \tag{15}$$

From both (13) and (14), we obtain

$$\dot{E}_i = -2F_i \tag{16}$$

for both holonomic and nonholonomic mobile robots. Since $k_{xi}>0$, $k_{yi}>0$ and $k_{\theta i}>0$, E_i tends to decrease, and hence the self-deployment motion of R_i is Lyapunov asymptotically stable. However, in practice, all member robots are moving. As indicated by the definition of the artificial potential energy on R_i (6), the motion of R_i is coupled with that of other objects in its neighborhood. The changing positions of other objects in R_i neighborhood result in the changing equilibrium position of R_i. Therefore, at each moment, R_i moves towards its new equilibrium position. However, this change will slow down as the robots spread out, and R_i deployment motion will converge, as we observe from the simulations. As a future work, we will study the analytical relationship between the global convergence at the multi-robot system level and the local convergence at the individual robot level.

The influence of the dynamic change in the multi-robot topological structure on the operational efficiency of deployment will also be investigated. Though the concept of topological neighborhood reduces the computational and communication complexities of the multi-robot deployment problem by coupling the deployment motion of each member robot with that of only its one-hop topological neighbors defined by Delaunay Triangulation, once the system topology changes, however, intensive communication and computation are required to update the Delaunay Triangulation of the whole multi-robot system. We will seek to maintain the global topologic optimality using local topologic adjustment instead of global reorganization.

6. Acknowledgements

Special thanks are given to Prof. Jindong Tan from the Department of Electrical and Computer Engineering, Michigan Technological University and Prof. Xin Wang from the Department of Electrical and Computer Engineering, State University of New York at Stony Brook for their suggestions to this work.

7. References

Bishop, B. E. (2007). On the use of capability functions for cooperative objective coverage in robot swarms. *2007 IEEE International Conference on Robotics and Automation.*

Borenstein, J.; Everett, H. R.; Feng, L. & Wehe, D. (1997). Mobile robot positioning: sensors and techniques. *Journal of Robotic Systems,* Vol.14, No.4, 231-249.

Butler, Z. & Rus, D. (2003). Event-based motion control for mobile sensor networks, *IEEE Pervasive Computing,* Vol.2, No.4, 34–43.

Cortes, J.; Martinez, S.; Karatas, T. & Bullo, F. (2004). Coverage control for mobile sensing networks, *IEEE Transactions on Robotics and Automation,* Vol.20, No.2, 243–255.

Fan, W. H.; Liu, Y. H.; Wang, F. & Cai, X. P. (2005). Multi-robot formation control using potential field for mobile ad-hoc networks. *2005 IEEE International Conference on Robotics and Biomimetics.*

Goldstein, H. (1980). *Classical Mechanics, 2nd edition,* Addison-Wesley, Reading, MA.

Howard, A.; Mataric, M. J.; Sukhatme, G. S. (2002). Mobile sensor network deployment using potential fields: a distributed, scalable solution to the area coverage problem. *Proceedings of the 6th International Conference on Distributed Autonomous Robotic Systems.*

Jiang, Q. (2006). An improved algorithm for coordination control multi-agent system based on r-limited Voronoi partitions. *Proceedings of the 2006 IEEE International Conference on Automation Science and Engineering.*

Jenkin, M. & Dudek, G. (2000). The paparazzi problem. Proceedings of the 2000 IEEE/RSJ International Conference on Intelligent Robots and Systems.

Ji, M. & Egerstedt, M. (2007). Distributed coordination control of multiagent systems while preserving connectedness. *IEEE Transactions on Robotics,* Vol.23, No.4, 693-703.

Jung, B. & Sukhatme, G. S. (2002). Tracking targets using multiple robots: the effect of environment occlusion. *Autonomous Robots,* Vol.13, 191-205.

Kerr, W.; Spears, D.; Spears, W. & Thayer, D. (2005). Two formal gas models for multi-agent sweeping and obstacle avoidance. *Lecture Notes in Artificial Intelligence,* Vol.3228, 111-130.

Lam, M. & Liu, Y. (2006). ISOGRID: an efficient algorithm for coverage enhancement in mobile sensor networks. *Proceedings of the 2006 IEEE/RSJ International Conference on Intelligent Robots and Systems.*

Pac, M. R.; Erkmen, A. M. & Erkmen, I. (2006). Scalable self-deployment of mobile sensor networks: a fluid dynamics approach. *Proceedings of the 2006 IEEE/RSJ International Conference on Intelligent Robotics and Systems.*

Parker, L. (1999). Cooperative robotics for multi-target observation. *Intelligent Automation and Soft Computing,* Vol.5, No.1, 5-19.

Parker, L. (2002). Distributed algorithms for multi-robot observation of multiple moving targets. *Autonomous Robots,* Vol.12, 231-255.

Preparata, F. & Shamos, M. I. (1985). *Computational Geometry: An Introduction,* Springer-Verlag, New York.

Poduri, S.; Sukhatme, G. S. (2004). Constrained coverage for mobile sensor networks. *Proceedings of 2004 IEEE International Conference on Robotics and Automation.*

Popa, D. O.; H. E. Stephanou; Helm, C. & Sanderson, A. C. (2004). Robotic deployment of sensor networks using potential fields. *Proceedings of the 2004 IEEE International Conference on Robotics & Automation.*

Reif, J. H. & Wang, H. (1999). Social potential fields: a distributed behavioral control for autonomous robots. *Robotics and Autonomous Systems,* Vol.27, 171-194.

Sander, P. V.; Peleshchuk, D. & Grosz, B. J. (2002). A scalable, distributed algorithm for efficient task allocation. *Proceedings of the first international joint conference on Autonomous agents and multiagent systems.*

Schwager, M.; Slotine, J. & Rus, D. (2007). Decentralized, adaptive control for coverage with networked robots. *2007 IEEE International Conference on Robotics and Automation.*

Tan, J.; Xi, N.; Sheng, W. & Xiao, J. (2004). Modeling multiple robot systems for area coverage and cooperation. *Proceedings of the 2004 IEEE International Conference on Robotics & Automation.*

Tan, J. (2005). A scalable graph model and coordination algorithms for multi-robot systems. *Proceedings of 2005 IEEE/ASME International Conference on Advanced Intelligent Mechatronics.*

Winfield, A. (2000). Distributed sensing and data collection via broken ad hoc wireless connected networks of mobile robots, In: *Distributed Autonomous Robotic Systems 4,* Parker, L.; Bekey, G. & Barhen, J., Springer-Verlag.

Zhou, Y. & Tan, J. (2008). Deployment of multi-robot systems under the nonholonomic constraint. *IEEE/ASME International Conference on Advanced Intelligent Mechatronics.*

Zhou, Y. (2008). A distributed self-deployment algorithm suitable for multiple nonholonomic mobile robots. *Proceedings of the ASME 2008 International Design Engineering Technical Conferences & Computers and Information in Engineering Conference.*

15

Controlling a Finger-arm Robot to Emulate the Motion of the Human Upper Limb by Regulating Finger Manipulability

Jian HUANG[1], Masayuki HARA[2] and Tetsuro YABUTA[3]
[1]*Kinki University,*
[2]*Swiss Federal Institute of Technology (EPFL),*
[3]*Yokohama National University,*
[1,3]*Japan*
[2]*Swiss*

1. Introduction

The human upper limb possesses a high degree of freedom (DOF) and its redundant structure permits greater flexibility in various dexterous manipulations. The simplest structure of a multifingered robot arm is constructed by fixing a robot finger onto the end effector of a robot arm. A robot with such a structure is also called a macro-micro manipulator (Nagai & Yoshigawa, 1994, 1995; Yoshikawa, et al. 1993). Similar to the human upper limb, the finger-arm robot exhibits a high redundancy. The movement of the robots with such high redundancies creates the problem of how to determine the numerous DOFs of its joints.

Controlling a robot with a high degree of redundancy is a fundamental problem in the field of robotics. A large number of studies have been published on the methodology for determining the redundant DOFs of a robot. Avoidance control of kinematics singularity (Nakamura & Hanafusa, 1986; Furusho & Usui 1989) and obstacle collision avoidance (Khatib, 1986; Maciejewski & Klein, 1985; Loeff & Soni, 1975; Guo & Hsia, 1993; Glass et. Al, 1995) by using redundant DOFs has been mostly investigated. In order to realize desired solutions for the above mentioned problems, methods involving null space (Vannoy & Xiao, 2004) and the criterion function (Kim & Kholsa, 1992; Ma & Nechev, 1995; Ma et al, 1996) have been typically applied.

The finger-arm robot is unlike conventional redundant manipulators. The finger is usually lightweight and has a small link size as compared to the arm. Therefore, it is inappropriate to directly apply the methods developed for controlling a redundant manipulator to the finger-arm robot. To achieve the dexterity like the human hand-arm, a lightweight finger should be actively moved whereas the arm cooperate the movement of the finger, which will greatly improve the performance of a robot (Khatib, 1995; Melchiorri & Salisbury, 1995). The human hand-arm system exhibits similar features. The human hand is obviously lighter, smaller and more sensitive as compared to the arm. The hand-arm coordination is well organized by the central nervous system so as to generate a natural motion. The motivation of this study is to develop a control method emulating a natural movement similar to that of a human upper limb.

Inspired by the human hand-arm movement, a motion control algorithm of a finger-arm robot has been proposed in our study based on the concept of using manipulability of the finger. An effective motion can be generated using the proposed method rather than merely calculating a geometric path for a kinematics solution or optimizing certain dynamic criteria by using the robot's redundant DOFs.
In our study, a heuristic motion control method for a finger-arm robot is firstly proposed. Using the heuristic method, the motion of the arm is heuristically determined by the manipulability of the finger (Huang et al., 2006, Quan et al., 2006). The arm moves to cooperate with the finger's movement in order to maintain finger's manipulability at a desired level. Thus, complicated motions of the finger-arm can be simply divided into separate motions for the arm and for the finger.
However, from the viewpoint of manipulability regulation, the performance of this heuristic method is unsatisfactory, especially when the finger's manipulability is actively required. To improve the property of manipulability regulation, a control of algorithm that employs the steepest ascent method to actively modulate the finger's manipulability is also proposed. Using the steepest ascent method, manipulability of the finger can be immediately increased when it drops below a given reference. As a result, the finger is robust to its singularity. We performed several experiments to demonstrate the effectiveness of the proposed methods.

2. The finger-arm robot and kinematics

2.1 Overview of the system

In this study, a robot finger with three compact motors (Yasukawa Co.) and a robot arm with six DOFs (PA-10, Mitsubishi Heavy Industry Co.) are used as shown in Fig.1. The finger robot is fixed onto the end effector of the robot arm. Such a finger-arm robot has 9 DOFs, whereas a task to be completed in the 3D space of the robot's base coordinate Σ_b requires 6 DOFs. Therefore, three DOFs are redundant. The task to be completed in this study is to generate a motion to trace a desired curve with the fingertip. Since the size of the finger robot is comparatively smaller as compared to that of the manipulator, the finger robot easily reaches its limit.

2.2 Kinematics

The end-effector coordinate and arm base coordinate are set as Σ_t and Σ_b respectively, as shown in Fig.1. The joint angular velocity vector $\dot{\boldsymbol{\theta}} \in R^{9\times1}$ of the finger-arm robot is defined as follows:

$$\dot{\boldsymbol{\theta}} = \begin{bmatrix} \dot{\boldsymbol{\theta}}_a \\ \dot{\boldsymbol{\theta}}_f \end{bmatrix} \tag{1}$$

where $\dot{\boldsymbol{\theta}}_a \in R^{6\times1}$ and $\dot{\boldsymbol{\theta}}_f \in R^{3\times1}$ are the joint angular velocities of the arm and the finger, respectively. Of the arm's end-effector position and orientation $\boldsymbol{p}_t \in R^{6\times1}$ in Σ_b as well as of the fingertip position and orientation $\boldsymbol{p}_f \in R^{6\times1}$ in Σ_b are defined as follows:

$$\boldsymbol{p}_t = \begin{bmatrix} x_t & y_t & z_t & \alpha_t & \beta_t & \gamma_t \end{bmatrix}^T \tag{2}$$

$$\boldsymbol{p}_f = [x_f \quad y_f \quad z_f \quad \alpha_f \quad \beta_f \quad \gamma_f]^T \tag{3}$$

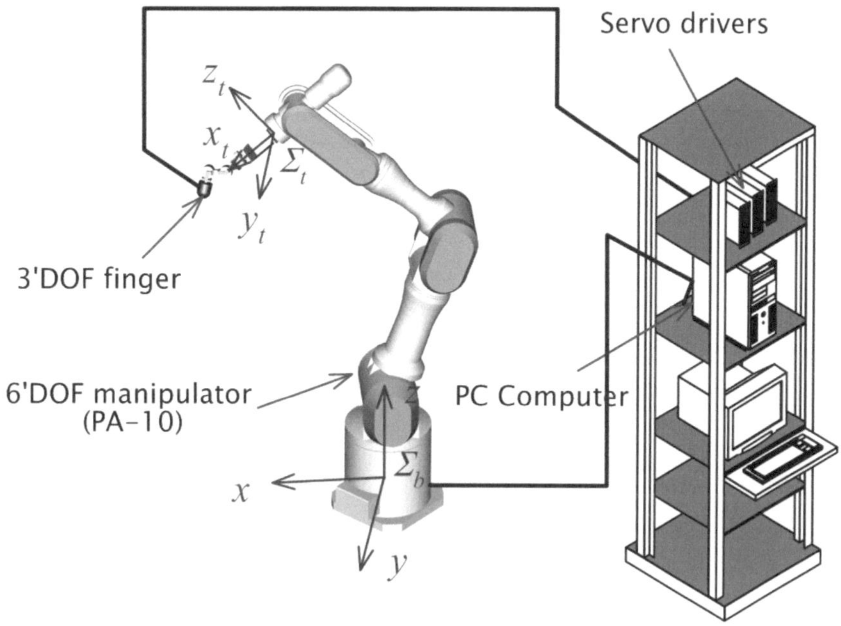

Fig. 1. Schematic diagram of the finger-arm robot

The relationship between its joint angular velocity $\dot{\boldsymbol{\theta}}$ and the fingertip velocity $\dot{\boldsymbol{p}}_f$ in Σ_b can be theoretically expressed as follows:

$$\dot{\boldsymbol{p}}_f = \boldsymbol{J} \cdot \dot{\boldsymbol{\theta}} \tag{4}$$

where $\boldsymbol{J} \in R^{6\times 9}$ is the Jacobian of the finger-arm robot. For the robot arm, we have

$$\dot{\boldsymbol{p}}_t = \boldsymbol{J}_a \cdot \dot{\boldsymbol{\theta}}_a \tag{5}$$

where $\boldsymbol{J}_a \in R^{6\times 6}$ is the Jacobian of the arm. For the finger, it is known that

$$^t\dot{\boldsymbol{p}}_f = \boldsymbol{J}_f \cdot \dot{\boldsymbol{\theta}}_f \tag{6}$$

where $J_f \in R^{3\times 3}$ is the Jacobian of the finger, and $^t\dot{\boldsymbol{p}}_f \in R^{3\times 1}$ is the fingertip velocity in coordinates Σ_t.

If a non-redundant robot is used, the Jacobian $\boldsymbol{J}$ is invertible. Subsequently, the joint angular velocity can be obtained as follows:

$$\dot{\boldsymbol{\theta}} = \boldsymbol{J}^{-1} \cdot \dot{\boldsymbol{p}} \tag{7}$$

However, because a redundant robot is used in our study, the Jacobian $\boldsymbol{J}$ in (4) is not a square matrix. Therefore, its inverse $\boldsymbol{J}^{-1}$ cannot be computed.

2.3 Manipulability of the finger

In robotics, manipulability is used as a criterion to describe the moving potential of a robot (Yoshikawa, 1985). Here, the manipulability W_f of the finger in Fig.1 is calculated as follows

$$W_f = \sqrt{\det(\boldsymbol{J}_f(\boldsymbol{\theta}_f)\cdot \boldsymbol{J}_f^T(\boldsymbol{\theta}_f))}$$
$$== l_2 l_3 \sin\theta_3 \left(l_1 + l_2 \sin\theta_2 + l_3 \sin(\theta_2+\theta_3)\right) \tag{8}$$

where $\boldsymbol{J}_f \in R^{3\times3}$ is the Jacobian of the finger, and the vector $\boldsymbol{\theta}_f = \begin{bmatrix}\theta_1 & \theta_2 & \theta_3\end{bmatrix}^T$ denotes the joint angles of the finger, l_i (i=1, 2, 3) is the ith link length of the finger.

When a human being requires his fingers and arm to perform a task, his limb is maneuvered such that his hand covers the largest possible range and the task is easily completed. In robotics, this property of the moving potential is referred to as manipulability.

3. Methods of motion control

3.1 The Heuristic Method (HM)

We assume T is the control cycle. As shown in Fig.2, at time $t=kT$, (k=0, 1, 2, $\cdots$) the fingertip moves along a desired trajectory $\boldsymbol{p}_d(k) \in R^{3\times1}$ in the arm's base coordinates Σ_b.

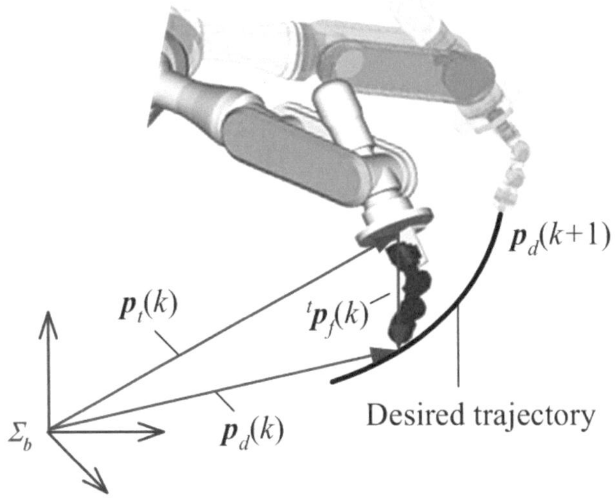

Fig. 2. Position vectors of the finger and the arm

The position of the arm's end-effector in Σ_b is $\boldsymbol{p}_t(k)$ and the fingertip position in Σ_t is ${}^t\boldsymbol{p}_f(k)$. Thus, we have

$$\boldsymbol{p}_d(k) = \boldsymbol{s}_1 \boldsymbol{p}_t(k) + \boldsymbol{R}_t \cdot {}^t\boldsymbol{p}_f(k) \tag{9}$$

where $\boldsymbol{R}_t \in R^{3\times3}$ is the rotation matrix of the arm in Σ_b, and $\boldsymbol{s}_1 \in R^{3\times6}$ is a constant matrix given as follows:

$$\boldsymbol{s}_1 = \begin{bmatrix} 1 & 0 & 0 & 0 & 0 & 0 \\ 0 & 1 & 0 & 0 & 0 & 0 \\ 0 & 0 & 1 & 0 & 0 & 0 \end{bmatrix} \tag{10}$$

If we assume that the orientation of the arm remains unchanged, the rotation matrix $\boldsymbol{R}$ is a constant. Thus, from (9), we get

$$\Delta \boldsymbol{p}_d(k) = \boldsymbol{s}_1 \Delta \boldsymbol{p}_t(k) + \boldsymbol{R}_t \cdot \Delta^t \boldsymbol{p}_f(k) \tag{11}$$

Where

$$\begin{aligned} \Delta \boldsymbol{p}_d(k) &= \boldsymbol{p}_d(k) - \boldsymbol{p}_d(k-1) \\ \Delta \boldsymbol{p}_t(k) &= \boldsymbol{p}_t(k) - \boldsymbol{p}_t(k-1) \\ \Delta^t \boldsymbol{p}_f(k) &= {}^t\boldsymbol{p}_f(k) - {}^t\boldsymbol{p}_f(k-1) \end{aligned} \tag{12}$$

Using (6), (11) can be expressed as

$$\begin{aligned} \Delta \boldsymbol{p}_d(k) &= \boldsymbol{s}_1 \Delta \boldsymbol{p}_t(k) + \boldsymbol{R}_t T \cdot {}^t\dot{\boldsymbol{p}}_f(k) \\ &= \boldsymbol{s}_1 \Delta \boldsymbol{p}_t(k) + T\boldsymbol{R}_t \boldsymbol{J}_f \dot{\boldsymbol{\theta}}_f(k) \end{aligned} \tag{13}$$

When the manipulability W_f is higher than a given reference manipulability W_{fr}, the movement of the arm is unnecessary. According to (13), we have

$$\Delta \boldsymbol{p}_t(k) = 0 \tag{14}$$

Subsequently, we get

$$\Delta \boldsymbol{p}_d(k) = T\boldsymbol{R}_t \boldsymbol{J}_f \dot{\boldsymbol{\theta}}_f(k) \tag{15}$$

Equation (15) expresses that only the finger moves with a joint angular velocity $\dot{\boldsymbol{\theta}}_f$ to trace the desired trajectory. Changing the joint angles of the finger results in a change in $\boldsymbol{J}_f$. As a result, two possibilities of W_f can be considered.

1. If W_f is still higher than W_{fr}, the finger will keep moving and tracing the desired trajectory, while the arm maintains its previous position.
2. If W_f falls below W_{fr}, moving the arm becomes necessary.

Further, if $\Delta \boldsymbol{p}_d(k)$ is theoretically assumed to be completed only by moving the arm, then from (13), we get

$$\Delta \boldsymbol{p}_d(k) = \boldsymbol{s}_1 \Delta \boldsymbol{p}_t(k), \qquad W_f(k) < W_{fr} \tag{16}$$

Thus,

$$\dot{\boldsymbol{\theta}}_f(k) = 0 \tag{17}$$

Equation (17) indicates that the finger stops moving. Hence, the manipulability W_f will remain unchanged as follows:

$$\Delta \boldsymbol{W}_f = 0 \tag{18}$$

Based on (16), we can say that moving the arm instead of moving the finger can theoretically prevent any further decrease in W_f. However, switching control between the arm and the finger by (15)~(17) result in an instant change in velocity.

In this study, in order to achieve the smooth movement of the arm, a desired position $\boldsymbol{p}_{td}(k)$ of the arm at time $t=kT$ is generated by

$$\boldsymbol{p}_{td}(k) = \boldsymbol{p}_t(k-1) + \Delta \boldsymbol{p}_{td}(k) \tag{19}$$

where

$$\Delta \boldsymbol{p}_{td}(k) = A(W_f)\boldsymbol{s}_1^T \Delta \vec{\boldsymbol{p}}_d(k) \tag{20}$$

Here, $A(W_f)$ is a scalar parameter related to the manipulability of the finger. $\Delta \vec{\boldsymbol{p}}_d(k)$ is an unit motion vector in the direction of the desired trajectory and can be computed as follows:

$$\Delta \vec{\boldsymbol{p}}_d(k) = \Delta \boldsymbol{p}_d(k) / |\Delta \boldsymbol{p}_d(k)| \tag{21}$$

In order to move the arm without suddenly changing its velocity, the parameter $A(W_f)$ is heuristically determined by

$$A(W_f) = \begin{cases} 0 & W_f(k) \geq W_{fr} \\ K_a(W_{fr} - W_f(k)) & W_f(k) < W_{fr} \end{cases} \tag{22}$$

where K_a is a selected coefficient.

Compared to (16), the heuristic method shown in (22) yields a smooth motion profile such that the arm moves without any instant change in velocity. Therefore, the finger can also move smoothly. Furthermore, when the generated movement of the arm given in (19) and (20) is larger than the necessary change $\Delta \boldsymbol{p}_d(k)$ of the desired trajectory given by (16), i.e.

$$|\boldsymbol{s}_1 \Delta \boldsymbol{p}_{td}(k)| > |\Delta \boldsymbol{p}_d(k)| \tag{23}$$

the finger will move in a direction such that the manipulability W_f increases. Therefore, we have

$$\Delta W_f \geq 0 \tag{24}$$

In reality, the arm will either not move or move very slowly when $\Delta W_f = W_f - W_{fr}$ is very small because of the friction it experiences at the joint motors and gears. This implies that W_f will probably keep decreasing for a short period. However, further drop of W_f will be definitely prevented as an integral effectiveness with the assist movement of the arm.

The proposed control block diagram of the heuristic method is shown in Fig.3 where, $\boldsymbol{\Lambda}_f$ represents the kinematics of the finger; $\boldsymbol{\Lambda}_a$, the kinematics of the arm; $\boldsymbol{J}_a$, the Jacobian of the arm; $\dot{\boldsymbol{\theta}}_a \in R^{6\times1}$, the joint velocity of the arm; and $\boldsymbol{G}_f(z)$ and $\boldsymbol{G}_a(z)$, the PID controllers of the finger and the arm, respectively. $\boldsymbol{G}_f(z)$ is defined as:

$$\boldsymbol{G}_f(z) = \boldsymbol{K}_P^f + \boldsymbol{K}_I^f \frac{z}{z-1} + \boldsymbol{K}_D^f(1 - z^{-1}) \tag{25}$$

where $\boldsymbol{K}_P^f \in R^{3\times3}$, $\boldsymbol{K}_I^f \in R^{3\times3}$ and $\boldsymbol{K}_D^f \in R^{3\times3}$ are the given gains of the PID controller. $\boldsymbol{G}_a(z)$ is given as follows:

$$\boldsymbol{G}_f(z) = \boldsymbol{K}_P^f + \boldsymbol{K}_I^f \frac{z}{z-1} + \boldsymbol{K}_D^f(1 - z^{-1}) \tag{26}$$

where $\boldsymbol{K}_P^a \in R^{6\times6}$, $\boldsymbol{K}_I^a \in R^{6\times6}$ and $\boldsymbol{K}_D^a \in R^{6\times6}$ are the given gains of the PID controller.

As shown in Fig.3, when time $t=kT$, based on the finger's manipulability W_f given by (8), the desired position $\boldsymbol{p}_{td}(k)$ of the end-effector of the arm is calculated from (19)~(22).

Subsequently, the desired position ${}^{t}\boldsymbol{p}_{fd}(k)$ of the finger can be computed by using (9). The obtained $\boldsymbol{p}_{td}(k)$ and ${}^{t}\boldsymbol{p}_{fd}(k)$are fed as input to each servo loop so as to generate the expected motion.

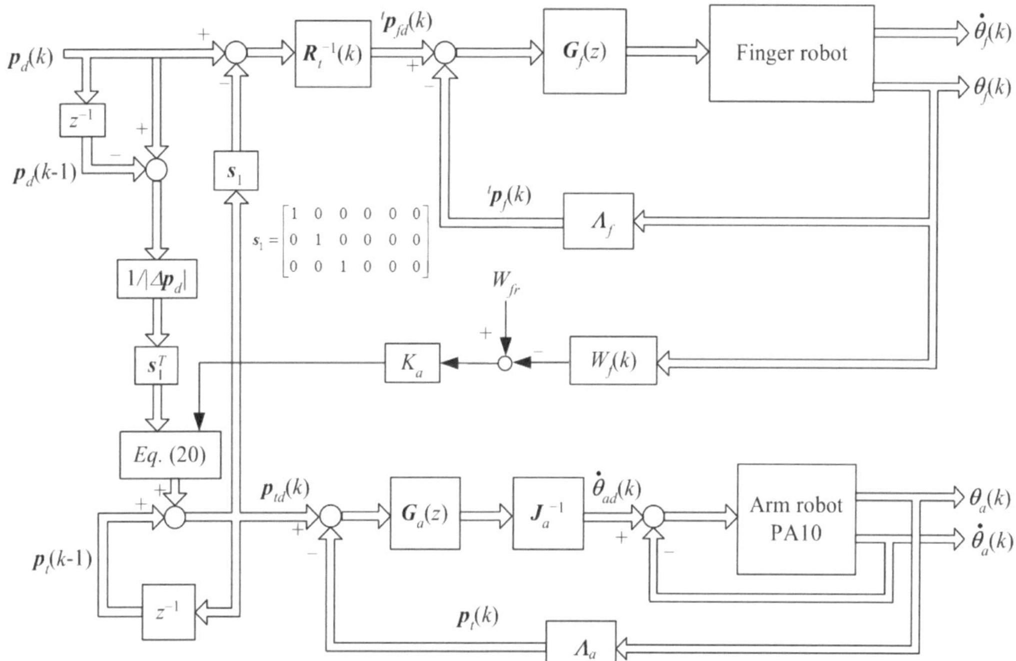

Fig. 3. Control block diagram of the heuristic method

3.2 The Steepest Ascent Method (SAM)

As described above, the basic concept of the heuristic method is to cooperatively move the arm to an expected position by using (19)~(22). The disadvantage of the heuristic method is that the finger's manipulability can not be directly increased. Therefore, it is inappropriate to apply the heuristic method to a task wherein the active modulation of the manipulability of the finger is strongly required. In order to effectively regulate the manipulability of the finger, employing a steepest ascent method is also attempted in our study. The most important feature of this method is that the manipulability of the finger W_f will increase rapidly once it is smaller than the reference value W_{fr}. The details of the algorithm are provided below.

When W_f is higher than W_{fr}, which is similar to the heuristic method, the movement of the arm is unnecessary. Therefore, only the finger moves to trace the desired trajectory as (15)

Whenever the manipulability of the finger reduces to a level smaller than the given reference W_{fr}, the movement of the arm is triggered. At this time, we apply the steepest ascent method to modulate the manipulability W_f of the finger by

$$\boldsymbol{\theta}_{fd}(k) = \boldsymbol{\theta}_{fd}(k-1) + \lambda \frac{\partial W_f}{\partial \boldsymbol{\theta}_f} \tag{27}$$

where λ is the gain coefficient. According to (8), for the finger robot, we have

$$\begin{cases} \dfrac{\partial W_f}{\partial \theta_1} = 0 \\ \dfrac{\partial W_f}{\partial \theta_2} = l_2 l_3 \sin\theta_3 \left(l_2 \cos\theta_2 + l_3 \cos(\theta_2 + \theta_3) \right) \\ \dfrac{\partial W_f}{\partial \theta_3} = l_2 l_3 \left(l_1 \cos\theta_3 + l_2 \cos\theta_3 \sin\theta_2 + l_3 \sin(\theta_2 + 2\theta_3) \right) \end{cases} \tag{28}$$

Then, the fingertip position ${}^t\boldsymbol{p}_f(k)$ in the arm's end effector coordinates Σ_t can be computed from the kinematics expressed as

$${}^t\boldsymbol{p}_f(k) = \Lambda_f(\boldsymbol{\theta}) \tag{29}$$

where $\boldsymbol{\Lambda}_f$ represents the kinematics of the finger robot. Therefore, the desired position $\boldsymbol{p}_{td}(k)$ of the arm's end effector in Σ_b can be computed by

$$\boldsymbol{p}_{td}(k) = \boldsymbol{s}_1^T(\boldsymbol{p}_d(k) - \boldsymbol{R}_t \cdot {}^t\boldsymbol{p}_f(k)) \,. \tag{30}$$

Equations (27)~(30) expresses the arm movement that must be performed once W_f decreases below W_{fr}. To achieve this effect, the arm must move to the desired position specified by (30). Therefore, the finger's joint angle is primarily maintained by the steepest ascent method. As compared to the heuristic method, the manipulability of the finger robot will increase immediately by using the steepest ascent method once it reduces to a level smaller

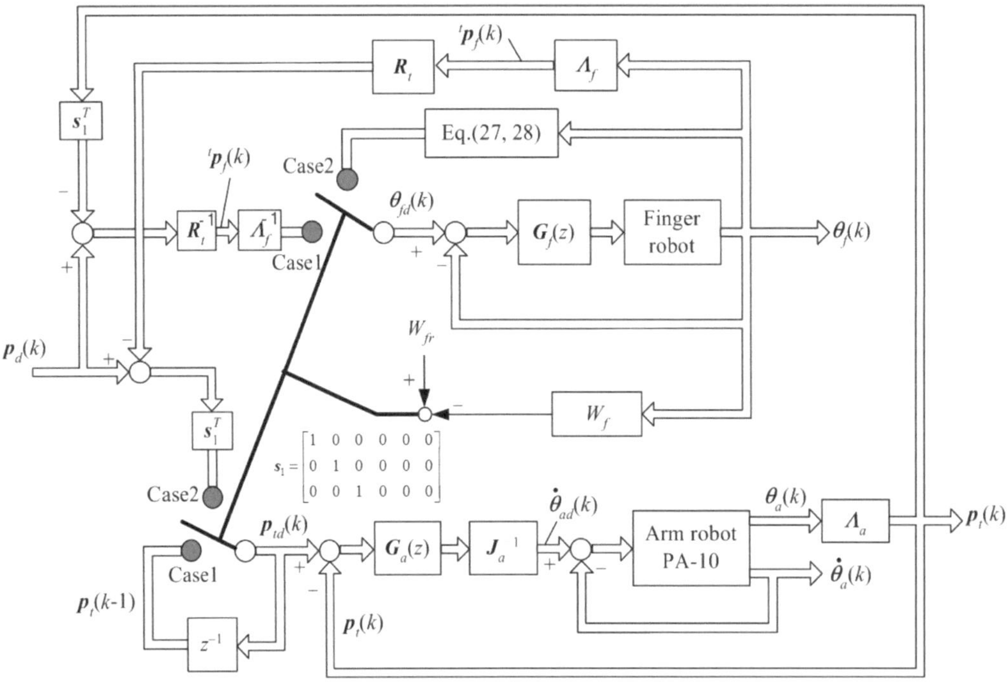

Fig. 4. Control block diagram of applying the steepest ascent method
(Case 1: $W_f > W_{fr}$, Case 2: $W_f \geq W_{fr}$)

than W_{fr}. Moreover, in this study, the steepest ascent method given by (27)~(30) will be applied once it is triggered, and it will be performed until the finger's manipulability reaches an upper threshold W_{ft}. The proposed control block diagram of the steepest ascent method is shown in Fig. 4.

4. Experimental results

To demonstrate the effectiveness of the proposed methods, some experiments were made.

4.1 Motion control using HM

Two distinct desired trajectories are charted out. Since the orientation of the fingertip with respect to the given curve is not specified in the experiment, the orientation of the arm's end-effector is determined as a constant vector. The other related control parameters are listed in Table I.

Control sampling interval, $T = 0.005$ s, Reference manipulability, $W_{fr} = 0.00018$,
Control parameters of the arm robot: Initial orientation of the fingertip in Σ_b: $\alpha_0 = 0$ [rad], $\beta_0 = 1.047$ [rad], $\gamma_0 = 0$ [rad],
Control parameters of the arm robot: $\boldsymbol{K}_P^a = \mathrm{diag}[5.0 \;\; 5.0 \;\; 5.0 \;\; 2.0 \;\; 2.0 \;\; 2.0]$ [1/s], $\boldsymbol{K}_I^a = \mathrm{diag}[0.5 \;\; 0.5 \;\; 0.5 \;\; 0.2 \;\; 0.2 \;\; 0.2]$ [1/s], $\boldsymbol{K}_D^a = \mathrm{diag}[0.1 \;\; 0.1 \;\; 0.1 \;\; 0.1 \;\; 0.1 \;\; 0.1]$ [1/s],
Control parameters of the finger robot: $\boldsymbol{K}_P^f = \mathrm{diag}[1.5\times10^4 \;\; 1.5\times10^4 \;\; 1.5\times10^4]$ [Nm/rad], $\boldsymbol{K}_I^f = \mathrm{diag}[0.4\times10^3 \;\; 0.4\times10^3 \;\; 0.4\times10^3]$ [Nm/rad], $\boldsymbol{K}_D^f = \mathrm{diag}[0.1\times10^3 \;\; 0.1\times10^3 \;\; 0.1\times10^3]$ [Nm/rad],
The total time taken to complete the task is 30 s, Thus, N=6000.

Table 1. Parameters of the experiments using HM.

(A) Tracing a Three Dimensional Sinusoidal Curve

When the amplitude of the desired sinusoidal curve is approximately equal to the total link length of the finger, the manipulability of the finger will fall to a very small value and the tracing task would not be completed without the assist movement of the arm. In this experiment, a three dimensional sinusoidal curve with an amplitude of 0.1[m] is given in Σ_b by

$$\begin{cases} x = 0.709 + 0.1\sin(2\pi k / N) \\ y = 0.079 + 0.3k/(N\sqrt{2}) \\ z = 0.794 - 0.3k/(N\sqrt{2}) \end{cases} \qquad (k = 0,1,2,\cdots,N) \tag{31}$$

where N is the total sampling number.

The obtained positions of the fingertip and the arm's end-effector are shown in Fig.5(a), and the results indicate that the arm also moves along a trajectory similar to that of the given sinusoidal curve to assist the finger to accurately trace the desired curve. The manipulability W_f and magnitude of the arm's velocity $|\dot{\boldsymbol{p}}_t|$ are drawn in Fig.5(b). In this figure, W_f of the finger is higher than W_{fr} during its starting period. Thus, only the finger moves to draw the curve, while the velocity of the arm is almost zero. Once W_f falls below W_{fr}, the velocity $|\dot{\boldsymbol{p}}_t|$ of the arm moves to augment movement of the finger. As a result, the manipulability of the finger increases.

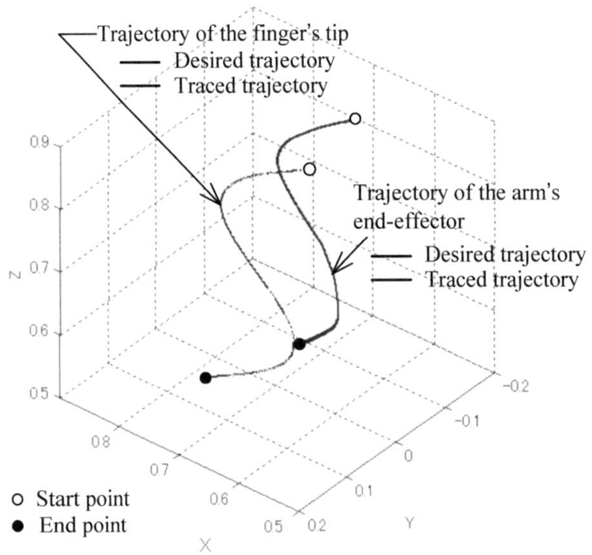

(a) Positions of the arm's end effctor and the fingertip in Σ_b

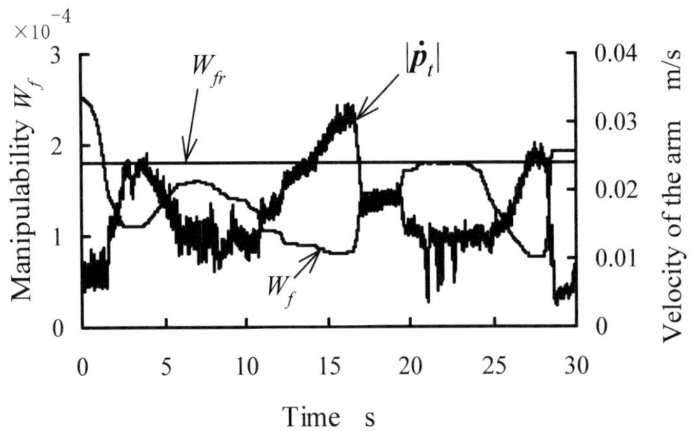

(b) Manipulability of the finger

Fig. 5. Results of tracing a 3D trajectory using the heuristic method

(B) Tracing a Free Hand Figure

The task of tracing a free-hand figure is also completed. This free-hand figure is composed of a small triangle (with edge lengths of 0.03[m], 0.04[m] and 0.05[m]), a small circle (with a diameter of 0.06[m]) and a longer straight line (with a length of 0.3[m]) to link the circle and the triangle together.

When a human draws such a figure, he will naturally move his finger primarily to draw the delicate part of the figure while he moves his arm to maintain the desired moving potential of his finger. As compared to the arm movement, hand movement consumes less energy because it has a small inertia.

The control parameters used in this experiment are same as listed in Table I, but W_{fr} is set to 0.0002. The positions of the arm's end-effector and the fingertip are shown in Fig.6(a), and the result of W_f is shown in Fig.6(b).

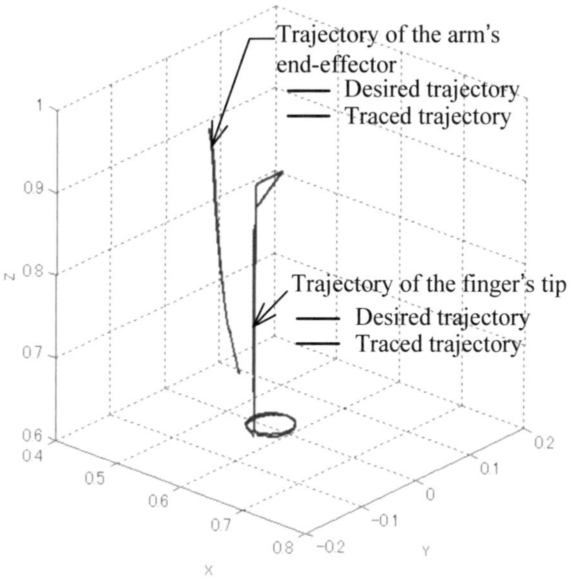

(a) Positions of the arm end effctor and the fingertip in Σ_b

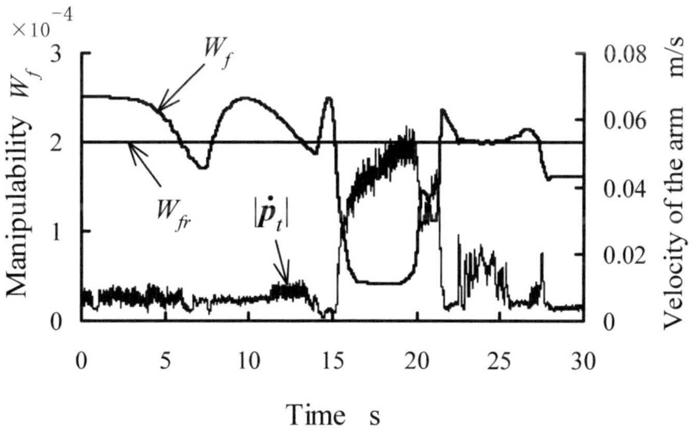

(b) Manipulability of the finger

Fig. 6. Results of tracing a free hand trajectory using the heuristic method

Unlike the movements in experiment I shown in Fig.5(a), the robot finger plays a dominant role in tracing the delicate parts of the triangle and the circle, while the arm moves along the straight line so as to maintain the desired value of the moving potential of the finger. As shown in Fig.6(b), the manipulability W_f of the finger is higher in the initial stages of drawing the triangle. However, the value of W_f falls gradually when the finger begins to move along the straight line toward the circle. To improve the W_f, $|\dot{\boldsymbol{p}}_t|$ of the arm increases resulting in an increase in W_f, as shown in Fig.6(b). When the finger reaches near the position where the circle needs to be traced, the arm stops moving and the finger traces the circle. Thus, the proposed method can naturally segregate the complicated motion of the finger-arm robot into two separate motions of the arm and the finger as in the case of a human being.

4.2 Motion control using SAM

To demonstrate the effectiveness of the steepest ascent method, a few experiments are performed. The control parameters used in the experiments are listed in Table II.

Control sampling interval: $T = 0.005$ s, Curve width: $L = 0.30$ [m], A=0.05[m],
Initial position and orientation of the fingertip in Σ_b: $x_0 = 0.381$ [m], $y_0 = 0.025$ [m], $z_0 = 1.043$ [m], $\alpha_0 = 0$ [rad], $\beta_0 = 1.047$ [rad], $\gamma_0 = 0$ [rad],
Control parameters of the arm robot: $\boldsymbol{K}_P^a = \mathrm{diag}[8.0\ \ 8.0\ \ 8.0\ \ 24.0\ \ 24.0\ \ 24.0]$ [1/s], $\boldsymbol{K}_I^a = \mathrm{diag}[1.0\ \ 1.0\ \ 1.0\ \ 1.0\ \ 1.0\ \ 1.0]$ [1/s], $\boldsymbol{K}_D^a = \mathrm{diag}[0.01\ \ 0.01\ \ 0.01\ \ 0.01\ \ 0.01\ \ 0.01]$ [1/s],
Control parameters of the finger robot: $\boldsymbol{K}_P^f = \mathrm{diag}[60.0\ \ 60.0\ \ 60.0]$ [Nm/rad], $\boldsymbol{K}_I^f = \mathrm{diag}[0.2\ \ 0.2\ \ 0.2]$ [Nm/rad], $\boldsymbol{K}_D^f = \mathrm{diag}[4.0\ \ 4.0\ \ 4.0]$ [Nm/rad],
Total sampling number: N=6000.

Table 2. Parameters of the experiments using SAM

(A) Tracing a Three Dimensional Sinusoid Curve

For comparing the heuristic method and the steepest ascent method, as shown in Fig.7, experiments are performed for tracing a three dimensional sinusoid curve amplitudes with the fingertip. The sinusoid curve is given as

$$\begin{cases} x = x_0 + Lk/N \\ y = y_0 \qquad\qquad\qquad k = 0,1,2,\cdots,N \\ z = z_0 + A\sin(2\pi k/N) \end{cases} \tag{32}$$

where (x_0, y_0, z_0) is the coordinate of the initial position; L, the curve width along the x axis; A, the amplitude of the sinusoid curve; k, the sampling count; and N, the maximum sampling number.

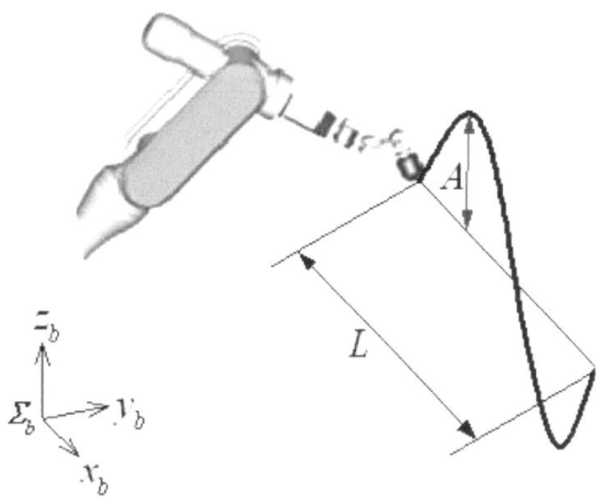

Fig. 7. Tracing a given sinusoid curve with the fingertip

The results obtained from the experiments by applying both the heuristic method and the steepest ascent method to trace the given sinusoid curve are shown in Fig.8(a) and Fig.8(b), respectively. The fingertip position p_f obtained by using both the heuristic method and the steepest ascent method is almost same as the desired p_{fd}, as shown in Fig.8(a) and Fig.8(b).

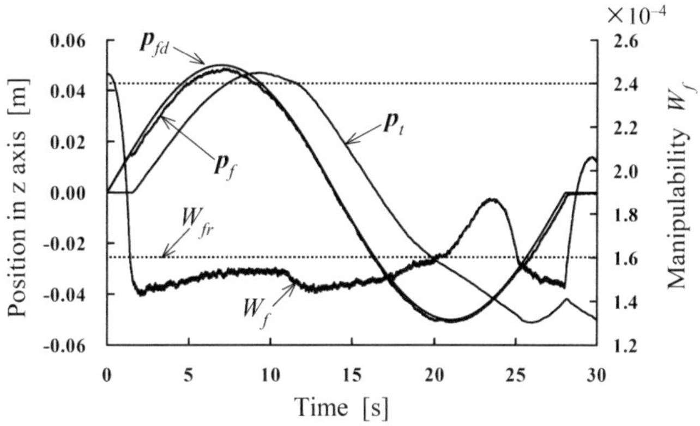

(a) Results obtained by heuristic method (W_{fr} = 1.6×10^{-4})

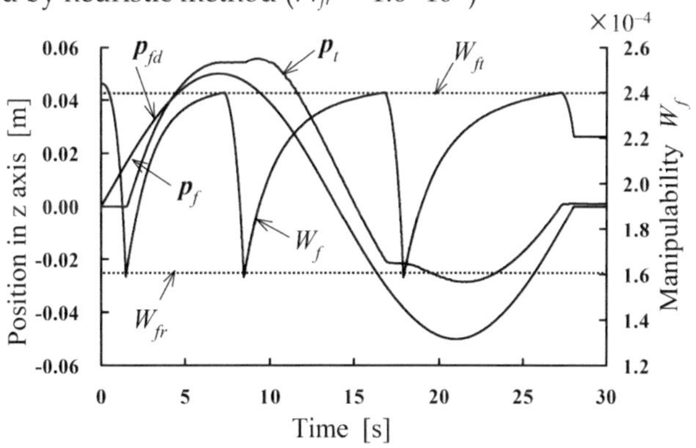

(b) Results obtained by the steepest ascent method (λ=5, W_{fr} =1.6×10^{-4}, W_{ft} =2.4×10^{-4})

Fig. 8. Results of tracing a sinusoid curve

However, the manipulability W_f of the finger in Fig.8(a), which is obtained by using the heuristic method increases gradually. This is because the basic concept of the heuristic method is to move the arm to a position where the finger's manipulability will not decrease further, instead of directly increasing the manipulability of the finger.

In contrast to the results of the heuristic method, the manipulability W_f of the finger increases significantly and is maintained above W_{fr} for almost the entire duration as indicated by Fig.8(b). This is because, once W_f drops below W_{fr}, the arm moves in the steepest direction which directly provides a moving potential to the finger.

(B) Influence of the Gain Efficient λ

Figure 8(b) indicates that the steepest ascent method effectively modulates W_f above W_{fr} by moving the arm in an efficient manner. In fact, the gain coefficient λ of the steepest ascent method given in (27) determines the speed at which W_f is modulated. To investigate the influence of λ on manipulability regulation, we conduct a few experiments using different values of λ. The results obtained with different values of λ (= 1, 5, and 10) are shown in Fig.9(a), Fig.9(b), and Fig.9(c), respectively.

In all the cases, the arm instantly moves once W_f reduces below W_{fr}. When we set λ = 1, W_f gradually increases and will finally reach the upper limit W_{ft} as shown in Fig.9(a). However, with a larger value of λ, as shown in Fig.9(b) and Fig.9(c), the arm will generate a fast and strong response for a quick movement. Therefore, W_{fr} increases noticeably and rapidly reaches it upper limit W_{ft}.

5. Discussions

We know that the hand-arm system of a human being has a high redundancy. Even the simplest hand movement requires a series of complicated computations that occur in the CNS. The type of coordinate mapping required and the manner of performing the computations to convert information from exterior space to joint space has perplexed scientists for about half a century. The experimental results of the endpoint stiffness of the human upper limb revealed that the shape and orientation of the stiffness varied proportionally with the location of the hand in the exterior work space (Hogan, et al. 1987). The mechanism of this phenomenon was theoretically summarized and the concept of stiffness ellipsoid was proposed (Mussa-Ivaldi, 1985). Further investigations on hand impedance were also conducted (Tsuji et al. 1988, 1994; Gomi rt al. 1997,1998) and same results revealed that circular ellipsoids are shown almost in front of the body centre, whereas narrow ellipsoids are shown at a greater distance from the body. In fact, the circular shape of the stiffness ellipsoid suggests that the stiffness is uniform in all directions in front of the body centre, whereas the narrow shape of the stiffness ellipsoid represents direction-dependant stiffness present at a greater distance from the body.

In the discussion on the manipulability measure of a robot end-effector, manipulability ellipsoid was first proposed theoretically in a Euclidian space (Yoshikawa, 1985). By using the manipulability theory, we can easily understand why the shape or orientation of the stiffness depends on the hand position. Since a position in front of body centre provides a higher manipulability, locating the hand at that position would provide a higher moving potential in order to easily deal with unexpected situations. Human beings have a natural tendency to move their hands and arms in a position that provides a higher manipulability.

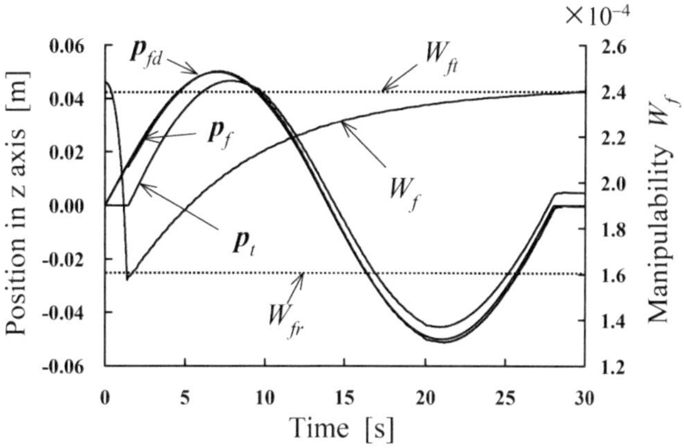

(a) When $\lambda = 1$

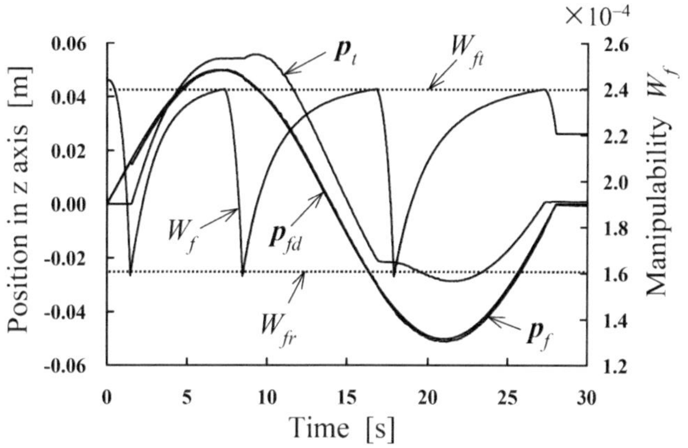

(b) When $\lambda = 5$

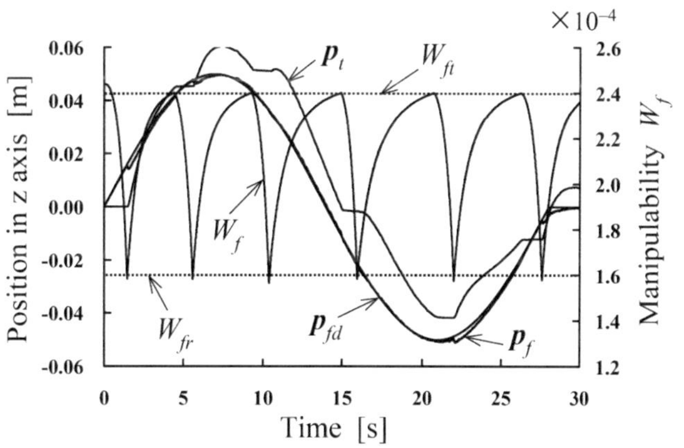

(c) When $\lambda = 10$

Fig. 9. Influence of the gain coefficient on SAM. ($W_{fr} = 1.6\times10^{-4}$, $W_{ft} = 2.4\times10^{-4}$)

This study attempts to propose a motion control for a redundant robot according to the manipulability of the finger inspired by the human hand-arm movement instead of merely calculating a geometric path for determining a kinematics solution.

6. Conclusions

Human beings always adopt the policy of locating their hands in front of the body center. In fact, locating the hand at this position would provide a higher moving potential in order to easily deal with unexpected situations. By using the manipulability theory, we can easily understand this motion policy. As compared to the reported studies, this study attempts to propose a motion control technique for a redundant robot according to its manipulability based on human hand-arm movement instead of merely calculating a geometric path for determining a kinematics solution.

By using the proposed methods of HM and SAM, the finger robot primarily moves, whereas the arm moves only to augment the finger's movement when the finger manipulability reduces below a given reference value. The performance of the finger becomes robust to its singularity by using the proposed steepest ascent method. The experimental results also reveal that the gain coefficient of the SAM plays an important role in response to the change of the finger's manipulability.

7. Future works

In this study, we proposed a motion control to complete unconstrained movement for a finger-arm robot using the manipulability of the finger. A new method of impedance control combined with the proposed method using the manipulability of the finger will be developed for the finger-arm robot to complete a contact task in our future study.

8. References

Furusho, J. & Usui, H. (1989). A control method for manipulator with redundant, *Trans. JSME (C)*, Vol.55, No.514, 1391-1398.

Guo, Z. Y. & Hsia, T. C. (1993). Joint Trajectory Generation for Redundant Robots in an Environment With Obstacles, *J. of Robotics System*, Vol.10(2), 199-215.

Glass, K.; Colbaugh, R.; Lim, D. & Seraji, H. (1995). Real-Time Collision Avoidance for Redundant Manipulators, *IEEE Trans. on Robotics and Automation*, Vol.11(3), 448-457.

Gomi, H & Kawato, M. (1997). Human arm stiffness and equilibrium-point trajectory during multi-joint movement, *Bio. Cybern.*, Vol.76, 163-171.

Gomi, H & Osu, R. (1998). Task dependant viscoelasticity of the human multijoint arm and its spatial characteristics for interaction with environments, *J. of Neuroscience*, Vol.18, 8965-8978.

Hogan, N.; Bizzi, E.; Mussa-Ivaldi, F.A. & Flash, T. (1987). Controlling multijoint motor behavior, *Exerc. Sports Sci. Rev.*, Vol.15, 153-190.

Huang, J.; Quan, B. T.; Harada, M. & Yabuta, T. (2006). Emulating the Motion of a Human Upper Limb: Controlling a Finger-arm Robot by using the Manipulability of its Finger, *Proc. IEEE International Conf. on Robotics and Biomimetics*, 607-612.

Khatib, O., (1986). Real-time obstacle avoidance for manipulators and mobile robots, *Int. J. of Robotics Research*, Vol.5, No.1, 90-98.

Khatib, O. (1995). Inertial Properties in Robotic Manipulation: An Object- Level Framework, *Int. J. of Robotics Research*, Vol.14, No.1, 19-36.

Kim, J. O. & Kholsa, P. K. (1992). Real Time Obstacle Avoidance Using Harmonic Potential Functions, *IEEE Trans. on Robot and Automation*, Vol. 8(3), 338-349.

Loeff, L. A. & Soni, A. H. (1975). An algorithm for computer guidance of a manipulator in between obstacles, *Trans. ASME, J. of Eng. for Industry*, Vol.97, No.3, 836-842.

Ma, S. & Nenchev, D. (1995). Real-time Dynamic Redundancy Control of Redundant Manipulators, *J. of Robotics Society of Japan*, Vol.13(5), 704- 710.

Ma, S.; Hirose, S. & Yoshinada, H. (1996). Efficient Redundancy Control of Redundant Manipulators, *J. of Robotics Society of Japan*, Vol.14(5), 703-709.

Maciejewski, A. A. & Klein, C. A. (1985). Obstacle avoidance for kinematically redundant manipulators in dynamically varying environments, *Int. J. of Robotics Research*, Vol.4, No.3, 109-117.

Melchiorri, C. & Salisbury, J. (1990). Exploiting the Redundancy of A Hand arm Robotic System, *MIT A.I. Memo* No.1261.

Mussa-Ivaldi, F.A.; Hogan, N. & Bizzi, E. (1985). Neural and geometric factors suberving arm posture, *J. Neuroscience*, Vol.5, 2732-2734.

Nagai, K. & Yoshikawa, T. (1994). Impedance control of redundant macro micro manipulator, *J. of Robotics Society of Japan*, Vol.12, No.5, 766-772.

Nagai, K. & Yoshikawa, T. (1995). Grasping and manipulation by robotic arm/ multifingered0hand mechanisms, *JRSJ*, Vol.13, No.7, 994-1005.

Nakamura, Y. & Hanafusa, H. (1986). Inverse kinematics solutions with singularity robustness for robot manipulator control, *Trans. ASME, J. Dyn. Syst., Meas. & Control*, Vol.108, No.4, 163-171.

Quan, B. T.; Huang, J.; Harada, M. & Yabuta, T. (2006). Control of a macro-micro robot system using manipulability of the micro robot, *JSME International Journal (C)*, Vol.49, No.3, 897-904.

Tsuiji, T.; Ito, K.; Nagamachi, M. & Ikemoto, T. (1988). Impedance Regulation in muculo-motor control system and the manipulation ability of the end point, *Trans. SICE*, Vol.24, No.4, 385-392.

Tsuiji, T.; Goto, K.; Ito, K. & Nagamachi, M. (1994). Estimation of human hand impedance during maintenance of posture, *Trans. SICE*, Vol.30, No.3, 319-328.

Vannoy, J. & Xiao, J. (2004). Real-Time Adaptive Trajectory-Optimized Manipulator Motion Planning, *Proc. IEEE/RSJ, Int. Conf. on IROS*, 497-502.

Yoshikawa, T. (1985). Manipulability of robotic mechanisms, *Int. J. of Robotics Research*, Vol.4, No.2, 3-9.

Yoshikawa, T.; Hosoda, K & Doi, T. (1993). Quasi-Static Trajectory Tracking Control of Flexible Manipulator by Macro-Micro Manipulator System, *J. of Robotics Society of Japan*, Vol.11(1), 140-147.

16

Elbow Prosthesis for Partial or Total Upper Limb Replacements

Federico Casolo
Politecnico di Milano,
Department of Mechanics
Italy

1. Introduction

While the prosthetic systems for hands amputation are fairly common, the active prostheses for the other joints of the upper limb are rarer. In addition to commercial reasons undoubtedly related to the low number of trans-humeral amputations than the hands amputations and to the exiguous number of shoulder disarticulation, technical reasons are also crucial. Among which, the electro-mechanical complexity of the system and, above all, the difficulty to design simple and "natural" motion control systems. Nevertheless active elbow and shoulder prosthetic joints can be very important for some upper limb amputees for the everyday life, for working or for other social reasons, in particular if the amputation is bilateral.

The upper limb prostheses can by functional or can have only cosmetic purposes. The functional prosthetic arms can have passive or active features.

The active artificial upper limbs can be divided into prostheses powered by body energy and prostheses powered by external energy. To the first group belong systems actuated by the motion of other body joints and systems moved by means of the arm muscles through their tendons(Chilress et al, 2001). The last, intuitive and theoretically easy to control, were somewhat popular in central Europe in mid 20th century, but they have been virtually abandoned because of critical problem both biological and mechanical.

On the contrary the upper limbs systems that transmit to the prostheses, through cables, the displacement of other body compartments are still common, and some special applications of this method are still under development. They are mostly suggested to active subjects, with adequate mobility of the other upper body structures, which want to feel a direct complete control of their artificial joints.

However, at present, most of artificial upper limbs are equipped by joints powered by external energy. This energy is in most cases electric and is stored in portable batteries but there are also examples of pressure energy stored in little gas tank mounted on the arm. An important requirement for the systems that use extra body energy is their autonomy: the stored energy must last at least for a whole day. Electrical prosthetic systems can benefit of the constant improvement of battery performance due to the rapid evolution of very popular potable products - e.g. cell. phones and computer -.

2. Active artificial arms motion control

Unlike to the body powered system which are directly driven by the patient and controlled by means of his senses, the ones using extra body power require a more complex approach to the arm motion control. At present one of the most important sources of control are the e.m.g. signal of residual muscles of the subjects; the muscles used can be inside the stump or external to it. The e.m.g. signals are normally used when the independent active axes are not more than three, but generally they are actuated in series because the co-activation of two or more specific muscles generate too much mental stress for the subject: only very few and young trained subjects demonstrated to be able to drive two joints simultaneously. Therefore, in general only one degree of freedom at a time is activated by the e.m.g. signal of two muscles, one for the joint flexion and one for the extension, while previously another signal is used to select the joint to be moved. The signal to operate the serial selector can be obtained or acting on proper belts or by means of another e.m.g. source. The sequential activation of the joints produces un-physiological and ungainly movements, therefore for multi axes systems other procedures have been implemented.

The purpose is to replicate the natural approach to the arm movement in which the subject governs the hand, thinking only very marginally to the joints of the arm. In fact, our approach to the artificial arm motion only requires the knowledge of position where the amputee want to place the artificial hand, all the other factors, such as joint trajectories and motors control are automatically managed by the system and are transparent to the subject (Camposaragna et al., 2001). Therefore the arm driving problem is made of two distinct parts: to interpret the patient will and to produce the arm movement, the first of which is the most difficult. In order to interpret the patient will, many parameters can contribute alone or combined to the aim, such as: e.m.g and b.c.i.(brain computer interface) signals, besides movement of other body parts. At present a research on the most effective composition of signals collectable from the patient to interpret his/her request of hand movement is under development at Politecnico di Milano, where a new b.c.i. helmet have been recently produced. Until now we are using only the head displacement to move the artificial arm and two e.m.g. signals to open and close the hand. We already tested two algorithms to drive the arm though the head movements, only one of which, during the action, requires that the patient directs continuously the head toward the instantaneous hand target. This technique have been also implemented in a virtual reality system designed to: identify the patients eligible to the use of the artificial limb, to train the subjects to drive the prosthesis before it is available, to personalize the driving algorithms parameters and to certify the progress in the learning process (Camposaragna et al, 2001).

The same procedure is used for multi axes arm systems addressed to shoulder disarticulated subjects, as well as to trans-humeral amputees: in the latter case together with the head kinematics also the stump position must be collected. The virtual reality exercises requires exactly the same input used by the real prosthetic arm, that includes the head kinematics, the e.m.g. signals to open or close the hand and, if the forearm is present, also its kinematics. To calculate the kinematics of each monitored body segment, the last version of our system is equipped with two Inertial Motion Tracker which includes a triaxial accelerometer, a triaxial gyroscope and a gravitometer (fig. 1).

The arm mathematical model used for the computer simulation take into account the electro-mechanical characteristics of the system, included the motors torque vs. velocity ratio and the whole system mechanical efficiency.

This allows training the subjects to develop the driving skills for his definitive artificial arm by means of a simulated arm which respond to the commands just like the prosthesis does.

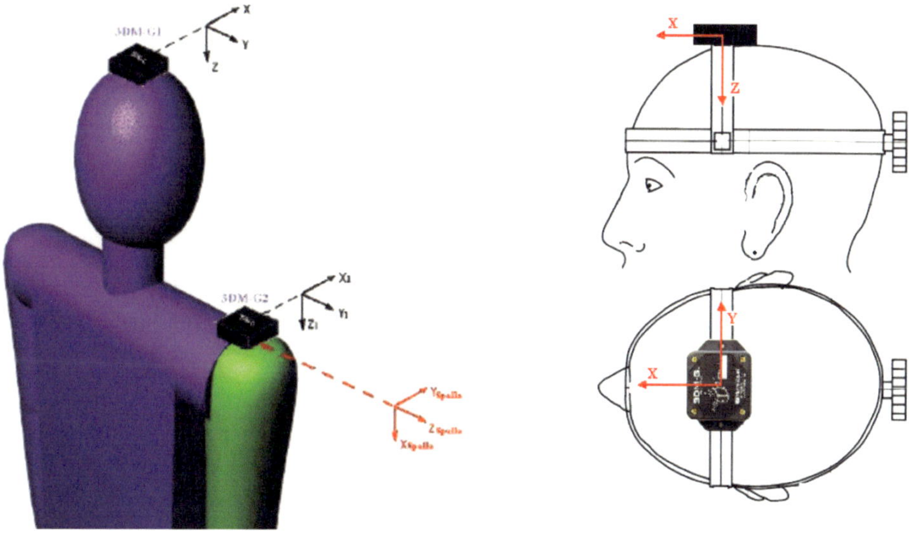

Fig. 1. Location of the Inertial Motion Trackers, with their reference axes

When the problem to extract, from the signal collected from the amputee, the kinematics of the hand target is solved, one of the main steps of the work is completed.
Next point is to define how to fill the gap between hand and target: the easiest solution would be to set the velocity inversely proportional to their distance, but with this procedure the average velocity is too slow, thus an exponential negative function have been implemented in order to increase the average velocity while keeping a low velocity near the target.
The next step is the solution of the forward dynamic problem of the arm, that is to calculate the required motor torque and to verify if it is compatible with the characteristic diagram of the motor (Legnani et al 1996). If negative, the dynamic requests are decreased in order to fulfil the motor limits.
The control system is modular: the main board calculates the target and arm kinematics, and send to the board of each motor - which is generally brushless - the setting points of velocity and position.
To decrease the computational load of the arm main board, the need of solving the trigonometric functions has been avoided by pre compiling a look-up table containing, for each square centimetre of the working volume of the hand, the corresponding angular joint coordinate of each motor of the limb. I2C protocol has been implemented for boards communication. The I2C bus physically consists of 2 active wires and a ground connection. The active wires, the serial data line, and the serial clock line, are both bidirectional (Nxp, 2007). The motors instantaneous velocity and position are calculated from the hall sensors signals. With our present approach the motor is controlled directly in velocity (fig.2) in order to be able to force also acceleration and deceleration, while the external ring is only used to check whether the position set point has been reached.

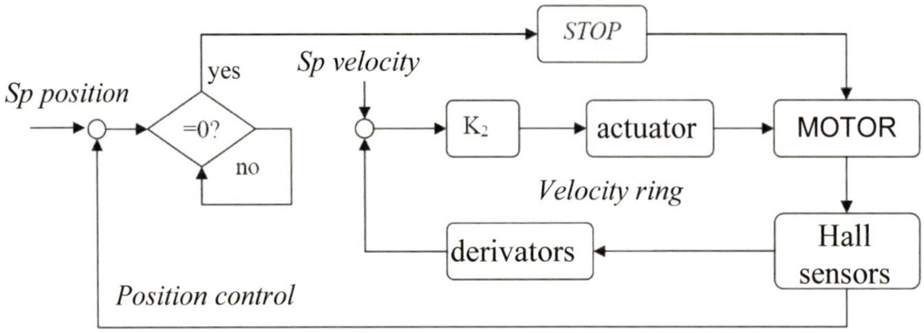

Fig. 2. Schema of the brushless motor control

3. Prosthetic active joint efficiency: an example of elbow transmission design

The system mechanical efficiency is a key factor of the whole project because it is strictly related to the torque required to the motor and consequently to the size and the weight of motor, of transmission and of the battery package. Also motor and transmission noise is, in general, inversely proportional to the mechanical efficiency. Most elbow prostheses of the past generation were quite inefficient because they employ small and fast DC motors coupled with transmissions made of long chains of gears. Each couple of gears waist energy for friction and the whole chain efficiency - product of the efficiencies of all the couple of gears - can assume very little values, even lover than 15%. This is the case of the elbow joint (fig.3a) that our group -MSUM- used as reference for the design of new elbows (Fig.3b, 3c).

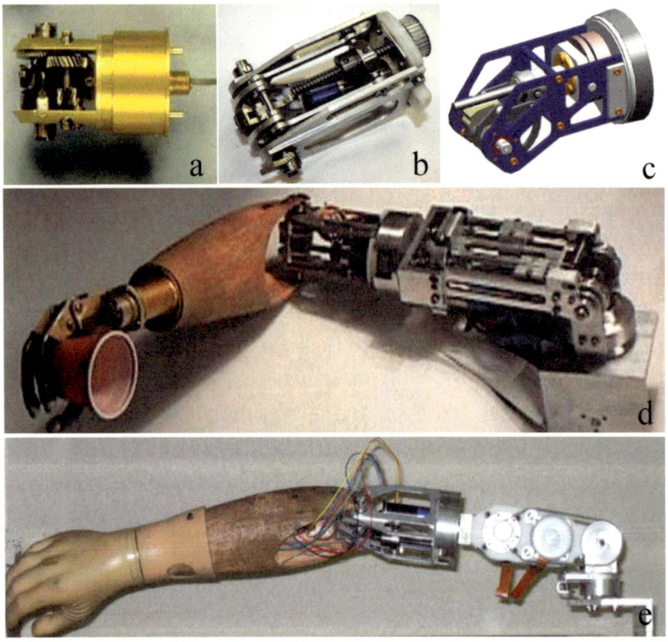

Fig. 3. a) Old elbow with a long chain of gears; b) MSUM elbow1; c) MSUM elbow2; d) Full arm with MSUM-shoulder1, MSUM-elbow2 and Otto-bock hand; e) Full arm with MSUM-shoulder2, MSUM-elbow1 with Otto-bock hand and wrist.

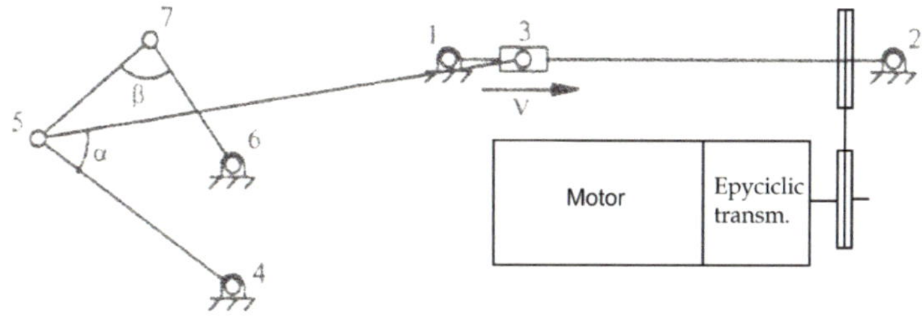

Fig. 4. Ebow1 with epicyclic transmission.

One method to increase the mechanical efficiency is to substitute, when possible, the gears by linkages with ball bearing and by screw-ball transmissions. The first elbow prosthesis designed by our MSUM group and used by patients with satisfaction, is made of a ball screw axis acting on a slider crank connected to a four bars linkage. The ball screw axis is connected to a brushless motor through an epicyclical transmission and a belt and pulley system. The measured overall mechanical efficiency of the system was increased up to 64%. In the second model (Fig.3c and Fig.4) the screwball axis is oscillating and is directly connected to a pancake brushless motor. This architecture raise the efficiency to over 80%. As an example, below are reported the main steps of the design and analysis procedure for the latter elbow system (Fig.5a).

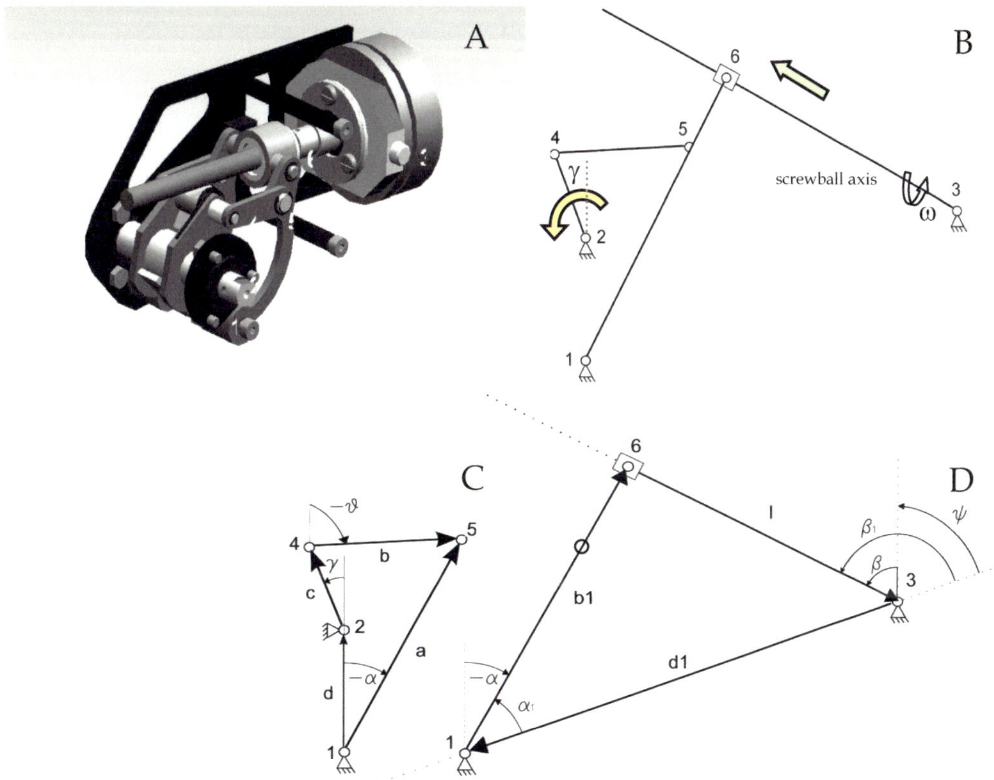

Fig. 5. a) MSUM elbow2 joint ; b) kinematic model of the full joint ; breakdown of the in two mechanisms in series with their vector representation : c) four bar linkage mechanism and d) slider-crank mechanism with the oscillating bar that is also the axis of the pancake motor.

Considering only the four bars linkage (fig. 5 c) its kinematics can be analyzed as follows:

$$a \cdot e^{i\alpha} - b \cdot e^{i\theta} - c \cdot e^{i\gamma} - d = 0$$

$$\begin{cases} b\cos\theta = a\cos\alpha - c\cos\gamma - d \\ b\sin\theta = a\sin\alpha - c\sin\gamma \end{cases}$$

$$a^2 - b^2 + c^2 + d^2 - 2ac\cos\gamma\cos a - 2ac\sin\gamma\sin a - 2ad\cos a + 2cd\cos\gamma = 0$$

and grouping to simplify the expressions

$$\begin{cases} A = -2ac\sin\gamma \\ B = -2ad - 2ac\cos\gamma \\ C = a^2 - b^2 + c^2 + d^2 + 2cd\cos\gamma \\ D = \sqrt{A^2 + B^2 - C^2} \end{cases}$$

we obtain, with few passages, the expressions of α and θ as function of γ

$$\alpha = \arcsin\left(-\frac{AC - BD}{A^2 + B^2}\right)$$

$$\theta = \arctan\left(\frac{c\sin\gamma - a\sin\alpha}{d + c\cos\gamma - a\cos\alpha}\right)$$

Similarly, considering only the slider crank oscillating system (Fig.5d):

$$b_1 \cdot e^{i\alpha_1} - l \cdot e^{i\beta_1} - d_1 = 0$$

$$\begin{cases} l = \sqrt{b_1^2 + d_1^2 - 2b_1 d_1 \cos\alpha_1} \\ \sin\beta_1 = {b_1 \sin\alpha_1}/{l} \end{cases}$$

$$l = \frac{p_{screw}\theta_m}{2\pi}$$

where l is the translation of the nut screw, θ_m is the ball scerw axis rotation angle and p_{screw} is the pitch of the ball-screw axis.

Therefore the transmission ratio(τ_{elbow}) of the complete mechanism can be written as:

$$\dot{\gamma} = \dot{\alpha}\frac{a \cdot \sin(\alpha - \theta)}{c \cdot \sin(\gamma - \theta)} \qquad \dot{\alpha} = -\frac{\dot{l}}{b_1 \cdot \sin(\alpha_1 - \beta_1)}$$

$$\tau_{elbow} = \frac{\dot{\gamma}}{\omega_m} = \frac{p_{screw} a \cdot \sin(\alpha - \theta)}{2\pi b_1 c \cdot \sin(\gamma - \theta)\sin(\alpha - \beta)}$$

τ_{elbow} expression shows that the mechanism is, by his nature, not homocinetic but the motion regularity can be improved by minimizing the following functions:

$$F_1 = \frac{\max(\tau) - \min(\tau)}{\max(\tau)}; \qquad F_2 = \max\left|\frac{\partial\tau}{\partial\gamma}\right|$$

obtaining, in this case, an average value of τ_{elbow} approximately of 0,0075 with a fluctuation within 15%. Nevertheless the software of the elbow joint driver can take into account the dependency of τ_{elbow} by γ.

The geometric parameters of the systems affect the transmission features and the size of the artificial joint. The influence of the variation of each single parameter can be analyzed separately (fig. 3) in order to choose the best fitting geometry of the system besides a tolerable fluctuation of the transmission ratio.

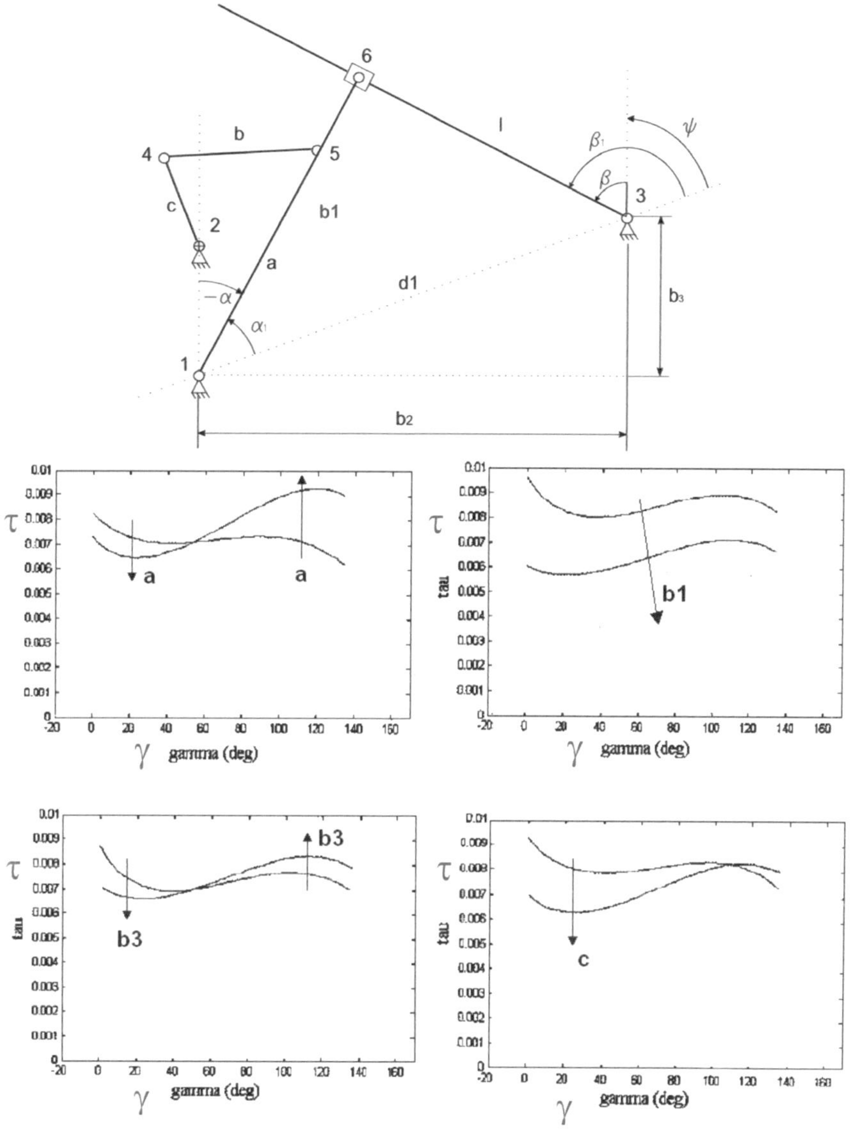

Fig. 6. Transmission ratio τ vs. elbow joint rotation γ: influence of parameters a, b_1, b_3, c

Another important parameter that must be taken into account for a good design of bars linkages is the transmission angle, which is the lower angle between two consecutive bars.
The optimum for this angle is 90° but it varies during the mechanism motion: in general it must never assume values lower than 40° in order to keep a good efficiency during the working phase and not to amplify the mechanism clearance (Magnani et al 1986). The

following graphs (Fig. 7) show that the elbow joint, previously designed to fulfil the transmission ratio and the size requirement, is also within the limits for all the transmission angles of the linkage.

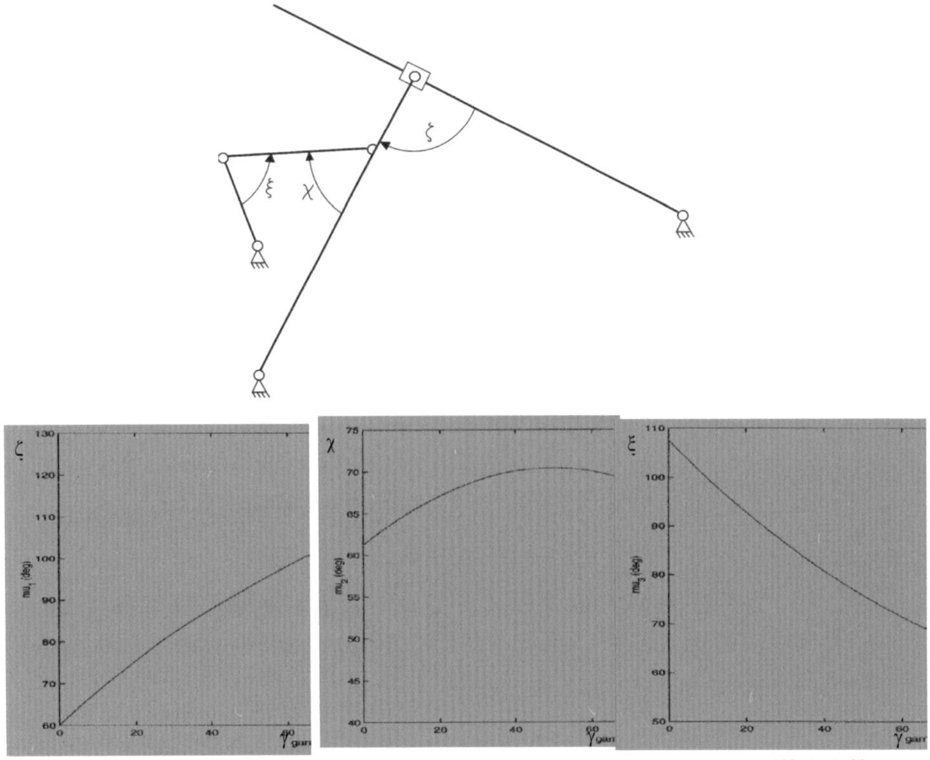

60°<ζ<130° 45°<χ<71° 60°<ξ<110°

Fig. 7. transmission angles ζ , χ and ξ of the analyzed mechanism vs. the elbow rotation γ

3.1 Extra d.o.f. for the artificial elbow

A prosthetic arm for subjects disarticulated at the shoulder is, for the most advanced prototypes, equipped with an active shoulder joint with two d.o.f.(Fig.3 d, e). Therefore, despite the natural humero-ulnar joint is almost a perfect revolute pair, a prosthetic elbow could benefit of an extra d.o.f. located proximally on the forearm and in series to the elbow flexion, in order to replace the effects of the lacking axial rotation of the humerus. This new active axis, which increases the weight and the system complexity, is only convenient for subjects with head and trunk limited mobility.

Only for some specific tasks of the arm, such as drinking, for which the hand orientation is important for the whole cup trajectory, the subject can benefit of the extra axis. The following figure shows that the extra axis (fig.5, images on the right) allows to lead the cup to the mouth maintaining its symmetry axis perfectly vertical (fig.3, red dots), but they also show that, if a certain tilt is accepted (yellow 20° and red dots 40°), it is also possible to drink with a full prosthetic arm mounting a single d.o.f. elbow(fig.5, images on the left). Furthermore, subjects with a good trunk mobility can easily adjust their head position to optimize the action.

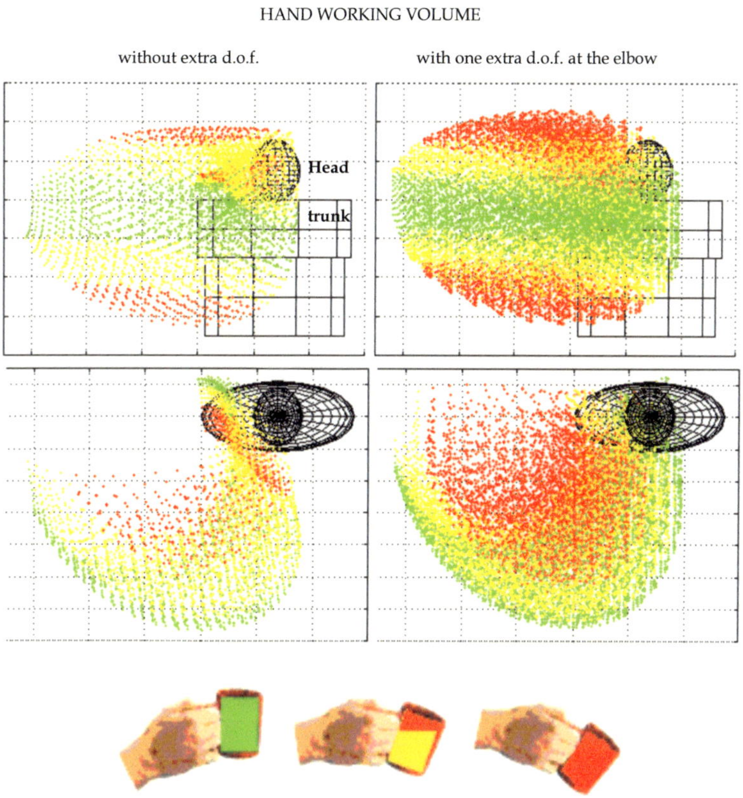

Fig. 5. Drinking task example: minimum inclination of a cup directed to the mouth following different trajectories for an artificial limb with two d.o.f. at the shoulder and one d.o.f. (left column) or two d.o.f. at the elbow (right column).

7. Conclusion

Externally powered prosthesis design requires preliminarily to solve the communication problem between the subject and the system: BCI is a promising technology but until now other signals collected for instance from head motion are used for driving purposes. Our research group designed modular systems for the whole upper limb replacement whose elbow can also be used separately for a transhumeral prosthetic arm. One approach to the design and the analysis of the performances of an electrically powered joint has been clarified referring to a real elbow joint developed by our group. Next improvement will concern the implementation of a recently developed mini hydrostatic transmission.

8. References

Camposaragna, M.; Casolo, F.; Cattaneo, B.; Lorenzi, F.(2001). Development of a computer simulation tool to design the motion control strategy of total artificial arm, *Computer simulation in Biomechanics* (p.165-172), Libreria Clup, ISBN887090, Milan

Chilress, D.S.; Weir, R.F.; Craig, W; Heckathorne, M.D. (2001). Cineplasty as a control input for externally powered prosthetic components. *J. of Rehabilitation Research and Development*, Vol.38, No.4, (July/August 2001)

Magnani, P.L ; Ruggieri G. (1986). *Meccanismi per machine automatiche.* UTET, ISBN8802039348, Milan

Legani, G.; Casolo, F.; Righettini, P.; Zappa, B.(1996). A homogeneous matrix approach to 3D kinematics and dynamics – part 2: Applications to chains of rigid bodies and…, *Mechanism & machine theory,* vol. 31, nº5, ISSN0094-114X Elsevier, Oxford

Nxp, B.V. (2007) *I2C bus specification & user manual* rev.03, June 2007 *ed.NSP,* the Netherlands

17

Fuzzy Control Strategy for Cooperative Non-holonomic Motion of Cybercars with Passengers Vibration Analysis

Francesco Maria Raimondi and Maurizio Melluso
Dipartimento di Ingegneria dell'Automazione e dei Sistemi, University of Palermo
Italy

1. Introduction

The cybercars are electric road wheeled non-holonomic vehicles with fully automated driving capabilities. They contribute to sustainable mobility and are employed as passenger vehicles. Non-holonomic mechanics describes the motion of the cybercar constrained by non-integrable constraints, i.e. constraints on the system velocities that do not arise from constraints on the configuration alone. First of all there are thus with dynamic non-holonomic constraints, i.e. constraints preserved by the basic Euler-Lagrange equations (Bloch, 2000; Melluso, 2007; Raimondi & Melluso, 2006-a). Of course, these constraints are not externally imposed on the system but rather are consequences of the equations of motion of the cybercar, and so it sometimes convenient to treat them as conservation laws rather than constraints per se. On the other hand, kinematic non-holonomic constraints are those imposed by kinematics, such as rolling constraints. The goal of the motion control of cybercars is to allow the automated vehicle to go from one terminal to another while staying on a defined trajectory and maintaining a set of performance criteria in terms of speeds, accelerations and jerks. There are many results concerning the issue of kinematic motion control for single car (Fierro & Lewis, 1997). The main idea behind the kinematic control algorithms is to define the velocity control inputs which stabilize the closed loop system. These works are based only on the steering kinematics and assume that there exists perfect velocity tracking, i.e. the control signal instantaneously affects the car velocities and this is not true. Other control researchers have target the problems of time varying trajectories tracking, regulating a single car to a desired position/orientation and incorporating the effects of the dynamical model to enhance the overall performance of the closed loop system. The works above are based on a backstepping approach, where the merging of kinematic and dynamic effects leads to the control torques applied to the motors of the wheels. A Fuzzy dynamic closed loop motion control for a single non-holonomic car based on backstepping approach and oriented to stability analysis of the motion errors has been developed by Raimondi & Melluso (2005). In Raimondi & Melluso (2006-b) and Raimondi & Melluso (2007-a) adaptive fuzzy motion control systems for single non-holonomic automated vehicles with unknown dynamic and kinematic parameters and Kalman's filter to localize the car have been presented. With regards to the problems of cooperative control of multiple cybercars, a number of techniques have been developed for omni-directional

(holonomic) wheeled cars (see Gerkey & Mataric, 2002; La Valle & Hutchinson, 1998). Decentralized algorithms have been dealt with for holonomic cars by Lumelsky & Harinarayan (1997). With regards to the cooperation of multiple non-holonomic cybercars, few results have been published. On this subject, an approach based on the definition of suitable functions of inverse kinematics to control the motion of a platoon of autonomous vehicles has been presented by Antonelli & Chiaverini (2006). The problem of controlling multiple non-holonomic vehicles by using fuzzy control so that they converge to a source has been studied by Driessen *et al*, (1999). However, since the cars do not have passengers on board, all the studies above do not consider the problem of the acceleration and jerk. For fully automated operation with passengers, a trajectory planning method that produces smooth trajectories with low acceleration is required. The jerk, i.e. the derivative of the acceleration, adversely affects the efficiency of the control algorithms and passengers comfort, so that it has to be reduced. Not many results have been published on this subject (Labakhua *et al.*, 2006; Panfeng *et al.*, 2007).
In this chapter a new closed loop fuzzy control system for non-holonomic motion of multiple cybercars in presence of passengers is proposed. The control strategy merges an innovative decentralized planning trajectory algorithm and a new fuzzy motion control law. About the cooperation, if the target position is fixed, then a number of cybercars has to reach the target one, without to come into collision with the other closest vehicles. The trajectories are planned as the desired time evolution for the position and orientation of some representative point of each cybercar. Forward trajectories are planned only, i.e. trajectories without manoeuvres. In other words all the cooperative cybercars should not stop, except, of course, at the initial and final position. Therefore circular trajectories with continuous curvature have been chosen. Since, for example, in airport the cybercars move in preferential roads without obstacles, the environment in which they move is considered free of obstacle. To ensure the trajectory tracking of all the cooperative cybercars, a new control strategy based on fuzzy inference system is proposed and developed. The fuzzy system generates the control torques for all the cybercars. The cybercars are still employed to transport passengers which are inevitable exposed to vibrations (Birlik & Sezgin, 2007). The acceleration is adopted as preferred measurement of the human vibration exposure. Therefore, with respect to other control theories, the parameters of the fuzzy controller developed in this chapter may be tuned with respect of the ISO 2631-1 standard, which proposes a comfort scale using a mean acceleration index.
This chapter is organized as follows. Section 2 presents the dynamical model of multiple cybercars which has to be employed to project the dynamic fuzzy control system. Also the acceleration model is formulated to develop a control strategy where the passenger comfort is ensured. Section 3 presents a new decentralized cooperative trajectory planner, where the aim is that all the cybercars must reach a target position without collisions between them. Section 4 presents the closed loop fuzzy motion control system, where the asymptotical stability of the motion errors given by the difference between the reference trajectory planned in Section 3 and the actual trajectory of each cybercar is proved by using the Lyapunov's theorem and the Barbalat's Lemma (Slotine & Li, 1991). The parameters of the fuzzy control law are investigated to ensure a good level comfort of the passengers. In this sense, the adjustment of the saturation values of the fuzzy dynamic control surfaces guarantees low values of the longitudinal, lateral accelerations and jerks. In Section 5 experimental tests in a Matlab environment are employed to confirm the effectiveness of the proposed motion control strategy. Some conclusions are drawn in Section 6.

2. Model formulation of multiple non-holonomic cybercars

Consider a system made up of r non-holonomic cybercars. A schematization of the system in open chain configuration is shown in Fig. 1. Now indicating with $\mathbf{q}_i(t) \in R^n$ the time varying coordinates of the position and orientation of the i-cybercar,

Fig. 1. System of r vehicles in open chain configuration

the complete system is subject to $r \times m$ non-holonomic constraints given by:

$$\boldsymbol{\rho}(\mathbf{q}_1(t), \mathbf{q}_2(t), ..., \mathbf{q}_r(t), \dot{\mathbf{q}}_1(t), \dot{\mathbf{q}}_2(t), ..., \dot{\mathbf{q}}_r(t)) = \mathbf{0}, \tag{1}$$

where:

$$\boldsymbol{\rho}(\mathbf{q}_1(t), \mathbf{q}_2(t), ..., \mathbf{q}_r(t), \dot{\mathbf{q}}_1(t), \dot{\mathbf{q}}_2(t), ..., \dot{\mathbf{q}}_r(t)) = \begin{bmatrix} \mathbf{A}_1(\mathbf{q}_1(t))\dot{\mathbf{q}}_1(t) \\ \mathbf{A}_2(\mathbf{q}_2(t))\dot{\mathbf{q}}_2(t) \\ \cdots \\ \mathbf{A}_\mathbf{r}(\mathbf{q}_r(t))\dot{\mathbf{q}}_r(t) \end{bmatrix} \tag{2}$$

and $\mathbf{A}_i(\mathbf{q}_i(t)) \in R^{m \times n}$ are the matrices of the non-holonomic constraints of the motion of each vehicle. Consider the following well known dynamic model (Fierro & Lewis , 1997) :

$$\mathbf{M}_i(\mathbf{q}_i(t))\ddot{\mathbf{q}}_i(t) + \mathbf{C}_i(\mathbf{q}_i(t), \dot{\mathbf{q}}_i(t))\dot{\mathbf{q}}_i(t) = \mathbf{E}_i(\mathbf{q}_i(t))\boldsymbol{\tau}_i(t) - \mathbf{A}_i^\mathbf{T}(\mathbf{q}_i(t))\boldsymbol{\lambda}_i, \quad i = 1 \ldots r, \tag{3}$$

where $\mathbf{M}_i(\mathbf{q}_i(t)) \in R^{n \times n}$ is a positive definite inertial matrix, $\mathbf{C}_i(\mathbf{q}_i(t), \dot{\mathbf{q}}_i(t)) \in R^{n \times n}$ is the centripetal Coriolis matrix, $\boldsymbol{\tau}_i(t) \in R^p$ is a vector of the torques applied to the right and left wheels of each vehicle, $\mathbf{E}_i(\mathbf{q}_i(t)) \in R^{n \times p}$ is a mapping between the torques above and the forces applied along the components of $\mathbf{q}_i(t)$, $\mathbf{A}_i(\mathbf{q}_i(t))$ are the matrices given by (2), while $\boldsymbol{\lambda}_i \in R^m$ is a vector of Lagrange multipliers. Let $\mathbf{S}_i(\mathbf{q}_i(t)) \in R^{n \times (n-m)}$ be a full rank matrix made up of a set of smooth and linearly independent vectors spanning the null space of $\mathbf{A}_i(\mathbf{q}_i(t))$, i.e.:

$$\mathbf{A}_i(\mathbf{q}_i(t))\mathbf{S}_i(\mathbf{q}_i(t)) = \mathbf{0}, i = 1 \ldots r. \tag{4}$$

Due to the non-holonomic constraints (1), it is possible to find velocity vectors $\mathbf{v}_i(t) \in R^{n-m}$, such that:

$$\dot{\mathbf{q}}_i(t) = \mathbf{S}_i(\mathbf{q}_i(t))\mathbf{v}_i(t), i = 1 \ldots r. \tag{5}$$

Referring to the i-vehicle shown in Fig. 2, let $P_0(x_{0i}, y_{0i})$ be the reference point of the motion and indicate the orientation with θ_i. Indicate with (X, Y) and with (X_c, Y_c) the ground and the body references respectively. Let r_i be the radius of the wheels. Indicate with b_i the distance from the center of the wheel to the longitudinal axis and with d_i the distance from the reference point to the mass center P_c. For the later description indicate with m_i and I_i the mass and the inertia of the *i*-vehicle respectively. In view of the previous considerations, the following model can be written:

$$\dot{\mathbf{q}}_i(t) = \mathbf{S}_i(\mathbf{q}_i(t))\mathbf{v}_i(t),$$
$$\mathbf{S}_i(\mathbf{q}_i(t)) = \begin{pmatrix} \cos\theta_i(t) & 0 \\ \sin\theta_i(t) & 0 \\ 0 & 1 \end{pmatrix}, \quad (6)$$
$$\mathbf{v}_i^{\mathbf{T}}(t) = [u_i(t) \;\; \omega_i(t)], \mathbf{q}_i^{\mathbf{T}}(t) = [x_{0i}(t) \;\; y_{0i}(t) \;\; \theta_i(t)],$$
$$i = 1 \ldots r,$$

where $u_i(t)$ and $\omega_i(t)$ are the linear and angular velocities of the *i*-cybercar of the cooperative system.

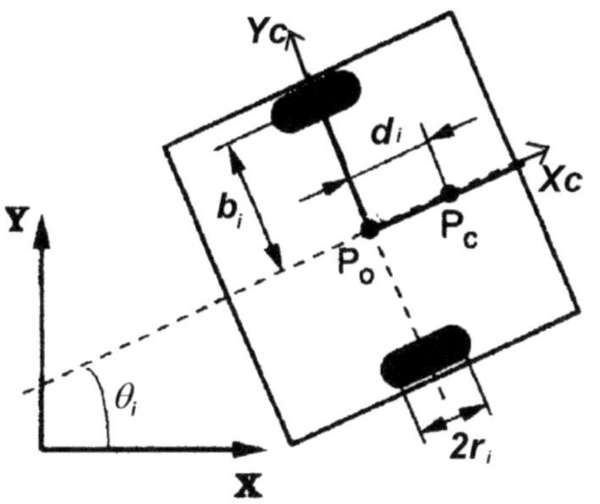

Fig. 2. Cybercar with reference systems

To design a control law which consider the ISO 2631-1, and, therefore, to analyze the vibration of the passengers during the motion, it is necessary to obtain the acceleration model. Let the acceleration vector of each cooperative vehicle be:

$$\mathbf{a}_{\mathbf{w}i}^{\mathbf{T}}(t) = [\hat{a}_{xi}(t) \;\; \hat{a}_{yi}(t)]$$
$$i = 1, \ldots, r \quad (7)$$

The accelerations $\hat{a}_{xi}(t)$ and $\hat{a}_{yi}(t)$ are the *longitudinal and lateral accelerations* respectively of the *i*-cybercar of the cooperative system. The accelerations (7) could be obtained through the following steps. By considering the kinematical model of a non-holonomic vehicle (cf. eq. 6), it is possible to calculate the accelerations along the axes of the ground reference (X,Y) (see Fig. 2) in function of the linear and angular velocities. It results:

$$a_{xi}(t) = \frac{d^2 x_{0i}(t)}{dt^2} = \ddot{x}_{0i}(t);$$
$$a_{yi}(t) = \frac{d^2 y_{0i}(t)}{dt^2} = \ddot{y}_{0i}(t). \tag{8}$$

Differentiating (6) leads to:

$$\begin{bmatrix} \ddot{x}_{0i}(t) \\ \ddot{y}_{0i}(t) \\ \ddot{\theta}_i(t) \end{bmatrix} = \begin{bmatrix} -\dot{\theta}_i(t)\sin\theta_i(t) & 0 \\ \dot{\theta}_i(t)\cos\theta_i(t) & 0 \\ 0 & 0 \end{bmatrix} \begin{bmatrix} u_i(t) \\ \omega_i(t) \end{bmatrix} +$$
$$+ \begin{bmatrix} \cos\theta_i(t) & 0 \\ \sin\theta_i(t) & 0 \\ 0 & 1 \end{bmatrix} \begin{bmatrix} \dot{u}_i(t) \\ \dot{\omega}_i(t) \end{bmatrix}, \tag{9}$$
$$i = 1 \ldots r.$$

Therefore the human body is subjected to forces along the X and Y axes. Now it is necessary to project the forces above along the axes of the body reference, i.e. X_c and Y_c, so that the lateral and longitudinal accelerations (7) can be calculated as:

$$\hat{a}_{xi}(t) = a_{xi}(t)\cos\theta_i(t) + a_{yi}(t)\cos\theta_i(t);$$
$$\hat{a}_{yi}(t) = a_{yi}(t)\cos\theta_i(t) - a_{xi}(t)\sin\theta_i(t). \tag{10}$$

After some calculations it results:

$$\hat{a}_{xi}(t) = -u_i(t)\dot{\theta}_i(\mathrm{t})\sin\theta_i(t)\cos\theta_i(t) + \dot{u}_i(\mathrm{t})\cos^2\theta_i(t) +$$
$$+ u_i(t)\dot{\theta}_i(\mathrm{t})\cos\theta_i(\mathrm{t})\sin\theta_i(t) + \dot{u}_i\sin^2\theta_i(t) = \dot{u}_i;$$
$$\hat{a}_{yi} = -u_i(t)\dot{\theta}_i(t)\sin\theta_i(t)\cos\theta_i(t) + \dot{u}_i(t)\cos^2\theta_i(t) +$$
$$+ u_i(t)\dot{\theta}_i(t)\cos\theta_i(t)\sin\theta_i(t) + \dot{u}_i(t)\sin\theta_i(t) = u_i\dot{\theta}_i. \tag{11}$$

Due to non-holonomic constraints given by (2), the components of $\mathbf{q}_i$ (*i=1…r*) vector are not independent. For this reason the dynamic model (3) cannot be employed directly for the motion control. A dynamic model in body fixed coordinates has to be used. Substituting the equation (6) into model (3) leads to:

$$\overline{\mathbf{M}}_i\dot{\mathbf{v}}_i(t) + \overline{\mathbf{V}}_{\mathbf{m}i}(\omega_i(t))\mathbf{v}_i(t) = \mathbf{S}_i^{\mathbf{T}}(\mathbf{q}_i(t))\mathbf{E}_i(\mathbf{q}_i(t))\boldsymbol{\tau}_i(t)$$
$$i = 1 \ldots r, \tag{12}$$

where:

$$\overline{\mathbf{M}}_i = \mathbf{S}_i^{\mathbf{T}}(\mathbf{q}_i(t))\mathbf{M}_i(\mathbf{q}_i(t))\mathbf{S}_i(\mathbf{q}_i(t)) = \begin{pmatrix} m_i & 0 \\ 0 & I_i \end{pmatrix},$$

$$\overline{\mathbf{V}}_{\mathbf{m}i}(\omega_i(t)) = \mathbf{S}_i^{\mathbf{T}}(\mathbf{q}_i(t))(\mathbf{M}_i(\mathbf{q}_i(t))\dot{\mathbf{S}}_i(\mathbf{q}_i(t)) + \mathbf{V}_{\mathbf{m}i}(\mathbf{q}_i(t))\mathbf{S}_i(\mathbf{q}_i(t))) = \begin{pmatrix} 0 & -d_i\omega_i m_i \\ d_i\omega_i m_i & 0 \end{pmatrix}, \tag{13}$$

$$i = 1 \ldots r.$$

3. Decentralized cooperative trajectory planning algorithm

In this paragraph a new decentralized trajectory planning algorithm is developed for the non-holonomic cooperative motion of the cybercars. The problem is the following. All the cybercars must reach a target position without collisions between them. Once the distances between the initial positions of the cybercars and the target are known, a decentralized algorithm permits to plane circular trajectories intersection-free. After communication to each cybercar of the initial position and target coordinates, the trajectory planner of each cybercar provide to plane a circular trajectory independently of each other. This means we have a decentralized planner. More precisely, indicate with $[x_{0i}(0) \quad y_{0i}(0) \quad \theta_i(0)]$ the initial position and orientation of each cybercar. An automated vehicle moving with constant linear and angular velocities, tracks a circular trajectory. Therefore a reference motion may be planned along a circumference that includes the initial coordinates above and the position of the target. With reference to Fig. 3, let $C_i(x_{oi}(0), y_{0i}(0), \theta_i(0)$ be the initial position of the i-cybercar. Let $B(x_T, y_T, \theta_T)$ the position and orientation of the target. Let Δx_i and Δy_i be the shiftings along the tangent and radial directions respectively of the i-vehicle. The i-cybercar moves from C_i to B along a circular trajectory of which the radius is equal to λ_i, while the center is located in D.

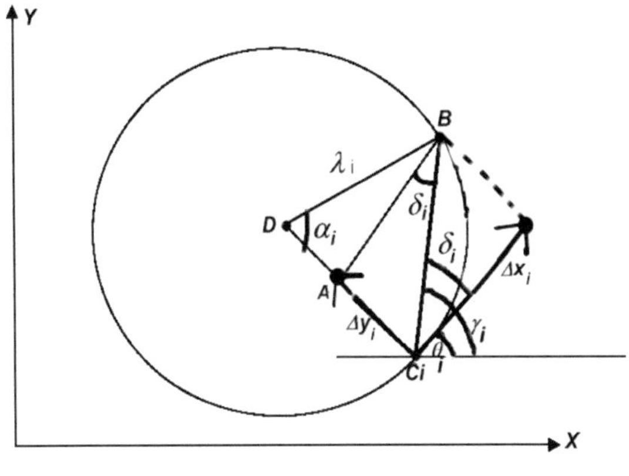

Fig. 3. Motion planning

It yields:

$$BC_i = \tilde{d}_i = \sqrt{(x_T - x_{0i}(0))^2 + (y_T - y_{0i}(0))^2},$$
$$\gamma_i = \arctan\left[(y_T - y_{0i}(0))/(x_T - x_{0i}(0))\right], \tag{14}$$
$$i = 1...r,$$

where γ_i is the angle between BC_i and the x axis. Consider the following angular relation (see Fig .3):

$$\delta_i = \gamma_i - \theta_i(0) =$$
$$= \arctan\left[(y_T - y_{0i}(0))/(x_T - x_{0i}(0))\right] - \theta_i(0). \tag{15}$$

The length of the line BA is equal to the distance Δx_i. Therefore, if we consider the triangle $C_i AB$, then:

$$\Delta x_i = \tilde{d}_i \cos \delta_i ; \; \Delta y_i = \tilde{d}_i \sin \delta_i . \tag{16}$$

The angular shifting α_i between C_i and B results as it follows:

$$\Delta x_i = \lambda_i \sin \alpha_i . \tag{17}$$

From observation of the triangle DAB, it results:

$$\lambda_i^2 = (\lambda_i - \Delta y_i)^2 + \Delta^2 x_i . \tag{18}$$

The solution of the equation (13) with respect to λ_i is:

$$\lambda_i = (\Delta^2 x_i + \Delta^2 y_i) / 2\Delta y_i . \tag{19}$$

Now the values of the reference angular (ω_{ri}) and linear (u_{ri}) velocities of each cooperative vehicle may be calculated as follows:

$$\begin{aligned} &\omega_{ri} = \frac{\alpha_i}{\Delta T}; \; u_{ri} = \omega_i \lambda_i ; \\ &i = 1 \ldots r, \end{aligned} \tag{20}$$

where ΔT is a fixed look-ahead time interval chosen by the designer. Let us consider multiple automated cybercars in an initial open chain configuration (cf. Fig. 1), i.e. collinear and with the same orientations given by Δx_i *(i=1,2)*. The algorithm above allows circular trajectories without intersections to be planned, so that the vehicles will avoid collisions while moving. Each trajectory is planned independently of the others. This means we have a decentralized cooperation of the vehicles. Fig. 4 shows an example, where two vehicles are considered in open chain configuration $C_1 - C_2$. One observes that the first vehicle of the open chain follows a circular trajectory from C_1 to the target *B* along Δx_1, while the second vehicle follows a circular trajectory from C_2 to the target one along Δx_2. The distance between C_1 and *B* is smaller than the distance between C_2 and *B*, so that, based on the equation (9), it is $d_1 < d_2$. Consequently, based on the equations (16) and (19), the radius of the circumference tracked by the first vehicle (i.e. λ_1) is smaller than the radius of the circumference tracked by the second vehicle (i.e. λ_2). Since the cybercars are initially collinear and have the same orientations, and the circumferences must be include both the initial positions of the vehicles and the target position, the trajectories are without intersections and the vehicles can reach the target without collisions. The method can be used for *r* cooperative vehicles in initial open chain $C_1 - C_2 - \ldots - C_r$, so that each vehicle can reach the target without coming into collision with other vehicles.
Note that the vehicles have to be in open chain configuration initially, i.e. collinear. If there is a vehicle which is not mutually collinear, it must reach a collinear position. On this subject, some studies have focused on modelling formations of non-holonomic vehicles (Bicho & Monteiro, 2003).

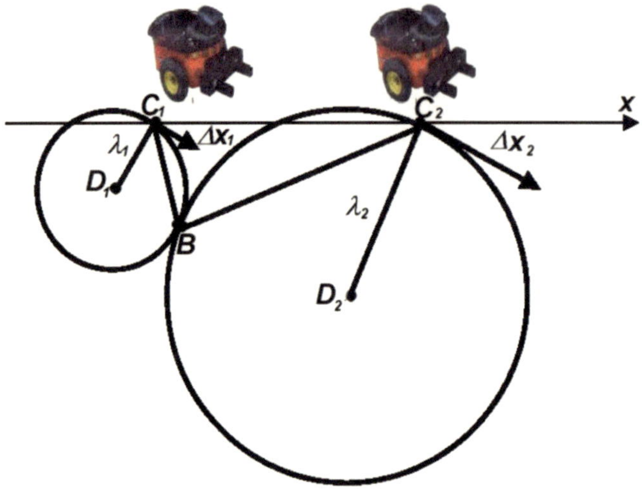

Fig. 4. Trajectory planning without collisions for multiple vehicles in initial open chain configuration

4. Fuzzy dynamic closed loop motion control for cooperative cybercars with passengers comfort

Consider the *i*-cybercar of the cooperative system (cf. Fig. 2).The kinematical model is given by (6), while the dynamical model is given by (12) and (13). Employing the values of linear and angular velocities given by (20) and using the kinematical model (6) lead to the following equations for the circular reference motion of each cybercar:

$$\begin{aligned} &\dot{x}_{ri}(t) = u_{ri}\cos\theta_{ri}(t); \dot{y}_{ri}(t) = u_{ri}\sin\theta_{ri}(t); \\ &\dot{\theta}_{ri}(t) = \omega_{ri}, \quad i = 1...r. \end{aligned} \tag{21}$$

Let the following vectors:

$$\begin{aligned} &\mathbf{q}_i^{\mathbf{T}}(t) = \left[x_{0i}(t)\ \ y_{0i}(t)\ \ \theta_i(t)\right], \\ &i = 1...r \end{aligned} \tag{22}$$

be the position and orientation of each cybercar. One defines the following motion errors between the planned circular reference trajectories the state variables given by (22) as it follows:

$$\mathbf{e}_i(t) = \begin{bmatrix} \cos\theta_i(t) & \sin\theta_i(t) & 0 \\ -\sin\theta_i(t) & \cos\theta_i(t) & 0 \\ 0 & 0 & 1 \end{bmatrix} \times \begin{bmatrix} x_{ri}(t) - x_{0i}(t) \\ y_{ri}(t) - y_{0i}(t) \\ \theta_{ri}(t) - \theta_i(t) \end{bmatrix} = \begin{bmatrix} e_{xi}(t) \\ e_{yi}(t) \\ e_{\theta i}(t) \end{bmatrix}, \quad i = 1...r. \tag{23}$$

The errors $e_{xi}(t)$ and $e_{yi}(t)$ are said to be the longitudinal and lateral motion errors respectively, while $e_{\theta i}(t)$ is the orientation error. The following velocity control laws are employed for each cybercar:

$$\mathbf{v}_{ci}(t) = \begin{bmatrix} u_{ci}(t) \\ \omega_{ci}(t) \end{bmatrix} = \begin{bmatrix} u_{ri} \cos e_{\theta i}(t) + K_{1i} e_{xi}(t) \\ \omega_{ri} + u_{ri}(K_{2i} e_{yi}(t) + K_{3i} \sin e_{\theta i}(t)) \end{bmatrix},$$
$$K_{1i}, K_{2i}, K_{3i} > 0, \tag{24}$$
$$i = 1 \ldots r.$$

Replacing (24) into (6) leads to the following closed loop mathematical model:

$$\dot{\mathbf{e}}_i(t) = \begin{bmatrix} (\omega_{ri} + u_{ri}(K_{2i} e_{yi}(t) + K_{3i} \sin e_{\theta i}(t))) e_{yi}(t) - K_{1i} e_{xi}(t) \\ -(\omega_{ri} + u_{ri}(K_{2i} e_{yi}(t) + K_{3i} \sin e_{\theta i}(t))) e_{xi}(t) + u_i \sin e_{\theta i}(t) \\ -u_{ri}(K_{2i} e_{yi}(t) + K_{3i} \sin e_{\theta i}(t)) \end{bmatrix}, \tag{25}$$
$$i = 1 \ldots r.$$

It is possible to formulate the following theorem.

Theorem 1. Consider the cooperative system (6), in closed loop with the velocity control laws (24). If the linear and angular reference velocities given by (21) are limited functions, then the equilibrium state of the closed loop model (25) is the origin of the state space and it is asymptotically stable.

Proof. From model (25), it is evident that, if $\dot{\mathbf{e}}_i(t) = 0$ (*i=1…r*), then $\mathbf{e}_i(t) = 0$ (*i=1…r*), so that the equilibrium point is the origin of the state space. Consider the following extended vector $\mathbf{e}(t) \in R^{(r \times n) \times 1}$ which contains the motion errors of all the cooperative vehicles:

$$\mathbf{e}^{\mathrm{T}}(t) = \begin{bmatrix} \mathbf{e}_1(t) & \mathbf{e}_2(t) & . & . & . & \mathbf{e}_r(t) \end{bmatrix}. \tag{26}$$

The following Lyapunov's function is chosen:

$$V(\mathbf{e}(t)) = \frac{1}{2} \sum_{i=1}^{r} K_{1i}(e_{xi}^2(t) + e_{yi}^2(t)) + 2 \sum_{i=1}^{r} \frac{K_{1i}}{K_{2i}}(1 - \cos e_{\theta i}(t)). \tag{27}$$

The function (27) is definite positive. By calculating the time derivative of the function (27) and substituting the equations (25) into result, it yields:

$$\dot{V}(\mathbf{e}(t)) = -\sum_{i=1}^{r} K_{1i} e_{xi}^2(t) - \sum_{i=1}^{r} u_{ri} \frac{2K_{1i}K_{3i}}{K_{2i}} \sin^2 e_{\theta i}(t). \tag{28}$$

The function (28) does not depend on lateral motion errors $e_{yi}(t)$ (*i=1…r*), so that it is equal to zero in correspondence of the inputs $\begin{bmatrix} 0 & e_{yi}(t) & 0 \end{bmatrix}$. Therefore the function (28) is semi-definite negative. The conclusion is that the components of the vector (26) are stable and bounded. Since all the motion errors and the reference velocities are bounded, the second time derivative of the function (28) is bounded, therefore Barbalat's lemma implies that the function (28) converges to zero when *t* diverges, so that the longitudinal motion errors and the orientation errors of all the cooperative vehicles converge to zero. From the second and third equations of system (20) it follows that:

$$\lim_{t\to\infty} \dot{e}_{yi}(t) = 0, \quad i = 1 \ldots r. \tag{29}$$

Therefore:

$$e_{yi}(\infty) = \overline{e}_{yi}, \quad i = 1 \ldots r. \tag{30}$$

where $\overline{e}_{yi}$ is a constant value. Since the orientation errors converge to zero, from the third equation of the system (25) it results:

$$-u_i K_{2i} \overline{e}_{yi} = 0, \quad i = 1 \ldots r. \tag{31}$$

It can be concluded that the lateral motion errors converge asymptotically to zero. Q.E.D.
By employing the kinematical control strategy (24), it is difficult to control directly the lateral and longitudinal accelerations which are responsible of harmful effects on the passengers. For this reason a fuzzy dynamical control strategy is developed below, where the properties of the fuzzy maps assures the Lyapunov's stability of the motion errors given by (25), while the saturation properties of the maps ones permit to control directly the maximum acceleration of each vehicle of the cooperative system during the motion. Let $\tilde{\boldsymbol{\eta}}_i(t)$ the time varying error between the velocity control laws given by (24) and the physical velocity $\mathbf{v}_i(t)$ of each cybercar (i.e. the solution of the differential equations (12)):

$$\tilde{\boldsymbol{\eta}}_i(t) = \begin{bmatrix} \tilde{\eta}_{1i}(t) \\ \tilde{\eta}_{2i}(t) \end{bmatrix} = [\mathbf{v}_{\mathbf{c}i}(t) - \mathbf{v}_i(t)], \quad i = 1 \ldots r. \tag{32}$$

The fuzzy inference mechanism is explained below. The inputs of the fuzzy system are the errors (32). The fuzzy rules for $\tilde{\eta}_{ji}$ (*j=1,2* and *i=1,2…r*) are the following:

1) if $\tilde{\eta}_{ji}$ is negative and $\dot{\tilde{\eta}}_{ji}$ is negative then the output Σ_{ji} has a negative big value;

2) if $\tilde{\eta}_{ji}$ is negative and $\dot{\tilde{\eta}}_{ji}$ is positive, then the output Σ_{ji} has a negative small value;

3) if $\tilde{\eta}_{ji}$ is positive and $\dot{\tilde{\eta}}_{ji}$ is negative, then the output Σ_{ji} has a positive small value;

4) if $\tilde{\eta}_{ji}$ is positive and $\dot{\tilde{\eta}}_{ji}$ is positive, then the output Σ_{ji} has a positive big value.

Now we assume the following dynamical control laws (Raimondi & Melluso, 2007-b):

$$\begin{aligned} &\mathbf{S}_i^T(\mathbf{q}_i(t))\mathbf{E}_i(\mathbf{q}_i(t))\boldsymbol{\tau}_i(t) = \\ &= \overline{\mathbf{M}}_i(\dot{\mathbf{v}}_{\mathbf{c}i}(t) + \dot{\mathbf{s}}_i(t)) + \overline{\mathbf{V}}_{\mathbf{m}i}(\omega_i(t))\mathbf{v}_i(t) = \begin{bmatrix} F_i(t) \\ T_i(t) \end{bmatrix}, \\ &i = i \ldots r \end{aligned} \tag{33}$$

where $\mathbf{S}_i(\mathbf{q}_i(t))$, $\overline{\mathbf{M}}_i$ and $\overline{\mathbf{V}}_{\mathbf{m}i}(\omega_i(t))$ are given by (6) and (13) respectively, $F_i(t)$ and $T_i(t)$ are the linear force and the angular momentum applied to the *i*-cybercar, while $\dot{\mathbf{s}}_i(t)$ is the output of the fuzzy inference system, so that it is:

$$\dot{\mathbf{s}}_i(t) = \boldsymbol{\Sigma}_i(\tilde{\boldsymbol{\eta}}_i(t), \dot{\tilde{\boldsymbol{\eta}}}_i(t)) = \begin{bmatrix} \Sigma_{1i}(\tilde{\eta}_{1i}(t), \dot{\tilde{\eta}}_{1i}(t)) \\ \Sigma_{2i}(\tilde{\eta}_{2i}(t), \dot{\tilde{\eta}}_{2i}(t)) \end{bmatrix}, \qquad i = 1 \ldots r \tag{34}$$

where $\boldsymbol{\Sigma}_{ji}$ (*j=1,2 i=1…r*) are the nonlinear input-output maps of the fuzzy inference system. The maps above depends on the choice of the input and output memberships. Figs 5 and 6 show the membership functions for obtaining Σ_{1i} (*i=1…r*).

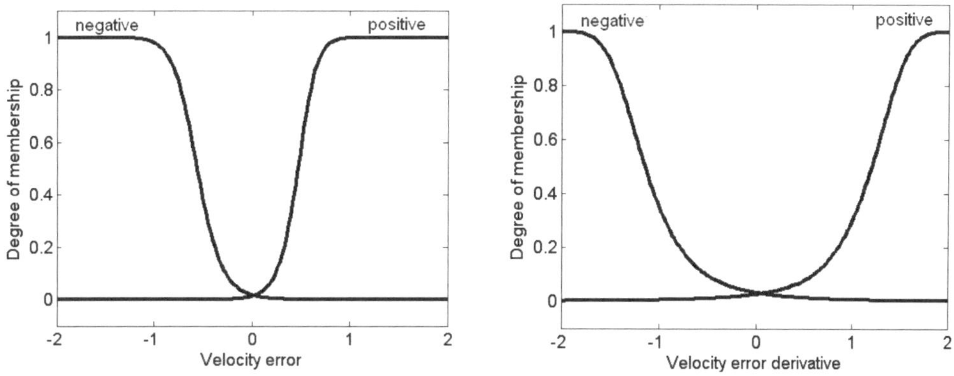

Fig. 5. Input membership functions

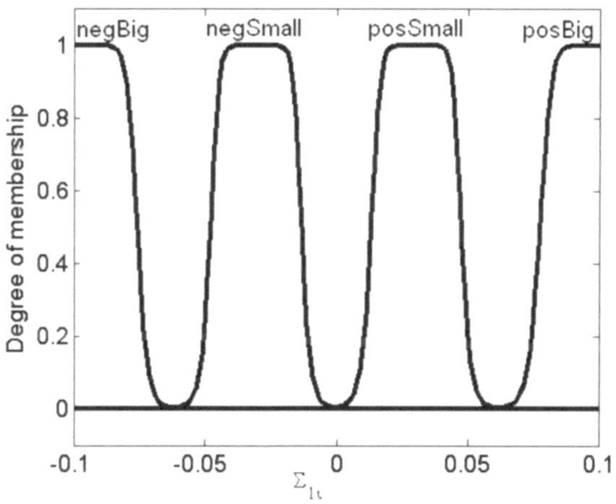

Fig. 6. Output membership function

The memberships employed for obtaining the control surfaces Σ_{2i} (*i=1…r*) have the same symmetry. The input values of the memberships of Fig. 5 are chosen in such a way to be in the range of the physical velocity errors obtained in our application. The outputs Σ_{ji} (*j=1,2*

i=1,2,…r) given by (34) represent the variations of the torques applied to the right and left wheels of each cooperative cybercar for positive or negative changes of the speed errors given by (32). Due to the choice of the form of the memberships functions, the outputs above are nonlinear uniformly continuous and bounded functions. They contribute to evaluating the acceleration or deceleration of the cybercars, to track the planned reference motion with good dynamic perfomances.
Assumption 1. The following properties for the nonlinear fuzzy maps given by (33) are assumed:

$$
\begin{aligned}
&1)\ \Sigma_{ji}(0,0)=0;\\
&2)\ \Sigma_{ji}(\eta_{ji}(t),\dot{\tilde{\eta}}_{ji}(t))=-\Sigma_{ji}(-\tilde{\eta}_{ji}(t),-\dot{\tilde{\eta}}_{ji}(t));\\
&3)\ \exists\delta_{ji}>0:\left|\Sigma_{ji}(\eta_{ji}(t),\dot{\tilde{\eta}}_{ji}(t))\right|\le\delta_{ji};\\
&4)\ \Sigma_{ji}(\eta_{ji}(t),0)=0\Rightarrow\tilde{\eta}_{ji}(t)=0;\\
&5)\ \tilde{\eta}_{ji}(t)\Sigma_{ji}(\tilde{\eta}_{ji}(t),\dot{\tilde{\eta}}_{ji}(t))>0;\\
&6)\ \begin{array}{l}0\le\tilde{\eta}_{ji}(t)(\Sigma_{ji}(\eta_{ji}(t),\dot{\tilde{\eta}}_{ji}(t))-\Sigma_{ji}(0,\dot{\tilde{\eta}}_{ji}(t)));\\ 0\le\dot{\tilde{\eta}}_{ji}(t)(\Sigma_{ji}(\eta_{ji}(t),\dot{\tilde{\eta}}_{ji}(t))-\Sigma_{ji}(\tilde{\eta}_{ji}(t),0)).\end{array}
\end{aligned}
\tag{35}
$$

The assumption above is a necessary condition to ensure that each cooperative cybercar can follow the planned trajectory; also by varying the values δ_{ij}, the acceleration saturation values can be changed, so that the motion of each cybercar can be controlled with low accelerations. Note that, if we choice different number or form of the membership functions, then the properties given by (35) cannot be guaranteed.
Substituting (33) into dynamical model (12) leads to:

$$
\begin{aligned}
&\dot{\tilde{\boldsymbol{\eta}}}_i(t)+\boldsymbol{\Sigma}_i(\tilde{\boldsymbol{\eta}}_i(t),\dot{\tilde{\boldsymbol{\eta}}}_i(t))=\mathbf{0},\\
&i=1\ldots r.
\end{aligned}
\tag{36}
$$

It can be written:

$$
\begin{aligned}
&\overline{\boldsymbol{\Sigma}}_i(\tilde{\boldsymbol{\eta}}_i(t),\dot{\tilde{\boldsymbol{\eta}}}_i(t))=\mathbf{0},\\
&i=1\ldots r,
\end{aligned}
\tag{37}
$$

so that:

$$
\begin{aligned}
&\dot{\tilde{\boldsymbol{\eta}}}_i(t)=\mathbf{f}_i(\tilde{\boldsymbol{\eta}}_i(t)),\\
&i=1\ldots r.
\end{aligned}
\tag{38}
$$

Considering equations (25) and (38) leads to the following closed loop model of the fuzzy dynamic control system:

$$\dot{\mathbf{e}}_i(t) = \begin{bmatrix} \dot{e}_{xi}(t) \\ \dot{e}_{yi}(t) \\ \dot{e}_{\theta i}(t) \end{bmatrix} = \begin{bmatrix} (\omega_{ri} + u_{ri}(K_{2i}e_{yi}(t) + K_{3i}\sin e_{\theta i}(t)))e_{yi}(t) - K_{1i}e_{xi}(t) \\ -(\omega_{ri} + u_{ri}(K_{2i}e_{yi}(t) + K_{3i}\sin(e_{\theta i})))e_{xi}(t) + u_{ri}\sin e_{\theta i}(t) \\ -u_{ri}(K_{2i}e_{yi}(t) + K_{3i}\sin e_{\theta i}(t)) \end{bmatrix},$$
$$\dot{\tilde{\boldsymbol{\eta}}}_i(t) = \mathbf{f}_i(\tilde{\boldsymbol{\eta}}_i(t)), \tag{39}$$
$$i = 1 \ldots r.$$

Now the following theorem may be formulated.

Theorem 2. Consider the closed loop system given by (39). Under the assumption 1, the equilibrium state of the modeled system is the origin of the state space and it is asymptotically stable.

Proof. Consider the following extended state vector:

$$\bar{\mathbf{e}}(t) = \begin{bmatrix} \bar{\mathbf{e}}_1(t) \\ \bar{\mathbf{e}}_2(t) \\ \cdot \\ \cdot \\ \cdot \\ \bar{\mathbf{e}}_r(t) \end{bmatrix}, \tag{40}$$

where:

$$\bar{\mathbf{e}}_i(t) = \begin{bmatrix} e_{xi}(t) \\ e_{yi}(t) \\ e_{\theta i}(t) \\ \tilde{\eta}_{1i}(t) \\ \tilde{\eta}_{2i}(t) \end{bmatrix} = \begin{bmatrix} \mathbf{e}_i(t) \\ \tilde{\boldsymbol{\eta}}_i(t) \end{bmatrix}, \tag{41}$$
$$i = 1 \ldots r.$$

From equations (39) it results:

$$\left[\dot{e}_{xi}(t)\,\dot{e}_{yi}(t)\,\dot{e}_{\theta i}(t)\right]^{\mathrm{T}} = \mathbf{0}, \quad \forall t; \quad i = 1 \ldots r, \tag{42}$$

so that:

$$\left[e_{xi}(t)\,e_{yi}(t)e_{\theta i}(t)\right]^{\mathrm{T}} = \mathbf{0}, \quad \forall t; i = 1 \ldots r. \tag{43}$$

Also, if $\dot{\tilde{\boldsymbol{\eta}}}_i(t) = \mathbf{0}$, then:

$$\boldsymbol{\Sigma}_i(\tilde{\boldsymbol{\eta}}_i(t), \mathbf{0}) = \mathbf{0}, \quad i = 1 \ldots r, \tag{44}$$

so that, based on the property 4 given in the Assumption 1, it can be said that:

$$\tilde{\boldsymbol{\eta}}_i(t) = \mathbf{0}, \quad \forall t, \quad i = 1...r. \tag{45}$$

This implies that:

$$\mathbf{0} = \mathbf{f}_i(\mathbf{0}), \quad i = 1...r. \tag{46}$$

Therefore the equilibrium point of model (39) is the origin of the state space. The following Lyapunov's function is chosen:

$$\begin{aligned} V(\overline{\mathbf{e}}(t)) = & \frac{1}{2}\sum_{i=1}^{r} K_{1i}(e_{xi}^2(t) + e_{yi}^2(t)) + 2\sum_{i=1}^{r} \frac{K_{1i}}{K_{2i}}(1 - \cos e_{\theta i}(t)) \\ & + \frac{1}{2}\sum_{i=1}^{r} \tilde{\boldsymbol{\eta}}_i^{\mathrm{T}}(t)\tilde{\boldsymbol{\eta}}_i(t) + +\sum_{j=1}^{2}\sum_{i=1}^{r} \int_0^{\tilde{\eta}_{ji}} \Sigma_{ji}(\xi_{ji}, 0) d\xi_{ji}, \\ & K_{1i}, K_{2i} > 0. \end{aligned} \tag{47}$$

The first, second and third terms of function (47) are always positive. Now, from the property 6 of the fuzzy maps it results:

$$0 \le \tilde{\eta}_{ji}(t)(\Sigma_{ji}(\eta_{ji}(t), \dot{\tilde{\eta}}_{ji}(t)) - \Sigma_{ji}(0, \dot{\tilde{\eta}}_{ji}(t))), \quad j = 1,2; \; i = 1,2,...r. \tag{48}$$

Consequently, if $\dot{\eta}_{ji}(t) = 0$, then it yields:

$$0 \le \tilde{\eta}_{ji}(t)(\Sigma_{ji}(\eta_{ji}(t), 0) - \Sigma_{ji}(0,0)), \tag{49}$$

so that, based on the property 1, one obtains:

$$0 \le \tilde{\eta}_{ji}(t)\Sigma_{ji}(\eta_{ji}(t), 0), \quad j = 1,2; \; i = 1,2...r. \tag{50}$$

It can be concluded that:

$$\sum_{j=1}^{2}\sum_{i=1}^{r} \int_0^{\tilde{\eta}_{ji}} \Sigma_{ji}(\xi_{ji}, 0) d\xi_{ji} > 0, \quad \forall \tilde{\eta}_{ji}, \quad j = 1,2; \; i = 1,2...r. \tag{51}$$

Therefore the function (47) is positive definite. Calculating the time derivative of the function (47) and replacing (39) into it lead to:

$$\dot{V}(\bar{\mathbf{e}}) = -\sum_{i=1}^{r} K_{1i} e_{xi}^2(t) - \sum_{i=1}^{r} u_{ri} \frac{2K_{1i}K_{3i}}{K_{2i}} \sin^2(e_{\theta i}(t)) - \sum_{i=1}^{r} \tilde{\boldsymbol{\eta}}_i^{\mathrm{T}}(t) \boldsymbol{\Sigma}_i(\tilde{\boldsymbol{\eta}}_i(t), \dot{\tilde{\boldsymbol{\eta}}}_i(t)) + \\ - \sum_{i=1}^{r} \boldsymbol{\Sigma}_i^{\mathrm{T}}(\tilde{\boldsymbol{\eta}}_i(t), \dot{\tilde{\boldsymbol{\eta}}}_i(t)) \boldsymbol{\Sigma}_i(\tilde{\eta}_i(t), \mathbf{0}) \; . \tag{52}$$

The first and second terms of (52) are negative. Based on the property 5 (see Assumption 1), the elements of the summation of the third term of (52) are positive numbers, so that the term above is negative. From property 5 and inequality (50) it yields:

$$\tilde{\eta}_{ji}^2(t) \Sigma_{ji}(\tilde{\eta}_{ji}(t), 0) \Sigma_{ji}(\tilde{\eta}_{ji}(t), \dot{\tilde{\eta}}_{ji}(t)) > 0, \\ j = 1,2; \;\; i = 1,2,\ldots,r, \tag{53}$$

so that, if $\Sigma_{ji}(\tilde{\eta}_{ji}(t), 0)$ is positive, then $\Sigma_{ji}(\tilde{\eta}_{ji}(t), \dot{\tilde{\eta}}_{ji}(t))$ is positive. Therefore the elements of the summation of the fourth term of function (52) are positive. Note that function (52) does not depend on the lateral error, therefore it can be concluded that the function above is negative semi-definite. Therefore the components of the vector (40) are bounded. Now it is possible to calculate the second time derivative of function (47). Based on the previous considerations, it is a bounded function, so that, by applying the Barbalat's lemma, it follows that:

$$\lim_{t \to \infty} \dot{V}(\bar{\mathbf{e}}) = 0. \tag{54}$$

From (52) it can be concluded that the errors $e_{xi}, e_{\theta i}, \tilde{\eta}_{1i}, \tilde{\eta}_{2i}$ (i=1,…,r) converge to zero. Replacing (24) into (32) leads to:

$$\tilde{\boldsymbol{\eta}}_i(t) = \begin{bmatrix} u_{ci}(t) - u_i(t) \\ \omega_{ci}(t) - \omega_i(t) \end{bmatrix} = \begin{bmatrix} \tilde{\eta}_{1i}(t) \\ \tilde{\eta}_{2i}(t) \end{bmatrix} = \\ = \begin{bmatrix} u_{ri} \cos(e_{\theta i}(t)) + K_{1i} e_{xi}(t) - u_i(t) \\ \omega_{ri} + K_{2i} u_{ri} e_{yi}(t) + K_{3i} u_{ri} \sin e_{\theta i}(t) - \omega_i(t) \end{bmatrix}, \\ i = 1 \ldots r. \tag{55}$$

Since the errors $e_{xi}, e_{\theta i}, \tilde{\eta}_{1i}, \tilde{\eta}_{2i}$ converge asymptotically to zero, the lateral errors e_{yi} of each cooperative cybercar converges asymptotically to zero. Q.E.D.

Fig. 7 illustrates the block scheme of the Fuzzy dynamical motion closed loop control system for a single cybercar.

With regards to the passengers comfort, several factor influence vibration discomfort in relation to passenger activities, e.g. seated posture, use of backrest. Passengers usually adopt their posture to attenuate the intensity of vibrations and jerks in order to perform their activities satisfactorily. However the transmission of vibrations on the human body is higher if a passenger uses armrest, backrest or places boot feet on the floor. Therefore attenuation of vibration exposure is a very important requirement of a motion control system for cybercars. There are various means by which the vibration may be expressed, such as

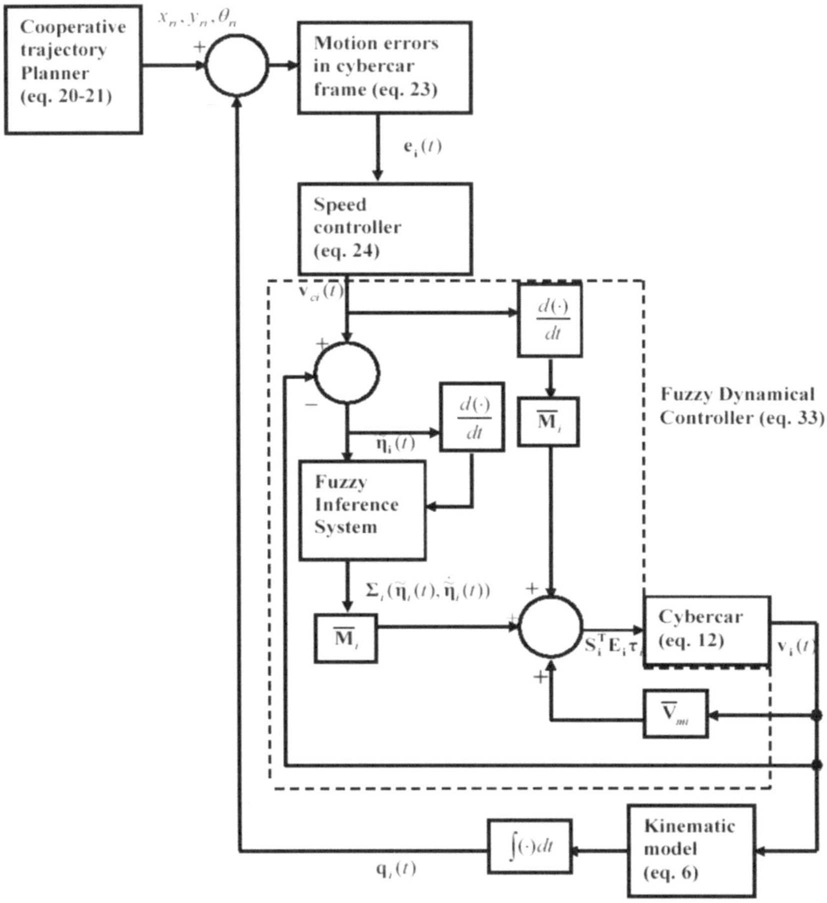

Fig. 7. Block scheme of the dynamical fuzzy control system for cybercar motion

displacement, velocity and acceleration. Of these physical quantities acceleration is generally adopted as preferred measured of quantifying the severity of human vibration exposure (Suzuki, 1998). From (10) it appears that the accelerations which cause vibrations on the human body depend on the curvature of the trajectory. Indicate with $\mathbf{a}_{\mathbf{wi}} \in R^2$ the following vector:

$$\mathbf{a}_{\mathbf{wi}}^{\mathbf{T}} = \left[\hat{a}_{xi}(t)\ \hat{a}_{yi}(t)\right], \quad i = 1 \ldots r, \tag{56}$$

where $\hat{a}_{xi}(t)$ and $\hat{a}_{yi}(t)$ are the accelerations given by (11). Indicate with $\overline{a}_{xi}$ and $\overline{a}_{yi}$ the r.m.s. values of the accelerations given by (56). Indicate the jerks with the following vector:

$$\mathbf{j}_i^{\mathbf{T}}(t) = [j_{xi}(t)\ j_{yi}(t)], \quad i = 1 \ldots r. \tag{57}$$

The components $j_{xi}(t)$ and $j_{yi}(t)$ are said to be the *lateral and longitudinal jerks* of each cybercar, i.e. the rate of change of the accelerations as it follows:

$$
\begin{aligned}
j_{xi}(t) &= \frac{d\hat{a}_{xi}(t)}{dt}, \\
j_{yi}(t) &= \frac{d\hat{a}_{yi}(t)}{dt}, \\
i &= 1 \ldots r.
\end{aligned}
\tag{58}
$$

Indicate with $\bar{j}_{xi}$ and $\bar{j}_{yi}$ the R.M.S. values of the jerks given by (58). The ISO 2631-1 Standard relates comfort of the passengers with the r.m.s. overall acceleration which causes vibrations acting on the human body defined as:

$$
\begin{aligned}
\tilde{a}_{wi} &= \sqrt{\alpha^2 \bar{a}_{xi}^2 + \beta^2 \bar{a}_{yi}^2 + \gamma^2 \bar{a}_{zi}^2}, \\
i &= 1 \ldots r,
\end{aligned}
\tag{59}
$$

where $\bar{a}_{xi}$ and $\bar{a}_{yi}$ are given by (56), $\bar{a}_{zi}$ is the acceleration on the z axis perpendicular to plane Xc,Yc (see Fig. 2), while α, β and γ are multiplying factors. Since each vehicle moves on the plane, the acceleration $\bar{a}_{zi}$ is equal to zero. Based on the ISO 2631-1, the relations between the values of the overall acceleration given by (58) and the passenger comfort of the *i*-cybercar are given by the Table 1 .

R.M.S. overall acceleration	*Passenger comfort level*
$\tilde{a}_{wi} < 0.315 m/s^2$	Not uncomfortable
$0.315 < \tilde{a}_{wi} < 0.63 m/s^2$	A little uncomfortable
$0.5 < \tilde{a}_{wi} < 1 m/s^2$	Fairy uncomfortable
$0.8 < \tilde{a}_{wi} < 1.6 m/s^2$	Uncomfortable
$1.25 < \tilde{a}_{wi} < 2.5 m/s^2$	Very uncomfortable
$\tilde{a}_{wi} > 2.5 m/s^2$	Extremely uncomfortable

Table 1. ISO 2631-1 Standard

By using our fuzzy approach, it is possible to obtain low values of the lateral and longitudinal accelerations in easy way. In fact the saturation values of the outputs of the fuzzy maps, i.e. the values δ_{ji} given by the third property of (35), represent a saturation of the linear and angular accelerations of the cooperative cybercars, so that they are degree of freedom and can be chosen by the designer. In particular, from equation (9) it can be seen that the accelerations along the axes of the ground reference depend on the linear acceleration $\dot{u}_i(t)$ and on the angular acceleration $\dot{\omega}_i(t)$. Now, during the motion, the values of $\dot{u}_i(t)$ and $\dot{\omega}_i(t)$ depend on the outputs of the dynamical fuzzy control law given by (33) and (34). Fig. 8 shows the typical fuzzy control surfaces obtained by using the fuzzy control law (33), where the acceleration given by (34) satisfies the properties of the assumption 1.

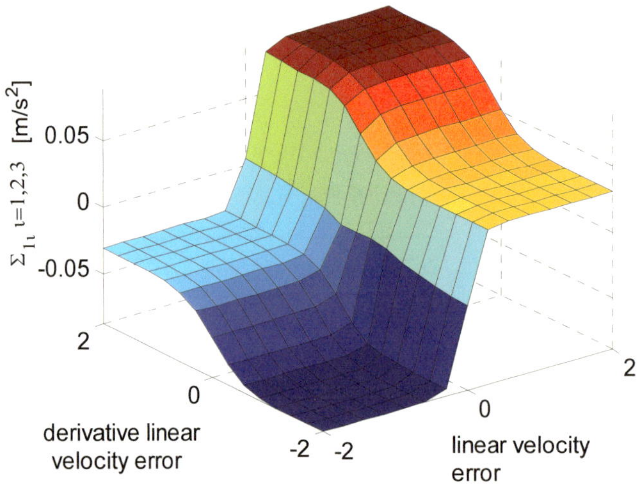

Fig. 8. Fuzzy mapping

The saturation property of the fuzzy map causes the accelerations above to be bounded, and so, after a few attempts, the designer can choose the ranges of the crisp output values of the fuzzy system in order to satisfy the ISO 2631-1 standard and to reduce the vibrations and the jerks. It is evident that the the output values of Fig. 8 fall within a small range, since a motion with low acceleration and jerk is desired. The simulation experiments described in the next section confirm the efficiency of our algorithm in terms of cooperation, stability of the fuzzy motion control system and passenger comfort.

5. Simulation experiments

In this performed simulations the efficiency of the cooperative fuzzy motion control law proposed and developed in this chapter and the good level comfort of the passengers during the motion of the cybercar is illustrated. The parameters of the cybercars have been chosen based on existing cybercars (McDonald & Voge, 2003). The weight of a cybercar is 300kg, the width is 1.45m, the height is 1.6m, while the length is 3.7m. Referring to Fig. 2, the kinematical parameters are chosen as:

$$r_i = 0.40m;\; b_i = 0.725m;\; i = 1,2,3. \tag{60}$$

The dynamical parameters are:

$$M_i = 300kg;\; d_i = 1.5m;\; i = 1,2,3. \tag{61}$$

The parameters of the speed control law (24) are given by:

$$K_{1i} = 25; K_{2i} = 20; \quad i = 1,2,3. \tag{62}$$

The reference trajectories of each cybercar were generated using the decentralized algorithm developed in Section 3, so that the initial motion error values are equal to zero. In fact the circumferences have been generated based on the distance between the initial position of the cybercars and the position of the target. Initially the vehicles are in open chain configuration along y-direction. The initial positions of the three cooperative cybercars are the following:

$$\begin{aligned} &x_{01}(t=0) = 2m;\; x_{02}(t=0) = 2m;\; x_{03}(t=0) = 2m; \\ &y_{01}(t=0) = 2m; y_{02}(t=0) = 3m; y_{03}(t=0) = 4m; \\ &\theta_{01}(t=0) = 1{,}74rad; \theta_{02}(t=0) = 1.74rad; \theta_{03}(t=0) = 1.74rad. \end{aligned} \tag{63}$$

All the generalized coordinates given by (63) refer to a ground reference whose origin is shown in Fig. 9. The position coordinates of the target with respect to the ground reference are:

$$\begin{bmatrix} x_T \\ y_T \end{bmatrix} = \begin{bmatrix} 1.7m \\ -2.8m \end{bmatrix}. \tag{64}$$

To analyze the performances in terms of passenger comforts, we compare for cases with reference to the parameters δ_{ij} given by the third property of (35) :

a. low values of the parameters $\delta_{ji} \; (i = 1,2;\; j = 1,2,3)$;

b. high values of the parameters $\delta_{ji} \; (i = 1,2;\; j = 1,2,3)$.

Case a- The absolute saturation values of the crisp outputs of the fuzzy control surfaces are:

$$\begin{aligned} &\delta_{1i} = \left|\Sigma_{1i\_sat}\right| = 0.1m/s^2; \\ &\delta_{2i} = \left|\Sigma_{2i\_sat}\right| = 0.1rad/s^2; \\ &i = 1,2,3. \end{aligned} \tag{65}$$

Fig. 9 shows the planar trajectories of the cybercars as planned by using the algorithm given in Section 3.

Initially the cybercars are in open chain configuration. The trajectories are intersections-free and therefore there are not collisions during the motion.

The graphs of Fig. 10 and 11 show the time evolutions of the velocity errors given by (32). Due to the dynamics of the cybercars, there are not perfect velocity tracking, i.e. the speed control laws (24) do not affect instantaneously the linear and angular velocities, but the errors converge to zero after some times.

The most significant graphs which illustrate the stability performances of the motion errors of each cybercar are drawn in Figs. 12 and 13, where the time evolutions of the longitudinal, lateral and orientation errors given by (23) are shown.

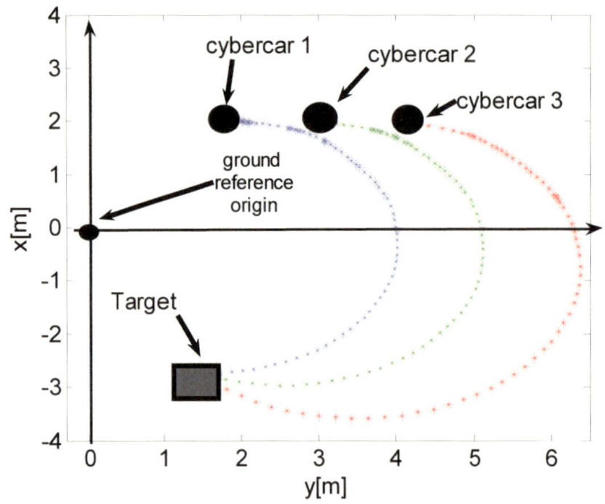

Fig. 9. Motion of three cooperative cybercars

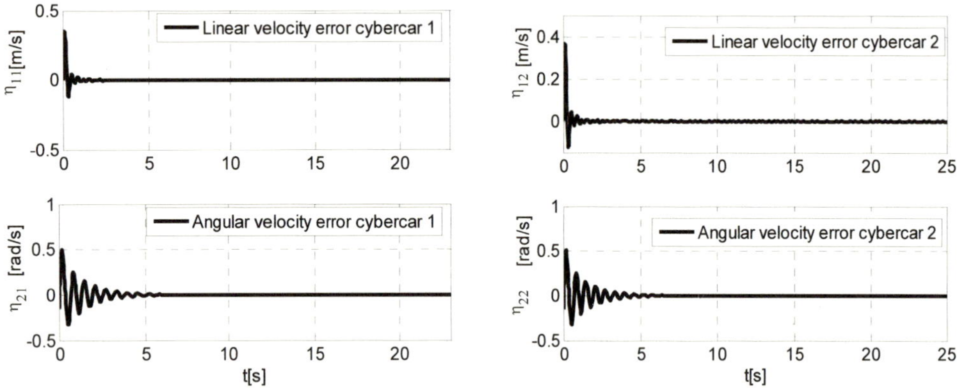

Fig. 10. Velocity errors of the cybercars 1 and 2

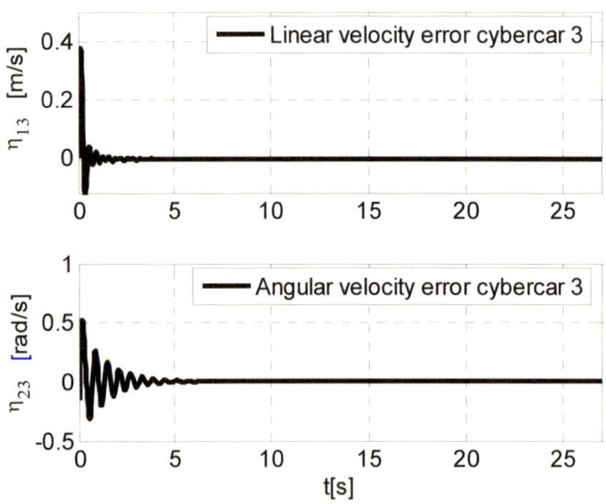

Fig. 11. Velocity errors of the cybercars 3

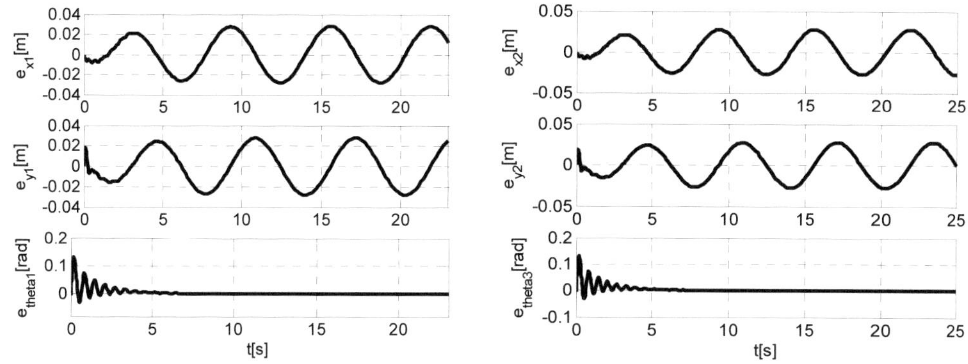

Fig. 12. Tracking errors of cybercars 1 and 2.

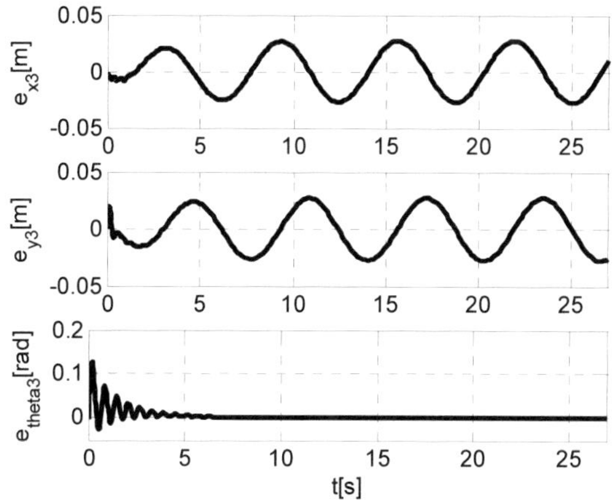

Fig. 13. Tracking errors of cybercar 3

Now we investigate on the passenger comforts with the saturation values given by (65). Figs. 14 and 15 shows the time evolution of the accelerations given by (56) which are responsible of vibrations on the human body, while Table 2 illustrates the r.m.s values of the same accelerations and the overall acceleration given by the mean index (59).

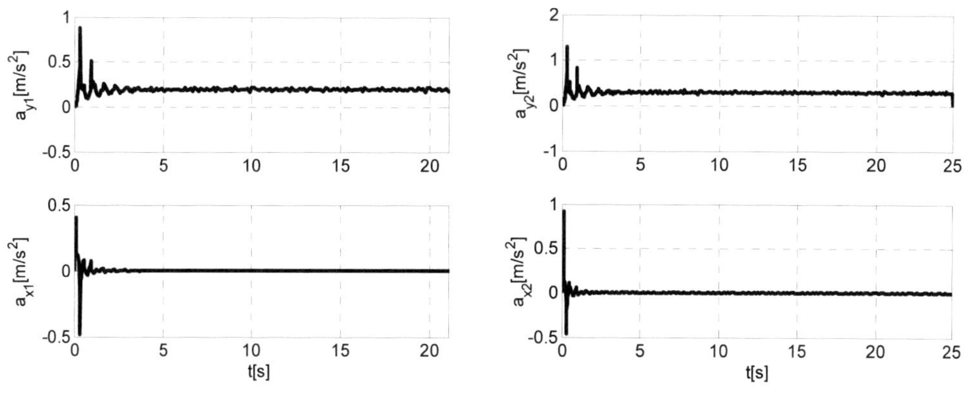

Fig. 14. Lateral and longitudinal accelerations of the cybercars 1 and 2.

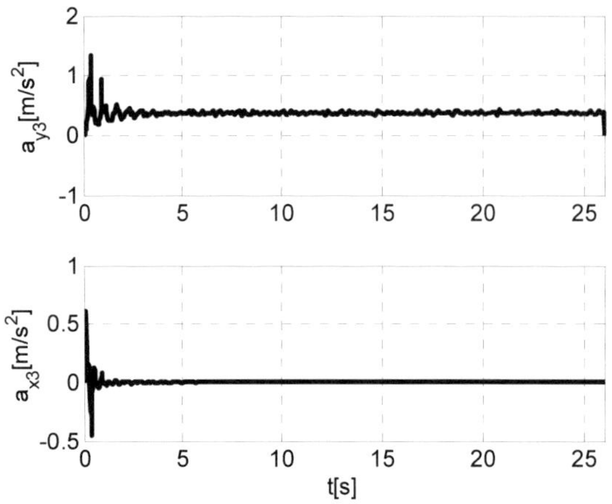

Fig. 15. Lateral and longitudinal accelerations of the cybercars 3.

Longitudinal acceleration	**Lateral acceleration**	**Overall Acceleration**
$\overline{a}_{x1} = 0.0787 m/s^2$	$\overline{a}_{y1} = 0.2034 m/s^2$	$\widetilde{a}_{w1} = 0.305 m/s^2$
$\overline{a}_{x2} = 0.0882\ m/s^2$	$\overline{a}_{y2} = 0.3097 m/s^2$	$\widetilde{a}_{w2} = 0.45 m/s^2$
$\overline{a}_{x3} = 0.0902 m/s^2$	$\overline{a}_{y3} = 0.3853 m/s^2$	$\widetilde{a}_{w3} = 0.554 m/s^2$

Table 2. r.m.s. and mean accelerations of all the cybercars

Note that the values of the r.m.s. overall accelerations are between "Not uncomfortable" and "A little uncomfortable" (see Table 1 and 2), so that the passengers comfort level is very good.

Case b- With respect to (65), the saturation values of the fuzzy control surfaces are increased as follows:

$$\begin{aligned} &\delta_{1i} = \left|\Sigma_{1i\_sat}\right| = 0.5 m/s^2; \\ &\delta_{2i} = \left|\Sigma_{2i\_sat}\right| = 0.5 rad/s^2; \\ &i = 1,2,3. \end{aligned} \tag{66}$$

The comfort of the passengers are also studied in this case. On this subject Figs. 16 and 17 illustrate the lateral and longitudinal accelerations of the cybercars. The r.m.s values of the accelerations above and the mean acceleration given by the index (59) are listed in Table 3, while the values of the jerks given by (58) in cases of low and high saturation values of the fuzzy control surfaces are illustrated in table 4.

Figures 16 and 17 and the results of the table 3 show that the overall accelerations are between "fairy uncomfortable" and "uncomfortable", so that the comfort of the passengers during the motion is bad. By the results shown in Tables 2-4 it is evident that, in case of low values saturation of the fuzzy control surfaces, the accelerations and the jerks are reduced, which means ride passengers comfort enhancement. Therefore the designer can be choice the parameters of the fuzzy controller to optimize the vibrations acting on the human body of the passengers.

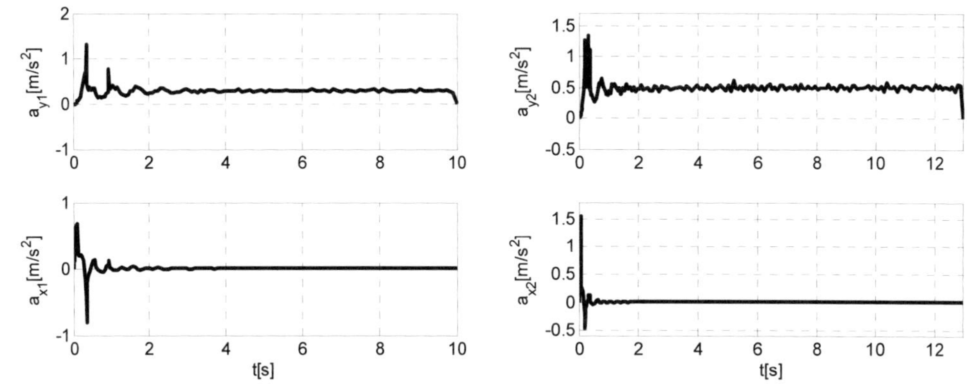

Fig. 16. Lateral and longitudinal accelerations of the cybercars 1 and 2

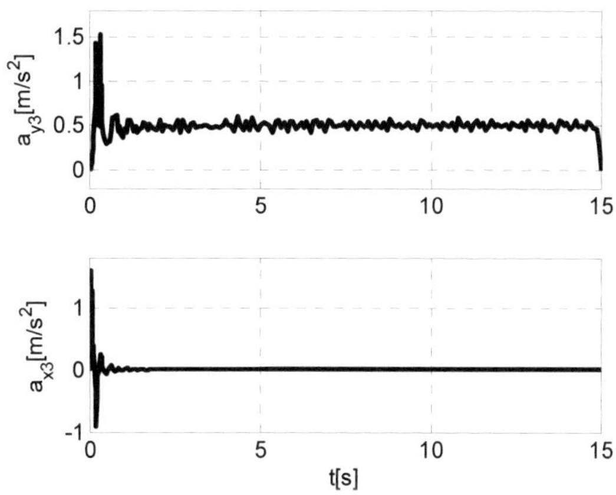

Fig. 17. Lateral and longitudinal accelerations of the cybercar 3

Longitudinal acceleration	Lateral acceleration	Overall Acceleration
$\overline{a}_{x1} = 0.1714 m/s^2$	$\overline{a}_{y1} = 0.3113 m/s^2$	$\widetilde{a}_{w1} = 0.497 m/s^2$
$\overline{a}_{x2} = 0.1903\, m/s^2$	$\overline{a}_{y2} = 0.4934 m/s^2$	$\widetilde{a}_{w2} = 0.54 m/s^2$
$\overline{a}_{x3} = 0.4559 m/s^2$	$\overline{a}_{y3} = 0.5673 m/s^2$	$\widetilde{a}_{w3} = 0.79 m/s^2$

Table 3. r.m.s. and mean accelerations of all the cybercars

Jerks with low saturation values of the fuzzy control surfaces (cfr. (65)).	Jerks with high saturation values of the fuzzy control surfaces (cfr. (66)).
$\bar{j}_{x1} = 0.1177 m/s^3$	$\bar{j}_{x1} = 1.1589 m/s^3$
$\bar{j}_{y1} = 0.0226 m/s^3$	$\bar{j}_{y1} = 0.4712 m/s^3$
$\bar{j}_{x2} = 0.0458 m/s^3$	$\bar{j}_{x2} = 2.6166 m/s^3$
$\bar{j}_{y2} = 0.0225 m/s^3$	$\bar{j}_{y2} = 0.5837 m/s^3$
$\bar{j}_{x3} = 0.1118 m/s^3$	$\bar{j}_{x3} = 2.2999 m/s^3$
$\bar{j}_{y3} = 0.0226 m/s^3$	$\bar{j}_{y3} = 0.5702 m/s^3$

Table 4. r.m.s. lateral and longitudinal jerks of all the cybercars

6. Conclusion

In this chapter a new fuzzy cooperative control algorithm for multiple fully automated cybercars, where the parameters of fuzzy controller may be tuned to obtain low vibrations on the body of the passengers, has been developed. A generalized mathematical model for multiple cybercars to project the fuzzy control system and an acceleration model to ensure the comfort of the passengers have been formulated. A new decentralized trajectory planner which guarantee the absence of collisions between the closest vehicles has been presented. A new fuzzy control strategy which stabilizes all the cooperative vehicles in the planned trajectories has been developed, where the asymptotical stability of the motion errors has been proved by using Lyapunov's theorem and Barbalat's lemma. Good passengers comfort levels during the motion has been ensured by tuning of the saturation of the fuzzy maps. In the simulation tests an example in case of motion control of three automated vehicles has been developed. Trajectories without intersections have been generated and, by choosing suitable input-output values of the fuzzy maps, the stability of the motion errors and very good passengers comfort levels based on ISO 2631-1 Standard have been obtained.

7. Aknowledgements

This work was realized with the contribution of the MIUR ex-60%.
All sections have been equally and jointly developed by the authors.

8. References

Antonelli, G. & Chiaverini, S. (2006). Kinematic Control of Platoons of Autonomous Vehicles. *IEEE Transactions on Robotics*, Vol. 22, Issue 6, pp. 1285-1292, ISSN 1552-3098.

Bicho, E. & Monteiro, S. (2003). Formation control of multiple mobile robots: a non-linear attractor dynamics approach, *Proceedings of IEEE International Conference on Intelligent Robots and Systems,* pp. 2016-2022, Vol. 2, ISBN 0-7803-7860-1.

Birlik, G. & Sezgin, O. C. (2007). Effect of Vibration on Trasportation System, In: *Springer Proceedings in Physics,* pp. 85-90, Springer Netherlands, ISBN 978-1-4020-5400-6.

Bloch, A.M. (2000). *Nonholonomic Mechanics and Control,* Springer Verlag, ISBN 0-387-95535-6.

Driessen, J.B.; Feddema, J.T. & Kwok, K.S. (1999). Decentralized Fuzzy Control of Multiple Nonholonomic Vehicles. *Journal of Intelligent and Robotic Systems,* Vol. 26, Issue 1, Kluwer Academics Publishers, pp. 65-70, ISSN 0921-0296.

Fierro, R. & Lewis J. (1997). Control of a Nonholonomic Mobile Robots: Backstepping Kinematics into Dynamics. *Journal of Robotics System,* Vol. 14, Issue 3, pp. 149-163, ISSN 0741-2223.

Gerkey, B.P. & Mataric, M.J. (2002). Sold!: Auction Methods for Multi-Robots Coordination. *IEEE Transactions on Robotics and Automation,* Vol. 18, Issue 5, pp. 758-768, ISSN 0882-4967.

La Valle, S.M. & Hutchinson, S.A. (1998). Optimal Motion Planning for Multiple Robots having Independent Goals. *IEEE Transactions on Robotics and Automation,* Vol. 14, Issue 6, pp. 912-925, ISSN 0882-4967.

Labakhua, L. ; Nunes, U. ; Rodrigues, R. & Leite, F.S. (2006). Smooth Trajectory Planning for Fully Automated Passengers Vehicles, *Proceedings of the 3th International Conference on Informatics in Control, Automation and Robotics (ICINCO),* pp. 89-96, ISBN 972-8865-60-0, Setubal (Portogal), INSTICC press.

Lumelsky, V.J. & Harinarayan, K.R. (1997). Decentralized Motion Planning for Multiple Mobile Robots: the Cocktail Party Model. *Autonomous Robots,* Vol. 4, Issue 1, pp. 121-135, ISSN 0929-5593.

McDonald, M. & Voge, T. (2003). Analysis of Potentials and Limitations of Cybernetic Transport Systems. *Report of TRG Institute,* University of Southampton (UK).

Melluso, M. (2007). Intelligent Fuzzy Lyapunov/EKF Control for Automatic Motion of Ground Vehicles with Non-Holonomic Constraints and Uncertainties. *PhD Thesis,* Dipartimento di Ingegneria dell'Automazione e dei Sistemi, University of Palermo (Italy).

Panfeng, H.; Kai, C.; Janping, Y. & Yangsheng, X. (2007). Motion Trajectory Planning of Space Manipulator for Joint Jerk Minimization, *Proceedings of IEEE International Conference on Mechatronics and Automation,* pp. 3543-3548, ISBN 1-4244-0828-8, Harbin (China).

Raimondi, F.M. & Melluso, M. (2005). A New Fuzzy Dynamic Controller for Autonomous Vehicles with Nonholonomic Constraints. *Robotics and Autonomous Systems,* Elsevier, Vol. 52, pp. 115-131, ISSN 0921-8890.

Raimondi, F.M. & Melluso, M. (2006-a). Fuzzy EKF Control for Wheeled Nonholonomic Vehicles, *Proceedings of 32th IEEE Conference on IECON 2006,* pp. 43-48, ISBN 1-4244-0136-4 , Paris (France).

Raimondi, F.M. & Melluso, M. (2006-b). Robust Trajectory Tracking Control and Localization based on Extended Kalman Filter for Nonholonomic Robots, In: *Advanced Technologies, Research-Development-Application,* edited by Bojan Lalic, pp. 699-724, Pro Literatur Verlag, ISBN 3-86611-197-5, Mammendorf (Germany).

Raimondi, F.M. & Melluso M. (2007-a). Fuzzy Adaptive EKF Motion Control for Non-Holonomic and Underactuated Cars with Parametric and Non-Parametric Uncertainties. *IET Control Theory and Applications,* Vol.1, Issue 5, pp. 1311-1321, ISSN 1751-8644.

Raimondi, F.M. & Melluso M. (2007-b). Fuzzy Cooperative Control of Automated Ground Passenger Vehicles, *Proceedings of IEEE Conference on ETFA 2007,* pp. 1364-1371, ISBN 1-4244-0826-1, Patras (Greece).

Slotine, J.J. & Li, W. (1991). *Applied Nonlinear Control,* Prentice Hall International Edition, ISBN 0-13-040890-5, London UK.

Suzuki, H. (1998). Research Trends of Riding Comfort Evaluation in Japan. *Proceedings of the Institution of Mechanical Engineers part F- Journal of Rail and Rapid Transit,* Vol. 212, Issue 1, pp. 61-72, ISSN 0954-4097.

18

Characteristics of Mechanical Noise during Motion Control Applications

Mehmet Emin Yüksekkaya, Ph.D.
College of Engineering
Usak University
Turkey

1. Introduction

Signal characteristics and processing are an important factor during today's digital world including the motion control and strain measurement applications. A digital signal is someone that can assume only a finite set of values is given for both the dependent and independent variables being analyzed (Smith, 2006). The independent variables are usually time or space; and the dependent variables are usually amplitudes. To use digital signal processing tools effectively, an analog signal must be converted into its digital representation in time space. In practice, this is implemented by using an analog-to-digital converter (A/D), which is an integral part of data acquisition (DAQ) cards (Vaseqhi, 2009). One of the most important parameters of an analog input system is the sampling rate at which the DAQ card samples an incoming signal. During the measurement and processing of the signal digitally, it is common to face noise problems interfacing the signals captured (Yuksekkaya, 1999; Chu & George, 1999; Kester, 2004). The noise could be coming from various sources with different characteristics and affecting the measurement systems. Once the signal is contaminated with the noise, the reading from the instruments will not be representing the actual situation of the physical phenomenon being captured. Therefore, it is an important area of practice to analyze the characteristics of the noise for any implementation of the signal analysis before constructing further refinements for data analysis. Furthermore, it would be a more practical to take some precautions in order to reduce the effects of the noise on the signal. Even it is possible to use some tools to decrease the effects of the noise to the signal ratio, it would be more practical to eliminate the noise as much as possible at the first hand. It is also evident from the industrial applications that the cost of initial investments for any noise elimination applications is cheaper than that of later investments.

Computer-based data acquisition systems using small computers have been successfully applied in many industrial applications including the motion control processes producing high performances at relatively low costs. As the investment cost of data processing systems decreases, it is getting more common to see a number of data acquisition systems implemented applications in our daily life. The benefits of a data acquisition system include: an improved analysis, accuracy and consistency, reduced analysis time and cost, and lower response time for an out-of-control situation regarding quality. It could be easily noticed that there would be a tremendous amount of noise superimposed on the signal coming from the measurement units. The noise could be coming from different sources depending on the application area. The main sources of the noise, however, are mechanical and electrical noises commonly found at the industrial applications (Yuksekkaya, 1999). Therefore, refinements are necessary for most of the times so that the noise problems could be eliminated from the signal in order to make an accurate measurement during the motion control.

During the industrial applications such as CNC controlled lathes and load cells taking the dynamic measurements, a considerable amount of mechanical noise could be superimposed to the signal from the ground due to the vibration of the buildings. The mechanical noise problem could damage the reading from the instruments due to the noise superimposed to the signal. In this text, an extensive analysis of the mechanical noise due to the building vibration has been analyzed and possible solutions to the problem discussed.

2. Diagnostics of noise in the signal

As stated, digital signal is a finite set of values in both the dependent and independent variables. One of the most important parameters of an analog input system is the rate at which the DAQ card samples an incoming signal. A fast sampling rate acquires more points in a given time. As a result, a better representation of the signal is formed. Sampling too slow may result in a poor representation of the signal. This may cause a misrepresentation of a signal, which is commonly known as an aliasing effect. In order to avoid aliasing effects, the *Nyquist Theorem* states that a signal must be acquired at the rate greater than twice the maximum frequency component in the signal acquired (Ramirez, 1985). Figure 1 indicates the basic divisions of different signal types. The most fundamental division is stationary and non-stationary signals. Stationary signals are characterized by average properties that do not vary with time and independent of the particular sample record used to determine them. The term "non-stationary" covers all signals that do not satisfy the requirements for stationary signals. Computer-based data acquisition systems using small computers have been successfully applied in many applications producing high performances at relatively low costs. The benefits of a data acquisition system include: an improved analysis, accuracy and consistency, reduced analysis time and cost, and lower response time for an out-of-control situation regarding quality. A typical data acquisition system consists of several parts: a signal conditioning module, a data acquisition hardware (A/D converter), analysis hardware, and data analysis software.

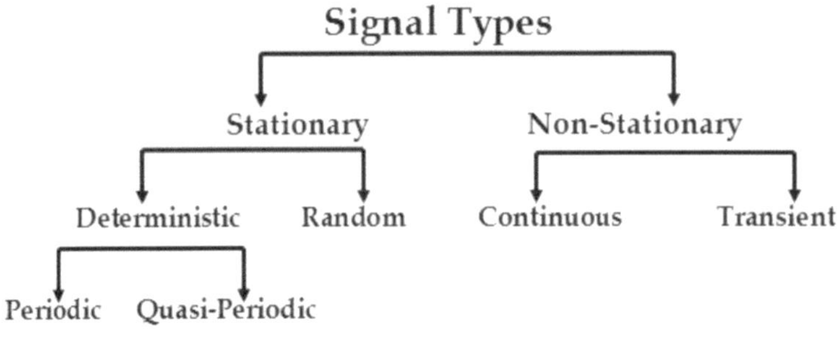

Fig. 1. Classification of signal types.

In any digitally working environment, most of the time, there were two types of main noise problems that needed to be solved in order to have a meaningful result as follows (Yuksekkaya & Oxenham 1999; Yuksekkaya et. al. 2008):

1) Electrical Noise:
    i) Static
        a) Fluorescent lamps
        b) Computer screens
        c) Others
    ii) Dynamic
        a) AC power lines
        b) Stepper motor
        c) Transformers
        d) Magnetic fields from other equipment
2) Mechanical Noise:
    i) Vibration from stepper motor
    ii) Vibration from building
    iii) Vibration from the other sources

2.1 Electrical noise

Tensile testing devices are a combination of a strain measurement unit and a stepper motor which drives the measurement unit. During the processing of a strain measurement, the location of measurement unit should be precisely located in order to have an accurate stress-stain reading from the instrument. Most of the time, strain measurements are taken in the presence of electrical and magnetic fields, which can superimpose electrical noise on the measurement signals. If the electrical noise is not controlled properly, the noise can lead to inaccurate results and incorrect interpretation of the signals coming from the strain gages as well as the inaccurate location data. In order to control the noise level and maximize the signal-to-noise ratio, it is first necessary to understand the types and characteristics of electrical noise as well as the sources of such noises. Without understanding the noise and its sources, it is impossible to apply the most effective noise-reduction methods on any particular instrumentation problem.

Virtually, every electrical device that generates, consumes, or transmits power is a potential source for causing noise in strain gages. In general, the higher the voltage or current level and the closer the circuit is to the electrical device, the greater the induced noise will be superimposed to the signal. A list of common sources of electrical noise could be found in any signal analysis textbooks and electrical noise from those sources could be categorized into two basic types, that is: electrostatic and magnetic noises (Croft et al., 2006; Agres, 2007). The characteristics of these two types of noise are different; and they require different noise reduction techniques in order to eliminate their effects on the signals. Most of the noise coming from outside may be eliminated by using shielded, twisted cables and eliminating the ground loops in the system (more than one connection of the system to the ground). Furthermore, electromagnetic noises could be eliminated by using a special designed apparatus named as *Faraday Cage* if the application requires it.

2.2 Mechanical noise

It would be practically possible to see that a strain measurement signal could be so sensitive that it would be continuously picking up mechanical noise from different sources such as from the buildings and from the stepper motors. It is necessary to analyze the building and stepper motor vibration sources separately. An extensive analysis of the mechanical noise coming from the building and potential solutions for the vibration sources are given as follows:

2.2.1 Building vibration

Laboratory measurement instruments could be located in either stationary or mobile laboratories depending on the type of measurements necessary to perform. Regardless of the location of the instruments, it is a known fact that the ground vibration will affect the instrument's reading if correct precautions will not be applied. A considerable amount of mechanical noise would be coming from the ground due to the vibration of the building. In order to eliminate, or at least minimize, the effect of the ground movement as much as possible, usually the testing instrument was mounted on the top of a heavy marble block that was supported by a spring-like material. In order to analyze the vibrating mechanical system, let us consider an object hanging from a spring as shown in Figure 2 (Halliday et al., 2007; Zill & Cullen, 2006).

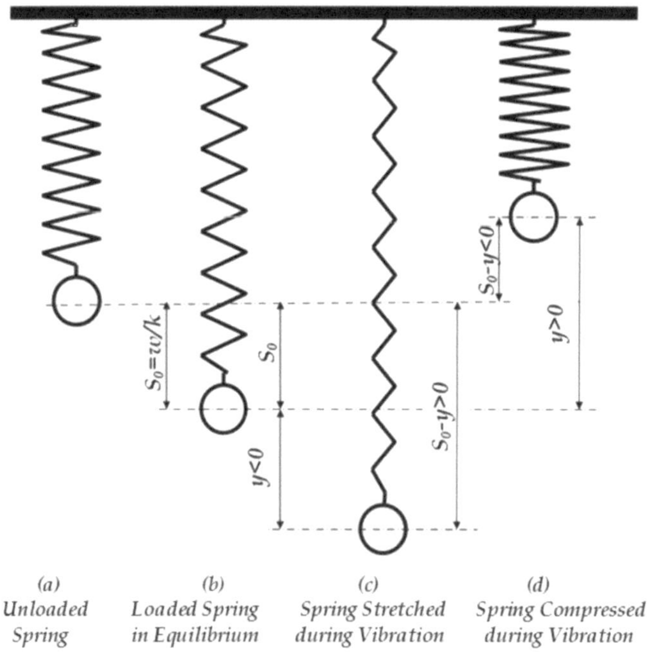

Fig. 2. Schematic of mass spring system.

The weight, w, of the object is the magnitude of the force of gravity acting on the measurement instrument. The mass, m, of the object is related to the weight of the object, w as follows:

$$w = mg \tag{1}$$

where g is the acceleration of the gravity. In order to keep complications to a minimum for the sake of analysis purposes, it was assumed that the spring obeys Hooke's law, that is, force is proportional to displacement. Namely,

$$f_s = ky \tag{2}$$

Where y is the displacement and k is the stiffness of the spring element. A realistic analysis of the vertical motion of the mass would take into account not only the elastic and gravitational forces, but also, the effects of the friction affecting the system and all other forces that act externally on the suspended mass. Considering the other forces diminishes the amplitude of the vibration. In order to keep the analysis as simple as possible, let us do not take them into the account now and talk about the details later.

Observing that $m = w/g$ and applying Newton's second law of motion, (*force* = *mass* * *acceleration*) to the system gives a very popular equation:

$$\frac{w}{g}\frac{d^2y}{dt^2} + ky = 0 \tag{3}$$

This is an equation of harmonic motion and its solution was discussed in almost every differential equation book (Halliday et al., 2007; Zill & Cullen, 2006). The solution for such a system is:

$$y = A\cos\sqrt{\frac{kg}{w}}t + B\sin\sqrt{\frac{kg}{w}}t \tag{4}$$

If the term $\sqrt{\frac{kg}{w}}$ is set to be ω, then Equation 4 can be written by using a periodic function, as used above, in order to have a more compact form as follows:

$$y = A\cos\omega t + B\sin\omega t \tag{5}$$

Regardless of the values of A and B, that is, regardless of how the system is set in motion, Equation 4 describes the periodic motion with the period of

$$2\pi\sqrt{\frac{w}{kg}} \tag{6}$$

or frequency of

$$\frac{1}{2\pi}\sqrt{\frac{kg}{w}} \tag{7}$$

Whether there is friction in the system or not, the quantity $\frac{1}{2\pi}\sqrt{\frac{kg}{w}}$ is called the *natural frequency*, (ω_n), of the system because this is the frequency at which the spring-mass system would vibrate naturally if no frictional or non-elastic forces other than gravity were present in a given system.

As mentioned earlier, the process by which free vibration diminishes in the amplitude is called the damping effect. If the damping effect presents in the system, the energy of the vibrating system will be dissipated by various mechanisms affecting the system, and often, more than one mechanism could be present in the system at the same time. In such a system, the damping force, f_D, is related to the velocity across the linear viscous damper by the following equation:

$$f_D = c\frac{dy}{dt} \tag{8}$$

where the constant, *c*, is the viscous damping coefficient and has the unit of *force*time/length*. It is important to get the correct damping factor for a given system. Unfortunately, unlike the stiffness of the spring, the damping coefficient cannot be calculated from the dimensions of the structure or the size of the structural elements. Therefore, it should be evaluated from the vibration experiments on actual structures in order to get a precise coefficient of damping ratio for minimizing the effects of mechanical vibration to the measurement instruments.

This system is usually called a mass-spring-damp system, and its governing equation can be written according to the Newton's second law of motion as follows:

$$\frac{w}{g}\frac{d^2y}{dt^2} + c\frac{dy}{dt}ky = p(t) \tag{9}$$

where $p(t)=p_0 sin(\omega t)$ is the function of the external force (from ground vibration, etc.) acting to the system. The nature of the free motion of the system will depend on the roots of the related characteristic equation of the second order differential equation given in the equation. The characteristic equation for this second order differential equation is given as follows:

$$-\frac{cg}{2w} \mp \frac{g}{2w}\sqrt{c^2 - \frac{4kw}{g}} \tag{10}$$

It is clear that *g*, *k*, and *w* are all positive quantities and *c* is a non-negative and real number. Therefore, the characteristic of the solution of this second order differential equation depends upon the term $\sqrt{c^2 - \frac{4kw}{g}}$. It is clear that there are three possibilities depending upon the values of *k*, *w*, and *c*, for the solution of this second order differential equation namely,

$$\sqrt{c^2 - \frac{4kw}{g}} \begin{cases} > 0 \\ = 0 \\ < 0 \end{cases} \tag{11}$$

If $\sqrt{c^2 - \frac{4kw}{g}} > 0$, there is a relatively large amount of friction, and, naturally enough, the system or its motion is said to be over-damped. In this case, the roots of the characteristic equation are real and unequal. The general solution is given by:

$$y = Ae^{m_1 t} + Be^{m_2 t} \tag{12}$$

where both of the roots of the second order differential equation, m_1 and m_2, are negative. Thus, y approaches zero as time increases indefinitely. If $\sqrt{c^2 - \frac{4kw}{g}} = 0$, it is at the borderline in which the roots of the characteristic equation are equal and real. In this case, the free motion can be expressed as follows:

$$y = Ae^{mt} + Bte^{mt} = (A + Bt)e^{mt} \tag{13}$$

From the equation $\sqrt{c^2 - \frac{4kw}{g}}$, critical damping can be defined as:

$$c_{cr} = \frac{2w}{g}\omega_n = 2m\omega_n \tag{14}$$

Then, the damping ratio, ζ, is defined as:

$$\zeta = \frac{c}{2m\omega_n} = \frac{c}{c_{cr}} \tag{15}$$

If $\sqrt{c^2 - \frac{4kw}{g}} < 0$, the motion is said to be under-damped. The roots of the characteristic equation are then the conjugate complex numbers given by:

$$m_1, m_2 = -\frac{cg}{2w} \mp i\sqrt{\frac{kg}{w} - \frac{c^2g^2}{4w^2}} \tag{16}$$

where $i = \sqrt{-1}$. Then, the general solution for the differential equation is given by:

$$y = Ae^{-\frac{cg}{2w}t}\cos\sqrt{\frac{kg}{w} - \frac{c^2g^2}{4w^2}}t + Be^{-\frac{cg}{2w}t}\sin\sqrt{\frac{kg}{w} - \frac{c^2g^2}{4w^2}}t \tag{17}$$

By defining some of the terms in the equation given above differently and using some trigonometric identities, the solution of the equation may be written in a more compact form as follows:

$$\sqrt{\frac{kg}{w} - \frac{c^2g^2}{4w^2}} = \omega_n\sqrt{1 - \zeta^2} = \omega_D \tag{18}$$

The differential equation is solved, subject to the initial conditions $y = y(0)$, and $\frac{dy}{dt} = dy/dt|_{y=0}$ The particular solution of such a system is given by:

$$y_p = \frac{p_0}{k\left[1 - \frac{\omega^2}{\omega_n^2}\right]^2 + \left[2\zeta\left(\frac{\omega}{\omega_n}\right)\right]^2}\left[\left(1 - \left(\frac{\omega}{\omega_n}\right)^2\right)\cos(\omega t) - \frac{2\zeta\omega}{\omega_n}\sin(\omega t)\right] \tag{19}$$

By setting $C = \frac{p_0\left(1 - \frac{\omega^2}{\omega_n^2}\right)}{k\left[1 - \frac{\omega^2}{\omega_n^2}\right]^2 + \left[2\zeta\left(\frac{\omega}{\omega_n}\right)\right]^2}$ and $D = \frac{-2p_0\zeta\frac{\omega}{\omega_n}}{k\left[1 - \frac{\omega^2}{\omega_n^2}\right]^2 + \left[2\zeta\left(\frac{\omega}{\omega_n}\right)\right]^2}$, the Equation 19 can be written as

$$y_p = C\sin(\omega t) + D\cos(\omega t) \tag{20}$$

However, the complete solution of Equation 9 consists of transient and steady parts given as follows:

$$y(t) = e^{-\zeta\omega_n t} \underbrace{(A\cos\omega_D + B\sin\omega_D}_{\text{transient}} + \underbrace{C\sin\omega t + D\cos\omega t}_{\text{steady state}} \tag{21}$$

where the constant A and B can be determined in terms of the initial displacement and initial velocity. The steady state deformation of the system, due to harmonic force given in Equation 19 can be rewritten as follows:

$$y(t) = y_0 \sin(\omega t - \phi) = \frac{p_0}{k} R_d \sin(\omega t - \phi) \tag{22}$$

where $y_0 = \sqrt{C^2 + D^2}$and $\phi = \tan^{-1}\left(\frac{D}{C}\right)$. Substituting for C and D gives deformation response factor, R_d, and phase angle, ϕ.

$$R_d = \frac{1}{\sqrt{\left[1-\frac{\omega^2}{\omega_n^2}\right]^2 + \left[2\zeta\left(\frac{\omega}{\omega_n}\right)\right]^2}} \tag{23}$$

and

$$\phi = \tan^{-1}\left(\frac{2\zeta\frac{\omega}{\omega_n}}{1-\frac{\omega^2}{\omega_n^2}}\right) \tag{24}$$

Differentiating Equation 22 gives the velocity as follows:

$$\frac{dy(t)}{dt} = \frac{p_0}{\sqrt{km}} R_v \cos(\omega t - \phi) \tag{25}$$

where R_v is the velocity response factor and related to R_d by

$$R_v = \frac{\omega}{\omega_n} R_d \tag{26}$$

After applying the basic definitions and solutions of the differential equation for a mass-spring-damper system, force transmission and vibration isolation can be taken into account as follows: Consider the mass-spring-damper system (The system is the instrument itself and any foundation making the total weight higher), shown in Figure 3 subjected to a harmonic force. The force transmitted to the base is given by:

$$f_T = f_s + f_D = ky(t) + c\frac{dy(t)}{dt} \tag{27}$$

Substituting Equation 22 for y (t) and Equation 25 for $\frac{dy(t)}{dt}$, and using Equation 26 give:

$$f_T(t) = (y_{st})_0 R_d [k\sin(\omega t - \phi) + c\omega\cos(\omega t - \phi)] \tag{28}$$

The maximum value of f_T (t) over t is:

$$(f_T)_0 = (y_{st})_0 R_d \sqrt{k^2 + c^2\omega^2} \tag{29}$$

which after using $(y_{st})_0 = \frac{p_0}{k}$ and $\zeta = \frac{c}{2m\omega_n}$ can be expressed as:

$$\frac{(f_T)_0}{p_0} = R_d \sqrt{1 + \left(2\zeta\frac{\omega}{\omega_n}\right)^2} \tag{30}$$

Substituting Equation 23 gives an equation for the ratio of the maximum transmitted force to the amplitude p_0 of the applied force, known as the transmissibility (*TR*) of the system for a mass-spring-damper application:

$$TR = \sqrt{\frac{1+\left(2\zeta\frac{\omega}{\omega_n}\right)^2}{\left[1-\frac{\omega^2}{\omega_n^2}\right]^2+\left[2\zeta\left(\frac{\omega}{\omega_n}\right)\right]^2}} \tag{31}$$

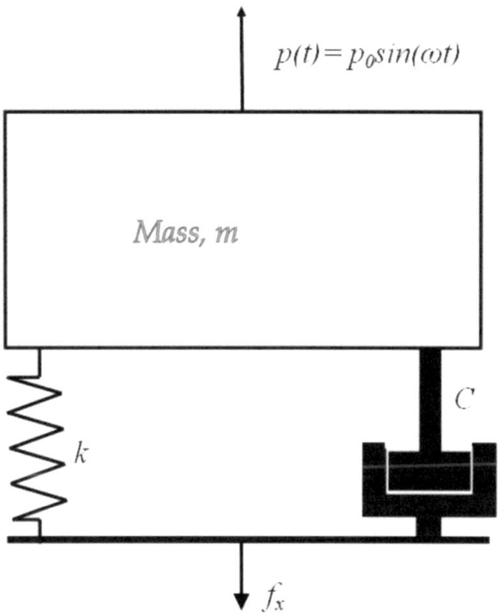

Fig. 3. Simple spring-mass-damper system

The transmissibility is plotted in Figure 4 as a function of the frequency ratio ω/ω_n, for several values of the damping ratio, ζ. For the transmitted force to be less than the applied force, the stiffness of the support system, and hence, the natural frequency, should be small enough so that the ratio of ω/ω_n should be bigger than $\sqrt{2}$ as seen in Figure 4. No damping is desired in the support system because, in this frequency range, damping increases the transmitted force. This implies a trade-off between a soft spring material to reduce the transmitted force and an acceptable static displacement. If the excitation frequency, ω, is much smaller than the natural frequency, ω_n, of the system, (*i.e.*: the mass is static while the ground beneath it is dynamic). This is the concept underlying isolation of a mass from a moving base by using a very flexible support system. For example, instruments or even buildings have been mounted on natural rubber bearings in order to isolate them from the ground-borne, vertical vibration (typically, with frequencies that range from 25 to 50 Hz) due to the rail traffic (Bozorgnia & Bertero, 2004; Chen & Lui, 2005). It would be also advisable to use rubber like material on the testing instruments in order to diminish the effects of the ground vibrations.

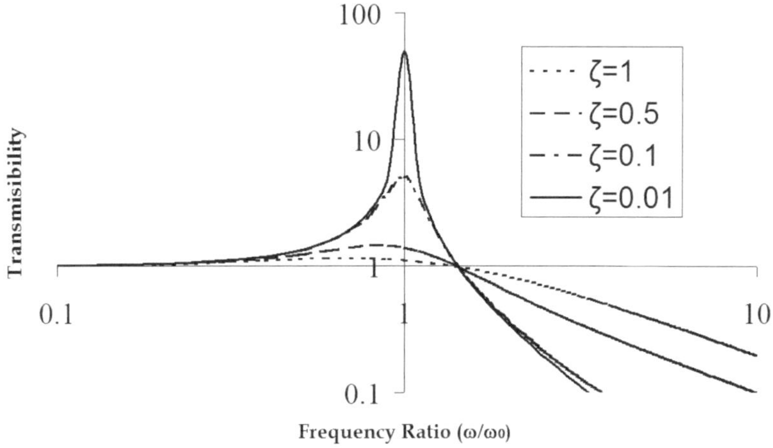

Fig. 4. Transmissibility for harmonic excitation for various damping factors (Axes are in logarithmic scale).

Figure 5 shows a typical building vibration effect on the acquired signal from an instrument. In order to reduce the amount of vibration that is transmitted to the instrument, natural rubber-like materials, such as tennis balls, are the appropriate choice. Figure 6 shows the effect of vibration dampers on the instrument. As seen from the graph, the usage of vibration damper reduces the effects of mechanical vibrations significantly. Some further improvements can be achieved by precisely calculating (and if necessary modifying) the stiffness of the insulation material taking into consideration the low damping coefficient or increasing the weight of instrument.

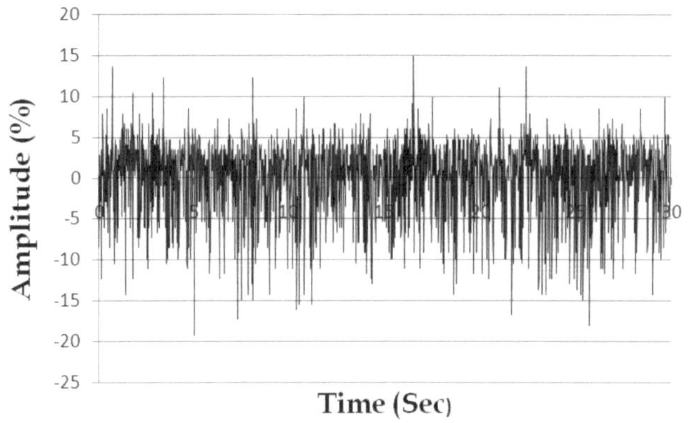

Fig. 5. Effects of building vibration on the acquired signal before using damper

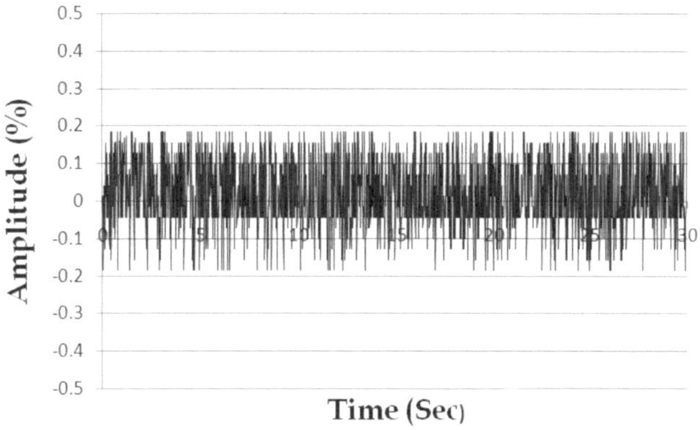

Fig. 6. Effects of building vibration on the acquired signal after using damper

In the analysis of the ground vibration, as seen, only vertical vibration of the building is considered. However, it is clearly known that buildings are exciting in three dimensions. As seen in the analysis, it was assumed that only vertical vibration had a significant effect on the data. This assumption may introduce some experimental errors into the measurement. However, a significant drop in the amplitude of the noise transmitted suggested that either the tennis balls were also eliminating some of the vibration effects coming from the other directions, or the vibration coming from the other directions did not have a significant effect on the signal. Therefore, vibration effects from other directions were not investigated further in this analysis.

In the engineering view of the problem, the ground vibration problem has a crucial effect on measurement instruments. Therefore, it would be advisable to take all of the necessary precautions in order to reduce the amount of transmitted ground vibration to the minimum level as much as possible by using damper systems which make the ratio of ω/ω_n is bigger than $\sqrt{2}$.

3. References

Agres, D. (2007). Dynamics of Frequency Estimation in the Frequency Domain, *IEEE Transactions on Instrumentation and Measurement,* Vol: 56, Issue:6, (December 2007) 2111-2118 ISSN: 0018-9456.

Bozorgnia, Y. & Bertero, V. V. (2004). *Earthquake Engineering: From Engineering Seismology to Performance-Based Engineering,* CRC Press LLC, ISBN: 0-203-48624-2, Boca Raton, FL.

Chen, W. F. & Lui, E. M. (2005). *Earthquake Engineering for Structural Design,* CRC Press Taylor & Francis Group, ISBN: 0-8493-7234-8, Boca Raton, FL.

Chu, E. & George, A., (1999). *Inside the FFT Black Box Serial and Parallel Fast Fourier Transform Algorithms,* CRC Press LLC, ISBN: 0-8493-0270-6 Boca Raton, FL.

Croft, A.; Davison, R. & Hargreaves, M. (2000). *Engineering Mathematics: A Foundation for Electronic, Electrical, Communications and System Engineers, Third Edition,* Pearson, Prentice Hall, ISBN:0-13-026858-5, Rugby, Warwickshire, UK.

Halliday, D; Resnick, R. & Walker, J. (2007). *Fundamentals of Physics Extended 7th Edition,* John Wiley & Sons, Inc., ISBN: 0-471-47062-7, Hobeken, NJ.

Kester, W., (2004). *The Data Conversion Handbook,* Elsevier, ISBN: 0-7506-7841-0, Oxford, UK.

Ramirez, R. W. (1985). *The FFT Fundamentals and Concepts,* Tektronix, Inc. ISBN: 0-13-314386-4, Cliff, NJ.

Smith, S. W. (2006). *The scientist & Engineer's Guide to Digital Signal Processing,* California Technical Publishing, ISBN: 0-9660176-3-3, San Diego, CA.

Vaseqhi, S. V, (2009). *Advanced Digital Signal Processing and Noise Reduction: Fourth Edition,* John Wiley & Sons Ltd., ISBN: 978-0-470-75406-1, West Sussex, UK.

Yuksekkaya, M. E. & Oxenham, W. (1999). Analysis of Mechanical and Electrical Noise Interfacing the Instrument During Data Acquisition: Development of a Machine for Assessing Surface Properties of Fibers, Textile, Fiber and Film Industry Technical Conference, Atlanta, GA.

Yuksekkaya, M. E. (1999). A Novel Technique for Assessing the Frictional Properties of Fibers, *Ph.D. Dissertation,* North Carolina State University, Raleigh, NC.

Yuksekkaya, M. E.; Oxenham, W. & Tercan, M. (2008). Analysis of Mechanical and Electrical Noise Interfacing the Instrument During data Acquisition for Measurement of Surface Properties of Textile Fibers, *IEEE Transactions on Instrumentation and Measurement,* Vol:57, No:12, (December 2008) 2885-2890 ISSN: 0018-9456.

Zill, D. G. & Cullen, M. R. (2006). *Advanced Engineering Mathematics, Third Edition,* Jones and Bartlett Publishers, ISBN: 13-978-0-7637-4591-2, Sudbury, MA.

19

Hybrid Magnetic Suspension Actuator for Precision Motion Control

Dengfeng Li[1] and Hector Gutierrez[2]
[1]*Laser Photonics, LLC*
[2]*Florida Institute of Technology*
USA

1. Introduction

The blooming development of semiconductor fabrication technologies, such as those used for manufacturing of micro electromechanical systems, has increased the interest to develop actuators that can provide micro scale or nano scale motion, with large range of travel range and fast dynamic response. For these applications, the main candidate technologies are magnetic suspension actuation (MSA) and piezoelectric actuation (PZT).

Direct piezoelectric actuation (i.e., actuation based on the strain of a piezo stack by the application of voltage or charge) has inherently a very limited range of travel. Although the displacement can be enlarged by mechanical amplification (Robbins, 1991) the range available for direct PZT actuation is at most in the order of a few hundred microns, and the precise control of PZT actuators involves other challenges such as hysteresis and drift. To achieve millimeter-range travel with piezoelectric actuation, ultrasonic motors or inchworm mechanisms have been devised. These devices can achieve excellent positioning performance but are limited in their force and torque load capacity, and have small tracking bandwidth.

MSA technologies are an interesting alternative to PZT in that they can provide substantially larger load capacity, fast response and large range of travel without introducing the complex microscopic phenomena associated with mechanical contact. The MSA devices based on attractive force can be found in works of Trump *et al* (Trump,1997; Kuo, 2003). All of these are small gap actuators which is not suitable for applications where the actuator is remote from the controlled target. More recently, large gap MSA devices which magnify the magnetic field to overcome the gap limitation can be found in (Lin, 2007) (Craig, 2007) and (Khamesee, 2005). However, it seems that the mass of the controlled object is required to be very small. When the mass and air gap become larger, a potential problem is the steep increase in required control effort (actuator current) may lead to heat dissipation problems and actuator core saturation. A novel hybrid MSA concept which has a great promise to realize large gap while avoiding the need of large currents and actuator saturation is presented in this paper. In the proposed design, a pair of permanent magnets in repulsive configuration is located coaxial with an attractive force electromagnet, as shown in Fig. 1. In this arrangement, zero current is needed to obtain large gap since passive force is used. Controlled attractive forces are used to provide stabilization of the controlled object.

This push-and-pull strategy is particularly useful when a large mass is to be suspended over a large range of travel, since the nominal gap is mainly dependent on the strength of the permanent magnets.
The dynamics of a MSA actuator are dominated by the nonlinear force-gap-current relationship. The most common controller design approach has been the use of linear control based on the linearization of the nonlinear dynamics around a nominal operation point. Controller performance deteriorates rapidly when the actuator position moves away from the nominal operation point. In (Trump, 1997) and (Ludwick, 1996), feedback linearization is successfully applied to a class of systems that can be described by the controllability canonical form. However, to achieve its ideal performance, feedback linearization requires perfect knowledge of the system's dynamics. In an application with large range of travel, parameter variation presents a considerable disturbance that has to be compensated by the controller. A robust nonlinear compensation scheme that synthesizes feedback linearization and disturbance estimation is presented in (Mittal, 1997). Variable structure control (VSC) is another good choice to provide robustness to external perturbations and model uncertainties. In (Gutierrez, 2005), a modified equivalent control approach is used to describe the dynamics of sliding mode. However, a reaching condition that uses a Lyapunov function approach ensures good transient performances only in systems with a single input since there is single sliding surface (Hung, 1993). In systems with multiple inputs there might be multiple sliding surfaces, and the transient performances on some sliding surfaces are typically unknown which might induce slow response or chattering. To obtain fast response and low chatter in a MSA application with long range of travel, an enhanced quasi-sliding mode control with reaching law is proposed and its performances are compared with the PID control with feedback linearization.

2. Actuator design and force model parameters identification

The proposed hybrid MSA is shown in Fig. 1, where two cylinders composed of four NdFeB magnets are placed in coaxial with a round electromagnetic coil. To reduce flux leakage, the permanent magnets are coated with ferrous sleeves. The nominal levitation gap can be modified by changing the length of the permanent magnet stack. The axial force in the HMSA is therefore composed of two components: a repulsive force exerted by the permanent magnets, and an attractive force exerted by the coil.

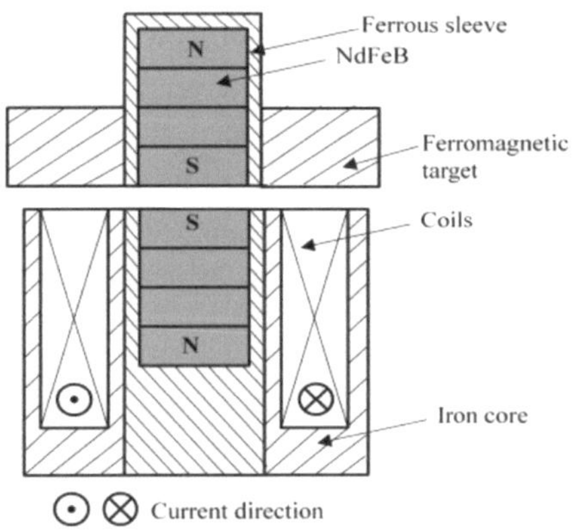

Fig. 1. Hybrid Magnetic Suspension Actuator

The flux lines shown in Fig. 2 demonstrate that the flux leakage can be neglected due to the high permeability of ferrous sleeve. It's reasonable to assume that both the permanent magnet stack have constant flux. The magnetomotive force of each permanent magnet stack is (Chikazumi,2005):

$$\mathcal{F} = \frac{B_r}{\mu} L$$

in which B_r is the residual induction of the permanent magnet, L is the length of the stack and μ is permeability of NdFeB. When the values are B_r=12000 Gauss,L=31.75mm,$\mu=1.15\mu_0$, the magnetomotive force of permanent magnet is 26364 ampere-turn.

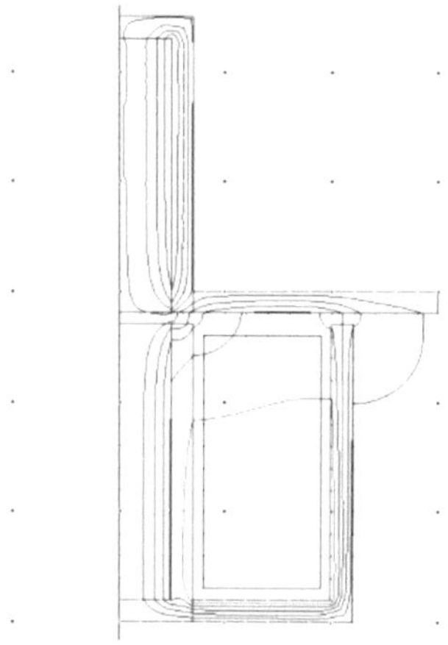

Fig. 2. Flux Lines in Hybrid Magnetic Suspension Actuator

Since the allowed current passing through the coils is designed to be less than 2A and the total number of turns is 270, the NI value of the coil is 540 ampere-turn which is far less than the magnetomotive force of the permanent magnet. This ensures that the moving permanent magnet will not be demagnetized by the field of the coils. The parameters of the proposed HMSA are listed in Table 1.

Diameter of NdFeB (mm)	Thickness of NdFeB (mm)	Residual induction (Gauss)	Thickness of sleeve (mm)	Length of Coil winding (mm)	Coil Diameter (mm)	Number of turns
9.47	7.93	12000	2.25	30.48	0.5	270

Table 1. Parameters of Hybrid Magnetic Suspension Actuator

In order to calculate the force of permanent magnet, a parametric simulation is performed using Ansoft Maxwell as shown in Fig. 3. Then a zero-current test was performed to characterize the force exerted by the permanent magnets. A one-DOF experimental setup to determine force model parameter is shown in Fig. 4, where a ferromagnetic target is mounted at the end of an aluminum beam supported on a flexure. The HMSA is mounted on a clamp which is attached to the base plate. The gap is sensed by a capacitance gauge with an operating range of 500 to 1500μm, and the nominal levitation gap is determined by the relative location of the permanent magnets. The force of permanent magnet is measured by a load cell mounted on a clamp. By adjusting the screw, the air gap is gradually increased from 500 to 1000μm without applying current in the coils. The force corresponding to each gap value is recorded and plotted in Fig. 3 to compare with the simulation data. One can see that in the working range of our selected capacitive sensor the force gap relationship can be approximately treated as linear for simplicity.

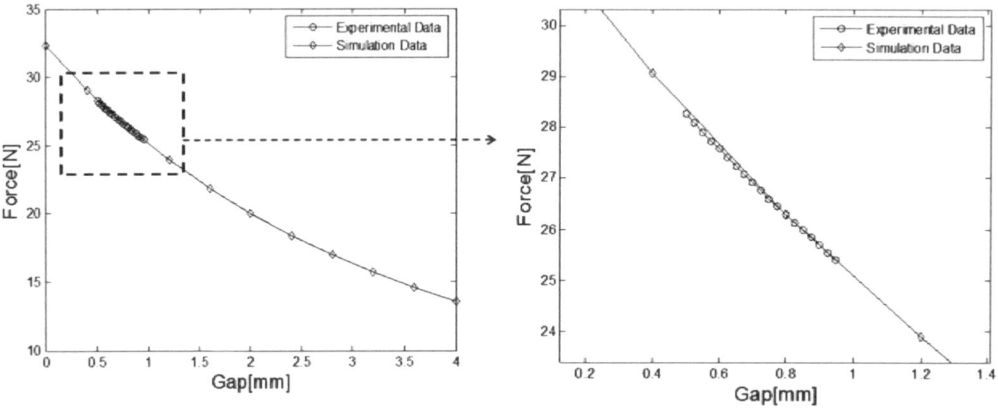

Fig. 3. Experimental and Simulation Repulsive Force Exerted by the Permanent Magnets in the HMSA with zero Current (Left) full view (Right) zoom-in view

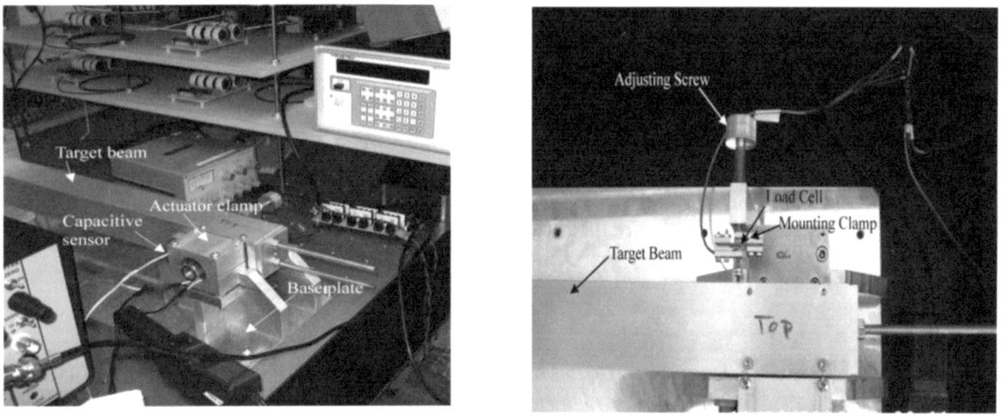

Fig. 4. One DOF Experimental Setup (Left) front view (Right) top view

It is well known that the force produced by attractive-type MSA actuators is proportional to the current squared and inversely proportional to the gap squared. Several methods to find force parameters of attractive-type MSA actuators have been described (Oliveira,1999; Mittal,1997; Lin,1993). This paper takes into account the force of permanent magnet discussed above in the force model of HMSA. Thus, the model for the axial force generated by the HMSA is proposed as:

$$F = K_1 g + K_2 - K_3 \frac{I^2}{(g + x_0)^2} \tag{1}$$

where g is the air gap, I is the coil current ,K_1 and K_2 are constants related to the force exerted by the permanent magnets, K_3 and x_0 are constants related to the electromagnetic attractive force. K_1 and K_2 were first calculated by optimal polynomial fit using the experimental data shown in Fig. 3. After this, DC currents were applied to the HMSA coil, and the corresponding forces and gaps were measured. To cover the same gap range as used for the zero-current measurements, nine data sets were acquired over nine intervals covering the proposed range. Denoting the measured gap, force and current as g_{MEA},F_{MEA} and I_{MEA} respectively, the parameter identification problem can be converted to a multidimensional unconstrained nonlinear optimization problem given by:

$$min \sum \left(F_{MEA} - \left(K_1 g_{MEA} + K_2 - K_3 \frac{{I_{MEA}}^2}{(g_{MEA} + x_0)^2} \right) \right)^2 \quad (2)$$

The initial guesses for K_1 and K_2 were chosen close as the values calculated from Fig. 3. The model parameters in (1) were found to be $K_1=-6523.7N/m$, $K_2=29.588N$, $K_3=7.0306\times10^{-6}Nm^2/A^2$, and $x_0=1.2443\times10^{-3}m$. The experimentally measured force-gap-current relationship is shown on Fig. 5.

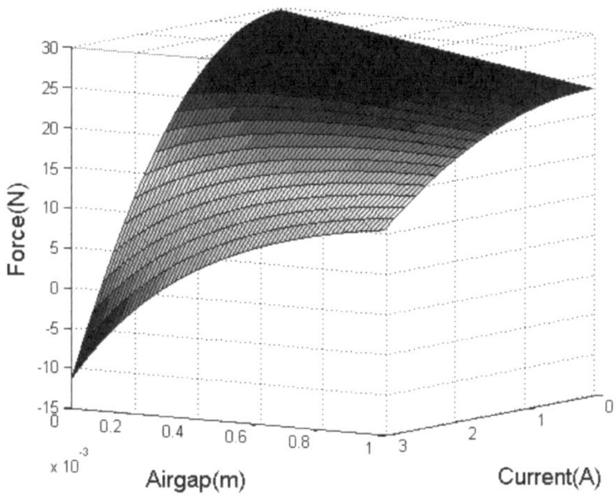

Fig. 5. Experimental Force-Gap-Current Relationship in the HMSA

3. Experimental system modelling and control design

Fig. 6 shows the block diagram of our one-DOF motion control system using HMSA. The position feedback signal is conditioned and used by a DSP board for control computation. The calculated control effort is fed to a servo amplifier to produce the desired current and drive the HMSA. The mass and stiffness of the beam is $M=1.2kg$ and $K_f=2000N/m$ respectively. Assuming the initial gap is g_0 and the discrete state vector is $x(k)= [x_1(k)\ x_2(k)]^T$ with sampling time Δt, the state space equations are:

$$\begin{aligned} x_1(k+1) &= x_2(k)\Delta t + x_1(k) \\ x_2(k+1) &= \frac{1}{M}\left(K_1 x_1(k) + K_2 - K_3 \frac{I^2(k)}{(x_1(k)+x_0)^2} + K_f\left(\frac{g_0}{2} - x_1(k)\right)\right)\Delta t + x_2(k) \end{aligned} \quad (3)$$

Equation (3) can be converted to the controllability canonical form with bounded uncertainty:

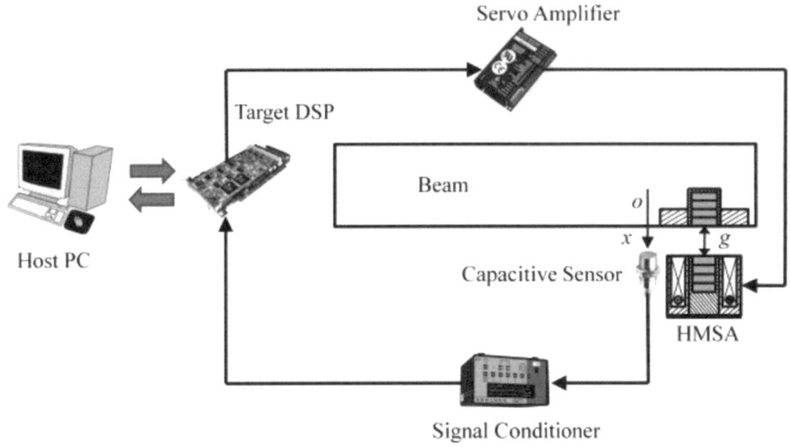

Fig. 6. Experimental One DOF Motion Control System

$$\begin{aligned} x_1(k+1) &= x_2(k)\Delta t + x_1(k) \\ x_2(k+1) &= \big(f(x_1(k)) + g(x_1(k))u(k)\big)\Delta t + x_2(k) \end{aligned} \tag{4}$$

Where

$$\begin{aligned} u(k) &= I^2(k) \\ f(x_1(k)) &= (K_1 - K_f)x_1(k) + K_2 + \frac{1}{2}K_f g_0 \\ g(x_1(k)) &= -\frac{K_3}{(x_1(k)+x_0)^2} \end{aligned}$$

and $G_1 \leq g\ (x_1\ (k)) \leq G_2 < 0$ is a bounded uncertain control gain. If the reference trajectory is $x_r(k) = [x_{1r}(k)\ x_{2r}(k)]^T$, and the error vector is $e\ (k) = [x_1\ (k) - x_{1r}(k)\ x_2\ (k) - x_{2r}(k)]^T$, a sliding surface useful for tracking can be written as:

$$s(k) = c^T e(k), \text{ where } c = [\lambda \quad 1]^T \text{ and } \lambda > 0. \tag{5}$$

Our goal is to find a digital control law $u\ (k)$ that can quickly drive the state trajectory to (5) and keep it there within a specified boundary $\{e\ (k) \|\ s\ (e) < \varepsilon, \varepsilon > 0\}$. Instead of the conventional equivalent control approach, we propose an alternate method based on specifying a reaching law with linear rate:

$$s(k+1) - s(k) = -qs(k)\Delta t - sgn\big(s(k)\big)\varepsilon\Delta t, q > 0 \tag{6}$$

From Equation (5):

$$\begin{aligned} s(k+1) - s(k) = c^T\big(e(k+1) - e(k)\big) = x_2(k+1) - x_2(k) \\ -\big(x_{2r}(k+1) - x_{2r}(k)\big) + \lambda\Big(x_1(k+1) - x_1(k) - \big(x_{1r}(k+1) - x_{1r}(k)\big)\Big) \end{aligned} \tag{7}$$

From Equation (4):

$$x_2(k+1) - x_2(k) = \big(f(x_1(k)) + g(x_1(k))u(k)\big)\Delta t \tag{8}$$

Substituting (6) and (8) into (7), a quasi-sliding mode control law can be defined as:

$$u(k) = \frac{\hat{u}(k) - s(k)q - sgn\left(s(k)\right)\varepsilon}{g\left(x_1(k)\right)} \tag{9}$$

where

$$\begin{aligned}\hat{u}(k) = -f\left(x_1(k)\right) + \frac{1}{\Delta t}\left(x_{2r}(k+1) - x_{2r}(k)\right) \\ - \frac{\lambda}{\Delta t}\left(x_1(k+1) - x_1(k) - \left(x_{1r}(k+1) - x_{1r}(k)\right)\right)\end{aligned} \tag{10}$$

To ensure finite reaching time, the following sliding condition must be satisfied:

$$s(k+1) - s(k) < -\eta sgn\left(s(k)\right)\Delta t,\ \eta > 0 \tag{11}$$

Comparing (11) and (6) yields:

$$\varepsilon \geq |\xi(D+\eta) - s(k)q| + (\xi - 1)|\hat{u}(k)| \tag{12}$$

where $\xi = (G_2/G_1)^{1/2}$ is the gain margin (Slotine,1987). It has been demonstrated in (Gao, 1993) that the time required to reach the sliding surface from an arbitrary initial state is given by:

$$T = \frac{1}{q} ln\left(1 + \frac{q|s|}{\varepsilon}\right) \tag{13}$$

which shows that for a given value of q the reaching time is reduced by increasing ε. However, a higher value of ε also leads to more chatter, as can be seen from the control law (9). To reduce chatter, we replace the switching function in (6), (9) and (11) with a saturation function, and introduce a boundary layer ϕ. The quasi-sliding control law is finally given by:

$$u(k) = \frac{\hat{u}(k) - s(k)q - sat\left(\frac{s(k)}{\phi}\right)\varepsilon}{g\left(x_1(k)\right)} \tag{14}$$

Both (9) and (14) show that full state feedback is required to calculate the control action. In a practical real-time implementation of the HMSA, only position feedback would be available. A high-gain state observer is used to estimate velocity from position measurements.
Given the nominal model $\hat{\phi}\left(\hat{x}_1(k), u(k)\right) = f\left(\hat{x}_1(k)\right) + g\left(\hat{x}_1(k)\right)u(k)$, the perturbations on $\hat{\phi}$ can be modelled as $\phi_d\left(\hat{x}_1(k), x_1(k)\right)$ so that

$$\phi\left(x_1(k), u(k)\right) = \hat{\phi}\left(\hat{x}_1(k), u(k)\right) + \phi_d\left(\hat{x}_1(k), x_1(k)\right).$$

As a result, (8) can be rewritten as:

$$\begin{aligned}x_1(k+1) &= x_2(k)\Delta t + x_1(k) \\ x_2(k+1) &= \phi\left(x_1(k), u(k)\right)\Delta t + x_2(k) \\ y(k) &= x_1(k)\end{aligned} \tag{15}$$

where $\hat{x}_1(k)$ is the estimated position and $y(k)$ is the measured position. A velocity observer for (19) can be given by:

$$
\begin{aligned}
\hat{x}_1(k+1) &= \left(\hat{x}_2(k) + h_1\left(y - \hat{x}_1(k)\right)\right)\Delta t + \hat{x}_1(k) \\
\hat{x}_2(k+1) &= \left(\hat{\phi}\left(\hat{x}_1(k), u(k)\right) + h_2\left(y - \hat{x}_1(k)\right)\right)\Delta t + \hat{x}_2(k) \\
\hat{v}(k) &= \hat{x}_2(k)
\end{aligned}
\tag{16}
$$

where $\hat{v}(k)$ is the estimated velocity, h_1 and h_2 are observer gains. Subtracting (16) from (15) shows the dynamics of the estimation error:

$$
\begin{aligned}
\tilde{x}_1(k+1) &= -\left(h_1\tilde{x}_1(k) + \tilde{x}_2(k)\right)\Delta t + \tilde{x}_1(k) \\
\tilde{x}_2(k+1) &= -\left(h_2\tilde{x}_1(k) + \phi_d\left(\tilde{x}_1(k), x_1(k)\right)\right)\Delta t + \tilde{x}_2(k)
\end{aligned}
\tag{17}
$$

A block diagram of the proposed observer-based quasi-sliding mode control is shown on Fig. 7.

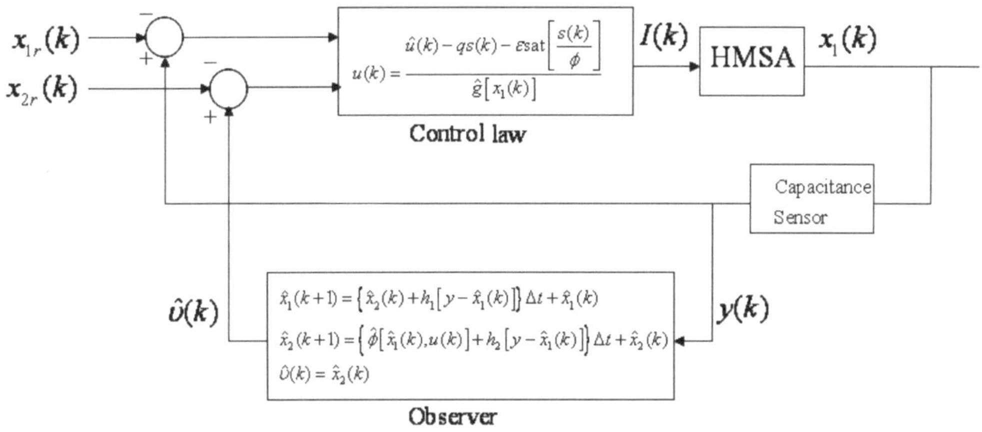

Fig. 7. Observer based Quasi Sliding Mode Control of the proposed HMSA

4. Experimental results

Several experiments were carried out to test both positioning and tracking performances of the proposed controller. All control algorithms were implemented on a TI TMS320C67 DSP computer, at a sampling rate of 2 kHz.

4.1 Consecutive step response

To demonstrate the positioning capabilities of the proposed HMSA, 10μm consecutive step commands are used. The target is stabilized at its initial gap of 1190μm and then multiple step commands are executed. To obtain large gap measurement capability, the capacitive sensor is set at coarse measurement mode. Fig. 8 shows the experimental positioning performance of the proposed quasi-sliding control with linear reaching rate. Fig. 9 shows the experimental result of the conventional PID control with feedback linearization. The performances of both controllers in terms of the overshoot, settling time and the steady state error are listed in Table 2.

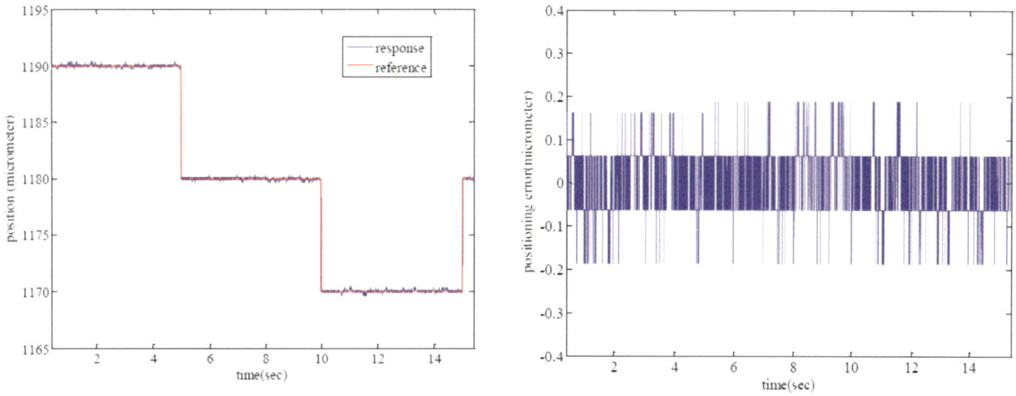

Fig. 8. Positioning Performances of Quai Sliding-Mode Control with Linear Reaching Law (Left) 10μm consecutive step response (Right) positioning error

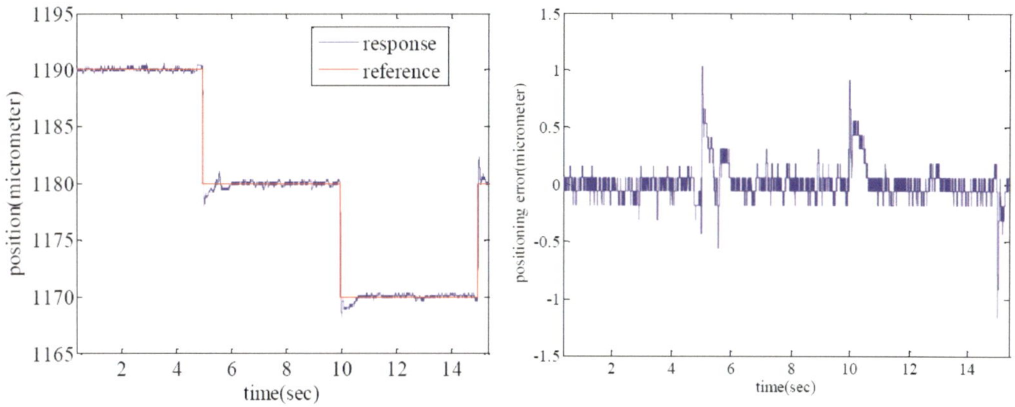

Fig. 9. Positioning Performances of PID Control with feedback linearization (Left) 10μm consecutive step response (Right) positioning error

Controllers / indexes	Quasi-sliding mode control with linear reaching law	PID with feedback linearization
overshoot	none	1μm
settling time	0.02s	0.7s
steady state error	±200nm	±200nm

Table 2. Performances for 10μm consecutive step responses

4.2 Tracking response

To test the tracking performance of the proposed HMSA actuator, it is commanded to follow a trapezoidal trajectory starting from 1000μm with amplitude of 400nm. The sliding control law (14) is implemented and the results are shown on Fig. 10. To exhibit more precise motion control capability, the capacitive sensor is set at fine measurement mode which gives tracking error of ±60nm.

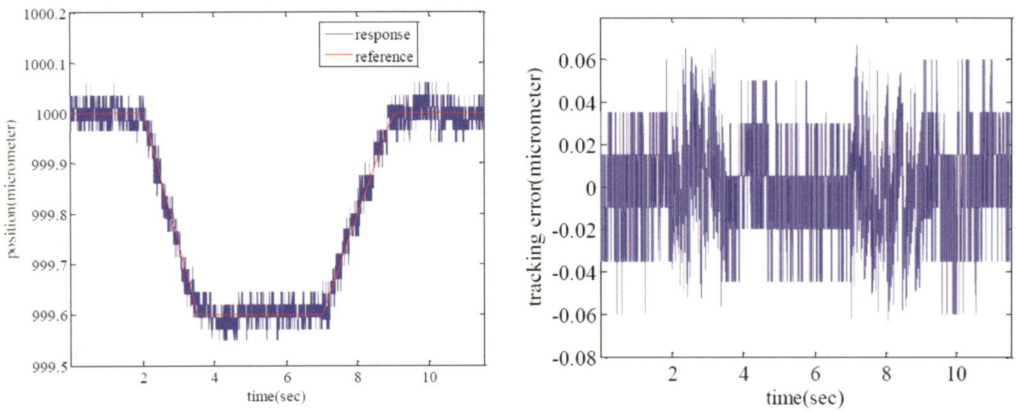

Fig. 10. Tracking Performances of Quasi Sliding-Mode Control with Linear Reaching Law (Left) 400nm trapezoidal tracking (Right) tracking error

5. Conclusion

A novel 1-DOF hybrid magnetic suspension actuator that has a great promise to realize large gap while avoiding large levitation currents and actuator saturation has been presented. In the proposed design, a pair of permanent magnets in repulsive configuration is located coaxial with an attractive force electromagnet. Controlled attractive forces are used to provide stabilization. The passive push-active pull strategy offers substantial advantages over other designs described in the literature when a large load is to be accurately suspended over a large range of travel. The proposed actuator can be used modularly to control multiple axes of motion in a multi-DOF positioning application that requires millimeter-range travel with fast response and sub-micron accuracy.

The force model parameters of the proposed actuator were identified using a multidimensional unconstrained nonlinear optimization method. To control the designed magnetic suspension system, a quasi-sliding mode controller (QSM) has been presented, in which a linear reaching law is used instead of the typical Lyapunov function approach. Since full knowledge of the state vector is required, a nonlinear high-gain observer was also designed and implemented. For comparison, a PID control with standard feedback linearization is also implemented for the magnetic suspension system. Several experiments were performed to demonstrate both the positioning and tracking capabilities of the proposed actuator. A fast, stable response with low chatter has been achieved via QSM. In a positioning control, the proposed method demonstrates faster response and better transient performance when compared with those of standard feedback linearization. In a tracking control, it also achieves a minimal positioning error of ±60nm which is very useful for systems in which steady-state accuracy is critical.

6. References

Chikazumi, S. (2005). *Physics of Ferromagnetism*. Oxford University Press, pp. 19-21.

Craig, D. and Khamesee, M. B. (2007). Motion Control of a Large Gap Magnetic Suspension System for Microrobotic Manipulation. *J. Phys. D: Appl. Phys.* 40, pp. 3277-3285.

Gao, W., and Hung, J. C. (1993). Variable Structure Control of Nonlinear Systems: A New Approach. *IEEE Trans. Industrial Electronics*, vol. 40, No.1, pp.45–55.

Gutierrez, H. M. and Ro, P. I. (2005). Magnetic Servo Levitation by Sliding-Mode Control of Nonaffine Systems with Algebraic Input Invertibility. *IEEE Trans. Industrial Electronics*, vol. 52, No.5, pp.1449–1455.

Hung, J. Y., Gao, W., and Hung, J. C. (1993). Variable Structure Control: A Survey. *IEEE Trans. Industrial Electronics*, vol. 40, No.1, pp.2–22.

Khamesee, M. B. and Shameli, E. (2005). Regulation Technique for a large gap magnetic field for 3D non-contact manipulation. *The Elsevier Journal of Mechatronics*, 15, pp. 1073-1087.

Kuo, S-K, Shan, X., and Menq, C-H. (2003). Large Travel Ultra Precision Motion Control of a Magnetic-Suspension Stage. *IEEE Trans. Mechatronics*, vol. 8, No.3, pp.334–341.

Lin, C. E. and Jou, H. L.. (1993). Force Model Identification for Magnetic Suspension Systems via Magnetic Field Measurement. *IEEE Trans. Instrumentation and Measurement*, vol. 42, No.3, pp.767–771.

Lin, F-J, Teng, L-T and Shieh, P-H. (2007). Intelligent Sliding-Mode Control Using RBFN for Magnetic Levitation System. *IEEE Trans. Industrial Electronics*, vol. 54, No.3, pp.1752–1762.

Ludwick, S. J., Trumper, D.L., and Holmes, M. L.. (1996). Modeling and Control of a Six Degree-of-Freedom Magnetic/Fluidic Motion Control Stage. *IEEE Trans. Control Systems Technology*, vol. 4, No.5, pp.553–564, September.

Mittal, S. and Menq, C-H. (1997). Precision Motion Control of a Magnetic Suspension Actuator Using a Robust Nonlinear Compensation Scheme. *IEEE Trans. Mechatronics*, vol. 2, No.4, pp.268–280.

Oliveira, V. A., Costa, E. F., and Vargas, J. B. (1999). Digital Implementation of a Magnetic Suspension Control System for Laboratory Experiments. *IEEE Trans. Education*, vol. 42, No.4, pp.315–322.

Robbins, W. P., Polla, D. L., and Glumac, D. E. (1991). High-Displacement Piezoelectric Actuator Utilizing a Meander-Line Geometry-Part I: Experimental Characterization. *IEEE Trans. Ultrasonics, Ferroelectrics, and Frequency Control*, vol. 38, No.5, pp.454–460.

Slotine, J.J. and Li, W. (1987). *Applied Nonlinear Control*. Englewood Cliffs, NJ Prentice Hall.

Trumper D. L., Olson, S. M., and Subrahmanyan, P. K.. (1997). Linearizing Control of Magnetic Suspension Systems. *IEEE Trans. Control Systems Technology*, vol. 5, No.4, pp.427-438.

20

Three Degrees-of-Freedom Hybrid Stage With Dual Actuators and Its Precision Motion Control

Yonmook Park
Mechatronics and Manufacturing Technology Center, Samsung Electronics Co., Ltd.
Republic of Korea

1. Introduction

Nanotechnology can be defined as the research, development and processing of materials, devices and systems in which structure on a dimension of less than several hundreds of nanometer is essential to obtain the required functional performance. There are currently two different approaches to nanotechnology. The first approach is called engineering nanotechnology. This approach involves using classical mechanical and electrical engineering principles to build structures with tolerances at levels approaching a nanometer. The other approach is called molecular nanotechnology. This approach is concerned with self-assembled machines. Due to the startling progress of nanotechnology, we can make high density memories and flat panel display panels with the help of nano-positioning systems. Furthermore, it allows us to manipulate a molecule with scanning probe microscopes.

One of the largest challenges in the field of nanotechnology is precision motion control of macroscopic stages. Especially, precision motion control on the nanometer level that delivers precision position stability with high bandwidth is a very important issue for industrial and scientific applications that especially include the lithography and inspection of integrated circuit patterns (Lee & Kim, 1997; Kwon et al., 2001; Pahk et al., 2001) and the fabrication and operation of high-density magnetic data storage devices (Chung et al., 2000; Kim & Lee, 2004; Lee & Kim, 2004; Du et al., 2005).

In general, traditional stages are designed to operate with respect to a number of kinematic constraints, which are assemblages of mechanical parts and need to be compounded in order to perform multiple degrees-of-freedom (DOF) motion (Shan et al., 2002). These kinematic constraints provide contact friction and error accumulation (Awabdy et al., 1998), and thus, it is very difficult to implement a high-performance motion control system with high bandwidth as well as a precision multiple DOF stage that is capable of large travel with nanometer position stability.

For stages only using coarse actuators such as linear motor or hydraulic actuator, there are nonlinear friction in low-speed motion and resonance mode in high-frequency motion. Thus, it is very difficult to achieve precision motion control performance with only coarse actuators although they provide large travel. One of the methods to overcome their limitation is adopting fine actuators such as piezoelectric actuator or voice coil motor (VCM). But, the travel of fine actuators is about several hundreds of micron meters to several millimeters, which is the limitation of them. Thus, if we design a hybrid actuation system with coarse and fine actuators, we can utilise their advantages and mutually compensate their drawbacks. So far, there have been several studies for the design of hybrid actuation systems. For instance, ultraprecision dual-servo systems have been proposed by Lee & Kim, 1997, Kwon et al., 2001, and Pahk et al., 2001 for lithography steppers. Dual-stage actuation systems consisting of a VCM and a microactuator have been developed by Fan et al., 1995 and Li & Horowitz, 2001 to obtain high servo bandwidth and perform the disturbance rejection. A novel control design that aims to achieve a low-hump sensitivity function for a dual-stage system in hard disk drives has been studied by Du et al., 2005. Note that, according to Du et al., 2005, a servo control system with low-hump sensitivity function is able to reduce the contribution from disturbance to a system. Fundamental control designs of dual-stage hard disk drive systems have been presented by Chung et al., 2000, Kim & Lee, 2004 and Lee & Kim, 2004, and performance enhancement methods of dual-stage servo systems have been proposed by Wu et al., 2002 and Li et al., 2003.

This chapter presents a three DOF precision hybrid stage that can move and align an object on it for the measurement of its three-dimensional image using the confocal scanning microscope (CSM). The CSM can observe a sub-micron meter-sized material due to its fine resolution and has a three-dimensional surface profiling capability. The hybrid stage consists of two individually operating x-y-θ stages, called the coarse stage and the fine stage. The coarse stage is driven by the three linear motors, and the fine stage is driven by the four VCMs. The coarse and fine stages are not mechanically interconnected and can be controlled independently.

For control of the hybrid stage, the author proposes a precision motion controller in this chapter. The precision motion controller consists of a position and velocity control loop, an anti-windup compensator to eliminate the windup problem that occurs in the controller, a generator of optimal force to optimally control the fine stage, a precision position determiner to determine the exact position of the fine stage and a perturbation observer that can observe the perturbation of the fine stage and compensate it. Note that, in this chapter, the exact position of the fine stage means the centre of the fine stage that is precisely determined by considering the orientation angle of the fine stage. The performances of the precision motion controller are evaluated by experiment.

The remainder of this chapter is organized as follows. In Section 2, the system overview of the hybrid stage is described. In Section 3, the hybrid stage control method is presented. In Section 4, the experimental results of the hybrid stage motion control are given. Finally, some concluding remarks are given in Section 5.

2. Hybrid stage with dual actuators

2.1 Overview

Fig. 1 shows the schematic of the hybrid stage presented in this paper. The objective of the hybrid stage is to move and align an object on it for the measurement of its three-dimensional image using the CSM. The CSM has a capability of the optical sectioning and can generate three-dimensional surface profile. The measurement principle of the CSM is based on the fact that only light reflected from the focal point of the objective lens contributes to the image, whereas all diffusely scattered light beams are filtered out by a pinhole. This creates a focused two-dimensional image of all object points that are located during the scanning process in the focal plane, similar to the contour lines of a map. Scanning the whole samples with an automatically varying focal plane results in a highly resolved and enlarged image of the corresponding surface section. The vertical and horizontal resolutions of the CSM are 30 and 140 nm, respectively.

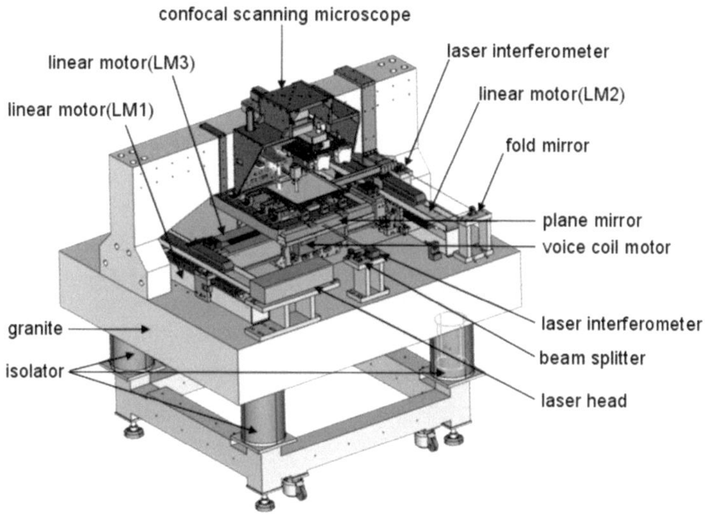

Fig. 1. Schematic of the hybrid stage

The hybrid stage consists of two individually operating x-y-θ stages, called the coarse stage and the fine stage. The coarse stage produces the initial movement of an object, and the fine stage provides the final alignment of the object. Since the laser interferometer with 0.31 nm resolution is used as position sensor in the fine stage, the precision control for the final alignment of an object can be possible. The moving parts of the hybrid stage are sustained by air bearings so that they can float on the base plate without mechanical contact. The material of the base plate is granite, and the base plate is connected with an isolator that can suppress internal and external vibrations.

2.2 Coarse stage

The schematic of the coarse stage is shown in Fig. 2. The coarse stage is driven by the three linear motors, and uses the 14 air bearings as guide and the three linear encoders as position sensor. The linear motor can be moved by the following Lorentz force

$$F_{LM}(t) = \oint i_{LM}(t) dl_{LM} \times B_{LM} \tag{1}$$

where $F_{LM}(t)$, $i_{LM}(t)$, $l_{LM}(t)$ and B_{LM} are force, current, coil length and flux density of the linear motor, respectively. A three-phase linear motor with the force of 233 N is used. The three linear motors are mechanically linked in an H-shaped rigid frame and can generate the x-y-θ motion of the coarse stage. Specifically, the stators of the two linear motors LM_1 and LM_2 are fixed to the base plate and parallel to each other. And the sliders of the two linear motors LM_1 and LM_2 are connected by the stator of the linear motor LM_3 that floats on the base plate. Thus, the movements of the three linear motors LM_1, LM_2, and LM_3 determine the x-y-θ motion of the coarse stage. The mass of each individual linear motor is 15 kg. The position sensor of the coarse stage is a linear encoder with the resolution of 5 nm. The coarse stage offers a large workspace of 500×500 mm^2.

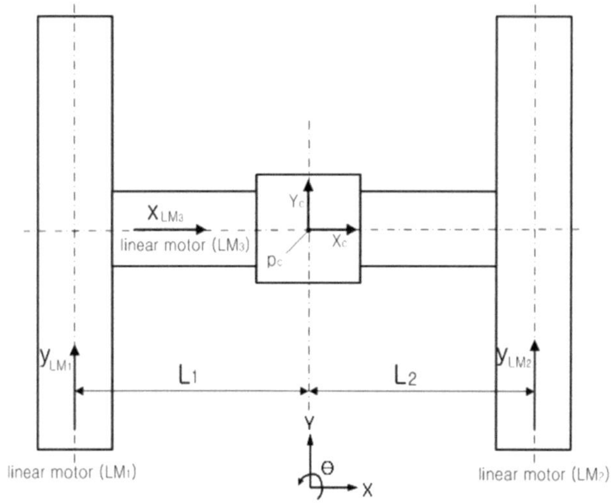

Fig. 2. Configuration of the coarse stage

Next, the kinematics of the coarse stage is derived. For the coarse stage shown in Fig. 2, let the vector $p_c(t)$ be given by

$$p_c(t) = \left[x_c(t) \quad y_c(t) \quad \theta_c(t) \right]^T \tag{2}$$

where $x_c(t)$ and $y_c(t)$ are the X_c and Y_c positions of the coarse stage, respectively, and $\theta_c(t)$ is the orientation angle of the coarse stage. Let the vector $p_{LM}(t)$ be given by

$$p_{LM}(t) = \left[x_{LM_3}(t) \quad y_{LM_1}(t) \quad y_{LM_2}(t) \right]^T \tag{3}$$

where $x_{LM_3}(t)$, $y_{LM_1}(t)$ and $y_{LM_2}(t)$ are the displacements of the three linear motors LM_3, LM_1 and LM_2, respectively. In (3), $x_{LM_3}(t)$, $y_{LM_1}(t)$ and $y_{LM_2}(t)$ are measured by the linear encoder. Under the assumption that the orientation angle $\theta_c(t)$ in (2) is very small, the vector $p_c(t)$ in (2) can be determined by the following equation

$$p_c(t) = H_c(t) p_{LM}(t)^T \tag{4}$$

where

$$H_c(t) = \begin{bmatrix} 1 & 0 & 0 \\ 0 & \dfrac{L_2 - x_{LM_3}(t)}{L_1 + L_2} & \dfrac{L_1 + x_{LM_3}(t)}{L_1 + L_2} \\ 0 & -\dfrac{1}{L_1 + L_2} & \dfrac{1}{L_1 + L_2} \end{bmatrix} \tag{5}$$

In (5), L_1 and L_2 are the distances from the centers of the linear motors LM_1 and LM_2 to the center of the coarse stage, respectively. In the sequel, the position and orientation angle of the coarse stage are represented as in (4).

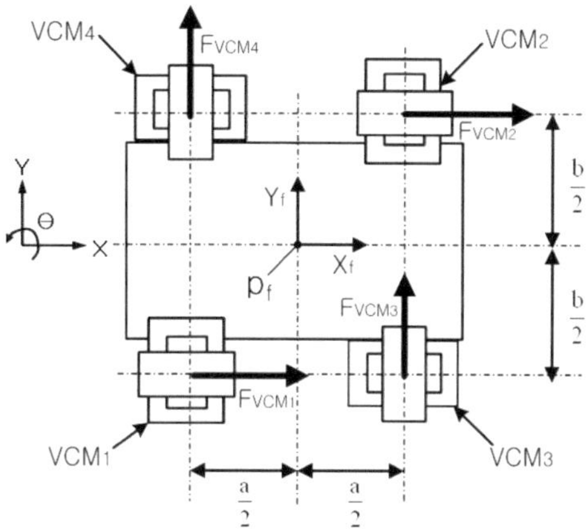

Fig. 3. Configuration of the fine stage

2.3 Fine stage

The configuration of the fine stage is shown in Fig. 3. The fine stage is driven by the four VCMs, and uses the four air bearings as guide and a laser interferometer as position sensor. The four VCMs lie on the same plane so that the tilting forces that cause the roll and pitch motions of the fine stage are negligible. The VCM can be moved by the following Lorentz force

$$F_{VCM}(t) = \oint i_{VCM}(t) dl_{VCM} \times B_{VCM} \tag{6}$$

where $F_{VCM}(t)$, $i_{VCM}(t)$, l_{VCM} and B_{VCM} are force, current, coil length and flux density of VCM, respectively. The four VCMs generate the x-y-θ motion of the fine stage by the following equation

$$AF_{VCM}(t) = u_f(t) \tag{7}$$

where

$$A = \begin{bmatrix} 1 & 1 & 0 & 0 \\ 0 & 0 & 1 & 1 \\ \frac{b}{2} & -\frac{b}{2} & \frac{a}{2} & -\frac{a}{2} \end{bmatrix} \tag{8}$$

$$F_{VCM}(t) = \begin{bmatrix} F_{VCM_1}(t) & F_{VCM_2}(t) & F_{VCM_3}(t) & F_{VCM_4}(t) \end{bmatrix}^T \tag{9}$$

and

$$u_f(t) = \begin{bmatrix} -F_{f_x}(t) & -F_{f_y}(t) & T_{f_\theta}(t) \end{bmatrix}^T \tag{10}$$

In (8)-(10), a is the distance from the centre of VCM_1 to the center of VCM_3 (or from the center of VCM_2 to the center of VCM_4), b is the distance from the center of VCM_1 to the center of VCM_4 (or from the centre of VCM_2 to the center of VCM_3), $F_{VCM_i}(t)$, $i=1,\cdots,4$ are the forces of VCM_i, $i=1,\cdots,4$, respectively, $F_{f_x}(t)$ and $F_{f_y}(t)$ are the X-axis and Y-axis control forces for the fine stage, respectively, and $T_{f_\theta}(t)$ is the control torque for the fine stage.

If we apply the current to coils of VCM_1 and VCM_2, the fine stage is driven in the X-axis direction. Similarly, we apply the current to coils of VCM_3 and VCM_4 for a driving in the Y-axis direction. In addition, the fine stage is driven in the θ direction if we make proper current and apply it to each coil. The VCM has the force of 220 N, and the mass of the fine stage is 36.5 kg. The position sensor of the fine stage is a laser interferometer with the resolution of 0.31 nm. The laser interferometer measures the x-y-θ motion of the fine stage by projecting laser beams onto the L-shaped plane mirror attached on top of the fine stage. The workspace of the fine stage is 5 × 5 mm^2, and the range of the orientation angle of the fine stage is 0.05 deg.

Now, the kinematics of the fine stage is derived. Let the vector $p_f(t)$ be given by

$$p_f(t)=\left[x_f(t) \quad y_f(t) \quad \theta_f(t)\right]^T \tag{11}$$

where $x_f(t)$ and $y_f(t)$ are the X_f and Y_f positions of the fine stage, respectively, and $u_f(t)$ is the orientation angle of the fine stage. Let the vector $p_{VCM}(t)$ be given by

$$p_{VCM}(t)=\left[x_{VCM}(t) \quad y_{VCM}(t) \quad \theta_{VCM}(t)\right]^T \tag{12}$$

where $x_{VCM}(t)$ and $y_{VCM}(t)$ are the X-axis and Y-axis displacements of VCM, respectively, and $u_{VCM}(t)$ is the orientation angle of VCM, which is equal to $u_f(t)$. In (12), $x_{VCM}(t)$, $y_{VCM}(t)$ and $\theta_{VCM}(t)$ are measured by the laser interferometer. Since the orientation angle $\theta_{VCM}(t)$ in (12) is not very small compared with $\theta_c(t)$ in (2), we should consider $\theta_{VCM}(t)$ to determine $x_f(t)$ and $y_f(t)$ in (11) and obtain the following equations by lengthy calculation.

$$\begin{aligned} x_f(t)=&[(x_{VCM}(t)+r)\cos\theta_{VCM}(t)-r]\cos\theta_{VCM}(t)\\ &-[(y_{VCM}(t)-r)\cos\theta_{VCM}(t)+r]\sin\theta_{VCM}(t) \end{aligned} \tag{13}$$

$$\begin{aligned} y_f(t)=&[(x_{VCM}(t)+r)\cos\theta_{VCM}(t)-r]\sin\theta_{VCM}(t)\\ &+[(y_{VCM}(t)-r)\cos\theta_{VCM}(t)+r]\cos\theta_{VCM}(t) \end{aligned} \tag{14}$$

where r is the distance from the centre of the fine stage to the L-shaped plane mirror attached on top of the fine stage. Consequently, the position of the fine stage can be precisely determined by (13) and (14).

Note that, as shown in Fig. 4, the VCM consists of magnet, yokes and coil. The magnet and yoke of VCM are fixed on the fine stage. On the other hand, the coil of VCM is fixed on the coarse stage. In addition, the magnet sticks to the yokes and does not come into contact with the coil. Thus, the coarse and fine stages are not mechanically interconnected and can be controlled independently.

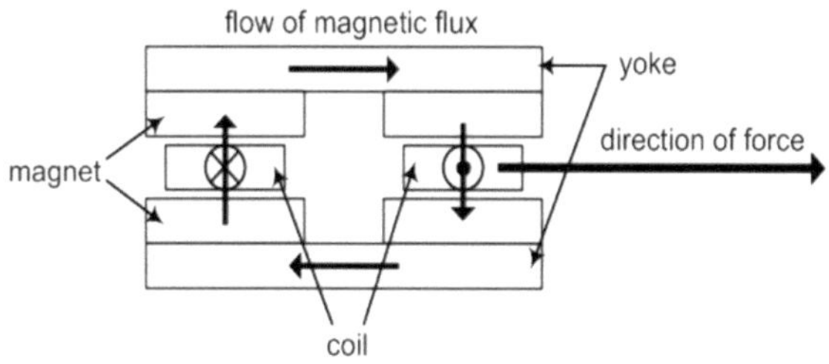

Fig. 4. Configuration of the voice coil motor

3. Precision motion control of hybrid stage with dual actuators

3.1 Overview

This section presents a precision motion control method of the x-y-θ motion of the hybrid stage. The motion control performance of the hybrid stage mainly depends on the motion control performance of the fine stage because the fine stage accomplishes the final alignment of an object. Therefore in this section the attention is focused on the precision motion control of the fine stage.

The block diagram of the hybrid stage control system is shown in Fig. 5. The coarse and fine stages are independently controlled under the common reference command. Let the reference command of the hybrid stage be given by

$$p_{ref}(t) = \left[x_{ref}(t) \quad y_{ref}(t) \quad \theta_{ref}(t) \right]^T \tag{15}$$

where $x_{ref}(t)$, $y_{ref}(t)$ and $\theta_{ref}(t)$ are the X-axis position reference command, the Y-axis position reference command and the orientation angle reference command of the hybrid stage, respectively.

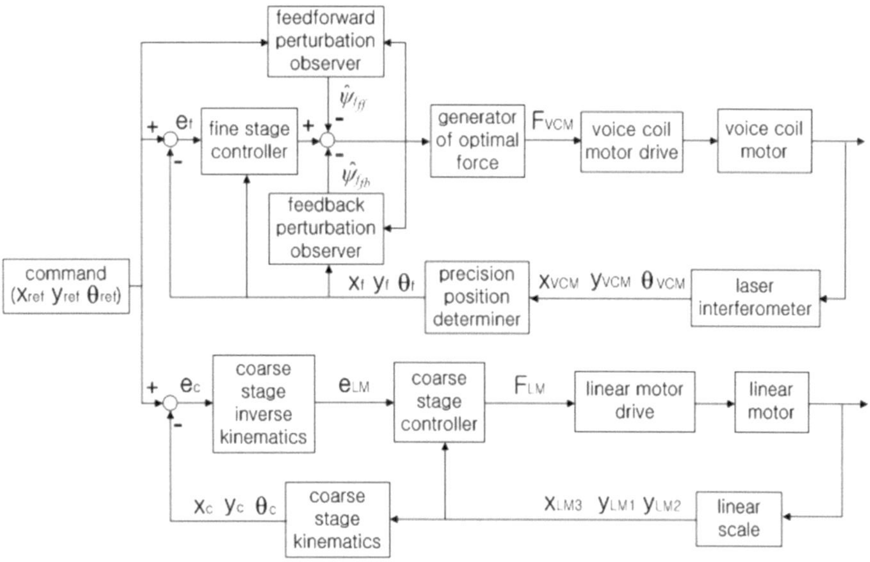

Fig. 5. Block diagram of the hybrid stage control system

3.2 Coarse stage control system

The error vector $e_c(t)$ for the coarse stage is defined as

$$e_c(t) = p_{ref}(t) - p_c(t) \tag{16}$$

where $p_{ref}(t)$ and $p_c(t)$ are defined in (15) and (2). Now the author explains each component of the coarse stage control system. First, the coarse stage controller consists of a position control loop, a velocity control loop and an antiwindup compensator. Specifically, as shown in Fig. 6, the position control loop has a proportional controller, and the velocity control loop has a proportional and integral controller. In addition, the velocity control loop is combined with an anti-windup compensator based on Bohn & Atherton, 1995 in order to eliminate the windup problem caused by the integral controller. In Fig. 6, $k_i(t)$, $i = 1,\cdots,4$ are positive scalars and s is the Laplace operator. The coarse stage controller generates the three control inputs $F_{LM_1}(t)$, $F_{LM_2}(t)$ and $F_{LM_3}(t)$ for control of the x-y-θ motion of the coarse stage where $F_{LM_1}(t)$, $F_{LM_2}(t)$ and $F_{LM_3}(t)$ are the control forces for the three linear motors LM_1, LM_2 and LM_3, respectively.

Second, the coarse stage kinematics implies the transformation of $p_{LM}(t)$ in (3) into $p_c(t)$ in (2) by (4). Finally, the coarse stage inverse kinematics represents the transformation of $e_c(t)$ in (16) into the error vector $e_{LM}(t)$, given by

$$e_{LM}(t) = H_c^{-1} e_c(t) = H_c^{-1}(p_{ref}(t) - p_c(t)) = H_c^{-1} p_{ref}(t) - p_{LM}(t) \tag{17}$$

where $H_c(t)$ is defined in (5) and its inverse matrix is given by

$$H_c^{-1}(t) = \begin{bmatrix} 1 & 0 & 0 \\ 0 & 1 & -L_1 - x_{LM_3}(t) \\ 0 & 1 & L_2 - x_{LM_3}(t) \end{bmatrix} \tag{18}$$

3.3 Fine stage control system

The error vector $e_f(t)$ for the fine stage is defined as

$$e_f(t) = p_{ref}(t) - p_f \tag{19}$$

where $p_{ref}(t)$ and $p_f(t)$ are defined in (15) and (11). The author describes each component of the fine stage control system. First, as shown in Fig. 6, the fine stage controller has the same structure that the coarse stage controller has. The fine stage controller produces the three control inputs $F_{f_x}(t)$, $F_{f_y}(t)$ and $T_{f_\theta}(t)$ for control of the x-y-θ motion of the fine stage.

Second, the precision position determiner means the equations of (13) and (14). Third, the generator of optimal force is proposed to make the optimal forces of the four VCMs. As shown in (7), after designing the three control inputs $F_{f_x}(t)$, $F_{f_y}(t)$ and $T_{f_\theta}(t)$, we should determine the four forces $F_{VCM_i}(t)$, $i = 1,\cdots,4$ of the four VCMs. In this case, (7) has infinitely many solutions for the four forces $F_{VCM_i}(t)$, $i = 1,\cdots,4$ because it is underdetermined with three equations in four unknowns. Among many solutions to the above problem, the author presents a meaningful solution to (7) by considering a least squares problem. Before deriving a meaningful solution to (7), the following definition is given.

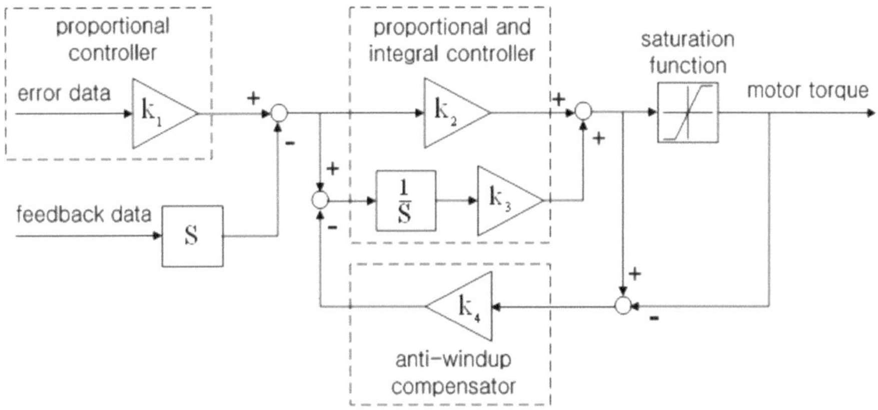

Fig. 6. Block diagram of the coarse stage and fine stage controllers

Definition 1 (Leon, 1995): Let $A \in R^{m\times n}$ have the rank of $q < n$. Then the singular value decomposition of A is given by

$$A = U\Sigma V^T = \begin{bmatrix} U_1 & U_2 \end{bmatrix}^T \begin{bmatrix} \Sigma_1 & 0_{q\times(n-q)} \\ 0_{(m-q)\times q} & 0_{(m-q)\times(n-q)} \end{bmatrix} \begin{bmatrix} V_1^T \\ V_2^T \end{bmatrix} = U_1 \Sigma_1 V_1^T \tag{20}$$

where $U = \begin{bmatrix} U_1 & U_2 \end{bmatrix} \in R^{m\times m}$ and $V = \begin{bmatrix} V_1 & V_2 \end{bmatrix} \in R^{n\times n}$ are orthogonal matrices with $U_1 \in R^{m\times q}$, $U_2 \in R^{m\times(m-q)}$, $V_1 \in R^{n\times q}$ and $V_2 \in R^{n\times(n-q)}$. Moreover, $0_{m\times n}$ denotes the $m \times n$ zero matrix and $\Sigma_1 \in R^{q\times q}$ is a diagonal matrix given by

$$\Sigma_1 = \begin{bmatrix} \sigma_1 & 0 & 0 & \cdots & 0 \\ 0 & \sigma_2 & 0 & \cdots & 0 \\ 0 & 0 & \sigma_3 & \cdots & 0 \\ \vdots & \vdots & \vdots & \ddots & \vdots \\ 0 & 0 & 0 & \cdots & \sigma_q \end{bmatrix} \tag{21}$$

with the entries satisfying

$$\sigma_1 \geq \sigma_1 \geq \cdots \geq \sigma_q > 0 \tag{22}$$

Then the following theorem shows that the singular value decomposition provides the key to solve the least squares problem for design of the optimal forces of the four VCMs.

Theorem 1: Consider the equation (7). Let the singular value decomposition of A be $U\Sigma V^T$ and define

$$A^+ = V\Sigma^+ U^T \tag{23}$$

where A^+ denotes the pseudo-inverse of A. Then the following is a solution to (7):

$$F_{VCM}(t) = A^+ u_f(t) = \begin{bmatrix} \frac{1}{2} & 0 & \frac{b}{a^2+b^2} \\ \frac{1}{2} & 0 & -\frac{b}{a^2+b^2} \\ 0 & \frac{1}{2} & \frac{a}{a^2+b^2} \\ 0 & \frac{1}{2} & -\frac{a}{a^2+b^2} \end{bmatrix} \begin{bmatrix} -F_{f_x}(t) \\ -F_{f_y}(t) \\ T_{f_\theta}(t) \end{bmatrix} \tag{24}$$

Moreover, if $h(t)$ is any other solution to (7), then we can guarantee

$$\|F_{VCM}(t)\|_2 < \|h(t)\|_2 \tag{25}$$

where $\|\cdot\|_2$ denotes the Euclidean norm.

Proof: Let $F_{VCM}(t) \in R^4$ and define

$$j(t) = U^T u_f(t) = \begin{bmatrix} j_1(t) \\ j_2(t) \end{bmatrix} \tag{26}$$

$$k(t) = V^T F_{VCM}(t) = \begin{bmatrix} k_1(t) \\ k_2(t) \end{bmatrix} \tag{27}$$

From the definition of singular value decomposition and (26) and (27), we can obtain

$$\begin{aligned} \|u_f(t) - AF_{VCM}(t)\|_2^2 &= \|U^T\|_2^2 \|u_f(t) - AF_{VCM}(t)\|_2^2 = \|U^T u_f(t) - U^T(U\Sigma V^T F_{VCM}(t))\|_2^2 \\ &= \|U^T u_f(t) - \Sigma V^T F_{VCM}(t)\|_2^2 = \|j(t) - \Sigma V^T F_{VCM}(t)\|_2^2 = \|j(t) - \Sigma k(t)\|_2^2 \\ &= \left\| \begin{bmatrix} j_1(t) \\ j_2(t) \end{bmatrix} - \begin{bmatrix} \Sigma_1 & 0 \\ 0 & 0 \end{bmatrix} \begin{bmatrix} k_1(t) \\ k_2(t) \end{bmatrix} \right\|_2^2 = \left\| \begin{bmatrix} j_1(t) - \Sigma_1 k_1(t) \\ j_2(t) \end{bmatrix} \right\|_2^2 \\ &= \|j_1(t) - \Sigma_1 k_1(t)\|_2^2 + \|j_2(t)\|_2^2 \end{aligned} \tag{28}$$

Since $j_2(t)$ is independent of $F_{VCM}(t)$, it follows that $\|u_f(t) - AF_{VCM}(t)\|_2^2$ will be minimal if and only if $\|j_1(t) - \Sigma_1 k_1(t)\|_2 = 0$. Furthermore, $\|u_f(t) - AF_{VCM}(t)\|_2^2$ will be zero if and only if $\|j_1(t) - \Sigma_1 k_1(t)\|_2 = 0$ and $\|j_2(t)\|_2 = 0$. Thus, $F_{VCM}(t)$ becomes a solution to (7) if and only if $F_{VCM}(t) = Vk(t)$ and $j_2(t) = 0$ where $k(t)$ is a vector of the form

$$k(t) = \begin{bmatrix} k_1(t) \\ k_2(t) \end{bmatrix} = \begin{bmatrix} \Sigma_1^{-1} j_1(t) \\ k_2(t) \end{bmatrix} \tag{29}$$

Especially, $F_{VCM}(t) = A^+ u_f(t)$ is a solution to (7) because

$$F_{VCM}(t) = Vk(t) = V\begin{bmatrix}\Sigma_1^{-1}j_1(t)\\0\end{bmatrix} = V\begin{bmatrix}\Sigma_1^{-1} & 0\\0 & 0\end{bmatrix}\begin{bmatrix}j_1(t)\\j_2(t)\end{bmatrix} = V\Sigma^+ j(t) = V\Sigma^+U^T u_f(t)$$
$$= A^+u_f(t) = \begin{bmatrix}\frac{1}{2} & 0 & \frac{b}{a^2+b^2}\\ \frac{1}{2} & 0 & -\frac{b}{a^2+b^2}\\ 0 & \frac{1}{2} & \frac{a}{a^2+b^2}\\ 0 & \frac{1}{2} & -\frac{a}{a^2+b^2}\end{bmatrix}\begin{bmatrix}-F_{f_x}(t)\\-F_{f_y}(t)\\T_{f_\theta}(t)\end{bmatrix} \tag{30}$$

Next, if $h(t)$ is any other solution to (7), $h(t)$ must be of the form

$$h(t) = Vk(t) = V\begin{bmatrix}\Sigma_1^{-1}j_1(t)\\k_2(t)\end{bmatrix} \tag{31}$$

where $k_2(t) \neq 0$. Then, for $h(t)$ of (31), we can obtain the following result:

$$\begin{aligned}\|h(t)\|_2^2 &= \|Vk(t)\|_2^2 = \|k(t)\|_2^2 = \|\Sigma_1^{-1}j_1(t)\|_2^2 + \|k_2(t)\|_2^2\\ &> \|\Sigma_1^{-1}j_1(t)\|_2^2 = \|F_{VCM}(t)\|_2^2\end{aligned} \tag{32}$$

This completes the proof. □

Physically, $F_{VCM}(t)$ in (24) is the minimal norm solution to (7) such that the condition of (25) holds for any other solution to (7). Therefore, if we use $F_{VCM}(t)$ in (24), we can achieve the optimal performance in the sense of the control effort.

Finally, the author presents the feedforward and feedback perturbation observers by extending the study of Kwon et al., 2001. Specifically, the perturbation applied to the nominal dynamics of the fine stage can be expressed by

$$\psi_f(t) = H_{f_n}\ddot{p}_f(t) + B_{f_n}\dot{p}_f(t) - u_f(t) \tag{33}$$

where H_{f_n} and B_{f_n} are the nominal inertia matrix and the nominal viscous damping coefficient of the fine stage, respectively, $u_f(t)$ is the control input of the fine stage defined in (10). Since one-step delay in signals is inevitable for the causality between input and output in practice, the perturbation observer is presented as follows

$$\hat{\psi}_f(t) = D_f\left\{H_{f_n}\ddot{p}_f(t-t_c) + B_{f_n}\dot{p}_f(t-t_c)\right\} - u_f(t-t_c) \tag{34}$$

where D_f is a diagonal matrix with scalar elements that plays the role of approximating $\hat{\psi}_f(t)$ to the real perturbation of $\psi_f(t)$, and t_c is the control interval. If we apply $\hat{\psi}_f(t)$ in (34) to the nominal dynamics of the fine stage in order to compensate the perturbation, the nominal dynamic equation of the fine stage can be changed to

$$H_{f_n}\ddot{p}_f(t) + B_{f_n}\dot{p}_f(t) = \eta_f(t) + \tilde{\psi}_f(t) \tag{35}$$

where $\eta_f(t) = u_f(t) + \hat{\psi}_f(t)$ is the tracking control input and $\tilde{\psi}_f(t) = \psi_f(t) - \hat{\psi}_f(t)$ is the perturbation compensation error. Since the reference command can be utilized in tracking control, the author presents the feedforward perturbation observer as follows

$$\hat{\psi}_{f_{ff}}(t) = D_{f_{ff}} \left\{ H_{f_n} \ddot{p}_{ref}(t - t_c) + B_{f_n} \dot{p}_{ref}(t - t_c) \right\} - u_f(t - t_c) \tag{36}$$

where $D_{f_{ff}}$ is a diagonal matrix with scalar elements. Also, the residue of the perturbation, given by $\Delta\tilde{\psi}_f(t) = \psi_f(t) - \hat{\psi}_{f_{ff}}(t)$, is compensated by the following feedback perturbation observer

$$\begin{aligned} \hat{\psi}_{f_{fb}}(t) &= \Delta\tilde{\psi}_f(t - t_c) \\ &= D_{f_{fb}} \left\{ H_{f_n} \ddot{p}_f(t - t_c) + B_{f_n} \dot{p}_f(t - t_c) \right\} - D_{f_{ff}} \left\{ H_{f_n} \ddot{p}_{ref}(t - 2t_c) + B_{f_n} \dot{p}_{ref}(t - 2t_c) \right\} \\ &\quad - u_f(t - t_c) + u_f(t - 2t_c) \end{aligned} \tag{37}$$

where $D_{f_{fb}}$ is a diagonal matrix with scalar elements.

It is remarkable that the perturbation observers presented in this section are the generalization of the perturbation observers developed by Kwon et al., 2001 because their study can be regarded as a special case of the proposed method with $D_{f_{ff}} = \text{diag}[1 \ \ 1 \ \ 1]$ and $D_{f_{fb}} = \text{diag}[1 \ \ 1 \ \ 1]$ where diag means the diagonal matrix. Because of $D_{f_{ff}}$ and $D_{f_{fb}}$, we have the extra freedom of designing the perturbation observers in real application. The perturbation observers presented in this section cannot help using delayed information because the current perturbation is monitored in discrete time with one-step delay. Using delayed information may result in bandwidth degradation and it is inevitable.

With a similar manner presented by Kwon et al., 2001, if we assume that the fine stage is time invariant during a control interval, the full state is available, the change of external disturbances during the control intervals is bounded, and the nominal inertia matrix of the fine stage H_{f_n} satisfies the following condition for the real inertia matrix of the fine stage $H_f(k)$ for all samples k

$$0 < H_{f_n} < 2H_f(k) \tag{38}$$

then the perturbation compensation error $\tilde{\psi}_f(t)$ in (35) is well bounded in a sufficiently small value.

4. Experimental results

As the hybrid stage control platform, the author uses the dSPACE system that features a power PC processor and is directly connected to all dSPACE I/O boards. The dSPACE system is an efficient and reliable engineering tool to develop and test control systems, and is in widespread use in many automotive industries. The graphical user interface software is programmed in order to control the hybrid stage by using the dSPACE system. By the graphical user interface software, we can give the target position command and target orientation angle command to the hybrid stage and can set all control parameters of the precision motion controller.

Then the performances of the precision motion controller of the hybrid stage are evaluated by experiment. The update rate of the dSPACE system is set to be 1 kHz. The author initially decides the gains k_1, k_2 and k_3 of the coarse stage and fine stage controllers by adopting the Ziegler–Nichols method (Ogata, 1996), which is very useful to select the control gains of a proportional, integral and derivative-type controller for complex dynamic systems in practice, and then further tunes these gains in order to obtain a desired control performance in terms of the step response. Also, the author designs the gain k_4 of the anti-windup compensator by an experimental method such that we make the overshot appearing in the step response, caused by windup, as small as possible. In the sequel, the undamped natural frequencies of the X-axis and Y-axis motions are decided to lie approximately at 117.909 and 118.448 rad/s, respectively. Also, the damping ratios of the X-axis and Y-axis motions are decided to lie approximately at 0.590 and 0.595, respectively.

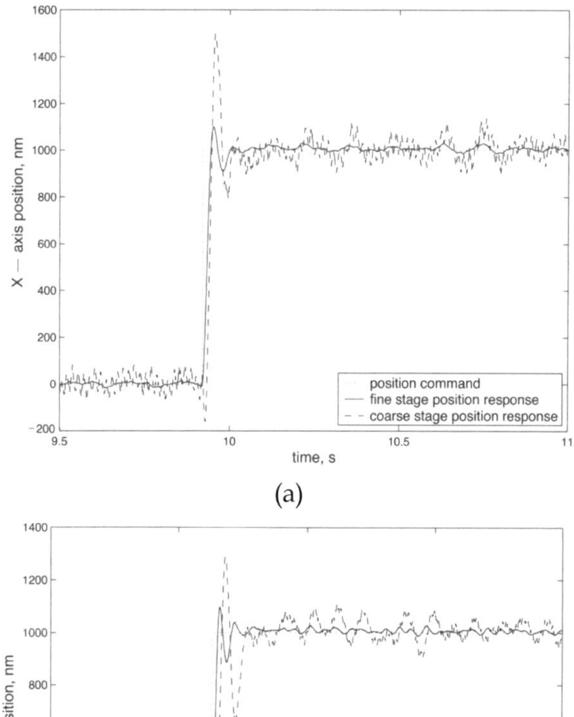

(a)

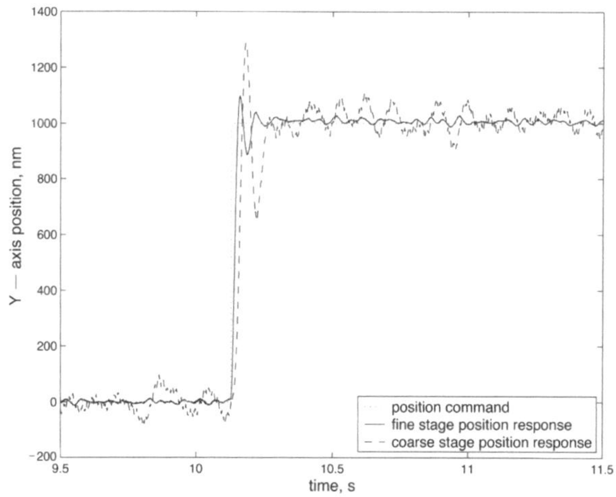

(b)

Fig. 7. Step responses of the hybrid stage by a step input of 1000 nm magnitude: (a) X-axis step response, (b) Y-axis step response

Fig. 7 shows the experimental result of the X-axis and Y-axis step responses of the hybrid stage by a step input of 1000 nm magnitude. From Fig. 7, we see that the maximum overshoots of the X-axis and Y-axis motions are 10.07% and 9.76%, respectively, the delay times of the X-axis and Y-axis motions are 0.014 and 0.014 s, respectively, the rise times of the X-axis and Y-axis motions are 0.026 and 0.025 s, respectively, the peak times of the X-axis and Y-axis motions are 0.033 and 0.033 s, respectively, and the 5% settling times of the X-axis and Y-axis motions are 0.071 and 0.072 s, respectively. Also, Fig. 7 demonstrates that the hybrid stage effectively responds to a step input of 1000 nm magnitude in the X-axis and Y-axis motions. Specifically, when the coarse and fine stages are operated for a step input of 1000 nm magnitude, the X-axis and Y-axis steady-state errors of the fine stage after 0.071 and 0.072 s rising periods remain within 50 nm, respectively. On the other hand, the X-axis and Y-axis steady-state errors of the coarse stage reach more than 100 and 80 nm although the time elapses 0.08 and 0.12 s after the step input, respectively. Therefore it is concluded that the hybrid stage has remarkable advantages in terms of the response time and positioning accuracy. Although there is no direct contact between the coarse and fine stages, the motion errors or vibrations of the coarse stage may lead to variations of the interaction force between the two stages. These variations present themselves as disturbances to the fine stage, and the resolution of the fine stage degrades unless the bandwidth of the fine stage is substantially higher than that of the coarse stage. This observation explains the fluctuation of the fine stage at steady state in Fig. 7.

Note that $F_{VCM}(t)$ in (24) is the minimal norm solution to (7). In order to demonstrate the result, the Euclidean norm histories of current inputs of the four VCMs is shown in Fig. 8. As shown in Fig. 8, the generator of optimal force yields the control effort of about 0.05 A. On the other hand, we need the control effort of about 2.8 A if we do not use the generator of optimal force. In this case, we can save the control effort significantly for the operation of the four VCMs by adopting the generator of optimal force.

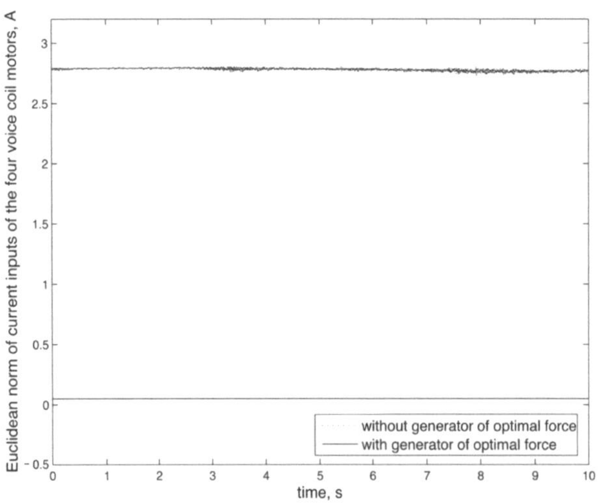

Fig. 8. Euclidean norm histories of current inputs of the four VCMs

Since the perturbation gives rise to a vibration of the hybrid stage in practice, the author determines the gains of the perturbation observers by the experimental method such that we make the position stability of the hybrid stage as small as possible. Then Fig. 9 shows the experimental result for the X-axis and Y-axis position stabilities of the hybrid stage. From Fig. 9, the X-axis and Y-axis position stabilities by the perturbation observers are about ±10 nm, respectively. On the other hand, the X-axis and Y-axis position stabilities are about ±30 nm, respectively, if we do not use the perturbation observers. Consequently, we see that the perturbation observers have the function of observing the perturbation and compensating it effectively about 66%.

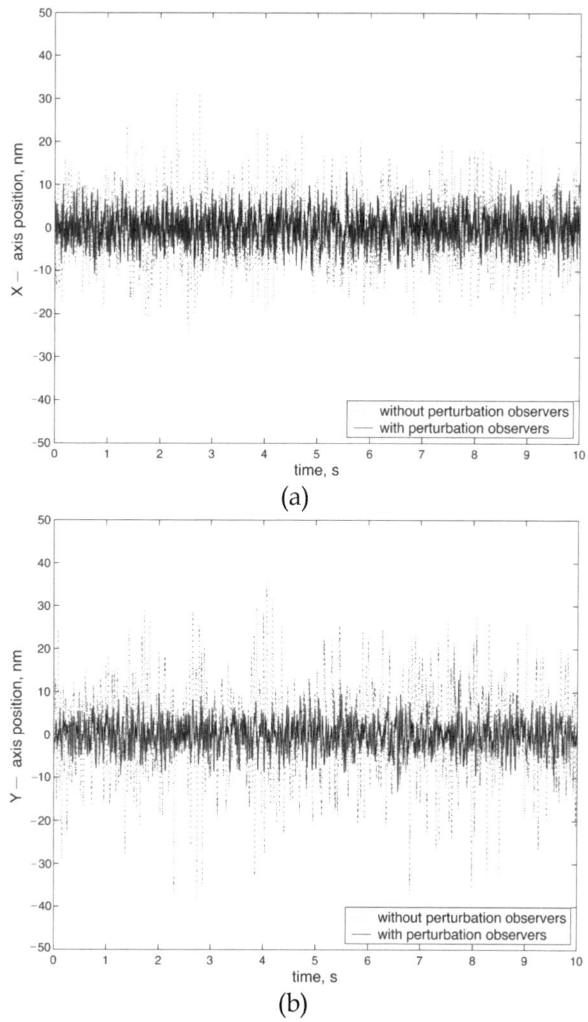

Fig. 9. Position stabilities of the hybrid stage: (a) X-axis position stability, (b) Y-axis position stability

Now the X-axis and Y-axis incremental step responses and orientation angle responses of the hybrid stage are evaluated by applying some step input to the hybrid stage. Specifically, the X-axis and Y-axis target positions are increased by 10 nm from 0 to 50 nm, and then decreased by 10 nm from 50 to 10 nm, respectively. Note that the coarse stage is on operation when the author conducts the X-axis and Y-axis incremental step responses and orientation angle responses.

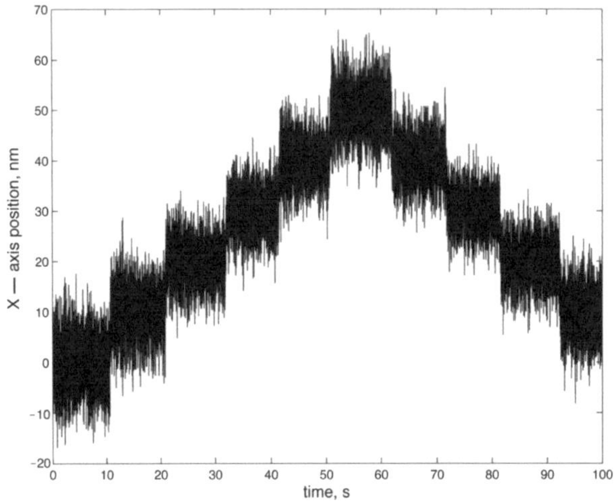

(a)

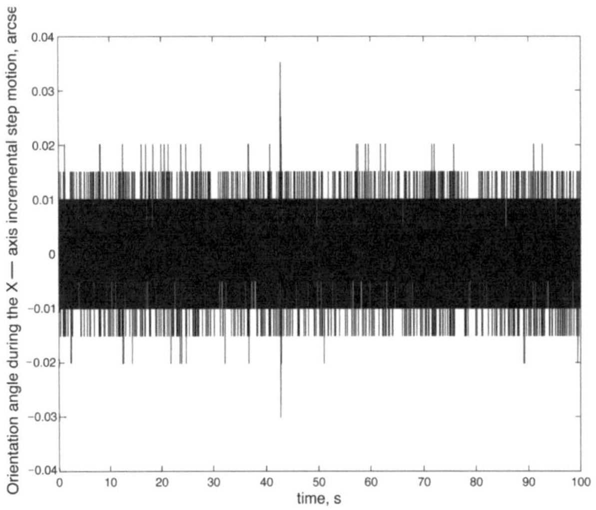

(b)

Fig. 10. X-axis incremental step response and orientation angle response of the hybrid stage: (a) X-axis incremental step response, (b) Orientation angle response during the X-axis incremental step motion

Then Fig. 10 shows the experimental results of the X-axis incremental step response and the orientation angle response during the X-axis incremental step motion, and Fig. 11 shows the experimental results of the Y-axis incremental step response and the orientation angle response during the Y-axis incremental step motion. From Figs. 10 and 11, we see that the resolutions of the X-axis and Y-axis motions are about ±10 nm, respectively, and the fluctuations of the orientation angle during the X-axis and Y-axis incremental step motions are about ±0.02 and ±0.04 arcsec, respectively. Note that 1 arcsec is equal to 1/3600 deg. Finally, the X-axis and Y-axis bidirectional repeatabilities of the hybrid stage are tested. Note that the repeatability is the error between a number of successive attempts to move the machine to the same position (Slocum, 1992). And the bidirectional repeatability is the repeatability achieved when the target position is approached from two different directions

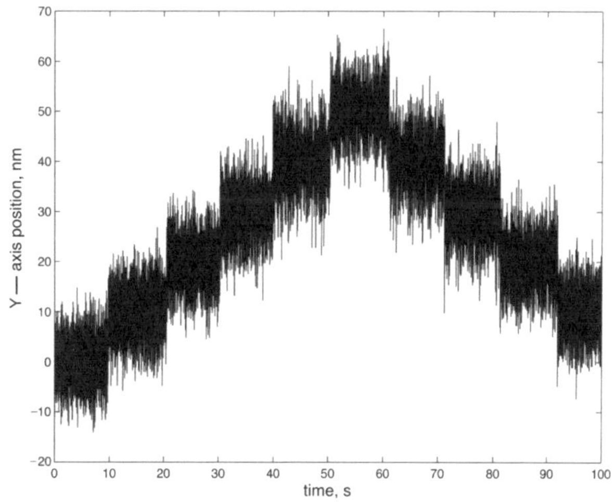

(a)

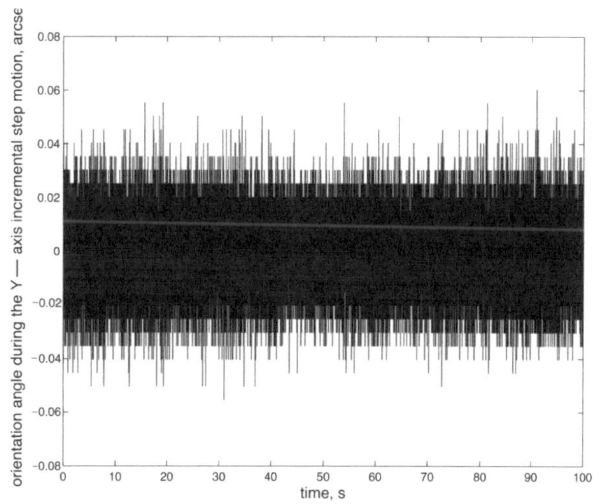

(b)

Fig. 11. Y-axis incremental step response and orientation angle response of the hybrid stage: (a) Y-axis incremental step response, (b) Orientation angle response during the Y-axis incremental step motion

(Slocum, 1992). In order to evaluate the X-axis bidirectional repeatability, the X-axis target position is increased by 40 mm from 0 to 400 mm, and then decreased by 40 mm from 400 to 0 mm. Also, in order to evaluate the Y-axis bidirectional repeatability, the Y-axis target position is increased by 30 mm from 0 to 300 mm, and then decreased by 30 mm from 300 to 0 mm. Then Fig. 12 shows the experimental results of the X-axis and Y-axis bidirectional repeatabilities. As shown in Fig. 12, the X-axis and Y-axis bidirectional repeatabilities are about 48.9 and 40.7 nm(6σ), respectively.

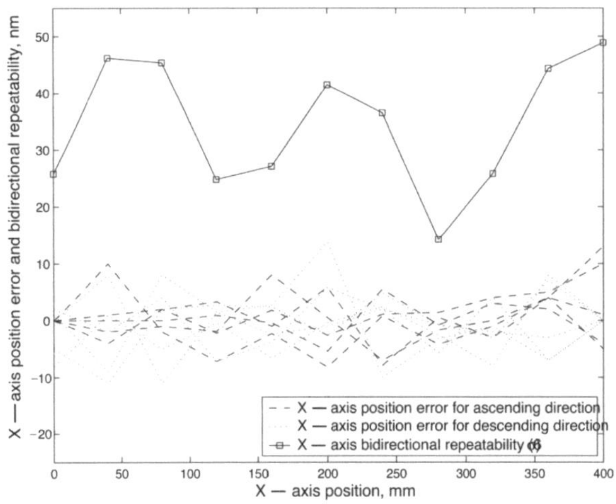

(a)

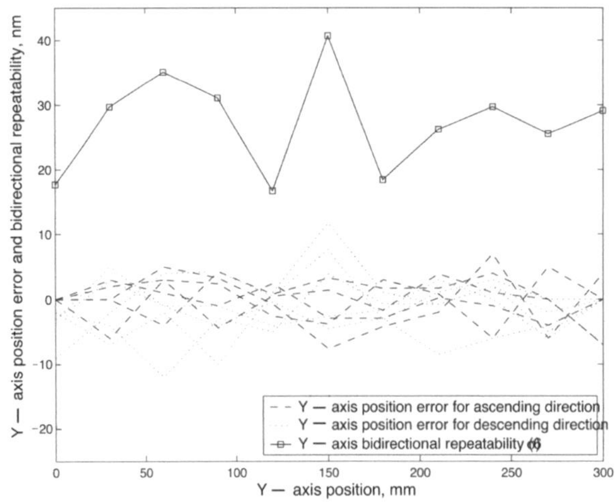

(b)

Fig. 12. Bidirectional repeatabilities of the hybrid stage: (a) X-axis bidirectional repeatability, (b) Y-axis bidirectional repeatability

5. Conclusion

In this chapter, the author presented a three degrees-of-freedom precision hybrid stage that can move and align an object on it for the measurement of its three-dimensional image using the confocal scanning microscope. Since the hybrid stage consists of two individually operating x-y-θ stages, it has not only a long operation travel but also a fine position stability. In order to control the hybrid stage, the author proposed a precision motion controller. The author evaluated the performances of the precision motion controller by experiment with a hardware setup. The experimental results showed that the precision motion controller provided the hybrid stage with desirable advantages in terms of the response time, positioning accuracy, control effort and perturbation compensation.

6. References

Awabdy, B.A.; Shin, W.-C. & Auslander, D.M. (1998). Nanometer positioning of a linear motion stage under static loads. *IEEE/ASME Trans. Mechatron.*, 3, 2, 113-119

Bohn, C. & Atherton, D.P. (1995). An analysis package comparing PID anti-windup strategies. *IEEE Control Syst. Mag.*, 15, 2, 34-40

Chung, C.C.; Seo, C.W. & Lee, S.H. (2000). Two degree-of-freedom dual-stage actuator controller design for hard disk drives. *IEEE Trans. Magn.*, 36, 5, 2255-2257

Du, C.; Guo, G. & Wu, D. (2005). Low-hump sensitivity function design for dual-stage HDD systems with different microactuators. *IEE Proc., Control Theory Appl.*, 152, 6, 655-661

Fan, L.S.; Ottesen, H.H.; Reiley, T.C. & Wood, R.W. (1995). Magnetic recording head positioning at very high track densities using a microactuator-based two-stage servo system. *IEEE Trans. Ind. Electron.*, 42, 3, 222-233

Kim, Y.-H. & Lee, S.-H. (2004). An approach to dual-stage servo design in computer disk drives. *IEEE Trans. Control Syst. Technol.*, 12, 1, 12-20

Kwon, S.; Chung, W.K. & Youm, Y. (2001). On the coarse/fine dual-stage manipulators with robust perturbation compensator, *Proc. IEEE Int. Conf. Robot. Autom.*, pp. 121-126, Seoul, Korea, May 2001

Lee, C.-W. & Kim, S.-W. (1997). An ultraprecision stage for alignment of wafers in advanced microlithography. *Precision Eng.*, 21, 2-3, 113-122

Lee, S.-H. & Kim, Y.-H. (2004). Minimum destructive interference design of dual-stage control systems for hard disk drives. *IEEE Trans. Control Syst. Technol.*, 12, 4, 517-531

Leon, S.J. (1995). *Linear Algebra With Application*, Prentice Hall, New Jersey

Li, Y. & Horowitz, R. (2001). Mechatronics of electrostatic microactuators for computer disk drive dual-stage servo system. *IEEE/ASME Trans. Mechatron.*, 6, 2, 111-121

Li, Y.; Horowitz, R. & Evans, R. (2003). Vibration control of a PZT actuated suspension dual-stage servo system using a PZT sensor. *IEEE Trans. Magn.*, 39, 2, 932-937

Ogata, K. (1996). *Modern Control Engineering*, Prentice Hall, New Jersey

Pahk, H.J.; Lee, D.S. & Park, J.H. (2001). Ultra precision positioning system for servo motor-piezo actuator using the dual servo loop and digital filter implementation. *Int. J. Mach. Tools Manuf.*, 41, 1, 51-63

Shan, X.; Kuo, S.-K.; Zhang, J. & Menq, C.-H. (2002). Ultra precision motion control of a multiple degrees of freedom magnetic suspension stage. *IEEE/ASME Trans. Mechatron.*, 7, 1, 67-78

Slocum, A.H. (1992). *Precision Machine Design*, Prentice Hall, New Jersey

Wu, D.; Guo, G. & Chong, T.C. (2002). Midfrequency disturbance suspension via microactuator in dual-stage HDDs. *IEEE Trans. Magn.*, 38, 5, 2189-2191

21

FPGA-Realization of a Motion Control IC for X-Y Table

Ying-Shieh Kung and Ting-Yu Tai
Southern Taiwan University
Taiwan

1. Introduction

The development of a compact and high performance motion controller for precision X-Y table, CNC machine etc. has been a popular field in literature (Goto et al., 1996; Wang & Lee, 1999; Hanafi et al., 2003). In position control of X-Y table, there are two approaches to be considered. One is semi closed-loop control and the other is full closed-loop control. The full closed-loop control with feed-backed by a linear encoder as the table position signal has a better positioning performance than the semi closed-loop control that a rotary encoder attached to PMSM is feed-backed as the position signal. However, to develop a motion control IC for X-Y table, the fixed-point digital signal processor (DSP) and FPGA provide two possible solutions in this issue. Compared with FPGA, DSP suffers from a long period of development and exhausts many resources of the CPU (Zhou et al., 2004).

For the progress of VLSI technology, the FPGA has been widely investigated due to its programmable hard-wired feature, fast time-to-market, shorter design cycle, embedding processor, low power consumption and higher density for implementing digital control system (Monmasson & Cirstea, 2007; Naouar et al., 2007; Jung & Kim, 2007). FPGA provides a compromise between the special-purpose ASIC (application specified integrated circuit) hardware and general-purpose processors (Wei et al., 2005). Therefore, using an FPGA to form a compact, low-cast and high performance servo system for precision machine has become an important issue. However, in many researches, the FPGA is merely used to realize the hardware part of the overall control system. Recently, fuzzy control has been successfully demonstrated in industrial control field (Sanchez-Solano et al., 2007; Kung & Tsai, 2007).Compared with other nonlinear approaches, FC has two main advantages, as follows: (1) FC has a special non-linear structure that is universal for various or uncertainty plants. (2) the formulation of fuzzy control rule can be easily achieved by control engineering knowledge, such as dynamic response characteristics, and it doesn't require a mathematical model of controlled plant. In literature, Li et al. (2003) utilized an FPGA to implement autonomous fuzzy behavior control on mobile robot. Lin et al. (2005) presented a fuzzy sliding-mode control for a linear induction motor drive based on FPGA. But, due to the fuzzy inference mechanism module adopts parallel processing circuits, it consumes much more FPGA resources; therefore limited fuzzy rules are used in their proposed method. To solve this problem, a FSM joined by a multiplier, an adder, a LUT (Look-up table), some comparators and registers are proposed to model the FC algorithm of the

PMSM drive system. Then a VHDL is adopted to describe the circuit of the FSM (Hsu et al., 1996). Due to the FSM belongs to the sequential processing method; the FPGA resources usage can be greatly reduced. Further, in recent years, an embedded processor IP and an application IP can now be developed and downloaded into FPGA to construct a SoPC environment (Altera, 2004), allowing the users to design a SoPC module by mixing hardware and software in one FPGA chip (Hall & Hamblen, 2004). The circuits required fast processing but fixed computation are suitable to be implemented by hardware in FPGA, and the heavy computation or complicated processing can be realized by software in FPGA (Kung et al., 2004; Kung & Shu, 2005). The results of the software/hardware co-design increase the programmability, flexibility of the designed digital system, enhance the system performance by parallel processing and reduce the development time.

To exploit the advantages, a motion control IC for X-Y table based on the new-generation FPGA technology is developed in this study and shown in Fig.1 (Kung et al., 2006), which the scheme of position/speed/current vector control of two PMSMs can be realized by hardware in FPGA, and the motion trajectory for X-Y table can be realized by software using Nios II embedded processor. Hence, all functionalities, which are based on software/hardware co-design, required to construct a full closed-loop control for X-Y table can be integrated and implemented in one FPGA chip. In addition, the FPGA resources usage can be greatly reduced by using the FSM in the control algorithm design. Herein, the Altera Stratix II EP2S60F672C5ES (Altera, 2008), which has 48,352 ALUTs (Adaptive Look-UP Tables), maximum 718 user I/O pins, total 2,544,192 RAM bits, and a Nios II embedded processor which has a 32-bit configurable CPU core, 16 M byte Flash memory, 1 M byte SRAM and 16 M byte SDRAM, are used. Finally, an experimental system included by an FPGA experimental board, two inverters, two sets of A/D converter and an X-Y table, is set up to verify the correctness and effectiveness of the proposed FPGA-based motion control IC.

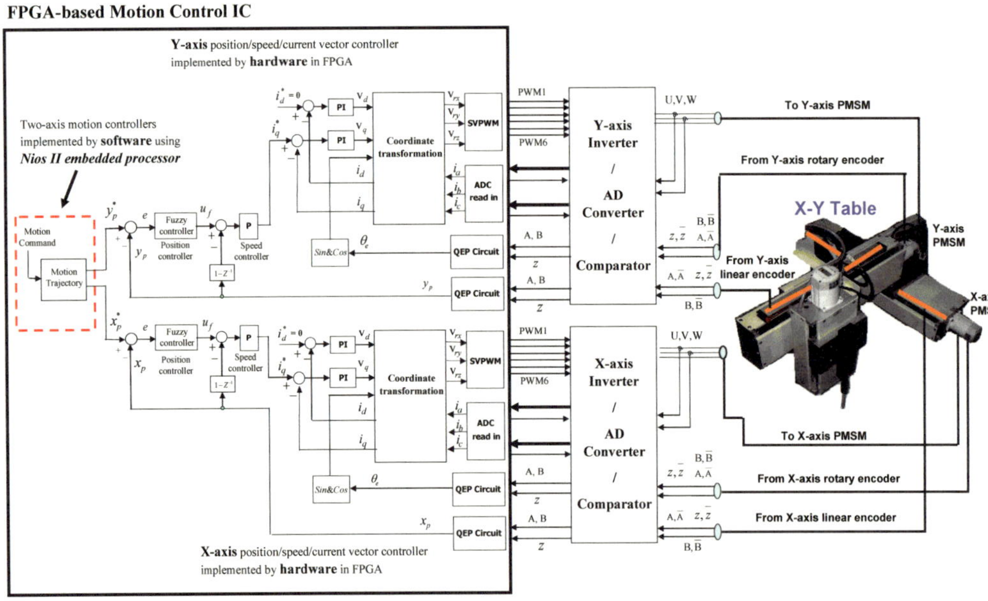

Fig. 1. The architecture of the FPGA-based motion control system for X-Y table

2. System description and controller design of X-Y table

The X-Y table is driven by two PMSMs which the current, speed and position loop in each PMSM drive adopts vector control, P control and fuzzy control, respectively The architecture of the proposed FPGA-based motion control IC for X-Y table is shown in Fig. 1. The modeling of PMSM, the fuzzy control algorithm and the motion trajectory planning are introduced as follows:

2.1 Mathematical model of PMSM and current vector controller

The typical mathematical model of a PMSM is described, in two-axis d-q synchronous rotating reference frame, as follows

$$\frac{di_d}{dt} = -\frac{R_s}{L_d} i_d + \omega_e \frac{L_q}{L_d} i_q + \frac{1}{L_d} v_d \tag{1}$$

$$\frac{di_q}{dt} = -\omega_e \frac{L_d}{L_q} i_d - \frac{R_s}{L_q} i_q - \omega_e \frac{\lambda_f}{L_q} + \frac{1}{L_q} v_q \tag{2}$$

where v_d, v_q are the d and q axis voltages; i_d, i_q, are the d and q axis currents, R_s is the phase winding resistance; L_d, L_q are the d and q axis inductance; ω_e is the rotating speed of magnet flux; λ_f is the permanent magnet flux linkage.
The current loop control of PMSM drive in Fig.1 is based on a vector control approach. That is, if the i_d is controlled to 0 in Fig.1, the PMSM will be decoupled and controlling a PMSM like to control a DC motor. Therefore, after decoupling, the torque of PMSM can be written as the following equation,

$$T_e = \frac{3P}{4} \lambda_f i_q \underline{\underline{\Delta}} K_t i_q \tag{3}$$

with

$$K_t = \frac{3P}{4} \lambda_f \tag{4}$$

Finally, considering the mechanical load with linear table, the overall dynamic equation of linear table system is obtained by

$$T_e - T_L = J_m \frac{2\pi}{r} \frac{d^2 s_p}{dt^2} + B_m \frac{2\pi}{r} \frac{ds_p}{dt} \tag{5}$$

where T_e is the motor torque, K_t is force constant, J_m is the inertial value, B_m is damping ratio, T_L is the external torque, s_p represents the displacement of X-axis or Y-axis table and r is the lead of the ball screw.
The current loop of the PMSM drive for X- or Y-table in Fig.1 includes two PI controllers, coordinate transformations of Clark, Modified inverse Clark, Park, inverse Park, SVPWM (Space Vector Pulse Width Muldulation), pulse signal detection of the encoder etc. The coordination transformation of the PMSM in Fig. 1 can be described in synchronous rotating reference frame. Figure 2 is the coordination system in rotating motor which includes

stationary *a-b-c* frame, stationary α-β frame and synchronously rotating *d-q* frame. Further, the formulations among three coordination systems are presented as follows.

1. *Clarke*: stationary *a-b-c* frame to stationary α-β frame.

$$\begin{bmatrix} i_\alpha \\ i_\beta \end{bmatrix} = \begin{bmatrix} \frac{2}{3} & \frac{-1}{3} & \frac{-1}{3} \\ 0 & \frac{1}{\sqrt{3}} & \frac{-1}{\sqrt{3}} \end{bmatrix} \begin{bmatrix} i_a \\ i_b \\ i_c \end{bmatrix} \tag{6}$$

2. Modified $Clarke^{-1}$: stationary α-β frame to stationary *a-b-c* frame.

$$\begin{bmatrix} v_a \\ v_b \\ v_c \end{bmatrix} = \begin{bmatrix} 1 & 0 \\ \frac{-1}{2} & \frac{\sqrt{3}}{2} \\ \frac{-1}{2} & \frac{-\sqrt{3}}{2} \end{bmatrix} \begin{bmatrix} v_\beta \\ v_\alpha \end{bmatrix} \tag{7}$$

3. *Park* : stationary α-β frame to rotating *d-q* frame.

$$\begin{bmatrix} i_d \\ i_q \end{bmatrix} = \begin{bmatrix} \cos\theta_e & \sin\theta_e \\ -\sin\theta_e & \cos\theta_e \end{bmatrix} \begin{bmatrix} i_\alpha \\ i_\beta \end{bmatrix} \tag{8}$$

4. $Park^{-1}$: rotating *d-q* frame to stationary α-β frame.

$$\begin{bmatrix} v_\alpha \\ v_\beta \end{bmatrix} = \begin{bmatrix} \cos\theta_e & -\sin\theta_e \\ \sin\theta_e & \cos\theta_e \end{bmatrix} \begin{bmatrix} v_d \\ v_q \end{bmatrix} \tag{9}$$

where θ_e is the electrical angle.

In Fig. 1, two digital *PI* controllers are presented in the current loop of PMSM. For the example in *d* frame, the formulation is shown as follows.

$$e_d(k) = i_d^*(k) - i_d(k) \tag{10}$$

$$v_{p\_d}(k) = k_{p\_d}\, e_d(k) \tag{11}$$

$$v_{i\_d}(k) = v_{i\_d}(k-1) + k_{i\_d}\, e_d(k-1) \tag{12}$$

$$v_d(k) = v_{p\_d}(k) + v_{i\_d}(k) \tag{13}$$

the e_d is the error between current command and measured current. The k_{p_d}, k_{i_d} are *P* controller gain and *I* controller gain, respectively. The $v_{p\_d}(k)$, $v_{i\_d}(k)$, $v_d(k)$ are the output of *P* controller only, *I* controller only and the *PI* controller, respectively. Similarity, the formulation of *PI* controller in *q* frame is the same.

2.2 Fuzzy controller (FC) for position control loop

The position controllers in X-axis and Y-axis table of Fig. 1 adopt fuzzy controller, which includes fuzzification, fuzzy rules, inference mechanism and defuzzification. Herein, an FC design method for X-axis and Y-axis table is presented. At first, position error and its error change, e , de are defined by

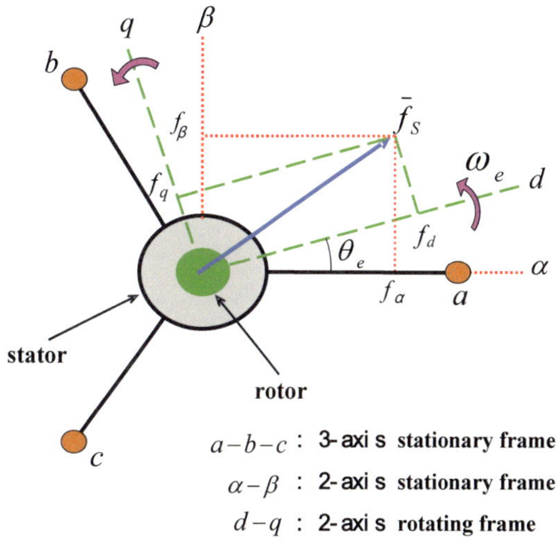

Fig. 2. Transformation between stationary axes and rotating axes

$$e(k) = s_p^*(k) - s_p(k) \tag{14}$$

$$de(k) = e(k) - e(k-1) \tag{15}$$

The K_{er} and K_{der} are the gains of the input variables e and de, respectively, as well as u_f is the output variables of the FC. The design procedure of the FC is as follows:

a. Take the E and dE as the input linguist variables, which are defined by $\{A_0, A_1, A_2, A_3, A_4, A_5, A_6\}$ and $\{B_0, B_1, B_2, B_3, B_4, B_5, B_6\}$, respectively. Each linguist value of E and dE are based on the symmetrical triangular membership function which is shown in Fig.3. The symmetrical triangular membership function are determined uniquely by three real numbers $\xi_1 \le \xi_2 \le \xi_3$, if one fixes $f(\xi_1) = f(\xi_3) = 0$ and $f(\xi_2) = 1$. With respect to the universe of discourse of [-6.6], the numbers for these linguistic values are selected as follows:

$$A_0{=}B_0: \{-6,-6,-4\},\ A_1{=}B_1: \{-6,-4,-2\},\ A_2{=}B_2: \{-4,-2,0\},\ A_3{=}B_3: \{-2,0,2\},$$

$$A_4{=}B_4: \{0,2,4\},\ A_5{=}B_5: \{2,4,6\},\ A_6{=}B_6: \{4,6,6\} \tag{16}$$

b. Compute the membership degree of e and de. Figure 3 shows that the only two linguistic values are excited (resulting in a non-zero membership) in any input value, and the membership degree $\mu_{A_i(e)}$ can be derived, in which the error e is located between e_i and e_{i+1}, two linguist values of A_i and A_{i+1} are excited, and the membership degree is obtained by

$$\mu_{A_i}(e) = \frac{e_{i+1} - e}{2} \quad \text{and} \quad \mu_{A_{i+1}}(e) = 1 - \mu_{A_i}(e) \tag{17}$$

where $e_{i+1} \underline{\underline{\Delta}} -6 + 2*(i+1)$. Similar results can be obtained in computing the membership degree $\mu_{B_j(de)}$.

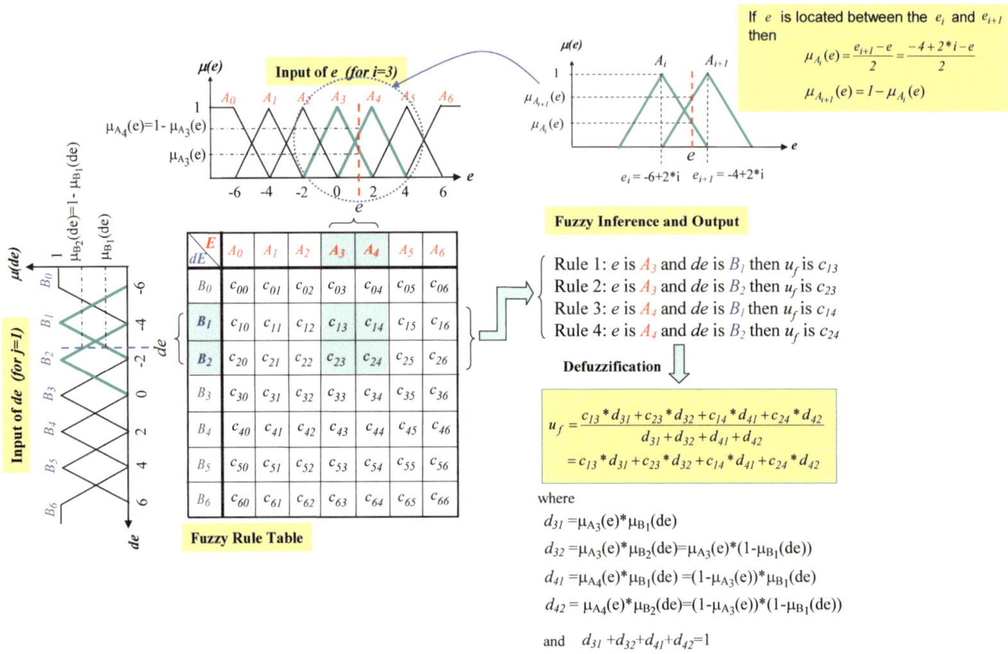

Fig. 3. Fuzzification, fuzzy rule table, fuzzy inference and defuzzication

c. Select the initial fuzzy control rules by referring to the dynamic response characteristics (Liaw et al., 1999), such as,

$$IF\ \ e\ \ is\ A_i\ \ and\ \ \Delta e\ is\ B_j\ \ THEN\ \ u_f\ \ is\ \ c_{j,i} \tag{18}$$

where i and j = 0~6, A_i and B_j are fuzzy number, and $c_{j,i}$ is real number. The graph of fuzzification and fuzzy rule table is shown in Fig. 3.

d. Construct the fuzzy system $u_f(e,de)$ by using the singleton fuzzifier, product-inference rule, and central average defuzzifier method. Although there are total 49 fuzzy rules in Fig. 3 will be inferred, actually only 4 fuzzy rules can be effectively excited to generate a non-zero output. Therefore, if the error e is located between e_i and e_{i+1}, and the error change de is located between de_j and de_{j+1}, only four linguistic values A_i, A_{i+1}, B_j, B_{j+1} and corresponding consequent values $c_{j,i}$, $c_{j+1,i}$, $c_{j,i+1}$, $c_{j+1,i+1}$ can be excited, and the (18) can be replaced by the following expression:

$$u_f(e,de) = \frac{\sum_{n=i}^{i+1}\sum_{m=j}^{j+1} c_{m,n}[\mu_{A_n}(e)*\mu_{B_m}(de)]}{\sum_{n=i}^{i+1}\sum_{m=j}^{j+1}\mu_{A_n}(e)*\mu_{B_m}(de)} \triangleq \sum_{n=i}^{i+1}\sum_{m=j}^{j+1} c_{m,n} * d_{n,m} \tag{19}$$

where $d_{n,m} \triangleq \mu_{A_n}(e)*\mu_{B_m}(de)$. And those $c_{m,n}$ denote the consequent parameters of the fuzzy system.

2.3 Motion trajectory planning of X-Y table

The point-to-point, circular and window motion trajectories are usually considered to evaluate the motion performance for X-Y table.

a. In point-to-point motion trajectory, for smoothly running of the table, it is designed with the trapezoidal velocity profile and its formulation is shown as follows.

$$s(t)=\begin{cases}\frac{1}{2}At^2+s_0 & 0\le t\le t_a\\ v_m(t-t_a)+s(t_a) & t_a\le t\le t_d\\ -\frac{1}{2}A(t-t_d)^2+v_m(t-t_d)+s(t_d) & t_d\le t\le t_s\end{cases} \tag{20}$$

Where $0<t<t_a$ is at the acceleration region, $t_a<t<t_d$ is at the constant velocity region, and $t_d<t<t_s$ is at the deceleration region. The s represents the position command in X-axis or Y-axis table; A is the acceleration/deceleration value; s_0 is the initial position; v_m is the maximum velocity; t_a, t_d and t_s represents the end time of the acceleration region, the start time of the deceleration region and the end time of the trapezoidal motion, respectively.

b. In circular motion trajectory, it is computed by

$$x_i=r\sin(\theta_i) \tag{21}$$

$$y_i=r\cos(\theta_i) \tag{22}$$

with $\theta_i=\theta_{i-1}+\Delta\theta$. Where $\Delta\theta$, r, x_i, y_i are angle increment, radius, X-axis trajectory command and Y-axis trajectory command, respectively.

c. The window motion trajectory is shown in Fig.4. The formulation is derived as follows:

$$\text{a-trajectory}: x_i=x_{i-1}, y_i=S+y_{i-1} \tag{23}$$

$$\text{b-trajectory}: (\theta_i:\frac{6}{4}\pi\rightarrow 2\pi, \text{and } \theta_i=\theta_{i-1}+\Delta\theta)$$

$$x_i=O_{x1}+r\cos(\theta_i), y_i=O_{y1}+r\sin(\theta_i) \tag{24}$$

$$\text{c-trajectory}: x_i=S+x_{i-1}, y_i=y_{i-1} \tag{25}$$

$$\text{d-trajectory}: (\theta_i:\pi\rightarrow\frac{6}{4}\pi, \text{and } \theta_i=\theta_{i-1}+\Delta\theta)$$

$$x_i=O_{x2}+r\cos(\theta_i), y_i=O_{y2}+r\sin(\theta_i) \tag{26}$$

$$\text{e-trajectory}: x_i=x_{i-1}, y_i=-S+y_{i-1} \tag{27}$$

$$\text{f-trajectory}: (\theta_i:\frac{1}{2}\pi\rightarrow\pi, \text{and } \theta_i=\theta_{i-1}+\Delta\theta)$$

$$x_i = O_{x3} + r\cos(\theta_i), y_i = O_{y3} + r\sin(\theta_i) \quad (28)$$

$$\text{g-trajectory} : x_i = -S + x_{i-1}, y_i = y_{i-1} \quad (29)$$

$$\text{h-trajectory} : (\theta_i : 0 \to \frac{1}{2}\pi, \text{and } \theta_i = \theta_{i-1} + \Delta\theta)$$

$$x_i = O_{x4} + r\cos(\theta_i), y_i = O_{y4} + r\sin(\theta_i) \quad (30)$$

$$\text{i-trajectory} : x_i = x_{i-1}, y_i = S + y_{i-1} \quad (31)$$

where S, $\Delta\theta$, x_i, y_i are position increment, angle increment, X-axis trajectory command and Y-axis trajectory command, respectively. In addition, the (O_{x1}, O_{y1}), (O_{x2}, O_{y2}), (O_{x3}, O_{y3}), (O_{x4}, O_{y4}) are arc center of b-, d-, f-, and h-trajectory in the Fig. 4 and r is the radius. The motion speed of the table is determined by $\Delta\theta$.

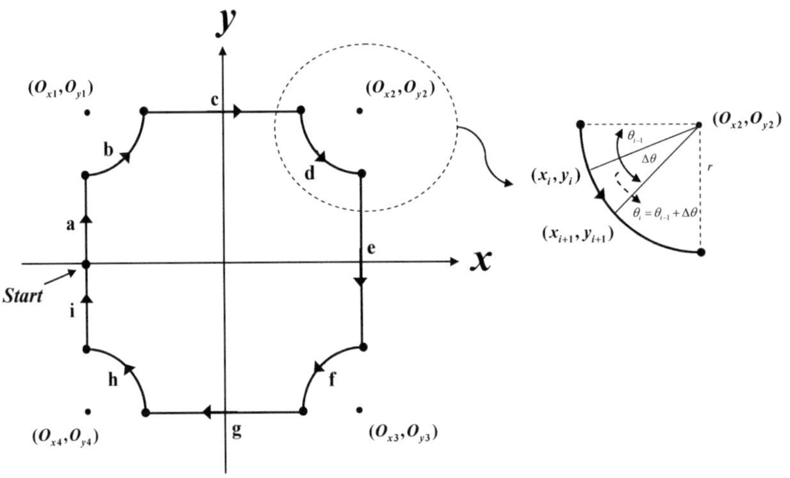

Fig. 4. Window motion trajectory

3. Design of an FPGA-based motion control IC for X-Y table

The architecture of the proposed FPGA-based motion control IC for X-Y table is shown in Fig. 1, in which the motion trajectory is implemented by software using Nios II embedded processor and the current vector controller, the position and speed controller for two PMSMs are implemented by hardware in FPGA chip. However, in this section, we firstly introduce the concept of finite state machine (FSM). Then use FSM to design the complicated control algorithm, such as the FC and the vector controller in PMSM drive.

3.1 Finite state machine (FSM)

To reduce the use of the FPGA resource, FSM is adopted to describe the complicated control algorithm. Herein, the computation of a sum of product (SOP) shown below is taken as a case study to present the advantage of FSM.

$$Y = a_1 * x_1 + a_2 * x_2 + a_3 * x_3 \tag{32}$$

Two kinds of design method that one is parallel processing method and the other is FSM method are introduced to realize the the computation of SOP. In the former method, the designed SOP circuit is shown in Fig. 5(a), and it will operate continuously and simultaneously. The circuit needs 2 adders and 3 multipliers, but only one clock time can complete the overall computation. Although the parallel processing method has fast computation ability, it consumes much more FPGA resources. To reduce the resource usage in FPGA, the designed SOP circuit adopted by using the FSM method is proposed and shown in Fig. 5(b), which uses one adder, one multiplier and manipulates 5 steps (or 5 clocks time) machine to carry out the overall computation of SOP. Although the FSM method needs more operation time (if one clock time is 40ns, the 5 clocks needs 0.2 μ s) than the parallel processing method in executing SOP circuit, it doesn't loss any computation power. Therefore, the more complicated computation in algorithm, the more FPGA resources can be economized if the FSM is applied. Further, VHDL code to implement the computation of SOP is shown in Fig.6

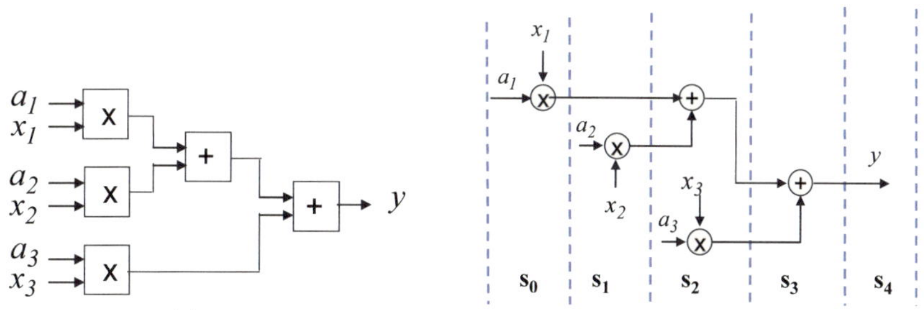

(a) (b)

Fig. 5. Computation of SOP by using (a) parallel operation (b) FSM operation

```
LIBRARY IEEE;
USE IEEE.std_logic_1164.all;
USE IEEE.std_logic_arith.all;
USE IEEE.std_logic_signed.all;
LIBRARY lpm;
USE lpm.LPM_COMPONENTS.ALL;
ENTITY SOP IS
    port(CLK_40n              :IN  STD_LOGIC;
        A1,A2,A3,X1,X2,X3     :IN  STD_LOGIC_VECTOR(11 downto 0);
        Y                     :OUT STD_LOGIC_VECTOR(23 downto 0)
        );
END matrix;
ARCHITECTURE SOP_arch OF SOP IS
    SIGNAL mula,mulb          :STD_LOGIC_VECTOR(11 downto 0);
    SIGNAL mulr               :STD_LOGIC_VECTOR(23 downto 0);
    SIGNAL adda,addb,addr     :STD_LOGIC_VECTOR(23 downto 0);
    SIGNAL CNT                :STD_LOGIC_VECTOR(7 downto 0);
BEGIN
multiplier: lpm_mult
  generic map(LPM_WIDTHA=>12,LPM_WIDTHB=>12,LPM_WIDTHS=>12,LPM_WI
            DTHP=>24,LPM_REPRESENTATION=>"signed",LPM_PIPELINE=>1)
  port map(dataa=> mula,datab=> mulb,clock=> clk,result=> mulr);

adder: lpm_add_sub
  generic map(lpm_width=>24,LPM_REPRESENTATION=>"signed",lpm_pipeline=>1)
  port map(dataa=>adda,datab=>addb,clock=> clk,result=>addr);
```

(a)

```
GEN: block
BEGIN
  PROCESS(CLK_40n)
    BEGIN
      IF CLK_40n'EVENT and CLK_40n='1' THEN
        CNT<=CNT+1;
        IF CNT=X"00" THEN
          mula  <= A1;
          mulb  <= X1;
        ELSIF CNT=X"01" THEN
          adda  <= mulr;
          mula  <= A2;
          mulb  <= X2;
        ELSIF CNT=X"02" THEN
          addb  <= mulr;
          mula  <= A3;
          mulb  <= X3;
        ELSIF CNT=X"03" THEN
          adda  <= addr;
          addb  <= mulr;
        ELSIF CNT=X"04" THEN
          Y     <= addr;
          CNT <= X"00";
        END IF;
      END IF;
  END PROCESS;
END BLOCK GEN;
END SOP_arch;
```

(a) continue

Fig. 6. VHDL code to implement the computation of SOP

3.2 Design of an FPGA-based motion control IC for X-Y table

The internal architecture of the proposed FPGA-based motion control IC for X-Y table is shown in Fig. 7. The FPGA is used by Altera Stratix II EP2S60 and a Nios II embedded processor can be downloaded into FPGA to construct an SoPC environment. The Altera Stratix II EP2S60 has 48,352 ALUTs (Adaptive Look-UP Tables), maximum 718 user I/O pins, total 2,544,192 RAM bits, and Nios II embedded processor is a 32-bit configurable CPU core, 16 M byte Flash memory, 1 M byte SRAM and 16 M byte SDRAM. A custom software development kit (SDK) consists of a compiled library of software routines for the SoPC design, a Make-file for rebuilding the library, and C header files containing structures for each peripheral. The motion control IC, which is designed in this SoPC environment, comprises a Nios II embedded processor IP and an application IP. The application IP implemented by hardware is adopted to realize two position/speed/current vector controllers of PMSMs and two QEP circuits of linear encoder. The circuit of each current vector controller includes a current controller and coordinate transformation (CCCT), SVPWM generation, QEP detection and transformation, ADC interface, etc. The speed loop uses P controller and the position loop adopts FC. The sampling frequency of the position control loop is designed with 2kHz. The frequency divider generates 50 Mhz (Clk), 25 Mhz (Clk-sp), 16 kHz (Clk-cur), and 2 kHz (Clk-po) clock to supply all circuits in Fig. 7.

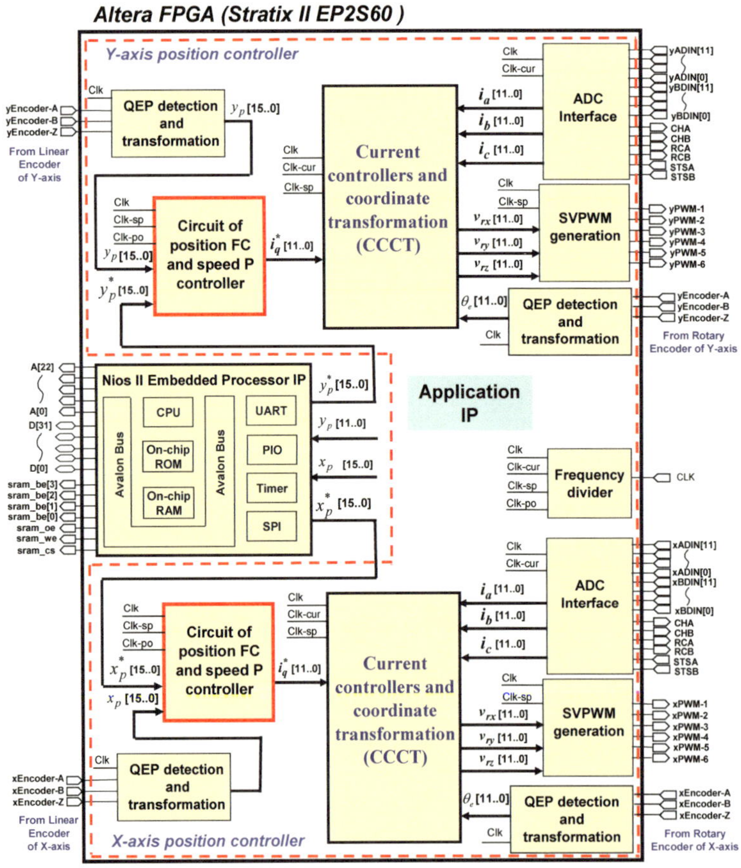

Fig. 7. Internal circuit of the proposed FPGA-based motion control IC

The internal circuit of CCCT performs the function of two PI controllers, table look-up for *sin/cos* function and the coordinate transformation for Clark, Park, inverse Park, modified inverse Clarke. The CCCT circuit designed by FSM is shown in Fig. 8, which uses one adder, one multiplier, an one-bit left shifter, a look-up-table and manipulates 24 steps machine to carry out the overall computation. The data type is 12-bit length with Q11 format and 2's complement operation. In Fig. 8, steps s0~s1 is for the look-up *sin/cos* table; steps s2~s5 and s5~s8 are for the transformation of Clarke and Park, respectively; steps s9~s14 is for the computation of d-axis and q-axis PI controller; and steps s15~s19 and s20~s23 represent the transformation of the inverse Park and the modified inverse Clarke, respectively. The operation of each step in FPGA can be completed within 40ns (25 MHz clock); therefore total 24 steps need 0.96 μs operation time. Although the FSM method needs more operation time than the parallel processing method in executing CCCT circuit, it doesn't loss any control performance in overall system because the 0.96 μs operation time is much less than the designed sampling interval, 62.5 μs (16 kHz) of current control loop in Fig. 1. To prevent numerical overflow and alleviate windup phenomenon, the output values of I controller and PI controller are both limited within a specific range.

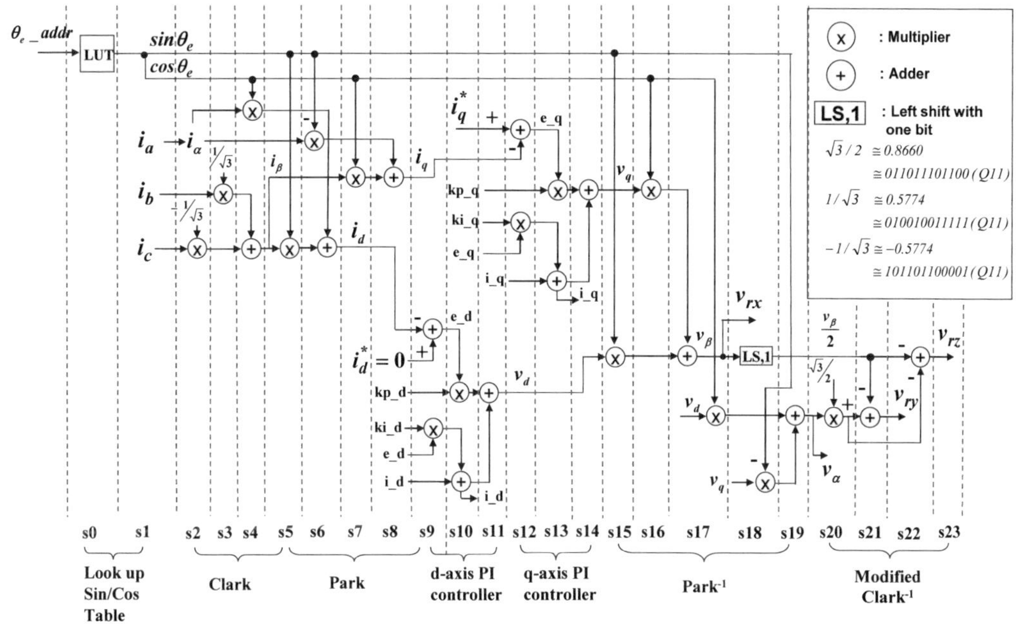

Fig. 8. Designed CCCT circuit in Fig. 7

An FSM is employed to model the FC of the position loop and P controller of the speed loop in PMLSM and shown in Fig. 9, which uses one adder, one multiplier, a look-up table, comparators, registers, etc. and manipulates 23 steps machine to carry out the overall computation. With exception of the data type in reference model are 24-bits, others data type are designed with 12-bits length, 2's complement and Q11 format. Although the algorithm of FC is highly complexity, the FSM can give a very adequate modeling and easily be described by VHDL. Furthermore, steps s_0~s_2 are for the computation of speed, position error and error change; steps s_3~s_6 execute the function of the fuzzification; s_7 describes the look-up table and s_8~s_{16} defuzzification; and steps s_{17}~s_{22} execute the computation of speed and current command output. The SD is the section determination of *e* and *de*, and its flow chart of circuit design is shown in Fig.10. And the RS,1 represents the right shift function with one bit. The operation of each step in Fig.9 can be completed within 40ns (25 MHz clock) in FPGA; therefore total 23 steps need 0.92μs operation time. It doesn't loss any control performance in the overall system because the operation time with 0.92μs is much less than the sampling interval, 500 μs (2 kHz), of the position control loop in Fig.1.

In Figure 7, with exception of the CCCT circuit, others circuit design, like SVPWM and QEP, are presented in Fig. 11(a) and 11(b), respectively. The SVPWM circuit is designed to be 12 kHz frequency and 1μs dead-band. The circuit of the *QEP* module is shown in Fig.11(b), which consists of two digital filters, a decoder and an up-down counter. The filter is used for reducing the noise effect of the input signals *PA* and *PB*. The pulse count signal *PLS* and the rotating direction signal *DIR* are obtained using the filtered signals through the decoder circuit. The *PLS* signal is a four times frequency pulses of the input signals *PA* or *PB*. The *QEP* value can be obtained using *PLS* and *DIR* signals through a directional up-down counter.

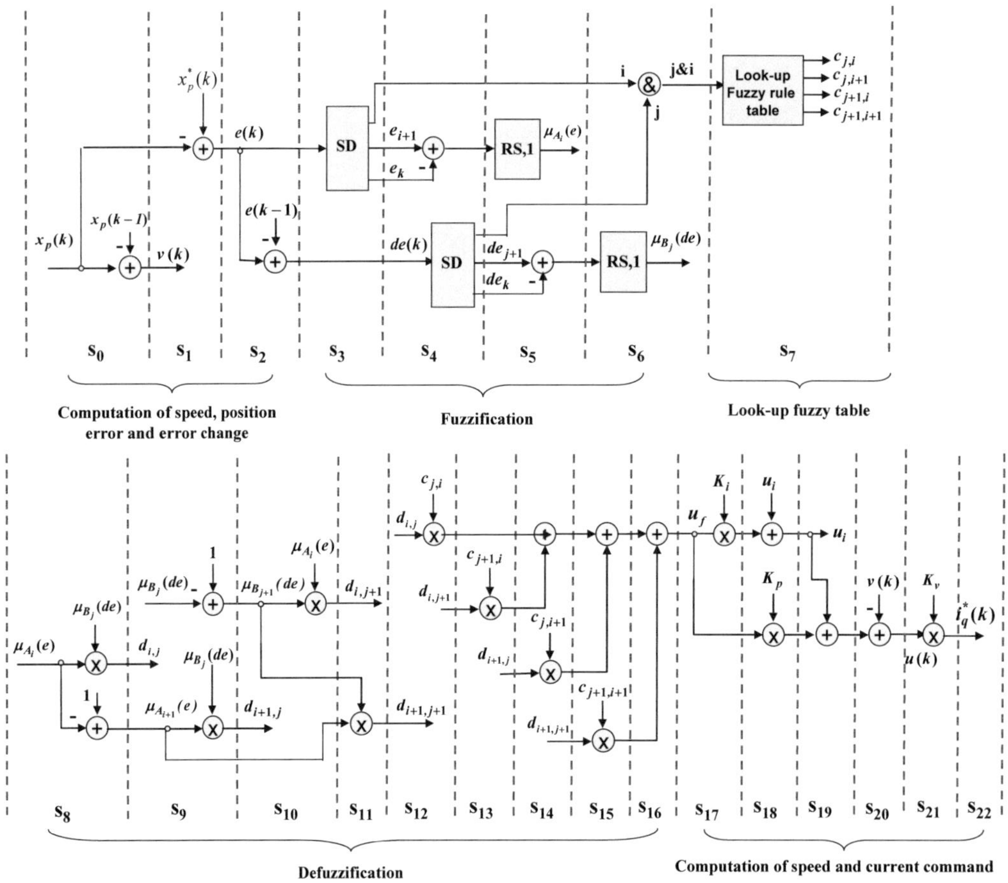

Fig. 9. State diagram of an FSM for describing the FC in position loop and P controller in speed loop

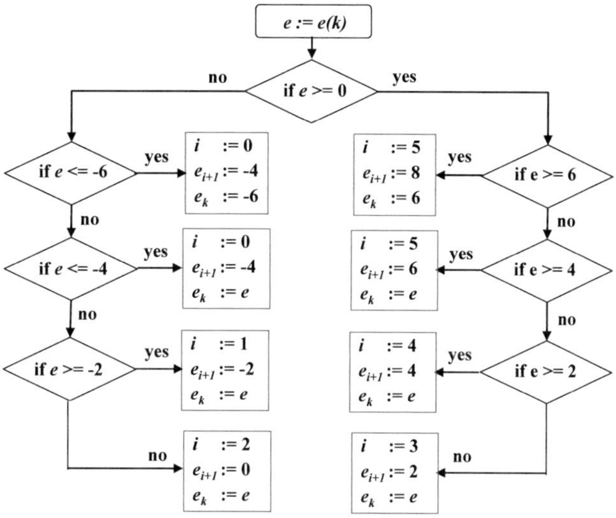

Fig. 10. Section determination in Fig. 9

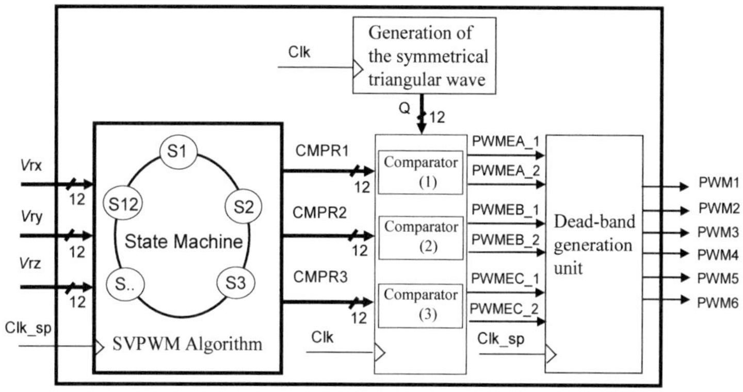

(a)

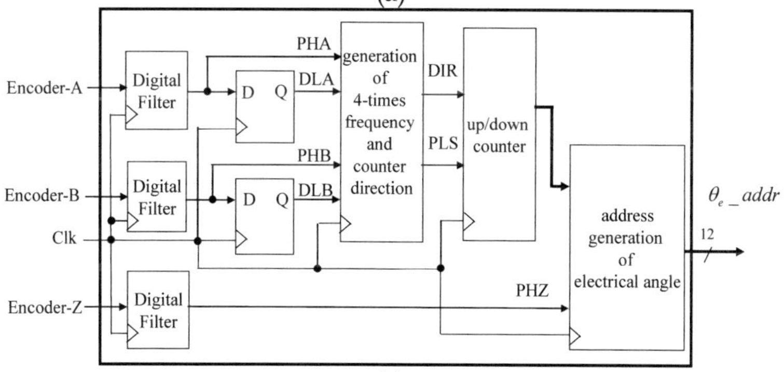

(b)

Fig. 11. Block diagram of (a) SVPWM circuit (b) QEP circuit

The Nios II embedded processor IP is depicted to perform the function of the motion trajectory for X-Y table in software. Figure 12 illustrates the flow charts of the main program and the interrupt service routine (ISR), where the interrupt interval is designed with 2ms.

All programs are coded in the C programming language in Fig.10. Then, through the complier and linker operation in the Nios II IDE (Integrated Development Environment), the execution code is produced and can be downloaded to the external Flash or SDRAM via JTAG interface. Using the C language to develop the control algorithm has the portable merit and is easier to transfer the mature code from the other processor to the Nios II embedded processor. Finally, Table 1 shows the FPGA utility of the proposed motion control IC and the overall circuits included a Nios II embedded processor IP (5,059 ALUTs and 78,592 RAM bits) and an application IP (10,196 ALUTs and 102,400 RAM bits), use 31.5% ALUTs resource and 7.1% RAM resource of Stratix II EP2S60.

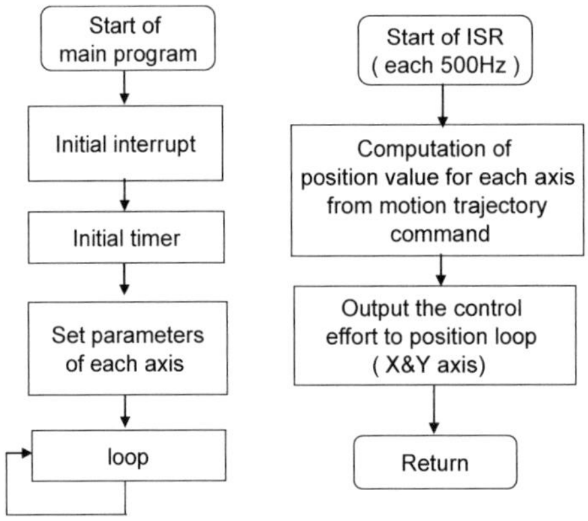

Fig. 12. Flow chart of the main and ISR program in Nios II embedded processor

IP	Module circuit	Logic gate (ALUTs)	Memory (Bits)
Nios II embedded processor IP		5,059	78,592
Application IP (for X-axis and Y-axis)	Position fuzzy controller and speed P controller	1,943 × 2	0
	Current controller and coordinate transformation (CCCT)	649 × 2	49,152 × 2
	SVPWM generation	1,220 × 2	0
	ADC interface	123 × 2	0
	QEP detection and transformation	79 × 4	0
	others	1,005 × 2	2,048 × 2
Total		**15,255**	**180,992**

Table 1. Utility evaluation of a motion contron IC for X-Y table in FPGA

4. Experiments and results

The overall experimental system is depicted in Fig. 1. This system includes an FPGA experimental board, two sets of voltage source IGBT inverter and an X-Y table which is driven by two PMSMs and two ball-screws. The power, rating, voltage, current and rating speed of PMSM are 200W, 92V, 1.6A and 3000rpm, respectively. A 2500 ppr rotary encoder attached to PMSM is used to measure the motor's electrical angle. Two linear encoders with 5μm resolution are mounted on the X-axis and Y-axis table as a position sensor. Each ball-

screw has 5mm lead. The inverter has 6 sets of IGBT type power transistors. The collector-emitter voltage of the IGBT is rating 600V, the gate-emitter voltage is rating ±12V, and the collector current in DC is rating 25A and in short time (1ms) is 50A. The photo-IC, Toshiba TLP250, is used for gate driving circuit of IGBT. Input signals of the inverter are PWM signals from FPGA chip. The FPGA-Altera Stratix II EP2S60 in Fig.1 is used to develop a full digital motion controller for X-Y table. The motion trajectory are implemented by software using Nios II embedded processor, and the two axis position/speed/current vector controller are implemental by hardware in FPGA. In the experimental system, the PWM switching frequency of inverter is designed with 12k Hz, dead-band is 1μs, and the sampling frequency in current loop and position loop of the PMSM are designed with 16kHz and 500Hz, respectively. The motion control algorithms are coded by C language.
In experiment, the position step response and the motion trajectory control are used to evaluate the dynamic performance of the proposed system. In the experiment of the step response, the results of X-axis and Y-axis table under 10 mm amplitude and 0.5Hz square wave command are shown in Fig. 13. The rising time, overshoot and steady-state value in Fig. 13(a) are 110ms, 14% and near 0mm, and in Fig. 13(b) are 90ms, 15% and near 0mm. It reveals that the mass carried in X-axis table is heavier than those in Y-axis table. In the experiment of the motion trajectory tracking, one-dimensional trapezoidal motion trajectory, two-dimensional circular and window motion trajectory are tested and its experimental tracking results are shown in Figs. 14 ~ 16. In one-dimensional motion trajectory, the trapezoidal velocity profile is considered which the acceleration and deceleration is designed with $500mm/s^2$, maximum speed is 125mm/s, and the overall displacement is designed with moving from 0 mm to 100 mm position. The trajectory tracking results in each axis corresponding with the aforementioned input commands is shown in Fig. 14. It can be seen that the motion of X-axis and Y-axis table can give a perfect tracking with command target both in position or speed trajectory. Further, in two-dimensional motion trajectory, the circular motion trajectory control with center (60, 60) mm and radius 50mm is evaluated and the tracking errors are the maximum $\pm$ 0.55 mm in X-axis, and $\pm$ 0.75 mm in Y-axis in Fig. 15. The window motion trajectory designed as Fig.4 and its experimental result is shown in Fig. 16, which also shows the tracking errors maximum $\pm$ 0.5 mm in X-axis, and $\pm$ 0.9 mm in Y-axis. Therefore, from the experimental results of Figs. 13~16, it demonstrates that the proposed FPGA-based motion controller IC for X-Y table is effective and correct.

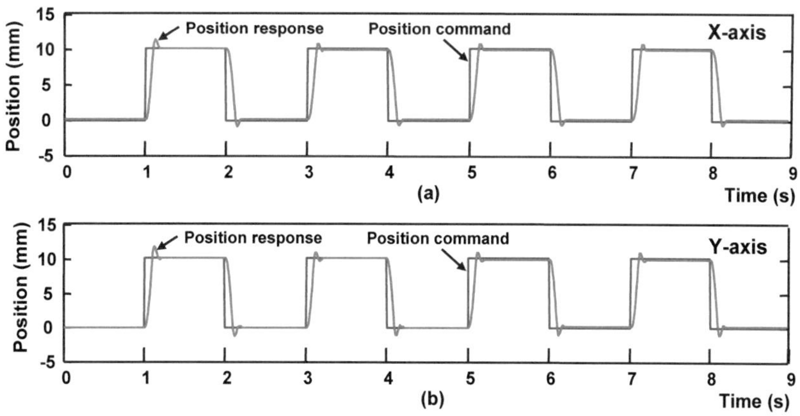

Fig. 13. Step response for (a) X-axis table (b) Y-axis table

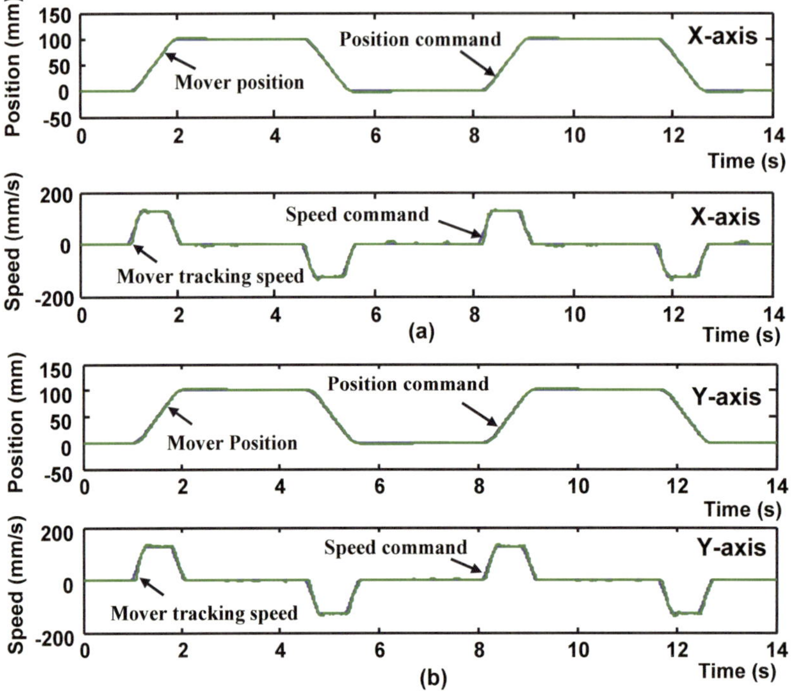

Fig. 14. (a) Position and speed tracking response in X-axis and in (b) Y-axis table

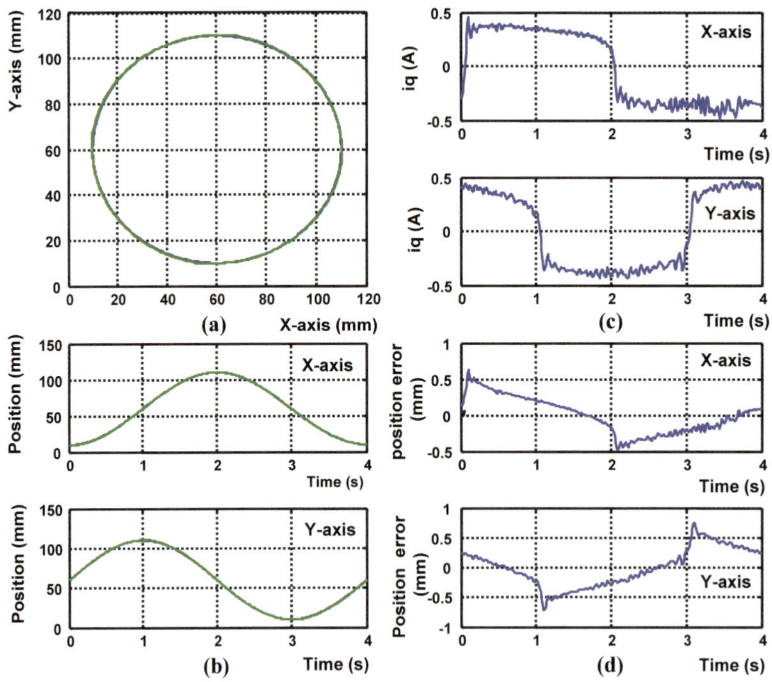

Fig. 15. Response of the circular trajectory (a) circular trajectory response (b) response for X- and Y- axis (c) control effort (d) tracking error

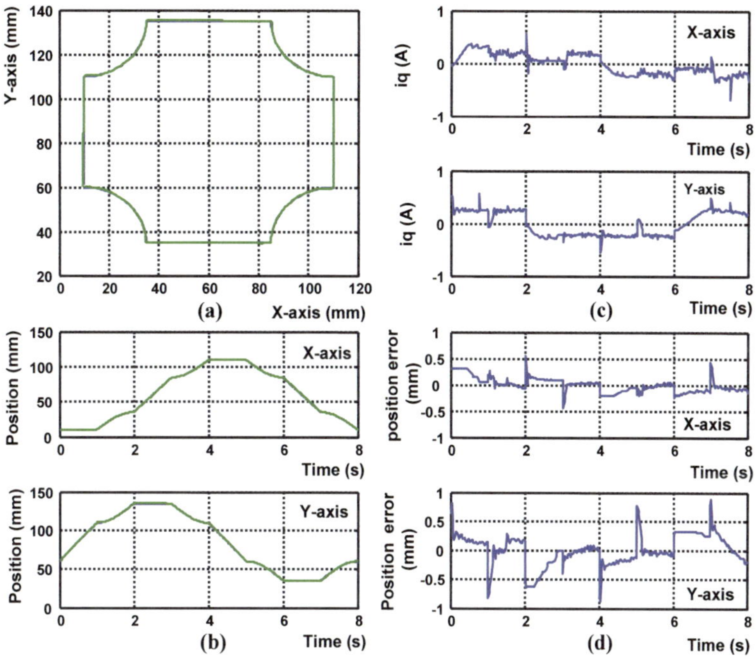

Fig. 16. Response of the window trajectory (a) window trajectory response (b) response for X- and Y- axis (c) control effort (d) tracking error

5. Conclusion

This study successfully presents a motion control IC for X-Y table based on novel FPGA technology. The works herein are summarized as follows.

1. The functionalities required to build a fully digital motion controller of X-Y table, such as the two current vector controllers, two speed P controllers, and two position fuzzy controllers and one motion trajectory planning, have been integrated in one FPGA chip.
2. An FSM joined by one multiplier, one adder, one LUT, or some comparators and registers has been employed to model the overall FC algorithm and the CCCT in vector control of the PMSM, such that it not only is easily implemented by VHDL but also can reduce the FPGA resources usage.
3. The software/hardware co-design technology under SoPC environment has been successfully applied to the motion controller of X-Y table.

However, the experimental results by step response, point-to-point, window and circular motion trajectory tracking, has been revealed that the software/hardware co-design technology with the parallel processing well in the motion control system of X-Y table.

6. References

Altera Corporation, (2004). *SOPC World.*

Altera (2008): www.altera.com.

Goto, S.; Nakamura M. & Kyura, N. (1996). Accurate contour control of mechatronic servo systems using gaussian networks, *IEEE Trans. Ind. Electron.*, vol.43, no. 4, pp. 469-476.

Hall, T.S. & Hamblen, J.O. (2004). System-on-a-programmable-chip development platforms in the classroom, *IEEE Trans. on Education*, vol.47, no.4, pp.502-507.

Hanafi, D.; Tordon, M. & Katupitiya, J. (2003). An active axis control system for a conventional CNC machine, *Proceedings of IEEE/ASME International Conference on Advanced Intelligent Mechatronics*, pp. 1188-1193.

Hsu, Y.C.; Tsai, K.F.; Liu, J.T. & Lin, E.S. (1996) *VHDL modeling for digital design synthesis*, KLUWER ACADEMIC PUBLISHERS, TOPPAN COMPANY (S) PTE LTD.

Jung, S. & Kim, S.S. (2007). Hardware implementation of a real-time neural network controller with a DSP and an FPGA for nonlinear systems, *IEEE Trans. Ind. Electron.*, vol. 54, no. 1, pp. 265-271.

Kung, Y.S.; Huang, P.G. & Chen, C.W. (2004). Development of a SOPC for PMSM drives, *Proceedings of the IEEE International Midwest Symposium on Circuits and Systems*, vol. II, pp. II-329~II-332.

Kung, Y.S. & Shu, G.S. (2005). Design and Implementation of a Control IC for Vertical Articulated Robot Arm using SOPC Technology, *Proceeding of IEEE International Conference on Mechatronics*, pp. 532~536.

Kung, Y.S.; Tseng, K.H. & Tai, F.Y. (2006) FPGA-based servo control IC for X-Y table, *Proceedings of the IEEE International Conference on Industrial Technology*, pp. 2913-2918.

Kung, Y.S. & Tsai, M.H. (2007). FPGA-based speed control IC for PMSM drive with adaptive fuzzy control, *IEEE Trans. on Power Electronics*, vol. 22, no. 6, pp. 2476-2486.

Li, T.S.; Chang S.J. & Chen, Y.X. (2003). Implementation of Human-like Driving Skills by Autonomous Fuzzy Behavior Control on an FPGA-based Car-like Mobile Robot, *IEEE Trans. Ind. Electro.*, vol. 50, no.5, pp. 867-880.

Liaw, C.M. & Kung, Y.S. (1999) Fuzzy control with reference model-following response, *Fuzzy Theory Systems: Techniques and Applications*, vol.1, pp. 129-158.

Lin, F.J.; Wang, D.H. & Huang, P.K. (2005). FPGA-based fuzzy sliding mode control for a linear induction motor drive, *IEE Proc.- Electr. Power Application*, vol. 152, no.5, pp. 1137-1148.

Monmasson, E. & Cirstea, M.N. (2007). FPGA design methodology for industrial control systems - a review, *IEEE Trans. on Industrial Electronics*, Vol. 54, No. 4, pp.1824-1842.

Naouar, M.W.; Monmasson, E.; Naassani, A.A.; Slama-Belkhodja, I. & Patin, N. (2007). FPGA-based current controllers for AC machine drives - a review, *IEEE Trans. Ind. Electron.*, vol. 54, no.4, pp.1907-1925.

Sanchez-Solano, S.; Cabrera, A.J.; Baturone, I.; Moreno-Velo, F.J. & Brox, M. (2007). FPGA implementation of embedded fuzzy controllers for robotic applications, *IEEE Trans. Ind. Electron.*, vol. 54, no. 4, pp. 1937-1945.

Wang, G.J. & Lee, T.J. (1999). Neural-network cross-coupled control system with application on circular tracking of linear motor X-Y table, *International Joint Conference on Neural Networks*, pp. 2194-2199.

Wei, R.; Gao, X.H.; Jin, M.H.; Liu, Y.W.; Liu, H.; Seitz, N.; Gruber, R. & Hirzinger, G. (2005). FPGA based Hardware Architecture for HIT/DLR Hand, *Proceeding of the IEEE/RSJ International Conference on Intelligent Robots and System*, pp. 523~528.

Zhou, Z.; Li, T.; Takahahi, T. & Ho, E. (2004). FPGA realization of a high-performance servo controller for PMSM, *Proceeding of the 9th IEEE Application Power Electronics conference and Exposition*, vol.3, pp. 1604-1609.

22

A Long-Stroke Planar Actuator with Multiple Degrees of Freedom by Minimum Number of Polyphase Currents

Yasuhito Ueda and Hiroyuki Ohsaki
The University of Tokyo
Japan

1. Introduction

There are a large number of drive systems employing numerous actuators in industry. As such, the performances of these actuators require constant improvement in terms of higher speed and precision, miniaturization, and lower energy consumption. In addition, most of these drive systems need a design that permits MDOF (Multi-Degree-Of-Freedom) motion. Motion controls allowing MDOF have been practically realized by using stacked multiple actuators (Ebihara et al., 2005). However, there are problems in attempting to improve the performance of these types of drive systems such as a larger and more complicated structure, fluctuation of the center of gravity, and Abbe errors in position measurement due to the multiple-moving parts. In order to eliminate these problems, MDOF actuators—which have only a single moving part, but are capable of being directly driven with MDOF—are emerging technologies for future applications (Ebihara et al., 2005).

Most MDOF actuators can be classified into two prominent types: a planar actuator that can drive in two-degree-of-freedom (2-DOF) translational directions; and, a spherical actuator that can drive in 2-DOF rotational directions. As for forms of the driving forces, electromagnetic, piezoelectric, magnetostrictive, and electrostatic types of MDOF actuators have been proposed (Ueda, 2009). Most models are of the electromagnetic actuator type because of mightiness and good controllability of the driving forces (Ueda, 2009).

This study deals with electromagnetic planar actuators, which have a mover capable of traveling over large translational displacements in a plane. The electromagnetic planar actuators that have been proposed can be classified by their drive principle such as stepping, direct-current, induction, and synchronous types. Synchronous planar actuators are expected to offer extremely good controllability of the motion controls, and furthermore in attempting to position a mover precisely, the mover should have no problematic wire in order to avoid heat generation at the mover and tension due to the wire that often deteriorates the drive characteristics. Therefore, synchronous planar actuators with a permanent-magnet mover have been studied actively. However, the movable area tends to be quite narrow due to the use of conventional magnetic circuits for the MDOF drives, which are spatially separated from one another as shown in Fig. 1 (Kim & Trumper, 1998), unless the planar actuator has a large number of armature coils (Jeon et al., 2007).

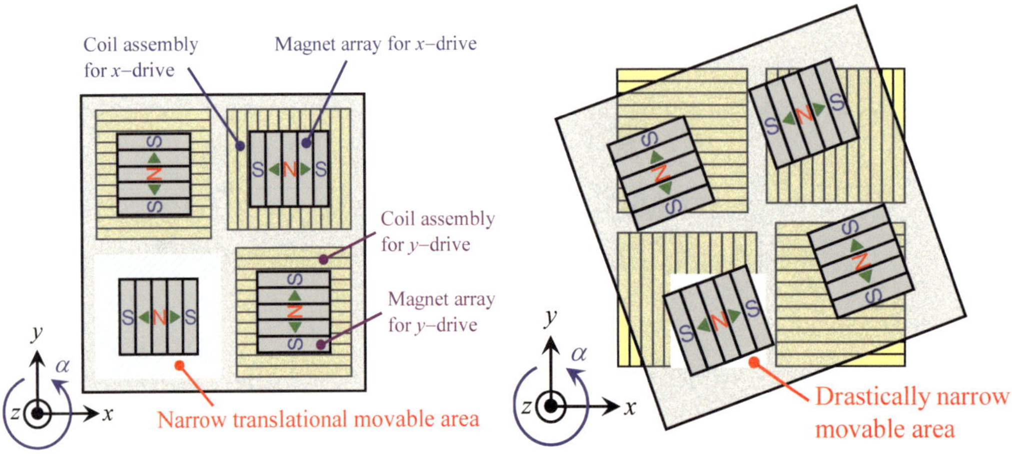

(a) When not displaced in α. (b) When displaced in α.

Fig. 1. Movable area of a prior planar actuator with spatially separated magnetic circuits

With this in mind, this study is aimed at designing high-performance planar actuators that have the following drive performances:

- decoupled control for 3-DOF (Three-Degree-Of-Freedom) motions on a plane.
- wide movable area that can be extended regardless of the number of armature coils.
- ease of mover miniaturization.
- no problematic wiring that can negatively influence drive performance.
- small number of armature currents to control.

Next, we propose a design for a novel synchronous planar actuator having spatially superimposed magnetic circuits for the 3-DOF drives. The magnetic circuits are a combination of a two-dimensional (2-D) Halbach permanent-magnet mover, and mutually overlapped stationary polyphase armature conductors. The movable area can be easily extended by increasing the length of the armature conductors, regardless of their number. However, independently controlling MDOF driving forces by means of superimposed magnetic circuits is very difficult and an extremely important issue in this study. This paper demonstrates a design for a planar actuator that enables MDOF driving forces to be controlled by using spatially superimposed magnetic circuits resulting from minimum number of armature currents.

First, based on the results of a numerical analysis of the driving forces, we design a decoupled control law for the 3-DOF driving forces on a plane by using two polyphase armature currents (Ueda & Ohsaki, 2008a; Ueda & Ohsaki, 2009). We experimentally demonstrate that the 3-DOF motions of the mover can be independently controlled by using two polyphase armature currents, and, therefore the planar actuator has the widest movable area of all planar actuators that have only two polyphase armature conductors.

Second, in order to further improve drive characteristics, the planar actuator is theoretically redesigned so that the mover can be stably levitated and the 3-DOF motions above a plane can be controlled by using three polyphase currents (Ueda & Ohsaki, 2008b).

2. 3-DOF long-stroke planar actuator

The fundamental structure of the 3-DOF planar actuator dealt with in this study is shown in Fig. 2. The planar actuator has a mover consisting of a 2-D Halbach permanent-magnet array and six stationary overlapped armature conductors, which are arranged in two layers of three.

2.1 Fundamental structure

2.1.1 2-D Halbach permanent-magnet mover

A Halbach permanent-magnet array generates a high-intensity magnetic field with a quasi-sinusoidal distribution along the array direction on one side of the magnet array (Halbach, 1986). Therefore, applying Halbach permanent-magnet arrays to coreless electric machines both enhances their driving forces and diminishes their force ripples (Zhu & Howe, 2001). The mover, with which this study deals, consists of five-pole-and-nine-segment Halbach permanent-magnet arrays arranged two-dimensionally in the x_l– and y_l–directions as shown in Fig. 3. Each permanent-magnet component was designed so as to maximize the ratio of the driving force of the planar actuator to the total mass of the mover at a nominal gap between the mover and stator, which is 0.5 mm (Ohsaki et al., 2003). The flux density B_z is distributed quasi-sinusoidally in the x_l– and y_l–directions, and can be approximately expressed near the mover center as follows:

$$B_z(x_l, y_l, z) = B_{zm}(z)\cos\left(\frac{\pi}{\tau_{PM}}x_l\right)\cos\left(\frac{\pi}{\tau_{PM}}y_l\right) \tag{1}$$

where B_{zm} is a maximum flux density on a plane distant from the mover surface in the z–direction, and τ_{PM} is pole-pitch length in the x_l– and y_l–directions. Equation (1) can be rewritten by utilizing the x_m–y_m coordinate, fixed with the mover, as follows:

$$B_z(x_m, y_m, z) = \frac{B_{zm}(z)}{2}\left(\cos\left(\frac{\pi}{\tau}x_m\right) + \cos\left(\frac{\pi}{\tau}y_m\right)\right). \tag{2}$$

Equation (2) indicates that the permanent-magnet mover generates multipole magnetic fields superimposed spatially with the same lengths of pole pitch τ in the x_m– and y_m–directions. Pole pitch τ in the x_m– and y_m–directions can be expressed by the pole pitch τ_{PM} in the x_l– and y_l–directions as follows:

$$\tau = \tau_{PM} / \sqrt{2}. \tag{3}$$

2.1.2 Stationary overlapped armature conductors

There are two pairs of three stationary armature conductors arranged on two layers on a double-layered printed circuit board, for the x– and y–directional drives. All the armature conductors are meander-shaped and are designed so that pitch length of the meander shape is τ, corresponding to the pole-pitch length of the permanent-magnet mover in the x_m– and y_m–directions. The three armature conductors arranged on each layer are mutually distant at $2\tau / 3$ intervals, and supply the three-phase alternating currents for the x– or y–directional drives. Between the two layers, there is a thin insulating layer that creates different gap lengths; the gap length between the mover and the armature conductors for the x–directional drive is different from that between the mover and the armature conductors for the y–directional drive. So, it is extremely important to make the insulating layer as thin as possible.

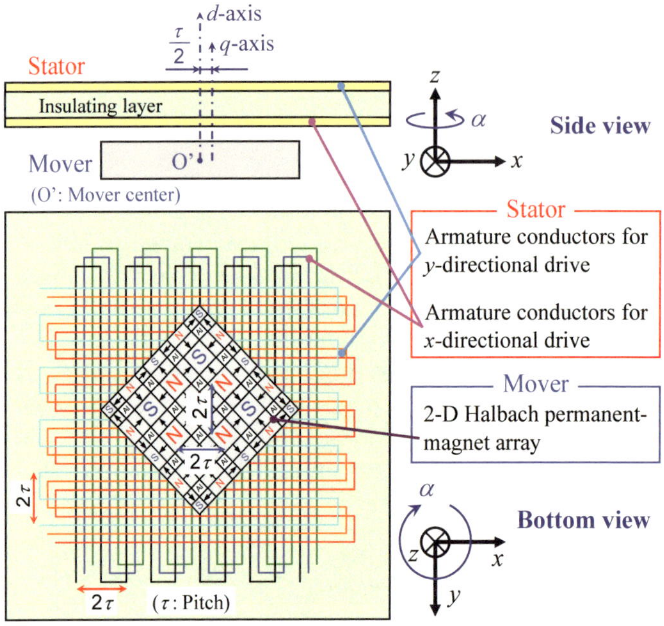

Fig. 2. Fundamental structure of planar actuator dealt with in this study

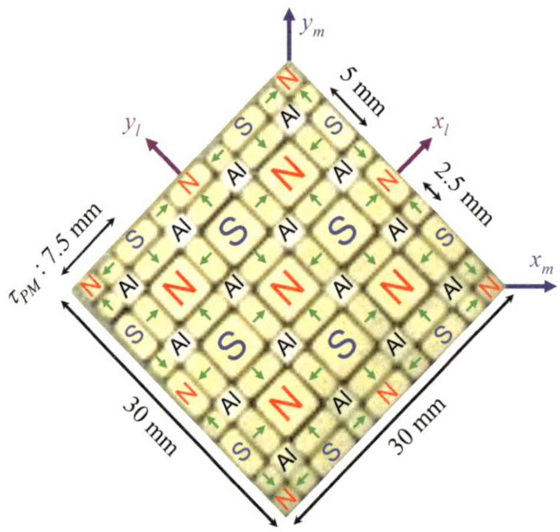

Fig. 3. Dimension and magnetization of permanent-magnet mover

All the armature conductors are arranged over the stator, and so are always subjected to the magnetic field for the x– and y–directional drives generated by the permanent-magnet mover regardless of the mover position on the stator. Therefore, the mover can travel over the wide stator area. Furthermore, lengthening all the armature conductors extends the movable area without increasing the number of armature conductors, thus the power-supply system does not become complicated.

2.1.3 Drive principle

When the y_m– and x_m–axes, fixed with the mover, are parallel to the armature conductors for the x– and y–directional drives as shown in Fig. 2, the mover generates a quasi-sinusoidal flux-density distribution with pole pitch τ, corresponding to the meander-shaped pitch in length, in the x– and y–directions as shown in Eq. (2). Therefore, supplying two sources of three-phase alternating current to the two pairs of three armature conductors forms two magnetic circuits, as in linear synchronous motors, in the x– and y–directions, and consequently generates driving forces in the x– and y–directions. Although not all the magnetic circuits formed over the stator are mutually-separated, exciting the armature conductors for the x– and y–directional drives independently generates driving forces in the x– and y–directions, respectively, because of their mutually-orthogonal directions.

Displacing the α–position (yaw position) and expressing the rotational position around the z–axis, not only influences the translational forces, it but also generates torque around the z–axis because of the broken symmetry of the spatial distribution of the translational forces acting on the mover. So it is extremely important to simultaneously control not only the x– and y–motions, but also the α–motions. Therefore, the characteristics of the translational forces and torque for the α–positions need to be investigated in detail.

2.2 Static force characteristics

2.2.1 Analytical model

The driving forces acting on the mover of the planar actuator can be calculated from armature current $\boldsymbol{i}$ and flux density $\boldsymbol{B}$ using the Lorentz force equation $\boldsymbol{F}_L = \boldsymbol{i} \times \boldsymbol{B}$. Figure 4 shows the configuration of the mover and stator of the analysis model, and shows that the mover is displaced in the α–direction and an armature current $\boldsymbol{i}_{jk}$ is supplied to an armature conductor l_{jk}, where j (= x or y) and k (= u, v, or w) express the driving direction and phase name of the three-phase currents, respectively. When a line element dl_{jk}, which is a small part of the armature conductor l_{jk} is in the flux density $\boldsymbol{B}$, translational force $\boldsymbol{F}$ and torque $\boldsymbol{T}$ acting on the mover can be expressed as follows:

$$\boldsymbol{F} = -\sum_{j,k} \int_{l_{jk}} \left(\boldsymbol{i}_{jk} \times \boldsymbol{B} \right) dl_{jk} \tag{4}$$

$$\boldsymbol{T} = -\sum_{j,k} \int_{l_{jk}} \left(\boldsymbol{r}_{jk} - \boldsymbol{r}_m \right) \times \left(\boldsymbol{i}_{jk} \times \boldsymbol{B} \right) dl_{jk} \tag{5}$$

where $\boldsymbol{r}_{jk}$ and $\boldsymbol{r}_m$ are position vectors of the line element dl_{jk} and the mover center O′ with respect to the stationary coordinate $x_s y_s z_s$. From these equations, if the armature currents $\boldsymbol{i}_{jk}$ and flux density $\boldsymbol{B}$ are constant, the translational forces $\boldsymbol{F}$ and torques $\boldsymbol{T}$ are proportional to the square and cube of a side of the mover, respectively, because the lengths of the integration passes in Eqs. (4) and (5) are proportional to the square of a side of the mover. On the other hand, if mass density ρ of the mover is constant, the mass and inertia-tensor elements of the mover are proportional to the third and fifth powers of a side of the mover, respectively. With this in mind, we can see that acceleration in the translational and rotational directions becomes twice and four times, respectively, when all sides of the mover become half.

In this analytical model, two pairs of three-phase currents (i_{ju}, i_{jv}, and i_{jw}) are given as follows:

$$i_{ju} = I_j \sin\left(\theta_{sj}\right) \tag{6}$$

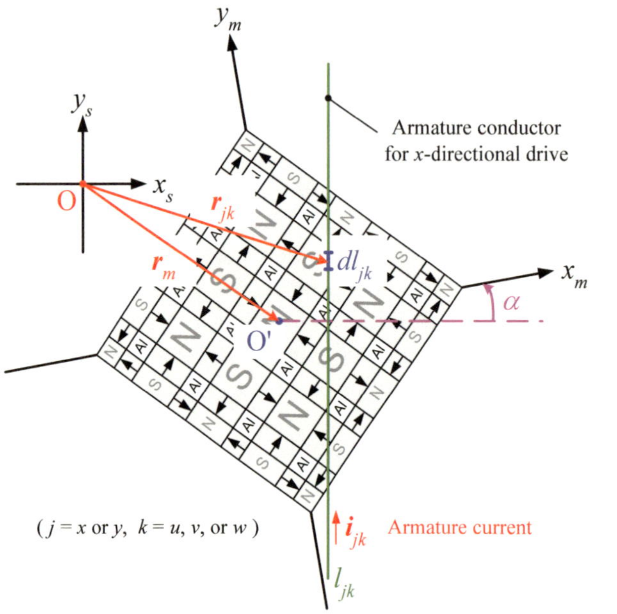

Fig. 4. Configuration of mover and stator in the analysis model

$$i_{jv} = I_j \sin\left(\theta_{sj} + \frac{2\pi}{3}\right) \tag{7}$$

$$i_{jw} = I_j \sin\left(\theta_{sj} + \frac{4\pi}{3}\right) \tag{8}$$

where I_j and θ_{sj} are the amplitude and the phase of the three-phase currents, respectively. A phase difference between magnetic fields generated by the armature conductors and the mover can be defined and expressed with the following equation:

$$\theta_{dj} = \theta_{sj} - \frac{\pi}{\tau} j_s \tag{9}$$

where x_s and y_s are the x– and y–positions of the mover center O′ with respect to the stationary coordinate $x_s y_s z_s$. Controlling the amplitude I_j and phase θ_{sj} of the armature conductors controls the 2-D mover motions.

2.2.2 Numerical analysis results

Exciting the armature conductors for the x– and y–directional drives generates the x– and y–directional driving forces using the same principle, and so this analysis deals with the armature conductors for only the x–directional drive.

When the x_m– and y_m–axes are parallel to the stationary x_s– and y_s–axes, respectively, the yaw angle α is defined to be 0 deg. Figure 5 shows a numerical analysis result of the translational force F_x and torque T_z for the phase difference θ_{dx} with the following analysis conditions:

- amplitude of the armature currents: I_x = 2 A, I_y = 0 A
- phase of the armature currents: θ_{sx} = –180 ~ 180 deg
- mover position: x_s = 0 mm
- yaw angle: α = 10 deg
- flux density due to the magnet mover: measurement result for air gap = 0.5 mm.

Figure 5 indicates that the translational force F_x and torque T_z can be expressed as sinusoidal functions with respect to the phase difference θ_{dx}. Equations (4) and (5) also indicate the translational force F_x and torque T_z are proportional to the amplitude of the armature currents, and therefore can be expressed as follows:

$$F_x = K_{Fx}(\alpha) I_x \sin\theta_{dx} \tag{10}$$

$$T_z = K_{Tx}(\alpha) I_x \cos\theta_{dx}. \tag{11}$$

These equations indicate that phases of the armature currents generating the translational force F_x and torque T_z differ by 90 deg. Furthermore, the translational forces F_x are proportional to the armature currents generating a 90-deg phase-lead magnetic field with respect to the magnetic field due to the magnet mover (θ_{dx} = 90 deg), and the torques T_z are proportional to the armature currents generating the same-phase magnetic field (θ_{dx} = 0 deg).

The system constants K_{Fx} and K_{Tx} depend on the yaw angle α. The system constants K_F and K_T can be calculated by fitting the analysis results of the driving forces to Eqs. (10) and (11). Figure 6 shows the calculation results of the system constants K_{Fx} and K_{Tx}. The same driving forces can be generated every 90 deg in the α–direction because of the symmetric structure of the permanent-magnet mover. Figure 6 shows a period of the system constants K_{Fx} and K_{Tx} in the α–direction. Figure 6 indicates that K_{Fx} is maximum and K_{Tx} = 0 N·m/A when the yaw angle α = 0 deg, and so the translational force F_x is maximum and the torque T_z is not generated. When the mover is displaced in the α–direction, K_{Fx} becomes smaller and $K_{Tx} \neq 0$ N·m/A. The driving forces can be generated in the range within the yaw angle α = ±30 deg because of $K_{Fx} \neq 0$ N/A and $K_{Tx} \neq 0$ N·m/A. The movable area in the α–direction is widest of all planar actuator having only two pairs of polyphase conductors. Therefore, this planar actuator is suitable for 3-DOF long-stroke planar manipulation using only a few conductors.

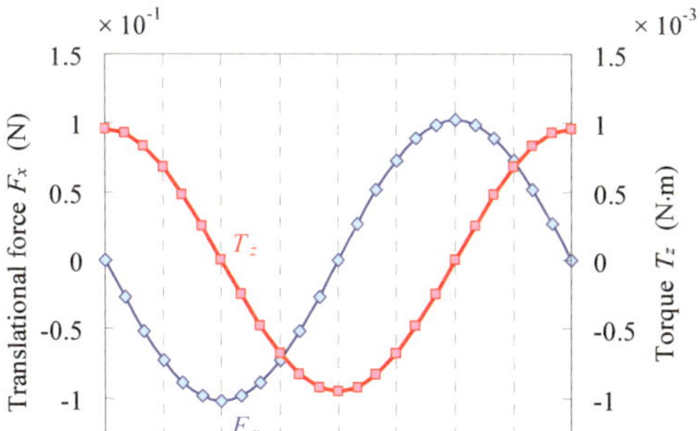

Fig. 5. Translational force F_x and torque T_z vs. phase difference θ_{dx}

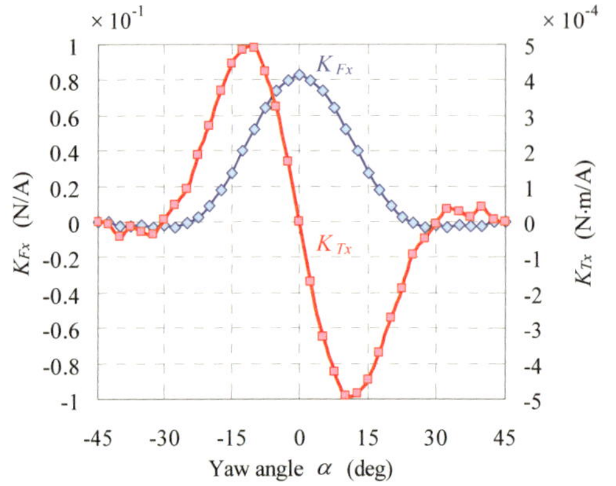

Fig. 6. System constants K_{Fx} and K_{Tx} vs. yaw angle α

2.3 Decoupled control for 3-DOF motions

2.3.1 DQ decomposition

DQ decomposition in conventional rotary machines separates armature-current components that are generating torques from those that are not (Fitzgerald et al., 1990). From preciously shown analysis results, DQ decomposition in this planar actuator separates the armature-current components, generating the translational forces F_x, F_y or torques T_z.

The direct axis (d–axis) and quadrature axis (q–axis) are attached to the mover, and move together with the mover. In a phasor diagram, conventionally, the d–axis is aligned with the magnetic field axis because of the permanent-magnet mover, and the q–axis leads the d–axis by 90 deg. In other words, the d–axis current is intended to generate the same-phase magnetic field as that resulting from the permanent-magnet mover, and the q–axis current is intended to generate a 90-deg phase-lead magnetic field. Figure 7 shows phasor diagrams for the relation between the dq–frame and uvw–frame. The u–, v–, and w–phase currents generate magnetic fields with definite phases, which are out of phase from one another by 120 deg. The α'–axis current generates the same-phase magnetic field as that from the

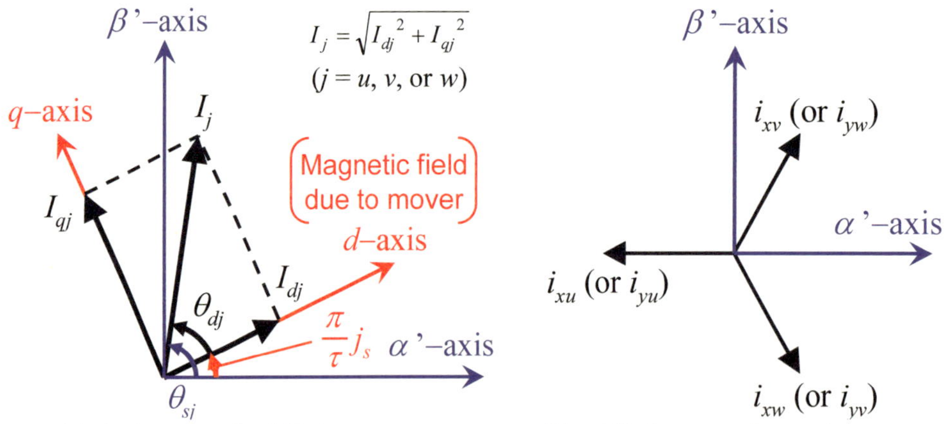

(a) dq–frame and $\alpha'\beta'$–frame. (b) $\alpha'\beta'$–frame and uvw–frame.

Fig. 7. Phasor diagrams showing relation between dq–frame and uvw–frame permanent-magnet mover when the mover center is at the origin of the stationary coordinate $x_s y_s z_s$. The β'–axis leads the α'–axis by 90 deg. Figure 7 indicates that the d– and q–axes lead the α'– and β'–axes by $(\pi j_s/\tau)$, respectively, which is proportional to the mover positions j_s (= x_s or y_s). The armature currents having the amplitude I_j and phase θ_{sj} can be decomposed to the d–axis currents I_{dj} and q–axis currents I_{qj} as follows:

$$I_{dj} = I_j \sin\theta_{dj} \tag{12}$$

$$I_{qj} = I_j \cos\theta_{dj}\,. \tag{13}$$

From Eqs. (10)–(13), the translational forces F_x, F_y and torques T_z resulting from supplying the two pairs of three-phase currents to the armature conductors for the x– and y–directional drives, can be expressed with the d– and q–axis currents I_{dj}, I_{qj} as follows:

$$F_x = K_{Fx}(\alpha) I_{qx} \tag{14}$$

$$F_y = K_{Fy}(\alpha) I_{qy} \tag{15}$$

$$T_z = K_{Tx}(\alpha) I_{dx} + K_{Ty}(\alpha) I_{dy}\,. \tag{16}$$

Equations (14)–(16) indicate that the translational forces F_x, F_y and torques T_z are proportional to the q– and d–axis currents I_{qj}, I_{dj}, respectively. The system constants for the y–directional drive K_{Fy}, K_{Ty} are slightly different from those for the x–directional drive K_{Fx}, K_{Tx} because of difference in the air gaps between the mover and armature conductors for the x– or y–directional drives. As mentioned above, the driving forces from the armature currents can be simply described.

2.3.2 3-DOF force control

Supplying the appropriate d– and q–axis currents independently controls the translational forces F_x, F_y and torques T_z from Eqs. (14)–(16). The degrees of freedom for the armature-current controls are four (I_{dx}, I_{dy}, I_{qx}, and I_{qy}), and one larger than that for the mover motions, which is three (x, y, and α) as shown in Fig. 8. Controlling the two q–axis currents I_{qx}, I_{qy} is essential to controlling the two translational forces F_x, F_y. The two d–axis currents I_{dx}, I_{dy} generate the torques T_z, and, so the torque controls have redundancy. In fact, the d–axis current I_{dx} generates the torques T_z more efficiently than the d–axis current I_{dy} because of the air-gap difference ($K_{Tx} > K_{Ty}$). However, each armature current is limited by its own rating, power supply, and so on. In Fig. 8, I_c is the current limit of all the armature currents. So if large driving forces F_x, F_y, T_z are required, then armature currents should be supplied so as to optimally satisfy the references of the driving forces under the current limit for producing maximum performance. This study, however, focuses principally on the verification of decoupled 3-DOF motion controls for a long-stroke planar actuator, and, therefore deals with a simply decoupled control algorithm.

Figure 9 shows a control block diagram for the translational forces F_x, F_y and torques T_z in this study. In Fig. 9, x_{ref}, y_{ref}, and α_{ref} are references of the mover positions in the x–, y–, and α–directions, respectively. In order to decide the references for the driving forces F_x^*, F_y^*, T_z^*, feedback controls in each degree of freedom are performed with three different PID (Proportional-Integral-Derivative) algorithms. The PID parameters are determined so that settling time of the x–, y–, and α–directional drives is shorter than 0.5 s. In this study, references of the d– and q–axis currents I_{dj}^*, I_{qj}^* are calculated from the system constants K_{Fj}, K_{Tj} and the driving force references F_x^*, F_y^*, T_z^* as follows ($j = x$ or y);

$$I_{dj}^* = T_z^* / 2K_{Tj}(\alpha) \tag{17}$$

$$I_{qj}^* = F_j^* / K_{Fj}(\alpha). \tag{18}$$

The system constants K_{Fj}, K_{Tj} are calculated from the detected yaw angle α by interpolation of the analysis data shown in Fig. 6. As we can see from Fig. 7 (a), the references of the amplitude and phase of the three-phase currents I_j^* and θ_{sj}^* can be calculated from the current references in the dq–frame I_{dj}^*, I_{qj}^* and the mover positions in the x– and y–directions. Then, from Fig. 7, references of the three-phase currents i_{ju}^*, i_{jv}^*, i_{jw}^*, ($j = x$ or y) can be calculated from the amplitude I_j^* and phase θ_{sj}^*.

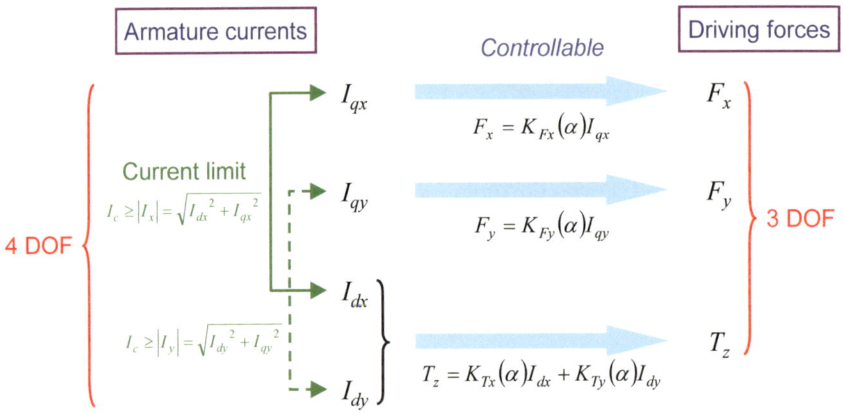

Fig. 8. Degrees of freedom for armature-current control and mover motion

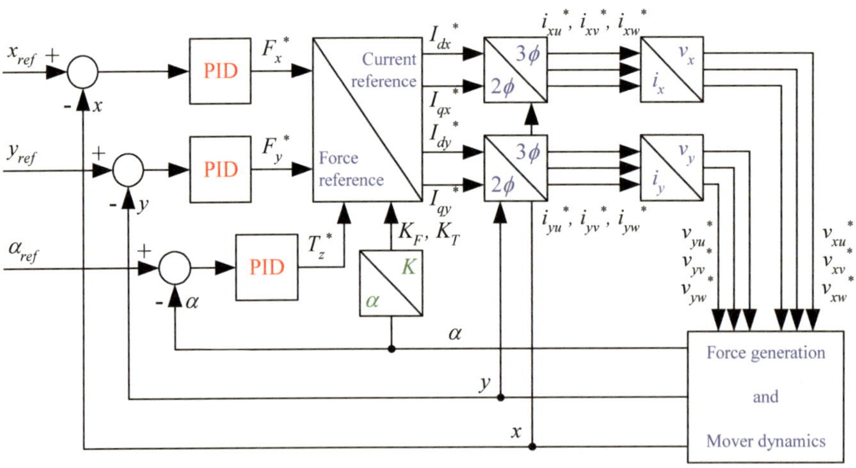

Fig. 9. Control block diagram for translational forces F_x, F_y and torques T_z

In this study, each-phase armature voltage v_{jk}^* is calculated from the resistances R and armature-current references i_{jk}^* as follows;

$$v_{jk}^* = Ri_{jk}^* . \tag{19}$$

Equation (19) does not consider back electromotive force, and causes errors between the armature currents and their references. In this study, the errors are compensated by the PID controls. As described above, the six armature-voltage references v_{jk}^* can be calculated from the driving force references F_x^*, F_y^*, T_z^*. Supplying the six armature voltages v_{jk}^* generates decoupled 3-DOF driving forces.

2.3.3 Experimental setup

In order to control the 3-DOF (x, y, and α) motions of the mover by position feedback, sensing the mover positions which are allowed anywhere on the wide stator area is extremely important. If multiple sensors are utilized for single-axis displacements, we require three or more sensors to detect the 3-DOF positions. The combination method for multiple sensors determines the measurement performance, for instance, detection accuracy and measurable area, and so is also extremely important. This study uses three single-axis laser-displacement sensors (laser triangulation) because of their long measurable area (several tens of mm) and high resolution (several μm).

When the mover is driven in 3-DOF directions, suspending and smoothly guiding the mover on a plane are also extremely important. This study uses ball bearings, 1-mm glass spheres, as the suspension and guide mechanism because they have smaller friction forces and are easily installed.

Figure 10 shows the configuration of the experimental system for the 3-DOF motion controls of the mover. The mover is supported by many ball bearings and guided on a plane having small frictional forces. At the same time, the mover faces the three-phase armature conductors for the x– and y–directional drives through 0.5-mm and 0.63-mm air gaps, respectively. The three laser-displacement sensors irradiate three points on three lateral sides of the mover, and detect displacements at the three points by triangulation. A personal computer (PC), running a Digital Signal Processor (DSP), inputs the three sensor signals and calculates the 3-DOF positions of the mover. The PC also calculates the six armature-voltage references based on a decoupled motion-control algorithm, and outputs the six voltages to the six armature conductors through the power amplifiers. Then, the armature currents and magnetic field, resulting from the magnet mover, interact and independently controlled driving forces are generated.

2.3.4 Experimental results

This subsection presents verification of the 3-DOF drive characteristics of the planar actuator ascertained by experimental study, and describes the experimental results under various conditions. First, in order to verify the decoupled 3-DOF motion controls of the planar actuator with superimposed magnetic circuits, drive tests of a mover with 3 DOF were performed. Second, the movable area in the α–direction was experimentally verified.

In this planar actuator, even if a large α–displacement of the mover occurs, the driving forces decrease less than conventional planar actuators because the magnetic circuits for the x–, y–, and α–directional drives are always formed. The planar actuator has a mover capable of infinitely-large translational motions on a plane by increasing the length of the armature conductors, in principle, as mentioned in Subsection 2.1.2.

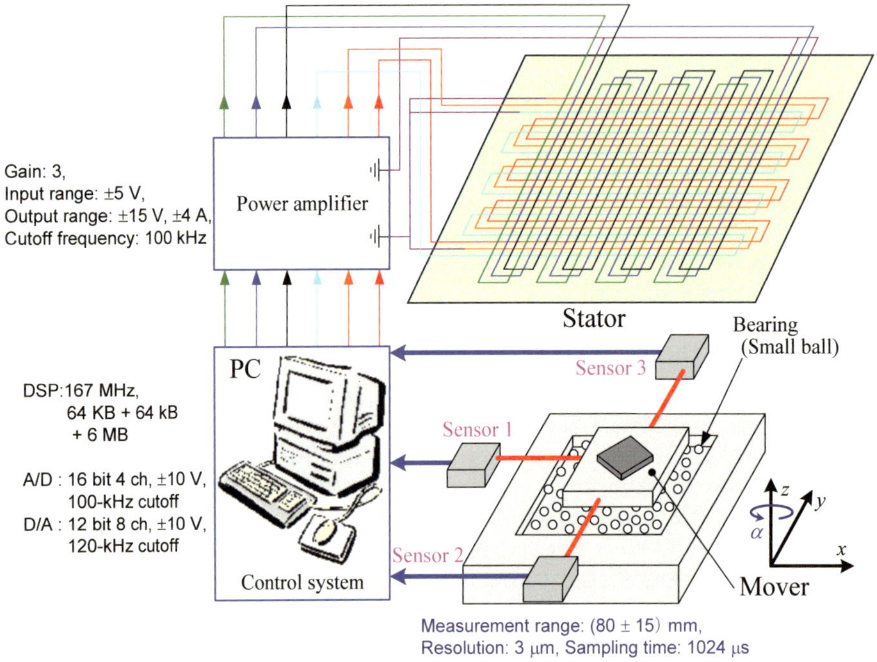

Fig. 10. Configuration of experimental system for 2-D drives

Next, in order to verify the long-stroke motion-control characteristics with 3 DOF, experimental results of the drive tests were described under the following four conditions:

(I) Simultaneous sine response for the x–, y– and α–directions (Ueda & Ohsaki, 2008a); Verification of decoupled motion controls in the x–, y– and α–directions, and an evaluation of follow-up controls are performed.

Figure 11 shows the experimental results of the mover motions when sine signals, which have a 10-mm amplitude and a 2-s period, are simultaneously given as the x–, y–, and α–position references. Figure 11 indicates that the mover can simultaneously track the three sine-reference positions over a wide movable area with less position error.

(II) α–directional drive at the yaw angle $\alpha \approx 0$ deg (Ueda & Ohsaki, 2009):

Whereas the translational forces in the x– and y–directions are maximum at the yaw angle $\alpha = 0$ deg, torque in the α–direction cannot be generated, as mentioned in Subsection 2.2.2. Therefore, controllability of the α–motion is presumed to deteriorate at the yaw angle $\alpha \approx 0$ deg. Then, in order to verify the motion-control characteristics at the yaw angle $\alpha \approx 0$ deg, ramp response for the α–direction was investigated in the same range.

Figure 12 shows the experimental results of the mover motions and armature currents for ramp response with the position references $(x_{ref}, y_{ref}) = (0, 0)$ and $\alpha_{ref} = -5 + t$ (deg), where t is time. Figure 12 indicates that the mover can travel in the α–direction in the range within the yaw angle $\alpha = \pm 5$ deg. When the yaw angle α closes to 0 deg, the large d–axis currents I_{dx} and I_{dy} are required to control the α–position. In the range within the yaw angle $\alpha = \pm 1$ deg, the α–position becomes more oscillatory. The current capacity of the power amplifiers is presumed to be the reason the d–axis currents are insufficient to control the α–motion, and consequently cause the oscillation in the α–position.

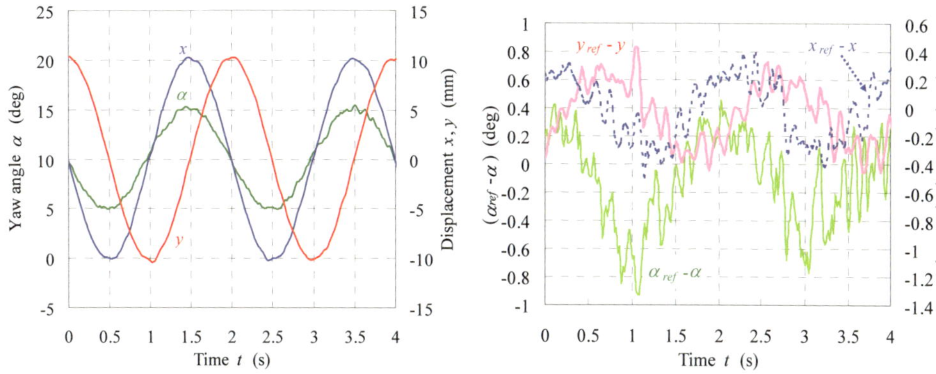

(a) Mover position. (b) Position error.

Fig. 11. Experimental results of simultaneous sine response for x–, y– and α–directions

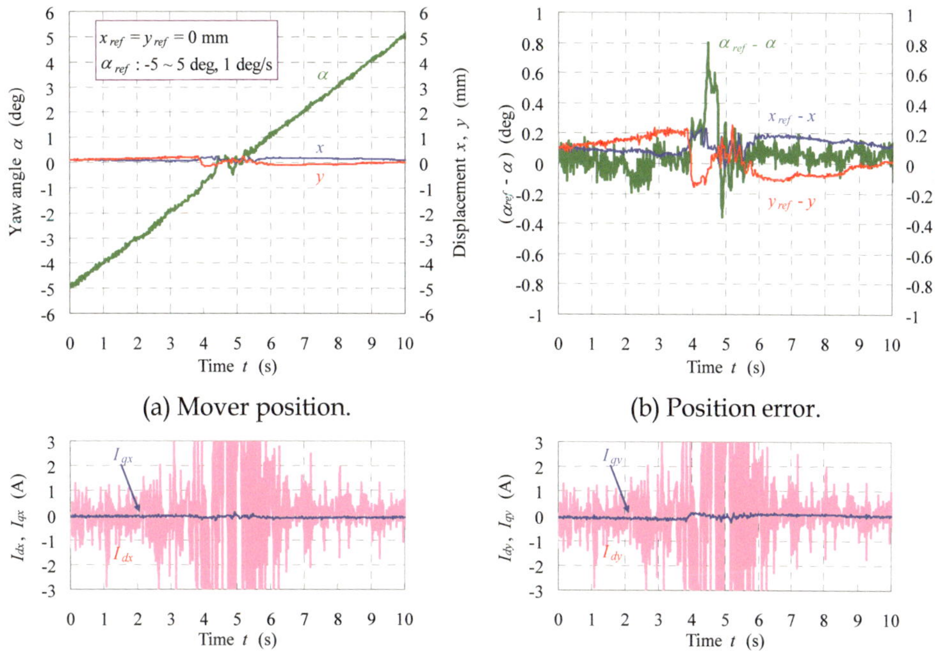

(a) Mover position. (b) Position error.

(c) Armature currents for the x–drive. (d) Armature currents for the y–drive.

Fig. 12. Experimental results of ramp response for α–direction at yaw angle $\alpha \approx 0$ deg

(III) α–directional drive at the yaw angle $\alpha > 20$ deg (Ueda & Ohsaki, 2009):
Increasing the yaw angle α can decrease the driving forces especially at the yaw angle $\alpha > 20$ deg. Then, in order to verify the movable area and motion-control characteristics at the yaw angle $\alpha > 20$ deg, ramp response for the α–direction was investigated in the same range.
Figure 13 shows the experimental results of the mover motions and armature currents for ramp response with the position references (x_{ref}, y_{ref}) = (0, 0) and $\alpha_{ref} = 18 + t$ (deg). Figure 13 indicates that the mover can travel in the α–direction at the yaw angle $\alpha < 26$ deg. When the yaw angle α becomes larger, the large q–axis currents I_{qx} and I_{qy} are required to control the

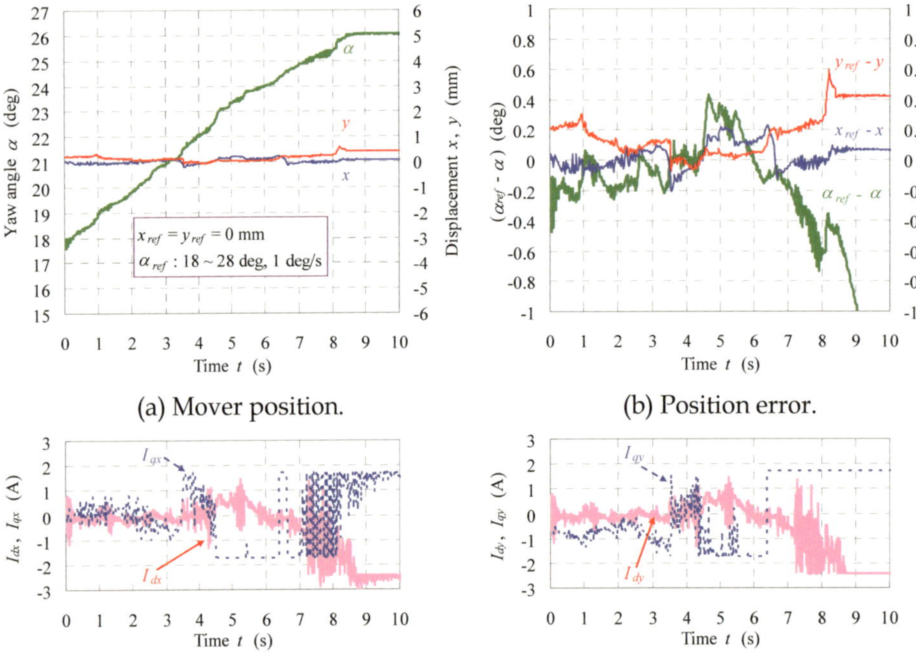

(a) Mover position. (b) Position error.

(c) Armature currents for the x–drive. (d) Armature currents for the y–drive.

Fig. 13. Experimental results of ramp response for α–direction at yaw angle $\alpha > 20$ deg

x– and y–positions. At the yaw angle $\alpha > 22$ deg (time $t > 4$ s), the q–axis currents I_{qx} and I_{qy} are limited by the maximum current = 1.7 A. At the yaw angle $\alpha > 26$ deg (time $t > 8$ s), the mover is stationary, and, therefore the driving forces required to overcome the friction forces between the mover and ball bearings cannot be generated.

(IV) 90-deg step response for the α–direction (Ueda & Ohsaki, 2009):

The driving forces of the planar actuator have a 90-deg periodicity for the α–direction because of the symmetric magnetized mover. Furthermore, the driving forces are generated in the range within the yaw angle $\alpha = \pm 26$ deg in the 90-deg periodicity. Then, in order to verify feasibility of periodic 90-deg stepping drives in the α–direction, a 90-deg step response was investigated.

Figure 14 shows experimental results of the mover motions and armature currents for 90-deg step response with the initial positions (x_{in}, y_{in}, α_{in}) = (0, 0, 10) and the position references (x_{ref}, y_{ref}, α_{ref}) = (0, 0, 100, 190, 280, or 370). Figure 14 indicates that a 90-deg stepping drive can be realized. The mover positions could not be detected in the range within 0.10 s < time t < 0.28 s due to geometrical problem in laser triangulation. At the yaw angle $\alpha > (14 + 90)$ deg, corresponding to time $t > 0.28$ s, the mover motions can be controlled because of the 90-deg periodicity in the yaw direction.

2.4 Summary of chapter 2

This chapter proposed a novel synchronous planar actuator having advantages in terms of the wide movable area of the magnet mover, which is independent of the number of armature conductors, presents the design for the experimental system for verification of the

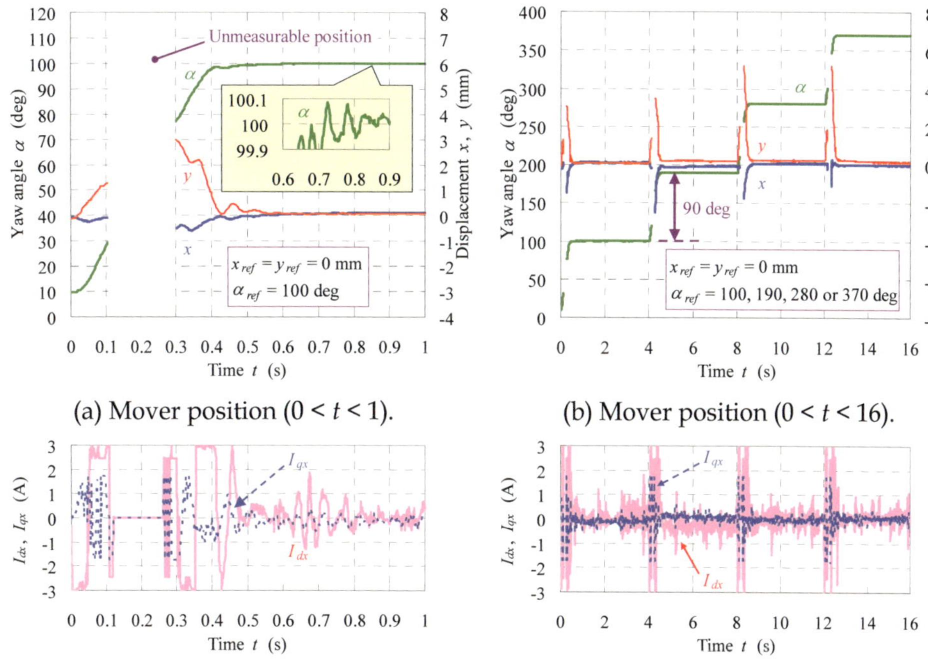

(a) Mover position ($0 < t < 1$).

(b) Mover position ($0 < t < 16$).

(c) Armature currents ($0 < t < 1$).

(d) Armature currents ($0 < t < 16$).

Fig. 14. Experimental results of 90-deg stepping response for α–direction

drive characteristics of this long-stroke planar actuator, and describes the experimental results of long-stroke 3-DOF motion controls. From these results, we successfully demonstrated that the 3-DOF motions of the mover can be independently controlled by two polyphase armature currents. The movable area in the x– and y–directions is infinitely wide in principle, and that in the α–direction is in the range within ±26 deg, meaning the planar actuator has the widest movable area of all planar actuators that have only two polyphase armature conductors.

3. Magnetically levitated MDOF planar actuator

This chapter presents a feasibility verification as to whether a planar actuator can magnetically suspend a mover, capable of 3-DOF motions on a plane, so as to further improve the drive performance of a planar actuator.

3.1 Actuator design

The proposed planar actuator has spatially superimposed magnetic circuits for the x–, y– and α–directions, which are its most important feature and enable the mover to travel over a wide movable area on a plane by exciting only two polyphase armature conductors. The magnetically levitated planar actuator is also designed so that all the magnetic circuits are mutually superimposed. In order to design the planar actuator, a numerical analysis of 6-DOF driving forces for 6-DOF mover positions is performed.

3.1.1 6-DOF force analysis

The driving forces, including the suspension forces, greatly depend on the size of the gap between the mover and armature conductors, and therefore this gap needs to be precisely controlled. Generally, reducing this gap increases the driving forces. If the mover is located below the stator, attraction forces to the stator are required to suspend the mover. However, the attraction forces are increased by reducing the gap, which makes the vertical motions of the mover unstable. Conversely, if the mover is located above the stator, repulsion forces from the stator are required to suspend the mover. The repulsion forces are increased by reducing the gap, and so the vertical motions are stable. Therefore, in this study, the mover of the magnetically levitated planar actuator is positioned on the stator.

Figure 15 shows the analytical model for the driving forces. In this figure, the mover and polyphase armature conductors for the x–direction only are shown. A moving 2-D Halbach permanent-magnet array has the same structure as shown in Fig. 3, and four-pole-and-seven-segment magnetization with pole-pitch length τ_{PM} = 3 mm along the x_l– and y_l–directions. Its dimensions are 11 mm × 11 mm × 2 mm, which are almost two-fifths the size of the magnet-array dimension shown in Fig. 3. The ultimate miniaturization of the permanent-magnet mover enables higher accelerations to be generated using the same armature currents and flux density as given in Subsection 2.2.1.

In the mover motions, there are 3-DOF rotations. However, this analysis deals with the rotations around only one axis (x_m, y_m, or z_m). The rotational angles around the x_m–, y_m–, and z_m–axes are referred to as roll angle γ, pitch angle β, and yaw angle α, respectively. The driving forces acting on the mover can be calculated from the Lorentz force law with the same equations as Eqs. (4) and (5).

Figure 16 shows the analysis results of the driving forces F_x, F_z, T_x, T_y, T_z for the yaw angle α when the d– and q–axis currents for the x–directional drive are supplied (I_{dx} = 1 A, or I_{qx} = 1 A), the air gap between the mover bottom and armature conductors is 0.5 mm, and

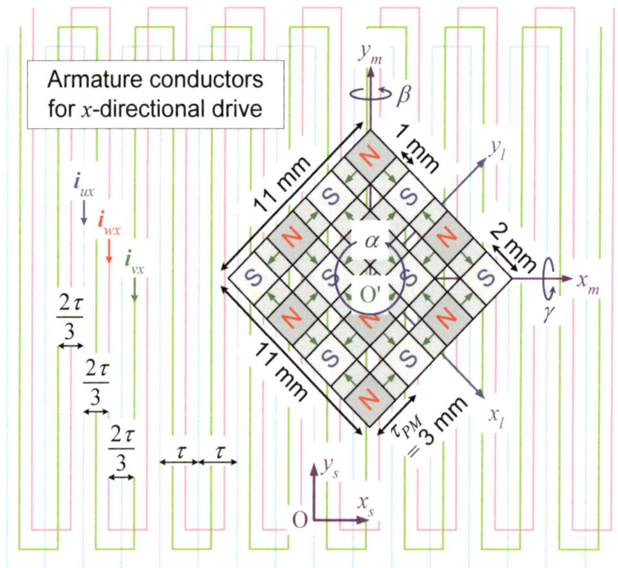

Fig. 15. Analytical model for 6-DOF driving forces

the pitch and roll positions are not displaced ($\beta = \gamma = 0$ deg). Figure 16 indicates that the d–axis current generates the translational forces F_z and torques T_z, and the q–axis current generates the translational forces F_x and torques T_x, T_y. The translational forces F_x, F_z and torques T_y are almost constant, and the torques T_x and T_z are proportional to the yaw angle α when the yaw angle $\alpha \approx 0$ deg. Because of the symmetric magnetization of the mover, the same driving forces can be generated every 180 deg.

From the analysis results shown above and others for the pitch angle β and the roll angle γ, the driving forces F_x, F_y, F_z, T_x, T_y, T_z can be expressed from the d– and q–axis currents I_{dx}, I_{qx}, I_{dy}, I_{qy} as follows:

$$\begin{bmatrix} F_x \\ F_y \\ F_z \\ T_x \\ T_y \\ T_z \end{bmatrix} = \begin{bmatrix} \\ \\ \boldsymbol{K}_{FT}(\alpha, \beta, \gamma) \\ 6 \times 4 \text{ matrix} \\ \\ \\ \end{bmatrix} \begin{bmatrix} I_{dx} \\ I_{qx} \\ I_{dy} \\ I_{qy} \end{bmatrix} \approx \begin{bmatrix} 0 & K_F & 0 & 0 \\ 0 & 0 & 0 & K_F \\ -K_F & 0 & -K_F & 0 \\ K_{TP}\gamma & K_{TP}\alpha & K_{TP}\gamma & K_{TC} \\ K_{TP}\beta & -K_{TC} & K_{TP}\beta & K_{TP}\alpha \\ -K_{TP}\alpha & K_{TP}\gamma & -K_{TP}\alpha & K_{TP}\beta \end{bmatrix} \begin{bmatrix} I_{dx} \\ I_{qx} \\ I_{dy} \\ I_{qy} \end{bmatrix} \tag{20}$$

where $\boldsymbol{K}_{FT}$ is a 6×4 matrix and all elements of the matrix nonlinearly depend on the yaw angle α, pitch angle β, and roll angle γ. In this study, the pitch and roll displacements of the mover are assumed to be very small ($\beta \approx 0$ deg and $\gamma \approx 0$ deg) because of small air gap (less than 1 mm) between the mover and stator, and in the range, all elements of $\boldsymbol{K}_{FT}$ almost linearly depend on the pitch and roll displacements. Furthermore, if the yaw displacements are assumed also to be very small ($\alpha \approx 0$ deg), all elements of $\boldsymbol{K}_{FT}$ almost linearly depend on the yaw displacements, and the system-constant matrix $\boldsymbol{K}_{FT}$ is expressed approximately as shown in Eq. (20).

In Eq. (20), K_{FC}, K_{TC}, and K_{TP} are constant (in this analysis, for a 0.5-mm air gap, $K_{FC} \approx 17$ mN, $K_{TC} \approx 12$ mN·mm, and $K_{TP} \approx 4.5$ mN·mm). Equation (20) also indicates that the driving forces due to the d–axis currents I_{dx} and I_{dy} are equal because of the symmetry of the actuator.

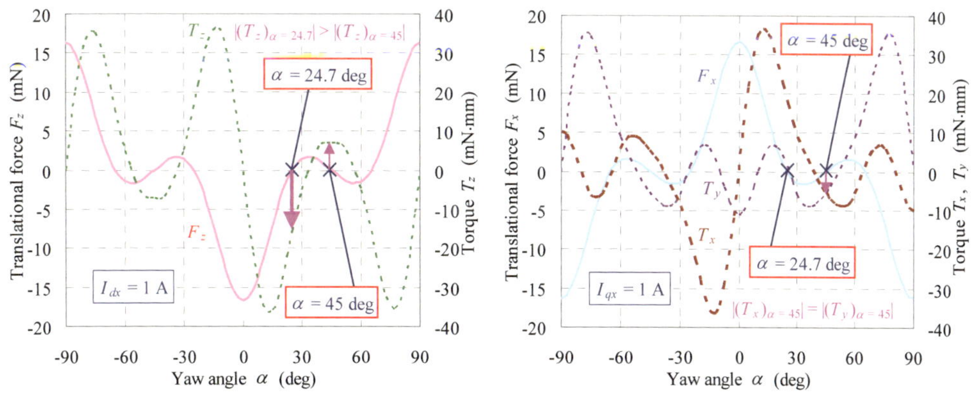

(a) Driving forces from I_{dx} = 1 A. (b) Driving forces from I_{qx} = 1 A.

Fig. 16. Driving forces for yaw angle α at pitch and roll angles $\beta = \gamma = 0$ deg

Therefore, even if the two currents I_{dx} and I_{dy} are controlled, only 1-DOF driving forces can be controlled in the range within $\alpha \approx 0$ deg, $\beta \approx 0$ deg, and $\gamma \approx 0$ deg. Therefore, controlling the four armature currents in the dq–frame controls the 3-DOF motions of the mover (for instance, x–, y– and z–motions, or x–, y–, and α–motions). In order to realize both 3-DOF motion controls on a plane and magnetic suspension, the planar actuator needs to be redesigned.

3.1.2 Conceptual design

In order to suspend the mover, suspension forces that balance the force of gravity need to be generated. Equation (20) indicates that negative d–axis currents (I_{dx}, I_{dy} < 0) generate suspension forces ($F_z > 0$). Figure 17 shows schematic views of when the d–axis currents are supplied. Negative d–axis currents to actively control levitation forces (F_z > 0) always generate restoring torques against the β– and γ–displacements. The restoring torques stabilize the β– and γ–motions of the mover.

Equation (20) also shows that the q–axis currents I_{qx}, I_{qy} generate the translational forces F_x, F_y on a plane without vertical forces F_z. Therefore, the d– and q– axis currents I_{dx}, I_{qx}, I_{dy}, I_{qy} :

- independently control the translational forces F_x, F_y, F_z
- stabilize the pitch and roll motions.

However, the d–axis currents utilized to control the suspension forces F_z, generate yaw-directional torques proportional to the yaw angle α, that is, they generate instable yaw motions. Therefore, in order to realize both 3-DOF motion controls on a plane and magnetic suspension, a stabilization mechanism for the yaw motions is needed.

The torques acting on the mover depend on the relative yaw, pitch, and roll distances between the mover and the armature conductors, but relative pitch and roll distances should be always nearly equal to 0 deg in order to maintain a small air gap. In this study, new armature conductors with different relative distances in the yaw direction from the armature conductors for the x– and y–directional drives are introduced to control the yaw motion.

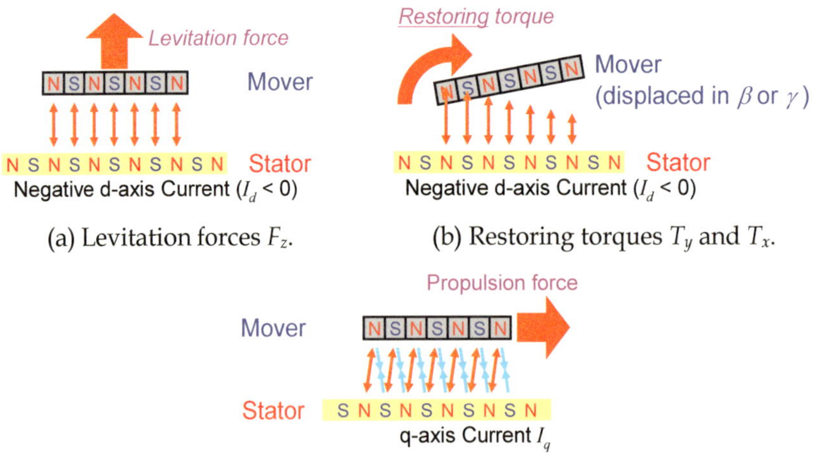

(a) Levitation forces F_z.

(b) Restoring torques T_y and T_x.

(c) Propulsion forces F_x and F_y.

Fig. 17. Conceptual design of a magnetically levitated planar actuator

Figure 16 indicates that the d–axis current generates translational forces F_z and torques T_z, and the q–axis current generates translational forces F_x, F_y and torques T_x, T_y when the pitch and roll positions are not displaced ($\beta = \gamma = 0$ deg). So, at least four kinds of the q–axis currents, that is, four pairs of polyphase currents are needed to actively control 6-DOF motions. Furthermore, Fig. 16 also indicates that the d– and q–axis currents generate only torques without translational forces when the relative yaw distance is 24.7 deg or 45 deg. As mentioned in Subsection 3.1.1, a magnitude of torque T_z resulting from the mover tiled by 24.7 deg is larger than that by 45 deg. Therefore in this study, the armature conductors are tilted by 24.7 deg in the yaw direction from the armature conductors for the x–directional drive, I term this arrangement "armature conductors for the α–directional drive." When the yaw angle of the mover $\alpha = 0$ deg, the d–axis currents for the α–directional drive $I_{d\alpha}$:

- generate only torques T_z
- without vertical forces F_z.

Therefore, the d–axis currents $I_{d\alpha}$ can separate the generation of the vertical forces F_z and torques T_z, and stabilize the yaw motion. To date, the d– and q–axis currents are generated by three-phase currents, but they can be also be generated by two-phase currents. In this study, a magnetically levitated planar actuator with three pairs of two-phase armature conductors is organized as shown in Fig. 18.

3.2 Dynamic behavior of mover

The mover has 3-DOF translational and rotational motions because there is no mechanical suspension mechanism. When the physical quantities of the mover motion are represented,

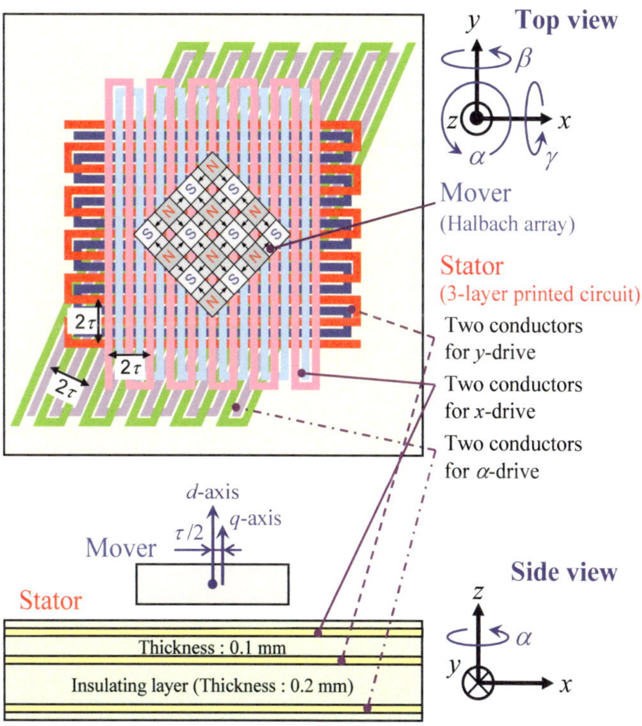

Fig. 18. Fundamental structure of magnetically levitated planar actuator

it is extremely important what coordinates are respected. The translational motions are often represented with respect to the stationary coordinate, and the rotational motions are often represented with respect to the mover coordinate. This section introduces an equation for the 6-DOF motions of the mover that describes the dynamic behavior.

The equation of the motion of the mover can be represented by the translational forces acting on the mover $\boldsymbol{F}_{sm} = [F_x \ F_y \ F_z]^T$ and torques around the mover center O′ $\boldsymbol{T}_{sm}' = [T_x' \ T_y' \ T_z']^T$ from mass M and inertia tensor $\boldsymbol{J}_m'$ of the mover as follows:

$$M\frac{d\boldsymbol{v}_{sm}}{dt} = \boldsymbol{F}_{sm} + \boldsymbol{F}_g \tag{21}$$

$$\boldsymbol{J}_m'\frac{d\boldsymbol{\omega}_{sm}'}{dt} = \boldsymbol{T}_{sm}' - \boldsymbol{\omega}_{sm}' \times \left(\boldsymbol{J}_m' \boldsymbol{\omega}_{sm}'\right) \tag{22}$$

where $\boldsymbol{v}_{sm} = [v_x \ v_y \ v_z]^T$ and $\boldsymbol{F}_g = [0 \ 0 \ -Mg]^T$ are velocity of the mover and the force of gravity acting on the mover, respectively.

Equations (21) and (22) represent 3-DOF translational and rotational motion equations of the mover, respectively. All variables in the translational and rotational motion equations are represented with respect to the stationary coordinate $x_s y_s z_s$ and mover coordinate $x_m y_m z_m$, respectively. The position $\boldsymbol{r}_{sm} = [x \ y \ z]^T$, with respect to the stationary coordinate $x_s y_s z_s$, and Euler angle $\boldsymbol{\phi} = [\alpha \ \beta \ \gamma]^T$, which is defined from α, β, and γ as orderly counterclockwise rotations about the z-, y- and x-axes, respectively, can be represented by the velocity $\boldsymbol{v}_{sm}$ and angular velocity $\boldsymbol{\omega}_{sm}'$, respectively, as follows:

$$\frac{d\boldsymbol{r}_{sm}}{dt} = \boldsymbol{v}_{sm} \tag{23}$$

$$\frac{d\boldsymbol{\varphi}}{dt} = \boldsymbol{R}_{\omega\varphi}\left(\boldsymbol{\varphi}\right)^{-1} \boldsymbol{\omega}_{sm}' \tag{24}$$

where $\boldsymbol{R}_{\omega\phi}$ is a 3×3 matrix and all elements of the matrix nonlinearly depend on the Euler angle $\boldsymbol{\phi}$. Equations (21)–(24) can represent dynamic behaviors of the mover with 6 DOF.

3.3 Planar motion control with stable magnetic levitation

This section discusses six-current controls to stably levitate the mover and actively control the x–, y–, z–, and α–motions. There are two important things for the motion controls:

- to generate independent translational forces F_x, F_y, and F_z with stable torques in the γ– and β–directions.
- to generate torques in the α–direction with less interference to translational forces F_x, F_y, and F_z.

This section first presents driving forces resulting from three pairs of two-phase armature currents, and then the driving force-control system.

3.3.1 Translational motion control

In this study, three pairs of two-phase currents $\boldsymbol{i}_j = [I_{1j} \ I_{2j}]^T$ ($j = x$, y, or α) are assumed to be supplied to the three pairs of two-phase armature conductors as shown in the following equations:

$$I_{1j} = -I_j \cos\left(\theta_{sj}\right) \tag{25}$$

$$I_{2j} = I_j \sin\left(\theta_{sj}\right) \tag{26}$$

Figure 19 shows phasor diagrams for the relation between the dq–frame and $\alpha'\beta'$–frame. The currents I_{1x} and I_{1y} generate the opposite-phase magnetic field to that resulting from the permanent-magnet mover when the mover position in the x– and y–directions $(x, y) = (x_s, y_s)$ and the Euler angle $\boldsymbol{\phi} = (0, 0, 0)$. The α'–axis are aligned to the opposite side of the current I_{1j} axis, and the β'–axis leads the α'–axis by 90 deg. The current $I_{1\alpha}$ generates a magnetic field that is tilted by φ = –24.7 deg around the α–direction from that caused by current I_{1x}. Bearing this in mind, the armature currents in the dq–frame I_{dj} and I_{qj} can be represented by the currents I_{1j} and I_{2j} as follows ($j = x, y$, or α):

$$\begin{bmatrix} I_{dj} \\ I_{qj} \end{bmatrix} = \begin{bmatrix} \cos\left(\pi j_s / \tau\right) & -\sin\left(\pi j_s / \tau\right) \\ \sin\left(\pi j_s / \tau\right) & \cos\left(\pi j_s / \tau\right) \end{bmatrix} \begin{bmatrix} I_{2j} \\ I_{1j} \end{bmatrix} \tag{27}$$

$$\alpha_s = x_s \cos\phi - y_s \sin\phi\,. \tag{28}$$

These pairs of d– and q–axis currents generate the translational forces $\boldsymbol{F}_{sm}$ and torques $\boldsymbol{T}_{sm}'$ as follows:

$$\begin{bmatrix} F_x \\ F_y \\ F_z \\ T_x' \\ T_y' \\ T_z' \end{bmatrix} = \begin{bmatrix} \\ \\ \boldsymbol{K}\left(\boldsymbol{r}_{sm}, \boldsymbol{\varphi}\right) \\ 6 \times 6 \text{ matrix} \\ \\ \\ \end{bmatrix} \begin{bmatrix} I_{dx} \\ I_{qx} \\ I_{dy} \\ I_{qy} \\ I_{d\alpha} \\ I_{q\alpha} \end{bmatrix} \approx \begin{bmatrix} 0 & & 0 & 0 & 0 & 0 \\ 0 & 0 & 0 & & 0 & 0 \\ & 0 & & 0 & 0 & 0 \\ & & & & & \\ & & & & & \\ & & & & & \end{bmatrix} \begin{bmatrix} I_{dx} \\ I_{qx} \\ I_{dy} \\ I_{qy} \\ I_{d\alpha} \\ I_{q\alpha} \end{bmatrix} \tag{29}$$

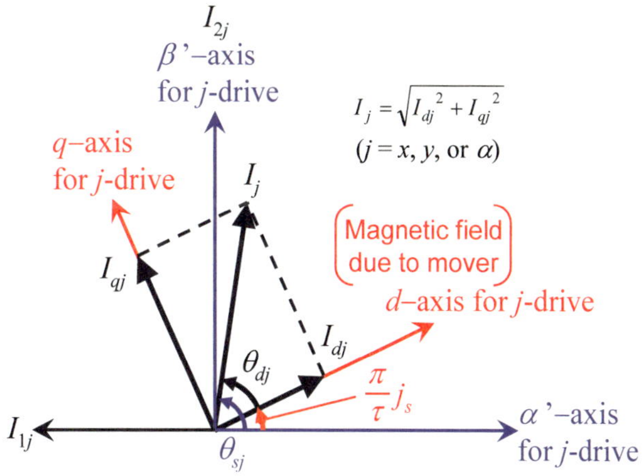

Fig. 19. Phasor diagram showing relation between dq–frame and $\alpha'\beta'$–frame

where $\boldsymbol{K}$ is a 6 × 6 matrix, and all elements of $\boldsymbol{K}$ depend on the mover position $\boldsymbol{r}_{sm}$ and Euler angle $\boldsymbol{\phi}$. Where Euler angle $\boldsymbol{\phi} \approx \boldsymbol{0}$, $\boldsymbol{K}$ can be approximated as shown in Eq. (29), and therefore 3-DOF translational forces F_x, F_y, and F_z can be independently controlled by two-phase currents $\boldsymbol{i}_x$ and $\boldsymbol{i}_y$.

In this study, references of the translational forces $\boldsymbol{F}_{sm}^* = [F_x^* \;\; F_y^* \;\; F_z^*]^T$ are determined from the mover positions $\boldsymbol{r}_{sm} = [x \;\; y \;\; z]^T$ and position references $\boldsymbol{r}_{sm}^* = [x^* \;\; y^* \;\; z^*]^T$ by three PID controls.

$$\boldsymbol{F}_{sm}^{\;*} = \boldsymbol{P}_F\left(\boldsymbol{r}_{sm}^{\;*} - \boldsymbol{r}_{sm}\right) - \boldsymbol{D}_F \frac{d\boldsymbol{r}_{sm}}{dt} - \boldsymbol{F}_g \tag{30}$$

where $\boldsymbol{P}_F = \mathrm{diag}(P_{Fx}, P_{Fy}, P_{Fz})$ and $\boldsymbol{D}_F = \mathrm{diag}(D_{Fx}, D_{Fy}, D_{Fz})$ are proportional and differential parameters, respectively. In this study, references of the armature currents $\boldsymbol{i}_x^*$ and $\boldsymbol{i}_y^*$ are calculated from those of the translational forces $\boldsymbol{F}_{sm}^*$ as follows:

$$\begin{bmatrix} I_{dx}^{\;*} \\ I_{qx}^{\;*} \end{bmatrix} = \begin{bmatrix} K_{11} & K_{12} \\ K_{31} & K_{32} \end{bmatrix}^{-1} \begin{bmatrix} F_x^{\;*} \\ F_z^{\;*}/2 \end{bmatrix} \tag{31}$$

$$\begin{bmatrix} I_{dy}^{\;*} \\ I_{qy}^{\;*} \end{bmatrix} = \begin{bmatrix} K_{23} & K_{24} \\ K_{33} & K_{34} \end{bmatrix}^{-1} \begin{bmatrix} F_y^{\;*} \\ F_z^{\;*}/2 \end{bmatrix} \tag{32}$$

Supplying the armature currents $\boldsymbol{i}_x$ and $\boldsymbol{i}_y$ equal to the references $\boldsymbol{i}_x^*$ and $\boldsymbol{i}_y^*$ generates the translational forces $\boldsymbol{F}_{sm}$ equal to the references $\boldsymbol{F}_{sm}^*$.

3.3.2 Rotational motion control

The armature currents $\boldsymbol{i}_x$ and $\boldsymbol{i}_y$ generate not only the translational forces $\boldsymbol{F}_{sm}$, but also the torques $\boldsymbol{T}_{sm}'$. Therefore, it is extremely important to investigate how the torques $\boldsymbol{T}_{sm}'$ resulting from the armature currents $\boldsymbol{i}_x$ and $\boldsymbol{i}_y$ influence the rotational motions of the mover. When the Euler angle $\boldsymbol{\phi} \approx \boldsymbol{0}$, the torques T_z', T_y', and T_x' are dominant on the Euler angle α, β, and γ, respectively. Next I performed a numerical analysis of the torque characteristics due to the armature currents for the x–directional drive when rotational motions with more than 1 DOF occur.

Furthermore, from analysis results of the 6-DOF driving forces, when rotational motions with more than 1 DOF occur in the range within –2 deg < α, β, and γ < 2 deg, a 6 × 4 submatrix of $\boldsymbol{K}$ is almost in agreement with $\boldsymbol{K}_{FT}$ in Eq. (20). Therefore, negative d–axis currents I_{dx}, I_{dy} that control the suspension forces F_z generate stable restoring torques T_y', T_x'. However, the q–axis currents that control the translational forces F_x, F_y generate torques T_z', T_y', T_x', which are not stable restoring torques. So next I performed a numerical analysis of the torque characteristics due to the armature currents for the α–directional drive.

Figure 20 shows the torques due to the armature conductors for the α–directional drive at $(\beta, \gamma) = (0, 0)$. When the Euler angles $(\beta, \gamma) = (0, 0)$, the d–axis current $I_{d\alpha}$ generates only the torque T_z' and the q–axis current $I_{q\alpha}$ generates only the torques T_y', T_x'. Therefore, the torques T_y' and T_x' cannot be independently controlled by the armature currents for the α–directional drive. When the Euler angle $\boldsymbol{\phi} \approx \boldsymbol{0}$ and angular velocity $\boldsymbol{\omega}_{ms}' \approx \boldsymbol{0}$, a linearized equation of the rotational motion can be obtained from Eqs. (22) and (24) as follows:

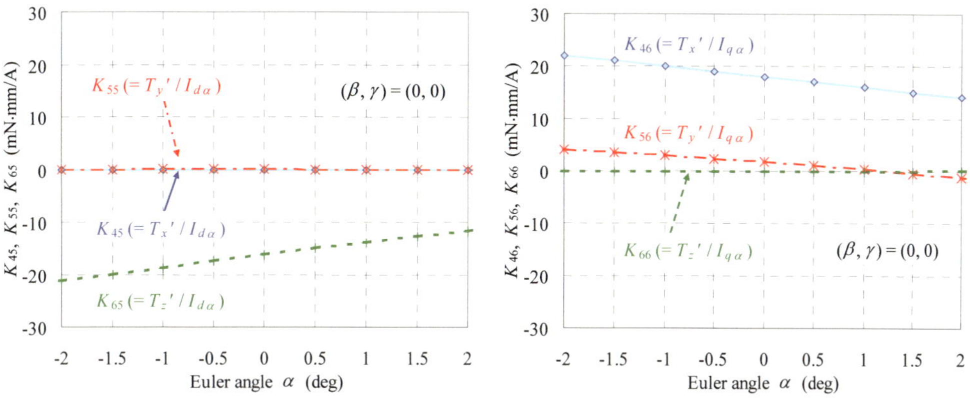

(a) K_{45}, K_{55}, and K_{65}. (b) K_{46}, K_{56}, and K_{66}.

Fig. 20. Analysis result of the torques from the armature conductors for the α–drive

$$\frac{d^2\boldsymbol{\varphi}}{dt^2} \approx \boldsymbol{R}_{\omega\varphi}\left(\boldsymbol{J}_m{}'\right)^{-1}\boldsymbol{T}_{sm}{}' = \boldsymbol{T}_E = \begin{bmatrix} T_\alpha & T_\beta & T_\gamma \end{bmatrix}^T. \tag{33}$$

In this study, $\boldsymbol{T}_E^*$, which is the reference of $\boldsymbol{T}_E$, is determined by a PD control from the Euler angle α and the reference α^* as follows:

$$T_\alpha^* = P_{T\alpha}\left(\alpha^* - \alpha\right) - D_{T\alpha}\frac{d\alpha}{dt} \tag{34}$$

where $P_{T\alpha}$ and $D_{T\alpha}$ are proportional and differential parameters, respectively. Then, the references T_β^* and T_γ^* are determined to be zero because of the suppression of the β– and γ–motions. The torque references $T_x'^*$ and $T_z'^*$ can be calculated from the reference $\boldsymbol{T}_E^*$ by Eq. (33). Then, the references of the armature currents for the α–directional drive $I_{d\alpha}^*$ and $I_{q\alpha}^*$ can be calculated for the torque references $T_x'^*$ and $T_z'^*$ as follows:

$$\begin{bmatrix} I_{d\alpha}{}^* \\ I_{q\alpha}{}^* \end{bmatrix} = \begin{bmatrix} K_{45} & K_{46} \\ K_{65} & K_{66} \end{bmatrix}^{-1}\left(\begin{bmatrix} T_x{}'^* \\ T_z{}'^* \end{bmatrix} - \begin{bmatrix} T_{xa}{}' \\ T_{za}{}' \end{bmatrix}\right) \tag{35}$$

where T_{xa}' and T_{za}' are torques due to the armature currents $\boldsymbol{i}_x$ and $\boldsymbol{i}_y$, and can be represented as follows:

$$\begin{bmatrix} T_{xa} \\ T_{za} \end{bmatrix} = \begin{bmatrix} K_{41} & K_{42} & K_{43} & K_{44} \\ K_{61} & K_{62} & K_{63} & K_{64} \end{bmatrix}\begin{bmatrix} I_{dx} \\ I_{qx} \\ I_{dy} \\ I_{qy} \end{bmatrix}. \tag{36}$$

Supplying the armature currents $\boldsymbol{i}_\alpha$ equal to the references $\boldsymbol{i}_\alpha^*$ generates $\boldsymbol{T}_E$ nearly equal to $\boldsymbol{T}_E^*$, and controls the rotational motions with less interference to the translational motions.

3.4 Numerical analysis of mover motion

Motion characteristics with 6 DOF can be obtained by solving Eqs. (21)–(24) using the Runge-Kutta method. In order to numerically solve the equations, it is necessary to calculate the driving forces $\boldsymbol{F}_{sm}$ and $\boldsymbol{T}_{sm}'$ at each time step. The calculation at each time step consists of an integration of Lorentz force acting on the line segments as shown in Eqs. (4) and (5), and so requires a lot of computation time. The flux density $\boldsymbol{B}$ acting on the armature conductors greatly depends on the mover position $\boldsymbol{r}_{sm}$ and Euler angle $\boldsymbol{\phi}$. Therefore, the driving forces $\boldsymbol{F}_{sm}$ and $\boldsymbol{T}_{sm}'$ are functions of the mover position $\boldsymbol{r}_{sm}$ and Euler angle $\boldsymbol{\phi}$. In this study, the system-constant matrix $\boldsymbol{K}$ was calculated and the data table of $\boldsymbol{K}$ was made before the motion analysis. Then, the system-constant matrix $\boldsymbol{K}$ is calculated from the mover position $\boldsymbol{r}_{sm}$ and Euler angle $\boldsymbol{\phi}$ by interpolating it with the data table at each time step. The analysis conditions are shown as follows:

- time step $dt = 0.2$ ms
- control period $t_c = 2$ ms
- initial position $\boldsymbol{r}_i = \boldsymbol{0}$
- initial Euler angle $\boldsymbol{\phi}_i = \boldsymbol{0}$.

When the z–position is zero, the mover is assumed to be on the stator. The proportional and differential parameters are determined so that the settling times in the x–, y–, z–, and α–motions are less than 1 s. In this analysis, to investigate the planar motion control and magnetic levitation, the following position reference is given:

- $x^* = 2\cos(\pi t)$ mm
- $y^* = 2\sin(\pi t)$ mm
- $z^* = 0.15$ mm
- Euler angle $\alpha^* = 0$ deg.

Figure 21 shows the analysis result of the mover motions under this analysis condition, and indicates that the mover can track the reference positions in the x– and y–directions, and be positioned in the z– and α–directions with suppression of the β– and γ–displacements. Therefore, mover motions can be controlled with stable magnetic levitation. The q–axis currents I_{qx} and I_{qy} used to control the translational forces F_x and F_y also generate simultaneously the torques T_y' and T_x', respectively. Therefore, displacement of the Euler angles β and γ slightly occurs.

3.5 Summary of chapter 3

This chapter presents a feasibility verification of a planar actuator with both 3-DOF planar motions and magnetic suspension of the mover in order to further improve performance. Then, based on a numerical analysis of the 6-DOF driving forces, a planar actuator having a mover positioned above a plane and magnetically levitated by only six currents and the six-current-control algorithm were conceptually designed. Furthermore, I validated the designed planar actuator by numerical analysis of the 6-DOF motions. The results obtained in this paper indicate the possibility of the realization of a high-performance MDOF planar actuator:

- decoupled 3-DOF motion control and magnetic levitation on a plane.
- wide movable area by a small number (six) of armature conductors.
- extendible movable area regardless of the number of armature conductors.
- small millimeter-sized mover.
- no problematic wiring to adversely affect drive performance.

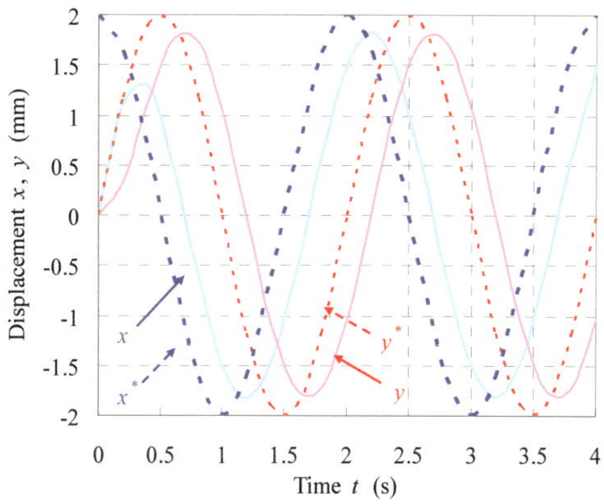

(a) Planar motions x, y.

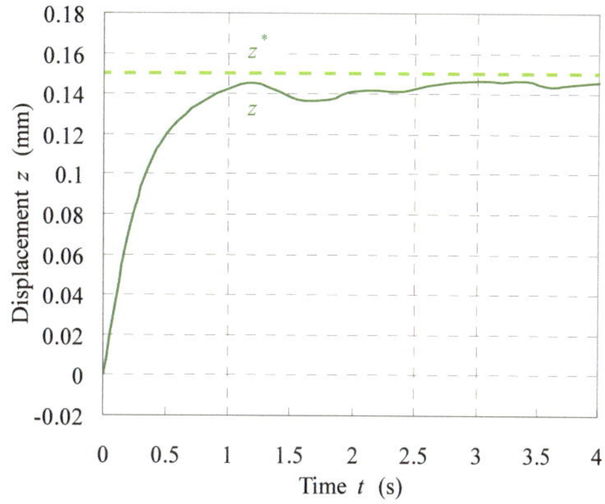

(b) Vertical motion z.

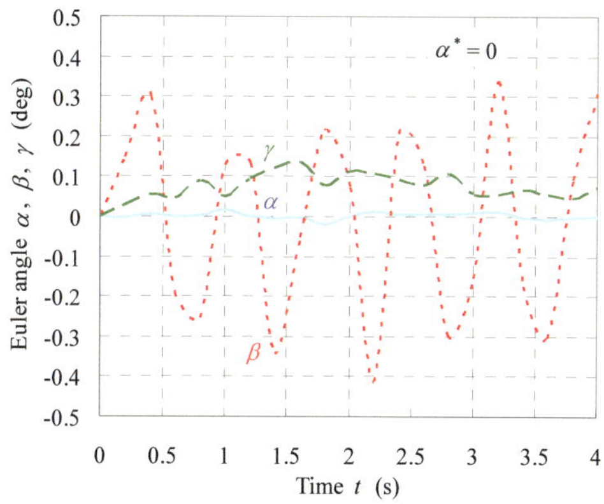

(c) Rotational motions α, β, γ.

Fig. 21. Analytically-obtained 6-DOF mover motions (Ueda & Ohsaki, 2008b)

4. Conclusion

This paper presents high-performance MDOF planar actuators with a permanent-magnet mover capable of traveling over a wide movable area on a plane, with just a small number of stationary armature conductors. The combination of the mover and stator can generate spatially superimposed magnetic fields for the MDOF drive, and therefore increasing the length of the armature conductors can easily expand the movable area regardless of the number of armature conductors. A planar actuator was conceptually designed and fabricated. The fabricated planar actuator can independently control the 3-DOF motions of the mover. Furthermore, in order to eliminate deterioration of the drive characteristics due to friction forces, the planar actuator was redesigned so that the mover could be stably levitated and the 3-DOF motions on a plane could be controlled. Then, the mover motion characteristics were successfully verified by means of a numerical analysis. Next, a small fabrication size was realized by integrating the permanent-magnet array and armature conductors for the MDOF drive. The planar actuator has the first millimeter-sized mover and would provide a significant starting point when used with small electromechanical components in an MDOF drive.

5. References

Ebihara, D.; Wakui, S.; Tsuchiya, J.; Inui, S.; Wakiwaka, H.; Ohsaki, H.; Yano, T.; Maeno, T.; Watanabe, T.; Hirata, K.; Dohmeki, H.; Ohashi, K.; Tomita, Y.; Maemura, A & Takemura, K. (2005). *Investigation of Possibility of Multi-Dimensional Drive System,* The Institute of Electrical Engineers of Japan, ISSN 0919-9195, MARUI-Kobunsha Corp.

Fitzgerald, A. E.; Kingsley, C. Jr. & Umans, S. D. (1990). *Electric Machinery,* McGraw Hill Higher Education, ISBN 978-0073660097

Halbach, K. (1986). Concepts for Insertion Devices that will Produce High-Quality Synchrotron Radiation. *Nuclear Instruments and Methods in Physics Research Section A,* Vol.246, No.1-3, May 1986, pp.77-81, ISSN 0168-9002

Jeon, J. W.; Caraiani, M.; Kim, Y. J.; Oh, H. S. & Kim, S. S. (2007). Development of Magnetic Levitated Stage for Wide Area Movements. *Proceedings of the International Conference on Electrical Machines and Systems*, pp.1486-1491, Korea, October 2007, Seoul

Kim, W. J. & Trumper, D. L. (1998). High-Precision Magnetic Levitation Stage for Photolithography. *Precision Engineering,* Vol.22, No.2, April 1998, pp.66-77, ISSN 0141-6359

Ohsaki, H.; Teramura, N.; Huang, X.; Tsuboi, Y. & Ootani, Y. (2003). Electromagnetic Characteristics of a Coreless Surface Motor Using Halbach Permanent Magnets. *Proceedings of the International Symposium on Linear Drives for Industry Applications,* pp.105-108, UK, September 2003, Birmingham

Ueda, Y. (2009). Design and Control of a High-Performance Multi-Degree-of-Freedom Planar Actuator. *Ph.D. Thesis at the University of Tokyo,* March 2009

Ueda, Y. & Ohsaki, H. (2008a). A Planar Actuator with a Small Mover Traveling over Large Yaw and Translational Displacements. *IEEE Transaction on Magnetics,* Vol.44, No.5, May 2008, pp.609-616, ISSN 0018-9464

Ueda, Y. & Ohsaki, H. (2008b). Six-Degree-of-Freedom Motion Analysis of a Planar Actuator with a Magnetically Levitated Mover by Six-Phase Current Controls. *IEEE Transaction on Magnetics,* Vol.44, No.11, November 2008, pp.4301-4304, ISSN 0018-9464

Ueda, Y. & Ohsaki, H. (2009). Compact Three-Degree-of-Freedom Planar Actuator with Only Six Currents Capable of Driving over Large displacements in Yaw Direction. *IEEJ Transaction on Industry Applications,* Vol.129, No.3, March 2009, pp.332-341, ISSN 1348-8163

Zhu, Z. O. & Howe, D. (2001). Halbach Permanent Magnet Machines and Applications: A Review. *IEE Proceedings-Electric Power Applications,* Vol.148, No.4, July 2001, pp.299-308, ISSN 1350-2352

23

Sensorless V/f Control of Permanent Magnet Synchronous Motors

Daniel Montesinos-Miracle[1], P. D. Chandana Perera[2], Samuel Galceran-Arellano[1] and Frede Blaabjerg[3]
[1]*Centre d'Innovació Tecnològica en Convertidors Estàtics i Accionaments, Departament d'Enginyeria Elèctrica, Universitat Politècnica de Catalunya*
[2]*Department of Electrical and Information Engineering, Faculty of Engineering, University of Ruhuna*
[3]*Institute of Energy Technology, Aalborg University*
[1]*Spain*
[2]*Sri Lanka*
[3]*Denmark*

1. Introduction

The increasing energy cost demands for more efficient motion control systems in domestic and industrial applications. Power electronics and control can contribute to increase the efficiency of present systems, but it can also be dealt with new efficient solutions for old applications.

Instead of using constant speed using variable speed drives in motion control applications the efficiency of the systems can be increased. The most common control methods used in drives are V/f control, vector control and direct torque control (DTC) [1]. In continuous running applications, a small increase in efficiency means a huge energy savings per year. These continuous running applications are mainly pumps, fans and compressors for heating, ventilating and air conditioning (HVAC) applications. In these applications where high dynamics are not required, a simple digital implementation of V/f control can be used instead of more complex vector or DTC with the same performance [2, 3].

The workhorse for these applications has been the induction motor for years. The induction motor is a well known motor, a cheap motor, and does not require position sensor to implement a low-cost control for this kind of applications.

But efficiency can be improved if the induction motor is substituted by a permanent magnet synchronous motor (PMSM) [4]. However, in permanent magnet synchronous motors, the stator currents have to be synchronized with the rotor permanent magnet in order to produce the required torque and not to lose synchronization. For this purpose a rotor position sensor is required. The need of a rotor position sensor increases the cost and reduces the reliability. Self synchronization can be achieved using damper windings [5], but due to cost, efficiency and high-cost, they are generally not implemented in PMSMs [6].

Therefore, it is necessary to develop new control strategies for PMSMs to avoid the use of the rotor position sensor. Because HVAC applications do not demand for a high

performance control, the V/f control strategy is suitable for these drives. However, even using a V/f control strategy for permanent magnet synchronous motors, there is a need in synchronization for stator currents with the rotor magnet position.
This chapter present a sensorless V/f control for permanent magnet synchronous motors.

2. Park equations of permanent magnet synchronous motors

The Park transformation is useful when modeling a PMSM, since there is no angle dependent terms appear in equations, providing easy analysis of the system [7].
With the electrical equations in Park variables, it is easy to obtain an expression for the motor generated torque as a function of electrical variables. This produced mechanical torque links the electrical world with the mechanical world, and completes the model of the system. The model equations of the permanent magnet synchronous motor in Park variables are:

$$\dot{i}_{ds}^{r} = -\frac{R_s}{L_d} i_{ds}^{r} + \omega_r \frac{L_q}{L_d} i_{qs}^{r} + \frac{v_{ds}^{r}}{L_d} \tag{1.1}$$

$$\dot{i}_{qs}^{r} = -\frac{R_s}{L_q} i_{qs}^{r} - \omega_r \frac{L_d}{L_q} i_{ds}^{r} - \frac{\lambda_m}{L_q}\omega_r + \frac{v_{qs}^{r}}{L_q} \tag{1.2}$$

$$\dot{\omega}_r = \frac{3}{2}\left(\frac{n}{2}\right)^2 \frac{1}{J_m}\lambda_m i_{qs}^{r} + \frac{3}{2}\left(\frac{n}{2}\right)^2 \frac{1}{J}\left(L_d - L_q\right) i_{ds}^{r} i_{qs}^{r} - \frac{B_m}{J_m}\omega_r - \frac{n}{2J}T_l \tag{1.3}$$

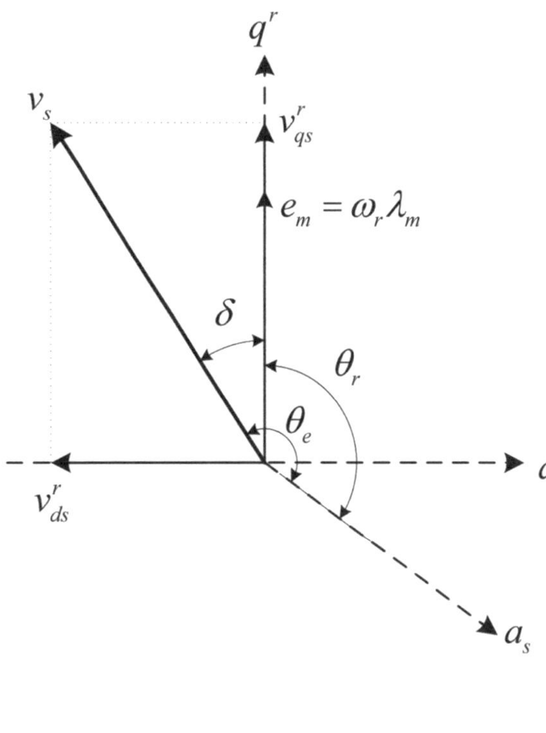

Fig. 1.1. Load angle

Sometimes, it is useful to define a new variable called load angle [8]. The load angle is the angle between the stator electrical applied voltage and the *emf* generated by rotor magnets when rotating, as can be seen in Figure 1.1. With this new equation, the system equations in state space variables are defined as

$$\dot{i}_{ds}^{r} = -\frac{R_s}{L_d} i_{ds}^{r} + \omega_r \frac{L_q}{L_d} i_{qs}^{r} - \frac{v_s \sin\delta}{L_d} \tag{1.4}$$

$$\dot{i}_{qs}^{r} = -\frac{R_s}{L_q} i_{qs}^{r} - \omega_r \frac{L_d}{L_q} i_{ds}^{r} - \frac{\lambda_m}{L_q}\omega_r + \frac{v_s \cos\delta}{L_q} \tag{1.5}$$

$$\dot{\omega}_r = \frac{3}{2}\left(\frac{n}{2}\right)^2 \frac{1}{J_m}\lambda_m i_{qs}^{r} + \frac{3}{2}\left(\frac{n}{2}\right)^2 \frac{1}{J}\left(L_d - L_q\right) i_{ds}^{r} i_{qs}^{r} - \frac{B_m}{J_m}\omega_r - \frac{n}{2J} T_l \tag{1.6}$$

$$\dot{\delta} = \omega_e - \omega_r \tag{1.7}$$

The above equations are the state space model, but this model contains non-linear terms. To analyze the stability of the system, a linear model must be obtained.

3. Stability analysis

The non-linear model can be linearized by substituting each variable as [8]

$$x_i = X_i + \Delta x_i \tag{1.8}$$

where x_i is the variable, X_i is the steady state value, and Δx_i is a perturbation from the steady state value. Then, the linearized system is

$$\Delta\dot{\mathbf{x}} = \mathbf{A}\left(\mathbf{X}\right)\Delta\mathbf{x} + \mathbf{B}\left(\mathbf{X}\right)\Delta\mathbf{u} \tag{1.9}$$

Applying this linearization technique to the state space model of the permanent magnet synchronous motor, the linearized model is obtained as

$$\begin{Bmatrix} \Delta\dot{i}_{ds}^{r} \\ \Delta\dot{i}_{qs}^{r} \\ \Delta\dot{\omega}_r \\ \Delta\dot{\delta} \end{Bmatrix} = \begin{pmatrix} -\frac{R_s}{L_d} & \frac{L_q\omega_{r0}}{L_d} & \frac{L_q I_{qs}^{r}}{L_d} & -\frac{V_s\cos\delta_0}{L_d} \\ -\frac{L_d\omega_{r0}}{L_q} & -\frac{R_s}{L_q} & -\frac{1}{L_q}\left(L_d I_{ds}^{r} + \lambda_m\right) & -\frac{V_s\sin\delta_0}{L_q} \\ \frac{3}{2}\left(\frac{n}{2}\right)^2 \frac{1}{J_m}\left(L_d - L_q\right) I_{qs}^{r} & \frac{3}{2}\left(\frac{n}{2}\right)^2 \frac{1}{J_m}\left(\lambda_m + \left(L_d - L_q\right) I_{ds}^{r}\right) & -\frac{B_m}{J_m} & 0 \\ 0 & 0 & -1 & 0 \end{pmatrix} \begin{Bmatrix} \Delta i_{ds}^{r} \\ \Delta i_{qs}^{r} \\ \Delta\omega_r \\ \Delta\delta \end{Bmatrix} + \begin{pmatrix} -\frac{\sin\delta_0}{L_d} & 0 & 0 \\ \frac{\cos\delta_0}{L_q} & 0 & 0 \\ 0 & 0 & -\frac{n}{2J_m} \\ 0 & 1 & 0 \end{pmatrix} \begin{Bmatrix} \Delta v_s \\ \Delta\omega_e \\ \Delta T_l \end{Bmatrix} \tag{1.10}$$

3.1. V/f open loop control

When the motor is operated at open loop V/f control, the applied voltage and frequency are constant, that is

$$\Delta v_s = 0 \tag{1.11}$$

$$\Delta \omega_e = 0 \tag{1.12}$$

The stability of the system is determined by the eigenvalues of the state matrix **A**(**X**). In open loop V/f control strategy with no-load, the motor produces no torque. Then, $I_{qs}^r = 0$. In order to minimize the losses, I_{ds}^r is also 0. In this case, the applied voltage must only compensate the *emf* in the *q* axis, $V_s = \omega_r \lambda_m = V_{qs}^r$ and $V_{ds}^r = 0$. Substituting this steady state conditions in the state matrix **A**(**X**) it is possible to obtain the stability characteristic of the system. Figure 1.2 shows the root locus diagram of the permanent magnet synchronous motor in an open loop V/f control at no-load as a function of stator frequency ω_e. For this figure the motor data can be found in Appendix A.

As seen in Figure 1.2, the motor becomes unstable above 100 Hz operation, i. e. half of the rated frequency. The most left poles are the named stator poles, and represent the fast electrical stator dynamics [9, 8]. The most right poles are named the rotor poles, and represent the slow mechanical dynamics. A poor coupling between the rotor and stator poles causes this instability [10, 11].

Figure 1.3 shows the dominant poles at different load levels. As seen, the stability characteristic is not modified with the load level.

Figure 1.4 shows this instability in a real system. As seen, at low frequency, the motor is stable (Figure 1.4(a)), but when increasing the frequency the motor becomes unstable (Figures 1.4(b) and 1.4(c)).

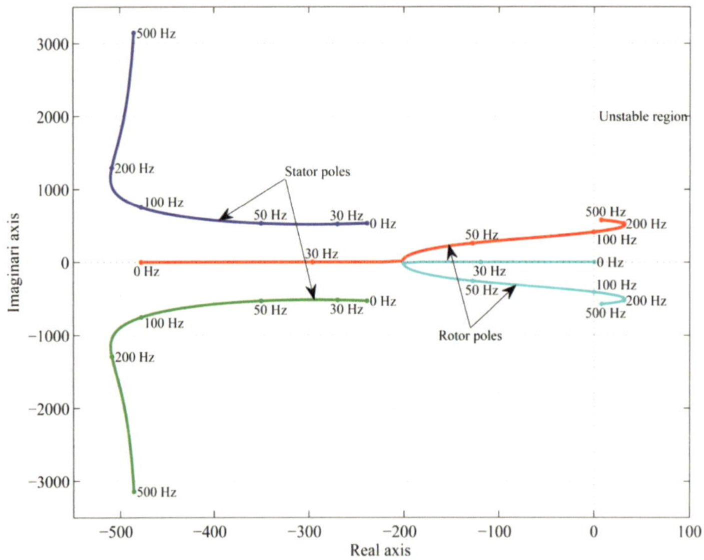

Fig. 1.2. Root locus of the permanent magnet synchronous motor operating at no-load in an open loop V/f control strategy

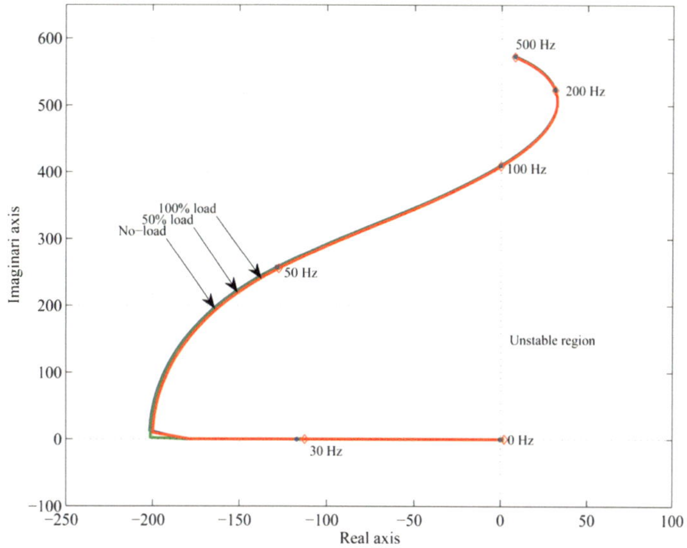

Fig. 1.3. Root locus of the permanent magnet synchronous motor operating at different load levels in an open loop V/f strategy

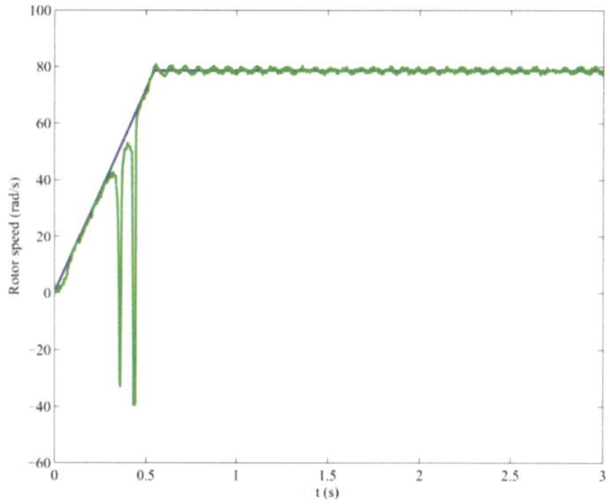

(a) Rotor speed at 50Hz stator frequency

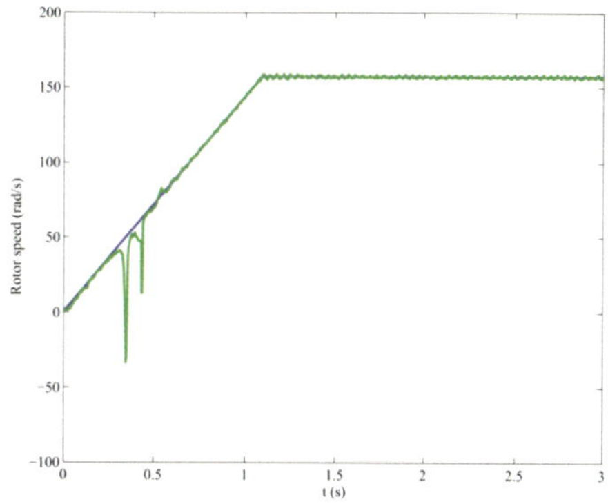

(b) Rotor speed at 100 Hz stator frequency

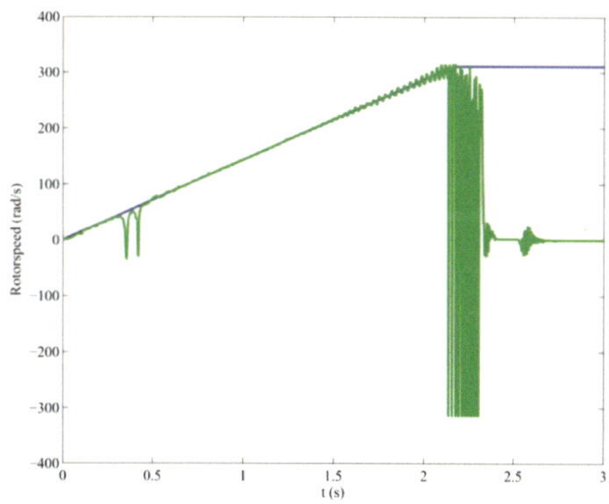

(c) Rotor speed at 200 Hz stator frequency

Fig. 1.4. Rotor speed at different stator frequencies in the open loop V/f control strategy

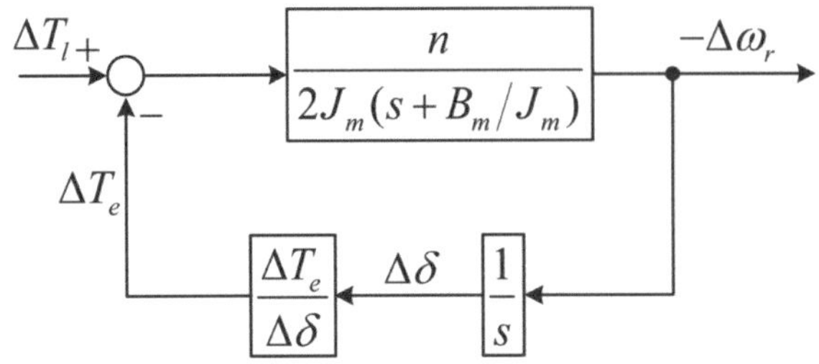

Fig. 1.5. Block diagram of the small signal model operating at V/f open loop control strategy

4. Stabilization of the V/f control in permanent magnet synchronous motor

The open loop V/f control does not assure the synchronization between stator currents and rotor position needed. If this synchronization is lost, the motor becomes unstable above a rotating frequency, as seen in section 1.3.

In order to operate the motor in a V/f control, a synchronization method is needed. The method proposed in this section is based on the stabilization of the operation of the permanent magnet synchronous motor in a V/f control strategy.

4.1 Small signal model of the permanent magnet synchronous motor

Assuming that the motor is operating in open loop V/f control, from the linearized model, the block diagram of Figure 1.5 can be obtained. The expression of $\frac{\Delta T_e}{\Delta\delta}$ is

$$\frac{\Delta T_e}{\Delta\delta} = \frac{3}{2}\frac{n}{2}\lambda_m \frac{\left(I^r_{ds} + \frac{\lambda_m}{L_d}\right)s^2 + \frac{R_s}{L_d}\left(\frac{\lambda_m}{L_d} - 2I^r_{ds}\right)s + \omega^2_{r0}\left(\frac{\lambda_m}{L_d} + I^r_{ds}\right) + \frac{R^2_s}{L^2_d}I^r_{ds}}{s^2 + \frac{2R_s}{L_d}s + \frac{R^2_s}{L^2_d} + \omega^2_{r0}} \tag{1.13}$$

In Figure 1.5, it can be seen that small load torque perturbations produce small rotating speed perturbations, through mechanical dynamics, that produce small perturbations of the load angle, that produce small produced torque perturbations.

It can be observed in equation (1.13) that, motor stability is not an explicit function of the torque in steady state, but, for a given voltage value, the current I^r_{ds} and therefore, coefficients in (1.13), are a function of torque. It must also be noted that, combinations of voltage and torque that gives equal values of I^r_{ds}, will have the same stability characteristic. In control methods that impose I^r_{ds}= 0, all the operating points will have the same stability characteristics [12].

From this model a simplification can be done supposing that the perturbations of T_e are linear with $\Delta\delta$ as [10],

$$\Delta T_e = K_e \Delta\delta \tag{1.14}$$

where K_e is the electromechanic spring constant. The electromagnetic torque in steady state T_{e0} is an expression of V_s, ω_r and δ_0. Now, the characteristic equation of the system operating in open loop is

$$1 + \frac{n}{2J_m\left(s + \frac{B_m}{J_m}\right)}\frac{K_e}{s} = 0 \tag{1.15}$$

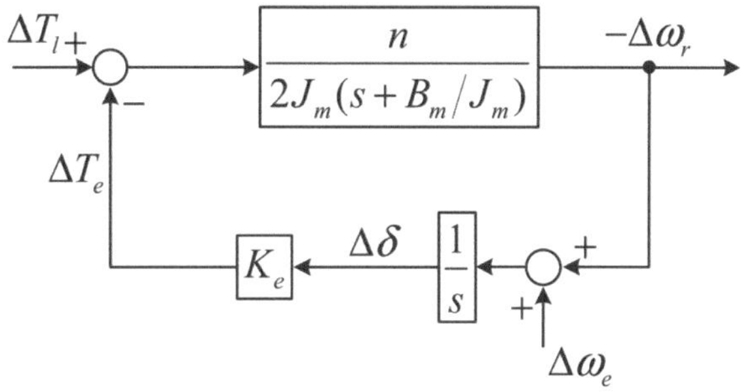

Fig. 1.6. Block diagram of the simplified small signal model operating at V/f control strategy

But looking at the model in (1.10), the perturbation of the load angle Δδ can be expressed as

$$\Delta\delta = \int (\Delta\omega_e - \Delta\omega_r)\, dt \tag{1.16}$$

and the model of Figure 1.6 can be obtained.

The instability of the permanent magnet synchronous motor described in section 1.3 is due to the low coupling between the electrical and mechanical modes. This instability shows a relatively small positive value of the dominant poles of the system. Therefore, a small damping must be added to the system to stabilize it [12]. In order to add this damping, it is only necessary to add to the torque a component proportional to the perturbation speed as

$$\Delta\omega_e = -K_v \frac{d\Delta\omega_r}{dt} \tag{1.17}$$

as seen in Figure 1.7. Now, the characteristic equation is

$$s^2 + \frac{2B_m + n + K_e K_v}{2J_m} s + \frac{n}{2J_m} K_e = 0 \tag{1.18}$$

and the stability characteristic of the system can be determined by K_v. This stabilization can be implemented measuring speed and extracting the perturbation, but then, a speed sensor is needed.

4.2 Stabilizing using power perturbations

Power perturbations can also be used to modulate the excitation frequency in order to add damping to the system and stabilize it. The power can be expressed as a steady state value plus a perturbation as

$$p_e = P_e + \Delta p_e = P_p + \frac{dW_e m}{dt} + \left(\frac{2}{n}\right)^2 \frac{J_m}{2} \frac{d}{dt}\omega_r^2 + \left(\frac{2}{n}\right)^2 B_m \omega_r^2 + \frac{2}{n}\omega_r T_l \tag{1.19}$$

where the first term are the loses and for small perturbations can be considered constant. The second term is the variation of the stored magnetic energy, and, however it is not constant, has a constant average value in an electrical rotation. Then, the power perturbations can be expressed as

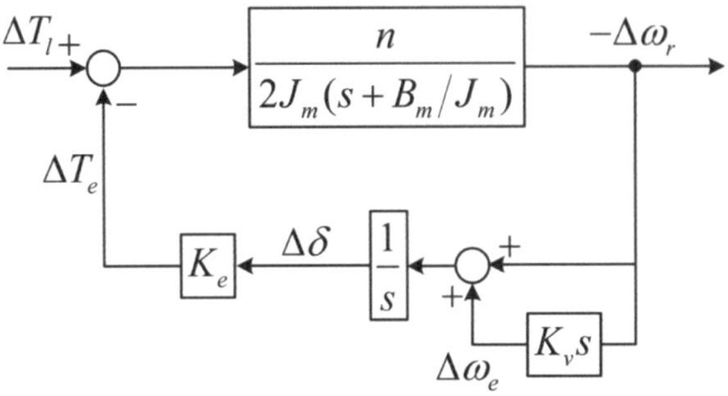

Fig. 1.7. Block diagram of the small signal model when the applied frequency is modulated with rotor speed perturbations

$$\Delta p_e = \left(\frac{2}{n}\right)^2 J_m \omega_{r0} \frac{d}{dt}\Delta\omega_r + 2\left(\frac{2}{n}\right)^2 B_m \omega_{r0}\Delta\omega_r + \frac{2}{n}T_{l0}\Delta\omega_r \tag{1.20}$$

Looking at equation (1.20) and at Figure 1.6, the excitation frequency can be modulated proportional to input power perturbations as

$$\Delta\omega_e = -K_p\Delta p_e = \\ -K_p\left(\left(\frac{2}{n}\right)^2 J_m\omega_{r0}\frac{d}{dt}(-\Delta\omega_r) + 2\left(\frac{2}{n}\right)^2 B_m\omega_{r0}(-\Delta\omega_r) + \frac{2}{n}T_{l0}(-\Delta\omega_r)\right) \tag{1.21}$$

and the block diagram of Figure 1.8 can be obtained. With this modulation technique, the characteristic equation of the system is

$$s^2 + \left(\frac{B_m}{J_m} + \frac{2K_e\omega_{r0}K_p}{n}\right)s + \frac{K_e}{2J_m}\left(n + 4\frac{2}{n}B_m K_p\omega_{r0} + 2T_{l0}K_p\right) = 0 \tag{1.22}$$

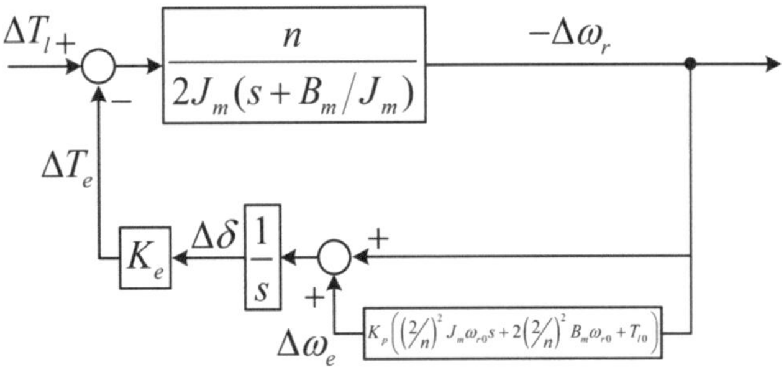

Fig. 1.8. Block diagram of the simplified smal signal model operating at V/f control where the applied frequency is modulated with input power perturbations

and the poles of the system are

$$s = -\left(\frac{B_m}{2J_m} + \frac{K_e\omega_{r0}K_p}{n}\right) \pm \sqrt{\left(\frac{B_m}{2J_m} + \frac{K_e\omega_{r0}K_p}{n}\right)^2 - \frac{K_e}{2J_m}\left(n + 4\frac{2}{n}B_m K_p\omega_{r0} + 2T_{l0}K_p\right)} \tag{1.23}$$

It can be seen that the damping of the system can be determined by the value of K_p in a steady state operation point. But, observing equation (1.20), it is possible to see that imposing a constant value of K_p, it is not possible to fix the poles of the system in the whole speed operating range. It is possible to almost fix the poles of the system not fixing K_p, but fixing the product $K_p\omega_{r0} = C_p$.

The frequency modulation can also be done using the perturbations of the DC link current [10]. Supposing that the DC link voltage is constant, power can be expressed as

$$p_e = P_e + \Delta p_e = V_{DC}\left(I_{DC} + \Delta i_{DC}\right) = P_p + \frac{dW_em}{dt} + \left(\frac{2}{n}\right)^2 \frac{J_m}{2}\frac{d}{dt}\omega_r^2 + \left(\frac{2}{n}\right)^2 B_m\omega_r^2 + \frac{2}{n}\omega_r T_l \tag{1.24}$$

if power at transistor level are not considered. Then, the modulation of the exciting frequency can be done as

$$\Delta\omega_e = -K_i\Delta i_{DC} = \frac{1}{V_{DC}}\left(\left(\frac{2}{n}\right)^2 J_m\omega_{r0}\frac{d}{dt}\left(-\Delta\omega_r\right) + 2\left(\frac{2}{n}\right)^2 B_m\omega_{r0}\left(-\Delta\omega_r\right) + \frac{2}{n}T_{l0}\left(-\Delta\omega_r\right)\right) \tag{1.25}$$

But, must be noted that

$$K_i = K_p V_{DC} \tag{1.26}$$

giving the same stability characteristics using input power perturbations and DC link current perturbations.

4.3 Stability verification

In the model expressed in (1.2), a new equation that describes the modulation of the excitation frequency must be added as

$$\Delta\omega_e = -K_p\Delta p_e \tag{1.27}$$

where input power perturbations are computed using a first order filter as

$$\Delta p_e = \frac{s}{s + \frac{1}{\tau_h}} p_e \tag{1.28}$$

where τ_h is high-pass filter time constant. Then, the fifth differential equation is

$$\Delta\dot{\omega}_e + \frac{\Delta\omega_e}{\tau_h} = -K_p\dot{p}_e \tag{1.29}$$

The input power can be calculated using the Park's voltages and currents as

$$p_e = \frac{2}{3}\left(v_{ds}^r i_{ds}^r + v_{qs}^r i_{qs}^r\right) \tag{1.30}$$

Supposing that the voltage is constant, i.e. no perturbation is applied, the time derivative of the power can be expressed as

$$\dot{p}_e = \frac{2}{3}V_s\left(-\dot{i}^r_{ds}\sin\delta + \dot{i}^r_{qs}\cos\delta - \left(i^r_{ds}\cos\delta + i^r_{qs}\sin\delta\right)\dot{\delta}\right) \tag{1.31}$$

To obtain the fifth system equation, equation (1.31) must be substituted in (1.29). Substituting time derivative expressions of currents and load angle, one can obtain the fifth differential equation as

$$\begin{aligned}\Delta\dot{\omega}_e = &\frac{3}{2}K_pV_s\left(\frac{\omega_r L_d\cos\delta}{L_q} - \frac{R_s\sin\delta}{L_d} + (\omega_e-\omega_r)\cos\delta\right)i^r_{ds}\\ &+\frac{3}{2}K_pV_s\left(\frac{\omega_r L_q\sin\delta}{L_d} + \frac{R_s\cos\delta}{L_q} + (\omega_e-\omega_r)\sin\delta\right)i^r_{qs}\\ &+\frac{3}{2}K_pV_s\frac{\lambda_m\cos\delta}{L_q}\omega_r\\ &-\frac{3}{2}K_pV_s\left(\frac{V_s\sin^2\delta}{L_d} - \frac{V_s\cos^2\delta}{L_q}\right)\\ &-\frac{1}{\tau_h}\Delta\omega_e\end{aligned} \tag{1.32}$$

Linearizing the new system, one can obtain the stability characteristics of the system with the modulation of the frequency using input power perturbations. The linearized system can be seen in Appendix B. Figure 1.9 shows the root locus as a function of K_p. As seen in Figure 1.9, the motor is stable for a range of values of K_p. The value of the K_p determines, as previously seen, the stability characteristics of the system. In this case, the cut-off frequency of the high-pass filter used to extract the input power perturbations is 2.5 Hz. The time constant τ_h is 0.0637 s. That locates the fifth pole of the system at s =-16. As seen in Figure 1.9, the rotor poles moves from the unstable region to stable region as K_p increases, but the stator poles moves from stable region to unstable region, giving the limits of the constant K_p.

As said before, in order to maintain the stability characteristics of the motor in the whole frequency range, the product $C_p = K_p\omega_{r0}$ is maintained constant. In this case, C_p =12.5664, giving a value of K_p =0.01 at rated speed.

The root locus of the motor as a function of excitation frequency can be seen in Figure 1.10. As seen in Figure 1.10 the poles of the system remains now in the stable region in the whole frequency range. It can also be seen in Figure 1.10 that the stability characteristics of the system is almost constant for frequencies above 100 Hz. This is because of the constant $C_p = K_p\omega_{r0}$ product.

5. Implementation of the stabilized V/f control

As seen in Figure 1.11, the implementation of the V/f control of PMSM with stabilization loop has two main parts. First the computation of power perturbations, and second, the computation of the voltage applied to the motor.

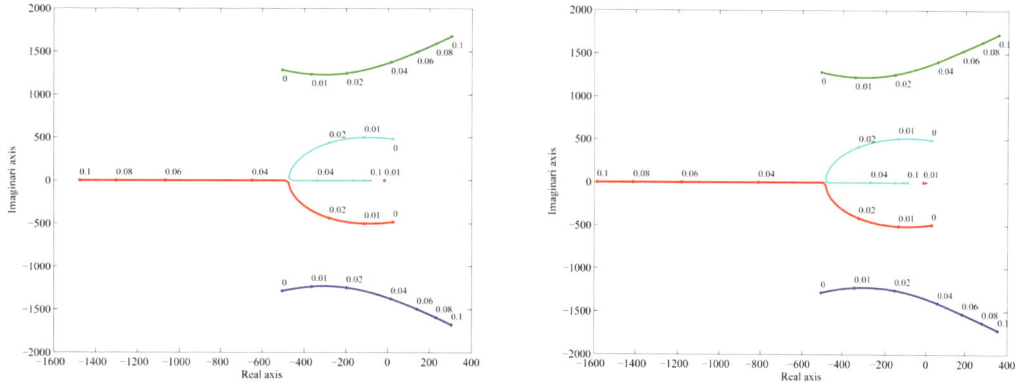

(a) Root locus of the system as a function of K_p operating the motor at no-load and rated speed

(b) Root locus of the system as a function of K_p operating the motor at full-load and rated speed

Fig. 1.9. Root locus of the stabilized motor as a function of K_p

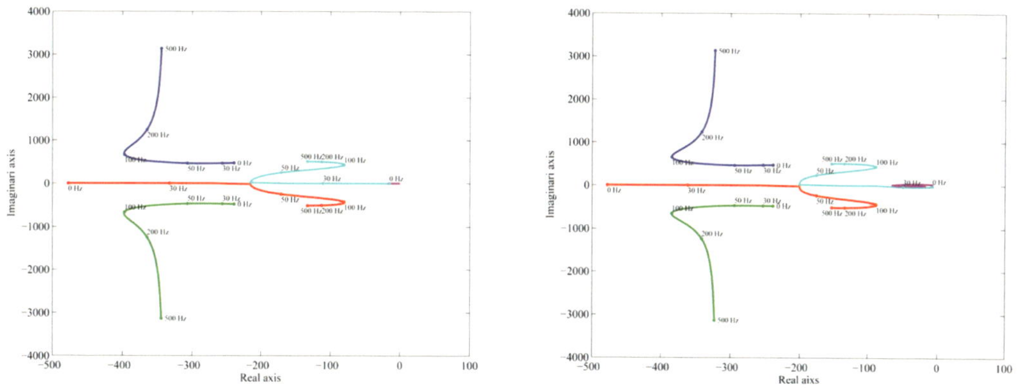

(a) Root locus of the system as a function of stator frequency when operating the motor at no-load

(b) Root locus of the system as a function of stator frequency when operating the motor at full-load

Fig. 1.10. Root locus of the stabilized motor as a function of stator frequency

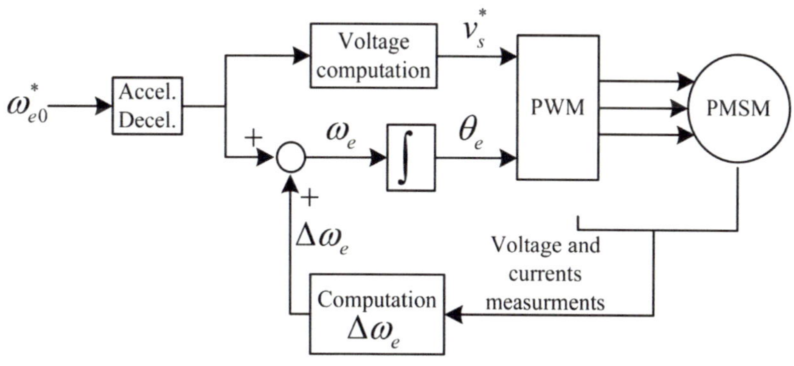

Fig. 1.11. Block diagram of the V/f control of PMSM with stabilization loop

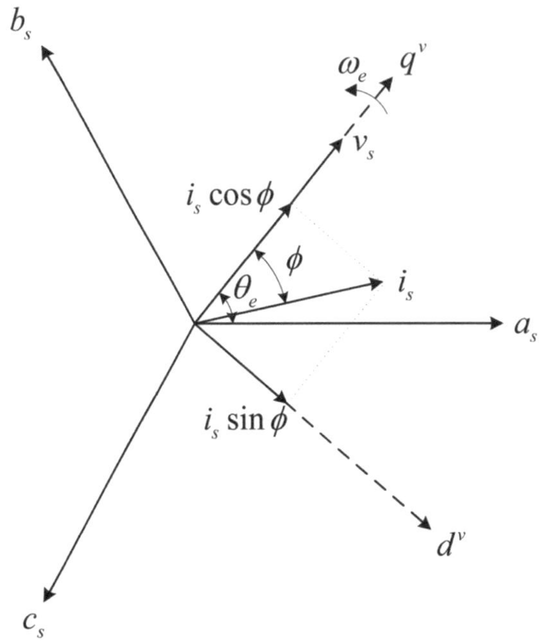

Fig. 1.12. Voltage and current vectors in the stator synchronous reference

5.1 Computation of power perturbations

The power delivered to the motor can be computed as

$$p_e = \frac{3}{2} v_s i_s \cos\phi \tag{1.33}$$

Observing Figure 1.12, $i_s \cos\phi$ can be computed as

$$i_{qs}^v = i_s \cos\phi = \frac{2}{3}\left(i_{as}\cos\theta_e + i_{bs}\cos\left(\theta_e - \frac{2\pi}{3}\right) - (i_{as}+i_{bs})\cos\left(\theta_e + \frac{2\pi}{3}\right)\right) \tag{1.34}$$

where the angle θ_e is the electrical angle of the applied voltage, and it is known. In (1.33), the v_s is the calculated voltage.

The input power perturbations can be easily obtained by a first order high-pass filter as

$$\Delta p_e = \frac{s}{s + \frac{1}{\tau_h}} p_e \tag{1.35}$$

5.2 Magnitude of the voltage vector

In the V/f control, the magnitude of the voltage vector is selected to maintain the motor flux constant, maintaining the relation V/f constant [13]. The steady state equations of the permanent magnet synchronous motor are

$$V_{ds}^{r} = R_s I_{ds}^{r} - \omega_{r0} L_q I_{qs}^{r} = R_s I_{ds}^{r} - \omega_{r0} \lambda_{ds}^{r} \tag{1.36}$$

$$V_{qs}^{r} = R_s I_{qs}^{r} + \omega_{r0} L_d I_{ds}^{r} + \omega_{r0} \lambda_m = R_s I_{qs}^{r} + \omega_{r0} \lambda_{qs}^{r} \tag{1.37}$$

If the resistive voltage droop is considered small

$$V_{ds}^{r} \approx -\omega_{r0} \lambda_{ds}^{r} \tag{1.38}$$

$$V_{qs}^{r} \approx \omega_{r0} \lambda_{qs}^{r} \tag{1.39}$$

The magnitude of the voltage vector is

$$V_s = \sqrt{\left(V_{ds}^{r}\right)^2 + \left(V_{qs}^{r}\right)^2} \approx \omega_{r0} \sqrt{\left(\lambda_{ds}^{r}\right)^2 + \left(\lambda_{qs}^{r}\right)^2} \tag{1.40}$$

and, then,

$$\frac{V_s}{\omega_{r0}} \approx \lambda_s \tag{1.41}$$

But, if the motor is operating at low speed or at high load level, the resistive voltage drop can not be neglected. When the motor operating at low speed or at high load level, maintaining the relation $\frac{V_s}{\omega_{r0}}$ constant does not maintains the flux constant, diminishing motor performance [14, 15]. In super-high-speed PMSM, with very low inductance, the stator resistance cannot be neglected even at high speeds [16]. Therefore, it is necessary to compensate the resistive voltage droop in the applied voltage. In HVAC applications, high-efficiency operation is desired. Low performance V / f control is ideal for efficiency-optimized control. This efficiency optimization can be achieved by means of controlling motor flux [17].

Voltage vector can be expressed adding the resistive voltage drop and the rotor permanent magnet induced voltage, as seen in Figure 1.13. The voltage vector can be expressed as

$$V_s = BC + CO = I_s R_s \cos\phi_0 + \sqrt{E_s^2 + I_s^2 R_s^2 \cos^2\phi_0 - I_s^2 R_s^2} \tag{1.42}$$

The vector E_s is the stator flux induced voltage. The stator flux is normally chosen to be equal as the rotor flux. Then,

$$E_s = \omega_{r0} \lambda_m \tag{1.43}$$

Even expression (1.42) is in steady state, the value of the voltage magnitude can be computed instantaneously as

$$v_s^* = R_s i_s \cos\phi + \sqrt{\left(\omega_{r0}\lambda_m\right)^2 + \left(R_s i_s \cos\phi\right)^2 - \left(R_s i_s\right)^2} \tag{1.44}$$

where i_s can be computed as

$$i_s = \sqrt{\left(i_{ds}^{s}\right)^2 + \left(i_{qs}^{s}\right)^2} = \sqrt{\frac{1}{3}\left(i_{as} + 2i_{bs}\right)^2 + \left(i_{as}\right)^2} \tag{1.45}$$

6. Experimental results

Figure 1.14 shows the block diagram of the implementation of the stabilizing loop. In i_s and $i_s \cos\phi$ computation, some ripple can be present. This ripple can affect the performance at low speed, deteriorating the voltage applied. Therefore, it is then necessary to filter these values.
Figure 1.15 shows the variation of the variables in a no-load ramp up and then adding torque step change. As it can be seen, with the stabilizing loop, the motor operation is stable in the whole frequency range. Once the rated speed is reached, at time 4 s, a rated torque

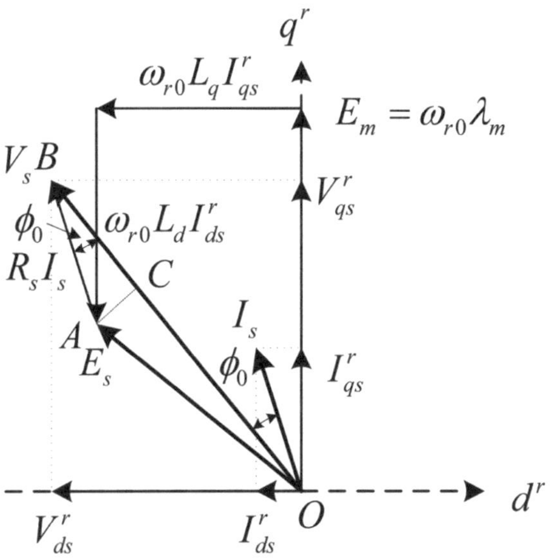

Fig. 1.13. Steady state vector diagram

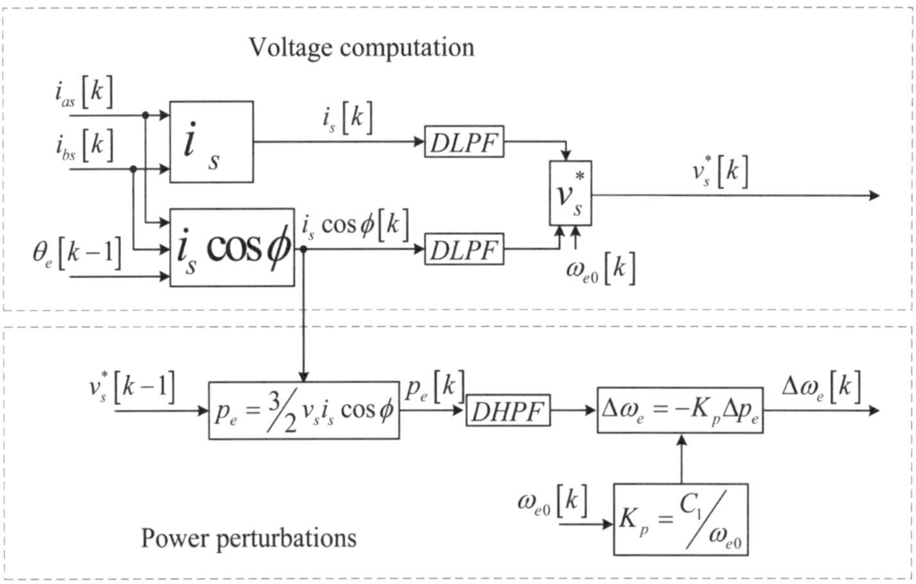

Fig. 1.14. Block diagram of the implementation of the stabilizing loop

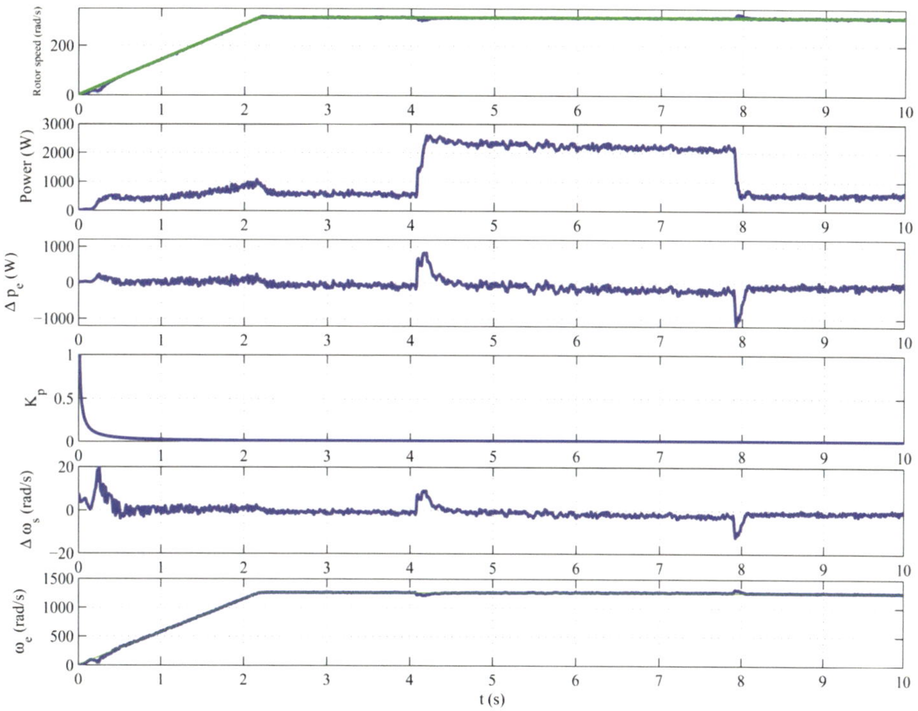

Fig. 1.15. Rotor speed, input power power perturbation, K_p, frequency perturbations $\Delta\omega_e$ and excitation frequency of the motor when operating at rated speed (200 Hz) with a load step.

step is applied, generating an input power perturbation. The stabilization loop compensates this perturbation, reducing the excitation frequency, and maintaining synchronism. At time 8 s, the torque is released, generating an input power perturbation. In this case, the stabilizing loop reacts increasing excitation frequency, to not to loose synchronization. In these torque steps, the variation of the rotation speed is less than 5 % of the rated speed during less than 200 ms. This performances are good enough for HVAC applications.

The variation of rotor speed when the stabilizing loop is removed can be seen in Figure 1.16. At time 7.5 s, the K_p is made zero, removing the stabilizing loop. Instantaneously, the motor loses synchronization.

The effects of the voltage vector resistive drop compensation can be seen in Figure 1.17. The motor is ramped up to a 40 Hz excitation frequency, including the resistive voltage drop compensation. Then, at time 7 s, this resistive voltage drop compensation is released, making R_s zero. Then voltage drop is no more compensated, and the motor loses synchronization, because not enough flux is created to maintain the rotor synchronized with the stator applied currents.

7. Conclusions

V/f control strategy for permanent magnet synchronous motors can be useful for HVAC applications, where not high performance is required. Permanent magnet synchronous motors have efficiency advantages over the induction motor. But open loop V/f control is not stable in the whole frequency range. As demonstrated, the V/f control strategy becomes

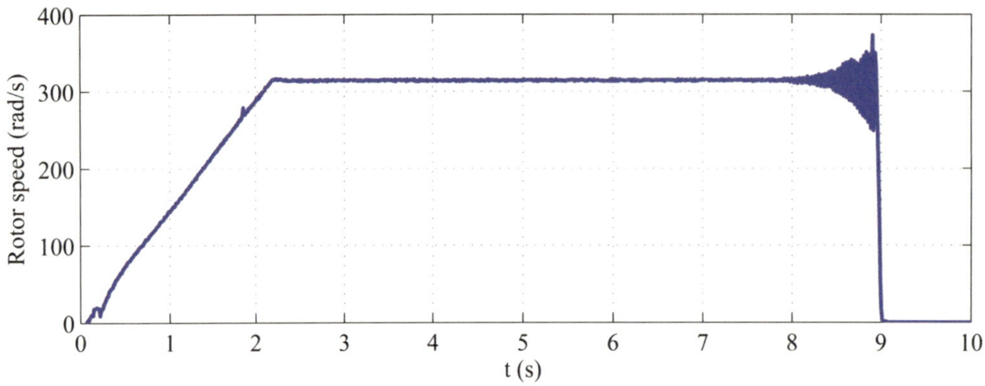

(a) Rotor speed

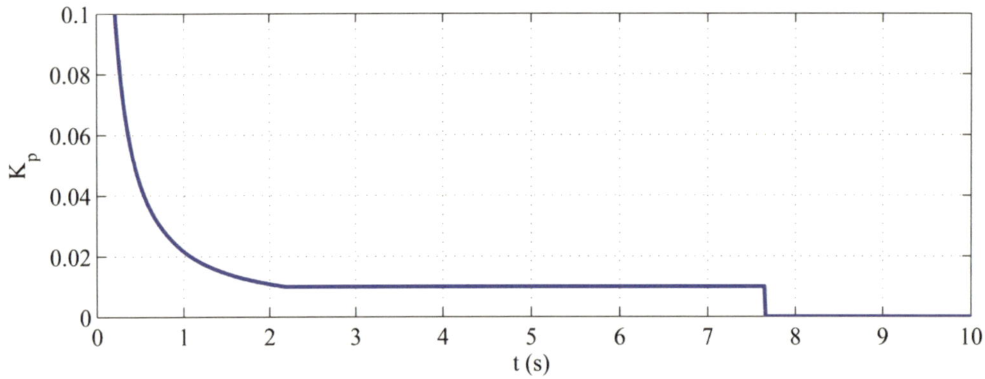

(b) K_p

Fig. 1.16. Rotor speed variation when $K_p = 0$

unstable, even at no-load, from a certain excitation frequency lower than the rated speed. Then, it is necessary to have a stabilizing loop in the system.

This stabilizing loop can be implemented by means of an speed measurement system, increasing cost, and complexity. The objective of this work is to develop a sensorless stabilizing loop. The presented strategy uses input power perturbations to stabilize the system. After adding the stabilizing loop, the V/f operation of the permanent magnet synchronous motor is stable for all the frequency range, and for any load torque applied to the motor.

Future research includes the estimation of the initial rotor position. Here, the motor is started from a known position, but for real applications, the rotor can be at any position. The stabilizing method developed uses some motor parameters. The variation of this parameters with temperature or even aging, must be studied. Increasing speed in PMSM over the rated speed means field weakening. For the operation above rated speed in HVAC applications is of interest, this method must be studied.

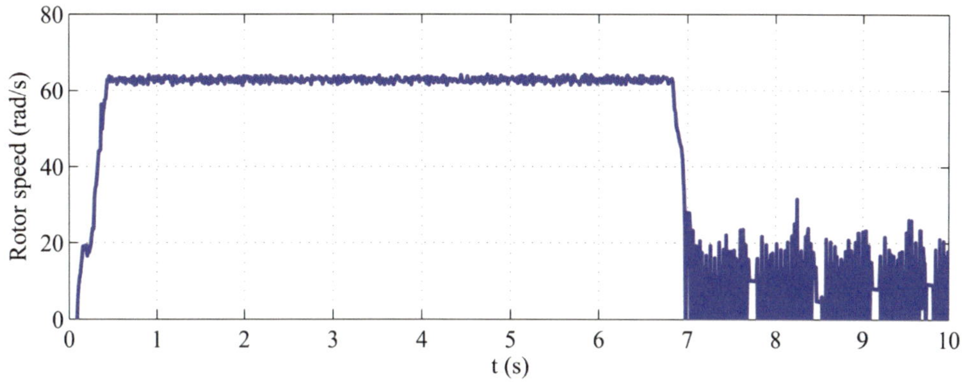

(a) Rotor speed at excitation frequency of 40 Hz

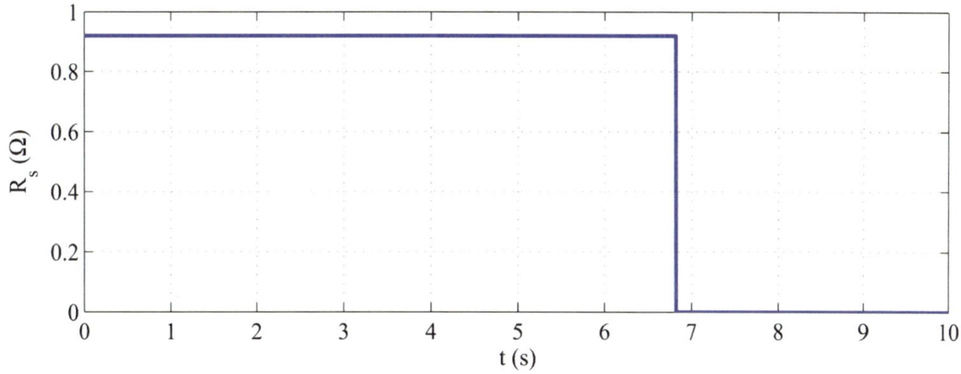

(b) R_s value used in computation. When it is made to zero, no voltage drop compensation is done

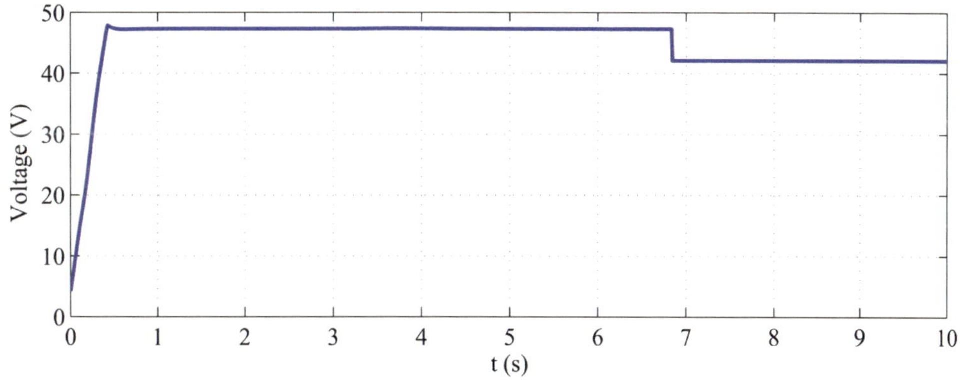

(c) Applied stator voltage

Fig. 1.17. Variation of rotor speed when there is no resistive voltage drop compensation at low speed (40 Hz)

A. Motor parameters

Model	MAVILOR BLS 115 4/400
Pole number (n)	8
Rated power	2,2 kW
Rated speed	3000 rpm
Rated frequency	200 Hz
Rated torque	8,1 Nm
Rated phase to phase voltage	400 V(rms)
Rated phase current	5,71 A(rms)
Stator resistance per phase (R_s)	0,92 Ω
d-axis inductance (L_d)	1,925 mH
q-axis inductance (L_q)	1,925 mH
Rotor permanent magnet flux (λ_m)	0,1674 V s rad^{-1}
Inertia of the mechanical system (J_m)	0,9724×10^{-3} kg m^2
Viscous friction coefficient (B_m)	1,3671×10^{-6} Nm s rad^{-1}

B. Linearization of the stabilized system

In order to analyze the stability of the system under frequency modulation, the model must be linearized. The linearized system has the form

$$\Delta\dot{x} = \mathbf{A}_1(\mathbf{X})\Delta\mathbf{x} + \mathbf{B}_1(\mathbf{X})\Delta T_l \tag{B.1}$$

where $\Delta\mathbf{x}$ is the state variables vector as

$$\Delta\mathbf{x} = \left\{\Delta i^r_{ds} \quad \Delta i^r_{qs} \quad \Delta\omega_r \quad \Delta\delta \quad \Delta\omega_e\right\}^T \tag{B.2}$$

and the B_1 matrix is

$$\mathbf{B}_1(\mathbf{X}) = \begin{pmatrix}0 & 0 & \frac{-n}{2J_m} & 0 & 0\end{pmatrix}^T \tag{B.3}$$

The elements of the state matrix $\mathbf{A}_1$ are

$$A_{11} = -\frac{R_s}{L_d}, A_{12} = \frac{L_q\omega_{r0}}{L_d}, A_{13} = \frac{L_q I^r_{qs}}{L_d}, A_{14} = -\frac{V_s\cos\delta_0}{L_d}, A_{15} = 0$$

$$A_{21} = -\frac{L_d\omega_{r0}}{L_q}, A_{22} = -\frac{R_s}{L_q}, A_{23} = -\frac{1}{L_q}\left(L_d I^r_{ds} + \lambda_m\right), A_{24} = -\frac{V_s\sin\delta_0}{L_q}, A_{25} = 0$$

$$A_{31} = \frac{3}{2}\left(\frac{n}{2}\right)^2\frac{1}{J_m}\left(L_d - L_q\right)I^r_{qs}, A_{32} = \frac{3}{2}\left(\frac{n}{2}\right)^2\frac{1}{J_m}\left(\lambda_m + \left(L_d - L_q\right)I^r_{ds}\right),$$

$$A_{33} = -\frac{B_m}{J_m}, A_{34} = 0, A_{35} = 0$$

$$A_{41} = 0, A_{42} = 0, A_{43} = -1, A_{44} = 0, A_{45} = 1$$

$$A_{51} = \frac{3}{2}K_pV_s\left(\frac{L_d\omega_{r0}\cos\delta_0}{L_q} - \frac{R_s\sin\delta_0}{L_d}\right), A_{52} = \frac{3}{2}K_pV_s\left(\frac{R_s\cos\delta_0}{L_q} + \frac{L_q\omega_{r0}\sin\delta_0}{L_d}\right),$$

$$A_{53} = \frac{3}{2}K_pV_s\left(\left(\frac{L_q}{L_d} - 1\right)I^r_{qs}\sin\delta_0 + \left(\left(\frac{L_d}{L_q} - 1\right)I^r_{ds} + \frac{\lambda_m}{L_q}\right)\cos\delta_0\right),$$

$$A_{54} = \frac{3}{2}K_pV_s\left(\left(\frac{1}{L_q} - \frac{1}{L_d}\right)V_s\sin 2\delta_0 + \frac{1}{L_d}\left(L_q\omega_{r0}I^r_{qs} - R_sI^r_{ds}\right)\cos\delta_0 - \right)$$

$$\left(\frac{1}{L_q}\left(R_sI^r_{qs} + \left(L_dI^r_{ds} + \lambda_m\right)\omega_{r0}\right)\sin\delta_0\right),$$

$$A_{55} = \frac{3}{2}K_pV_s\left(I^r_{ds}\cos\delta_0 + I^r_{qs}\sin\delta_0\right) - \frac{1}{\tau_h}$$

The eigenvalues of the state matrix can be used to analyze the stability characteristics of the system.

8. References

[1] P. Vas, Sensorless vector control and direct torque control. Oxford University Press, 1998.

[2] J. Itoh, N. Nomura, and hiroshi Ohsawa, "A comparison between *V/f* control and position-sensorless vector control for the permanent magnet synchronous motor," in *Proc. Power Conversion Conf.*, vol. 3, Osaka, Japan, Apr. 2002, pp. 1310-1315.

[3] P. D. Chandana Perera, "Sensorless control of permanent-magnet synchronous motor drives," Ph.D. dissertation, Institute of energy technology. Aalborg university, 2002.

[4] G. R. Slemon, *Electrical machines for drives.* IEEE Press, 1997, ch. 2 in Power electronics and variabale frequency drives. Tecnology and applications, pp. 36-76.

[5] T. M. Jahns, *Variable frequency permanent magnet AC machine drives.* IEEE Press, 1997, ch. 6 in Power electronics and variabale frequency drives. Tecnology and applications, pp. 277-325.

[6] B. K. Bose, Ed., Power electronics and variable frequency drives. Technology and applications. IEEE Press, 1997.

[7] D. W. Novotny and T. A. Lipo, *Vector control and dynamics of AC drives*, P. Hammond, T. J. E. Miller, and T. Kenjo, Eds. Oxford University Press, 2000.

[8] P. C. Krause, O. Wasynczuk, and S. D. Sudhoff, *Analysis of electrical machinery and drive systems*, 2nd ed. IEEE Press, 2002.

[9] G. C. Verghese, J. H. Lang, and L. F. Casey, "Analysis of instability in electrical machines," *IEEE Trans. Ind. Applicat.*, vol. IA-22, no. 5, pp. 853-864, 1986.

[10] R. S. Colby and D. W. Novotny, "An efficiency-optimizing permanent-magnet synchronous motor drive," *IEEE Trans. Ind. Applicat.*, vol. 24, no. 3, pp. 462-469, May/Jun. 1988.

[11] P. D. Chandana Perera, F. Blaabjerg, J. K. Pedersen, and P. Thøgersen, "A sensorless, stable *V/f* control method for permanent-magnet synchronous motor drives," *IEEE Trans. Ind. Applicat.*, vol. 39, no. 3, pp. 783-791, May/Jun. 2003.

[12] D. Montesinos-Miracle, "Modelització i control d'accionaments elèctrics," Ph.D. dissertation, Technical University of Catalonia, Jul. 2008.

[13] B. K. Bose, *Modern power electronics and AC drives.* Prentice Hall PTR, 2002.

[14] A. Munoz-Garcia, T. A. Lipo, and D. W. Novotny, "A new induction motor v/f control method capable of high-performance regulation at low speeds," *IEEE Trans. Ind. Applicat.*, vol. 34, no. 4, pp. 813-821, 1998.

[15] M. P. Kazmierkowski, R. Krishnan, and F. Blaabjerg, Eds., *Control in power electronics*, ser. Academic Press series in engineering. Academic Press. Elsevier Science, 2002.

[16] L. Zhao, C. H. Ham, Q. Han, T. X. Wu, L. Zheng, K. B. Sundaram, J. Kapat, and L. Chow, "Design of optimal digital controller for stable super-high-speed permanent-magnet synchronous motor," *IEE Proceedings -Electric Power Applications*, vol. 153, no. 2, pp. 213-218, 2006.

[17] F. Abrahamsen, F. Blaabjerg, J. K. Pedersen, and P. B. Thoegersen, "Efficiency-optimized control of medium-size induction motor drives," *IEEE Trans. Ind. Applicat.*, vol. 37, no. 6, pp. 1761-1767, 2001.

24

Fuzzy Sliding Mode Control of a Ball Screw Driven Stage

Mohammad Shams[1] and Masoud Safdari[2]
[1]*MSc. In Manufacturing,*
[2]*PhD Student, Mechanical Engineering Dept, University of New Mexico*
[1]*Iran,*
[2]*USA*

1. Introduction

Nowadays by introduction of Nanotechnology as cutting edge of research, researchers are studying and investigating phenomenon in nanoscales. Almost all various tools required for this purpose are calling for next generation of positioning systems with far better precision than available counterparts. As instance processes such as ultra precision machining and micro assembly require high precision positioning systems. These processes have their applications in semiconductor, optical and precision metrology industries. Accordingly there are vast varieties of available positioning systems. Among them, fines pitch screws, differential screws, piezo electric positioning systems, linear motors and ball screw driven stages could be mentioned. Each system has its own advantages and disadvantages. For example generally speaking a piezoelectric positioning system has better positioning accuracy while the stroke, the velocity and the maximum thrust force of this system is more limited. Ball screw driven systems have been the subject of much attention because of their major advantages like large stroke, higher speeds and thrust force, availability and low cost. Despite the architecture of each system, in order to achieve high precision in positioning, it is important to resolve the performance characteristics of the system in micro and submicron scale. Nonlinear phenomena such as friction, backlash and compliance can reduce the positioning precision and tracking accuracy. Among them friction always plays a key role.

The nonlinear behavior of friction may lead to consequences like limit cycle and instability in the dynamic system. As a consequence such behavior in a control system demands for more complex hierarchy in its controller. In other words, nonlinear positioning systems are mostly less precise in positioning and less accurate in tracking. Two important consequences of friction in sliding systems are steady state error in positioning in positioning and lag in tracking. These two are direct effect of the Stick-Slip behavior of friction. To eliminate these deficiencies, the nonlinear effects of friction should be compensated. Therefore a behavioral model of friction is required to eliminate these effects.

Various researches have been carried out to characterize and model the friction in a positioning system. Dahl et al. proposed a dynamic model describing the spring like behavior during stiction (Dahl, 1968). One shortcomings of this model is that Stribeck effect had not been considered in this model. First attempt to incorporate Stribeck effect to Dahl model has been done by Bliman and Sorine (Bliman & Sorine, 1991). Canudas et al.

proposed a model that captures most of frictional behaviors such as hysteresis, spring - like behavior during stiction, Stribeck effect and variable break - away force observed in practice (Canudas et al., 1995).

In order to compensate the friction in a controlled positioning system, different types of controllers have been proposed. Ro et al. developed a PD control scheme with a nonlinear friction estimation algorithm based on Canudas friction model for submicrometer positioning and tracking (Ro et al., 2000). In addition, variable structure systems can also be employed to compensate consequences of friction in motion. For instance sliding mode control can be utilized. This type of controller can treat a variety of uncertainties as well as bounded external disturbances and satisfy system characteristics such as stability, elimination or decrease of chattering, linearization of nonlinear and order reduction in accordance with a design target. However sliding mode controllers suffer from chatter phenomena. In order to reduce the chattering, especially in the systems with some degrees of uncertainty, more sophisticated controlling schemes like boundary layer sliding mode controller and integral sliding mode control are proposed. These controlling schemes can compromise between small chattering and good tracking precision. Chen et al. compensated for friction in static and dynamic regimes in a ball screw driven stage by developing a sliding mode controller for each regime (Chen et al. 2004). Fuzzy logic controllers are another type of variable structure controllers which are more applicable to complex systems and ill-defined systems. Regarding the fact that the governing parameters of friction model are mostly hard to define and also these parameters are subject to change in various controlling environments, it is suitable to utilize variable structure controllers for the purpose of compensation.

Fuzzy sliding mode control is another control strategy that attracts a great deal of interest. Lee et al. developed a sliding mode controller based on fuzzy variable boundary layer with a control gain and boundary layer thickness as design parameters (Lee et al., 2001). Liang et al. proposed a two input single output (TISO) fuzzy sliding mode controller (FSMC) by developing complementary sliding mode controller (CSMC) (Liang &, 2003). Mihajlov et al. applied a sliding mode controller enhanced by fuzzy PI controller to an electro-hydraulic positioning system, whose model includes internal friction (Mihajlov1 et al., 2002).

In following chapter, the phenomena of backlash and friction are introduced firstly and then Canudas friction model (Canudas et al., 1995) is used to capture the frictional behaviors in a ball screw driven stage. Conventional sliding mode control with boundary layer, complementary sliding mode control and TISO fuzzy sliding mode control (Liang &, 2003) are applied to this dynamic system for the purpose of the compensation of friction effects. Finally the performances of various controllers are evaluated in the sense of submicrometer positioning, tracking accuracy and control input to the system.

The outline of this chapter is as follows. In section 2, backlash and friction are described and then friction model proposed by Canudas is briefly introduced. In section 3, model of dynamic system is explained. In section 4, theory of sliding mode control is explained at first and then CSMC and TISO FSMC are described. Sections 5 and 6 include designed parameters and computer simulation respectively. The chapter ends with conclusion and future work remarks in section 7.

2. Introducing system nonlinearities

Friction and backlash are two important nonlinearities which dominate performances of ball screw driven stages in the sense of positioning precision and tracking accuracy. The term friction indicates any speed or force-dependent energy dissipation in a machine. Backlash implies mechanical hysteresis and multibody impaction. An introduction to these topics is outlined below and then a model for friction is introduced.

2.1 Backlash

In conventional engineering terminology, backlash is the distance through which one part of connected machinery, as a gear, or screw, can be moved without moving the connected parts, resulting from looseness in fitting or from wear. In other word, backlash results in mechanical hysteresis. Basically the existence of backlash between the components of a machine like ball screw and gears reduces machine's accuracy. Two main features of backlash are mechanical hysteresis and the impact which is mostly important between two relatively hard surfaces in contact. In classic model backlash is considered as an unsymmetrical deadband around the equilibrium. In this model for the simplicity it is assumed that all of dynamic impacts are fully plastic and therefore exhausting in the backlash zone. Figure 1 illustrates the phenomenon of backlash, reaction versus action, based on classical modeling considerations. It is worth noting that the slope and the deadbands may differ in either side.

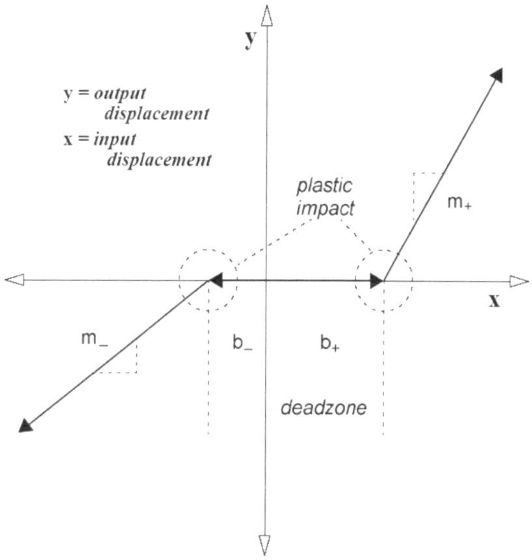

Fig. 1. Classical model of backlash. (de Marchi. 1998)

In modern model, both features of backlash which are mostly effective on the accuracy of the positioning in machine tool have been considered. In this model, the backlash is considered as a combination of deadband and an almost elastic impact between numbers of rigid bodies in contact. The impact can cause transient vibrations between rigid bodies (e.g. components made of hard metal). The existence of low frequency vibrations causes inaccuracy in positioning. The deadband in this model indicates the region in which there is not any considerable motion in output for a motion in input. In analogy with friction the deadband is similar to hysteresis caused by pre-sliding friction. The deadband is covering the point in which the velocity reduced to zero (the point of change in direction). Figure 2 illustrates the modern backlash model.

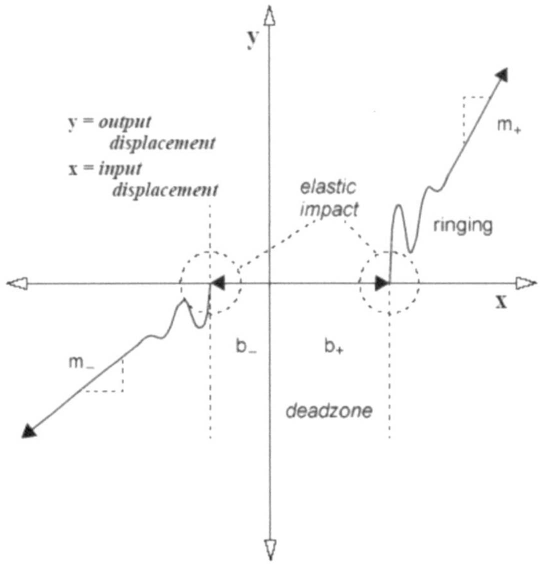

Fig. 2. Modern model of backlash. (de Marchi. 1998)

2.2 Friction

Friction force initiates from relative motion during contact and resists this motion. There are at least two different physical explanations for friction.

2.2.1 Classical friction model (static, dynamic and viscous)

Based on this model the mechanical clamping between surface asperities during contact is the origin of friction in case of static and dynamic friction. Thus the there should be a direct relation between the contact load (in vertical direction) and the friction force. However either the shape of contact surfaces or its effective contact area does not contribute to the friction. Additionally based on this model static friction force should be greater than dynamic friction force. Based on experimental results the coefficient of friction has been measured. It is also confirmed that the friction force is directly dependent to contact load. This model has been constructed based on the following assumptions:

- In each case, bulk and surface properties of the contacting pairs contribute to the friction.
- The amount of static friction force depends on the duration of contact.
- The amount of dynamic friction force is independent from sliding velocity of contact.
- A fraction of friction force is also dependent on the magnitude of adhesion force between the molecules on the contact surfaces.

Based on empirical investigations, the principle of Columb's friction represents friction as a simple mathematic relation:

$$f = n + t$$

Which f is the contact force, n is the normal force and $t < \mu n$ is the tangential shear force carried out by two surfaces in opposite direction of motion.

During sliding, $t = \mu_k n$ and in its threshold $t = \mu_s n$. Typically $\mu_s = \mu_k$ and the μ_s and μ_k are called the coefficients of static and dynamic friction respectively. The magnitude of μ_s and μ_k are taken to be constant for each pair of rubbing materials. The Columb's friction model investigated above provides a close approximation for clean, dry and geometrically regular surfaces in contact. The Columb's law of dynamic friction claims:

- The friction force is proportional to the contact load. Contact load is the magnitude of force acting along normal direction to the contact surfaces.
- For a constant load, the friction force is independent of the area of contact.
- The friction is independent of sliding velocity.
- The friction force is dependent to the nature of the pair of materials in contact.

During industrial revolution and introduction of high speed machines, a speed-dependent component of friction became evident, called viscous friction. Reynolds was the first person who investigated this type of friction which was related to the viscosity of a lubricant. The lubricants are typically non-Newtonian type and their viscosity increase with increasing shearing velocity. Figure 3 illustrates the friction model integrating static, dynamic and viscous types of friction.

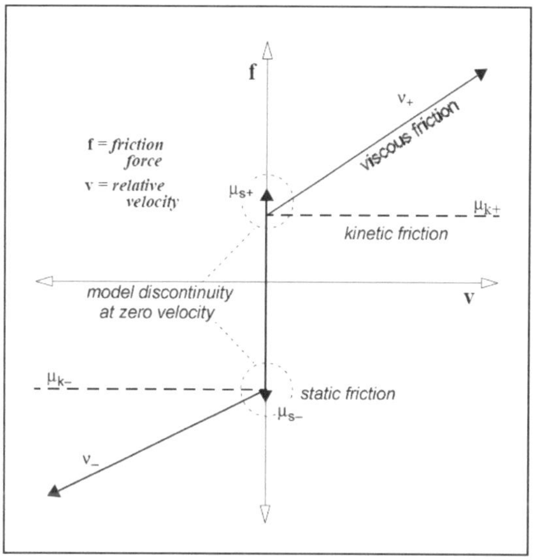

Fig. 3. Classical model of friction (static, dynamic and viscous). (de Marchi. 1998)

2.2.2 Modern model of friction (rolling friction, stribeck, stick-slip, hydraulic)

Contrasting classic model, modern model is more explicable via contact adhesion theories. Several reasons can be mentioned for this e.g. despite common sense; very smooth surfaces provide more friction than rougher surfaces. Researchers observed that actual area of contact is much less than apparent area. This difference that is due to surface asperities and the area of actual contact can be measured easily by measuring electrical conductivity between two surfaces in contact. In this model it is assumed that the magnitude of friction force is proportional to actual area of contact which is a fraction of apparent area. Since the actual area of contact is small, small amount of contact load causes enough pressure to weld contacting points to each other. So based on this model, frictional resistance is primarily

attributed to shearing of welds in contact points and secondarily to the stage of plowing the softer material by the harder one.
Based on experimental data gathered by Stribeck, the transition process from static friction stage to dynamic friction stage and finally to viscous stage is continuous. During measurement of friction on a spindle, he noted that the friction rapidly changes from its static value to a minimum dynamic value and then it will approach to its linear viscous friction. This transition that is a velocity dependent phenomenon is called Stribeck effect. This phenomenon is only observable over a very small velocity range and almost standstill or varies over a very long period of time during steady state operation. Stribeck friction sometimes is called negative friction which is because of existence of negative slope in the friction versus velocity curve. Later other researchers stated that negative friction is the prerequisite of the Stick-Slip phenomenon. This phenomenon originates from switching from static friction scheme to dynamic friction scheme between contacting surfaces. Finally disagreement on the dependence of friction force on velocity has been resolved as following:

- At very low speeds, friction force increases as speed increases.
- At medium speeds, (25 mm/s ~ 30 cm/s), friction force is almost independent of speed.
- At high speeds, friction force decreases as speed increases.

In very high speeds, it is accepted that the friction force decreases because of the separation of contact surfaces and therefore the molecular attraction and adhesion forces decreases. The modern friction model is illustrated in Fig 4 (de Marchi. 1998).

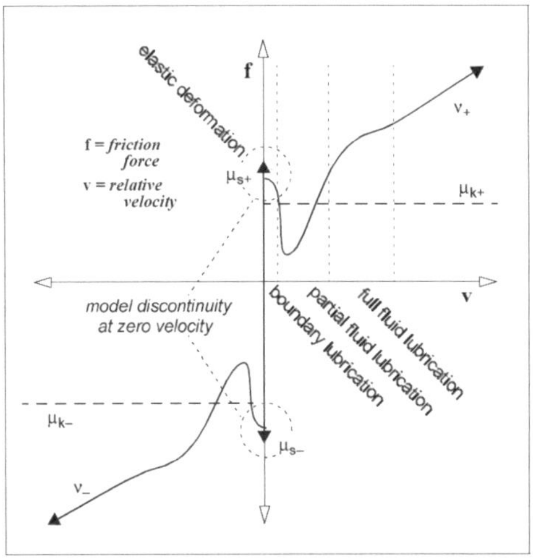

Fig. 4. Modern model of friction (static, Stribeck and viscous). (de Marchi. 1998)

2.3 Friction model

In order to achieve the desired precision in positioning and tracking, friction should be compensated. Therefore a frictional model capable to describe various behaviors of this nonlinear phenomenon should be used. Canudas (Canudas et al., 1995) proposed a model that considers the friction interface as elastic bristles which are attached to rigid solids (Fig.5). Relative velocity between the surfaces causes the bristles to be deformed and this deformation can be regarded as the source of friction. The governing equation of average bristle deflection can be stated as

$$\frac{dz}{dt} = v - \frac{|v|}{g(v)} z \tag{1}$$

where v is the relative velocity between two surfaces. The function $g(v)$ is positive and depends on factors such as material properties, lubrication and temperature and is inevitable not symmetrical. This function has direction dependent behavior. The correspondence of $g(v)$ to Stribeck effect is verified by the fact that $g(v)$ decreased monotonically by increasing speed. The friction force (torque) generated from bending of bristles and viscous friction can be stated as

$$F(\circ) = \sigma_0 z + \sigma_1 \frac{dz}{dt} + \sigma_2 v \tag{2}$$

where σ_0 is the stiffness, σ_1 is the damping coefficient that stands for the dissipation energy due to the deflection of bristles and σ_2 is viscous coefficient. The parametric equation of function $g(v)$ that can describe Stribeck effect is given by

$$\sigma_0 g(v) = F_C + (F_S - F_C) e^{-\left(\frac{v}{v}\right)^2} \tag{3}$$

where F_C is Coulomb friction force (torque), F_S is maximum amount of presliding friction force (torque) and v_s is the Stribeck velocity. The proposed model can be characterized by six parameters which are σ_0, σ_1, σ_2, F_C, F_S and v_s. This model can simulate most of the behaviors that are observed by friction in practice. These behaviors are not only attributed to the value of parameters but also are affected by nonlinear differential equation (1). Presliding displacement and variable break-away force are due to the system dynamics. The Stribeck shape of function $g(v)$ and model dynamics lead to hysteresis (Canudas et al., 1995).

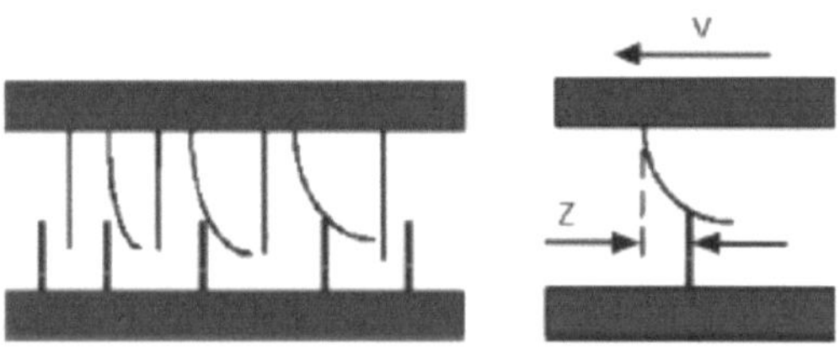

Fig. 5. Friction model

3. Positioning system modelling

Schematic diagram of a ball screw driven stage is illustrated in Fig.6. Rotational dynamics of the system can be described as follows:

$$\left(J_{rot} + m * l_p^2\right) \frac{\ddot{x}_1}{l_p} + F(\circ) = u(t) \tag{4}$$

where J_{rot} is the total rotational inertia of ball screw and motor, m is the moving masses, l_p is the lead pitch of the ball screw, $F(\circ)$ is the equivalent frictional torque of the ball screw, rotary and linear bearings and $u(t)$ is the input torque that is generated by a servo motor. The term $\frac{\ddot{x}_1}{l_p}$ is equivalent to ball screw rotational acceleration and also the term $m * l_p^2$ stands for rotational inertia of moving masses. Equation (4) can be simplified since the frictional torque of the ball screw dominates the system. In designing the controllers, the reduced form of equation (4) will be used.

$$J\ddot{\theta} + F(\circ) = u(t) \tag{5}$$

where J is the effective inertia of ball screw, motor and equivalent rotational inertia of moving masses. The significant parameters of the system are listed in Table 1.

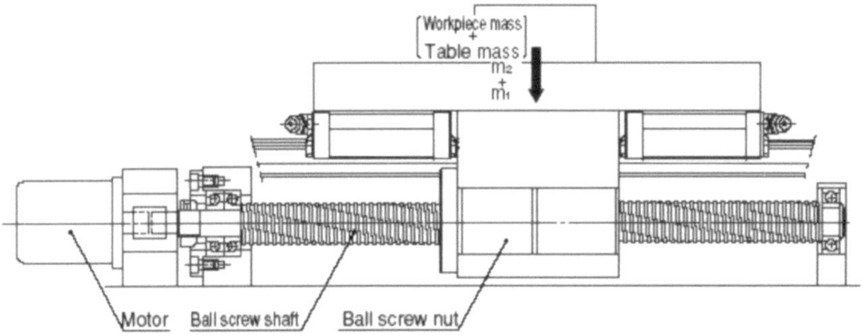

Fig. 6. Schematic of a ball screw driven stage

Moment of inertia of ball screw and motor	$2.12*10^{-4}$ (Kg.m²)
Maximum mass of table and work	62.4 (Kg)
Lead pitch	1.5923 (mm/Rad)

Table 1. System parameters

4. Control design

4.1 Uncertainties and control approaches

From engineering control viewpoint the inaccuracy of implemented model for a dynamic system can be categorized into these two comprehensive groups:

- Parameter uncertainties: the uncertainty or variation of the governing parameters of the dynamic model of the system.
- Unstructured uncertainties or unmodeled dynamics: This originates from the undetermined dynamic model of the system.

The existence of model inaccuracy can have strong adverse effect on nonlinear control systems. To safeguard the system from these harmful effects, adaptive and robust control techniques are widely used. If the system is suffering from parameter uncertainties it is possible to apply online or off-line parameter estimation techniques. As instance, adaptive control schemes can be regarded as a control system with on-line parameter estimation. But

existence of unknown and time varying effective parameters and lack of sufficient information from dynamic behavior of the system, limit the application of adaptive methods.
Variety of control approaches are devised to deal with the effect of unmodeled dynamics which one simple and important of them is sliding mode control. Because of the robustness, the sliding mode control is known as an effective control scheme of the systems with unstructured uncertainties. The only shortcoming of sliding mode controller is its vulnerability to chattering effect. However several compensating techniques are devised to increase the precision of tracking and also to decrease chattering effect. As instance, integral sliding mode controller, time varying boundary layer sliding mode controller and the combination of sliding controller and fuzzy controllers could be mentioned. As seen above, dynamic equation of the ball screw driven stage has frictional term and since there is no exact model for friction and friction model parameters are not known exactly and may be changed, it seems that sliding mode control could be a suitable control scheme for overcoming the adverse effect of these uncertainties. Sliding mode control is described in following subsections.

4.2 Sliding mode control

Considering the nth-order SISO nonlinear system described by

$$y^{(n)} = f(y,\dot{y},...,y^{(n-1)},t) + u \tag{6}$$

Assume that f is not known precisely but can be written as $f = \hat{f} + \Delta f$, where $\hat{f}$ is the nominal term and $\hat{f}$ is the uncertain term that is bounded by known function $\tilde{F}$, i.e.,

$$|\Delta f(\mathrm{x},t)| \le \tilde{F}(\mathrm{x},t)$$

where $\mathrm{x} = [x_1, x_2, ..., x_n]^T = [y, \dot{y}, ..., y^{(n-1)}]^T$. The control objective is to find a continuous sliding control u such that the output y of (6) will approximately track a reference signal, y_r, which is assumed to be nth-order continuously differentiable and all of its derivatives are uniformly bounded. The tracking error is given by:

$$e(t) = y(t) - y_r(t) \tag{7}$$

By defining the generalized error

$$s = (\frac{d}{dt} + \lambda)^n \xi \tag{8}$$

where $\xi = \int^t e(\tau)d\tau$ and $\lambda > 0$, nth-order tracking problem is converted to 1st-order stabilization problem in variable s. s is equivalent to time variable surface in system state space R^n defined by $s(x,t) = 0$ It is easy to show that by using the control input

$$u = \hat{u} + v_c \tag{9}$$

where

$$\hat{u} = -\hat{f} + x_d^{(n)} - \sum_{k=0}^{n-1} \binom{n+1}{k+1} e^{(k)} \lambda^{n-k} - \lambda^{n+1} \xi \tag{10}$$

and

$$v_c = -K * sat(\frac{s}{\phi}) \tag{11}$$

where ϕ is boundary layer thickness, $K = \tilde{F}(x,t) + \eta$ and η is the positive constant, the reaching condition (12) in the feedback control system can be satisfied.

$$\frac{1}{2}\frac{d}{dt}s^2 \le -\eta|s| \quad \forall|s| \ge \phi \tag{12}$$

This equation indicates that the squared distance to surface s, that measured by s^2, decreases along all system trajectories. $s(t)$ verifying (12) is known as a sliding surface and the system's behavior once on the surface is called sliding mode. Despite the presence of model imprecision and disturbance, s^2 remains a Lyapunov - like function of closed loop system. In addition, this control input ensures that starting from any initial state, the error trajectories will be in boundary layer $|s| \le \phi$ after finite time. Whenever $|s| \le \phi$, the guaranteed tracking precision will be

$$|e^{(i)}| \le (2\lambda)^i \varepsilon \tag{13}$$

where $\varepsilon = \phi / \lambda^{n-1}$ is known as boundary layer width and is shown in Fig. 7 for the case $n = 2$ (Slotine & Li, 1991).

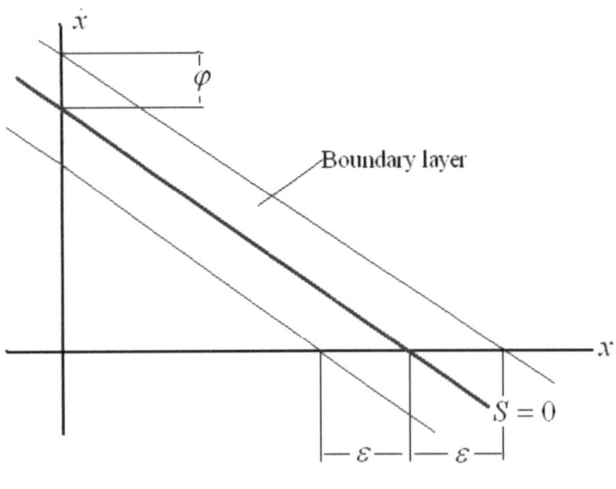

Fig. 7. The boundary layer for the case n=2

4.3 Complementary sliding mode control

Now an additional complementary transformation to the generalized error is defined as

$$s_c = (\frac{d}{dt} + \lambda)^{n-1}(\frac{d}{dt} - \lambda)\xi \tag{14}$$

The control input is given by (9) with $\hat{u}$ is defined in (10) and v_c is replaced by

$$v_c = -K * sat(\frac{s+s_c}{\phi}) \tag{15}$$

By applying this control input to the system, any initial state off the boundary layer will move to the boundary layer $|s+s_c| \le \phi$ in finite time and tracking precision will be

$$|e^{(i)}| \le \frac{1}{2}(2\lambda)^i \varepsilon \tag{16}$$

whenever the system state will be in boundary layer. It is obvious that the tracking precision is at least 50% better than that of in conventional sliding mode control (Liang &, 2003).

4.4 Fuzzy sliding mode control

Based on the definitions of generalized errors, a fuzzy sliding mode controller can be constructed. The fuzzy control input is in the form of

$$u = \hat{u} + K_f(s,s_c) \tag{17}$$

where $\hat{u}$ is given in (10) and $K_f(s,s_c)$ is the output of fuzzy controller. The fuzzy rules are in the form of:

$$Rule^i : if\ s = LS^i\ and\ s_c = LS_c^i, then\ u = LU^i$$

where LS^i and LS_c^i are the linguistic values of s and s_c, respectively. By appropriate tuning the parameters of fuzzy controller, a high performance controller can be obtained and no further information about uncertainties of the system will be needed. It is worth to point out that because the fuzzy controller is constructed on sliding mode controller which is robust, the stability of fuzzy controller would be guaranteed (Liang &, 2003).

5. Parameter design

The constant characteristic parameters of friction model in the ball screw driven stage, i.e. σ_0, σ_1, σ_2, F_C, F_S and v_s, parameters for evaluating $\hat{f}$ (i.e. $\hat{\sigma}_0$, $\hat{\sigma}_1$, $\hat{\sigma}_2$, $\hat{F}_C$, $\hat{F}_S$ and $\hat{v}_s$) and parameters for evaluating $\tilde{F}$ (i.e. $\tilde{\sigma}_0$, $\tilde{\sigma}_1$, $\tilde{\sigma}_2$, $\hat{F}_C$, $\hat{F}_S$ and $\hat{v}_s$) are listed in table 2. The parameters σ_0, σ_1, σ_2, and v_s are the same as reference (Canudas et al., 1995) while F_C and F_S are selected arbitrarily. The values of parameters of $\hat{f}$ are assumed to be uncertain enough in order to enable us validate our claims about robustness of the controllers.

σ_0	10^5	$\hat{\sigma}_0$	$2*10^5$	$\tilde{\sigma}_0$	10^5	Nm/Rad
σ_1	$\sqrt{10^5}$	$\hat{\sigma}_1$	$2*\sqrt{10^5}$	$\tilde{\sigma}_1$	$\sqrt{10^5}$	Nms/Rad
σ_2	0.4	$\hat{\sigma}_2$	1.2	$\tilde{\sigma}_2$	1	Nms/Rad
F_C	6.24	$\hat{F}_C$	6.24	$\hat{F}_C$	6.24	Nm
F_S	9.36	$\hat{F}_S$	9.36	$\hat{F}_S$	9.36	Nm
v_s	0.001	$\hat{v}_s$	0.005	$\hat{v}_s$	0.005	Rad/s

Table 2. Parameters of friction model

6. Computer simulation and analysis

In order to evaluate the performance of the proposed controllers in control of ball screw positioning system some simulations are performed in Matlab/ Simulink and Matlab/ Fuzzy toolbox and tracking error and control input to the system are compared when the system tracks various desired signals. We take $\lambda = 150$, $\eta = 120$ and $\phi = 0.1$ in simulation of the system when conventional sliding mode controller is applied. In simulation of the system when complementary sliding mode controller is applied, the constants parameters are taken as $\lambda = 150$, $\eta = 20$ and $\phi = 0.1$. λ equals to 150 in simulation of the system with fuzzy sliding mode controller. The membership functions of input variables s and s_c to the fuzzy controller are shown in Fig. 8. Membership functions of output variable $K_f(s,s_c)$ of the fuzzy controller are shown in Fig. 9. We used Mamdani method for fuzzy inference system, fuzzifier type was triangular and defuzzifier type was centroid.

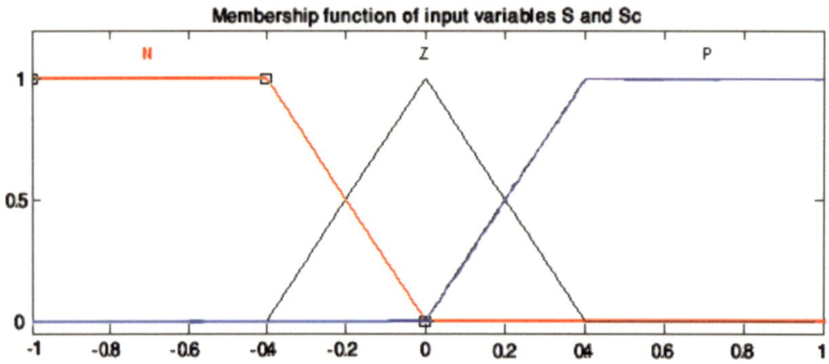

Fig. 8. Membership functions of input variables s and s_c

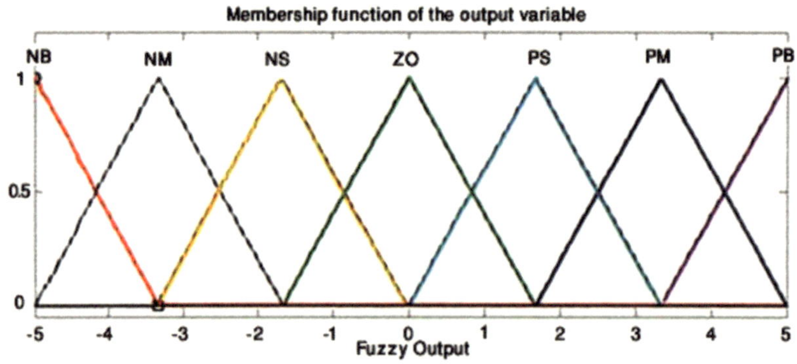

Fig. 9. Membership functions of output variable of the fuzzy controller

We develop the rule base of the fuzzy controller which s and s_c are its input variables and is the output variable in the form of

1. If " s " is "N" and " s_c " is "P" then "Fuzzy Output" is "PB".
2. If " s " is "N" and " s_c " is "Z" then "Fuzzy Output" is "PS".

3. If " s " is "N" and " s_c " is "N" then "Fuzzy Output" is "ZO".
4. If " s " is "Z" and " s_c " is "P" then "Fuzzy Output" is "PS".
5. If " s " is "Z" and " s_c " is "Z" then "Fuzzy Output" is "ZO".
6. If " s " is "Z" and " s_c " is "N" then "Fuzzy Output" is "NS".
7. If " s " is "P" and " s_c " is "P" then "Fuzzy Output" is "ZO".
8. If " s " is "P" and " s_c " is "Z" then "Fuzzy Output" is "NS".
9. If " s " is "P" and " s_c " is "N" then "Fuzzy Output" is "NB".

where P stands for positive; N stands for negative; PB stands for positive big; PS stands for positive small; both Z and ZO stand for zero; NB stands for negative big, and NS stands for negative small. The scaling factors of input and output variables are shown in Fig. 10.

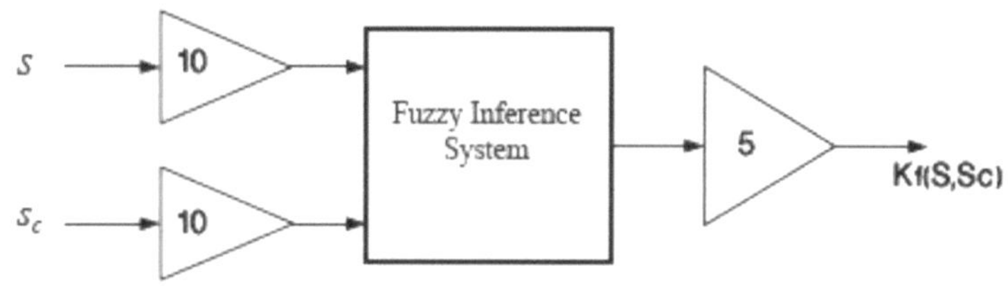

Fig. 10. Scaling factor of input and output variables of fuzzy controller

The response of the proposed controllers to positioning signal 1 (mm) is shown in Fig. 11. As seen in this figure, SMC has the maximum positioning error and FSMC has the minimum one. Also steady state error exists in the response of the SMC to this signal, but CSMC and FSMC are capable of eliminating this error in their responses. The control inputs of the SMC and CSMC are not zero after reaching the desired positions that can be attributed to spring like behavior of friction. But FSMC is capable of setting the control input to zero. But since the gains of fuzzy controller are not suitable for very small amount of error, an oscillating behavior is observed in control input.

The responses of the proposed controllers for various sinusoidal inputs with different amplitudes and frequencies are shown in Fig. 12 to Fig. 14. The results indicate that tracking error exists when SMC is applied to the system dynamics and control input to the system undergoes to fluctuations and high frequency chatter in some cases. Also this controller can not response properly to high frequency reference signal and its performance suffers from large tracking error, fluctuations and oscillations in control input. When CSMC is applied to the system, the tracking error becomes smaller in comparison with SMC. Its peaks occurred in the points where the direction of the motion is changed. Fluctuations and chatters in control inputs are also observed in this case but frequency of chattering is lower. High frequency reference signal can not be tracked properly in this case because of large tracking error and large values of fluctuations in the control input. FSMC is capable of eliminating tracking error except for where the direction of motion is changed and the occurred error is damped in finite time (less than 0.1 (s)). Control inputs are more smooth in this case and do not include fluctuations and chattering. The response of the controller is more desirable to high frequency reference signal. Tracking error does not grow drastically in this case and control input does not suffer from fluctuations and oscillations.

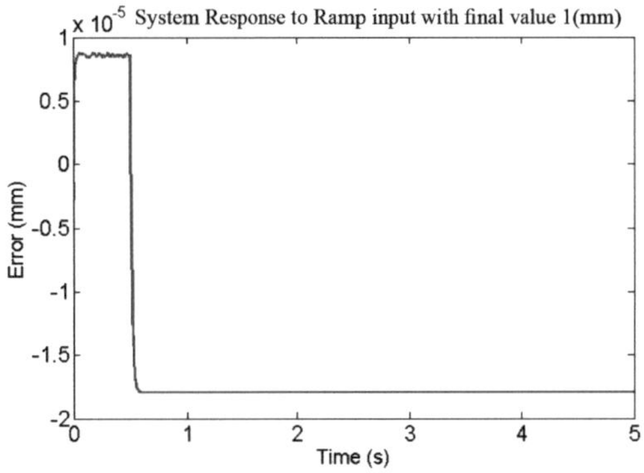

(a)

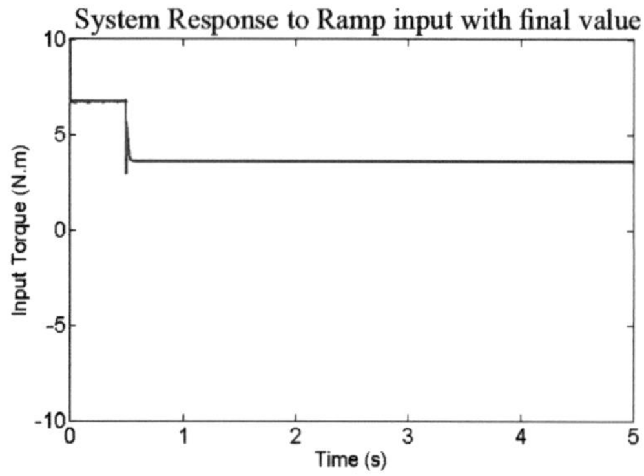

(b)

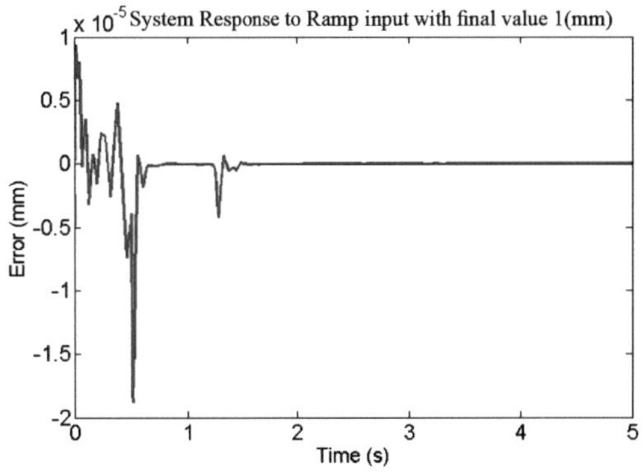

(c)

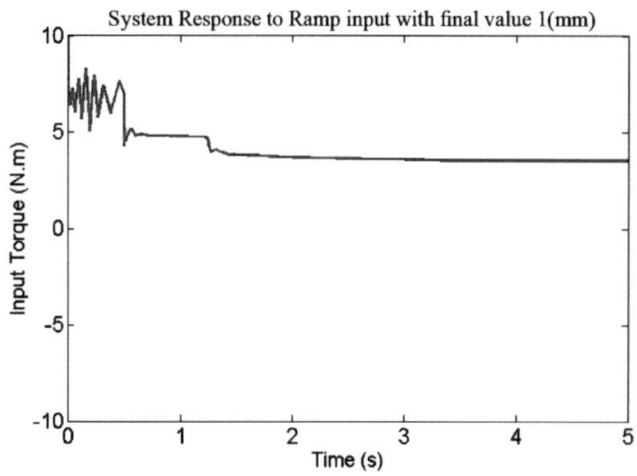

(d)

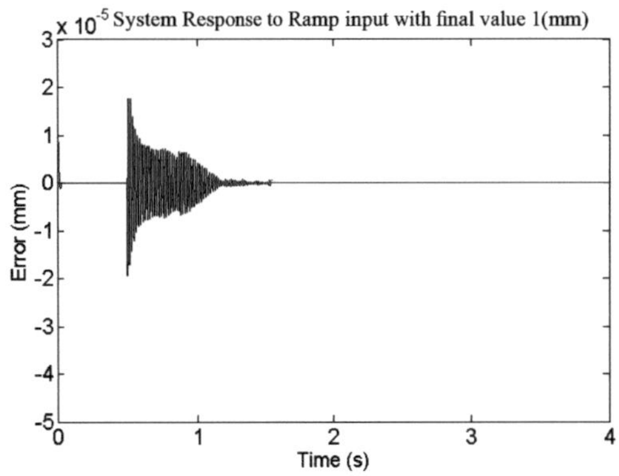

(e)

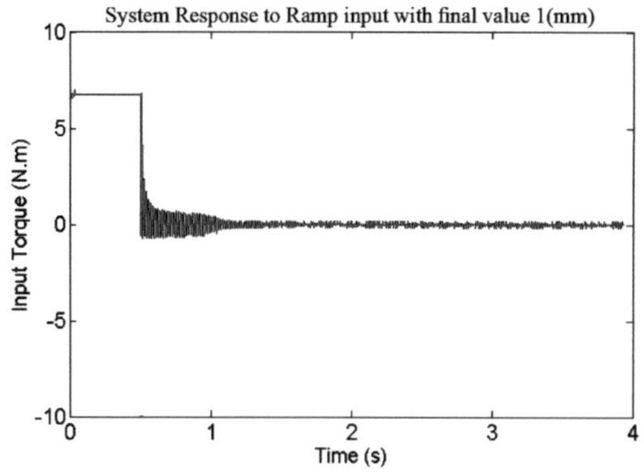

(f)

Fig. 11. Positioning response to step input with final value 1(mm). (a) Positioning error of SMC, (b) Control input of SMC, (c) positioning error of CSMC, (d) Control input of CSMC, (e) positioning error of FSMC, (f) Control input of FSMC.

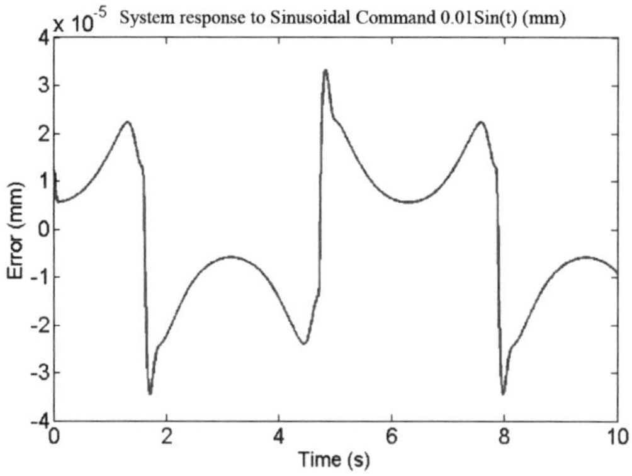

(a)

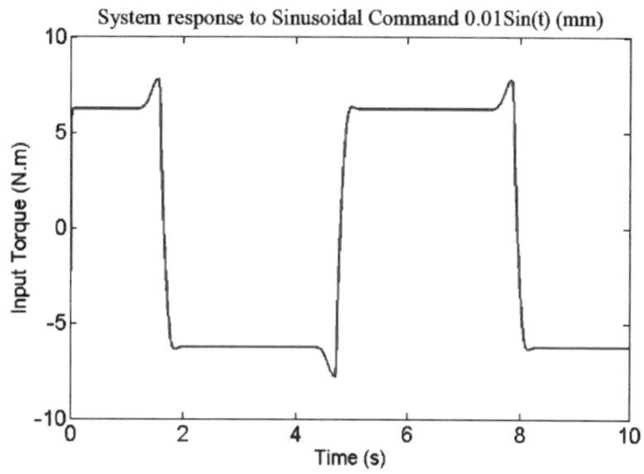

(b)

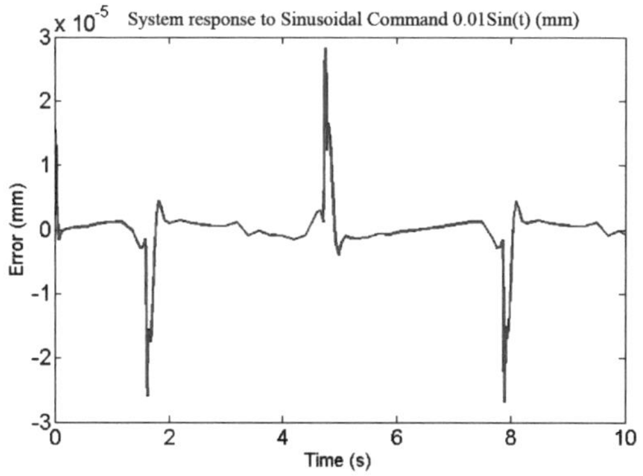

(c)

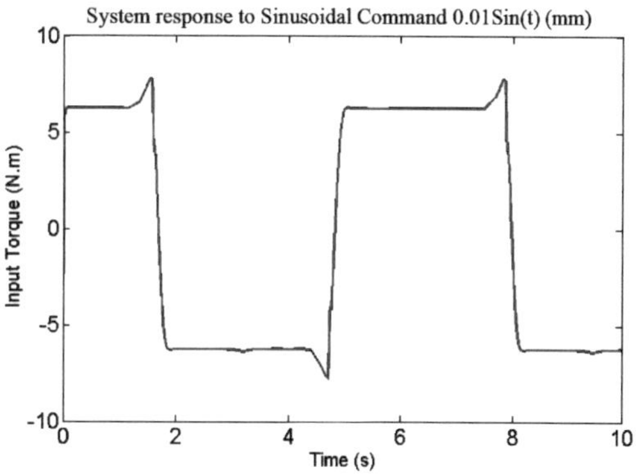

(d)

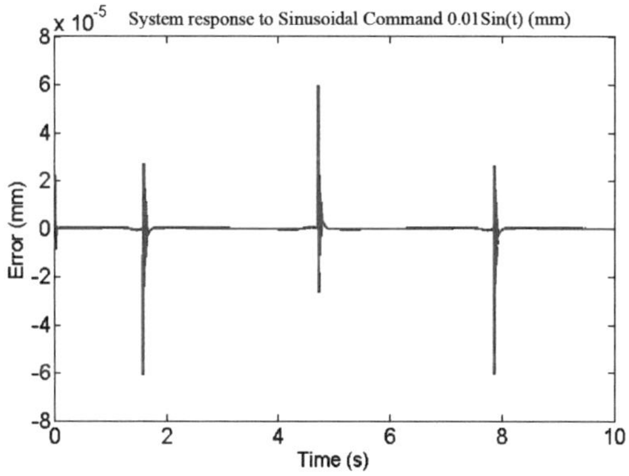

(e)

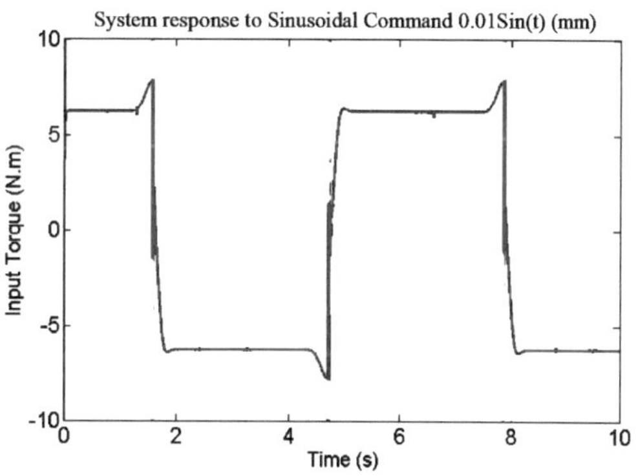

(f)

Fig. 12. Tracking response to sinusoidal input 0.01Sin(t) (mm). (a) Tracking error of SMC, (b) Control input of SMC, (c) Tracking error of CSMC, (d) Control input of CSMC, (e) Tracking error of FSMC, (f) Control input of FSMC

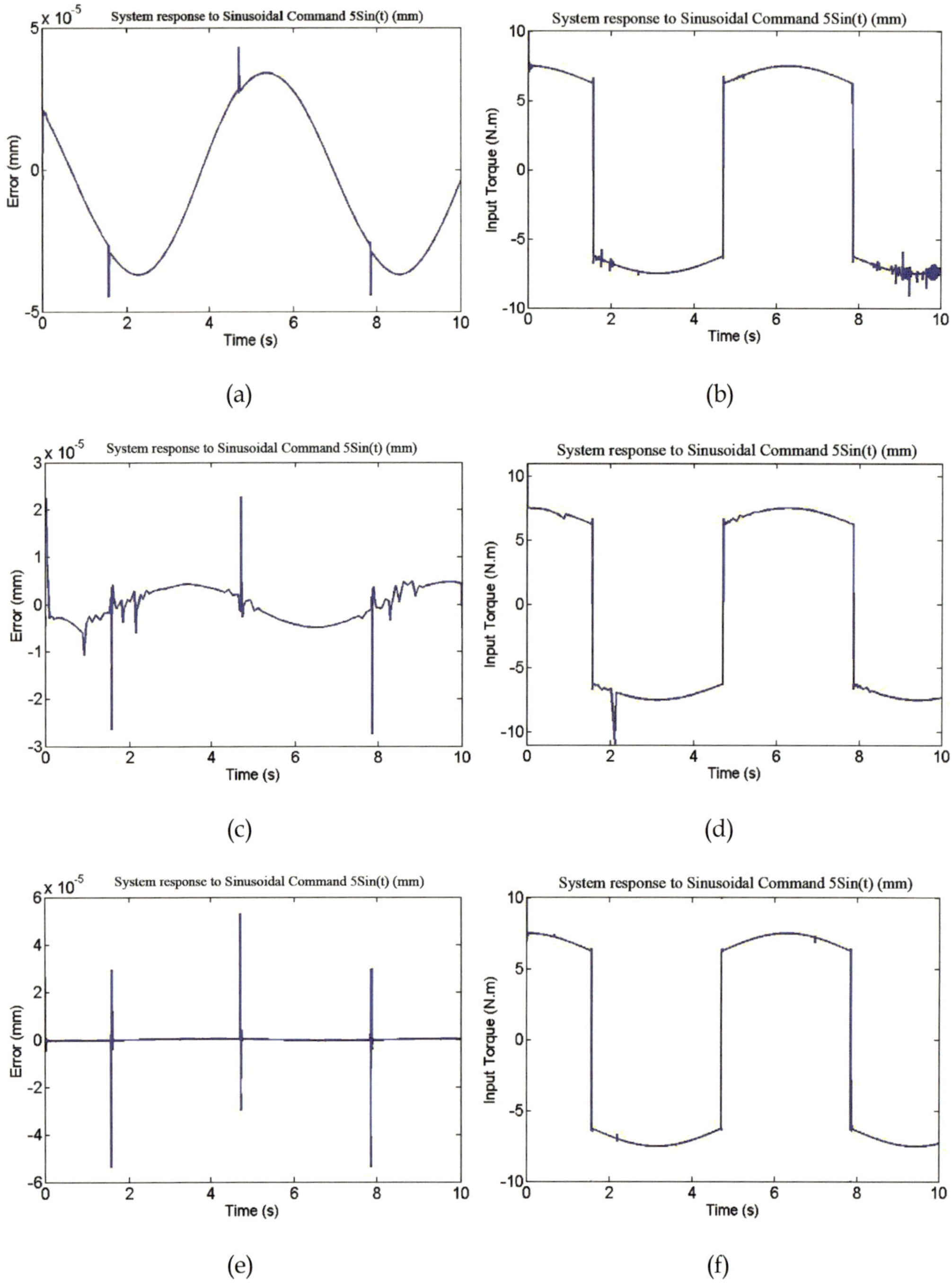

Fig. 13. Tracking response to sinusoidal input 5Sin(t) (mm). (a) Tracking error of SMC, (b) Control input of SMC, (c) Tracking error of CSMC, (d) Control input of CSMC, (e) Tracking error of FSMC, (f) Control input of FSMC

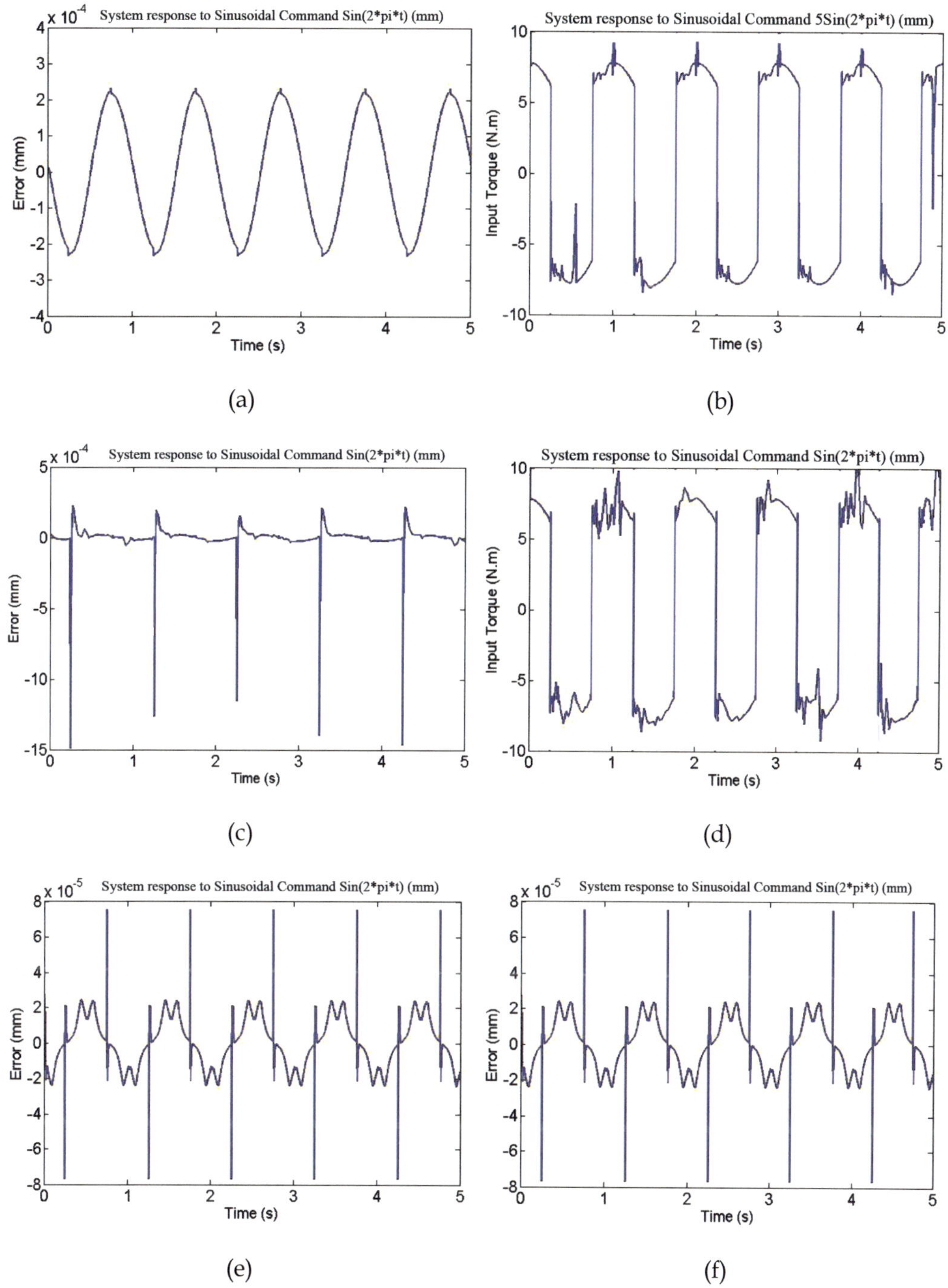

Fig. 14. Tracking response to sinusoidal input Sin(2*pi*t) (mm). (a) Tracking error of SMC, (b) Control input of SMC, (c) Tracking error of CSMC, (d) Control input of CSMC, (e) Tracking error of FSMC, (f) Control input of FSMC

7. Verification

7.1 Introducing Double Ball Bar (DBB)

DBB is consisting from two precise balls that each ball is 25.4 mm in diameter with geometric spherical error of less than 0.05 μm. These two balls are connected together by a low thermal expansion coefficient bar made of Invar. This bar is equipped with a measurement system with 0.1 μm resolution (Fig 15). Each ball magnetically attracted by the support and it is in contact with supporting surface in three points. One of these supports is attached to the machine table and the other one is fastened to the machine spindle. Once DBB assembled correctly, the machine is commanded to move in a circular path. The variations in the distance between the centers of the two balls is measured via the measurement system and corresponding analog signal is converted to digital signal and then its corresponding numerical value is calculated by computer. The variations of mentioned distance is identical to deviation from ideal circular path (Fig 16). Usually the data acquisition rate is such that the computer detects motion error ΔR in every 0.2 degree. So in every full rotation 1800 numeric values is sampled.

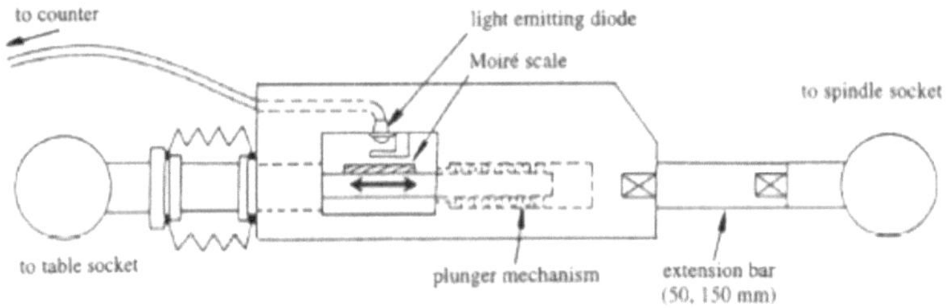

Fig. 15. Schematic of Double Ball Bar. (Kakino et al., 1993)

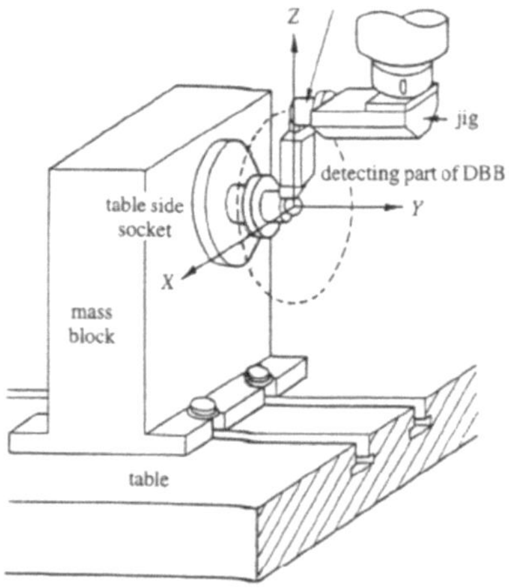

Fig. 16. Setup for accuracy inspection of a machine tool in vertical plane by Double Ball Bar. (Kakino et al., 1993)

7.2 Errors in CNC machines

In accuracy inspection of a machine tool by means of DBB, the source of the error in each CNC machine is divided into two main categories.

- Errors of which the magnitude is dependent to the feedrate and direction of the motion e.g. errors originated from stick motion, stick-slip behavior and vibration in the hydraulic valves and chains.
- Errors of which the magnitude is dependent to the position e.g. cyclic error of the ball screw and linear encoder, squareness errors between two axes and straightness errors of guide ways.

Most of the errors mentioned above are only practically observable but the effect of stick motion and stick-slip can also be shown in simulations. In the following paragraphs these error will be introduced.

- Stick motion error: This error is originated from lag in the response of control system when the direction of motion is reversed. During reversing the direction of motion, the velocity of the axis decreases to zero and hence static friction becomes the dominant frictional force in the system. Since break-away force is greater than dynamic friction force, the slider stays in stick condition and hence the tracking error increases. By increasing the magnitude of tracking error the controller provides relatively larger control input to the actuator to conquer break- away force and the axis again starts to catch up the reference signal. Fig 17 illustrates the slider in the stick condition for a short time after the direction of motion is reversed. The effect of this error is shown in Fig 18 in a polar coordinate system and with proper magnification. The magnitude of this error is dependent upon the feedrate and the amplitude of the reference signal. As can be seen in the figures 12-14 (a,c,e) the presence of a peak in diagrams of tracking error of sinusoidal reference signal can be attributed to stick motion error.

In the standard ISO 230-4 that is related to the testing of the NC and CNC machines using DBB, it is mentioned that in a machine equipped with linear encoder, reversal error will be detected and compensated by the machine controller but the compensation needs a certain amount of time and therefore some peaks will outbreak in the reversal points. (Figure 19). (BS ISO 230-4, 1996)

- Stick-slip: This phenomenon occurs in very low feedrate and is happening in a continuous sequence of stop and move motion in a short period of time. In a circular motion this phenomenon is observable in zero, 90, 180 and 270 degrees which feedrate of one axis reduces to zero. In Fig 20 illustrate the trace pattern

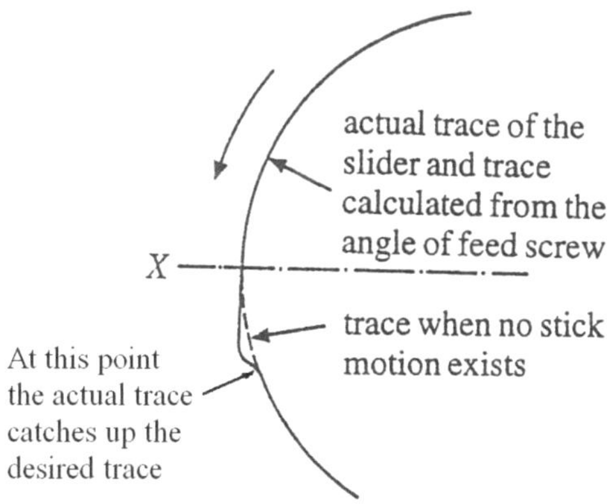

Fig. 17. Model of stick motion. (Kakino et al., 1993)

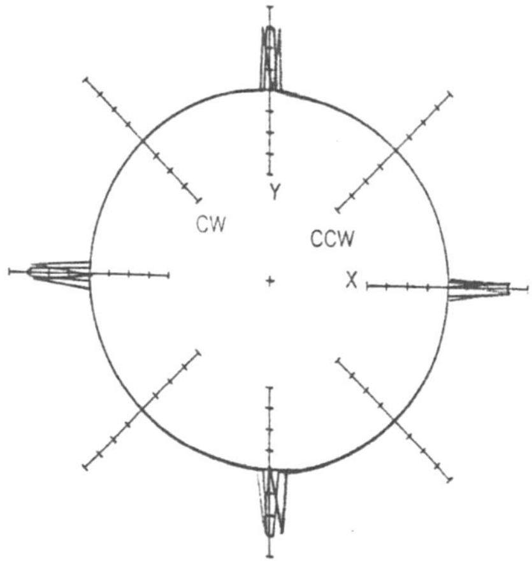

Fig. 18. Trace pattern caused by stick motion. (Kakino et al., 1993)

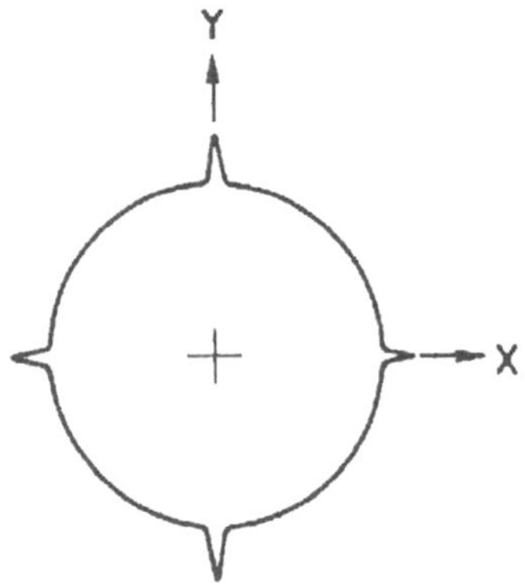

Fig. 19. Compensated reversal error. (BS ISO 230-4, 1996)

caused by stick-slip on a circular motion. It can be seen that there is a symmetric pattern around the axis which has this error. By increasing feedrate, this event decreases and then completely eliminated (Kakino et al., 1993). Since this phenomenon is considered in the assumed frictional model, its effect can not be seen in the simulations.

- In standard ISO 230-4 is mentioned that by increasing feedrate in circular motion the acceleration of the axis increases accordingly and the controller acts in such a way that the amplitude of the movement decreases at higher frequencies (Fig 21) (BS ISO 230-4, 1996). It is also obvious that a circular motion can be produced by combining two ball screw driven stages. The effect of increasing the frequencies of sinusoidal reference signals to each stage is that the diameter of resultant circular path becomes smaller than desired path. This observation also confirms the increase in the magnitude of error which can be seen in figure 14 (a,c,e). These figures are related to sine input to the system with 1mm amplitude and 1 second period. Hereby the magnitude of error increases as the frequency of the sine input increases and the magnitude of the error will be such that the amplitude of the resultant motion will be less than the desired value.

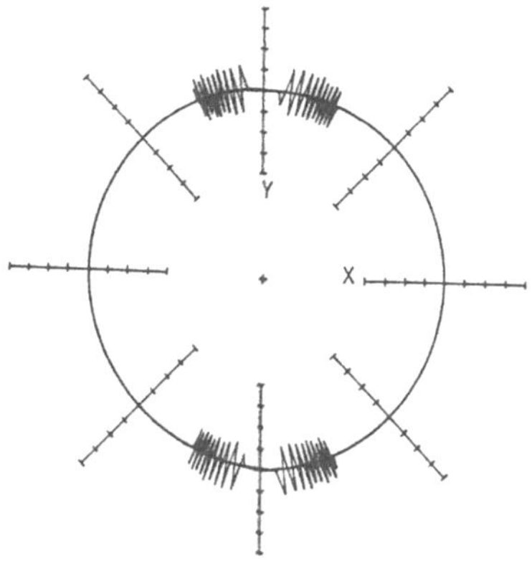

Fig. 20. Trace pattern caused by stick slip. (Kakino et al., 1993)

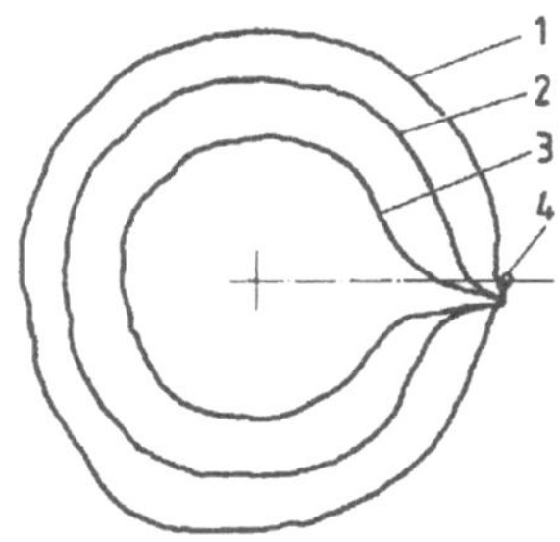

Fig. 21. Influence of acceleration of the axis on amplitude of resultant motion
(1- Low feedrate. 2- Medium feedrate. 3- High feedrate. 4- Starting and stop point). (BS ISO 230-4, 1996)

8. Conclusion

In recent years, due to the developments in nanotechnology, the demand for high-precision stages has been subjected to great attention. Systems to provide long-range and high-precision performance for positioning, tracking and contouring tasks have become stringent issues. Such systems are widely used in semiconductors, precision metrology, micro-assembly and micro- and nanoscale fabrication industries. Among the positioning systems, the ball screw driven systems have been widely used in academic researches and in the inferred industries. Several researches have been performed on precise positioning such as modeling friction, implementing different control strategies for compensating friction and achieving micron accuracy. In this chapter, the process of controlling of a single axis ball screw driven stage with micron positioning accuracy has been investigated.
In the first step, two nonlinear phenomena i.e. friction and backlash that can affect the accuracy of the designed stage are studied. Among them, friction has been the most dominant phenomena which significantly reduce the final achievable accuracy. In order to compensate the undesired effects of friction, it is necessary to provide its behavioral model. Among proposed models, Canudas model was selected that can explain frictional behavior of such positioning systems comprehensively and effectively.

In the next step, the theoretical description of a control system capable of compensating the undesired effects of friction is introduced. Since governing parameters of frictional model are not known accurately, the sliding mode control (SMC) technique is selected which is appropriate in the architecture of controllers with uncertainty. Subsequently a new type of sliding mode control approach known as complementary sliding mode (CSMC) and fuzzy sliding mode control (FSMC) was introduced and discussed.

Finally the described controllers are applied to nonlinear dynamics of the system and their performances were investigated by computer simulation for submicrometer positioning, tracking and control input. It is observed that not only FSMC could significantly improve the achievable accuracy of positioning and tracing, but also it is capable of smoothing inputs to the system and eliminating fluctuations. Also this type of controller has better response to high frequency sinusoidal reference signals. Finally the results have been validated by comparing them with those achieved from Double Ball Bar experimental results. Further investigation on evaluating the controllers' performances in the presence of other type of uncertainty such as inertia will be held in future.

9. References

Bliman, P.A & Sorine, M. (1991). *Friction modelling by hysteresis operators. Application to Dahl, Stiction, and Stribeck effects*, in Proc. Conf. Models of Hysteresis, Trento, Italy.

Canudas de Wit, C.; Olsson, H., Astrom, K.J. & Lischinsky, P. (1995). *A new model for control of systems with friction*, IEEE Transaction on Automatic Control, Vol 40, No.3.

Chen, C.L.; Jang, M.J. & Lin K.C. (2004). *Modeling and High Precision Control of a Ballscrew Driven Stage*, Precision Engineering, pp. 483-495.

Dahl, P. (1968). *A solid friction model*, Aerospace Corp., EI Segundo, CA, Tech, Rep. TOR-0158(3107-18).

Kakino, Y.; Ihara, Y.; Shinohara, A & Heidenhain, J. (1993). *Accuracy Inspection of NC Machine Tools by Double Ball Bar Method*, Hanser Gardner Publications, ISBN: 1569901600, Cincinnati, Ohio.

Liang, C.Y. & Su, J.P. (2003). *A New Approach to the Design of a Fuzzy Sliding Mode Controller*, Fuzzy Sets and Systems, pp. 111–124.

Lee, H.; Kim, E.; Kang, H.J. & Park, M. (2001). *A new sliding-mode control with fuzzy boundary layer*, Fuzzy Sets and Systems, pp. 135–143.

Marchi, J.A. (1998). *Modeling of Dynamic Friction, Impact Backlash and Elastic Compliance Nonlinearities in Machine Tools, With Application to Asymmetric Viscous and Kinetic Friction Identification*, A Thesis Submitted to the Graduate Faculty of Rensselaer Polytechnic Institute in Partial Fulfillment of the Requirements for the Degree of Doctor of Philosophy. Rensselaer Polytechnic Institute. Troy, New York.

Slotine, J.E. & Li. W.P. (1991). *Applied Nonlinear Control*, Prentice Hall, ISBN: 0130408905.

Mihajlov1, M.; Nikolić, V. & Antić, D. (2002). *Position control of an electro hydraulic servo system using sliding mode control enhanced by fuzzy PI controller*, Mechanical Engineering, Vol.1, pp.1217-1230.

Ro, P.I.; Shim, W. & Jeong, S. (2000). *Robust Friction Compensation for Submicrometer Positioning and Tracking for a Ball-Screw-Driven Slide System*, Precision Engineering, pp. 160–173

(1996). *Test code for machine tools - Part 4: Circular tests for numerically controlled machine tools*, BS ISO 230-4, 12 pages.

25

Dynamic Modeling and Performance Trade-offs in Flexure-based Positioning and Alignment Systems[1]

Vijay Shilpiekandula and Kamal Youcef-Toumi
Mechatronics Research Laboratory, Department of Mechanical Engineering
Massachusetts Institute of Technology, Cambridge, MA 02139, USA

1. Introduction

Precision positioning and alignment are critical to an emerging class of small-scale manufacturing and numerous motion control applications. The drive for better performance steers design and control effort into achieving high tolerances and stringent specifications in terms of parameters, such as resolution, range, load-capacity, and bandwidth. Examples of applications needing precision positioning and alignment include (i) high-bandwidth steering of mirrors in telecommunication applications [1], (ii) tool-sample alignment in stamping applications such as imprint lithography [2, 3] and micro-contact printing [4], and (iii) alignment of optically flat surfaces brought in close proximity to characterize fields and forces on small-scales, such as the Casimir force [5, 6].

A widely used set of designs for precision applications described above involve compliant mechanisms based on slender beam modules, also referred to as flexures [7, 8]. The advantages flexures offer are mainly smooth elastic motion without non-linearities such as friction or backlash [9]. Flexure-based mechanisms such as the diaphragm flexure involve the payload suspended on a radial or tangential arrangement of flexural beams. Various forms of such flexures have appeared over the past few decades for applications such as MEMS mirrors, and in angle alignment and guidance applications [4, 10, 11]. Analysis of the statics and dynamics of flexure-based mechanisms have been extensively studied [9, 12].

While flexure-based engineering designs have been around for many decades [13], designing them for dynamic performance has sought little attention. Few publications [14, 15] have appeared in this context. The design for dynamical performance of flexures in the context of mechanical advantage is detailed in [14]. A finite-element approach based on Euler-Bernoulli beam bending theory is formulated for analyzing dynamics in [15] and optimizing the design space for precision flexure-based applications in [16].

We build on the work presented in the literature and integrate models that can enhance the accuracy in predicting the dynamics of a given flexure-based design by including the effects of distributed mass and compliance of the flexures covering shear and rotational effects. These effects are shown to dominate at small flexure lengths. Further, we use the models to characterize design space parameters such as range, load-capacity, and bandwidth. While most of the current literature in flexure-based designs focuses on static values of performance variables such as angular position, or acceleration, we present a state-space approach for characterizing the bounds on these variables in the frequency domain. This is critical for ensuring that performance requirements are met within the usually large bandwidths of operation, an example application being fast steering of mirrors in telecommunications [1].

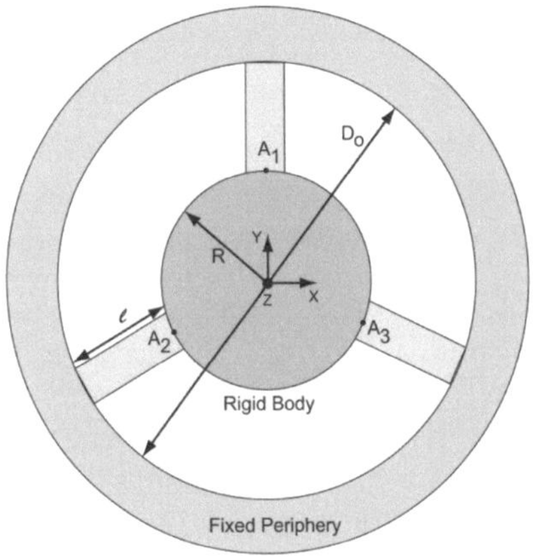

Fig. 1. A diaphragm flexure as a parallel kinematic mechanism with a central rigid mass connected by $n = 3$ flexural beam units to the ground. The dimension $D_0 = 2\ell + 2R$ is referred in this chapter as the footprint of the mechanism. The **Z** axis is shown pointing out of the page.

This chapter is organized as follows. In Section 2 we assemble lumped parameter models for a diaphragm flexure design. Section 3 covers the closed-form characterization of the design space from the dynamic models. Non-dimensional design plots and the efficacy of the models in capturing shear effects at small flexure lengths are addressed in this section. A state-space approach is used for characterizing key performance variables in Section 4. This section taps into multi-input multi-output (MIMO) analysis tools to develop a framework for mapping design requirements over to the state-space. The effects of manufacturing errors are studied in the context of decoupling and asymmetry in this section. Finally, we conclude with a summary of the contributions of the work.

2. Dynamic modeling

Our goal is to capture the out-of-plane behavior, i.e. the vertical translation, pitch, and roll degrees of freedom of diaphragm flexures used in precision angle alignment mechanisms. In this section, we assemble dynamic models for a class of diaphragm flexures - namely, those applying radial constraints on a central rigid mass via flexural beam units. We derive lumped parameter models representing the mass and stiffness of the diaphragm flexure. The applications of compliant mechanisms using the simple flexural beam units studied here span multiple scales from MEMS to meso-scale systems. In all these applications, it is desirable to develop accurate models since the mechanisms constitute the plant in the overall closed-loop control system.

2.1 Modeling flexural beam

To model bending of the flexural beam unit, shown in Fig. 2, we use a Timoshenko beam [17] model since a simple Euler-Bernoulli beam model cannot capture the effects of shear and rotational inertia. As will be shown later, these effects become significant for short beams, which are widely used in compliant mechanisms spanning multiple length scales. To model torsion of the flexural element, we use St. Venant's torsion formulation assuming that (i) the effects of restrained warping are negligible and (ii) bending and torsion are decoupled. Further, we assume that the deflections of the flexural beam element are small (an order of magnitude smaller than the beam thickness) and hence, neglect the effects of axial stretching and the resultant stress stiffening along the length of the beam element.

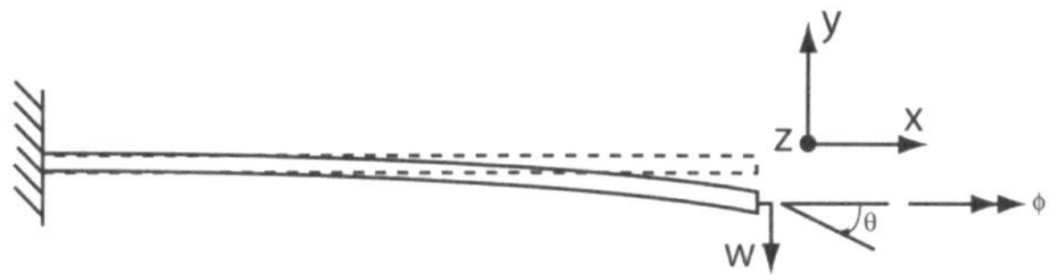

Fig. 2. Schematic diagram showing a flexural beam element with deflection $\omega(x,\ t)$, slope $\theta(x,\ t)$, and twist about the X-axis by an angle $\phi(x,\ t)$.

Under the above-mentioned assumptions, the distributed parameter model for the beam is well-documented in the literature [17] as being depicted by a set of partial differential equations in the deflection $w(x,\ t)$, slope $\theta(x,\ t)$, and angle of twist $\phi(x,\ t)$ listed in Section **A. 1** of the Appendix.

The infinite-dimensional behavior governed by the set of partial differential equations can be approximated to that arising for a one-element model using the method of assumed modes [18]. By this method, the infinite-dimensional behavior of the mechanism is approximated to a finite-series made up of spatially varying mode shape functions (or trial functions) with temporally varying mode amplitudes [19]. Since a one-element model is used for the beam, the distributed properties of the beam are lumped to the node at the guided end of the beam; the fixed node of the beam has no lumped mass or stiffness. Hence, from the three displacements assumed for the guided end of the beam, a three-DOF lumped parameter model can be derived.[2]

The detailed application of the assumed modes method to the set of partial differential equations governing the motion of the Timoshenko beam can be found in parts in [20] and [21]. The key results used in this work are highlighted here. Under the geometric boundary conditions of (i) one end $x = 0$ of the flexural beam being grounded and (ii) the other end $x = \ell$ subject to generalized displacements $\mathbf{V}(t) = [w(\ell,\ t)\ \theta(\ell,\ t)\ \phi(\ell,\ t)]^T$, (where $w(\ell,\ t)$ is the vertical deflection, $\theta(\ell,\ t)$ the slope, and $\phi(\ell,\ t)$ is the angle of twist), the corresponding 3×3 matrices – mass $\mathbf{M}_f$ and stiffness $\mathbf{K}_f$ are as given below:

---

[2] A many-element model can be used for each flexural beam to develop higher-order lumped parameter models, if desired.

$$\mathbf{M}_f = \begin{bmatrix} M_1 & M_2 & 0 \\ M_2 & M_3 & 0 \\ 0 & 0 & M_4 \end{bmatrix}; \quad \mathbf{K}_f = \begin{bmatrix} K_1 & K_2 & 0 \\ K_2 & K_3 & 0 \\ 0 & 0 & K_4 \end{bmatrix} \tag{1}$$

where the matrix values depend on material properties and geometry, and are tabulated in Section **A. 1** of the Appendix. Zeroes in either matrix result from the decoupling assumed between bending and torsion. The lumped mass and stiffness matrices are used as building blocks for assembling dynamic models of mechanisms involving flexural beam units. Note that we need to restrict these matrix building blocks to parallel kinematic configurations since the geometric boundary conditions corresponding to $x = 0$ have been assumed to be all zero. Formulations for serial kinematic configurations can be developed by altering this set of geometric boundary conditions [20].

2.2 Assembling global lumped parameter model

Here, we formulate the dynamics of parallel kinematic mechanisms that contain a rigid body connected to the ground through a multitude of flexural beam units. We integrate the lumped parameter model for the flexural beam in Section 2 with rigid body dynamical models using appropriate transformations to obtain the global model [15]. These transformations are chosen to ensure the continuity of nodal displacements at the interface between the rigid body and the flexures.

Consider a parallel kinematic mechanism with a central rigid circular disk centered at the origin and parallel to the horizontal **XY** plane of the cartesian **XYZ** space, as shown in Fig. 1. In the rest position, the principal axes of the disk **X'**, **Y'**, and **Z'** coincide with the cartesians axes **X**, **Y**, and **Z**, respectively. Let the disk be of radius R, thickness T, mass M_R, and moments of inertia J_{Rxx} and J_{Ryy} about the X and Y axes respectively. A number n of slender beam flexures, each of width W, thickness H, and length ℓ, are in the **XY** plane connecting every peripheral point P_i to the ground. The coordinates of P_i in the **X'Y'** plane are $(R\cos\alpha_I, R\sin\alpha_i)$ with angles $\alpha_i \in [0, 2\pi)$ for $i = 1, 2, 3\ldots\ n$.

Since the beams provide high axial (and hence in-plane **XY**) stiffness and low out-of-plane stiffness, we expect that the dominant modes correspond to the out-of-plane motion, namely vertical deflection, pitch, and roll. We hence assume that the out-of-plane motion of the disk is decoupled from the in-plane motion, i.e. the center of the disk always moves only vertically. For small vertical deflection $z(t)$ of the center of mass, and small angular rotations $\theta_x(t)$ and $\theta_y(t)$ about the **X** and **Y** axes respectively, the principal plane **X'Y'** of the disk moves out of the **XY** plane to the one depicted by

$$\mathbf{Z}_p(t) = \theta_y(t)\mathbf{X} + \theta_x(t)\mathbf{Y} + z(t) \tag{2}$$

For continuity of displacement at each of the nodes P_i, Eq. (2) can be used to show that the end-displacements $\mathbf{V}_i(t)$ of every i^{th} flexure are related to the global generalized (rigid body) displacements $\mathbf{V}_R(t)$ as follows:

$$\mathbf{V}_i(t) = \begin{Bmatrix} w_i(\ell, t) \\ \theta_i(\ell, t) \\ \phi_i(\ell, t) \end{Bmatrix} = \Re \begin{Bmatrix} z(t) \\ \theta_x(t) \\ \theta_y(t) \end{Bmatrix} = \Re \mathbf{V}_R(t) \tag{3}$$

where the transformation matrix $\Re$=

$$\begin{bmatrix} 1 & Rsin\alpha_i & Rcos\alpha_i \\ 0 & -sin\alpha_i & -cos\alpha_i \\ 0 & -cos\alpha_i & -sin\alpha_i \end{bmatrix} \tag{4}$$

2.3 Dynamics

Based on the mass and stiffness properties of the individual flexural beam units connected to the central rigid body, we need to derive the mass **M** and stiffness **K** properties of the assembly.

By formulating the Lagrangian of the assembly in terms of the rigid body displacements $\mathbf{V}_R(t)$, we develop the lumped mass and stiffness matrices of the overall parallel kinematic mechanism as follows:

$$\mathbf{M} = \sum_{i=1}^{n} \Re^T \mathbf{M}_{fi} \Re + \mathbf{M}_R; \;\; \mathbf{K} = \sum_{i=1}^{n} \Re^T \mathbf{K}_{fi} \Re \tag{5}$$

where$\mathbf{M}_{fi}$ and $\mathbf{K}_{fi}$ are the lumped mass and stiffness matrices, respectively, of the individual flexure building blocks given in Section 2.1, and Tables 4 and 5; $\mathbf{M}_R$ is the mass matrix of the rigid body and is given by:

$$\begin{bmatrix} M & 0 & 0 \\ 0 & J_{Rxx} & 0 \\ 0 & 0 & J_{Ryy} \end{bmatrix} \tag{6}$$

The equations of motion of the lumped parameter representation, for the free response case, is in the form given below:

$$\mathbf{M}\ddot{\mathbf{V}}_R + \mathbf{B}\dot{\mathbf{V}}_R + \mathbf{K}\mathbf{V}_R = \mathbf{0} \tag{7}$$

Note that we have not presented the modeling of damping matrix **B** in this chapter. Models such as proportional damping, given by $\mathbf{B} = b_m\mathbf{M}+b_k\mathbf{K}$, are widely used in the literature [22], where b_m and b_k are constants that depend on material properties and are experimentally determined from sine-sweep frequency response measurements. For the design of active or passive damping in flexure mechanisms, a survey and foam-based methods are detailed in [23].

3. Dynamic performance of diaphragm flexures

In this section, we use the dynamic models developed from Section 2.3 to examine (i) the influence of geometric arrangement of flexures on coupling between the global generalized displacements or modes, (ii) the best bandwidth possible for a given foot-print of a symmetric diaphragm flexure mechanism, and (iii) the performance trade-offs between parameters such as range, bandwidth, and load-capacity for the same.

3.1 Coupling

From the equations of motion of the diaphragm flexure derived in Section 2.3, the geometrical layout of flexural constraints that allow for static and dynamic decoupling of the three DOFs can be determined. For static decoupling, the off-diagonal terms in the overall stiffness matrix **K** should be zero. For dynamic decoupling, the off-diagonal terms in both the mass **M** and stiffness **K** matrices should be zero.

Static and dynamic decoupling is desirable, for instance, when the diaphragm flexure mechanism is controlled to vertically position the central rigid mass while ensuring low error motions in the other DOFs, namely pitch and roll. Stable decoupled systems tend to be more amenable to low error motions even under open-loop control. It should, however, be noted that perfect decoupling cannot be achieved in practice owing to non-uniformities arising from manufacturing or material properties. Nonetheless, designing a compliant mechanism to be as close to a decoupled dynamic system as possible is desirable [24]. Here, we examine conditions under which such decoupling is possible for a diaphragm flexure mechanism.

Substituting the values of $\Re$ from Eq. (4) into Eqs. (5) and (5), and given Eqs. (1) and (6), the conditions necessary for the off-diagonal terms in the global mass **M** and stiffness **K** matrices to be zero are as follows:

$$\sum_{i=1}^{n} cos\alpha_i = 0; \quad \sum_{i=1}^{n} sin\alpha_i = 0; \quad \sum_{i=1}^{n} sin2\alpha_i = 0; \tag{8}$$

Hence, the geometric arrangement of a number $n \geq 3$ of flexures around the central rigid mass allows for the overall mechanism to be close to being statically and dynamically decoupled if the above conditions are satisfied. Note here that each individual flexure of the mechanism has its vertical deflection and slope coupled (both statically and dynamically), it is only the parallel combination of three or more of them that allow for the decoupling between the global modes to occur. Some possible design layouts that satisfy Eq. (8) are discussed in Section **A. 2** of the appendix. The effect of deviations from perfect symmetry on design requirements such as maximum deflection, velocity, and acceleration are studied in Section 4.

3.2 Natural frequencies

The best -3 dB bandwidth possible for a closed-loop system depends on many factors, including the natural frequencies or poles of the open-loop plant. Fig. 3 shows the plots of undamped natural frequencies of the first three modes of the symmetric diaphragm flexure of Fig. 1. The plots show the variation corresponding to diaphragm flexure configurations with flexural beam length, ℓ, varied in the range of about 0.5 in to 3 in, while keeping the footprint $2\ell+2R$ at a constant value of 7 in. This constant footprint is chosen as a scaling factor for the length dimension and will be used in Section 3.3 to normalize all lengths in the design to formulate a non-dimensional study.

The plots of Fig. 1 show values of the undamped natural frequencies obtained from models based on St. Venant's torsion theory and one of two distinct beam bending theories – either (i) Timoshenko beam bending theory, or (ii) Euler Bernoulli beam bending theory. As explained earlier, the former beam bending theory accounts for shear and rotational effects versus, while the latter does not. In the plots of the figure, the frequency values obtained from a commercial FEA package are superimposed for comparison of the chosen models.

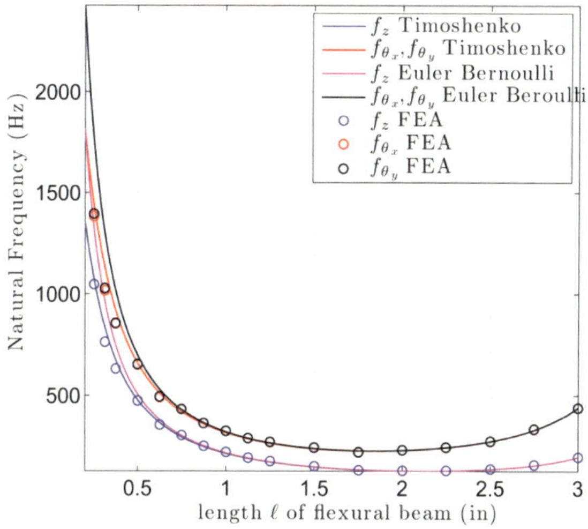

Fig. 3. Plot of undamped natural frequencies of first three modes of the symmetric diaphragm flexure of Fig. 1 for flexure beam length ℓ varying from about 0.2 in to 3 in, keeping footprint $D_0 = 2\ell + 2R$ at a constant value of $7in$. Other parameter values used in the simulation are: beam width $W = 0.75$ in, beam thickness $H = 0.1$ in, central rigid disk thickness $T = 2.5$ in, poisson's ratio $\nu = 0.33$, elastic modulus $E = 69$ GPa, density $\rho = 2700\ \frac{kg}{m^3}$.

The trends observed for the variation of natural frequencies for small flexural beam lengths is as expected, since small beam lengths result in large stiffness. Since the footprint is maintained constant, smaller beam lengths also imply large radius of the central disk and hence larger moving mass. However, the cubic dependence of stiffness on beam length dominates over the square dependence of mass on radius of the disk; hence the large natural frequencies at short beam lengths. For large beam lengths, the radius of the central disk is small, and hence the moving mass.[3] This effect is marginally larger than the loss in stiffness and hence the slight increase in natural frequency at large beam lengths.

For flexural beam lengths smaller than the shear approximation length factor $c \approx 0.6$ in, the Timoshenko model matches the trend from the FEA data better than the Euler-Bernoulli model. This confrms the prediction that shear effects dominate at small beam lengths and agrees with similar observations supporting the Timoshenko beam bending models for depicting the natural frequencies of short AFM cantilevers in [25].

Closed-form expressions for the natural frequency of the first three-dominant modes in the decoupled case, for large flexure lengths, are presented in Table 1. These expressions can be used as part of formulating an optimization problem, or to gain useful insights from parametric dependencies in designing a precision angular alignment setup based on diaphragm flexures.

[3] At the length scale of the diaphragm flexure discussed here, the moving mass is mainly composed of the central rigid disk. The lumped mass of the flexural beam is small at this length scale. However, it can be higher in other length scales, as in the case of a torsional MEMS mirror.

ω_z	$=$	$\sqrt{\dfrac{\frac{3EWH^3}{L^3}}{m+\frac{39}{35}m_f}}$
$\omega_{\theta_x,\theta_y}$	$=$	$\sqrt{\dfrac{\frac{3EWH^3}{L^3}(L^2+3R^2+3LR)+\frac{GWH^3}{2L^3}}{m(\frac{3R^2+T^2}{12})+m_f(\frac{L^2+39R^2+11LR}{70}+\frac{W^2+H^2}{24})}}$

Table 1. Closed-form expressions for natural frequencies of first three modes of diaphragm flexure.

3.3 Performance trade-offs

The design space for utilizing flexure-based precision angular alignment mechanisms can be characterizing in terms of key parameters such as the range, payload capacity, and bandwidth. Fig. 4 shows the variation of the key non-dimensionalized performance parameters as a function of the non-dimensional flexural beam length ℓ for all diaphragm flexures with a constant footprint of $D = 2\ell + 2R$. The performance parameters plotted in the figure are (i) the natural frequencies of the first three modes, namely deflection z and the two rotations θ_x and θ_y, (ii) the maximum load-capacity, F_{max}, defined as the load that causes the resultant axial stress in the flexural beams to reach the yield strength, σ_Y, of the material within a safety factor η, and (iii) the maximum vertical deflection δ_{max}, i.e. range under a given load. The normalization factors used for non-dimensionalizing the parameters are tabulated in Table 3, where ρ and E are the density and elastic modulus, respectively, of the material constituting the diaphragm flexure σ.

σ	F_{max}	K
$6\frac{Fl}{bh^2}$	$\frac{3EI}{l^3}$	$\frac{1}{6}\sigma_{max}\frac{bh^2}{l}$
$\frac{3}{2}\frac{Fl}{bh^2}$	$\frac{24EI}{l^3}$	$\frac{2}{3}\sigma_{max}\frac{bh^2}{l}$
$\frac{3}{2}\frac{Fl}{bh^2}$	$\frac{12EI}{l^3}$	$\frac{2}{3}\sigma_{max}\frac{bh^2}{l}$

Table 2. Flexural Building blocks Comparison

Parameter	Normalization Factor
ℓ	$D = 2\ell + 2R$
$f_z, f_{\theta_x}, f_{\theta_y}$	$f_0 = \frac{1}{4\pi}\sqrt{\frac{E}{\rho D^2}}$
F_{max}	$F_0 = \frac{1}{6}\frac{WH^2}{D}\frac{100\sigma_Y}{\eta}$
δ_{max}	$\delta_0 = \frac{10^6 F_0}{ED}$

Table 3. Normalization factors used for Design Parameters in Fig. 4.

The trade-off between load-capacity and range at different flexural beam lengths is evident from the figure. Small beam lengths allow for large load capacity and low range, whereas large beam lengths allow for low load capacity and large range. Natural frequencies are relatively low for intermediate beam lengths. The trade-off between natural frequencies and range is evident for small beam lengths, since stiffer beams have smaller deflections. The compromise between natural frequency and range is relatively mild at large beam lengths. An extension of the dynamic performance tradeoff characterization to the case of $n = 6$ flexural beams is presented in Fig. 5. From kinematic exact-contraint theory, it is known that only three constraints are needed to curb the three in-plane degrees of freedom of the rigid body. Thus for the case $n = 6$, three of the remaining constraints are redundant.

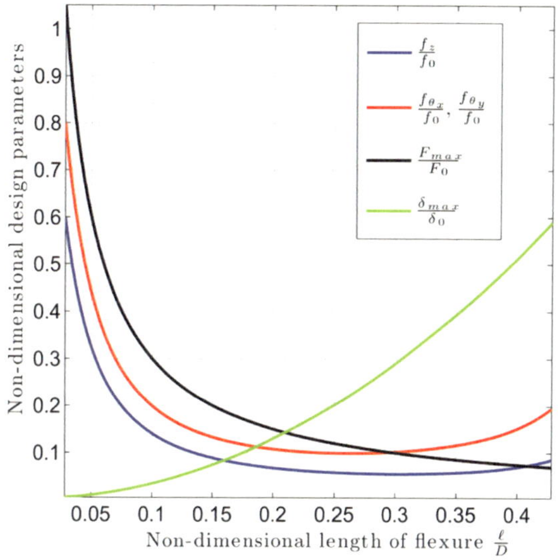

Fig. 4. Non-dimensional plots capturing the key dynamic performance parameters for diaphragm flexures of Fig. 3 with $n = 3$ flexural beams of the same footprint D, but different flexure lengths ranging in the approximate range 0-0.45 D. The parameters of interest are (i) the natural frequency of first three modes f_z, f_{θ_x}, and f_{θ_y}, all normalized by f_0, (ii) load capacity F_{max} normalized by F_0 and (iii) static vertical range δ_{max} normalized by δ_0.

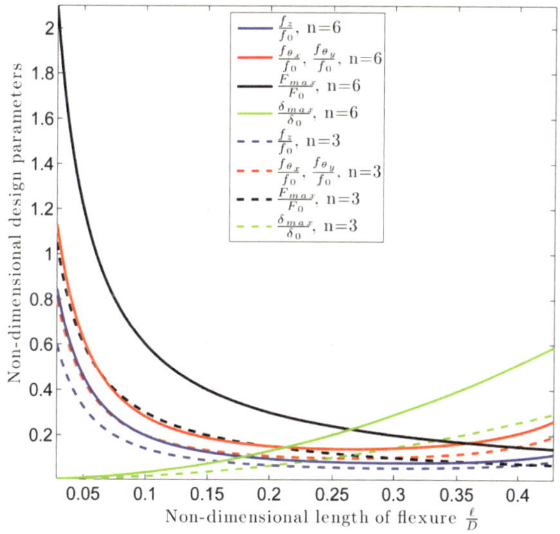

Fig. 5. Design parameter variations (shown as solid lines) captured for the case of double the number of flexures, i.e. $n = 6$, as compared to the case of $n = 3$ (shown as dashed lines) captured in Fig. 4.

The redundancy comes with added features which are desirable and facilitate improving the performance as follows. Since the load is distributed between larger number of flexures, the load-capacity envelope in this case is pushed higher to double that of the $n = 3$ case. The stiffnesses sum up since the flexural beams are arranged in a parallel combination. The resultant range is hence unchanged from that of the case of $n = 3$. The added slight benefit is that there is no reduction in the natural frequencies. Instead, there is an almost $\sqrt{2}$ increase in the natural frequency of all the first three modes. This increase can be explained from the doubling of stiffness and constancy of mass, since the added mass of the flexures is small compared to the rigid body here. Therefore, increasing the number of flexural beams offers improved performance in terms of range, load-capacity, and natural frequencies.

4. State-space performance analysis

Characterization of parameters such as maximum displacement (or range), maximum velocity and acceleration is critical for precision motion control applications. A procedure for identifying the static ($\omega = 0$) values of these parameters is developed for precision control of ball-screw drives in [26]. In this section, we present a state-space approach for determining the design performance parameters - maximum values of deflection (or range), velocity, and acceleration that are possible not only for static ($\omega = 0$) but for a large range of operating frequencies.

We follow the approach for the case of a diaphragm flexure used for precision angle alignment. To account for the case when symmetry cannot be ensured for the diaphragm flexure, we assume a coupled multi-input multi-output (MIMO) model, as against a collection of independent single-input single-output (SISO) models. We focus our analyses to parameters such as maximum vertical and angular displacement (range), velocity, and acceleration. The presented approach can be extended to map other design parameters to the state-space. Further, while the ideas presented here are general and applicable to the case when state or output feedback control is used as well, we focus our analysis on just the open-loop system.

4.1 State space formulation

We begin with a state vector $\mathbf{x}$ containing the generalized coordinates depicting the equations of motion of the system. One choice of state variables could be the generalized displacements and their first-order derivatives.

$$\dot{\mathbf{x}}_{m\times 1} = \mathbf{A}_{m\times m}\mathbf{x}_{m\times 1} + \mathbf{B}_{m\times r}\mathbf{u}_{r\times 1} \tag{9}$$

$$\mathbf{y}_{p\times 1} = \mathbf{C}_{p\times m}\mathbf{x}_{m\times 1} + \mathbf{D}_{p\times r}\mathbf{u}_{r\times 1} \tag{10}$$

The goal here is to find the maximum values of displacements, velocities, and acceleration for any set of inputs (which can be oriented in any direction in the input space). That means we need to compute the upper bounds on the amplification of a scalar component x_i, which is derived as:

$$x_i = \mathbf{E}_{i\,1\times m}\mathbf{x}_{m\times 1} \tag{11}$$

where the ith element of the row vector $\mathbf{E}_i$ is 1 and the rest of the m - 1 elements are zero. The component x_i can be any design variable, such as angular velocity, or vertical deflection of the diaphragm flexure.

For a chosen control law, in the Laplace domain, the following relations hold between the state vector $\mathbf{X}(s)_{m\times 1}$, its ith component $X_i(s)$, and the input vector $\mathbf{U}(s)$:

$$\mathbf{X}(s)_{m\times 1} = \mathbf{G}(s)_{m\times r}\mathbf{U}(s)_{r\times 1} \tag{12}$$

$$X_i(s)_{1\times 1} = \mathbf{E}_{i1\times m}\mathbf{G}(s)_{m\times r}\mathbf{U}(s)_{r\times 1} \tag{13}$$

The maximum amplification [27] of the component $X_i(s)$ for a given input $\mathbf{U}(s)$ can be expressed as the 2-induced (Euclidean) norm of the gain matrix $\mathbf{E}_{i1\times m}\mathbf{G}(s)_{m\times r}$. For the choice of $\mathbf{E}_i$, the gain matrix reduces to the ith row of $\mathbf{G}(s)$. Hence, its 2-induced norm reduces to a vector norm, and is given by its lone singular value. This singular value of $\mathbf{E}_{i1\times m}\mathbf{G}(s)_{m\times r}$ is always smaller than or equal to the singular values of the matrix $\mathbf{G}(s)$ and hence provides a tighter bound on the amplification of $X_i(s)$.

4.2 Application to diaphragm flexure

We now apply the above formulation to the case of the diaphragm flexure of Fig. 3 to derive tight upper bounds[4] on the amplification of state vector components, such as vertical deflection, or say, maximum angular velocity of the diaphragm flexure in a given control situation. We do not consider the feedback control problem here; however, the proposed method can be extended to that case as well.

The state vector $\mathbf{x}(t)$ and the input vector $\mathbf{u}(t)$ for a configuration with three linear actuators pushing down on the central rigid body at three locations $Q_i(Rcos\beta_i, Rsin\beta_i)$, for i = 1, 2, 3, are given as:

$$\mathbf{x}(t) = \begin{bmatrix} z(t) \\ \dot{z}(t) \\ \theta_x(t) \\ \dot{\theta}_x(t) \\ \theta_y(t) \\ \dot{\theta}_y(t) \end{bmatrix}; \quad \mathbf{u}(t) = \begin{bmatrix} F_1(t) \\ F_2(t) \\ F_3(t) \end{bmatrix}$$

The matrix $\mathbf{A}$ is assembled from components of $\mathbf{M}^{-1}\mathbf{K}$ from Eq. (5), and accounting for the derivative relationship pairs between the components of the state vector. By determining the force and moment components, the matrix $\mathbf{B}$ is given as

$$\begin{bmatrix} 0 & 0 & 0 \\ 1 & 1 & 1 \\ 0 & 0 & 0 \\ Rcos\beta_1 & Rcos\beta_2 & Rcos\beta_3 \\ 0 & 0 & 0 \\ -Rsin\beta_1 & -Rsin\beta_2 & -Rsin\beta_3 \end{bmatrix}$$

With the choice of $\mathbf{E}_i$ as described earlier, the maximum bound on each of the components of the state vector are found as shown in Fig. 6 for the diaphragm flexure of Fig. 3 containing three flexural beam units arranged symmetrically around the central rigid mass, and with

[4] The lower bound is zero, since for zero inputs, the components of the state vector are all zero.

linear actuators located at angles, $\beta_1 = 0$, $\beta_2 = \frac{2\pi}{3}$, $\beta_3 = \frac{4\pi}{3}$. The system is decoupled as seen from the variation of the singular values. With zero damping at the resonance peak, the maximum values of all variables assume exceedingly large values at the resonance frequency.

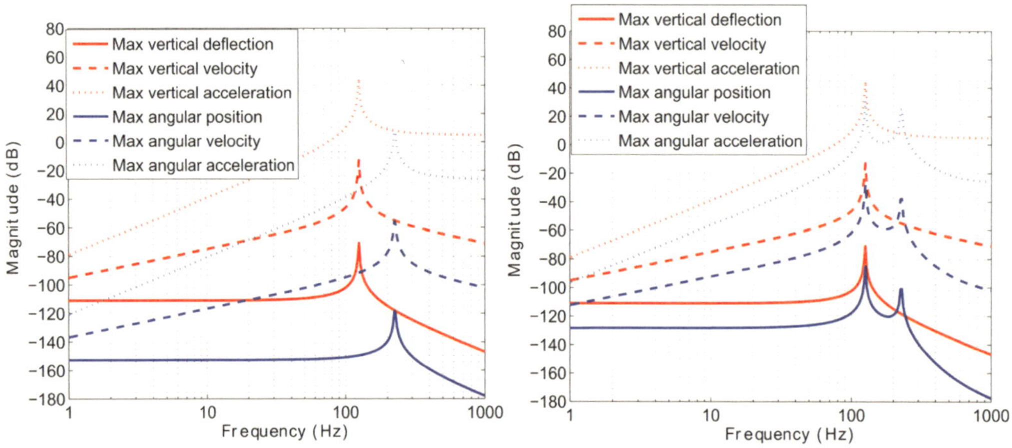

Fig. 6. The maximum amplification of states for an unit input vector (along any direction in the position space) is plotted for the case of three flexural beam units arranged symmetrically around the central rigid mass.

4.3 Effect of deviation from symmetry

The same analysis is repeated for the case when a 1° misalignment of one of the flexural beam units is caused by a manufacturing error. The system is now coupled, with deflection and angular position being dependent on each other, as seen from the two resonance peaks appearing in the variation of the singular values. This coupling implies that the angular position can assume exceedingly large values at a resonance frequency lower than that expected when perfect symmetry is ensured. The input directions that correspond to the maximum bound on a component $X_i(j\omega)$ at a chosen frequency ω lie along the right eigen vectors of the matrix $\mathbf{G}_i(j\omega)\mathbf{G}_i^H(j\omega)$, where $\mathbf{G}_i(j\omega)$ is the ith row of the matrix $\mathbf{G}(j\omega)$.

In summary, the benefits of using this approach for specifying the design performance variables are two fold - (i) it is applicable in case of deviations from perfect symmetry, allowing to analyze the effects of the deviations, and (ii) it gives the bounds not only for the static case (ω= 0) but also for the desired frequency range of interest. This approach can be incorporated into the design decision-making process, along with other important considerations, such as constraints imposed by physical limits, for example, saturation of the actuators, or limit stops in the path of a motion stage.

5. Conclusions

We have examined the need for diaphragm flexures in precision angular positioning applications. To accurately characterize the dynamics, we assembled lumped parameter models from mass and stiffness matrices for individual flexural building blocks and from the connected rigid body, as is done in typical finite element methods. Unlike previous

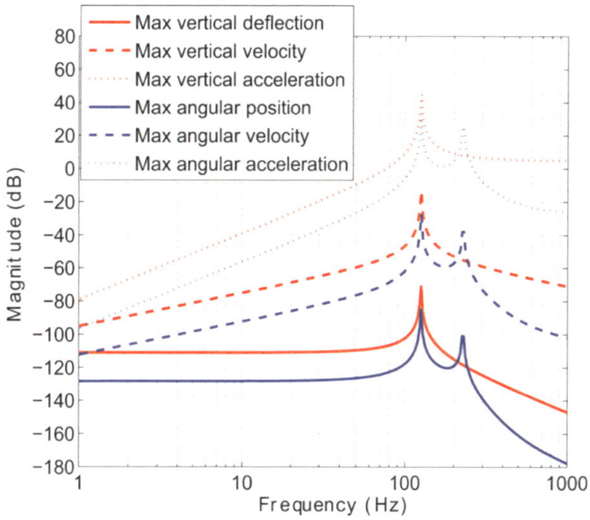

Fig. 7. The maximum amplification of states for an unit input vector (along any direction in the position space) is plotted for the case when a 1° misalignment of one of the flexural beam units is caused by a manufacturing error. Note that the system is now coupled as seen from the variation of the singular values.

works on flexure modeling, we found that Timoshenko beam bending models capture shear effects that dominate for short flexural beam lengths. We identified the key performance trade-offs in range, load-capacity, and natural frequencies of the first three modes of the diaphragm flexure. Redundancy in constraints was exploited to improve on load-capacity while ensuring the range or natural frequency requirements are met. Further, a control analysis based on singular value decomposition was formulated to capture the maximum values of performance variables such as linear or angular position and their derivatives. While perfect symmetry ensures decoupling between the modes, it was found that the amplification signature changes significantly in the presence of slight asymmetry caused by manufacturing errors. The dynamic modeling and state-space performance analysis detailed in this chapter are intended to serve as design tools for implementing high-precision motion control applications, and in particular, angular alignment based on diaphragm flexures.

APPENDIX

A.1 Distributed parameter model and approximation

The set of partial differential equations governing Timoshenko beam bending and St. Venant's torsion under the assumptions stated in Section 2.1 can be written in terms of the deflection $w(x, t)$, slope $\theta(x, t)$, and angle of twist $\phi(x, t)$ as follows [21, 20]:

$$EI_{yy}\frac{\partial^2\theta(x,t)}{\partial x^2} + \kappa AG\left\{\frac{\partial w(x,t)}{\partial x} - \theta(x,t)\right\} - \rho I_{yy}\frac{\partial^2\theta(x,t)}{\partial t^2} = 0 \tag{14}$$

$$\rho A\frac{\partial^2 w(x,t)}{\partial x^2} - \kappa AG\left\{\frac{\partial^2 w(x,t)}{\partial x^2} - \frac{\partial\theta(x,t)}{\partial x}\right\} = 0 \tag{15}$$

$$GJ_{xx}\frac{\partial^2\phi(x,t)}{\partial x^2}+I_{xx}\frac{\partial^2\phi(x,t)}{\partial t^2}=0 \quad (16)$$

where ρ, E, GJ_{xx} are the density, elastic modulus, and torsional rigidity, respectively; A, I_{yy}, and κ are the cross-sectional area, area moment of inertia about the neutral axis Y, and a geometry-dependent shear-factor, respectively. For a rectangular cross-section κ assumes a value of 0.833 [25].

The component values of the mass **M** and stiffness **K** matrices are listed in Tables 4 and 5. The parameters used in the tables are $\alpha=\left\{\frac{1}{1+2p}\right\}^2$ and $\beta=\frac{EI_{yy}}{6}\frac{c^2}{l^5}$, where $p=(\frac{c}{\ell})^2$, and $c=\sqrt{\frac{6EI_{yy}}{\kappa AG}}$ representing the length scale at which effects of shear dominate. For a flexural beam with rectangular cross-section of height H, c reduces to $\sqrt{1.1(1+\nu)}H$, where ν is the poisson's ratio of the material.

$$\begin{aligned}
M_1 &= \alpha m_f[\tfrac{13}{35}+\tfrac{7}{5}(\tfrac{c}{\ell})^2+\tfrac{4}{3}(\tfrac{c}{\ell})^4]+\tfrac{6}{5}\tfrac{\rho\alpha I_{yy}}{\ell}\\
M_2 &= -\alpha m_f\ell[\tfrac{11}{210}+\tfrac{11}{60}(\tfrac{c}{\ell})^2+\tfrac{1}{6}(\tfrac{c}{\ell})^4]\\
&\quad +\rho\alpha I_{yy}[\tfrac{1}{10}+(\tfrac{c}{\ell})^2]\\
M_3 &= \alpha m_f\ell^2[\tfrac{1}{105}+\tfrac{1}{30}\{(\tfrac{c}{\ell})^2+(\tfrac{c}{\ell})^4\}]\\
&\quad +\rho\alpha I_{yy}\ell[\tfrac{2}{15}+\tfrac{1}{3}(\tfrac{c}{\ell})^2+\tfrac{4}{3}(\tfrac{c}{\ell})^4]\\
M_4 &= \tfrac{I_{xx}}{3}
\end{aligned}$$

Table 4. Mass matrix component values

$$\begin{aligned}
K_1 &= \alpha\tfrac{12EI}{\ell^3}+144\alpha\beta\\
K_2 &= -\alpha\tfrac{6EI}{\ell^2}-72\alpha\beta\\
K_3 &= \alpha\tfrac{4EI}{\ell}[1+(\tfrac{c}{l})^2+(\tfrac{c}{l})^4]+36\alpha\beta\\
K_4 &= \tfrac{GJ_{xx}}{\ell}
\end{aligned}$$

Table 5. Stiffness matrix component values

A.2 Decoupled arrangements

To identify the designs of diaphragm flexures that are close to being statically and dynamically decoupled, numerical methods can be used to solve the conditions given in Eq. (8). A geometric interpretation of the first two conditions of Eq. (8) is presented as:

$$\sum_{i=1}^{n}(cos\alpha_i+jsin\alpha_i)=0; \quad (17)$$

where j is the imaginary number $\sqrt{-1}$. In the complex plane, the first two conditions of Eq. (8) hence represents a number n of unit vectors radiating from the origin and adding up to zero. An easy guess of a subset Γ_s of the solution space Γ is possible if we consider the case when the flexures are symmetrically arranged around the central rigid mass. Some possible solutions present in the symmetry solution subset Γ_s are as follows:

i. For odd values of n, $n \geq 3$, a possible solution subset is $\alpha_i=\varphi+\frac{(i-1)2\pi}{n}$ for i = 1, 2, 3, … n and $0 \leq \varphi < \frac{2\pi}{n}$.
ii. For even values of n, $n \geq 4$, there is an $\frac{n}{2}$ -fold symmetry, i.e. there are $\frac{n}{2}$ axes about which the flexural units are arranged symmetrically. A possible solution is one with all

unit vectors symmetrically arranged $\frac{2\pi}{n}$ apart, i.e. $\alpha_i = \varphi + \frac{(i-1)2\pi}{n}$ for i = 1, 2, 3, …n and $0 \leq \varphi < \frac{2\pi}{n}$. Another set can be obtained simply by rotating any two (or multiples of two) unit vectors adjacent to any axes of symmetry by the same angle, one in clockwise direction, and the other in counter-clockwise direction. This possible solution is illustrated in Fig. 8.

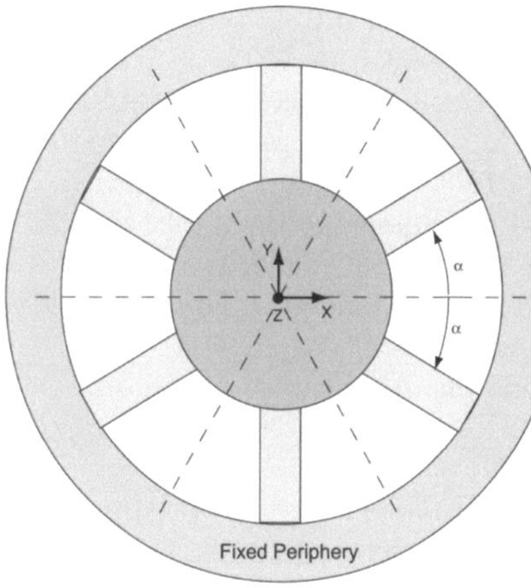

Fig. 8. A solution subset to the case when n = 6, corresponding to three-fold symmetry with angle between any two constraints being 2α = 60°. The dashed lines denote the axes of symmetry. If the flexure beams on either side of the horizontal axis of symmetry are brought symmetrically closer by an angle $\Delta\alpha$, they still satisfy the decoupling conditions of Eq. (8).

6. Acknowledgements

This work was supported by funding grants from the Manufacturing Systems and Technology program under the Singapore MIT Alliance. The first author would like to thank Ajay A. Deshpande and Mythili R. Vutukuru for their discussions on parts of the analysis.

7. References

[1] Loney GC, "High bandwidth steering mirror," US Patent 5,110,195, 1992.
[2] Choi BJ, Sreenivasan SV, Johnson S, et al, "Design of orientation stages for step and flash imprint lithography," Precision Engineering, Journal of the International Societies for Precision Engineering and Nanotechnology, vol. 25 (3), pp 192-199, July 2001.
[3] Choi KB, Lee JJ "Passive compliant wafer stage for single-step nano-imprint lithography,' Review of Scientific Instruments, 76 (7): Art. No. 075106 Jul 2005.
[4] Kendale AM, "Automation of soft lithographic microcontact printing," SM thesis, Department of Mechanical Engineering, Massachusetts Institute of Technology, Cambridge MA, 2002.
[5] Chan HB, Aksyuk VA, Kleiman RN, et al, "Quantum mechanical actuation of microelectromechanical systems by the Casimir force," Science, vol. 291 (5510): 1941-1944, Mar 2001.
[6] V. Shilpiekandula, "Progress through Mechanics: Small-scale Gaps," *Mechanics* (Publication of the American Academy of Mechanics), vol. 35, no. 9-10, pp. 3-6, Sep-Oct 2006.

[7] Slocum AH, "Precision machine design," Englewood Cliffs, NJ, Prentice Hall, 1992.
[8] Smith ST, "Flexures: elements of elastic mechanisms," Amsterdam, Gordon and Breach, 2000.
[9] Awtar S, Slocum AH, Sevincer E. "Characteristics of beam-based flexure modules," Journal of Mechanical Design, vol. 129 (6): 625-639 Jun 2007.
[10] V. Shilpiekandula and K. Youcef-Toumi, "Modeling and Control of a Programmable Filter for Separation of Biologically Active Molecules," In Proceedings of *American Control Conference*, pp. 394-399, June 2005.
[11] Awtar S and Slocum AH, "Design of Flexure Stages based on a Symmetric Diaphragm Flexure,' In Proceedings of American Society for Precision Engineering Annual Meeting, no. 1803, Norfolk, VA, 2005.
[12] Lobontiu N, "Compliant mechanisms: design of flexure hinges," Boca Raton, CRC Press, 2003; Howell LL, "Compliant mechanisms," New York , Wiley 2001.
[13] Jones RV, "Instruments and Experiences: Papers on Measurement and Instrument Design," John Wiley, 1988.
[14] Li Z, and S. Kota, "Dynamic Analysis of Compliant Mechanisms, "Proceedings of DETC, 27th Biannual Mechanisms and Robotics Conference," Sep 29 - Oct 2, Montreal, Canada, 2002.
[15] Choi KB, "Dynamics of a compliant mechanism based on flexure hinges," Proceedings of the Institution of Mechanical Engineers, Part-C, Journal of Mechanical Engineering Science 219 (2): 225-235 Feb 2005.
[16] Choi KB, Han CS, "Optimal design of a compliant mechanism with circular notch flexure hinges," Proceedings of the institution of Mechanical Engineers, Part-C, Journal of Mechanical Engineering Science 221 (3): 385-392 Mar 2007.
[17] Weaver W, Timoshenko SP, Young DH, "Vibration problems in engineering," New York, Wiley, 1990.
[18] Meirovitch L, "Computational Methods in Structural Dynamics," Sijthoff Noordhoff, Rockville MD, 1980.
[19] Dwivedy SK, Eberhard P. "Dynamic analysis of flexible manipulators, a literature review," Mechanism and Machine Theory, vol. 41 (7): 749-777, July 2006.
[20] Prezmieniecki JS, "Theory of matrix structural analysis," New York, McGraw-Hill, 1968.
[21] Ganesan N and Engels RC, "A dynamic finite element model for the Timoshenko beam," AIAA-1990-3547 Space Programs and Technologies Conference, Huntsville AL, Sep 1990.
[22] Inman DJ, "Engineering vibration," Englewood Cliffs, NJ, Prentice Hall,1994.
[23] Varanasi KK and Nayfeh SA, "Damping of Flexural Vibration Using Low-Density, Low-Wave-Speed Media," Journal of Sound and Vibration, vol. 292, pp. 402-414, 2006.
[24] Deo HV, Suh NP,"Mathematical transforms in design: Case study on feedback control of a customizable automotive suspension," CIRP Annals - Manufacturing Technology, vol. 53 (1), pp. 125-128, 2004.
[25] Hsu JC, Lee HL, Chang WJ, "Flexural vibration frequency of atomic force microscope cantilevers using the Timoshenko beam model," Nanotechnology, vol.18 (28): Art. No. 285503, July 2007.
[26] Varanasi KK, Nayfeh SA, "The Dynamics of Lead-Screw Drives: Low-Order Modeling and Experiments," Journal of Dynamic Systems, Measurement, and Control, Transactions of the ASME, vol. 126, pp. 388-396.
[27] Skogestad S and Postlethwaite I, "Multivariable feedback control: analysis and design," Hoboken, NJ, John Wiley, 2005.

26

Development of a LCD Photomask Based Desktop Manufacturing System

Ren C. Luo[1] and Jyh-Hwa Tzou[2]
[1]*Department of Electrical Engineering, National Taiwan University*
[2]*Department of Aeronautical Engineering, National Formosa University*
Taiwan, R. O. C.

1. Introduction

Desktop manufacturing refers to various rapid prototyping (RP) techniques where three-dimensional components are directly built, layer-by-layer, from a computer data description or computer aided design (CAD) file. Due to their ability and relative ease in transforming a conceptual design into a physical model, desktop manufacturing technologies have met escalating demand in the industry for shortening new product development cycle time. The layer-by-layer fabrication methodology also allows complex models to be made with ease.

In the desktop manufacturing processes, the geometry of the object to be manufactured can be obtained from computer aided design (CAD) model data, an existing object (through Reverse engineering) (Puntambenker et al., 1994) (Soboh et al., 1994) (Chua Chee Kai, 1994) or mathematical data (e.g., surface equations) (Ren C Luo & Yawei Ma, 1995). Prashant et al. (Prashant Kulkarni et al., 2000) had reviewed many process planning techniques in layer manufacturing. Most desktop manufacturing systems accept model data described in an intermediate file format called the .STL format. This file format approximates the original model geometry using a series of triangular facets.

After loading a .STL model, a slicing procedure is then applied to the tessellated model. In this process, the model is intersected with a set of horizontal planes to create a series of cross sections, or slices, comprised of contours that represent the material boundaries of the part to be generated. The contours are subsequently used to generate the NC tool paths for the desktop manufacturing system.

There are many commercial RP systems available on the market today such as InVision (3D Systems Corporation), Objet (Objet Geometries Ltd.), Perfactory (Envisiontec Inc.), stereolithography (SLA) (3D Systems Corporation) and fused deposition modeling (FDM) (Stratasys, Inc.). The InVision 3-D printer combines 3D Systems' multi-jet modeling (MJM) printing technology with an acrylic photopolymer model material. The Objet's Polyjet technology works by jetting photopolymer materials in ultra-thin layers (0.016 mm) onto a build tray, layer by layer, until the part is complete. The Perfactory RP system uses a photomonomer resin and a DLP projector to polymerize 3D finished parts. The stereolithography (SLA) and fused deposition modeling (FDM) are old RP processes. Both the laser beam of SLA system and the thermal extrusion head of FDM system generate 2D cross sectional areas using one-dimensional tool paths. The disadvantages of these systems are lower speed, the requirement for an expensive XYZ table and 3 axes motion control system. Another old RP process, Solid Ground Curing (SGC) (Cubital Ltd.), involves creating a temporary photomask of each layer, applying a thin coating of photopolymers and exposing the layer to a burst of ultraviolet light to cure it. Because the cross sections of one layer are cured at the same time, the SGC system has a faster build speed. But the mechanism of the SGC system is complex, and the price is very expensive.

The purpose of this paper is to develop a low cost desktop manufacturing system. We use a liquid crystal display (LCD) panel as a photomask(Luo, R. C. et al., 2000). With the bottom exposure method, the image of the LCD photomask is calculated from the sliced data. The desktop manufacturing system has the advantages of low cost, compactness and requiring no special physical support, making it suitable for use in an office.

S. C. Ventura et al. (S. C. Ventura et al., 1996) developed a Direct Photo Shaping process for the fabrication of functional ceramic components layer by layer. Each layer is photoimaged by a LCD or a digital light processing (DLP) projection system. Young, J. S. et al. (Young, J. S. et al. , 1999) described a novel device for producing 3D objects that has been developed using an LCD as a programmable, dynamic mask and visible light to initiate photopolymerization. Monneret, Serge et al. (Monneret, Serge et al. , 2002) presented a new process of microstereolithography to manufacture freeform solid 3D micro-components with outer dimensions in the millimeter size range. Huang, Y. -M. et al. (Huang, Y. -M. et al., 2003) analyzed the shrinkage deformation of the mask type stereolithography process. Jiang, C. -P. et al. (Jiang, C. -P. et al., 2005) developed a Masked Photopolymerization Rapid Prototyping (MPRP) system using LCD panel as dynamic mask with an upper exposure skill.

2. LCD photomask based desktop manufacturing system

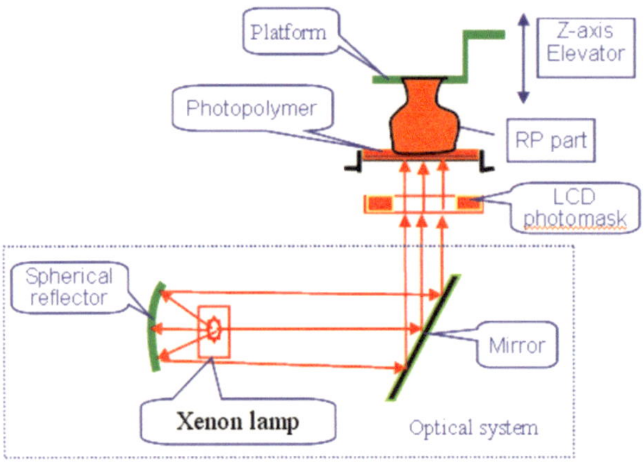

Fig. 1. Configuration of the LCD photomask based desktop manufacturing system

The LCD photomask based desktop manufacturing system structure is illustrated in Fig. 1. The hardware configuration of this system includes the LCD photomask, optical system, z-axis elevator and PC based control system. The optical system can generate parallel light passing through the LCD photomask to cure the photopolymer. The RP part is generated layer-by-layer and attached to a platform that rises as each successive layer is attached to the bottom-most face. The resin is deposited onto the transparent bottom plate. The platform and previously built structure are lowered into the resin, leaving a liquid film between the part and the bottom plate that has the correct thickness for the next layer. The new layer is formed beneath the platform by exposing the LCD photomask. After the layer is finished, the platform is raised, separating the layer from the bottom plate, filling and wiping the resin and the process is repeated until all layers are fabricated. The completed RP part is then removed from the platform, post cured and finished, if needed.

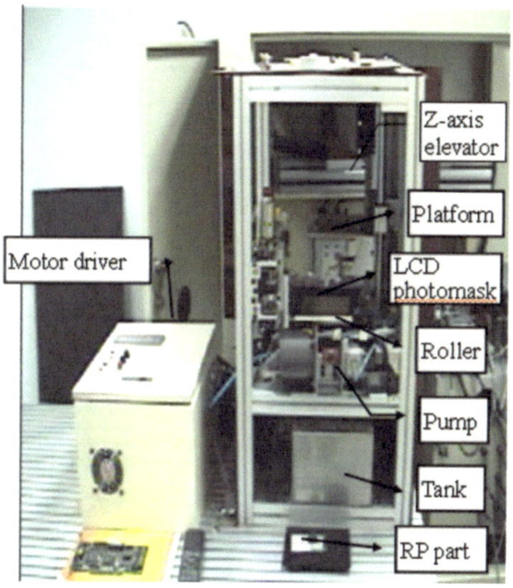

Fig. 2. Photo of the LCD photomask based desktop manufacturing system

The desktop manufacturing system architecture is illustrated in Fig. 2. The Z-axis elevator with a high precision ball screw is driven by an AC servomotor. A PC-based DSP motion controller is used to control the movement of the Z-axis elevator. The LCD photomask is connected to the VGA card of the computer.

The architecture of the proposed system includes five main components; 1) Data processing unit; 2) LCD photomask; 3) Optical system; 4) PC based DSP Motion Control System; 5) Z-axis elevator. The details of these components are as follows:

2.1 Data processing unit

The data processing unit performs the slicing procedure and the photomask generation process. The slicing procedure transforms the 3-D CAD model into a set of 2-D layer contours. According to this contour data, the photomask generation program exports the contour of each layer to the LCD photomask. Note that the region inside the contour is displayed in white color and the region outside the contour is displayed in black color.

2.2 LCD photomask

An LCD serves as a photomask which is used to display layer contours. The light source emits parallel light upwardly through the transparent portions of the LCD photomask to expose and solidify the entire layer at once. As shown in Fig. 3, a 14.1-inch TFT (Thin Film Transistor) LCD with 1024x768 pixels is used herein. Each pixel is 0.28 mm in both width and length, yielding a photomask resolution of under 0.28 mm. An insulating membrane is located below the LCD photomask to insulate the resin from ultraviolet and heat produced by the light source. Furthermore, the size of the RP part can be produced in this system is restricted by the size of the LCD panel used. If we can use larger LCD panel, the size of the RP part could be increased.

Fig. 3. The LCD photomask of the proposed system

2.3 Optical system:

The optical system strongly influences the system structure, forming method and building time for parts. The proposed system uses NAF-200N photo-curable liquid resin (Denken Engineering Co. Ltd., Japan), as the building material. NAF-200N solidifies under exposure to 680 nm visible light. A 275W xenon lamp serves as the light source. The optical spectrum of this xenon lamp, detected by a spectrometer, is shown in Fig. 4, where the spectral radiance is observed with maximum power at 736nm wavelength and a lot of energy at 680 nm. The experimental results confirm the ready solidification of NAF-200N under exposure to this xenon lamp. According to the experimental results, the layer thickness for one layer is 0.254mm. The curing time for one layer is 135 sec.

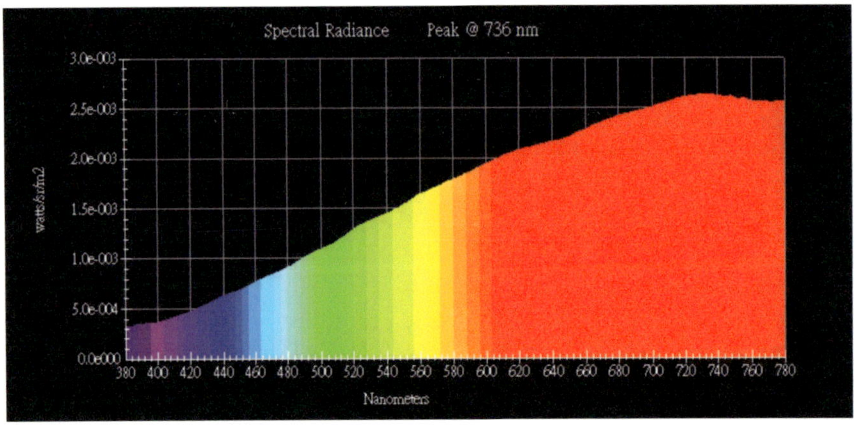

Fig. 4. The spectrogram of the 275W xenon lamp.

The optical system design is illustrated in Fig. 5(a). The actual structure is shown in Fig. 5(b).The proposed system uses plane-shaping instead of line-shaping. The visible light source is emitted from the bottom up to the part instead of from the top down in order to reduce the amount of resin wasted.

The ray tracing method is an important tool in geometrical optics. Matrix optics (Eugene Hecht, 2002) was used to design the RP machine optical system. A ray is described by its position and its angle with respect to the optical axis. The matrix form of several optical components can be shown as follows:

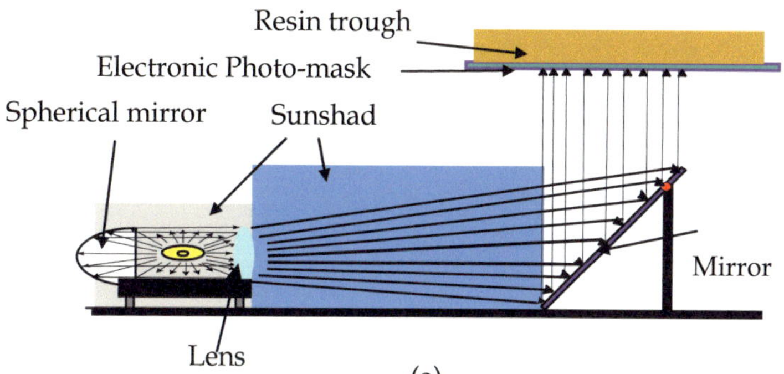

(a)

(b)

Fig. 5(a). The design of the curing light source; 5(b) Real structure of the curing light source

1. **Free - space propagation**

As shown in Fig. 6, a ray traversing a distance d is altered in accordance with $y_2 = y_1 + \theta_1 d$ and $\theta_2 = \theta_1$.

The ray-transfer matrix is $T = \begin{bmatrix} 1 & d \\ 0 & 1 \end{bmatrix}$, and $\begin{bmatrix} y_2 \\ \theta_2 \end{bmatrix} = \begin{bmatrix} 1 & d \\ 0 & 1 \end{bmatrix} \begin{bmatrix} y_1 \\ \theta_1 \end{bmatrix} = T \begin{bmatrix} y_1 \\ \theta_1 \end{bmatrix}$ (1)

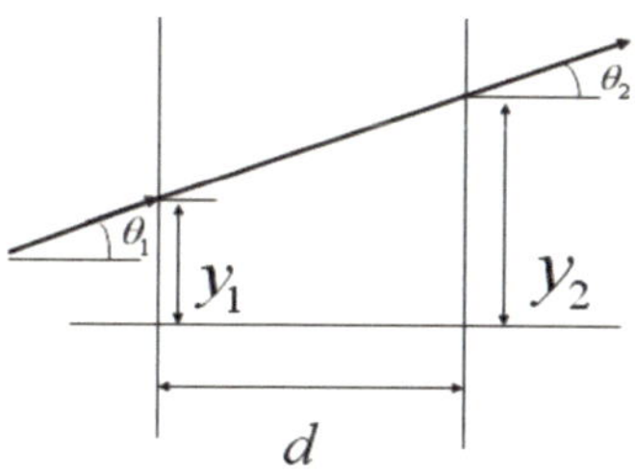

Fig. 6. Free - space propagation

2. **Transmission through a thin lens**

As shown in Fig. 7, the relation between θ_1 and θ_2 for paraxial rays transmitted through a thin lens of focal length f. Since the height remains unchanged $(y_2 = y_1)$, the refraction matrix of thin lens is $A = \begin{bmatrix} 1 & 0 \\ -\frac{1}{f} & 1 \end{bmatrix}$, and $\begin{bmatrix} y_2 \\ \theta_2 \end{bmatrix} = \begin{bmatrix} 1 & 0 \\ -\frac{1}{f} & 1 \end{bmatrix} \begin{bmatrix} y_1 \\ \theta_1 \end{bmatrix} = A \begin{bmatrix} y_1 \\ \theta_1 \end{bmatrix}$ (2)

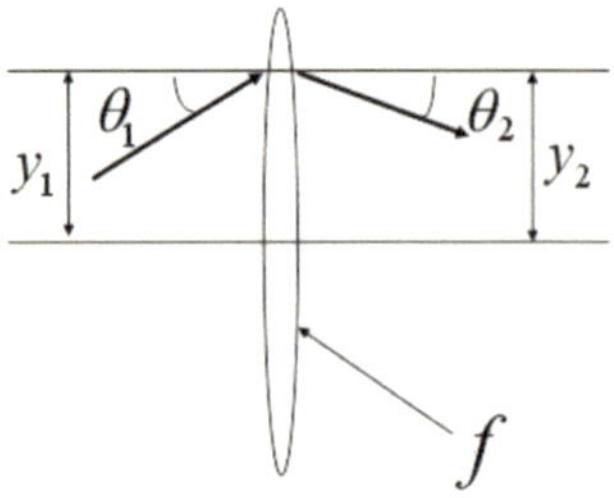

Fig. 7. Transmission through a thin lens

3. **Reflection from a planar mirror**

As shown in Fig. 8, the ray position is not altered $(y_2 = y_1)$, and we conclude that $\theta_2 = \theta_1$.

The ray-transfer matrix is therefore the identity matrix $R = \begin{bmatrix} 1 & 0 \\ 0 & 1 \end{bmatrix}$, and

$$\begin{bmatrix} y_2 \\ \theta_2 \end{bmatrix} = \begin{bmatrix} 1 & 0 \\ 0 & 1 \end{bmatrix} \begin{bmatrix} y_1 \\ \theta_1 \end{bmatrix} = R \begin{bmatrix} y_1 \\ \theta_1 \end{bmatrix} \quad (3)$$

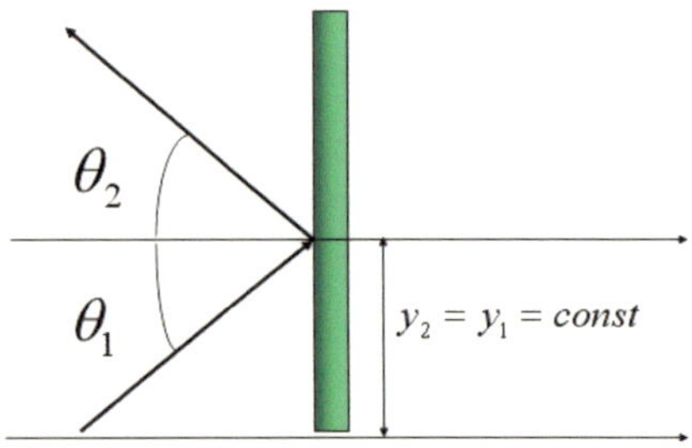

Fig. 8. Reflection from a planar mirror

4. **Reflection from a spherical mirror**

As shown in Fig. 9, the ray position is not altered $(y_2 = y_1)$. The reflection matrix of a spherical mirror is $S = \begin{bmatrix} 1 & 0 \\ -\frac{2}{R} & 1 \end{bmatrix}$, and $\begin{bmatrix} y_2 \\ \theta_2 \end{bmatrix} = \begin{bmatrix} 1 & 0 \\ -\frac{2}{R} & 1 \end{bmatrix} \begin{bmatrix} y_1 \\ \theta_1 \end{bmatrix} = S \begin{bmatrix} y_1 \\ \theta_1 \end{bmatrix}$ (4)

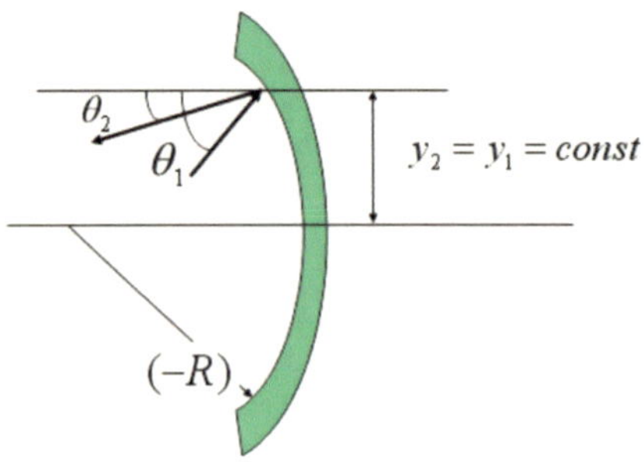

Fig. 9. Reflection from a spherical mirror

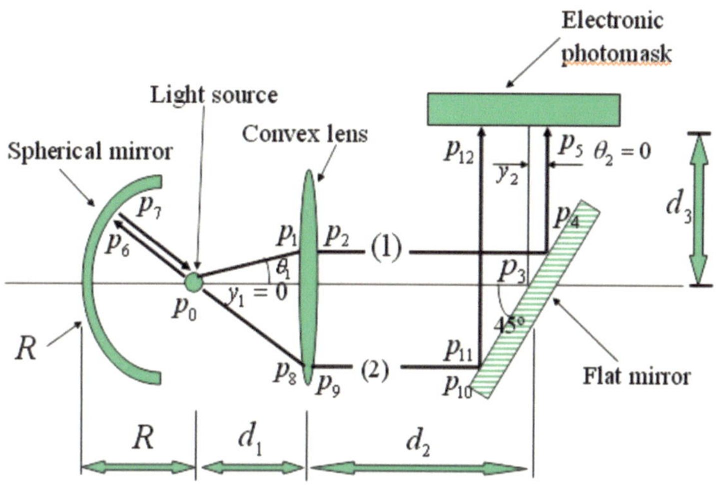

Fig. 10. The ray tracing diagram of the optical system of RP machine

The ray tracing diagram of the optical system of RP machine is shown in Fig. 10. The light source P_0 (xenon lamp) is located on the focus of the biconvex lens and on the center position of the spherical mirror. The ray emitted by light source P_0 can be divided into two parts, one [Ray trajectory (1)] transmits through the thin lens directly, and the other [Ray trajectory (2)] is reflected from the spherical mirror.

These two ray trajectories are discussed as follows:

- **Ray trajectory (1): $P_0P_1\ P_2\ P_3\ P_4\ P_5$**

The ray emitted from light source P_0, transmits through the thin lens. After reflecting the ray from the flat mirror, the reflected parallel light can be generated onto the LCD photomask position (P_5). The system matrix is then defined as:

$$M_1 = T_{54}R_{43}T_{32}A_{21}T_{10} = \begin{bmatrix} 1 & d_3 \\ 0 & 1 \end{bmatrix}\begin{bmatrix} 1 & 0 \\ 0 & 1 \end{bmatrix}\begin{bmatrix} 1 & d_2 \\ 0 & 1 \end{bmatrix}\begin{bmatrix} 1 & 0 \\ -\frac{1}{f} & 1 \end{bmatrix}\begin{bmatrix} 1 & d_1 \\ 0 & 1 \end{bmatrix} =$$

$$= \begin{bmatrix} 1-\frac{d_2}{f}-\frac{d_3}{f} & d_1+d_2\left(-\frac{d_1}{f}+1\right)-\frac{d_1 d_3}{f}+d_3 \\ -\frac{1}{f} & -\frac{d_1}{f}+1 \end{bmatrix} \tag{5}$$

Where:

T_{10} = The ray-transfer matrix is from P_0 to P_1

A_{21} = The thin lens refraction matrix is from P_1 to P_2

T_{32} = The ray-transfer matrix is from P_2 to P_3

R_{43} = The reflection matrix of a planar mirror is from P_3 to P_4

T_{54} = The ray-transfer matrix is from P_4 to P_5

Thus the ray at point P_5 on the LCD photomask position is given by:

$$\begin{bmatrix} y_2 \\ \theta_2 \end{bmatrix} = M_1 * \begin{bmatrix} y_1 \\ \theta_1 \end{bmatrix} = \begin{bmatrix} 1-\frac{d_2}{f}-\frac{d_3}{f} & d_1+d_2\left(-\frac{d_1}{f}+1\right)-\frac{d_1 d_3}{f}+d_3 \\ -\frac{1}{f} & -\frac{d_1}{f}+1 \end{bmatrix} \begin{bmatrix} y_1 \\ \theta_1 \end{bmatrix} \tag{6}$$

If d_1 = f and y_1 =0, equation (6) can be simplified as

$$\begin{bmatrix} y_2 \\ \theta_2 \end{bmatrix} = \begin{bmatrix} 1-\frac{d_2}{f}-\frac{d_3}{f} & f \\ -\frac{1}{f} & 0 \end{bmatrix} \begin{bmatrix} 0 \\ \theta_1 \end{bmatrix} \tag{7}$$

Consequently, $y_2 = f\ \theta_1$ and θ_2 =0 (8)

- **Ray trajectory (2): $P_0P_6\ P_7\ P_8\ P_9\ P_{10}\ P_{11}\ P_{12}$**

As shown in Fig.10, the ray emitted from light source P_0 is reflected by the spherical mirror. After the reflected ray is transferred through the thin lens and flat mirror, the parallel light can be generated onto the LCD photomask position (P_{12}). The system matrix is then defined as:

$$M_2 = T_{12\cdot 11} R_{11\cdot 10} T_{10\cdot 9} A_{98} T_{87} S_{76} T_{60} =$$

$$= \begin{bmatrix} 1 & d_3 \\ 0 & 1 \end{bmatrix} \begin{bmatrix} 1 & 0 \\ 0 & 1 \end{bmatrix} \begin{bmatrix} 1 & d_2 \\ 0 & 1 \end{bmatrix} \begin{bmatrix} 1 & 0 \\ -\frac{1}{f} & 1 \end{bmatrix} \begin{bmatrix} 1 & d_1+R \\ 0 & 1 \end{bmatrix} \begin{bmatrix} 1 & 0 \\ -\frac{2}{R} & 1 \end{bmatrix} \begin{bmatrix} 1 & R \\ 0 & 1 \end{bmatrix} =$$

$$= \begin{bmatrix} -1-\frac{2d_1}{R}+\frac{d_2}{f}+\frac{d_3}{f} & -d_1 \\ \frac{1}{f} & 0 \end{bmatrix} \tag{9}$$

Where:

T_{60} = The ray-transfer matrix is from P_0 to P_6

S_{76} = The reflection matrix from a spherical mirror is from P_6 to P_7

T_{87} = The ray-transfer matrix is from P_7 to P_8

A_{98} = The thin lens refraction matrix is from P_8 to P_9

$T_{10\cdot 9}$ = The ray-transfer matrix is from P_9 to P_{10}

$R_{11\cdot 10}$ = The reflection matrix of a planar mirror is from P_{10} to P_{11}

$T_{12\cdot 11}$ = The ray-transfer matrix is from P_{11} to P_{12}

Thus the ray at the point P_{12} on the LCD photomask position is given by:

$$\begin{bmatrix} y_2 \\ \theta_2 \end{bmatrix} = M_2 * \begin{bmatrix} y_1 \\ \theta_1 \end{bmatrix} = \begin{bmatrix} -1-\frac{2d_1}{R}+\frac{d_2}{f}+\frac{d_3}{f} & -d_1 \\ \frac{1}{f} & 0 \end{bmatrix} \begin{bmatrix} y_1 \\ \theta_1 \end{bmatrix} \tag{10}$$

If d_1 = f and y_1 =0, Eq. (10) can be simplified as

$$\begin{bmatrix} y_2 \\ \theta_2 \end{bmatrix} = \begin{bmatrix} 1 - \frac{d_2}{f} - \frac{d_3}{f} & f \\ -\frac{1}{f} & 0 \end{bmatrix} \begin{bmatrix} 0 \\ \theta_1 \end{bmatrix} \tag{11}$$

Consequently, $y_2 = f\ \theta_1$ and θ_2 =0 (12)

From the results of Eqs. (8) and (12), the ray angle θ_2 on the LCD photomask is independent of the incident ray angle θ_1 in two ray trajectories. The optical system emits parallel light upwardly (θ_2 =0) through the LCD photomask to expose and solidify the photo-curable resin.

The focal length of the biconvex lens and the radius of curvature of the spherical mirror are selected for 15 cm and 10 cm due to the machine space limitations, respectively. Because the high power xenon lamp could generate enough convection and radiation heat to affect the LCD photomask and resin, a sunshade is placed between the biconvex lens and flat mirror to reduce the heat transfer to the LCD photomask.

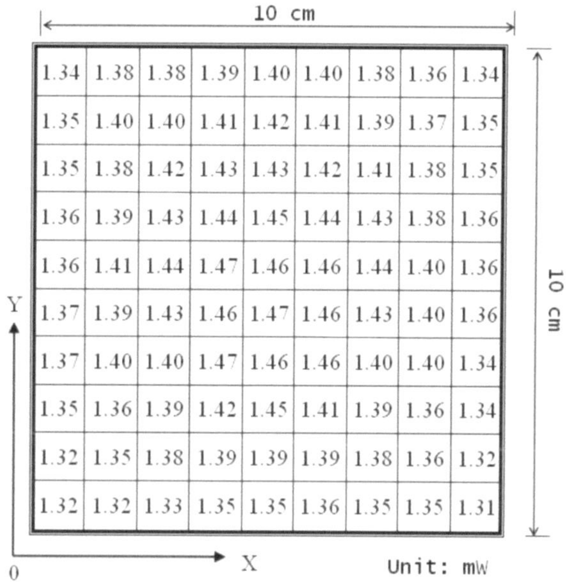

Fig. 11. The light power spreads on the LCD photomask

After constructing the optical system, measurement of the power that spreads onto the LCD photomask is necessary. This research uses an optical power meter to measure the light power that spreads onto the LCD photomask. The valid light area on the LCD photomask is 10 x 10 cm^2. This research divides the valid area into 100 equal parts. Each part is measured the light power using an optical power meter. The measured values are shown in Fig. 11. From these experimental results the light that spreads onto the LCD photomask is determined in good uniform. The light source uses 275W xenon lamp. The average light power through the LCD photomask is 1.43 mW.

3. Software design

3.1 The Bucket-Sorting algorithm

For most Rapid Prototyping systems, CAD models described in the .STL file format must be sliced into contours. An effective slicing algorithm is necessary for Rapid Prototyping systems. The simple approach is to intersect every facet with every slicing plane. This approach is time consuming. The Bucket sorting algorithm is used in the slicing pre-processing for search speed enhancement.

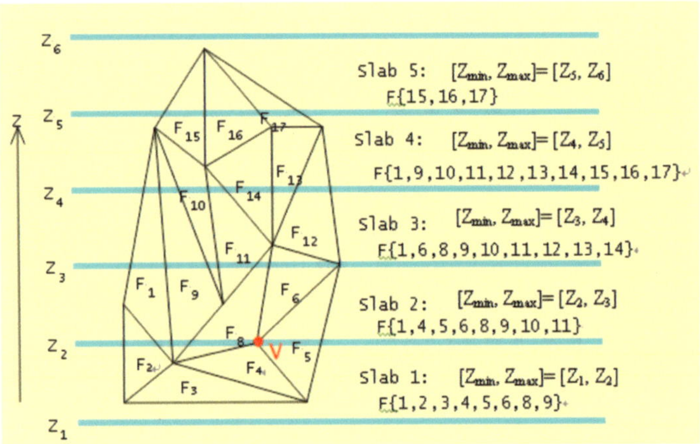

Fig. 12. The bucket-sorting algorithm for data sorting

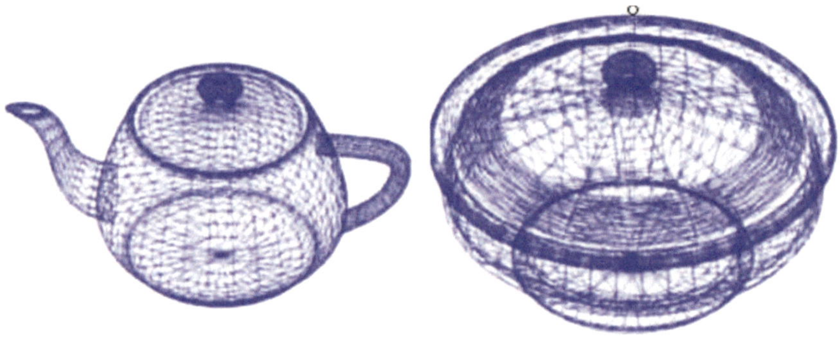

Fig. 13 (a).
Facet Number : 4096
Height range : 56.114 inch

Fig. 13 (b).
Facet Number : 5192
Height Range : 180.718 inch

Bucket number	1	5	10	20	30	40
Slicing time(sec)	49.251	13.309	8.633	6.409	5.709	5.388
Time ratio	100%	27%	17.5%	13%	11.6%	11%

Bucket number	50	60	70	80	90	100
Slicing time(sec)	5.238	5.257	5.268	5.287	5.558	5.598
Time ratio	10.6%	10.7%	10.7%	10.7%	11.3%	11.4%

Table 1. The slicing time (sec) for the different Bucket number [Fig. 13 (a).]

Bucket number	1	5	10	20	30	40
Slicing time(sec)	194.28	50.363	31.826	23.26	20.29	18.54
Time ratio	100%	25.9%	16.4%	12%	10.4%	9.5%

Bucket number	50	60	70	80	90	100
Slicing time(sec)	17.57	17.24	17.045	16.72	16.84	16.99
Time ratio	9.0%	8.9%	8.8%	8.6%	8.7%	8.8%

Table 2. The slicing time (sec) for the different Bucket number [Fig. 13 (b).]

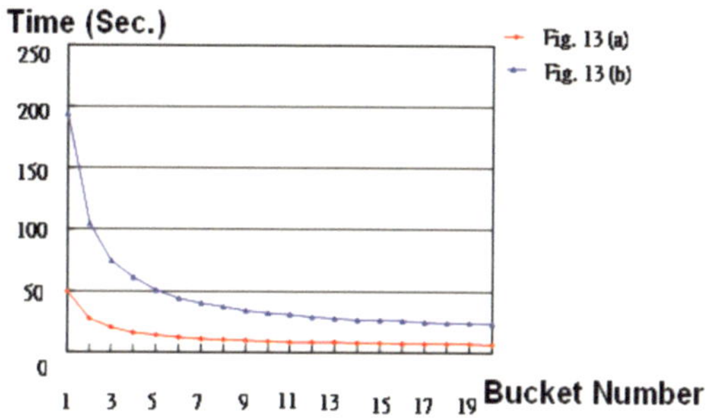

(a)

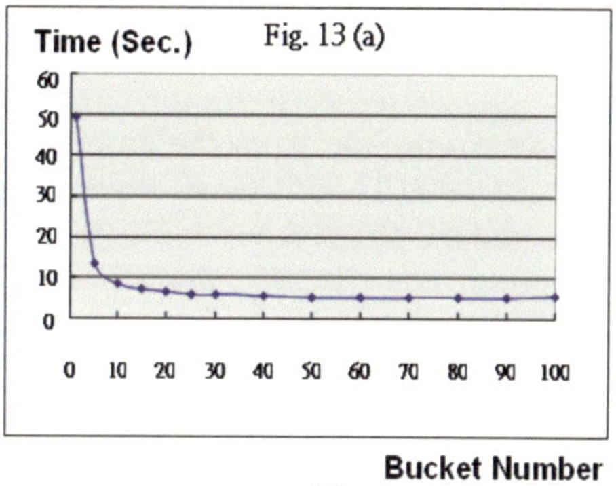

(b)

Fig. 14(a) The Slicing time for different Bucket number; Fig. 14(b) The Slicing time for different Bucket number

The bucket-sorting algorithm divides the spatial space into N subspaces. Searching for triangular data inside these smaller subspaces is often faster than browsing the entire space. In this study, the split space was also called a slab as illustrated in Fig.12. The slabs were generated by the defined maximum acceptable thickness and by the maximum and minimum Z-coordinate of the facets. Each slab was defined between a Z_{min} and a Z_{max} (Fig.12.), so that, when slicing at a specific height z, the specific slab was the one that included z within its limits [Z_{min}, Z_{max}]. A facet is assigned to a slab whenever one or more of its vertices fall within the slab's range. If a vertex has a Z value exactly equal to the boundary height between two slabs, that facet is assigned to both slabs. Fig.12 shows that the four facets (F_4,F_5,F_6,F_8) have a common vertex (V) and this vertex's Z-coordinate equals to the boundary height (Z_2) between two slabs (Slab 1 and slab 2). Consequently, these facets ((F_4,F_5,F_6, and F_8) are assigned to both slabs (slab1 & slab 2).

To implement the bucket-sorting algorithm, Figs. 13(a) and 13(b) are the input files. The Bucket number was changed to compare the slicing time. The results (slicing time) for different bucket numbers are shown in Table 1 [Fig. 13 (a).] and Table2 [Fig. 13 (b).]. If the bucket number is 1, the bucket-sorting algorithm was not used. From these results, the slicing time is greatly reduced using the bucket-sorting algorithm.
The results from Table 1 and Table2 are shown in Figs. 14(a) and 14(b). From these results, the slicing time is greatly reduced by the bucket-sorting algorithm. The slicing time ratio can be reduced by nearly 25% with 5 buckets. In Table 1, the slicing time ratio will reach the minimum value with 50 buckets. If the Bucket number is more than this certain value (50 for Table 1), the slicing time will increase. However, in Table2, the fastest slicing time will occur at 80 buckets. This optimum bucket number for slicing time is not a fixed value. It depends on the RP part's height, facet number and the size of the facets. In general, the optimum Bucket number value is 8~10. If the Bucket number is greater than 10, the slicing time will not be obviously decreased. This means that too many buckets are not useful for reducing the slicing time.

3.2 The LCD photomask display algorithm

The LCD photomask displays the cross-sectional contours of model layers and the optical system can project the light through the white areas of the photomask. The LCD photomask display algorithm is described as follows. The program fills with white color inside the contour, and fills with black color outside the contour. The light beam shines through the white areas to cure the resin. After curing one layer, the program will display the cross-sectional contour of the next layer in the LCD photomask. When all layers have been built, the program stops the RP machine and the physical part is finished.
We used Visual BASIC as the algorithm compiler. The program outputs display data to the LCD photomask to display the filled contours layer by layer. Fig. 15(a) shows a STL file which is created by Pro/Engineering 3D CAD software. Fig. 15(b) is the result after slicing the STL file. Fig. 16(a) and Fig. 16(b) show the cross-sectional contours filled with white color inside the contour.

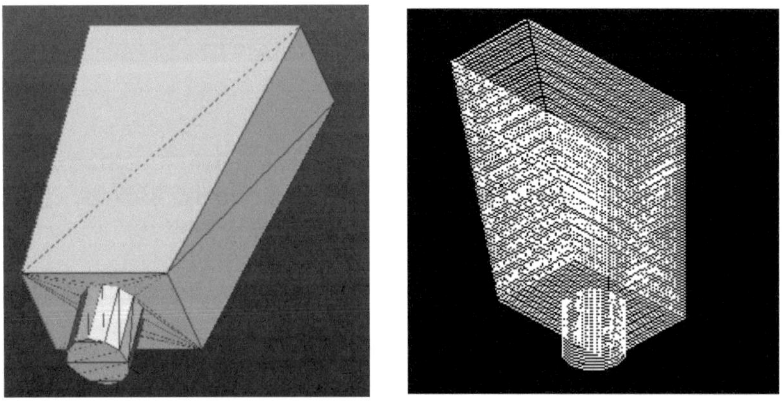

(a) (b)

Fig. 15(a) STL Model created by Pro/Engineering CAD software; Fig. 15(b) STL model sliced using the slicing process

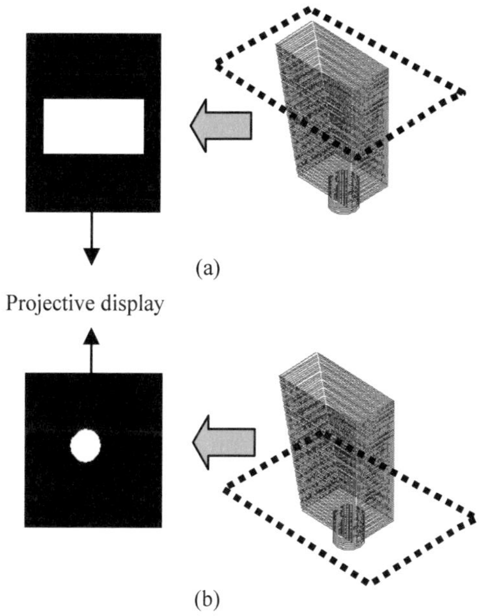

Fig. 16. LCD photomask shows the cross-sectional contours filled with white color

As described above, the proposed desktop manufacturing system uses photo-curable liquid resin NAF-200N to serve as the building material. This resin will solidify under exposure to 680 nm wavelength light. From the experimental results, the relation between the exposure time and hardened depth of the resin is illustrated in Table 3. In general, the proposed system uses uniform slicing and the layer thickness is set to 0.254mm so that the exposure time for each layer is 135 sec.

4. Experimental results

In order to compare the machining efficiency, the FDM 2000 RP machine (Stratasys Inc.) has been chosen for comparison. Although FDM 2000 is among the slowest system, it is the only RP system available in our laboratory. A 100mm x 100mm x 1mm thin plate is manufactured by the FDM 2000 and the proposed RP system. After finishing the slicing process, the tool path for FDM 2000 RP machine is shown in Fig. 17(a), and the LCD photomask display of the proposed RP system is shown in Fig. 17(b). The manufacturing time for building one layer is 582 seconds in FDM 2000. However, the manufacturing time for building one layer is 135 seconds in the proposed RP system. The machining efficiency of the proposed system is better than the FDM 2000 RP system. The accuracy of the proposed RP system is 0.4mm (0.015 inch), which is sufficient for real applications.

	1	2	3	4	5	6	7	8	9
Exposure time (sec)	15	30	45	60	75	90	105	120	**135**
Hardened thickness (mm)	0.010	0.015	0.030	0.050	0.06	0.100	0.150	0.200	**0.254**

Table 3. The experimental results: The exposure time relative to hardened depth of the resin

(a) (b)

Fig. 17(a) Tool path for FDM 2000 RP machine; Fig. 17(b) The LCD photomask display of the proposed RP system

4.1 Case 1

In Case 1, the 3D CAD model is illustrated in Fig. 18(a) and the STL model is shown in Fig. 18(b). The RP software reads the STL file first and proceeds with the slicing process. The sliced model is shown in Fig. 19(a).

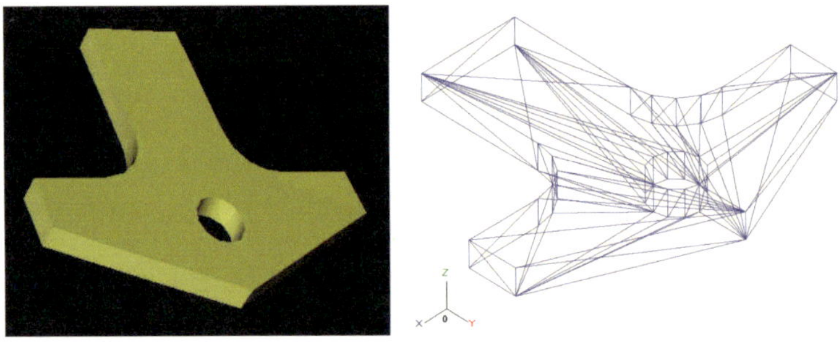

(a) (b)

Fig. 18(a) 3D CAD model of Case 1; Fig. 18(b) STL model of Case 1

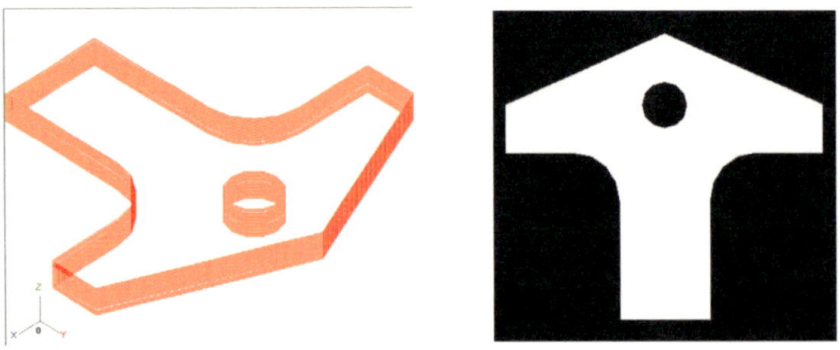

(a) (b)

Fig. 19(a) Illustration of sliced model; Fig. 19(b) Cross contour displayed in the LCD photomask

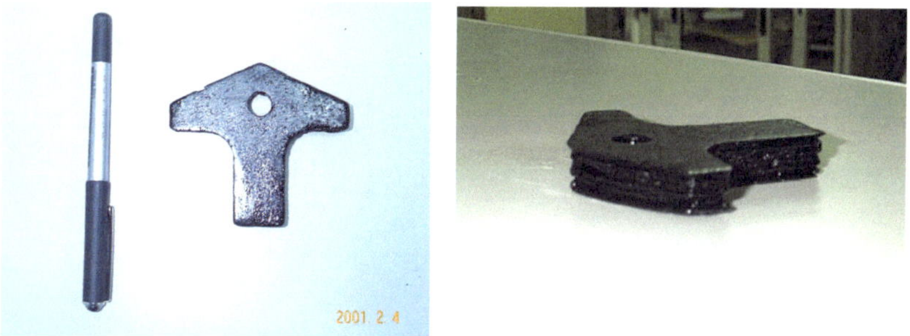

Fig. 20. RP part of Case 1 built by the proposed desktop manufacturing system

The LCD photomask displays the cross contour and the proposed desktop manufacturing system builds the physical model. The cross contour displayed in the LCD photomask is illustrated in Fig. 19(b). The physical part is built using the proposed system layer by layer. The finished RP part for Case 1 is shown in Fig. 20.

4.2 Case 2

In Case 2, the 3D CAD model is illustrated in Fig. 21(a) and the STL model is shown in Fig. 21(b). After slicing the STL file, the sliced model is illustrated in Fig. 22(a), and the LCD photomask cross contour is illustrated in Fig. 22(b). The physical part built using the proposed system is shown in Fig. 23.

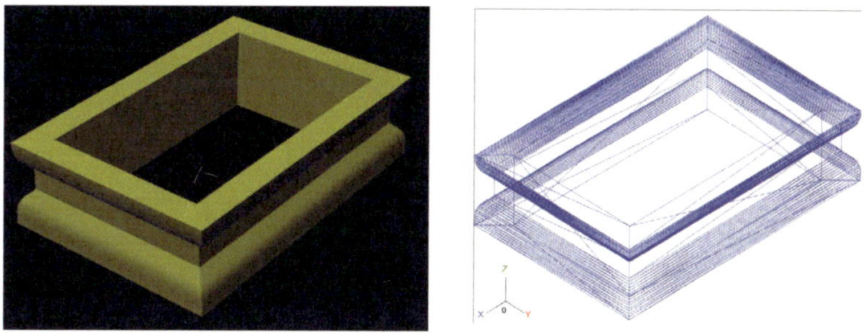

(a) (b)

Fig. 21(a). 3D CAD model of Case 2; Fig. 21(b) STL file of Case 2

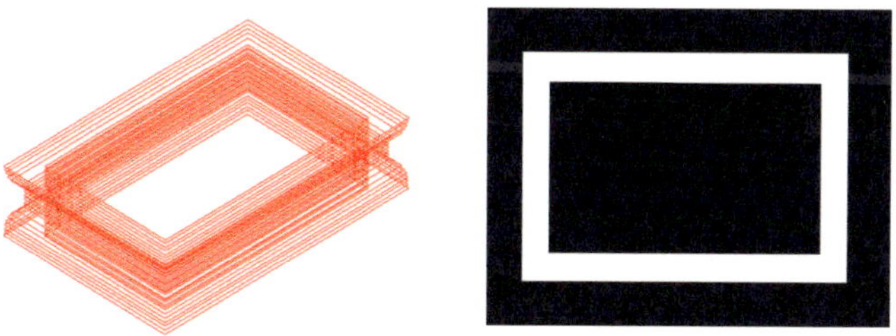

(a) (b)

Fig. 22(a). Sliced model of Case 2; Fig. 22(b) Cross contour in the LCD photomask

Fig. 23. RP part of Case 2 built by the proposed desktop manufacturing system

The experimental results of Case 1 and Case 2 are also shown in Table 4.

	Layer number	Layer thickness	Curing time of each layer	Total curing time
CASE 1	49	0.254 mm	135 sec	110.25 min
CASE 2	60	0.254 mm	135 sec	135 min

Table 4. The experimental results of Case 1 and Case 2

5. Conclusions

Desktop manufacturing system has been widely known as being able to fabricate 3D objects with complex geometric shapes. The purpose of this paper is to develop a LCD photomask based desktop manufacturing system. The main features of the proposed system are described as follows:

1. The LCD photomask is connected to the computer, to allow changing the cross sectional display of each layer quickly.
2. The software design includes a Slicing algorithm, LCD photomask display process, user interface and motion control program. The Bucket-sorting algorithm is used in the slicing pre-processing for search speed enhancement. The slicing time ratio can be reduced to nearly 25% with 5 buckets.
3. The hardware configuration of this architecture includes LCD photomask, optical system, z-axis elevator and PC based control system.
4. The Matrix optics is used to design the optical system.
5. The experimental results show that the proposed desktop manufacturing system can produce RP parts with good machining efficiency, but the surface roughness should be further improved.
6. The proposed desktop manufacturing system has the advantages of low cost, compactness, speed and accuracy, and requires no additional support, providing a valuable addition to the working office environment for designers at all levels and in all locations.

6. References

Bahaa E. A. Saleh and Malvin Carl Teich (1991). *Fundamentals of Photonics*, John Wiley & Sons, Inc.

Chua Chee Kai (1994). Three-dimensional rapid prototyping technologies and key development areas, *IEE Computing & Control Engineering Journal*, pp. 200 ~ 206.

Eugene Hecht, (2002). *OPTICS*, 4th edition, Pearson Education, Inc., publishing as Addison Wesley, San Francisco, CA.

Huang, Y.- M.; Jiang, C.-P. (2003). Numerical analysis of a mask type stereolithography process using a dynamic finite element method, *The International Journal of Advanced Manufacturing Technology*, Vol. 21, No. 9, pp 649-655.

Jiang, C. -P. *, J. -Y, Jeng, Y. -M., Huang and M. -J. Tsai (2005). Development of Masked Photo-polymerization Rapid Prototyping System using Liquid Crystal Display Panel, *Journal of the Chinese Institute of Industrial Engineers*, Vol. 22, No.1, pp76-81.

Luo, R. C., Tzou, J. H., and Lee, W. Z. (2000). The Development of LCD Panel Display Based Rapid Prototyping System for Advanced Manufacturing, *Proceedings* of the *2000 IEEE International Conference on Robotics and Automation*, San Francisco, CA, USA, Vol. 4, pp. 3083-3088.

Monneret, Serge; Le Gall, H.; Bade, V.; Devaux, F.; Mosset, A.; Lantz, E., (2002). Dynamic UV microstereolithography, *Applied Physics*, Vol. 6, No. 3, pp 213-218.

Prashant Kulkarni, Anne Marsan and Debasish dutta (2000). A review of process planning techniques in layer manufacturing, *Rapid Prototyping Journal*, Vol. 6, No. 1,pp. 18-35.

Puntambenker, N. Jablokow, A. and Sommer, H. (1994). A unified review of 3D model generation for reverse engineering, *Computer Integrated Manufacturing*, 7, pp259-269.

Ren C Luo and Yawei Ma (1995). A Slicing Algorithm for Rapid Prototyping and Manufacturing, *Proceedings of the IEEE international Conference on Robotics and Automation*, pp 2841-2846.

S. C. Ventura, S. C. Narang, S. Sharma, J. Stotts, C. Liu, S. Liu, L-H. Ho, D. Annavajjula, S.J. Lombardi, A. Hardy, M. Mangaudis, E. Chen, L. Groseclose (1996). A New SFF Process for Functional Ceramic Components, Proceedings of the Solid Freeform Fabrication, pp 327-334.

Soboh, T. Owen, J. Jaynes, C. Dekhil, M. and Henderson, T. (1995). Industrial inspection and reverse engineering, *Proceedings of the IEEE CAD-based Vision*, pp. 228-235.

Young, J. S. Fox, S. R. Anseth, K. S. (1999). A novel device for producing threedimensional objects, *Journal of Manufacturing Science & Engineering*. Vol. 121, No.3, pp 474-477.

http://www. 3dsystems.com/

http://www. objet.com/

http://www.envisiontec.de/index.htm

http://www.3dsystems.com/

http://www.stratasys.com/
http://www.cubital.com/

27

Positioning Systems for Bed Profiling in Hydraulics Physical Models

João Palma[1], Paulo Morais[1], Luís Guilherme[1] and Elsa Alves[2]
[1]*Scientific Instrumentation Centre,*
[2]*Department of Hydraulics and Environment,*
Laboratório Nacional de Engenharia Civil,
Portugal

1. Introduction

Physical hydraulic models are commonly used to study complex flow problems in the field of coastal engineering, estuaries, rivers, reservoirs and hydraulic structures. Laboratório Nacional de Engenharia Civil (LNEC) and other hydraulics research institutions developed some kinds of devices for recording bed profiles in flumes and hydraulic basins with varying types of bottom materials (sand, plastic, among others), most of which are based on servo-controlled positioning systems (Cardoso, 1964; Azevedo & Morais, 1989; Delft Hydraulics, 2005; HR Wallingford, 2006; Palma et al, 2008).

A common type of apparatus for this purpose consists of a vertical position controlled device driven by a small electric motor with a mechanical transmission actuating a rod as depicted in Fig. 1. At the lower end of the rod there is a probe which is the bottom sensor used for the tracking the rod point under water closely to the sediment bed. Sediments like sand or mud are often very weak materials and require contactless detection in order to prevent their destruction by the sensing probe itself. The vertical displacement of the moving elements out of water is converted into an electric signal representing depth.

The longitudinal profile recording is performed by the horizontal motion of the vertical servo controlled device, usually placed in a carriage or chariot, with rail or beam guidance; the horizontal (translation) movement may be done either by motorized or manual displacement.

A delicate part in this domain of application is the capability to detect very closely the bottom of a hydraulics flume or basin without effective contact and with good tracking capability, in order to avoid the penetration and destruction of the bed material. The problem will be addressed by appropriate design of multi-electrode conductivity probes taking into account also the need for compensating disturbance effects. Rough bed surface, variable granular sediments, water conductivity variations due to temperature and salinity, and also electromagnetic interference, are among the most disturbing causes that influence the sensing and servo-control performance.

In the following sections a number of problems and solutions for sediment bed profile tracking in hydraulics laboratory studies are discussed. An overview of solutions for vertical and horizontal motion control will be presented and discussed in the following, as regards both to the mechanical and the electrical drive subsystems. Experimental results from its application to a particular hydraulics study are shown. Particular attention is devoted to special phenomena and details that, not surprisingly, are decisive for achieving high-quality implementations.

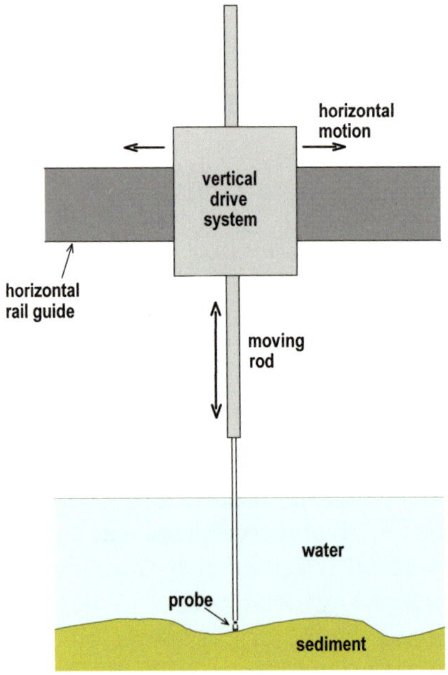

Fig. 1. Conceptual principle of a servo driven bottom follower.

2. Positioning and measurement apparatuses for bed profiling

2.1 Solutions for bed profiling instrumentation

The method that will be described in detail is based on the principle of a servo controlled vertical rod having a tip probe to detect the bottom by rapid electric conductivity variation. This method of transduction is used in several institutions (e.g. Delft Hydraulics, 2005; HR Wallingford, 2006). Its merits consist of providing high accuracy (reaching the order of tenths of a millimetre), good spatial resolution (in the order of the tip diameter that may be of a few millimetres) and the ability to track even rather inclined profiles (over 30º) as well as irregular ones.

The same type of instrument can also be used to measure varying water levels - a function denominated *limnimeter*. It can be used to track tide levels in laboratory basins, but the speed of response usually does not allow the tracking of waves.

In a brief review of other bed profiling methods ultrasonic transduction appears as an attractive alternative for depth measurement based on the transit time of acoustic pulses emitted from inside the liquid mass towards the bottom until the reception of an echo, as shown in Fig. 2a. This static technique has low maintenance requirements and has deserved major preference for batimetry studies in rivers or at the sea (e.g. Ernstsen et al, 2006) although it is identified with some difficulties in measuring very inclined or irregular bed surfaces. Laser profiling is also of an expensive kind of method that has good metrological characteristics (Yeh et al, 2009) but need careful flume or basin empting operations to leave dry surfaces which is not always acceptable.

The above techniques are applicable when the bottom is composed of sediment grain sizes several times less than the diameter of the tip probe or of the ultrasonic beam. For beds of granules or stones of a size up to several centimetres the profile can be tracked by a trolley lean rod with wheels at the end in contact with the bottom (see Fig. 2b) being dragged horizontally by the axis of rotation in the upper end (Rao & Rao, 2004). The depth is calculated by geometric relations as a function of the tilt angle, the wheel diameter, the rod length, the axis height and the horizontal coordinate. The measurement of inclination (Fraden, 1993) can be made by a potentiometer, an inclinometer or even by an optical pulse encoder.

The latter solution, which can be used both inside and outside water (amphibious method), can only move in one sense - dragging the rod with the wheels behind - thus requiring an automatic lift up mechanism or a manual procedure to allow the return motion. The spatial resolution is poor and the contact of the wheel with the bottom may also crush the sediment if it is not consistent enough.

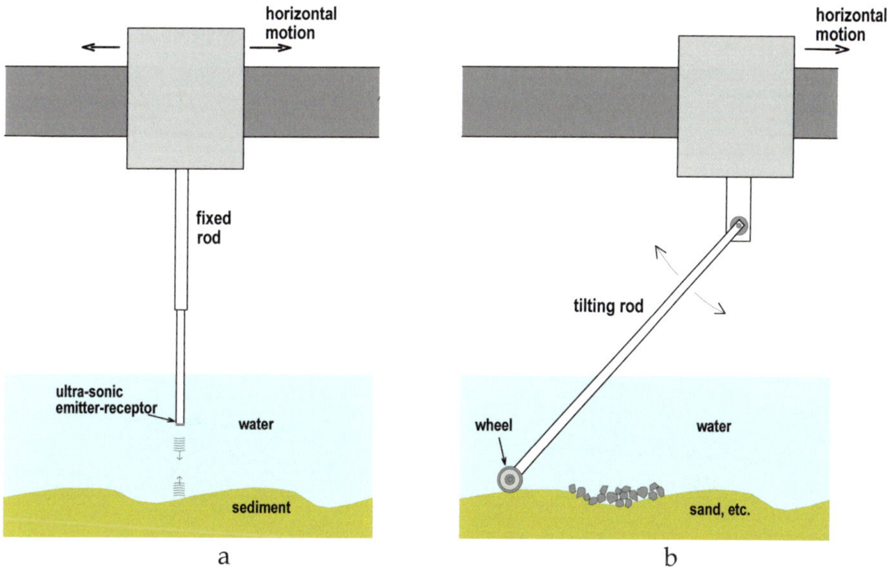

Fig. 2. Other bed profiling solutions: a) ultrasonic transit time; b) trolley with tilting rod.

There is another servo-controlled solution with a vertically moving rod like the one in Fig. 1, in which the bottom surface is detected by sensing a small force of contact applied to the sediment by a wheel placed at the lower end with a force cell (Delft Hydraulics, 2005). Although insensitive to changes in the conductivity of the water, this method is amphibious like the previous one, has poor spatial resolution and may not be used in weak sediment beds.

2.2 Horizontal motion of bed profilers

Different solutions may be conceived for diverse motion requirements in laboratory instrumentation positioning. All of them have advantages and disadvantages when taking into consideration factors like cost, maintenance, reliability and adequacy to the problem at hand. In Fig 3 shows several types of devices for linear movement.

The rack and pinion device (Fig. 3a) is one of the oldest that is still being used to provide precision slippage-less horizontal linear movement of bed followers. With proper accessories it can be set to eliminate any backlash. Due to a very high cost is inadequate for very large applications. It also requires additional devices to avoid damage in case of movement blockage. The rack can be used to drive a position measuring device like a potentiometer for small applications or an optical encoder.

The systems based on flexible elements to convert rotary motion into linear motion like the toothed belt or chain and sprocket driven systems (Fig 3b) or the cable driven systems (Fig. 3c) provide a low cost solution for large applications. Both types are much less precise than the rack and pinion device. Also, the cable driven systems are prone to slippage problems; an additional system is required in order to reliably measure position.

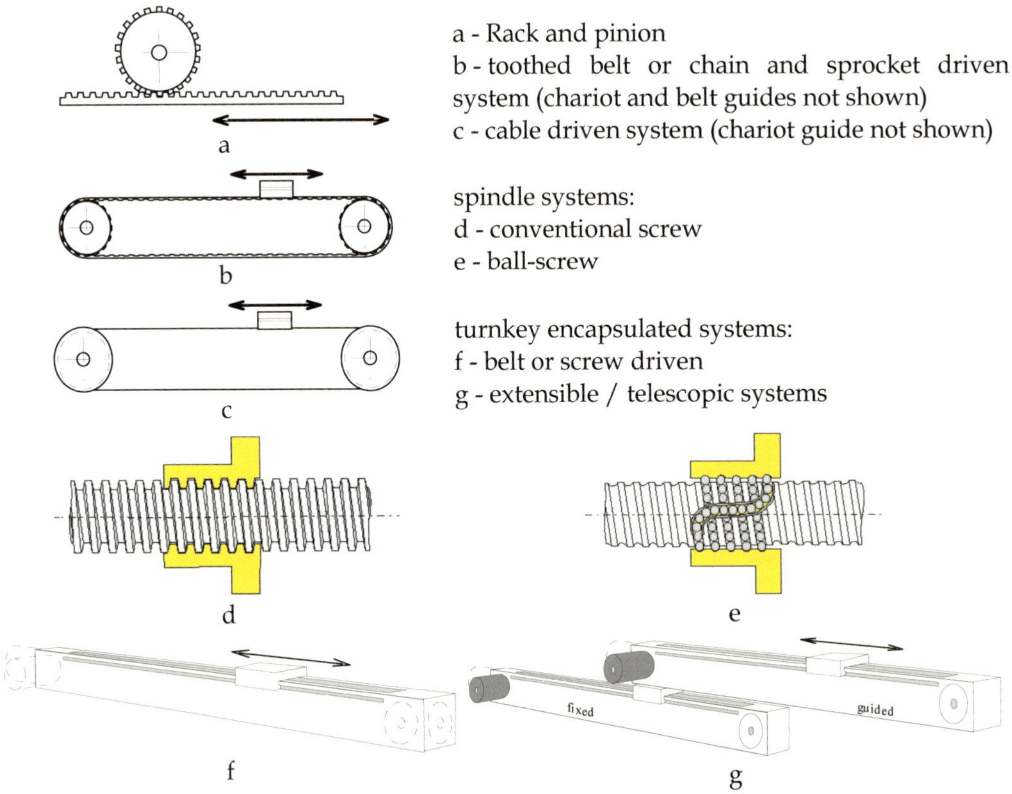

a - Rack and pinion
b - toothed belt or chain and sprocket driven system (chariot and belt guides not shown)
c - cable driven system (chariot guide not shown)

spindle systems:
d - conventional screw
e - ball-screw

turnkey encapsulated systems:
f - belt or screw driven
g - extensible / telescopic systems

Fig. 3. Typical solutions for horizontal movement of bed profiling devices

Spindle systems are primarily used to drive heavy applications or for particular cases where a very large dynamic performance is required with large accelerations and decelerations. They are primarily used to drive equipment for wave generation but nonetheless can be used to provide horizontal motion of bed following devices. The conventional spindle (Fig. 3d) is very cost effective but has backlash problems, higher friction and if not maintained properly very high wear in the short term. The ball-screw device avoids all of the main defects of the conventional screw device for a much higher initial cost. In the long run the ball-screw devices have proven much more cost effective for demanding applications under higher duty cycles.

In the last decade several suppliers of dedicated high performance positioning systems have entered the market of turnkey encapsulated solutions (Fig. 3f). These systems use the same type of devices already described in a compact ready to use format including chariot guides, electric drive motors, end travel switches and control hardware. Cost efficiency is the main criterion to consider in this type of solution. This type of system is compatible with complex situations requiring combination of two or more linear motions (Fig. 3g).
Other electromechanical motion schemes for hydraulics models include those for wave generation (both in flumes and in basins, and consisting of single paddle or multimachine arrangements (Palma et al, 2002), instrument chariot motion control, floodgates, tide gates, and repartition valves for controlling sea currents, each one with peculiar requirements.

2.3 Hydrodynamic interactions due to horizontal motion

Fluid flow patterns around an immersed body depend on the *Reynolds* number (Simiu, 1996; Blake, 1986). Taking as an example a cylindrical body, a very common shape for probe supporting rods, by increasing the flow velocity a number of different flow situations can be created, each situation being identified by a specific *Reynolds* number range (Fig. 4). For all cases it is assumed that the undisturbed flow is laminar. For very low values of *Reynolds* number ($Re < 1$) the flow remains attached to the cylinder through all periphery (Fig 4a). For $1 < Re < 30$ the flow remains symmetrical in shape but flow separation occurs downstream of the cylinder as shown in Fig. 4b giving origin to stationary vortices. For values of *Reynolds* number $40 < Re < 4000$ alternating vortices of symmetrical intensity are shed downstream of the cylinder forming the so called "*von Kárman* vortex trail" (Fig. 4c). For *Reynolds* number values above the last range the flow downstream of the cylinder became turbulent as shown in Fig. 4d.
The alternate shedding of vortices is responsible for forces acting on the cylinder that can be analyzed as the result of two fluctuating components, one due to drag is aligned with flow and another transverse to flow is due to lift (Naudasher, 1985; Blevins, 1977).
The fluctuating component due to lift has a frequency (*f*) that can be given by:

$$St = \frac{f\,d}{V} \tag{1}$$

where V is the velocity of the flow, d is a characteristic dimension of the body projected on a plane normal to mean flow velocity. The fluctuating component due to drag has a frequency that is double the transverse one.
The non-dimensional number St, usually known as *Strouhal* number, is a function of body cross-section geometry, fluid viscosity and flow velocity usually expressed through the *Reynolds* number which is given by the following expression

$$Re = \frac{V\,d}{\nu} \tag{2}$$

where ν is called the *kinematic viscosity*. A typical value for water kinematic viscosity is ν_{water}=0.01 cm^{2}s^{-1} at 20ºC (Simiu, 1996).
Considering the flow around a cylinder with circular cross-section, St is approximately constant and equal to 0.2 in the range of $30 < Re < 10^5$ (see Fig. 5) (Naudasher, 1985).
The fluctuating component due to lift can induce vibration in the body especially if the vortices shedding frequency is close to one of the natural frequencies of vibration of the

body that could lead to resonance problems. In either case the rod vibration can easily lead to bed following problems and excessive strain on the vertical movement control system.

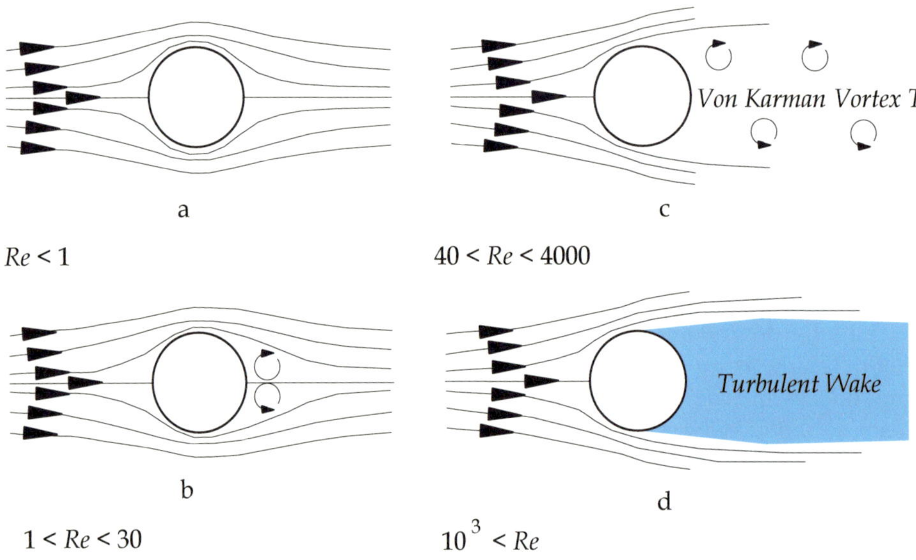

Fig. 4. Different types of flow around a cylinder as a function of the *Reynolds* number

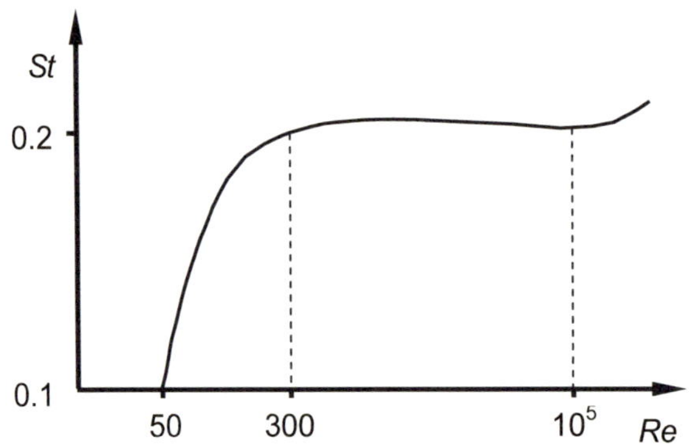

Fig. 5. *Strouhal* number as a function of *Reynolds* number for flow around a cylindrical body

When a body immersed in a fluid vibrates even when the fluid is stationary it gives origin to a non-stationary flow that generates additional forces. Those forces are in phase with the body displacement and can be seen as the effect of an *additional mass* also designated as *hydrodynamic mass*. As a direct consequence the immersed body natural frequencies became lower as the virtual fluid mass was directly attached to the vibrating body. The *hydrodynamic mass* is usually a function of the geometric characteristics of the body and fluid properties. In case of vibrating cylindrical bodies in water the *hydrodynamic mass* can be taken to be equal to the mass of the displaced volume of water (Naudasher, 1985; Simiu, 1996). These phenomena have to be taken into account in the mechanical design of the support structure and driving mechanisms of a bed profiler to work also under rapid horizontal displacement, especially when long immersed rods are required by the water depth and simultaneously light moving parts are necessary for obtaining high vertical accelerations.

2.4 Vertical motion of bed profilers

Vertical motion of bed profilers can be achieved by usage of some of the devices already described for horizontal motion with some restrictions. The major problem of vertical motion is related with dynamics, elimination of backlash problems and motion feedback to control system. This set of characteristics reduces the number of viable solutions to a smaller group as shown in Fig. 6.

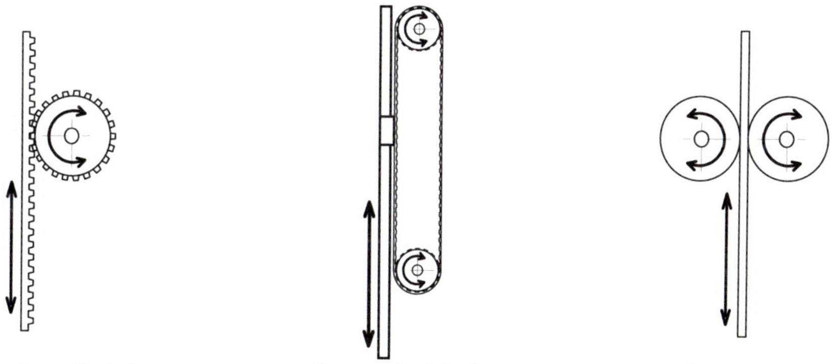

a - rack and pinion b - toothed belt c - friction wheel

Fig. 6. Typical solutions for vertical movement of probe of bed profiling devices (guide wheels, dumpers, brakes, backlash and vibration suppressors not shown)

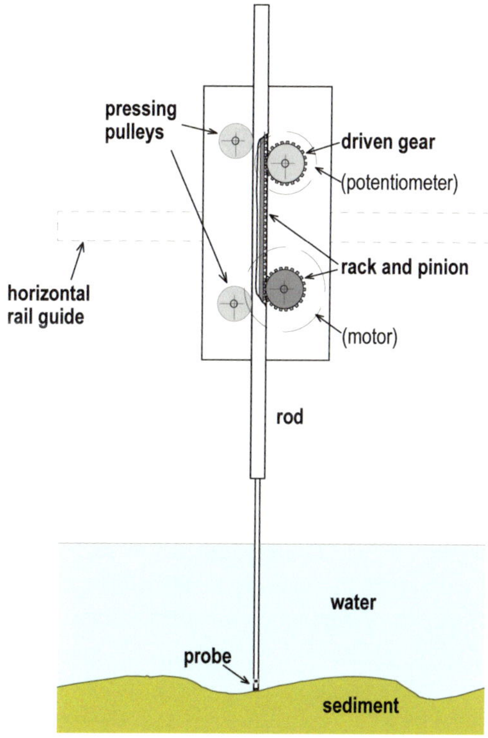

Fig. 7. Pictorial description of a complete mechanism for vertical motion of a bed profiler; the electric motor, the potentiometer and the pressing leaf spring are mounted on the back face; guide wheels are not shown.

The final solution adopted for the bed follower can be seen in the Fig. 7. The vertical motion is achieved by a rack and pinion device. Rod guidance and backlash reduction is provided by two pressing pulleys mounted on swivelling arms acted by a leaf spring and two fixed pulleys on the opposite side. The rack also drives a gear coupled to a multi-turn potentiometer to measure the vertical coordinate. Parking of the movable vertical rod is achieved by self-locking of the gear reductor coupled to the electric motor. A photo of the front side of the system built is presented in Fig. 13.

Some particular features of the electromechanical drive for this type of application result from the fact that it usually requires a micro machine of a few watts. Given the very low power involved some kind of linear electronic amplifier is preferred to the conventional switching converters that are inevitable for power ratings higher than a few tens of watts. This option is also a convenience in order to avoid harmonic interference in transduction circuits which are very sensitive and have to work relatively close to the power stage.

Linear motors could also apparently be used but are generally more expensive and disadvantageous in relation to the solutions with rotating motorization; in this case a gear reduction having self-braking capacity can avoid additional devices to counteract the effect of the gravity force which is required in a linear motor solution.

3. Bottom sensing and tracking control for a laboratory flume

3.1 Bottom sensing

Bed materials in hydraulic basins or flumes consist mostly of sediments with a variety of sizes which are not electrically conductive. Contactless bottom sensing can then be made with a tip probe located at the lower end of the vertical rod, with electrodes for sensing electric resistance.

Rough bed surface, variable granular sediments, water conductivity variations due to temperature and salinity, and also electromagnetic interference, are among the most disturbing causes that have a considerable influence on the sensing task and, at some extent, on servo-control performance.

The design of a tip sensing probe with three electrodes as shown in Fig. 8a solves two problems simultaneously: the transduction to an electric quantity indicating the proximity of the sediment surface and, additionally, the automatic compensation for disturbances that affect water conductivity. The distance from the probe to the bottom is transduced by the unbalance between impedances R_A and R_B. Under homogeneous temperature and salinity conditions R_A increases when the probe approaches the bottom narrowing the electric current field between A and C, while R_B remains constant.

Bridge circuits are widely used in instrumentation owing to its inherent merits for creating solutions with automatic self-compensation capability to certain disturbing effects (Fraden, 1993; Pallas-Areny & Webster 2001). The bridge circuit shown in Fig. 8b provides a fair compensation for water conductivity variations with temperature and salinity. In fact, conductivity variation affects simultaneously and proportionally R_A and R_B, still yielding an equilibrium condition:

$$R_1 R_A = R_2 R_B \rightarrow R_1 \alpha R_A = R_2 \alpha R_B \quad (3)$$

The bridge excitation is carried out in high frequency (10 kHz) ac voltage in order to avoid electrolytic polarisation of electrodes (see Fig. 8c). The bridge output is then dc decoupled, rectified, peak detected, amplified and filtered. The resulting voltage u_y vary nonlinearly with the distance y to the bottom, but may be linearised in a vicinity of less than 3 mm from the bottom surface (i.e. the working range) being suitable for linear control design techniques (D'Azzo & Houpis, 1981).

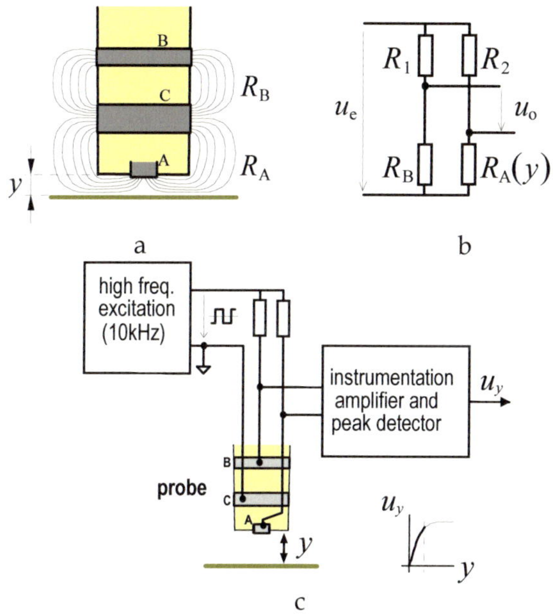

Fig. 8. a) Details of the bottom proximity sensor probe showing the fields of currents between electrodes; b) Bridge circuit for transducing the distance y to the bottom; c) block diagram showing excitation and signal conditioning stages.

In practice R_2 is a potentiometer used to adjust the equilibrium distance to the bottom. With the right compensation for conductivity variations by means of the third electrode that distance can be maintained reasonably constant without allowing the tip probe to go too close to the sediments with an added risk to get in touch with them in transients during the horizontal motion.

When the probe is close to the water surface a reciprocate effect occurs with R_B increasing while R_A remains constant. If this condition is explored the same instrument becomes a surface follower (or *limnimeter*). A similar phenomenon, but with catastrophic consequences, occurs in the opposite sense if the probe tip penetrates in the bottom sediment: in fact, resistance R_A increases again if the probe enters into the low conductive bed material which initiates an instability of the motion control as will be mentioned at the next section.

The sediment bed sensing function (or the water surface sensing function in a limnimeter) just described is required for motion control of the upright moving rod. A displacement transduction should be made additionally in order to acquire the vertical coordinate measurement at each point of the bed profile. A multi-turn linear potentiometer with a gear attached to a rack in the rod is a practical mean to produce a good resolution measurement.

The horizontal coordinate is easily obtained from an incremental encoder actuated by a system of two pulleys and a thin steel cable. Resolutions better than 1 mm are easy to obtain with a 500 ppr encoder in a pulley with a diameter of 120 mm. This method requires initial zeroing in order to produce absolute x coordinate values, by resetting the pulse count when the carriage is positioned at the conventional origin point.

3.2 Tracking control system

Very small direct current motors of about 12 V with gear reduction are typically used owing to their low cost, the simplicity of the necessary electronic drive circuits, and high acceleration capacity, as compared e.g. to stepper motors which could be the competitors at such low power ratings. DC motors having ironless rotor armature are especially interesting due to their extremely low moment of inertia: values of peak torque/moment of inertia over 100 krad/s^2 in the 10 W range are presently reported by manufacturers[1].

Older versions of servo controlled bed profile followers had greater stability problems due to the dc motor control method by direct armature voltage adjustment (Fig. 9a). In fact, the brush contact voltage drop of 1 to 2 V is quite significant in a 12 V motor giving rise to a strong nonlinearity, especially in the vicinity of zero speed. An alternative implementation made with a subordinate current control loop (Fig. 9b) allows overcoming the above difficulty and gives explicit torque control (Leonhard, 1996).

The approximate linear system model adopted for the servo control system is represented in the block diagrams of Fig. 10 with an inner loop for current control and an external loop for position control, without explicit speed regulation: $K_a\Phi$, R, τ_a, K_D and τ_m represent torque constant, rotor resistance, armature time constant, viscous damping coefficient and mechanical time constant, respectively, of the motor; the amplifier time constant, τ_e, in this case can be neglected; K and τ represent the equivalent gain and time constant, respectively, of the current control loop 1st order approximation. The effect of the gravity force is reported to the motor shaft as torque $T_g=M\,r\,g$ (M=linear moving mass, r=gear radius, g=gravity acceleration). The reference signal to the subordinate loop represents the reference value of current demanded by the position control loop and allows the application of a limiting device in order to protect the motor and the amplifier against over-currents (Fig. 10b).

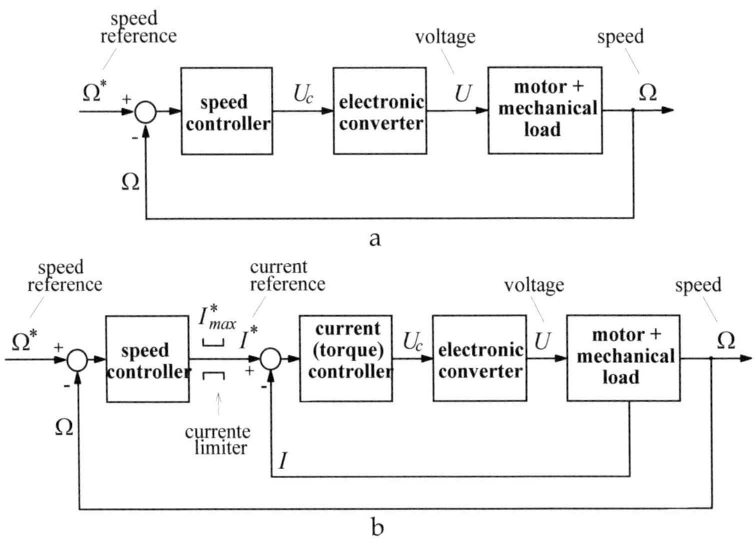

Fig. 9. Closed loop dc motor speed drive methods: a) with direct voltage adjustment; b) with a subordinate current control loop.

---

[1] E.g. www.maxonmotors.com, www.faulhaber.com

Lag-lead and PI structures were sought for the tracking controller C_p in order to get a compromise between error and response speed. Whenever possible simple controller structures have the important practical advantage of a reduced set of parameters to adjust. The PI controller type is also common in current control loops; however a C_i of proportional type can also be tolerated given that this is not a final controlled quantity.

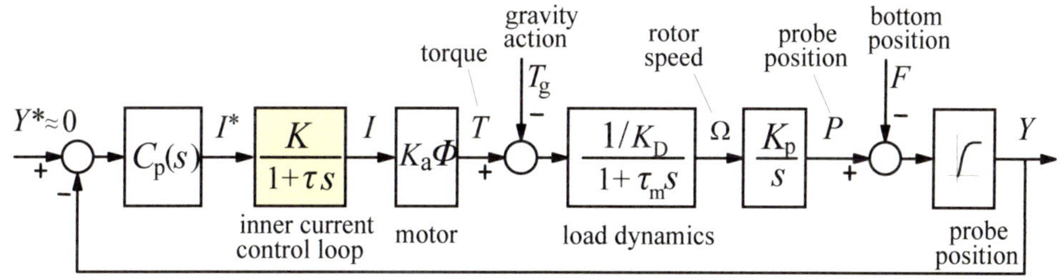

a

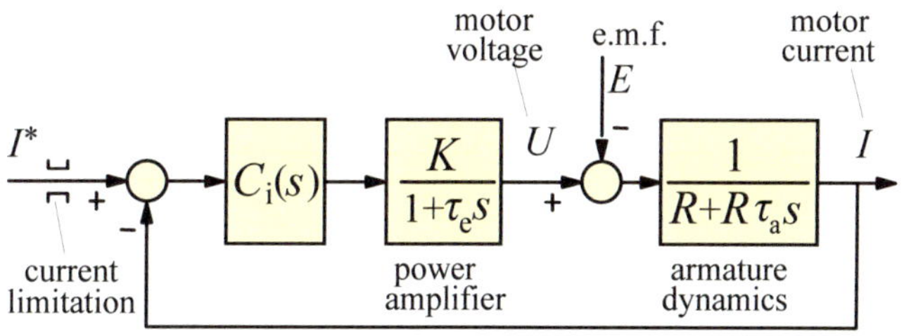

b

Fig. 10. a) Block diagram of the bottom tracking system. b) Detail of the inner current control loop.

A switching converter would be essential for higher power machines (Leonhard, 1996), due to efficiency requirements and also for the sake of semiconductor rating and heat dissipation. In this particular application, however, movable parts are relatively light thus allowing for a very low power motorization (approximately 10 W). Moreover, in many instrumentation applications with delicate transduction processes the interference caused by the harmonics generated by switching converters are a major concern. Thus a linear amplifier was preferred instead of a chopper given the almost insignificant heat losses involved in order to reduce the electromagnetic noise and interference in the sensing process.

The bed tracking control system was implemented with analogue electronics, as shown in Fig. 11. A current control scheme was adapted to an H-type power amplifier using a probe resistor R_p for current sensing and a triple operational amplifier arrangement (U_3,U_4,U_5): subtractor U_1 provides the current measurement signal from the voltage at R_p terminals, U_2 performs as the linear current source controller driving directly the left bridge arm and U_3 creates the inverted signal for the right bridge arm. The bridge is formed of BD 680 PNP and BD 681 NPN Darlington transistors. Current control was achieved with less than 2 milliseconds settling time. Overcurrent limitation is made by zener diodes coupled to the position controller (U_1 in Fig. 11) in order to prevent wind up when the PI function is set.

The system may be set to keep the probe at a constant distance of 1 to 3 mm from the bottom by adjusting the potentiometer R_2 (see Fig. 8b). Far from the linear region the output signal saturates (Fig. 8c), the maximum actuating torque is demanded and so is the current limitation in the inner motor control loop, corresponding to a situation of cruise travel.

Control design is based on the previous linear model where three major poles are found: a fast electrical one (P) due to the inner current control loop (with aprox. 1 ms time constant), a slow mechanical pole (P_m) of the motor and load model (with aprox. 40 ms time constant), and a pole at the origin resulting from the integration of speed that yields the position.

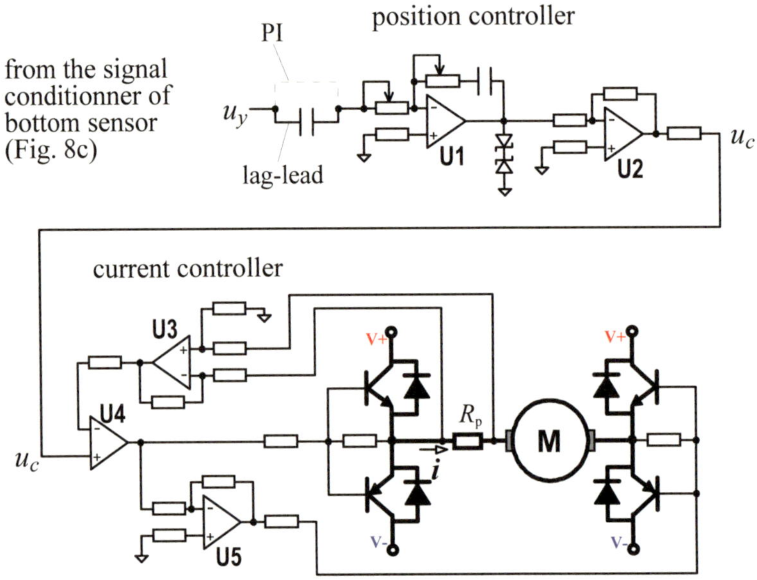

Fig. 11. Electronic power stage and controller circuit schematics.

The controller synthesis is described with the aid of the root locus sketched in Fig. 12. The controller consists of a pole P_c and a zero Z_c in a lag-lead arrangement. In fact the pole and zero placement may be chosen such that different emerging locus patterns can be found (see Fig. 12b) ranging from the lag-lead type to the PI type controller (when P_c is placed at 0). Stability is theoretically possible with both types and was experimentally confirmed. The PI controller, that gives no steady state error, generates two dominant complex poles that can be adjusted to produce a moderate overshoot. Neglecting the effect of pole P, given the dominance of P_m, the ITAE analytical criterion (D'Azzo & Houpis, 1981) was used as the guideline to obtain a global transfer function of the form shown in equation (4)

$$\frac{Y(s)}{Y^*(s)} = \frac{K\omega_o^3}{s^3 + 2\omega_o s_2 + 2\omega_o^2 s + \omega_o^3} \tag{4}$$

There is a high uncertainty in the gain of the linearised bottom sensor characteristic which implies a final experimental adjustment of the PI controller parameters - in practice, the reduced number of parameters allows uncomplicated empirical procedures for adjustment.

A faster speed response with still less overshoot was found with P_c slightly displaced from 0, i.e. in the lag-lead mode.

A more complex controller structure with one pole and two zeroes has been studied and tested in laboratory with good results, nevertheless the simpler versions are preferred for the sake of future adjustment needs.
The penetration of the probe tip inside the sediment mass is highly unwanted and should be avoided by all means: at a certain extent it is only lightly destructive of the bed materials, but if it goes deeper an instability occurs in the motion control reinforcing the downward movement due to the inverse variation of R_A (see section 3.1). Careful control adjustment or special logic detection schemes have to be implemented in order to avoid this kind of occurrence. Older versions of bed followers were significantly more prone to it.

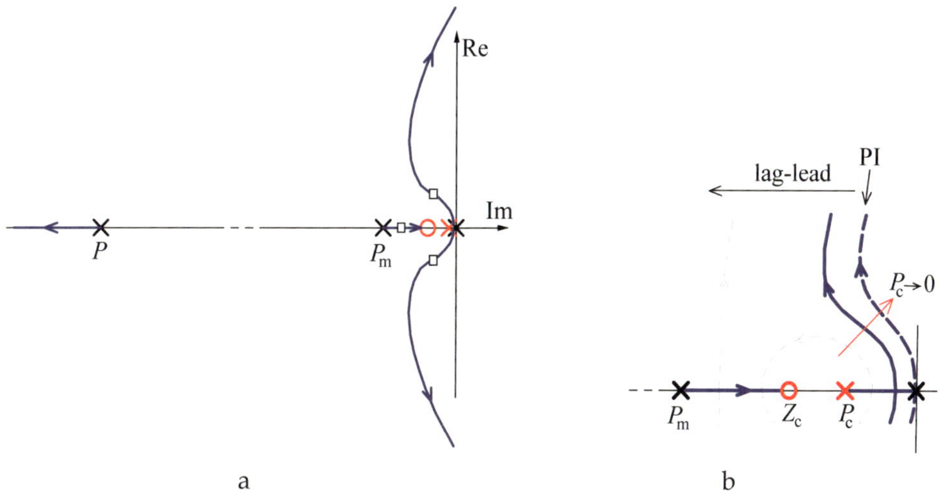

Fig. 12. Root locus sketch of the closed-loop system: a) general plan; b) enlarged view near the origin showing the effect of varying the controller pole P_c from lag-lead to PI.

4. Experimental application of a bed profiler

4.1 Computer command an data acquisition

This type of instrument has to be used in conjunction with PC based data acquisition and command equipment. In addition to the analogue part used for the vertical motion control of the profiler, a digital unit was developed (Palma et al, 2008) for converting the vertical coordinate analogue data, counting the pulses from the encoder for the horizontal coordinate, communicate with a computer to send data and state information as well as to receive commands. The horizontal coordinate data resolution is 1 mm and the vertical coordinate resolution is better than 0.1 mm, after the digital conversion.
Since the system may be at rest for long periods there are commands for (de-)energizing the power supply for amplifier and motor; the gear reductor coupled to the motor prevents gravity downwards movements after de-energisation. During rapid carriage travel (e.g. for return to origin) an order can be issued to suspend the bed tracking operation of the profiler and to commutate it for surface tracking. Fig. 13 shows a photo of the instrument apparatus in a carriage over the flume.

Fig. 13. Photograph showing the instrument block in a chariot over the flume.

4.2 Experimental results

The system as shown in Fig. 14 was used in a flume of the Department of Hydraulics and Environment of LNEC for the study of the sedimentation process in dam reservoirs, based on successive bed profile surveillances (González, 2007; Alves, 2008). The sedimentation results from de deposition of coarser sediments that form a delta and fine sediments (turbidity currents).

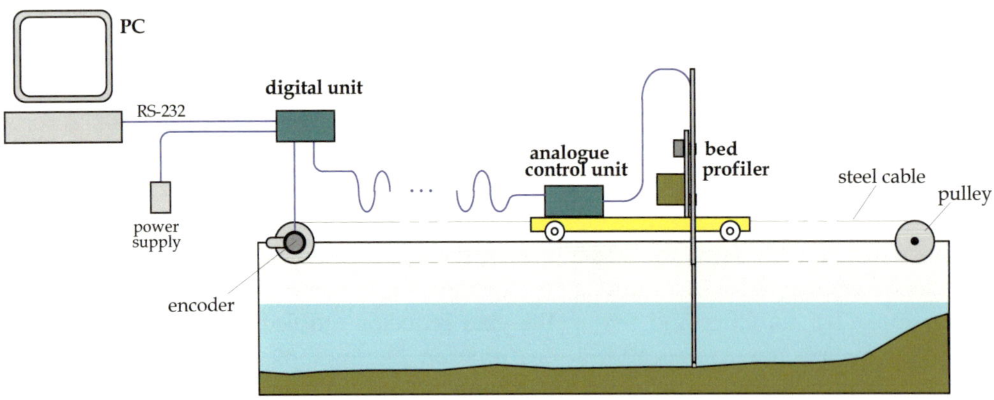

Fig. 14. Synoptic sketch of the bed profiler and the data acquisition system used in the flume model

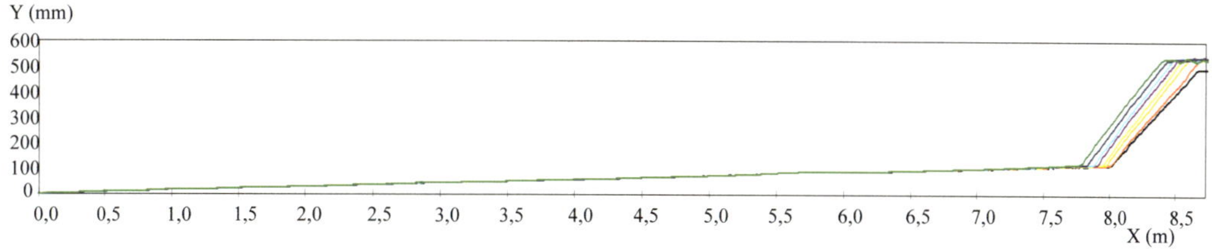

a

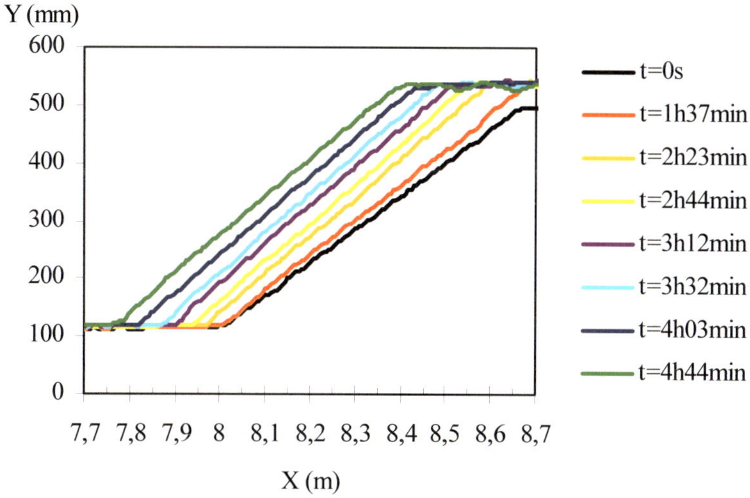

b

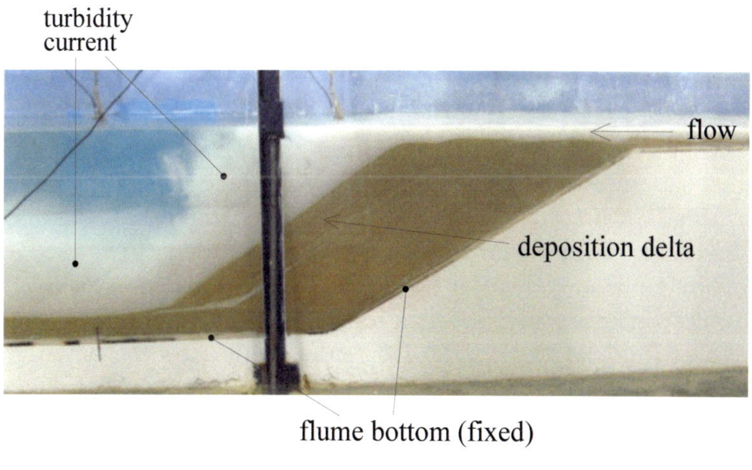

c

Fig. 15. a) Graphic view of the data acquired for whole flume length. b) Enlarged view of the profiles in the region of interest; c) Photo taken through the last sections of the transparent acrylic wall of the flume.

The flume has 10.60 m length, in which a segment of 8.7 m was used for this test, 0.30 m wide and 0.75 m maximum height, having a transparent acrylic wall on one side.
The shape of deposited sand after 4 h 44 min of test is shown in Fig. 15. The diagrams represented in Fig. 15a and 15b correspond to data collected from several profiles at given time intervals, and show the progress of the delta front edge. The shape of the delta can be seen in the photograph of Fig. 15c.
The study demonstrated that the equilibrium tilt angle of the front edge is approximately 33° and was attained about 2 h and 23 min after the test start.

5. Conclusion

Significant advancements had to be introduced in servo-controlled sediment bed profilers in order to obtain reliable solutions for these particularly demanding motion control tracking systems. Improvements in robustness and performance were achieved in the motion mechanisms as well as in the tracking control system in terms of higher stability, response speed and bottom sensing accuracy.
A number of mechanical refinements namely for backlash absorption, vibration reduction, guidance friction minimization and weight optimization have a great influence on the overall performance. The introduction of a subordinate control loop attenuates the nonlinearity effect of the collector-brush voltage drop, while simultaneously self-protects the motor and the amplifier against overcurrents. The bridge topology used in the amplifier for the current-source loop expands the capacity for producing high force peaks in transient acceleration/deceleration periods.
The penetration in the sediment by the probe tip, which occurred at times in the past, and was a critical kind of event, has been effectively avoided with the new design, owing to the servo controller improvements and to the re-shaping of the geometry of electrodes at the probe tip. The improvement in control performance is another important feature in order to avoid occasional destruction of sediments and to prevent instability. The addition of digital communication with a computer brings other advantages for command, diagnostics and data acquisition.
Experimental results demonstrated the appropriate performance of the proposed solution. After initial tests and adjustments the system has been used with success in a study of sedimentation caused by turbidity currents, for monitoring the progress of a relatively steep front edge of a delta with an approximate tilt angle of 33°.
The type of application described embeds the contribution of several disciplines. A number of particular details concerning physical measurements, mechanical design, electronics and electromechanical drives had to be tackled in order to develop the appropriate motion system.

6. References

Alves, E. (2008). *Sedimentation in Dam Reservoirs by Turbidity Currents*, PhD thesis, Technical University of Lisbon, IST

Azevedo, F. & Morais, P. (1989). *Mechanical Design of a 600 mm Bottom Follower* (in Portuguese), LNEC report 17/89-CPCE, Lisbon

Blake, W.K. (1986). *Mechanics of flow-induced sound and vibration,* Academic Group, ISBN 0121035026, Orlando

Blevins, R. D. (1977). *Flow-induced vibration,* Nostrand Reinhold, ISBN 894648233, New York

Cardoso, T. (1964). *A System for Automatic Recording of Bed Profiles* (in Portuguese), Traineeship Work Report, LNEC, Lisbon

D'Azzo & Houpis (1981). *Linear Control System Analysis and Design,* 2nd edition, McGraw-Hill, ISBN 0070161836, New York

Delft Hydraulics, *PV-09 Profile Indicator* and *PV-11 Amphibious Profile Indicator,* http://www.wldelft.nl/inst/intro/index.html

Ernstsen, V.; Noormets, R.; Winter; C.; Hebbeln, D.; Bartholomä, A.; Flemming, B. & Bartholdy, J. (2006). Quantification of dune dynamics during a tidal cycle in an inlet channel of the Danish Wadden Sea, *Geo-Marine Letters,* Vol. 26, Issue 3, pgs.151-163, Springer-Verlag, ISSN 1432-1157

Fraden, J. (1993). *AIP Handbook of Modern Sensors,* American Institute of Physics, ISBN 978-1563961083, New York

González, J. (2007). *Experimental Study of Reservoir Turbidity Currents. Development and Application of Data Processing Routines for Velocity and Bed Profiles,* LNEC report 104/2007-NRE, Lisbon

HR Wallingford, *2D Bed Profiling System for Physical Models,* www.hrwallingford.co.uk /Equipment/EQ001A_Bed_profiler_Jan06.pdf

Leonhard, W. (1996). *Control of Electrical Drives,* 2nd edition, Springer-Verlag, ISBN 3-540-59380-2, New York

Naudasher, E. (1985). *Engineering for Structures Subject to Flow-Induced Forces and Vibration,* LNEC, Intensive Course, Lisbon

Oumeraci, H. (2003). Role of Large-Scale Model Testing in Coastal Engineering - Selected Example Studies Performed in Gwk Hannover", *Proc. of the International Conference "Towards a Balanced Methodology in European Hydraulic Research",* pgs. 3.1-3.18, ISBN 9635111320, May, 2003, Budapest

Pallas-Areny, R., Webster, J. (2001). *Sensors and Signal Processing,* 2nd edition, Wiley, ISBN 978-0471545651, New York

Palma, J.; Carvalhal, F. & Morais, P. (2002). Distributed Control Architecture of a Multipaddle Wave Generator for Hydraulics Studies, *Proc. of the 7th International Conference on Electrical Machines, Converters and Systems* (ElectrIMACS), CD-ROM, pgs. 1-5, ISBN 2-921145-31-6, August, 2002, Montréal

Palma, J.; Guilherme, L. & Alves, E. (2008). Bed Profile Tracking and Data Acquisition System for Hydraulics Physical Models, *Proc. of the 14th IEEE Mediterranean Electrotechnical Conference* , pgs. 96 a 101, ISBN 978-1-4244-1633-2, May, 2008, IEEE, Ajaccio

Rao, S. & Rao, B. (2004). Stability of berm breakwater withreduced armor stone weight. *Ocean Engineering,* Vol. 31, Issues 11-12, pgs. 1577–1589, Elsevier, ISSN 0029-8018

Simiu, E.; Scanlan, R.H. (1996). *Wind effects on Structures – Fundamentals and Applications to Design,* John Wiley & Sons, ISBN 471121576, New York

Yeh, P.; Chang, C.; Henriksen, J.; Edge, B.; Chang, P.; Silver, A. & Vargas, A. (2009). Large-scale laboratory experiment on erosion of sand beds by moving circular vertical jets, *Ocean Engineering*, Vol. 36, Issues 3-4, pgs. 248–255, Elsevier, ISSN 0029-8018

28

Switching Control of Image Based Visual Servoing in an Eye-in-Hand System using Laser Pointer

Wen-Fang Xie[1], Zheng Li[1], Claude Perron[2] and Xiao-Wei Tu[1,2]
[1]*Department of Mechanical & Industrial Engineering, Concordia University,*
[2]*Aerospace Manufacturing Technology Centre, Institute for Aerospace Research,*
National Research Council Canada,
Canada

1. Introduction

To reduce the labor cost and increase the throughput in the manufacturing industry, there is an increasing demand for automated robotic manufacturing systems such as robotic assembly, bin picking, drilling and palletizing systems, which require accurate and fast robot positioning. The 3D machine vision system is normally used in a robotic manufacturing system to compensate for the robotic positioning errors due to unforeseen work environment and randomly-placed objects. The task of robot positioning using vision system is referred to as visual servoing which aims at controlling the pose of the robot's end effector relative to a target object or a set of target features (Hutchinson et al, 1996);(Corka, 1996); (Li et al, 2006); (Aouf et al, 2004). According to the features used as feedback in minimizing the positioning error, visual servoing is classified into three categories, Position Based Visual Servoing (PBVS) (Hutchinson et al, 1996); (DeManthon & Davis, 1995);(Wilson et al, 1996), Image Based Visual Servoing (IBVS) (Weiss et al, 1987); (Espiau, 1993); (Chaumette, 1998); (Wang & Cho, 2008) and Hybrid Visual Servoing (Malis et al, 1999).
Since IBVS was introduced in 1980, it has attracted the attention of many researchers and has been tremendously developed in recent years. The method is based on the principle that when the image feature error in 2-D image space is approaching zero, the kinematical error in Cartesian space approaches zero too (Hutchinson et al, 1996). In IBVS, the error for the controller is defined directly with respect to the image feature parameters (Weiss et al, 1987). Compared with PBVS, the advantages of IBVS are obvious. First, it is object model free, and robust to camera modeling and hand-eye calibration errors (Espiau, 1993). Second, the image feature point trajectories are controlled to move approximately along straight lines in the image plane. Hence, it is able to prevent the image features from leaving the FOV. However, the drawbacks of IBVS lie in the following aspects. Since the control law is merely designed in the image plane, the trajectory of the end effector in Cartesian space is not a straight line, and even odd in some cases. In other words, in order to reduce the image feature error to zero as soon as possible, unnecessary motions of end effector are performed. Moreover the system is stable only in a region around the desired position, and there may exist image singularities and image local minima (Chaumette, 1998) leading to IBVS failure.

The choice of the visual features is a key point to solve the problem of image singularities. Lots of studies have been done to find out the decoupled visual features with respect to the 6DOF of the robot. Such studies also ensure that the trajectory in Cartesian space is like a straight line (Tahri & Chaumette, 2005); (Pages et al, 2006); (Janbi-Sharifi & Fiocelli, 2004); (Krupa et al, 2003). In (Tahri & Chaumette, 2005), six image moment features were selected to design a decoupled control scheme. In (Pages et al, 2006), Pages et al. derived the image Jacobian matrix related to a laser spot as an image feature. The global convergences of the control law had been shown with a constant interaction matrix. However, the method needed the information of the planar object and only fit for the situation where the camera was located near the desired position. Another approach using a laser pointer in visual servoing was presented in (Krupa et al, 2003). Krupa et al. developed a vision system with stationary camera, which retrieved and positioned surgical instruments for operation. A laser pointer was used to project laser spot on the organ surface to control the depth. However, the servoing was only carried out in 3-DOF and the camera was motionless. Therefore, the system could not provide much flexibility in visual servoing in large scale environment.

Koichi Hashimoto et al. (Hashimoto & Noritsugu, 2000) introduced a method to solve the image local minima. The main idea was to define a potential function and to minimize it while controlling the robot. If the potential had local minima, the algorithm generated an artificial potential and then controlled the camera based on the artificial one. In (Kase et al, 1993), stereo based visual servoing was proposed to solve the depth estimation problem and calculate an exact image Jacobian matrix. However this kind of algorithm increased the computational cost. R. Mahony et al. (Mahony et al, 2002) introduced a method of choosing other types of image features instead of points for IBVS and focusing on the depth axis control. P.Y.Oh et al. (Oh & Allen, 2001) presented a partitioning DOF method for IBVS which used a 3-DOF robot with a 2-DOF pan tilt unit. The experimental results of tracking people were given. In (Corke & Hutchinson, 2001), another partitioned approach to visual servoing control was introduced, which decoupled the z-axis rotational and translational components of the control from the remaining DOF.

To overcome the aforementioned shortcomings of IBVS, some new approaches that integrate PBVS and IBVS methods have been developed (Gans et al, 2003); (Malis et al, 1999). The main idea is to use a hybrid of Cartesian and image space sensory feedback signals to control both Cartesian and image trajectories simultaneously. One example of such hybrid approach is 2.5D visual servoing (Malis et al, 1999), which was based on the estimation of the partial camera displacement. Recently, a hybrid motion control and planning strategy for image constraints avoidance was presented in (Deng et al, 2005); (Deng et al, 2003). This motion control part included a local switching control between the IBVS and PBVS for avoiding image singularity and image local minima. In addition, the planning strategy was composed of an artificial hybrid trajectory planner.

Inspired by the hybrid motion control and planning strategy, we proposed a new switch control approach to IBVS to overcome the shortcomings of IBVS. First, a laser pointer is adopted to realize on-line depth estimation to obtain image Jacobian matrix. Second, we added a laser point image feature to the chosen image features of the object. Based on the new image feature set, we can detect the object in the workspace even when the features of object are in FOV partially. Hence the available workspace is virtually enlarged to some extent. Furthermore, a set of imaginary target features are introduced so that a decoupled control scheme for IBVS can be designed. Third, we separated 3-DOF rotational motion from the translational motion to solve some image singularity problem such as 180 degree rotation around the optical axis (Chaumette, 1998) and image local minima in IBVS. This decoupled control strategy can make visual servoing system work over a large region around the desired position.

This switch control approach to IBVS with laser pointer is applied to a robotic assembly system. The system is composed of a 6-DOF robot, a camera mounted on the robot end effector, a simple off-the-shelf laser pointer rigidly linked to the camera and a vacuum pump for the object grasping. The whole algorithm consists of three steps. First the laser spot is driven onto a planar object. Since the laser pointer is mounted on the robot end effector, 3-DOF rotational motion of end effector can drive the object image features close to a set of imaginary image features so that the laser spot is projected on the object. Next, the image features of the object and laser spot are used to obtain the image Jacobian matrix, which is primarily used for controlling the end effector translational motion with respect to the object. Finally a constant image Jacobian at the desired camera configuration for IBVS is used to adjust the fine alignment so that the feature errors can reach to the image global minimum. The successful application of the proposed algorithm to an experimental robotic assembly system demonstrates the effectiveness of the proposed method.

The paper is organized as follows. In Section 2, the problem statement of visual servoing is introduced. A novel approach to switching control of IBVS with laser pointer is presented in Section 3. In Section 4, several experimental results are given to show the effectiveness of the proposed method. The concluding remarks are given in Section 5.

2. Problem statement

In this paper, we focus on an automated robotic assembly system which uses Eye-in-Hand architecture to perform visual servoing. In this system, the assembly task is to move the robot end effector together with a tool, such as a gripper or a vacuum pump, to approach a part with unknown pose, and then to grasp and assemble it to a main body fast and smoothly. Such an assembly task is a typical visual servoing control problem. Hence, IBVS is an appropriate method to achieve the task since all control input is computed in an image space without using the pose information.

Let $\dot{r} = \begin{bmatrix} {}^C v_C & {}^C \omega_C \end{bmatrix}^T = \begin{bmatrix} v_x & v_y & v_z & \omega_x & \omega_y & \omega_z \end{bmatrix}^T$ be a velocity screw of the camera. Define $f_i = \begin{bmatrix} x_i & y_i \end{bmatrix}^T, i = 1, 2, \cdots n$ as the image features and $\dot{f}_i = \begin{bmatrix} \dot{x}_i & \dot{y}_i \end{bmatrix}^T$ as the corresponding image feature velocity. Denote the desired image feature as $f_{di} = \begin{bmatrix} x_{di} & y_{di} \end{bmatrix}^T$ and $f_d = [f_{d1}\ f_{d2} \cdots f_{dn}]^T$ which are obtained by using a teaching by showing approach. In this paper, the teaching procedures are: (i) Move the robot end effector to a position where the pump can perfectly suck up the object; (ii) Move the robot in end effector frame to a new position where the whole object is in the FOV. Record this moving value as the constant transformation; (iii) Take an image as the target image for visual servoing. The four coplanar corners of the object are chosen as the target image features and the laser spot $f_d^l = \begin{bmatrix} x_d^l & y_d^l \end{bmatrix}^T$ is also selected as a target image feature in some cases.

Assume that the effective sizes of a pixel (s_x, s_y) are constant to simplify the visual servoing computation without loss of generality. The transformation between $\begin{bmatrix} x_i & y_i \end{bmatrix}^T$ and the pixel indices $\begin{bmatrix} u & v \end{bmatrix}^T$ depends only on camera intrinsic parameters.

In order to design the feedback control for robot based on the velocity of the feature points, we have the following relationship between the motion of image features and the physical motion of the camera:

$$\dot{f} = \mathbf{J}_{img}(f, Z)\dot{r} \tag{1}$$

where

$$\mathbf{J}_{img}(f, Z) = \begin{bmatrix} \mathbf{J}_{img}(f_1, Z_1) \\ \vdots \\ \mathbf{J}_{img}(f_n, Z_n) \end{bmatrix} \tag{2}$$

is the image Jacobian matrix, $Z = \begin{bmatrix} Z_1 & \cdots & Z_n \end{bmatrix}^T$ is the depth of each feature point, and $f = [f_1 \cdots f_n]^T$ is the image feature vector containing n features.
For each feature point (x_i, y_i), the image Jacobian matrix is represented as follows:

$$\mathbf{J}_{img}(f_i, Z_i) = \begin{bmatrix} \frac{\lambda}{Z_i} & 0 & -\frac{x_i}{Z_i} & -\frac{x_i y_i}{\lambda} & \frac{\lambda^2 + x_i^2}{\lambda} & -y_i \\ 0 & \frac{\lambda}{Z_i} & -\frac{y_i}{Z_i} & \frac{-\lambda^2 - y_i^2}{\lambda} & \frac{x_i y_i}{\lambda} & x_i \end{bmatrix} \tag{3}$$

where λ is the known focal length of the camera.
Equation (1) can be written as:

$$\dot{r} = \mathbf{J}_{img}^{+}(f, Z)\dot{f} \tag{4}$$

where $\mathbf{J}_{img}^{+}(f, Z)$ is the pseudo inverse of the image Jacobian. If the error function is defined as $e(f) = f - f_d$ and we impose $\dot{e}(f) = -Ke(f)$, a simple proportional control law is given by

$$\dot{r} = -K\mathbf{J}_{img}^{+}(f, Z)e(f) \tag{5}$$

where $\dot{r}$ is the camera velocity sent to the robot controller, K is the proportional gain which tunes the exponential convergence rate of f toward f_d.
It is assumed that the optical axis of the camera is coincident with the Z axis of the end effector. The motion of camera can be related to robot joint rates through the normal robot Jacobian and a fixed transformation between the motion of the camera and the end effector. When the image error function *e(f)* tends to zero, the kinematical error must also approach zero. However, an inappropriate choice of $\mathbf{J}_{img}^{+}(f, Z)$ may lead the system close to, or even cause a singularity of the Jacobian matrix, which may result in potential servoing failure (Malis et al, 1999).
The objective of IBVS in this paper is to control the end effector to approach an unknown object so that the image error function *e(f)* approaches zero while the trajectory in the Cartesian space is kept as short as possible. Meanwhile, the system should be designed with low cost and the visual servoing algorithm is developed with low computational load for real time robot control.

3. IBVS with laser system

In this section, a new approach to switching control of IBVS with laser pointer is presented to accomplish the aforementioned visual servoing tasks. This approach is designed to overcome the drawbacks of IBVS by installing an off-the-shelf laser pointer on the end effector for estimating the depth of the image features and separating the visual servoing procedures into several control stages.

(a) Robotic Eye-in-Hand System

The designed robotic Eye-in-Hand system configuration is shown in Figure 1, which is composed of a 6-DOF robot, a camera mounted on the robot end effector, a laser pointer rigidly linked to the camera and a vacuum pump for grasping object. In Figure 1, H denotes the transformation between two reference frames.

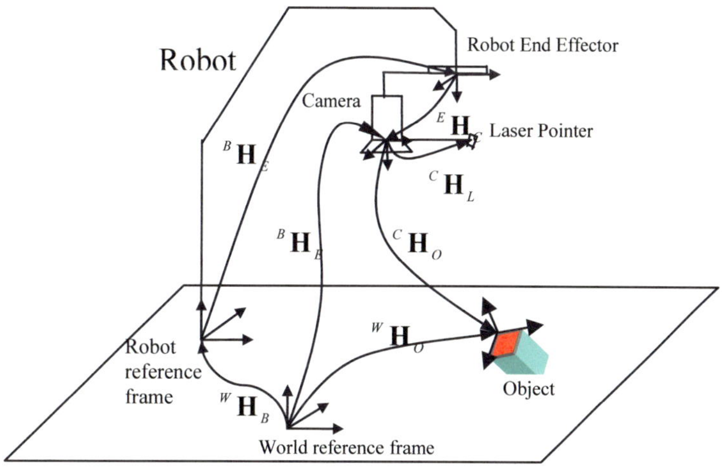

Fig. 1. Robotic Eye-in-Hand System Configuration

In traditional IBVS, since the control law is only designed in the image plane, unnecessary motions of the end effector are performed. In order to obtain the image Jacobian $\mathbf{J}_{img}(f,Z)$ which is a function of features and depths of object, depth estimation is needed for each feature point. A low cost laser pointer is thus adopted and installed on the end effector to measure the distance from the camera to the target for depth estimation. Through the laser triangulation method, the depth estimation can be achieved in a very short time. Moreover the laser spot can be chosen as an image feature that eases the image processing with low computational load for visual servoing.

(b) On-line Depth Estimation

In Equation (3), $\mathbf{J}_{img}(f,Z)$ is a function of features and their depths. Although one of the solutions to IBVS is to choose the target image features and their depths for a constant image Jacobian, it is proved to be stable only in a neighborhood of the desired position. Another solution is to estimate the depth of every feature point on-line. In this paper, a simple laser pointer based triangulation method is applied to estimate the depth on-line.

Assume that the laser beam is lying in the same plane with the camera optical axis. We use a camera-center frame $\{c\}$ with ${}^{c}z$ parallel to the optical axis. In this configuration, no matter how camera is moving, the trajectory of the laser spot in the image is a straight line passing

the principal point. If we consider the laser pointer as a camera and the laser beam is its optical axis, the straight line in the image is an epipolar line of the imaginary stereo configuration.
As shown in Figure 2, d denotes the horizontal distance between the laser beam and the optical axis of lens of the camera, and α is the angle between the laser beam and the horizontal line. Both of them are fixed and known when the laser pointer is installed. Point P is the intersecting point of the laser beam and the object surface. A function of depth Z_P with respect to pixel indices (u, v) (Trucco & Verri, 1998) is derived by applying the trigonometry. The depth Z_P calculated by Equation (6) can be used to approximate the depth of each feature point in the planar object surface under assumption that the size of object is small enough.

$$Z_p = \frac{d \sin \alpha}{\cos(\alpha - \beta)} \tag{6}$$

where β is the triangle inside the camera.

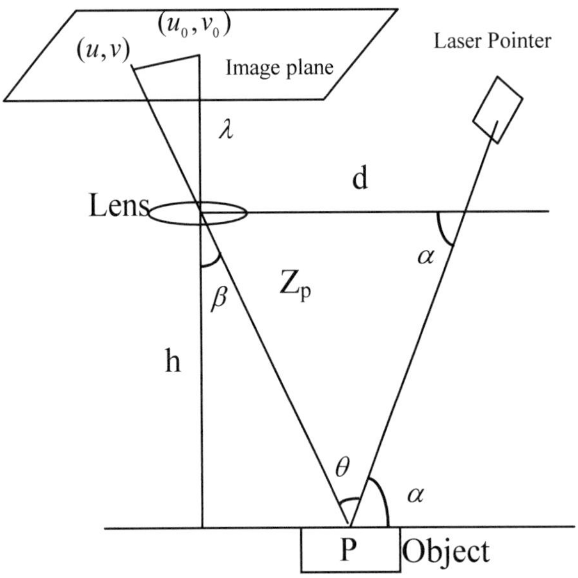

Fig. 2. Calculation of the Depth of a Point by Using Triangulation

Remark: It is noted that (6) is only valid when the camera and the laser are planar. In the robotic assembly system, the laser pointer is rigidly linked to the camera in a frame which is attached to the end effector of the robot. Inside the frame, the laser pointer is installed in a way that the laser line and object camera projections are coplanar. The possible generalization of this calibration phase to the case on no-coplanar laser and camera is under investigation.

(c) Switching Control of IBVS with Laser Pointer

The proposed algorithm is divided into three control stages which are a) driving the laser spot on the object, b) combining IBVS with laser spot for translational moving and c) adjusting the fine alignment. The block diagram of switching control system of IBVS with laser pointer is presented in Figure 3. The object is assumed to be stationary with respect to robot reference frame.

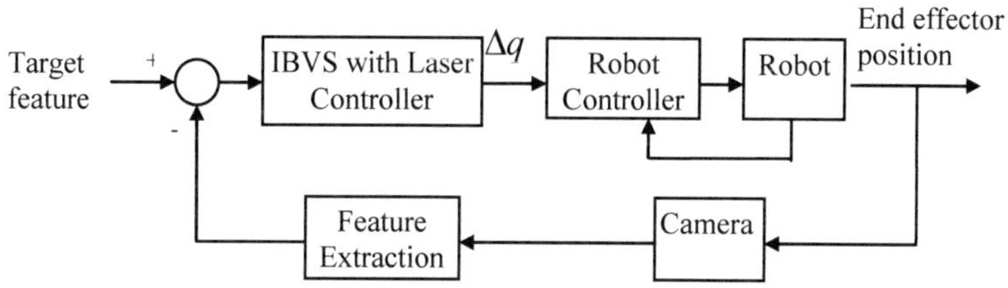

Fig. 3. Block diagram of IBVS with Laser control system

Stage 1: Driving the Laser Spot on the Object

To project the laser spot on the object, two kinds of situations need to be considered. One is that all features are in the FOV (Stage 1.A), and the other is that certain features are missing in the FOV (Stage 1.B). Both of them are discussed below in detail.

Stage 1.A: All Features in the FOV

When all image features are in the FOV, the control task is to drive the center of gravity of the object together with the image features of object near the current laser spot and the created imaginary image features by using 3DOF camera motion. Equation (1) can be decomposed into translational and rotational component parts as shown below

$$\dot{f} = \left[\mathbf{J}_{img}^{t}(f,Z) \quad \mathbf{J}_{img}^{r}(f) \right] \begin{bmatrix} {}^{C}v_{C} \\ {}^{C}\omega_{C} \end{bmatrix} \tag{7}$$

where $\mathbf{J}_{img}^{t}(f,Z)$ and $\mathbf{J}_{img}^{r}(f)$ are stacked by $\mathbf{J}_{img}^{t}(f_i,Z_i)$ and $\mathbf{J}_{img}^{r}(f_i)$ given by

$$\mathbf{J}_{img}^{t}(f_i,Z_i) = \begin{bmatrix} \dfrac{\lambda}{Z_i} & 0 & -\dfrac{x_i}{Z_i} \\ 0 & \dfrac{\lambda}{Z_i} & -\dfrac{y_i}{Z_i} \end{bmatrix} \tag{8}$$

$$\mathbf{J}_{img}^{r}(f_i) = \begin{bmatrix} -\dfrac{x_i y_i}{\lambda} & \dfrac{\lambda^2 + x_i^2}{\lambda} & -y_i \\ \dfrac{-\lambda^2 - y_i^2}{\lambda} & \dfrac{x_i y_i}{\lambda} & x_i \end{bmatrix} \tag{9}$$

It is noted that $\mathbf{J}_{img}^{t}$ is related to both features and their depths, and $\mathbf{J}_{img}^{r}$ is only a function of the image features. Since the laser pointer is mounted on the robot end effector, it is possible to control it by performing 3-DOF rotational motion. It is known that 2-DOF of rotational motion can drive a laser pointer to project its dot image on the target. Here the reason of using 3-DOF instead of 2-DOF is that the rotation along the camera Z axis can be used to solve some image singularity problems such as 180 deg rotation around the optical axis, which is well known case causing visual servoing failure as presented by F. Chaumette (Chaumette, 1998). To avoid this particular case, a set of imaginary target features are designed. Based on the target image features including the desired object features f_d and the laser spot f_d^l, a new imaginary image can be designed, which is shown in Figure 4. Let the distance between the target position of laser spot and the current laser spot be $d_l = \left\| f_d^l - f^l \right\|$. The imaginary object features f_{iof} are formed by shifting all target object features f_{di} by units d_l and adding the current laser spot f^l:

$$f_{iof} = [f_{d1} - d_l \quad f_{d2} - d_l \cdots f_{dn} - d_l \quad f^l]^T \tag{10}$$

It is assumed that the height of the object is relatively small. Hence there is no big discrepancy of the laser spot in the image plane when the laser spot moves from the workspace platform to the surface of object.

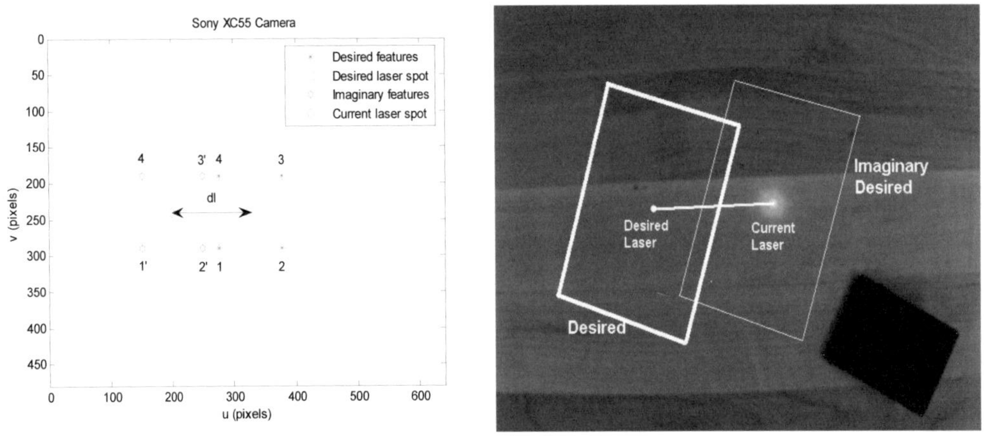

Fig. 4. Example of creating imaginary features, (a) in 2-D image space, (b) in 3-D Cartesian space

The center of gravity of object image features is served as an extra image feature which is defined as $f_{cg} = [(x_1 + \cdots + x_n)/n, (y_1 + \cdots + y_n)/n]^T$ in the image plane. The control goal is to minimize the error norm between the image features $f_{s1A} = [f_1 \quad f_2 \cdots f_n \quad f_{cg}]^T$ and the imaginary features, as presented below

$$\min\left\{ \left\| f_{s1A} - f_{iof} \right\| \right\} \tag{11}$$

From Equation (7), the relationship between the motion of the image features and the rotational DOF of camera ${}^C\omega_C$ is represented as

$$\dot{f}_{s1A} = \left[\mathbf{J}_{img}^t(f_{s1A}, Z_{s1A}) \quad \mathbf{J}_{img}^r(f_{s1A}) \right] \begin{bmatrix} {}^C v_C \\ {}^C \omega_C \end{bmatrix} \tag{12}$$

where $Z_{s1A} = [Z_1 \quad Z_2 \cdots Z_n \quad Z^l]^T$. The above equation can be written as:

$${}^C\omega_C = \mathbf{J}_{img}^{r\ +}(f_{S1A}) \left[\dot{f}_{s1A} - \mathbf{J}_{img}^t(f_{s1A}, Z_{s1A})\, {}^C v_C \right] \tag{13}$$

We set the translational DOF of camera motion as zero (${}^{C}v_{C} = 0$) and the following equation is obtained:

$$ {}^{C}\omega_{C} \cong \mathbf{J}^{r\;+}_{img}(f_{s1A})\left[\dot{f}_{s1A}\right] \tag{14} $$

Let the feature error defined as $e_{iof} = f_{s1A} - f_{iof}$. By imposing $\dot{e}_{iof} = -K_{l}e_{iof}$, one can design the proportional control law given by

$$ u_{s1A} = {}^{C}\omega_{C} = -K_{l}\mathbf{J}^{r\;+}_{img}(f_{s1A})e_{iof} \tag{15} $$

where ${}^{C}\omega_{C}$ is the camera angular velocity sent to the robot controller and K_{l} is the proportional gain.

Since we deliberately turn off the translational motion of camera, Equation (14) relating the image velocity to 3DOF camera rotational motion is approximately held. The proportional controller (15) cannot make the feature error e_{iof} approach zero exponentially. However, the current control task is only to drive the laser spot on the object and rotate the camera to make the image features of the object approach the imaginary features. The further tuning of visual servoing will be carried out in the next stages. Hence we adopt the switching rule to switch the controller to the second stage.

The switching rule is described as: when the error norm falls below a predetermined threshold value, the controller will switch from current stage to the second stage. The switching condition is given by

$$ \left\| f_{iof} - f_{s1A} \right\| \le f_{s1A\_0} \tag{16} $$

where f_{s1A_0} is the predetermined feature error threshold value.

Therefore, the controller (15) is expressed as follows:

$$ u_{s1A} = {}^{C}\omega_{C} = \begin{cases} -K_{l}\mathbf{J}^{r\;+}_{img}(f_{s1A})e_{iof} & \left\| f_{iof} - f_{s1A} \right\| > f_{s1A\_0} \\ u_{s2} & Otherwise \end{cases} \tag{17} $$

Notice that the function of this control law not only drives the laser spot on the object but also solves the image singularity problem. As mentioned in (Chaumette, 1998), a pure rotation of 180 deg around the optical axis leads to image singularity and causes a pure backward translational camera motion along the optical axis. In the proposed algorithm, the 3-DOF rotation of the camera is mainly executed in the first stage of control and the translational movement of camera is primarily executed in the second stage of control. Hence, the backward translational camera motion is avoided.

Stage 1.B: Features Partially seen in the FOV

When only partial object is in the FOV, some features are not available. In order to obtain all the features, we propose a strategy to control 2-DOF rotational motion of camera to project the laser spot on the centroid of the object in image plane till the whole object appears in the FOV. Hence the motion of the laser image feature $f^{l} = \left[x^{l} \quad y^{l}\right]^{T}$ is related to the camera motion as:

$$\dot{f}^l = \mathbf{J}^r_{imgXY}(f^l)\dot{r}_{xy} + \mathbf{J}_{img\_cz}(f^l)\dot{r}^T_{cz} \tag{18}$$

where $\dot{r}_{xy} = [\omega_x, \omega_y]^T$ represents 2-DOF rotational motion of camera, $\dot{r}_{cz} = \begin{bmatrix} v_x & v_y & v_z & \omega_z \end{bmatrix}^T$ is the rest of camera velocity screw, the image Jacobian matrices $\mathbf{J}^r_{imgXY}(f^l)$ and $\mathbf{J}_{img\_cz}(f^l)$ are defined respectively as:

$$\mathbf{J}^r_{imgXY}(f^l) = \begin{bmatrix} -\dfrac{x^l y^l}{\lambda} & \dfrac{\lambda^2 + x^{l2}}{\lambda} \\ \dfrac{-\lambda^2 - y^{l2}}{\lambda} & \dfrac{x^l y^l}{\lambda} \end{bmatrix} \tag{19}$$

$$\mathbf{J}_{img\_cz}(f^l, Z^l) = \begin{bmatrix} \dfrac{\lambda}{Z^l} & 0 & -\dfrac{x^l}{Z^l} & -y^l \\ 0 & \dfrac{\lambda}{Z^l} & -\dfrac{y^l}{Z^l} & x^l \end{bmatrix}.$$

Equation (18) can be written as:

$$\dot{r}_{xy} = \mathbf{J}^{r\,+}_{imgXY}(f^l)[\dot{f}^l - \mathbf{J}_{img\_cz}(f^l, Z^l)\dot{r}^T_{cz}]. \tag{20}$$

where $\mathbf{J}^{r\,+}_{imgXY}(f^l)$ is the pseudo-inverse of image Jacobian matrix $\mathbf{J}^r_{imgXY}(f^l)$.

As mentioned before, 2-DOF of rotational motion $\dot{r}_{xy} = [\omega_x \ \omega_y]^T$ allows the laser pointer to project its dot image close to the desired target. Thus the other elements of the camera velocity screw $\dot{r}_{cz} = \begin{bmatrix} v_x & v_y & v_z & \omega_z \end{bmatrix}^T$ are set to zero. Equation (20) becomes:

$$\dot{r}_{xy} \cong \mathbf{J}^{r\,+}_{imgXY}(f^l)\dot{f}^l \tag{21}$$

The centroid of the partial object in image is used as the desired laser image feature for the laser spot. To attain such centroid, one generally calculates the first order moments of the partial object image. Let R represents the region of the partial object in a binary image $I(k, j)$, which can be obtained by using fixed-level threshold. For a digital image, the moments of the region R are defined as:

$$m_{kj} = \sum_{(x,y)\in R} x^k y^j \quad k \geq 0, \qquad j \geq 0 \tag{22}$$

where (x, y) represent the row and column of a pixel in the region R respectively. According to the definition of moment, we have the area of the region R and the centroid of R as:

$$A_R = m_{00} \tag{23}$$

$$f_c = \begin{bmatrix} f_{xc} \\ f_{yc} \end{bmatrix} = \begin{bmatrix} \dfrac{m_{10}}{m_{00}} \\ \dfrac{m_{10}}{m_{00}} \end{bmatrix} \tag{24}$$

where f_c is the centroid of the partial object in the image.

Define the image feature error between the laser image feature and the centroid of partial object image feature as $e_c^l = [f^l - f_c]$. The proportional control law for 2-DOF of rotational motion is designed by imposing $\dot{e}_c^l = -K_{l2} e_c^l$:

$$u_{s1B} = \dot{r}_{xy} = \begin{cases} -K_{l2}\mathbf{J}_{imgXY}^{r+}(f^l)e_c^l & \left\| f^l - f_c \right\| > f_{s1B\_0} \\ u_{s1A} & Otherwise \end{cases} \tag{25}$$

where K_{l2} is the proportional gain and f_{s1B_0} is the predetermined feature error threshold value.

When the image feature error e_c^l enters into the range $\left\| f^l - f_c \right\| \le f_{s1B\_0}$, the laser spot is projected close to the planar object. In this paper, the end effector is assumed to be posed above the planar object and the FOV is relatively large compared with the image of the object. Hence, all the feature points enter into FOV when the laser spot is close to the desired laser image ($\left\| f^l - f_c \right\| \le f_{s1B\_0}$). Once all the feature points are obtained, the control law is switched to u_{s1A} in Stage 1.A.

Stage 2: Translational Moving of Camera

The key problem in switch control of Stage 2 is how to obtain the image Jacobian matrix of relating the motion of image features plus laser spot to the translational motion of camera. According to the derivation of traditional image Jacobian matrix, the target is supposed to be stationary. Hence the laser spot on the object can be considered as a stationary point adapting to image Jacobian matrix of traditional IBVS. Based on above scheme, the algorithm is presented in detail as follows.

Let $f_{s2} = [f_1 \cdots f_n \; f^l]^T$ represents n image features plus the laser image feature $f^l = [x^l \quad y^l]^T$. It is assumed that the object is small compared with the workplace. Hence the depth of the laser image close to the centroid of the object can be treated as a good approximation to the depth of all features. The modified relationship between the motion of the image features and the motion of camera is given by

$$\dot{f}_{s2} = \mathbf{J}_{img}(f_{s2}, Z^l)\dot{r} \tag{26}$$

where $\mathbf{J}_{img}(f_{s2}, Z^l)$ can also be decomposed into translational and rotational component parts as shown below

$$\mathbf{J}_{img}(f_{s2}, Z^l) = \left[\mathbf{J}_{img}^t(f_{s2}, Z^l) \quad \mathbf{J}_{img}^r(f_{s2}) \right] \tag{27}$$

where the two components are formed as:

$$\mathbf{J}_{img}^t(f_{s2}, Z^l) = \begin{bmatrix} \mathbf{J}_{img}^t(f_1, Z^l) \\ \vdots \\ \mathbf{J}_{img}^t(f_n, Z^l) \\ \mathbf{J}_{img}^t(f^l, Z^l) \end{bmatrix} \quad \mathbf{J}_{img}^r(f_{s2}) = \begin{bmatrix} \mathbf{J}_{img}^r(f_1) \\ \vdots \\ \mathbf{J}_{img}^r(f_n) \\ \mathbf{J}_{img}^r(f^l) \end{bmatrix} \tag{28}$$

Equation (26) can be written as:

$$\dot{f}_{s2} = \left[\mathbf{J}^{t}_{img}(f_{s2}, Z^{l}) \quad \mathbf{J}^{r}_{img}(f_{s2}) \right] \begin{bmatrix} {}^{C}v_{C} \\ {}^{C}\omega_{C} \end{bmatrix} = \mathbf{J}^{t}_{img}(f_{s2}, Z^{l})\,{}^{C}v_{C} + \mathbf{J}^{r}_{img}(f_{s2})\,{}^{C}\omega_{C} \tag{29}$$

And the translational motion of the camera is derived as:

$${}^{C}v_{C} = \mathbf{J}^{t\,+}_{img}(f_{s2}, Z^{l})\left[\dot{f}_{s2} - \mathbf{J}^{r}_{img}(f_{s2})\,{}^{C}\omega_{C} \right] \tag{30}$$

Set the rotational motion of the camera to zero. The above equation is rewritten as:

$${}^{C}v_{C} \cong \mathbf{J}^{t\,+}_{img}(f_{s2}, Z^{l})\,\dot{f}_{s2} \tag{31}$$

The control objective is to move the image features of object plus the laser spot image close to the target image features by using the translational motion of the camera. The target image features include four coplanar corner points of object image plus the desired laser spot f^{l}_{d} defined as $f_{s2\_D} = [f_{d1}\ f_{d2} \cdots f_{dn} \quad f^{l}_{d}]^{T}$. The error between them is defined as:

$$e_{s2} = f_{s2} - f_{s2\_D}$$

The translational motion of the camera is designed by imposing $\dot{e}_{s2} = -K_{s2}e_{s2}$:

$$u_{s2} = {}^{C}v_{C} = \begin{cases} -K_{s2}\mathbf{J}^{t\,+}_{img}(f_{s2}, Z^{l})e_{s2} & \left\| f_{s2} - f_{s2\_D} \right\| > f_{s2\_0} \\ u_{s3} & Otherwise \end{cases} \tag{32}$$

Where K_{s2} is the proportional gain and f_{s2_0} is the predetermined feature error threshold value.

The switching rule is described as: if the image feature error norm between the current image features and the desired image features falls below a threshold f_{s2_0} , the IBVS with laser pointer switches from the current stage to the third stage u_{s3}.

$$\left\| f_{s2} - f_{s2\_D} \right\| \le f_{s2\_0} \tag{33}$$

Since the translational motion of camera is approximately derived as (31), the proportional controller (32) can not drive the image feature e_{s2} to approach zero exponentially. However, the translational motion brings the image of object in the vicinity of target image features. A switching rule is thus set to switch from the current stage to the next one for the fine tuning of visual servoing. The threshold f_{s2_0} is directly related to the depth and affects the stability of the controller. Thus the selection of this threshold is crucial and it is normally set as relatively small pixels to maintain the stability of the system.

Stage 3: Adjusting the Fine Alignments

After applying the switching controllers in Stage 1 and 2, one can control the end effector to the neighbour of the target image features. It has been presented in (Chaumette, 1998) that the constant image Jacobian at the desired camera configuration can be used to reach the image global minimum. Hence, a constant image Jacobian matrix can be used in IBVS to adjust the fine alignment of end effector so that the pump can perfectly suck up the object. In this stage, the laser image feature is not considered as an image feature and the traditional IBVS with constant image Jacobian of target image features is applied.

Since the image features are close to their desired position in the image plane, as shown in (33), the depths of the points of image features are approximate to the desired ones (essentially planar object). The target features and their corresponding depths are applied to the traditional IBVS with constant image Jacobian matrix shown as:

$$\mathbf{J}_{img}(f_{id}, Z_{id}) = \begin{bmatrix} \frac{\lambda}{Z_{id}} & 0 & -\frac{x_{id}}{Z_{id}} & -\frac{x_{id}y_{id}}{\lambda} & \frac{\lambda^2 + x_{id}^2}{\lambda} & -y_{id} \\ 0 & \frac{\lambda}{Z_{id}} & -\frac{y_{id}}{Z_{id}} & \frac{-\lambda^2 - y_{id}^2}{\lambda} & \frac{x_{id}y_{id}}{\lambda} & x_{id} \end{bmatrix} \tag{34}$$

where f_{id} and Z_{id} are target features and their corresponding depths respectively.

The control goal of this stage is to control the pose of the end effector so that the image feature error between the current image features $f = [f_1 \cdots f_n]^T$ and the target image features $f_d = [f_{d1}\ f_{d2} \cdots f_{dn}]^T$ reaches the global minimum. The condition can be described as: if the feature error norm falls below a predetermined threshold, the whole IBVS with laser pointer will stop. The condition is presented as:

$$\|f_d - f\| \le f_{s3\_0} \tag{35}$$

where f_{s3_0} is a predetermined threshold value, $f = [f_1 \cdots f_n]^T$ is the image features and $f_d = [f_{d1}\ f_{d2} \cdots f_{dn}]^T$ is the desired feature.

The proportional controller is designed as:

$$u_{s3} = \begin{bmatrix} {}^{C}v_C \\ {}^{C}\omega_C \end{bmatrix} = \begin{cases} -\mathbf{J}^{+}_{img}(f_d, Z_d)(f - f_d) & \|f_d - f\| > f_{S3\_0} \\ Stop\ Seroving - Start\ grasping & Otherwise \end{cases} \tag{36}$$

where K_{s3} is the proportional gain and $Z_d = [Z_{1d} \quad \cdots \quad Z_{nd}]^T$ is the depth vector of each target feature point.

The threshold f_{S3_0} directly affects the accuracy of the robot end effector pose with respect to the object. The bigger value of threshold is chosen, the less accurate pose will be achieved. However, the positioning process time is reduced.

It is noticed that the proposed IBVS algorithm is derived from traditional IBVS. Therefore it inherits the advantages of IBVS, which does not need object model and is robust to the camera calibration and hand-eye calibration error. In addition, by using a laser pointer and the separate DOF method, the proposed switch algorithm decouples the rotational and translational motion control of the robotic end effector to overcome the inherent drawbacks of traditional IBVS such as image singularities and the image local minima.

4. Experimental results

The proposed IBVS with laser pointer has been tested on a robotic assembly system including an industrial robot Motoman UPJ with JRC controller, a laser pointer, and a PC based vision system including Matrox frame grabber and a Sony XC55 camera mounted on the robot. The robotic assembly system setup for testing IBVS with laser pointer is shown in Figure 5.

To verify the effectiveness of the proposed method, a plastic object shown in Figure 5 is chosen to be assembled in a metallic part. The four coplanar corners of the surface are selected as the target features. One of the advantages of the JRC controller in UPJ robot is that it accepts the position and orientation values and calculates the joint angle by itself, which eliminates the robot kinematical modeling error. However, the drawback of the JRC controller is that it cannot be used for real time control. In other words, when the designed controller generates a new position or orientation value and sends it to the JRC controller, it will not respond it until the previous position is reached in each iteration. With this limitation of hardware, we have to divide the calculated value into a serial of small increment by a constant factor in each step, and increase the sampling time as well. In the experiment, we chose constant factor as 50, and the sampling time as 2 seconds.

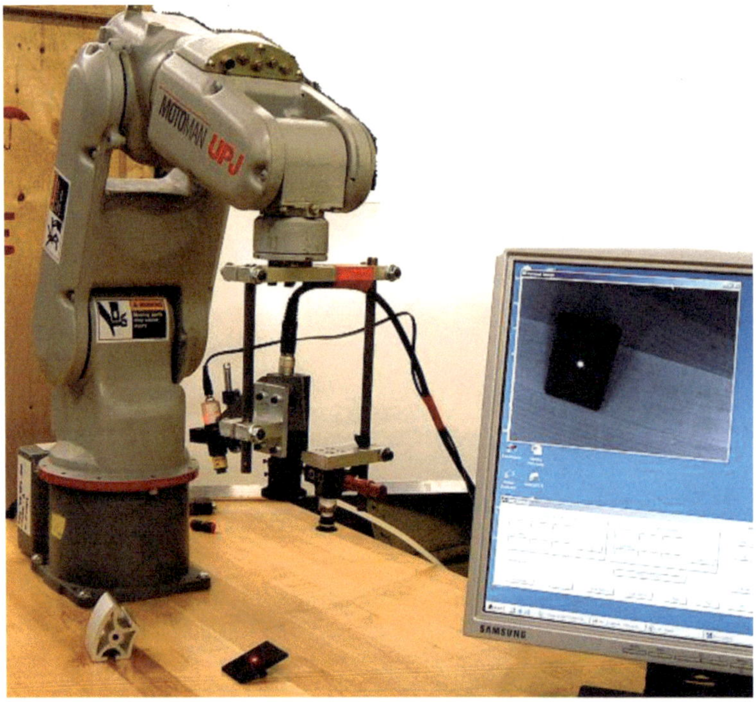

Fig. 5. Robotic assembly system setup and components to be assembled

The desired pose ${}^{E}\mathbf{H}_{O}$ is set as $[36 \;\; -66 \;\; 123 \;\; 0 \;\; 0 \;\; 13.7]$ (mm, deg) for teaching procedure. The predetermined threshold f_{s1A_0} in (16) is set as 10 pixels and f_{s1B_0} in (25) is set as 10 pixels. The f_{s2_0} in (32) is tuned as 10 pixels, which is approximate to 28mm in the experimental setup. f_{s3_0} in (35) is set as 3 pixels. All the proportional gains of control law are set as 1. When the condition of feature error norm falling below 3 pixels is met, the end effector is brought to the desired pose with respect to the plastic object. It should be noted that no matter where the object is located in the workplace, the desired pose of end effector with respect to the object is exactly the same. For any object position, there exists a fixed transformation between camera and end effector frame with this desired pose. Before the visual servoing starts, we need to teach the robot this fixed pose once. Therefore, commanding the robot end effector to move the predefined transformation vector in end effector frame perfectly lets the vacuum pump suck up the object. The assembled components shown in Figure 5 demonstrate the accuracy of the proposed method.

(a) With Good Calibration Value

The proposed algorithm is tested with good calibration values (Table 1). The parameters d and α for the fixed laser-camera configuration are adjusted to 8mm and 72 deg individually. The sequential pictures of successful assembly process are illustrated in Figure 6. The image trajectory obtained from the experiment is illustrated in Figure 7 (a). The rectangular with white edges represents the initial position and the black object shows the desired position. The object is posed manually with randomized initial position.

Parameters	Good calibration values	Bad calibration values
Principal point (pixel)	[326 228]	[384 288]
Focal length (mm)	6.02	7.2
Effective size (mm)	0.0074×0.0074	0.0074×0.0074
${}^{C}\mathbf{H}_{L}$ [X Y Z φ θ ψ] (mm, deg) measured by hand	[8 0 0 0 -18 0]	[6 0 0 0 -14 0]
${}^{E}\mathbf{H}_{C}$ [X Y Z φ θ ψ] (mm, deg) calculated by calibration	[-7 5 189 -2 0 -179]	[-20 20 210 5 5 180]

Table 1. Good and bad calibration values of system parameters

(b) With Bad Calibration Value

To test the robustness of the IBVS with laser pointer, camera calibration error is also added to intrinsic parameters with 20% deviation as shown in Table 1. The good calibration value and bad calibration value of transformation among camera reference, laser pointer frame and robot end effector frame are also shown in Table 1. The object position is the same as that of the experiment with good calibration value and the image trajectory resulted from experiment is shown in Figure 7 (b).

Although the image trajectory shown in Figure 7 (b) is distorted comparing with the trajectory presented in Figure 7 (a), the image features still converge to the desired position and the assembly task is successfully accomplished as well. Hence, the proposed algorithm is very robust to the camera calibration and hand-eye calibration error on the order of 20% deviation from nominal values.

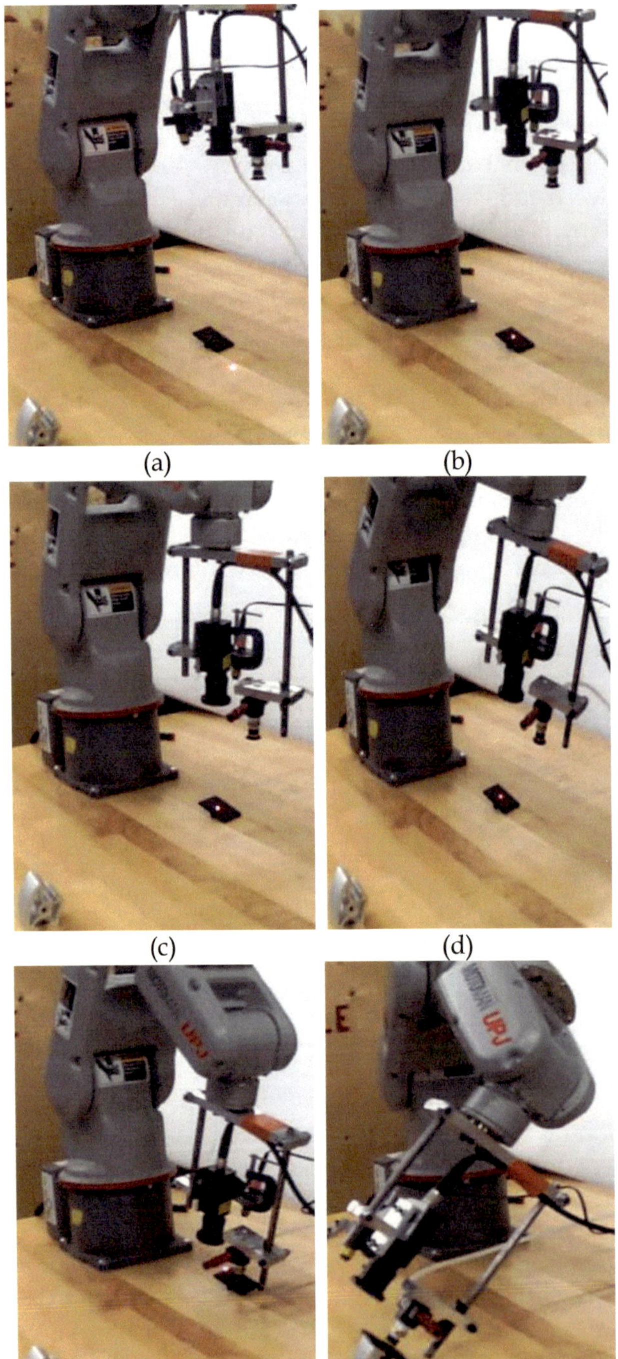

Fig. 6. Assembly sequence (a) Initial position (b) The end of stage 1, (c) The end of stage 2 (d) The end of stage 3 (ready to pick up the object) (e) Suck up the object (f) Assembly

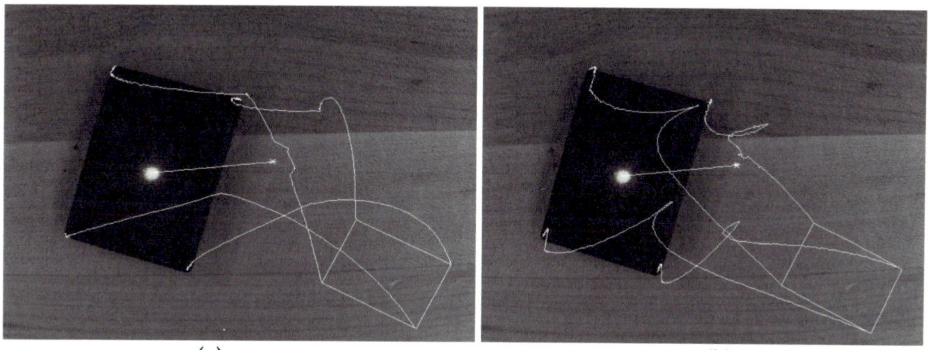

(a) (b)

Fig. 7. Image trajectory (a) with good calibration value (b) with bad calibration value

In order to test the depth estimation convergence, the experiments on the depth estimation are carried out under the conditions with good and bad calibration values. The experimental results are presented in Figure 8. Both results show that the estimated depths converge to the real distance and thus the depth estimation convergences are experimentally proved.

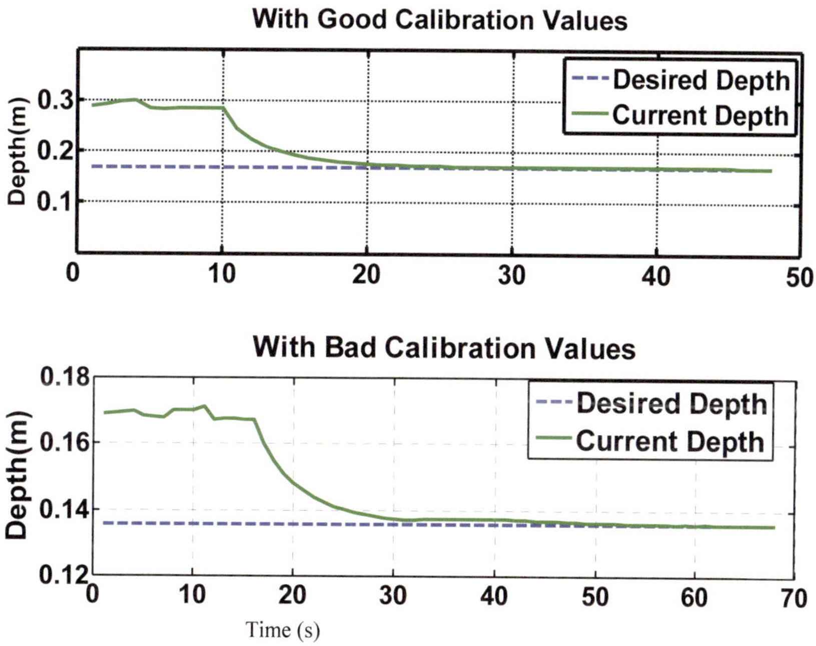

Fig. 8. Experimental results of depth estimation

5. Conclusion

In this paper, a new approach to image based visual servoing with laser pointer is developed. The laser pointer has been adopted and the triangular method has been used to estimate the depth between the camera and the object. The switch control of IBVS with laser pointer is decomposed into three stages to accomplish visual servoing tasks under various circumstances. The algorithm has been successfully applied in an experimental robotic assembly system. The experimental results verify the effectiveness of the proposed method and also validate the feasibility of applying the proposed method to industrial manufacturing systems. Future work includes testing the proposed method on a robot manipulator supporting real time control and conducting analytical convergence analysis of switch control algorithm.

6. References

N. Aouf, H. Rajabi, N. Rajabi, H. Alanbari and C. Perron (2004), "Visual object tracking by a camera mounted on a 6DOF industrial robot," IEEE Conference on Robotics, Automation and Mechatronics, vol. 1, pp. 213-218, Dec. 2004.

F. Chaumette (1998), "Potential problems of stability and convergence in image based and position-based visual servoing," in The Conference of Vision and Control, D. Kriegman, G.

G. Chesi, K. Hashimoto, D. Prattichizzo and A. Vicino (2004), "Keeping Features in the Field of View in Eye-In-Hand Visual Servoing: A Switching approach", IEEE Transactions on Robotics, Vol. 20, No. 5, pp. 908-913, Oct. 2004.

P. I. Corke and S. A. Hutchinson (2001), "A new partitioned approach to image-based visual servo control," IEEE Transactions on Robotics and Automation, vol. 17, no.4, pp. 507-515, Aug. 2001.

P. I. Corke (1996), "Robotics toolbox for MATLAB," IEEE Transactions on Robotics and Automation, vol. 3, no.1, pp. 24-32, 1996.

P. I. Corke (1996), "Visual Control of Robots: High Performance Visual Servoing," Research Studies Press, Australia, 1996.

D. F. DeMenthon and L.S. Davis (1995), "Model-based object pose in 25 lines of code," International Journal of Computer Vision, vol. 15, no. 1/2, pp. 123-142, 1995.

L. Deng, F. Janabi-Sharifi and W. J. Wilson (2005), "Hybrid Motion Control and Planning Strategies for Visual Servoing," IEEE Transactions on Industrial Electronics, vol. 52, no. 4, pp. 1024–1040, Aug. 2005.

L. Deng; W. J. Wilson and F. Janabi-Sharifi (2003), "Dynamic performance of the position-based visual servoing method in the cartesian and image spaces," in Proceedings, IEEE/RSJ International Conference on Intelligent Robots and Systems, vol. 1, pp. 510–515, 2003.

B. Espiau (1993), "Effect of camera calibration errors on visual servoing in robotics," in Experimental Robotics III: The 3rd International symposium on experimental robotics, Springer-Verlag, Lecture notes in control and information sciences, vol. 200, pp.182-192, 1993.

N. R. Gans and S. A. Hutchinson (2003), "An experimental study of hybrid switched system approaches to visual servoing," in Proceedings ICRA '03, IEEE International Conference on Robotics and Automation, vol. 3, pp. 3061–3068, Sept. 2003.

D. Hager, and A. Morse (1998), Eds. Berlin, Germany: Springer-Verlag, vol. 237, pp. 66-78. Lecture Notes in Control and Information Sciences, 1998.

K. Hashimoto and T. Noritsugu (2000), "Potential problems and switching control for visual servoing," in Proceedings, IEEE/RSJ International Conference on Intelligent Robots and System, vol. 1, pp. 423–428, Oct. 2000.

S. Hutchinson, G. D. Hager, and P. I. Corke (1996), "A tutorial on visual servo control," IEEE Transactions on Robotics and Automation, vol. 12, no. 5, pp. 651–670, Oct. 1996.

F. Janabi-Sharifi and M. Ficocelli (2004), "Formulation of Radiometric Feasibility Measures for Feature Selection and Planning in Visual Servoing", IEEE Transactions on Systems, Man and Cybernetics—Part B: Cybernetics, Vol. 34, No. 2, pp. 978-987, April 2004.

H. Kase, N. Maru, A. Nishikawa, S. Yamada and F. Miyazaki (1993), "Visual Servoing of the Manipulator using the Stereo Vision," in Proceedings of the IECON '93 IEEE International Conference on Industrial Electronics, Control, and Instrumentation, vol. 3, pp.1791 - 1796, 1993.

A. Krupa, J. Gangloff, C. Doignon, M. Mathelin. G. Morel, J. Leroy, L. Soler and J. Marescaux (2003), "Autonomous 3-D positioning of surgical instruments in robotized laparoscopic surgery using visual servoing," IEEE Transaction on Robotic and Automation, vol. 19, no.5, pp. 842-853, 2003.

Z. Li, W. F. Xie and N. Aouf (2006), "A Neural Network Based Hand-Eye Calibration Approach in Robotic Manufacturing Systems", CSME 2006, Calgary, May 2006.

R. Mahony, P. Corke and F. Chaumette (2002), "Choice of image features for depth-axis control in image based visual servo control," IEEE/RSJ International Conference on Intelligent Robots and Systems, vol.1, pp. 390–395, 2002.

E. Malis, F. Chaumette, and S. Boudet (1999), "2-1/2-D visual servoing (1999)," IEEE Transactions on Robotics and Automation, vol. 15, no.2, pp. 238–250, April 1999.

P. Y. Oh, and P. K. Allen (2001), "Visual Servoing by Partitioning Degrees of Freedom," IEEE Transactions on Robotics and Automation, vol. 17, no.1, pp. 1-17, 2001.

J. Pages, C. Collewet, F. Chaumette and J. Salvi (2006), "Optimizing Plane-to-Plane Positioning Tasks by Image-Based Visual Servoing and Structure light," IEEE Transactions on Robotics, Vol. 22, No.5, pp. 1000-1010, 2006.

O. Tahri and F. Chaumette (2005), "Point-Based and Region-Based Image Moments for Visual Servoing of Planar Objects", IEEE Transactions on Robotics, Vol. 21, No. 6, pp. 1116-1127, Dec. 2005.

E. Trucco and A. Verri (1998), "Introductory Techniques for 3-D computer Vision," Prentice Hall; 1998.

J. P. Wang and H. Cho (2008), "Micropeg and Hole Alignment Using Image Moments Based Visual Servoing Method", IEEE Transactions on Industrial Electronics, Vol. 55, No. 3, pp. 1286-1294, March 2008.

L. E. Weiss, A. C. Sanderson, and C. P. Neuman (1987), "Dynamic sensor based control of robots with visual feedback," IEEE Journal of Robotics and Automation, vol. 3, no.5, pp. 404–417, Oct. 1987.

W. J. Wilson, C. C. W. Hulls, and G. S. Bell (1996), "Relative end-effector control using Cartesian position-based visual servoing," IEEE Transactions on Robotics and Automation, vol. 12, no.5, pp. 684–696, Oct. 1996.

29

The Analysis and Optimization in Virtual Environment of the Mechatronic Tracking Systems used for Improving the Photovoltaic Conversion

Cătălin Alexandru and Claudiu Pozna
University Transilvania of Braşov
Romania

1. Introduction

The chapter presents researches in the field of increasing the efficiency of the solar energy conversion in electric energy, using tracking systems that change the position of the photovoltaic (PV) panel in order to maximize the solar radiation degree of use. From efficiency and safety point of view, we have selected a dual-axis equatorial tracking system, with two degrees of freedom. The both motions (daily and seasonal) are driven by rotary actuators, which integrate irreversible transmission for blocking the system in the intermediary positions, between actuatings.

The tracking system is approached in mechatronic concept, by integrating the electronic control system in the mechanical structure of the solar tracker. In this way, two main aspects are taken into consideration: optimizing the interaction between the mechatronic system components (the mechanic structure, and the control system); reducing the cost and time for the design process by replacing the traditional tests on hardware models, which are very expensive, with the testing in virtual environment. The study is performed by developing the virtual prototype of the mechatronic tracking system, which is a complex dynamical model. In fact, the virtual prototype is a control loop composed by the multi-body mechanical model connected with the dynamic model of the actuators and with the controller dynamical model. Using the virtual prototype, we are able to optimize the tracking mechanism, choose the appropriate actuators, and design the optimal controller.

The functionality of the tracking systems is evaluated from kinematic and dynamic point of view (i.e. motions, forces), as well as from energetic balance point of view. The key word of the design is the energetically efficiency. Using the tracking system, the PV panel follows the Sun and increase the collected energy, but the driving actuators consume a part of this energy. In order to analyze our solution, we will compare the PV system with tracking with a fixed panel, in standard testing conditions. One of the most important advantages of this kind of simulation is the possibility to perform measurements in any point and/or area of the system and for any parameter. This helps us to make quick decisions on any design changes without going through expensive prototype building and testing.

2. Actual knowledge stage & original contributions

The theme of the chapter belongs to a very important field: renewable sources for energy production - increasing the efficiency of the photovoltaic conversion. The researches in this field represent a priority at international level because provides viable alternatives to a series of major problems that humanity is facing: the limited and pollutant character of the fossil fuels, global warming or the greenhouse effect. The solution to these problems is the renewable energy, including the energy efficiency, the energy saving, and systems based on clean renewable energy sources, like sun, wind and water. The concept of sustainable development have been enounced for the first time in 1987, in the Brundland Commission Report, and subsequent adopted at the political level, so in the Conference for Development and Environment from Rio de Janeiro (1992) the participant countries have undertaken to develop national strategies for sustainable development - The Program Agenda 21.

The solar energy conversion is one of the most addressed topics in the fields of renewable energy systems. The present-day techniques allow converting the solar radiation in two basic forms of energy: thermal and electric energy. Likewise, there are additional fields that use the solar energy, for example the hydrogen technology. The technical solution for converting the solar energy in electricity is well-known: the photovoltaic (PV) conversion. The PV systems can deliver energy on large-scale to a competitive price is the conclusion of the European Commission for Energy, in the report "A Vision for Photovoltaic Technology for 2030 and Beyond" (2004). The report emphasizes as the development of advanced technologies in the photovoltaic area, and a European strong and competitive industry will support the strategic initiatives concerning to the security and the diversity of the electric energy sources.

The efficiency of the PV systems depends on the degree of use and conversion of the solar radiation. The energy balance refers to the surface that absorbs the incoming radiation and to the balance between energy inflow and energy outflow. The rate of useful energy leaving the absorber is given by the difference between the rate of incident radiation on absorber and the rate of energy loss from the absorber (Goswami et al., 2000; Tiwari, 2002). The degree of use of the solar radiation can be maximized by use of mechanical systems for the orientation of the PV panels in accordance with the paths of the Sun. Basically the tracking systems are mechanical systems that integrate mechanics, electronics, and information technology. These mechanisms are driven by rotary motors or linear actuators, which are controlled in order to ensure the optimal positioning of the panel relatively to the Sun position on the sky dome, on the entire period of the day (the daily motion, from East to West), and also depending on the season (the seasonal/elevation motion). The orientation of the photovoltaic panels may increase the efficiency of the conversion system from 20% up to 50% (Abdallah & Nijmeh, 2004; Canova et al., 2007; Hoffmann et al., 2008).

In literature, the increasing of the photovoltaic efficiency is approached mainly through the optimization of the conversion to the absorber level, and this because the subject is mainly specific for the chemical and electrical engineering. In our vision, this is an interdisciplinary field that concentrates the competences of three fundamental domains: chemical, electrical, and mechanical. The solar energy conversion field interferes with the mechanical engineering there where the optimization process of the conversion uses tracking systems based on mechatronic devices (the maximum degree of collecting is obtained when the solar radiation is normal on the active surface).

Having in view the operating principle, there are two fundamentals types of tracking systems: passive and active trackers. The passive trackers are based on thermal expansion of a Freon-based liquid from one edge of the tracker to another because of the heat sensitive working fluid (Clifford & Eastwood, 2004). The active trackers are based on electrically operated positioning drives, which need motors, gearboxes, mechanisms, couplings etc. Usually, the nowadays active tracking systems are based on planar or spatial linkages, gears, chain and belt transmissions.

The photovoltaic system with tracking is efficient if the following condition is achieved:

$$\varepsilon = (E_T - E_F) - E_C >> 0 \tag{1}$$

in which E_T is the electric energy produced by the photovoltaic panel with tracking, E_F - the energy produced by the same panel without tracking (fixed), and E_C - the energy consumption for orienting the PV panel. In the current conditions, the maximization of the efficiency parameter ε, by the optimization of the tracking system, became an important challenge in the modern research and technology.

In the design process of the tracking systems, the solar radiation represents the main input data. Interacting with atmospheric phenomena involving reflection, scattering, and absorption of radiation, the quantity of solar energy that reaches the earth's surface is reduced in intensity. The total solar radiation received at ground level includes two main components: direct solar radiation and diffuse radiation (Melіß, 1997). The solar radiation can be measured using traditional instruments, or can be digitally recorded with a data acquisition system. Within an EU funded project, a solar radiation atlas was realized for Europe (Scharmer & Greif, 2000). At the same time, there were developed large meteorological databases, such as Meteonorm - Global Meteorological Database for Engineers, Planners and Education. In addition, different models were developed for estimating the solar radiation. The traditional Angstrom's linear approach is based on measurements of sunshine duration, while relatively new methods are based on artificial neural networks - ANN (Tymvios et al., 2005). Different models for estimating the monthly mean solar radiation, including linear Angstrom-Prescot variation, quadratic equation, logarithmic variation, and exponential function, are comparatively analyzed by using as principal element the root mean square error (Salmi et al., 2007). A step by step procedure was developed for implementing an algorithm to calculate the solar irradiation, using both zenith and azimuth angles to describe sky element's position, for a surface that is tilted to any horizontal and vertical angle (Reda & Andreas, 2004).

Other papers refer to the computation of the yearly energy collection allowed by different tracking strategies. A theoretical analysis of different intervals of intermittent two-axis tracking of the Sun on the amount of annual energy received by solar panels was performed for estimating the solar radiation in standard sky condition (Koay & Bari, 1999). Using as input data the location latitude and commonly available values of monthly irradiation, a relation between the latitude of the chosen location and the most suitable tracker was established (Sorichetti & Perpinan, 2007). For comparing the energy capture between fixed tilt angle and sun trackers, in clear sky and mean sky conditions, the Moon-Spencer and the Aste models are frequently used (Canova et al., 2007). Specific software tools were also developed to simulate the energy yield of PV systems as a function of the ground cover ratio, for different tracking strategies (Narvarte & Lorenzo, 2007).

The active tracking systems contain mechanisms, which are driven by controlled motors - actuators. Regarding the control process, in literature, closed loop systems with photo sensors are traditionally used. The photo sensors are responsible for discrimination of the

Sun position and for sending electrical signals, proportional with the error, to the controller, which actuates the motors to track the Sun. Many authors have adopted this method as a basis in construction and design of such systems (Baltas et al., 1986; Dobon et al., 2003; Karimov et al., 2005). Although, the orientation based on the Sun detecting sensors, may introduce errors in detection of real sun position for variable weather conditions.
The alternative consists in the opened loop systems (Abdallah & Nijmeh, 2004; Alexandru & Pozna, 2008; Roth et al., 2004), which are based on mathematic algorithms/programs that may provide predefined parameters for the motors, depending on the sun positions on the sky dome (i.e. the astronomic movements of the Sun - Earth system). These positions can be precisely determined because they are functions of the solar angles that can be calculated for any local area. By using this control technique, based on predefined parameters, the errors introduced by the use of the sensors may be avoided.
Other solution is to incorporate some kind of Sun position sensor to check and calibrate automatically the astronomical control system. In addition, the tracking system can also be adjusted to provide maximum output energy, to self-trim it initially or self correct itself throughout its life (Sala et al., 2002). Such hybrid control system consists of a combination of opened loop tracking strategies, based on solar movement models, and closed loop strategies, using dynamic feedback controller (Rubio et al., 2007).
From the controller point of view, different control strategies are used, such as classical techniques as PID algorithm or more advanced strategy such as fuzzy logic controller - FLC. The design of a low cost two-axis solar tracker for obtaining a high precision positioning of the PV panel is made considering a control-board that is able to support different PID and FLC control strategies (Yazidi et al., 2006). An evolution of the fuzzy control concept is the fuzzy logic neural controller (FNLC), which allows the PV system to learn control rules (Chojnacki, 2005). A more complex controller incorporates the advantages of two alternate design techniques, namely the deadbeat regulator - for quick, rough control, and the LOG/LTR regulator - for soft final tracking (Rubio & Aracil, 1997). The first order Sugeno fuzzy inference system is used for modelling and designing the controller of an azimuth & elevation solar tracker (Alata et al., 2005); the estimation of the insolation incident on the two axis sun tracking system is determined by fuzzy IF-THEN rules.
The orientation principle of the photovoltaic panels is based on the input data referring to the position of the Sun on the sky dome. For the highest conversion efficiency, the sunrays have to fall normal on the receiver surface so the system must periodically modify its position in order to maintain this relation between the sunrays and the PV panel. The positions of the Sun on its path along the year represent input data for the design process of the tracking systems. The Earth describes along the year a rotational motion following an elliptical path around the Sun. During one day, the Earth also spins around its own axis describing a complete rotation that generates the sunrises and the sunsets. The variation of the altitude of the Sun on the celestial sphere during one year is determined by the precession motion, responsible for a declination of the Earth axis in consideration with the plane of the elliptic yearly path. In these conditions, for the design process of the tracking systems there are considered two rotational motions: the daily motion, and the yearly precession motion.
Consequently, there are two basic types of tracking systems (fig. 1): single-axis tracking systems (a), and dual-axis tracking systems (b, c). The single-axis tracking systems spins on their axis to track the Sun, facing east in the morning and west in the afternoon. The tilt angle of this axis equals the latitude angle of the loco because this axis has to be always

parallel with the polar axis. In consequence for this type of tracking system is necessary a seasonal tilt angle adjustment. The two-axis tracking systems combine two motions, so that they are able to follow very precisely the Sun path along the period of one year.

Depending on the relative position of the revolute axes, there are two types of dual-axis systems: polar (b), and azimuthal (c). For the polar trackers, there are two independent motions, because the daily motion is made rotating the PV panel around the polar axis. For the azimuthal trackers, the main motion is made by rotating the PV panel around the vertical axis, so that it is necessary to continuously combine the vertical rotation with an elevation motion around the horizontal axis.

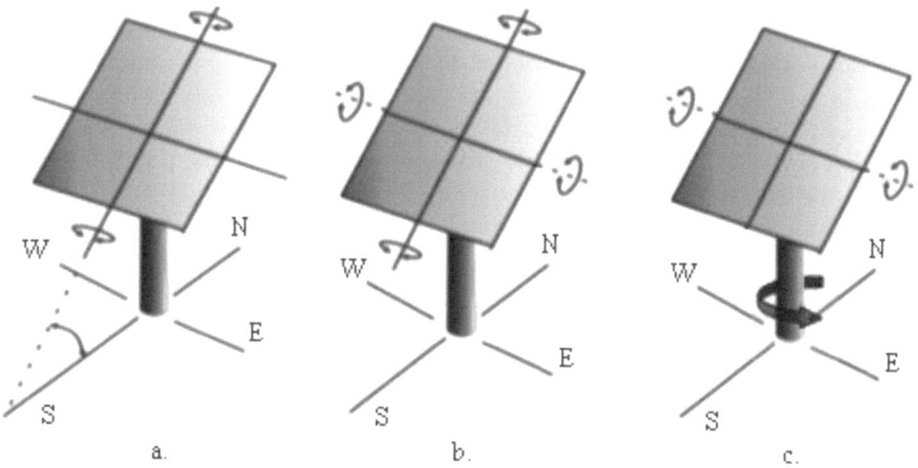

Fig. 1. Basic types of solar tracking systems

The research of the literature reveals the limits of the actual stage in the development of the mechanical device of the tracking systems. Since now there are no unitary modellings on structural, kinematical and dynamical aspects in designing the mechanical structure. At the same time, a general approaching for the conceptual design and the structural synthesis of these mechanisms is missing. Thus rises the necessity of a unitary modelling method of mechanisms, and in our opinion this method is the Multi Body Systems (MBS) method, which may facilitates the self-formulating algorithms (Garcia de Jalón & Bayo, 1994; Schiehlen, 1997). At the same time, in the reference literature, there were not identified compliant models of tracking systems that allow the precise simulation of the system behaviour with deformable components (bodies, joints); this is the reason for the missing of the studies concerning the vibratory and durability characteristics of the tracking systems.

On the other hand, the research is focused mostly on the quantity of the energy achieved by tracking and there is no evaluation on the energy consumption determined by the orientation (consumed energy by the motors/actuators). A possible cause of this case is due to the fact that the issue is not approached as an integrated assembly (mechanical device + actuating system and control). Thus, we propose the integration of the main components at the virtual prototype level, during the entire design process (i.e. the modelling in mechatronic concept), which allows at the same time the evaluation of the energy gain by orientation as well as the energetic consumption necessary for the motion.

In these terms, we can systematize the following scientific objectives of the paper: identifying & modelling the reference input data for the PV system design (i.e. the solar

radiation); conceptual design of the tracking mechanisms for identifying suitable solutions (from functional and constructive point of view); optimizing the tracking mechanisms from geometric point of view; in-detail modelling of the tracking mechanisms, including the friction and the deformability; developing the control strategies in order to obtain as much as possible incident radiation with minimum energy consumption; evaluating the vibratory characteristics; performing the energy balance (i.e. the energetic efficiency). In this way, the necessary conditions for approaching the functionality of the entire assembly (including the PV panel, the tracking mechanism, the motors, and the control system) are assured.
No less important there are the instruments (techniques) used for realizing the objectives, based on testing - simulation in the virtual environment (digital mock-up, virtual prototyping). This kind of approach is based on the design of the detailed digital models and the use of these in virtual experiments, by reproducing with the computer of the real phenomena. An important advantage of this process consists of the possibility to develop measurements in any point or zone and for any type of the parameter (motion, force, energy) that is not always possible to the experimental testing on the physical models (the missing of the adequate sensors or the misplace for these sensors, high temperatures etc.). The design process based on the virtual prototyping allows the fast evaluation of the shape, mounting and functions of the tracking systems, eliminating the large period amount necessary for physical prototyping, or very expensive design modifications.

3. Design tool - virtual prototyping platform

Traditional CAD/CAM/CAE practices used to evaluate the functional characteristics of the mechanical & mechatronic systems were focused on a concept referred to as art-to-part. Nearly all engineering software activity was oriented toward the design, development, and manufacturing of higher quality parts. Unfortunately, optimal part design rarely leads to optimal system design. The interaction of form, fit, function, and assembly of all parts in a product is a major contributor to overall product quality. The big opportunity to increase quality and reduce time and cost has now shifted to the system level (Ryan, 2001).
The system-focused approach involves the following software solutions: Digital Mock-Up (DMU) - to investigate product form and fit, Functional Virtual Prototyping (FVP) - to assess product function and operating performance, and Virtual Factory Simulation (VFS) - to assess assembly and manufacturability of the product. The integration of the system-focused tools provides a means for realizing the transition from physical to virtual prototyping, with all of the concomitant benefits.
The steps to create a virtual (software) prototype mirror the steps to build a physical (hardware) prototype (Haug et al., 1995). During the build phase, virtual prototypes are created of both the new product concept and any target products which may already exist in the market. The geometry and mass properties are obtained from component solid models. The structural, thermal and vibratory characteristics result from component finite element models or experimental tests.
One of the most important axioms for successful functional virtual prototyping is to simulate as test. Testing of hardware prototypes has traditionally involved both lab tests and field tests in various configurations, which are very expensive. With virtual prototyping, it is enough to create virtual equivalents of the lab and field tests.
To validate the virtual prototype, the physical and virtual models are tested identically, using the same testing and instrumentation procedures. The results are compared, and

design sensitivity analyses are performed to identify design parameters that have influence on the performance results that do not correspond. Afterwards, different changes on these parameters are realized in order to obtain an acceptable correlation.

Refining the virtual prototype involves the fidelity of the model. By replacing the rigid components with flexible counterparts, adding frictions, or representing the automatic systems that control the operating performance of the mechanical system can make the improvement of the virtual prototype.

The optimization of the virtual prototype is made with the following steps: parameterizing the model; defining the design variables; defining the objective functions for optimization and the design constraints; performing design studies and design of experiments; optimizing the model on the basis of the main design variables. Parameterizing the model simplifies changes to model because it helps to automatically size, relocate and orient bodies. Design variables allow creating independent parameters and tie modelling objects to them. Design study describes the ability to select a design variable, sweep that variable through a range of values and then simulate the motion behaviour of the various designs in order to understand the sensitivity of the overall system to these design variations.

The objective function of the optimization is a numerical quantification that distinguishes or rates candidate designs. The constraints are boundaries that directly or indirectly eliminate unacceptable designs; they often take the form of additional goals for the mechanism design. In general, the optimization problem is described as a problem to minimize or maximize the objective function over a selection of design variables, while satisfying various constraints on the design.

The analysis flow-chart of the tracking systems is shown in figure 2. The kinematic model of the tracking system contains the rigid parts, which are connected through geometric constraints (joints), and the specific geometric parameters; the input is made using kinematic restrictions (motion generators) that impose the motion of the driving elements (usually, the angular or linear positions). The aim of the kinematic analysis is to evaluate the relative motion between components, and to identify if the tracking system is able to generate the angular fields of the photovoltaic panel.

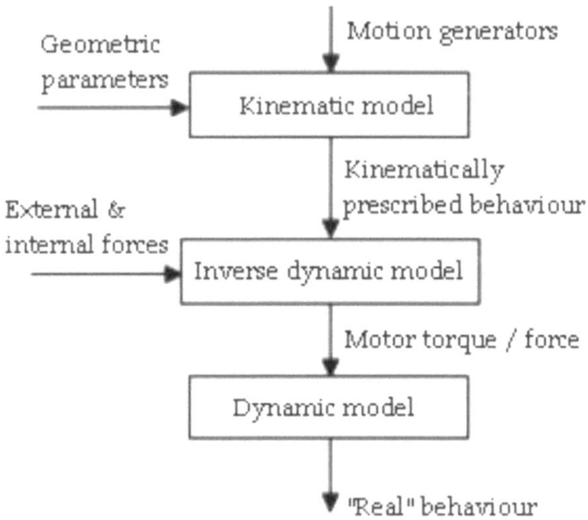

Fig. 2. The analysis flow-chart of the solar trackers

The inverse dynamic model includes the kinematic model and, in addition, the external and internal forces & torques. Basically, there are taken in consideration the mass forces, which depend on the geometric model and the material content for each mobile body, the friction forces, and the reactions in joints. In addition, the model can be completed with other external factors (perturbations), for example the wind or snow action. The aim of the inverse dynamic analysis is to determine the motor torque and/or the motor force applied by the driving element, in order to generate the cinematically-prescribed behaviour. The dynamic model of the tracking system (i.e. the virtual prototype) includes the components of the inverse dynamic model, but the input is made through the above-determined motor torques - forces; the goal is to evaluate the system's behaviour in real operating conditions.

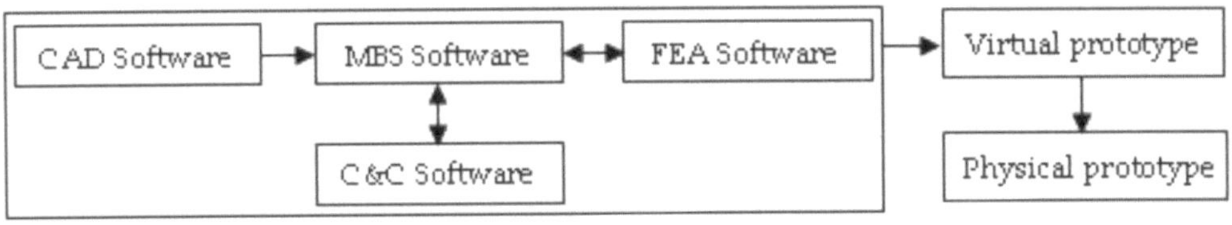

Fig. 3. Virtual (software) prototyping platform

Generally, the virtual prototyping platform, shown in figure 3, includes the following software products (Alexandru & Comsit, 2008): CAD - Computer Aided Design (ex. CATIA, PROENGINEER, SOLIDWORKS); MBS - MultiBody Systems (ex. ADAMS, SD-EXACT, DYMES); FEA - Finite Element Analysis (ex. NASTRAN, COSMOS, ANSYS); Command & Control (ex. MATLAB, EASY5). The MBS software, which is the main component of the platform, allows analyzing, optimizing, and simulating the system under real operating conditions. The CAD software is used for creating the geometric (solid) model of the mechanical system. This model contains data about the mass & inertia properties of the rigid parts. The part geometry can be exported from CAD to MBS using standard format files, such as IGS, DWG, DXF, STEP or PARASOLID. To import the geometry of the rigid parts, the MBS software reads the CAD file and converts the geometry into a set of MBS elements.

The FEA software is used for modelling flexible bodies in mechanical systems. Integrating flexibilities into model allows to capture inertial and compliance effects during simulations, to study deformations of the flexible components, and to predict loads with greater accuracy, therefore achieving more realistic results. The flexible body characteristics are defined in a finite element modelling output file (MNF - Modal Neutral File). The information in a MNF includes location of nodes and node connectivity, nodal mass and inertia, mode shapes, generalized mass and stiffness for modal shapes. The MBS model transmits to FEA the motion & load states in the mechanical system, which can be defined using a FEA Loads format file.

In the modern concept, the mechanical systems are approached as mechatronic systems, which integrate mechanics, electronics, and information technology. The mechatronic systems are built-up with some units with basic functions, which are made to interact between them in order to form a complex system with a given functionality. Integrating the control system in the mechanical model at the virtual prototype level, the mechanical designer and the controls designer can share the same model; they can also verify from one database the combined effects of a control system on a nonlinear, non-rigid model. In this way, the physical testing process is greatly simplified, and the risk of the control law being poorly matched to the physical (hardware) prototype is eliminated. The C&C (Command &

Control) software directly exchanges information with the MBS software: the output from MBS is input for C&C and vice-versa. The mechanical model and the control system communicate by passing state variables back and forth. The analysis process creates a closed loop in which the control inputs from the control application affect the MBS simulation, and the MBS outputs affect the control input levels (Alexandru, 2008).

In this approach, complex virtual prototypes for different mechanical & mechatronic systems (including the renewable energy systems) can be developed, which exactly replicate the structure (components) and the operating conditions of the real physical products. For this paper, in order to develop the virtual prototype of the photovoltaic tracking system, we have used a digital prototyping platform that integrates the following software solutions: CAD - CATIA (to create the solid model, which contains information about the mass & inertia properties of the parts), MBS – ADAMS/View (to analyze, and optimize the mechanical device), C&C - ADAMS/Controls and MATLAB/Simulink (to create the control block diagram, and to simulate the mechatronic system), FEA - ADAMS/AutoFlex (for identifying the eigenshapes, eigenfrequencies and equivalent stresses of the system).

4. Developing the virtual prototype of the tracking system

For identifying accurate and efficient mechanical configurations suitable for tracking systems, a structural synthesis method based on the multi-body systems theory has been developed. The conceptual design can be performed in the following stages (Comsit & Visa, 2007): identifying all possible graphs, by considering the space motion of the system, the type of joints, the number of bodies, and the degree of mobility; selecting the graphs that are admitting supplementary conditions imposed by the specific utilization field; transforming the selected graphs into mechanisms by mentioning the fixed body and the function of the other bodies, identifying the distinct graphs versions based on the preceding particularizations, transforming these graphs versions into mechanisms by mentioning the types of geometric constraints (ex. revolute joint - R, or translational joint - T).

The graphs of the multibody system are defined as features based on the modules, considering the number of bodies and the relationships between them (e.g. R, T, R-R, R-T, RR-RR, RR-RT and so on. In this way, a collection of possible structural schemes have been obtained. In order to select the structural solution for study, we applied specific techniques for product design such as multi-criteria analysis and morphological analysis (Pahl & Beitz, 1996). The multi-criteria analysis was performed using the following steps: selecting the possible variants in accordance with the structural synthesis; establishing the evaluation criteria and the weight coefficient for each criterion (the FRISCO formula); granting the importance note to a criterion and computing the product between the importance note and the weight coefficient in the consequences matrix.

The evaluation criteria of the solutions were referring to the tracking precision, the amplitude of the motion, the complexity of the system, the possibility for manufacturing and implementation. The final solution has been established based on the morphological analysis, the description of possible solutions being conducted by combinatorial procedures that associates the requirements to be met (parameters, functions, attributes). In this way, there is described the morphological table, which eliminates the irrational constructive or incompatible solutions.

Using this conceptual design method, we established the solution for the tracking system in study, which is a polar dual-axis mechanism, the basic scheme being shown in Fig. 1,b. The revolute axis of the daily motion is fixed and parallel with the polar axis. The both motions are directly driven by rotary motors, type MAXON ROBOTIS DX-117 (fig. 4). The motion is transmitted from the output shafts of the motors by using worm gears; these irreversible transmissions assure at the same time the blocking of the system in the stationary positions between actuatings, when the motors are stopped (in this way, there is no energy consumption in the stationary positions).

Fig. 4. The rotary motor MAXON DX-117

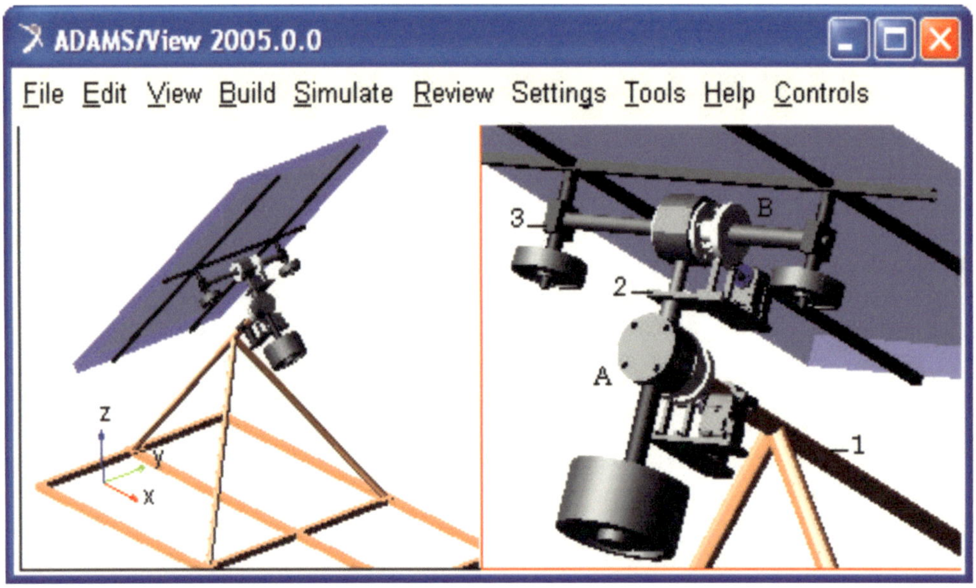

Fig. 5. The MBS mechanical model of the dual-axis tracking mechanism

The "multibody system" model of the mechanism, developed by using the MBS environment ADAMS/View of MSC Software, it is designed so that it has three parts, as follows (fig. 5): base/sustaining frame (1), on which there is disposed the support of the motor and the fixed axle of the revolute joint for the daily motion - A; intermediary element (2), which includes the mobile parts of the joint A, and the fixed part of the joint for the seasonal motion - B; support part (3), which contains the axle of the seasonal motion, the panel frame, and the PV panel.

For developing the constructive solution of the tracking system, we used the CAD software CATIA of Dassault Systems. The geometry transfer from CATIA to ADAMS was made using the STEP (Standard for the Exchange of Product Model Data) file format, through the ADAMS/Exchange interface. The STEP format describes the level of product through a specialized language (Express), which establishes the correspondence between the STEP file and the CAD model. The geometry of the PV panel corresponds to a VITOVOLT 200 panel, which has the following specifications: active surface - 1.26 m^2, weight - 15.5 Kg.

The virtual model takes into consideration the mass forces, the reaction in joints, and the joint frictions. The block diagram of the revolute joint is shown in figure 6, with the following notations: F_a, F_r – axial and radial joint reactions, T_r - bending moment, T_p - torque preload, T_f - frictional torque, R_n – friction arm, R_p – pin radius, R_b – bending reaction arm, μ - friction coefficient (static or dynamic), SW – switch block, MAG - magnitude block, ABS – absolute value block, FR - friction regime determination, ⊕ - summing junction, ⊗ - multiplication junction. Joint reactions, bending moment, and torque preload determine the frictional torque in a revolute joint. These force effects (one or more) can be turned-off by using switches. The joint reactions are converted into equivalent torques using the respective friction arm and pin radius. The joint bending moment is converted into an equivalent torque using pin radius divided by bending reaction arm.

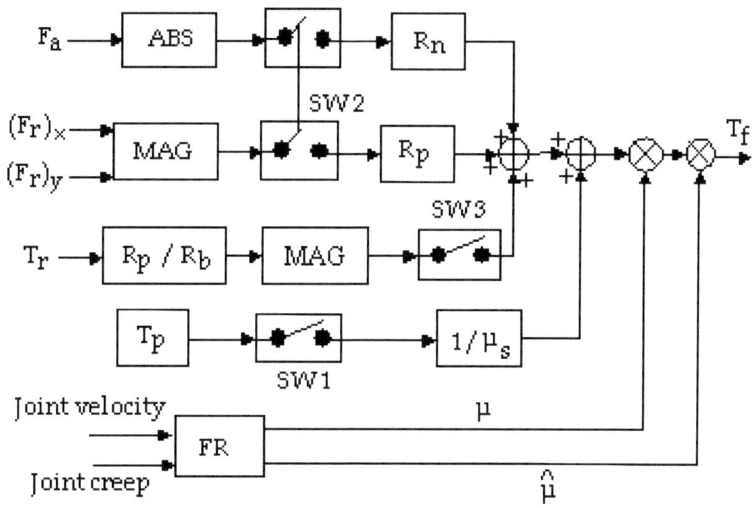

Fig. 6. The block diagram of the revolute joints

For simulating the real behaviour of the tracking system, in order to obtain more realistic results, we have developed the control system in the concurrent engineering concept, using ADAMS/Controls and MATLAB/Simulink. The solar tracker is an automated (controlled) system, which has as task the orientation of the PV panel (i.e. the effecter) on the imposed trajectory. Generally, the solar tracking system is composed by different subsystems, whose structure is shown in figure 7. In these terms, for the polar solar tracker in study (see fig. 5), there are the following subsystems: the DC rotary actuators, the mechanical transmissions, and the mechanism's structure.

The connecting scheme of the subsystems emphasis the following: the mechanism is a coupled system (each element has influence on the others); the transmissions are connected between the joints and the DC motors; the transmissions and the DC motors are uncoupled subsystems. In figure 7,b there is shown a section through a certain joint "k" (in this case, k=1, 2). The design problem can be formulated in the following ways: designing a control system which allows the displacement of the effecter on the imposed trajectory; because the imposed trajectory can be transformed, using the inverse kinematics, in n=2 imposed trajectories for the motor axes, the objective is to design a control system which allows the rotation of the motor axes on the imposed trajectory.

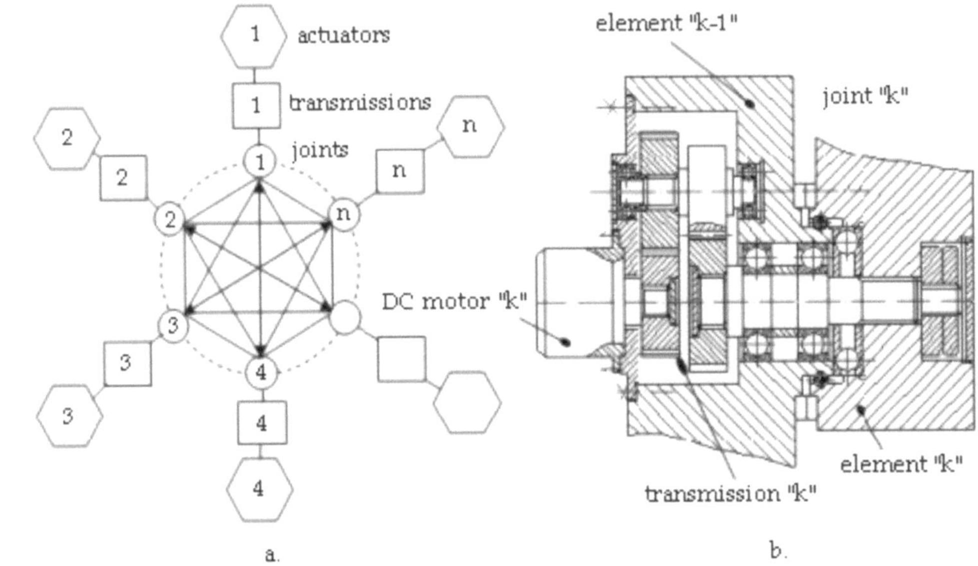

Fig. 7. The main subsystems of the solar tracker

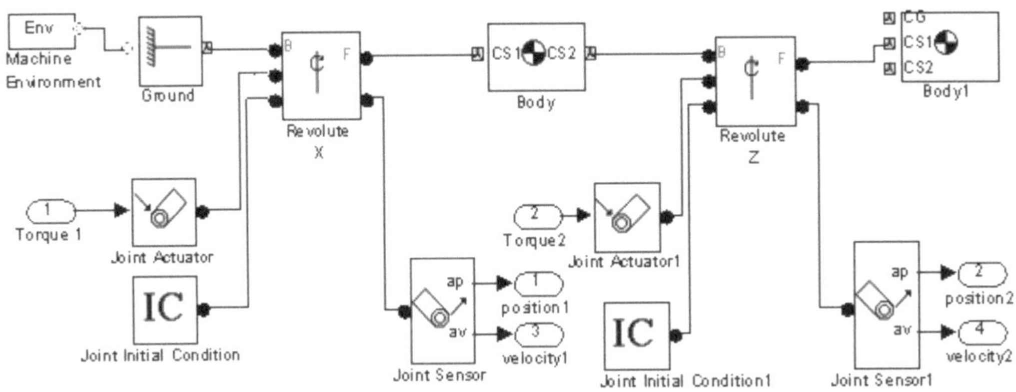

Fig. 8. The dynamic model of the solar tracker

The dynamic model of the polar tracking system is shown in figure 8, the MBS virtual prototype being the one in figure 5. The mechanical transmission is modeled with the following equations:

$$\begin{cases} q_{m1/2} = i_{1/2} \cdot q_{1/2} \\ T_{1/2} = \dfrac{1}{i_{1/2}} \cdot \tau_{1/2} \end{cases} \tag{2}$$

in which: $q_{1/2}$ is the angular position in the revolute joint; $q_{m1/2}$ – the angular position of the motor axis; $\tau_{1/2}$ – the torque in joint; $T_{1/2}$ – the motor torque; $i_{1/2}$ – the transmission ratio of the rotary motor.

The DC rotary motor, which is an electro-mechanical system, is modeled by the following relations (fig. 9):

$$V = R_a i + L\frac{di}{dt} + E\,,\, J\ddot{q}_m = K_M i - b\dot{q}_m - T\,,\, E = K_b\dot{q}_m \tag{3}$$

where V is the input voltage, T - the torque, R_a - the resistance, L - the inductance, i - the current, K_b - the contra-electromotor constant, q_m - the rotor position, $\dot{q}_m$ - the angular velocity of the rotor (ω_m), J - the inertial moment, K_M - the motor torque coefficient, b - the viscous friction; the values of the specific parameters are presented in table 1.

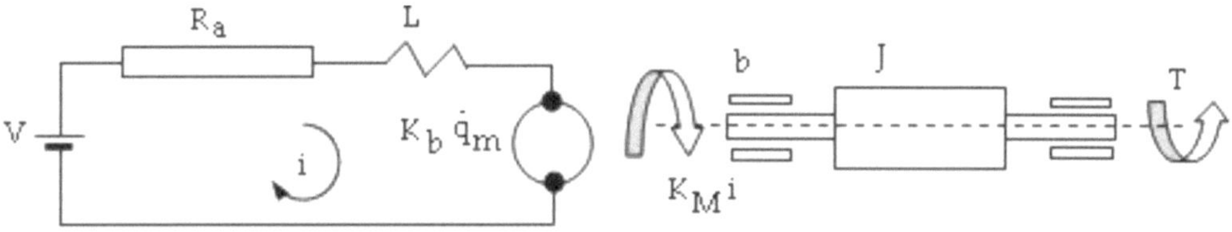

Fig. 9. The dynamic model of the DC motor

System	Parameter	1	2
Mechanical structure	m [kg]	0.7	0.5
	J [Nms²/rad]	1.8e-3	0.78e-3
	l [m]	0.5	0.35
	c [m]	0.25	0.125
Transmission	i	150	150
DC motor	J [Nms²/rad]	233e-6	
	L [H]	0.5	
	Ra [Ω]	0.8	
	b [Nms/rad]	0.1	
	K_M [Nm/amp]	176e-3	
	K_b [Vs/rad]	0.105	

Table 1. The numeric values of the motors' parameters

The structural block scheme of the tracking system is shown in figure 10. There can be identified the connections between subsystems, and the measures which define the communications. According with this model, the goal is to control the DC motors (the reference measures of the control problem are the revolution angles of the motor axes $q_{1/2}$), which are perturbed with the torques $T_{1/2}$. The torques computation is based on the dynamic model of the mechanical structure on which acts outside forces.

For detailing the previous diagram, we will focus on the DC motors subsystems, and will transform the equations (3) with the Laplace operator:

$$Q_m(s) = G_V(s)\cdot V(s) + G_T(s)\cdot T(s) \tag{4}$$

in which $\Omega_m(s)$ is the Laplace transform of ω_m, V(s) - the Laplace transform of V, T(s) - the Laplace transform of T, $G_V(s)$ & $G_T(s)$ - the transfer functions:

$$G_V(s) = \frac{K_M}{LJs^3 + (JRa + bL)s^2 + (bRa + K_b K_M)s}, \quad G_T(s) = \frac{Ls + Ra}{LJs^3 + (JRa + bL)s^2 + (bRa + K_b K_M)s} \tag{5}$$

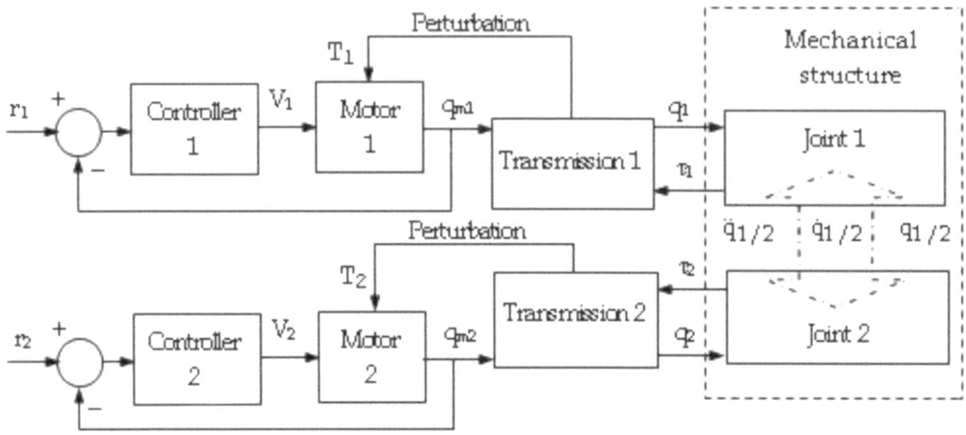

Fig. 10. The structural block scheme of the tracking system

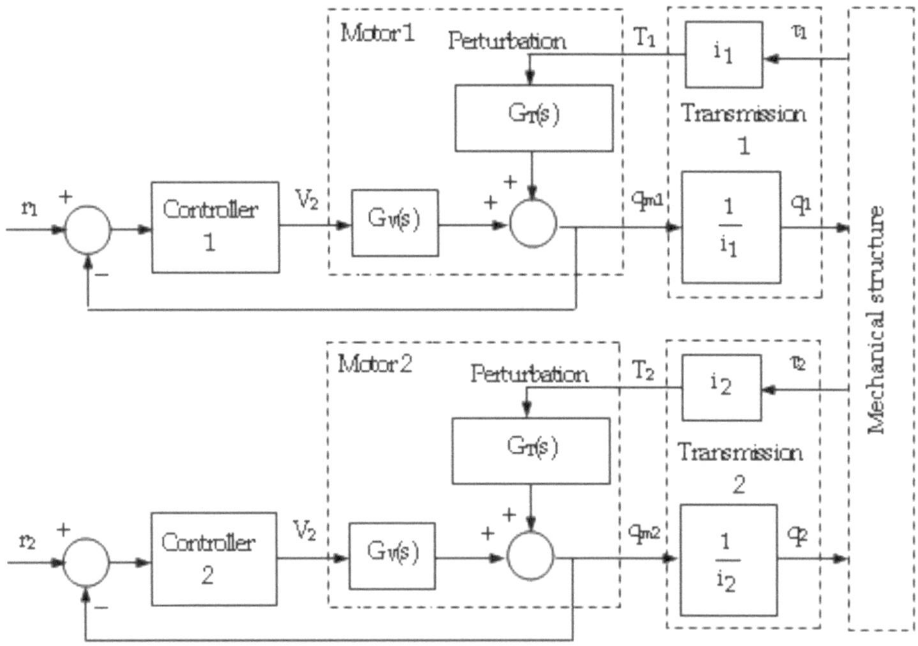

Fig. 11. The detailed control scheme

These relations lead to the detailed model shown in figure 11. Considering the numeric values from table 1, we obtained the following results for the transfer functions (G_V, and G_T respectively):

$$G_V(s) = \frac{1511}{s^3 + 430.8s^2 + 845.3s}, \quad G_T(s) = \frac{4291.84s + 6866.95}{s^3 + 430.8s^2 + 845.3s} \tag{6}$$

Using the mentioned control strategy (see figure 11) the plant transfer function is $G_V(s)$. From (6) we can extract the following observations: in order to obtain a zero error for the unperturbed control loop, because of the zero-pole, it is appropriate to use a proportional controller; in order to reject a constant perturbation we will use the diagram transformations presented in figure 12 and the final value theorem (7):

$$q_m(\infty) = \lim_{s \to 0} s \cdot \frac{1}{s} \cdot \frac{G_T}{1 + G_V G_C} \tag{7}$$

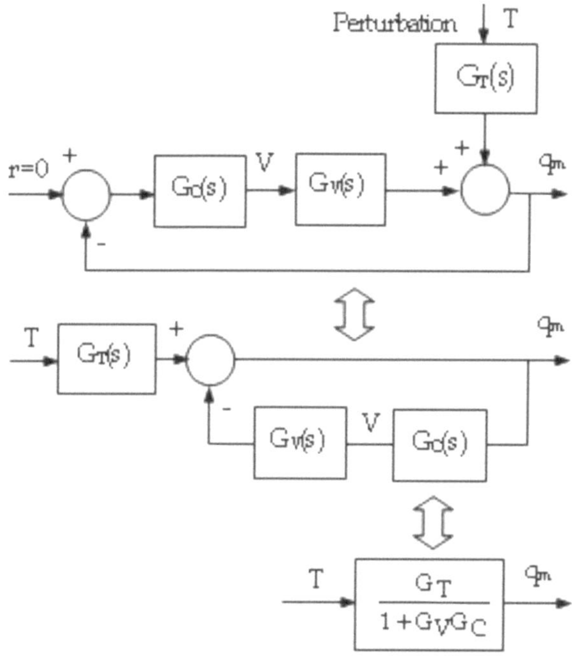

Fig. 12. The equivalent diagram

This result underlines that:

- using a proportional controller we can not reject the error caused by a constant perturbation:

$$q_m(\infty) = \frac{4.54}{K} \tag{8}$$

- for rejecting the perturbation effect ($q_m(\infty)=0$) we must use a controller which has a zero pole;
- the plant poles are:

$$\begin{cases} p_1 = 0 \\ p_2 = -1.9712 \\ p_3 = -428.8288 \end{cases} \tag{9}$$

- the step response of the control system is presented in figure 13.

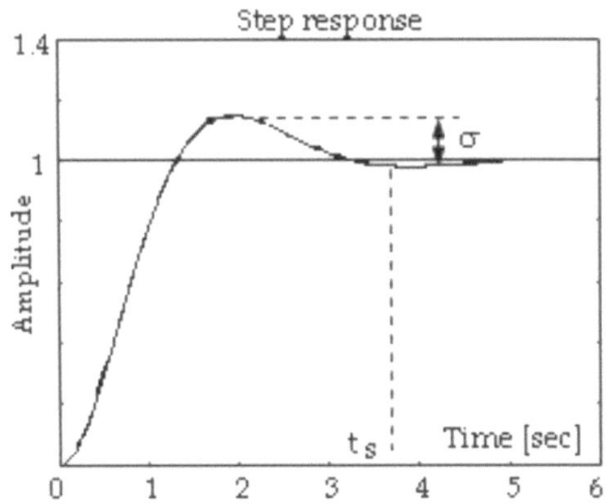

Fig. 13. The indicial response of the motor

Analyzing this result we obtain that the overshooting is $\sigma \cong 20\%$ and the transient regime duration is $t_s \cong 4s$. For the tracker control the recommended values are $\sigma < 5\%$, $t_s < 1s$. Meanwhile we will model the perturbations $T_{1/2}$ caused by the resistant torques in joints ($\tau_{1/2}$) with a step signal because these are bounded.

In conclusion:

- the tracking system is a complex system composed from three parts: the actuator, the gear transmission, and the mechanical structure of the tracker;
- the control strategy replace the use the actuator as the plant and consider the other mentioned subsystems perturbations source;
- the perturbations are bounded;
- for errors rejecting we will need a controller with a pole in origin;
- using the step response of the plant we have established the new transient response parameters.

For the controller synthesis in the state space we will adopt the following state vector:

$$x = [q \quad \dot{q} \quad i] \tag{10}$$

where q is the angular position, $\dot{q}$ - is the angular velocity, i - the current intensity.
Using the mentioned state vector, the DC motor dynamic model is:

$$\dot{x} = \begin{bmatrix} 0 & 1 & 0 \\ 0 & -\frac{b}{J} & \frac{K_M}{J} \\ 0 & -\frac{K_b}{L} & -\frac{R_a}{L} \end{bmatrix} \cdot x + \begin{bmatrix} 0 \\ 0 \\ \frac{1}{L} \end{bmatrix} \cdot u + \begin{bmatrix} 0 \\ -\frac{1}{J} \\ 0 \end{bmatrix} \cdot T \tag{11}$$
$$y = [1 \quad 0 \quad 0] \cdot x$$

and by replacing the numerical values for the parameters (see table 1) we obtain:

$$\dot{x} = \begin{bmatrix} 0 & 1 & 0 \\ 0 & -429.1845 & 755.3648 \\ 0 & -0.21 & -1.6 \end{bmatrix} \cdot x + \begin{bmatrix} 0 \\ 0 \\ 2 \end{bmatrix} \cdot u + \begin{bmatrix} 0 \\ -4291.8 \\ 0 \end{bmatrix} \cdot T \tag{12}$$
$$y = [1 \quad 0 \quad 0] \cdot x$$

We can split this model in two parts:

- the unperturbed part:

$$\dot{x} = \begin{bmatrix} 0 & 1 & 0 \\ 0 & -429.1845 & 755.3648 \\ 0 & -0.21 & -1.6 \end{bmatrix} \cdot x + \begin{bmatrix} 0 \\ 0 \\ 2 \end{bmatrix} \cdot u \qquad (13)$$
$$y = \begin{bmatrix} 1 & 0 & 0 \end{bmatrix} \cdot x$$

- the perturbation effect:

$$\dot{x} = \begin{bmatrix} 0 & 1 & 0 \\ 0 & -429.1845 & 755.3648 \\ 0 & -0.21 & -1.6 \end{bmatrix} \cdot x + \begin{bmatrix} 0 \\ -4291.8 \\ 0 \end{bmatrix} \cdot T \qquad (14)$$
$$y = \begin{bmatrix} 1 & 0 & 0 \end{bmatrix} \cdot x$$

The simulation of the result is realized using the scheme shown in figure 14. In the controller synthesis we have used the Ackerman algorithm. For the imposed transient response parameters we have obtained the following control gains: K=[5.9309 106.6441 -187.3923]; k_i = 16.94; the simulation of this result is shown in table 2.

Using this result, we can simulate the behaviour of the tracking system. In order to do this we must construct a block diagram (fig. 15) composed by the control loop and also by the perturbation source (the gear transmission and the tracker structure). In figure 15 we have used the following blocks: qd1/2 - the desired angular position of the tracker; motor 1/2 - the DC motors models; perturbation 1/2 - the perturbations models; mechanical device – the block that contain the mechanical structure of the tracking mechanism - see figure 8 (this block is used for determining the motor torques, and it works in inverse dynamics regime); ddq1/2 – the accelerations computation blocks. Using this simulation diagram we have obtained the results presented in table 3.

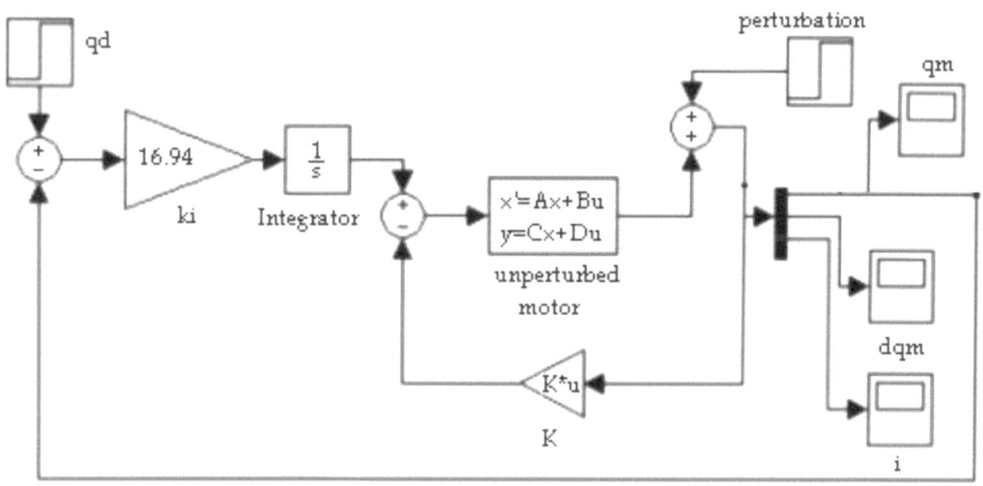

Fig. 14. The control scheme of the DC motor

Indicial response	Eliminating the perturbation
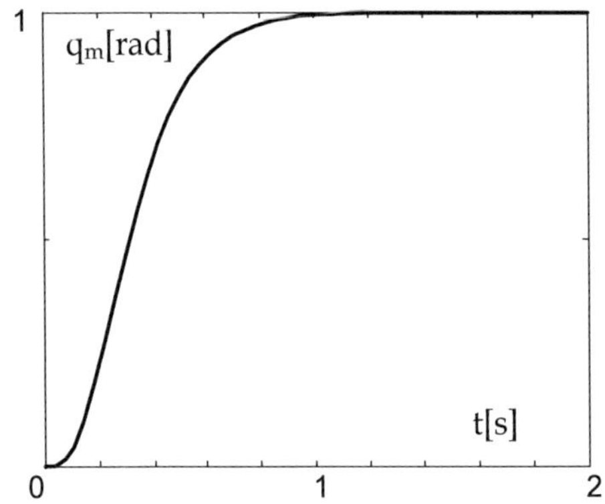	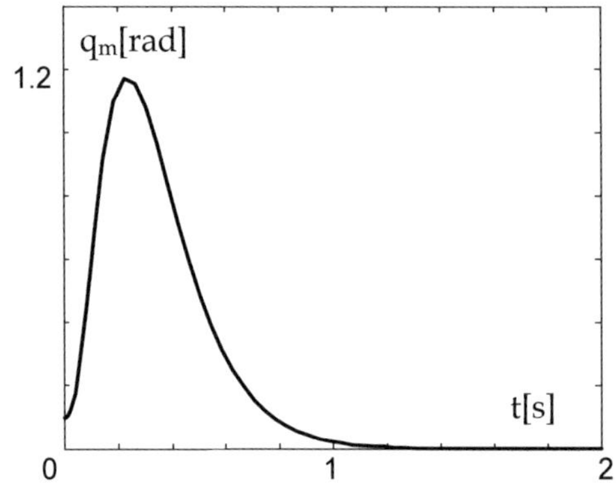
qd is a step signal and the perturbation is zero; the system react critical damped; the settling time is 1 sec	qd is zero; the perturbation is a step signal

Table 2. The simulation of the result for the DC motor

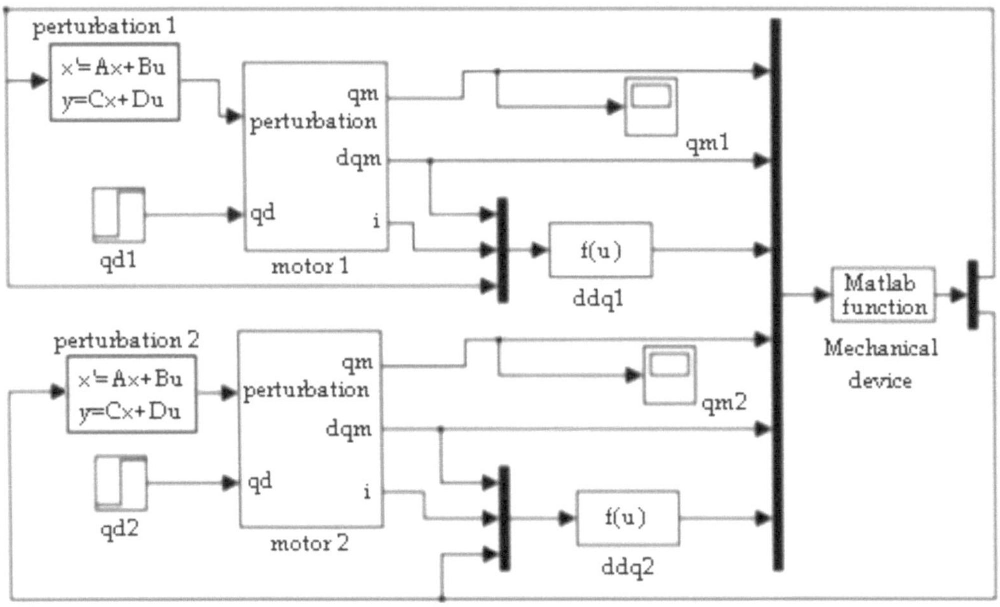

Fig. 15. The control diagram of the tracking mechanism (MATLAB)

DC motor 1	DC motor 2
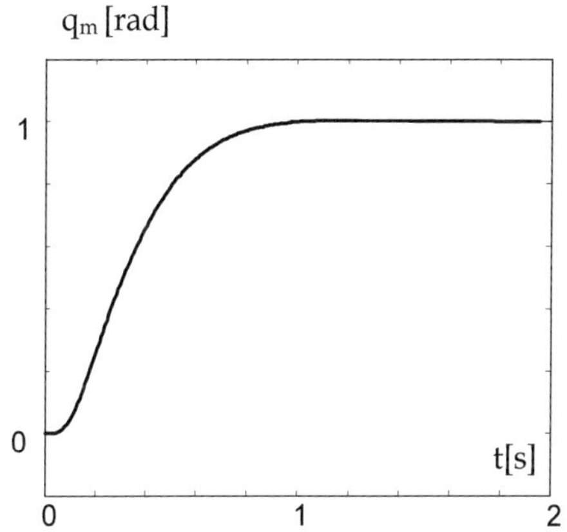	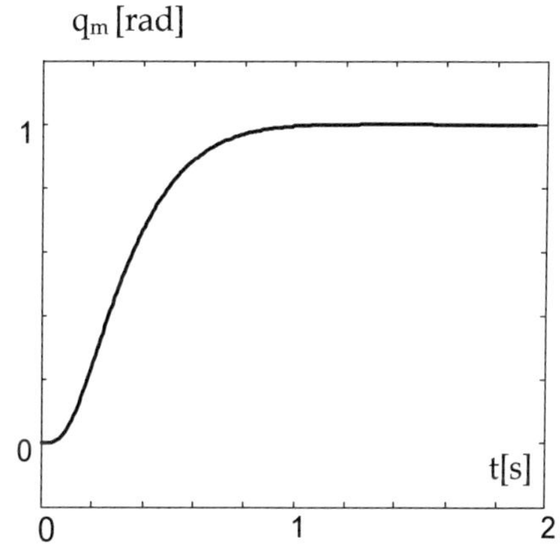
The desired angular position has been imposed with a step signal; the transient response has the settling time less then 1 sec; the settling error is zero	

Table 3. The angular positions of the DC motors

In the next step, for finishing the simulation diagram of the mechatronic tracking system in the concurrent engineering concept, in the control diagram there will be integrated the MBS model of the mechanical system, which it was developed in ADAMS/View (see fig. 5); this model replaces the MATLAB function (i.e. the mechanical device from figure 15). The objective is to control the angular positions (for the DC motors which drive the daily and seasonal motions), which are perturbed with the motor torques. For connecting the mechanical model (ADAMS) with the electronic control system (MATLAB), the input & output parameters have been defined. The daily & seasonal angles of the PV panel (i.e. the angular positions) represent the input parameters in the MBS mechanical model. The outputs transmitted to the controller are the motor torques generated by the DC motors.

For the input state variables, the run-time functions are 0.0 during each step of the simulation, because the variables will get their values from the control application. The run-time functions for these variables are VARVAL(daily_angle), and VARVAL(seasonal_angle), where VARVAL is a specific function that returns the value of the given variable. In other words, the input daily & seasonal angles get the values from the input variables. For the output state variables, the run-time functions return the sum of torques at locations.

The next step is for exporting the ADAMS plant files for the control application. The Plant Inputs refer the input state variables (daily & seasonal angles), and the Plant Outputs refer the output state variables (daily & seasonal motor torques). ADAMS/Controls save the input and output information in a specific file for MATLAB (*.m); it also generates a command file (*.cmd) and a dataset file (*.adm), which will be used during the simulation. With these files, the control diagram can be developed, in order to complete the link between the mechanical and actuating - control systems. For beginning, the *.m* file is loaded in the MATLAB command line, and then, for creating the MSC ADAMS block diagram, there is a special command, namely *adams_sys*. This builds a new model in Simulink (*adams_sys_.mdl*), which contains the MSC.Software S-Function block representing the non-linear MSC.ADAMS model (i.e. the mechanical system). The *ADAMS_sub* block is created based on the information from the *.m* file (Alexandru, 2008).

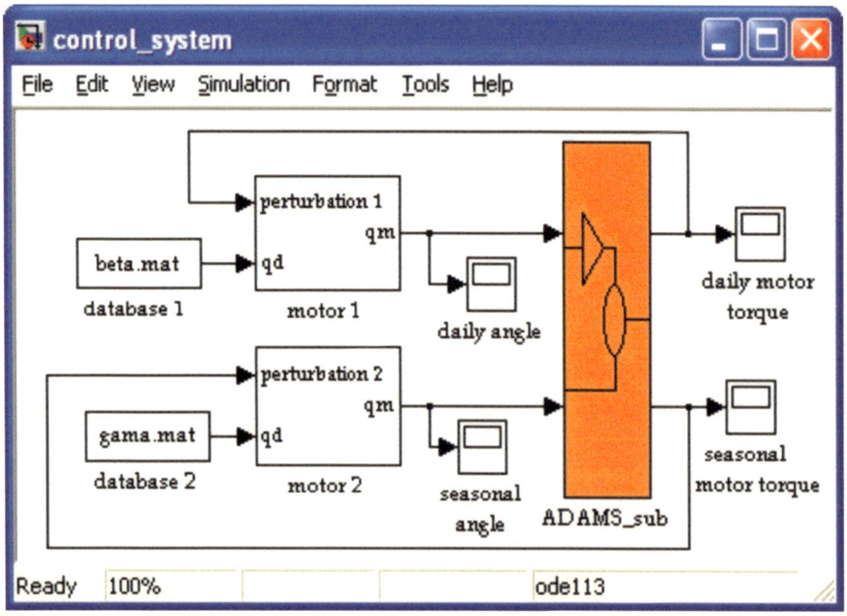

Fig. 16. The control system in the concurrent engineering concept (MATLAB + ADAMS)

Finally, the construction of the controls system block diagram (fig. 16) is made by integrating the ADAMS_sub block, the DC motors blocks, and the databases blocks containing the desired trajectory of the panel (for the daily & seasonal angles). To set the simulation parameters (e.g. the solver type, the communication interval, the number of communications per output step, the simulation & animation model), the MSC.ADAMS Plant Mask is used. In this way, ADAMS accepts the control inputs from MATLAB and integrates the mechanical model in response to them. At the same time, ADAMS provides the motor torques for MATLAB to integrate the control model.

5. Analyzing & optimizing the virtual prototype

The photovoltaic panel can be rotated without brakes during the day-light, or can be discontinuously driven (step-by-step motion), usually by rotating the panel with equal steps at every hour. Obviously, the maximum incident solar radiation is obtained for the continuous motion in the entire angular field (from -90° at sunrise, to +90° at sunset), but in this case the operating time of the system/motor is high. In our vision, the strategy for optimizing the motion law of the tracking system aims to reduce the angular field of the daily motion and the number of actuating operations, without significantly affecting the incoming solar energy, and to minimize the energy consumption for realizing the tracking.

The energy produced by the PV panel depends on the quantity of incident solar radiation, the active surface of the panel, as well as the panel's efficiency. The incident radiation, which is normal to the active surface, is given by the relation:

$$G_I = G_D \cdot \cos i \qquad (15)$$

where G_D is the direct terrestrial radiation, and i - the angle of incidence. The direct radiation is established using the next equation (Meliß, 1997):

$$G_D = G_0 \exp\left(-\frac{T_R}{0.9 + 9.4 \cdot \sin\alpha}\right) \qquad (16)$$

with the following components:

$$G_0 = \overline{G}_0 \cdot (1 + 0.0334 \cdot \cos x) \qquad (17)$$

$$x = 0.9856^\circ \cdot n - 2.72^\circ \qquad (18)$$

$$\alpha = \sin^{-1}(\sin\delta \cdot \sin\varphi + \cos\delta \cdot \cos\varphi \cdot \cos\omega) \qquad (19)$$

$$\omega = 15^\circ \cdot (12 - T) \qquad (20)$$

where G_0 is the extraterrestrial radiation, $\overline{G}_0$ - the solar constant (1367 W/m^2), n - the day number during a year, T_R - the distortion factor, which depends on the month and the geographic region, α - the solar altitude angle, δ - the solar declination, φ - the location latitude, ω - the solar hour angle, and T- the local time.
The angle of incidence is determined from the scalar product of the sunray vector and the normal vector on panel,

$$i = \cos^{-1}(\cos\beta \cdot \cos\beta^* \cdot \cos(\gamma - \gamma^*) + \sin\beta \cdot \sin\beta^*) \qquad (21)$$

$$\beta = \sin^{-1}(\cos\delta \cdot \sin\omega) \qquad (22)$$

$$\gamma = \sin^{-1}\left(\frac{\cos\alpha \cdot \cos\psi}{\cos\beta}\right) \qquad (23)$$

$$\psi = (\sin\omega)\cos^{-1}\left(\frac{\sin\alpha \cdot \sin\varphi - \sin\delta}{\cos\alpha \cdot \cos\varphi}\right) \qquad (24)$$

in which β and γ are the diurnal & seasonal angles of the sunray, β* and γ* - the daily & elevation angles of the panel, and the ψ - the azimuth angle.
It has been demonstrated that for every month there is one day whose irradiation is equal to the monthly average: it is the day in which the declination equals the mean declination of the month (Sorichetti & Perpinan, 2007). Due to this consideration, a noticeable facilitation is introduced in the computing calculation, considering just the mean days of each month instead of the 365 days of the year.
This work presents the exemplification for the summer solstice path of the Sun. The situation is similar for any period of the year: the energy gain is not affected by the variation of the radiation intensity because the gain is obtained from the difference between a tracked panel energy output and a non-tracked panel energy output. In these conditions, the numeric simulations were performed considering the Braşov geographic area, with the following specific input data: φ=45.5°, δ=23.45°, n=172, $T \in [5.579, 21.059]$, T_R=4.2, γ^*=22.05°.

For identifying the optimum angular field of the daily motion, we have considered the correlation between the motion amplitude and the local solar time, for obtaining symmetric revolute motions relative to the solar noon position (β*=0). The analysis has been performed for the following tracking cases: (a) β*∈[-90°, +90°], T∈[5.579, 21.059] - the maximum motion interval; (b) β*∈[-75°, +75°], T∈[6.869, 19.769]; (c) β*∈[-60°, +60°], T∈[8.159, 18.479]; (d) β*∈[-45°, +45°], T∈[9.449, 17.189]; (e) β*∈[-30°, +30°], T∈[10.739, 15.899]; (f) β*∈[-15°, +15°], T∈[12.029, 14.609]; (g) β*=0, T∈[5.579, 21.059] - the fixed (non-tracked) system. In this study, we considered that the PV panel in rotated without brakes (continuous motion).
The computations have been made using the above-presented mathematic model, obtaining in this way the incident radiation for the considered cases. Integrating the incident radiation curves (shown in figure 17), and taking into account the active surface (1.26 m^2) and the conversion efficiency of the panel (15%), we have obtained the quantity of electric energy produced by the PV system. Afterwards, the energy consumption for realizing the motion laws was determined by using the virtual prototype of the tracking system (see section 4).

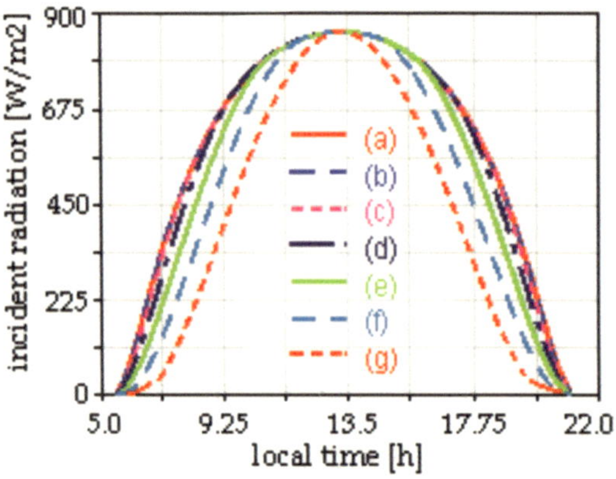

Fig. 17. The incident radiation curves

For the energy consumption, there is also considered the return of the tracking mechanism in the initial position (east/sunshine) after the sunset, which is performed with 10 degrees per minute (0.016667 hours). In this way, the energy balance was performed, the results being systematized in the table 4 (the energy gain is computed relative to the fixed panel case). Analyzing these results, we consider that the optimal angular field for the daily motion of the panel, in the summer solstice day, is β*∈[-60°, +60°], in which there is the maximum energy gain.

β*	E_T [Wh/day]	E_C [Wh/day]	Gain [%]
[-90°, +90°]	1740.88	44.09	37.87
[-75°, +75°]	1739.40	37.01	38.32
[-60°, +60°]	1726.44	17.64	38.84
[-45°, +45°]	1685.62	12.77	35.92
[-30°, +30°]	1598.91	5.84	29.44
[-15°, +15°]	1449.61	1.49	17.66
0°	1230.73	-	-

Table 4. The energy balance for different angular fields of the daily motion

Afterwards, in the optimal angular field, we have evaluated different step-by-step tracking strategies. The objective is to minimize the operating time, which is important for the durability & reliability of the tracking system. In these terms, we have developed - analyzed six tracking cases, depending on the number of steps (in consequence, the step dimension - $\Delta\beta^*$) for realizing the optimum angular field: 12 steps ($\Delta\beta^*=10°$), 10 steps ($\Delta\beta^*=12°$), 8 steps ($\Delta\beta^*=15°$), 6 steps ($\Delta\beta^*=20°$), 4 steps ($\Delta\beta^*=30°$), 2 steps ($\Delta\beta^*=60°$); in each case, the angular velocity for the motion steps is 10 degrees per minute.

One of the most important problems in the step-by-step tracking is to identify the optimum actuating time, in which the motion step has to be performed. In paper, the solution to this problem is obtained by developing an algorithm based on the following phases: the optimal angular field was segmented into the intermediary positions, depending on the step dimension for each case (e.g. for 4 steps there are the following positions: $\beta^* = \{\pm 60°, \pm 30°, 0°\}$), and the incident radiation curves have been consecutively obtained considering the panel fixed in these positions during the day-light; analyzing these curves, we have identified the moment in which the value of the incident radiation for a certain position *k* become smaller than the value in the next position *k+1*; in this moment, the motion step is performed; the analysis continues with the next pair of positions *k+1* and *k+2*, and so on. For example, in the figure 18 there are presented the diagrams for the 4-steps tracking case. Because there is a symmetrical motion relative to the noon position (T=13.319 - local time), the actuating moments are also symmetrically disposed (I-IV, II-III).

For the considered step-by-step tracking cases, the results of the energy balance are systematized in the table 5. According with these results, by using the above-presented algorithm for configuring the step-by-step orientation, we obtained values closed-by the

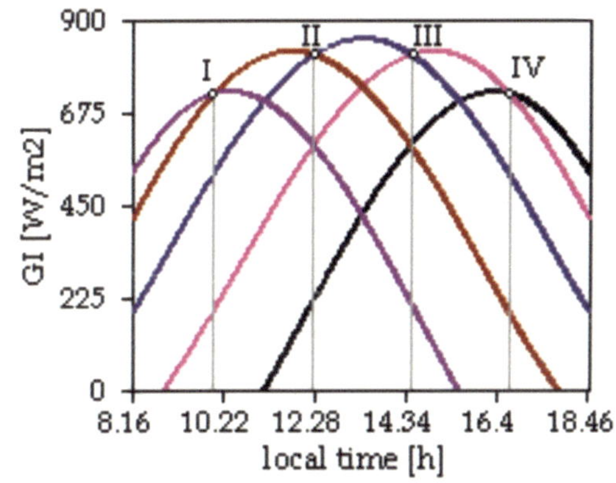

Fig. 18. The optimum actuating time

No. of steps	E_T [Wh/day]	E_C [Wh/day]	Gain [%]
12	1722.19	20.78	38.24
10	1719.72	19.99	38.11
8	1718.35	19.47	38.04
6	1715.39	19.00	37.84
4	1706.92	18.23	37.21
2	1670.97	17.65	34.34

Table 5. The energy balance for the step-by-step tracking cases

continuous motion case, and this demonstrates the viability of the adopted optimization strategy. As we can see, the energy consumptions for realizing the step-by-step motion laws are greater than the energy consumption for the continuous motion, and this because of the over-shootings that appear when the motor is turned-on/off.
Similar studies have been performed for different periods/days, obtaining in this way the optimum motion laws during the year.
The final aspect for this study is to refine the tracking system by adding flexible bodies. The modelling of the mechanical structure of the tracking system with finite elements has two main objectives: identifying the eigenshapes and eigenfrequencies of the system, which are useful to avoid the resonance phenomenon due to the action of the external dynamic loads (as wind or earthquake); identifying the maximum deformabilities and the equivalent stresses, depending on the action of the external factors and on the position of the structure.
In paper, the analysis of the structure is made for different values of the daily position of the panel, in the above-obtained optimum field, $\beta\in[-60^o, +60^o]$, the elevation being fixed at 22.05° (for the summer solstice day, in Braşov geographic area). The external load is materialized by forces and pressures which are produced by a 90 km/h wind speed which is acting on the structure; this speed generates a pressure of 1200 MPa on the panel's surface.
For creating the finite element model of the tracking system, we used the specific ADAMS/AutoFlex module, which automatically mesh the geometries imported from CAD system. In these terms, analyzing the refined virtual prototype of the tracking system, we obtained the following results: the variation of the eigenfrequencies with the daily position angle is insignificant from quantitative point of view; the maximum values of the equivalent stresses (von Misses), which appear in the rotational joints of the structure, are less than the admissible resistance values (90...120 MPa), and this assures the mechanical resistance condition (the maximum value is 71 MPa); the maximum value of the deformability is small (< 0.08mm), and this ensures proper functioning of the system. Among the specific results, in figure 19 there are shown the variations of the von Misses equivalent stresses (a) and deformations (b) fields, as well as the eigenfrequencies (c), for the daily angle β^*=-60°.

6. Conclusions

The application is a relevant example regarding the implementation of the virtual prototyping tools in the design process of the photovoltaic tracking systems. One of the most important advantages of this kind of simulation is the possibility to perform virtual measurements in any point or area of the tracking system, and for any parameter (for example motion, force, and/or energy). Using the virtual prototyping platform, we are able to optimize the mechanical structure of the tracking mechanism, choose the appropriate actuators, design the optimal controller, optimize the motion law, and perform the energy balance of the photovoltaic system. In this way, we are much better equipped to manage the risks inherent in the product development cycle.
Connecting the electronic control system and the mechanical device at the virtual prototype level (i.e. the concurrent engineering concept), the physical testing process is greatly simplified, and the risk of the control law being poorly matched to the real tracking system is eliminated. At the same time, integrating the finite element model in the multibody system analysis, we can quickly build a parametric flexible body representation of a component, analyze the system, make changes to the flexible body and evaluate the effect of the changes, all within the MBS environment.

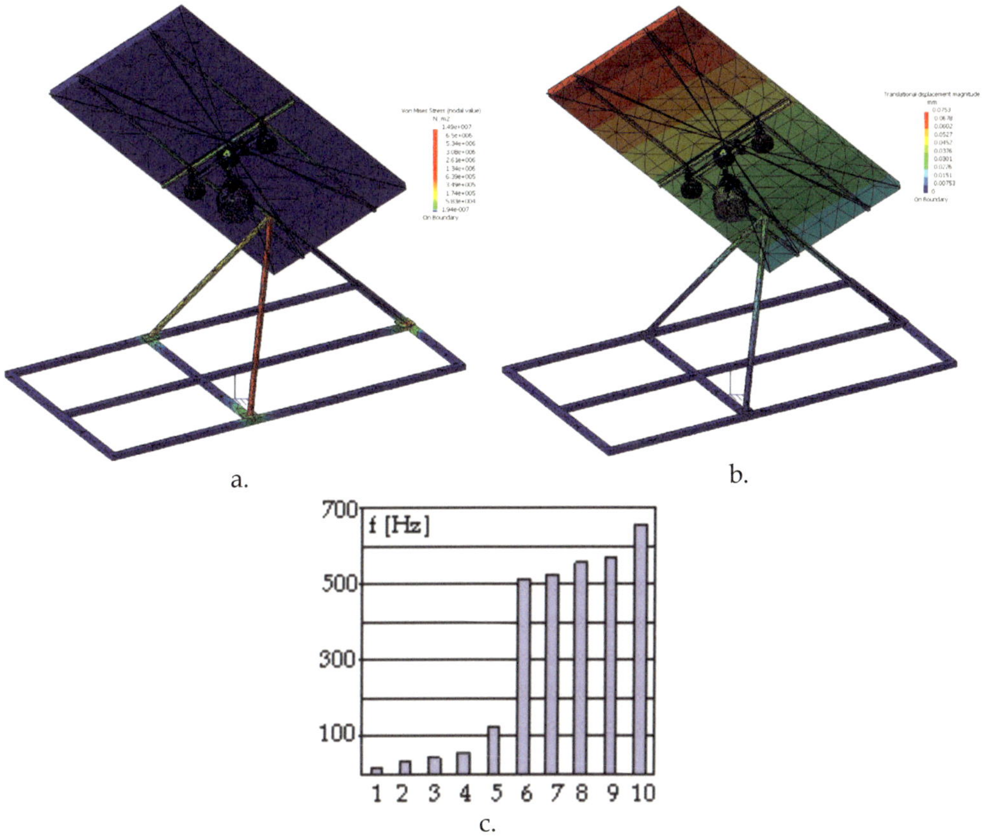

Fig. 19. Specific results of the finite element analysis (flexible multibody system)

The optimization strategy of the motion law, which is based-on the minimization of the angular field for the daily motion and the minimization of the number of actuating operations, leads to an efficient PV system, without developing expensive hardware prototypes. Thus, the behavioural performance predictions are obtained much earlier in the design cycle of the tracking systems, thereby allowing more effective and cost efficient design changes and reducing overall risk substantially.

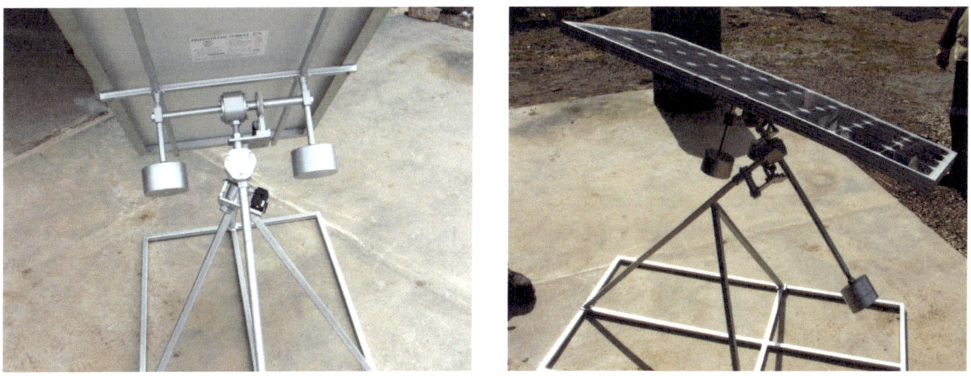

Fig. 20. The physical prototype of the polar tracking system

The tracking system has been manufactured (fig. 20) and it will be tested in the Centre Product Design for Sustainable Development from Transilvania University of Braşov, creating a real perspective for the research in the field. This will allow a relevant comparison between the virtual prototype analysis and the data achieved by measurements; the results of the comparative analysis will be presented in a future paper.

7. References

Abdallah, S. & Nijmeh, S. (2004). Two-axes sun tracking with PLC control. *Energy Conversion and Management*, Vol. 45, No. 11-12, 1931-1939, ISSN 0196-8904

Alata, M.; Al-Nimr, M.A. & Qaroush, Y. (2005). Developing a multipurpose sun tracking system using fuzzy control. *Energy Conversion and Management*, Vol. 46, No. 7-8, 1229-1245, ISSN 0196-8904

Alexandru, C. (2008). The mechatronic model of a photovoltaic tracking system. *International Review on Modelling and Simulations - IREMOS*, Vol. 0, No. 0, 64-74, ISSN 1974-9821

Alexandru, C. & Comsit, M. (2008). The energy balance of the photovoltaic tracking systems using virtual prototyping platform, *Proceedings of the 5th International Conference on the European Electricity Market - EEM*, pp. 1-6, ISBN 978-1-4244-1743-8, Lisbon, May 2008, IEEE/PES

Alexandru, C. & Pozna, C. (2007). Different tracking strategies for optimizing the energetic efficiency of a photovoltaic system, *Proceedings of the 16-th IEEE International Conference on Automation, Quality and Testing, Robotics - AQTR/THETA*, Vol. 3, pp. 434-439, ISBN 978-1-4244-2576-1, Cluj-Napoca, May 2008, IEEE/TTTC

Baltas, P.; Tortoreli, M. & Russel, P. (1986). Evaluation of power output for fixed and step tracking photovoltaics. *Solar Energy*, Vol. 37, No. 2, 147-163, ISSN 0038-092X

Canova, A.; Giaccone, L. & Spertino, F. (2007). Sun tracking for capture improvement, *Proceedings of the 22nd European Photovoltaic Solar Energy Conference - EUPVSEC*, pp. 3053-3058, ISBN 3-936338-22-1, Milan, September 2007, WIP-Renewable Energies

Chojnacki, J.A. (2005). Application of fuzzy logic neural network controllers in PV systems, *Proceedings of the 20th European Photovoltaic Solar Energy Conference - EUPVSEC*, pp. 2269-2272, ISBN 3-936338-19-1, Barcelona, June 2005, WIP-Renewable Energies

Clifford, M.J. & Eastwood, D. (2004). Design of a novel passive tracker. *Solar Energy*, Vol. 77, No. 3, 269-280, ISSN 0038-092X

Comsit, M. & Visa, I. (2007). Design of the linkages-type tracking mechanisms by using MBS method, *Proceedings of the 12th IFToMM World Congress*, pp. 582-587, ISBN 7-89492-107-6, Besancon, June 2007, IFToMM

Dobon, F.; Lugo, A. & Monedero, J. (2003). First results of the tetra-track system and control electronics, *Proceedings of the 3rd World Conference on Photovoltaic Energy Conversion - WCPEC*, pp. 2050 - 2053 , ISBN 4-9901816-0-3, Osaka, May 2003, IEEE

Garcia de Jalón, J. & Bayo, E. (1994). *Kinematic and Dynamic Simulation of Multibody Systems*, Springer-Verlag, ISBN 0-387-94096-0, New-York

Goswami, Y.; Kreith, F. & Kreider, J. (2000). *Principles of Solar Engineering*, Taylor & Francis, ISBN 1-56032-714-6, Philadelphia

Haug, E.; Choi, K.; Kuhl, J. & Vargo, J. (1995). Virtual prototyping for design of mechanical systems. *Journal of Mechanical Design*, Vo. 117, No. 63, 63-70, ISSN 1050-0472

Hoffmann, A.; Frindt, H.; Spinnler, M.; Wolf, J.; Sattelmayer, T. & Hartkopf, T. (2008). A systematic study on potentials of PV tracking modes, *Proceedings of the 23rd European Photovoltaic Solar Energy Conference - EUPVSEC*, pp. 3378-3383, ISBN 3-936338-24-8, Valencia, September 2008, WIP-Renewable Energies

Karimov, K.; Saqib, M. & Akhter, P. (2005). A simple photovoltaic tracking system. *Solar Energy*, Vol. 87, No. 1-4, 49-59, ISSN 0038-092X

Koay, C.W. & Bari, S. (1999). Intermittent tracking of flat plate collectors, *Proceedings of the World Renewable Energy Congress - WREC*, pp. 79-82, ISBN 9-8399-9481-6, Kuala Lumpur, June 1999, World Renewable Energy Network

Meliß, M. (1997). *Regenerative Energiequellen*, Springer-Verlag, ISBN 3-540-63218-2, Berlin

Narvarte, L. & Lorenzo, E. (2007). Tracking gains and ground cover ratio, *Proceedings of the 22nd European Photovoltaic Solar Energy Conference – EUPVSEC*, pp. 3153-3156, ISBN 3-9363-3822-1, Milan, September 2007, WIP-Renewable Energies

Pahl, G. & Beitz, W. (1996). *Engineering Desing*, Springer-Verlag, ISBN 3-5401-9917-9, London

Reda, I. & Andreas, A. (2004). Solar position algorithm for solar radiation applications. *Solar Energy*, Vol. 76, No. 5, 577-589, ISSN 0038-092X

Roth, P.; Georgiev, A. & Boudinov, H. (2004). Design and construction of a system for sun-tracking. *Renewable Energy*, Vol. 29, No. 3, 393-402, ISSN 0960-1481

Rubio, F.R. & Aracil, J. (1997). Design of a combined tracking control system. *Control Engineering Practice*, Vol. 5, No. 1, 23-31, ISSN 0967-0661

Rubio, F.R.; Ortega, M. & Gordillo, F. (2007). New control strategy for sun tracking. *Energy Conversion and Management*, Vol. 48, No. 7, 2174-2184, ISSN 0196-8904

Ryan, R. (2001). *Functional Virtual Prototyping*, Mechanical Dynamics Inc. Press, Michigan

Sala, G.; Pachon, D. & Anton, I. (2002). *Test, Rating and Specification of PV Concentrator Components and Systems*, Instituto de Energia Solar, Madrid

Salmi, M.; Chegaar, M. & Mialhe, P. (2007). Modèles d'estimation de l'irradiation solaire globale au sol. *Revue Internationale d'Heliotechnique Energie - Environment*, No. 35, 19-24, ISSN 0378-1305

Scharmer, K. & Greif, J. (2000). *The European Solar Radiation Atlas*, Ecole des Mines, Paris

Schiehlen, W.O. (1997). Multibody systems - roots and perspectives. *Multibody Systems Dynamics*, Vol. 1, No. 2, 149-188, ISSN 1384-5640

Sorichetti, R. & Perpinan, O. (2007). PV solar tracking systems analysis, *Proceedings of the 22nd European Photovoltaic Solar Energy Conference – EUPVSEC*, pp. 246-252, ISBN 3-9363-3822-1, Milan, September 2007, WIP-Renewable Energies

Tiwari, G.N. (2002). *Solar Energy: Fundamentals, Design, Modeling and Applications*, Alpha Science Int. Ltd., ISBN 0-8493-2409-2, Pangbourne

Tymvios, F.S.; Jacovides, C.P. & Scouteli, C. (2005). Comparative study of Angstrom's and artificial neural networks methodologies in estimating global solar radiation. *Solar Energy*, Vol. 78, No. 6, 752-762, ISSN 0038-092X

Yazidi, A.; Betin, F. & Notton, G. (2006). Low cost two-axis solar tracker with high precision positioning, *Proceedings of the 1st International Symposium on Environment Identities & Mediterrance Area*, pp. 211-216, ISBN 1-4244-0231-X, Corte-Ajaccio, July 2006, IEEE